1st International Conference in Physical Science and Advance Materials (PAM 2019)

IOP Conference Series: Materials Science and Engineering
Volume 757

Istanbul, Turkey
20 - 25 September 2019

ISBN: 978-1-7138-4702-1
ISSN: 1757-8981

Printed from e-media with permission by:

Curran Associates, Inc.
57 Morehouse Lane
Red Hook, NY 12571

Some format issues inherent in the e-media version may also appear in this print version.

Institute of Physics
Dirac House, Temple Back
Bristol BS1 6BE UK

Phone: 44 1 17 929 7481
Fax: 44 1 17 920 0979

techtracking@iop.org

Additional copies of this publication are available from:

Curran Associates, Inc.
57 Morehouse Lane
Red Hook, NY 12571 USA
Phone: 845-758-0400
Fax: 845-758-2633
Email: curran@proceedings.com
Web: www.proceedings.com

TABLE OF CONTENTS

IOP Conf. Series: Materials Science and Engineering **757** (2020) 011001 doi:10.1088/1757-899X/757/1/011001

PREFACE

" 1st. International Conference in Physical Science & Advance Materials PAM 2019" which be held on September 20-25, 2019 in Istanbul /Turkey in Sarissa Hotel, Beylikduzu Under the care of Council of Arab Economic Unity , the Arab Union of Trainers, the Lehaye University , Biruni University and Nişantaşı University with participation of the Iraqi Inventors and Innovators Society, member of International Federation of Inventors Association (IFIA) .More than 100 participants were joined the conference from , Asia ,Europe and Africa. The conference covered most of its topics through its sessions .Which started by a special lecture for each session . Posters activity were arranged too . The end ceremony showed a presented certificate for each participants .Our aim is to reflect the cooperation between ,scientists, researchers and students from all the world in the way to drive the physical science and material science ,material engineering to be in the development mode .And du to its scientific benefit we are encouraged to start our arrangement to held the second conference – PAM 2020 .

IOP Conf. Series: Materials Science and Engineering **757** (2020) 011002 doi:10.1088/1757-899X/757/1/011002

Peer review statement

All papers published in this volume of *IOP Conference Series: Materials Science and Engineering* have been peer reviewed through processes administered by the proceedings Editors. Reviews were conducted by expert referees to the professional and scientific standards expected of a proceedings journal published by IOP Publishing.

IOP Conf. Series: Materials Science and Engineering **757** (2020) 012001 doi:10.1088/1757-899X/757/1/012001

Studying the effect of Polyamide 6 Nanofibers on Physical and Chemical Properties of Unsaturated Polyester Reinforced by Glass fibers Composites

B M Al dabbagh1 , H J Kadhim2 and R M Salih1
1Applied Physics Branch, Applied Sciences department, University of Technology, IRAQ
2Babylon University, Material Engineering collage, IRAQ
1957000b@gmail.com1, hanajawad21@yahoo.com2, ranamahdi1@gmail.com1

Abstract.

In this study, the hybrid reinforcement materials of (glass fibers and polyamide 6 nanofibers) with 15% weight fraction added to unsaturated polyester (UPS) . The physical properties (density , bending strength, impact strength) and chemical resistance of UPS reinforced with of hybrid reinforcement (GF+PA6 nanofibers)with differen ratios: (0.99:0.01 ,0.98:0.02, 0.97:0.03, &0.96:0.04) wt.% by 15% weight fraction were studied. Results proved that, the impact strength increased from 0.5 to 1.5 K.J/m2, while the bending strength increases from (2 to 14.5) MPa with adding of (0.96:0.04) wt.% of reinforcement materials by 15% weight fraction.The weight loss of (UPS/GF) decreased from (1 to 0.1) with (0.04) wt. % of PA6 Nanofibers for 30 days and it stable at this value for 40 days, this is refers to increases of chemical resistance of samples in H2SO4 acid. The density of (UPS /GF) decreased from (1.29 g/cm3) to (1.01) g/cm3 for density of (UPS/19.96GF+0.04 PA6 Nanofibers).

Keywords : Electrospinning technique ,water purification p, olystyrene nanofibers contact angle, hydrophobiIntroduction .

1. Introduction

Nano science and technology have a relatively broad area of research . It focuses on things that are measured in nanometers. Combining active nanomaterials with polymers is an influential way to obtaining improve performances and get new employment for the lightweight material implementations. The resulting polymer nanocomposites interest, from advantages of matrix phase and nano-reinforcement phase [1]

Fibre/polymeric matrix composite materials utilized in a many implementations as civil , aerospace, automotive and other.[2]

Hybrid reinforcement phase used in many applications for superior physical and chemical properties, as : Detmoni et.al .1914 , studying the effect of silica micro particles on the flexural strength, bulk density and modulus of glass-fibre composites, he noted there are an enhancement of these properties.[3]

2008, Fu. et.al , studying the particle size and particle/matrix interface adhesion on mechanical properties of glass-polymer composites , they noted there are an improvement raises from adding of nanoparticles to these composites.[4]

Jo lio et.al 2016, manufactured and tested of mechanical properties , manufacturing parameters and particle loading of hybrid epoxy/carbon fiber composites with poly-di-allyldimethyl ammonium chloride-functionalized nanoparticle and micro silica . They noted there are an homogenous dispersion, diminution of the density and enhancement of the mechanical properties in the hybrid carbon fibers composites.[5] Brittle matrix fracture for long since represent the greatest problems in

IOP Conf. Series: Materials Science and Engineering **757** (2020) 012001 doi:10.1088/1757-899X/757/1/012001

fiber-reinforced composites. Newly, the combining of electrospun nanofibers is an effective method for enhancing the mechanical properties of these composites, without process problems and negatively affecting other properties particularly , mechanical properties [6-8].
Cristina M. et. al 2018, studying the effect of different electrospun nanofibers on mechanical properties of carbon/fibers epoxy composites. They noted , there are enhancement in flexural strength and mechanical energy of these composites [9].

2. **Experimental**:

2.1 Materials

1- Unsaturated polyester (UPS) , with density (1.19 g/cm3) , Sigma Aldridge company used as a matrix materials, and ethyl methyl ketone used as a hardener of polyester risen .
2- Fiber glass specifications (EG) with density (2.5 g/cm3) , China production used as the first type of a reinforcement materials .
3- Electrospun nanofibers of polyamide 6 with diameter about 100 nm , prepared by electrospinning technique and used as a second type of a reinforcement materials.
'Figures 1. and 2'. show the reinforcement materials .

.

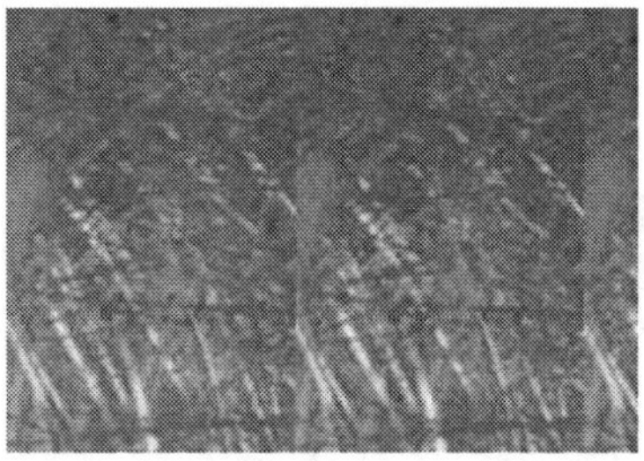

Figure 1: Photograph image of glass fibers

Figure 2:SEM Image of polyamide 6 Nanofibers

2.2 Methods

1- Electrospinning technique for prepared polyamide 6 nanofibers is used. 0.18% concentration of polyamide 6 dissolve in formic acid prepared by magnetic stirrer tool , it mix for 6 hr for homogeneous solution. It remain as stable solution for 3 hr for removing the bubbles from it.
2- It pumped through electrospinning machine for prepared the nanofibers under (25 kV power supply , 14 cm electrospinning gab, 0.48 mm needle diameter, and 0.5 ml/hr flaw rate).
3- Hand casting method used for prepared all the samples, (10*10) cm2 mold of casting prepared for fabricated the sample, then it cut according to the standard measurements of each test as in table 1.
4- Different weight ratios of a reinforcement material prepared for composites and nanocomposites fabricated as in table 2.

Table 1: standard dimensions of each test sample

No. of test	Test	Dimensions (length * width * thickness)
1	Density	5 cm*5cm *1
2	Bending strength	12.7cm * 1.2 cm* 1 cm
3	Impact strength	7.5 cm * 1 cm * 0.5 cm
4	Weight loss	1 cm * 1 cm * 1 cm

Table 2: Weight ratios of samples

No. of test	UPE	Glass Fiber	PA 6 nanofibers
1	100 %	0	0
2	85 %	15%	0
3	85%	14.99	0.01
4	85%	14.98	0.02
5	85%	14.97	0.03
6	85%	14.96	0.04

3. Result and Discussion

3.1 Bending Strength Test

'Figure 3'. Show the relationship between the bending strength and different ratios of reinforcement of composites and nanocomposites. Results prove the bending strength increases after adding of PA6 nanofibers, it is arise from 2 MPa for UPS to 14.5 MPa for UPS reinforced by (14.96 + 0.04)% wt.% of (GF+PA6 nanofibers) respectively. This is because transfer of composites behaviour from brittle to ductile materials which leads to more toughness of these nanocomposites. This is agreement with Cristina M. et.al [9]

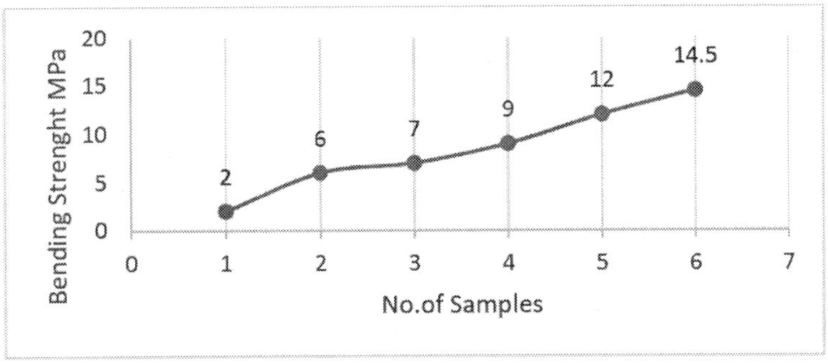

Figure 3: Bending strength of Nanocomposites with Deferent Ratios of a Reinforcement Material

3.2 Density

'Figure 4'. Show the density of composites under deferent ratios of reinforcement materials as in table 2. Density Results refers to increase of microcomposite density with adding the glass fibers with ratio 15% wt.%, this is because the glass fibers have more density than unsaturated polyester according to the rule of mixture of composites.

On the other hand , the density decreases after adding the PA 6 nanofibers , and it continue with decreasing by increasing of PA6 nanofibers weight ratio . It decreases from 1.5 g/cm3 with (0 wt%)of PA6 nanofibers to 1.01 g/cm3 with (0.04) wt.% of PA6 nanofibers . This is because the nanofibers is merged significantly with matrix material [10-11]

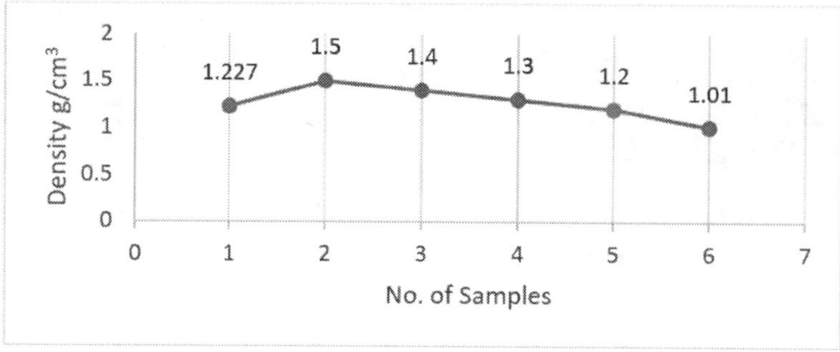

Figure 4: The Density of Composites with Different Ratios of a Reinforcement Material

3.3 Impact strength Test

'Figure 5.' Show the relationship between the impact strength and different ratios of reinforcement of composites and nanocomposites. Results prove the impact strength increases after adding of PA6 nanofibers, it is arise from 0.5 MPa for UPS to 1.5 KJ/m2 for UPS reinforced by (14.96 + 0.04)% wt.% of (GF+PA6 nanofibers) respectively. This is because Polymer nanofabric interleaving increased the threshold impact force (the force to cause initiation of impact damage) [13-15]

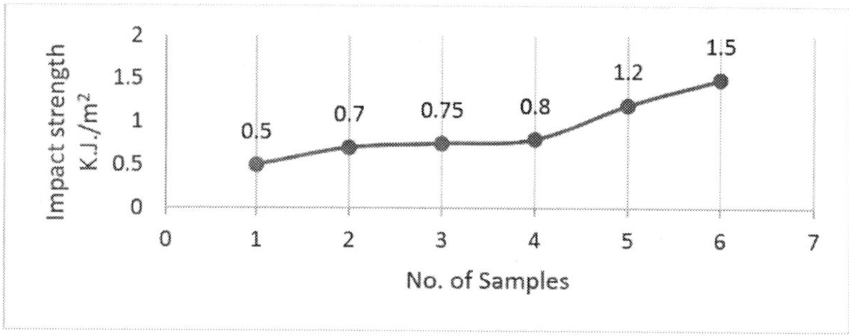

Figure 5: Impact strength of Nanocomposites with Deferent Ratios of a Reinforcement Material

3.4 Weight loss Test

'Figure 6'. Show the relationship between the weight loss of composites and nanocomposites with different ratios of reinforcement materials exposed to H2SO4 acid for 30-40 days with period rate 10 day for each test. Results proved the weight of UPS sample decreases with increasing the immersion time in the acid, this is because the degradation of polymer chain. Also , adding of GF leads to lower weight loss than UPS, This is due to a glass fiber barrier that prevents the acid from reaching the polymer chains and reduces its degradation. While with adding of the PA6 nanofibers leads to more reducing of weight loss of nanocomposites and preventing it from degradation this because it has more homogenous and lower voids . As well as the nanofibers prevent the polymer from degradation by its immerged dramatically in polymer chain this is leads to higher chemical resistance than other samples.[16]

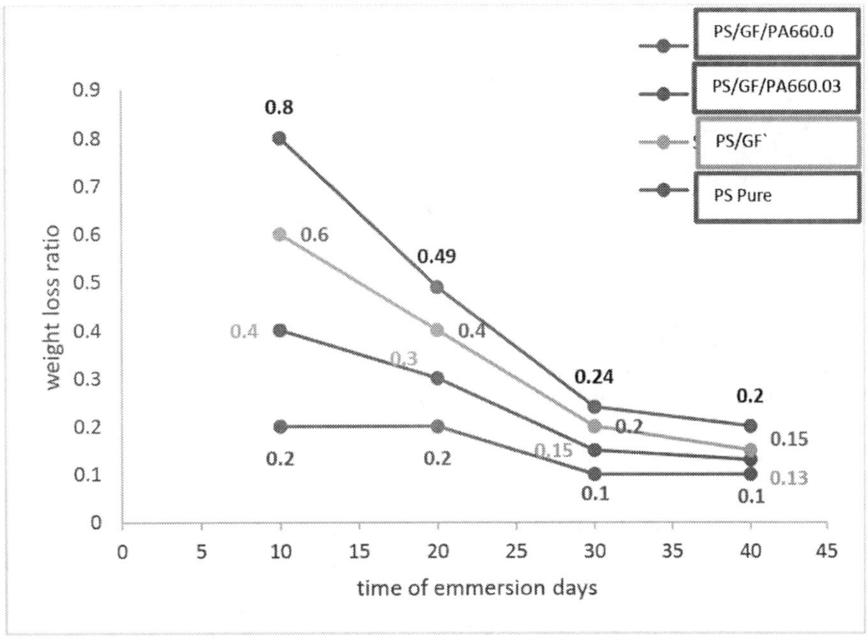

Figure 6: Weight loss ratio of Nanocomposites immersed in H_2SO_4 acid for 30-40 days with period time 10 day for each test

4. Conclusion

We conclude that the adding of PA6 Nanofibers leads to improve the mechanical and chemical properties of nanocomposites, it leads to increasing of impact and bending strength, while it leads to decrease of density. In addition, it leads to enhancement of chemical resistance via H2SO4 acid.

Acknowledgements

The authors extend their appreciation to the Technology University /Applied science department /materials branch, Nano technology research center and Babylon University / Material engineering collage for perform this search.

IOP Conf. Series: Materials Science and Engineering **757** (2020) 012001 doi:10.1088/1757-899X/757/1/012001

References:

[1] Bin Li, Nazanin E., Guan G. and Weidong S., J. Nano. Mat. MDPI, 1-11 (2016)
[2] J. Njuguna , K. Pielichowski, and J. Fan Ser. in Comp. Sci. and Eng. Woodhead, 472-539 (2012)
[3] Anine C. Detomia R. Maria Santosa S. Luiz Moni R. Filhoa C. Coelho M. Tulio H. Fabrizio S. Mater Des. Sci. Direct Else. 55 463–470L (2014)
[4] Shao-Y. Xi-Qiao F. Bernd L. Yiu-Wing M. Comp. B Sci. Direct Else. 39 933–961(2008)
[5] Ju L. Luciano M. Vieira T. Panzera, A. Marco A. and Fabrizio S. J. Comp. Mat. SAGE, 6 783-795 (2017)
[6] Kinloch, J. Taylor, A.. J. Mater. Sci. Springer, 37 433–460. (2002)
[7] Bucknall, C.B. J. Poly. Sci. Elsevier , 16 358 – 359 (1978)
[8] Girard-Reydet E. Vicard, V. Pascault, J. Appl. Poly. Sci. WILEY, 16 2433–2445. (1997)
[9] Cristina M. , Miren B. , Nieves M. , Ana P. Jon M. Jorge G. Jose M. , Estíbaliz A. and Jose L. Poly. MDPI 10. 1190. (2018)
[10] Aldabbagh B. and Alshimary H. J. Meta. & Mat. Sci. and Eng. Transtellar, 6 278-2524 (2016)
[11] Jintendra K., Hitoshi T. Antonio N. and Hyun K. Springer. (2015)
[12] Karthikeyans,P. Babu B. Siva K. , and Sabarinathan C. , Digest J. Nano. and Bios. MDPI, 11 625 – 632 (2016)
[13] Hamed S. Mohamad F. and Giangiacomo M. , Appl. Sci. MDPI, 8 .2406 (2018)
[14] Paul A. Shivalingappa L. Kunigal Sh. J. Comp. St. Sci. Direct Else. 92 1432–1439 (2010)
[15] Poveda, Roland L. and Nikhil G., Springer Briefs ebook , 77- 95 (2015)

IOP Conf. Series: Materials Science and Engineering **757** (2020) 012002 doi:10.1088/1757-899X/757/1/012002

Mechanical properties of nanocomposites for building materials

R M Salih* and Sh H Ahmad
Department of Applied Sciences, University of Technology, Baghdad, Iraq
*Corresponding author email: ranamahdi1@gmail.com

Abstract

This work aims to evaluate some mechanical properties (bending behavior, compressive strength and impact strength) for polymeric composites made of a polymer blend (epoxy+ polyester) as a matrix, with a hybrid reinforcement of both rock wool and nano clay. Hand lay-up technique was used to manufacture the castings, and epoxy and polyester were mixed in different weight ratios, starting from (50:50) up to (90:10) % of epoxy and polyester respectively. Optimum mixing ratio (OMR) was decided upon carrying out impact test on the specimens, and was then reinforced with nanoclay (Kaolite) in weight fractions (5 and 7.5)%,. These ratios were then hybridized with rockwool, added in a volume fraction (10%). Mechanical properties (bending behavior, compressive strength, and impact strength) were studied for these specimens. The results show that the blend reinforced with nano clay in a weight fraction (7.5)% and hybridized with rockwool volume fraction (10%) had the lowest deflection and the highest value of Young's modulus (1889.49)MPa in bending, while blend reinforced with 5% nanoclay and rockwool (10%) showed the highest value of compressive strength (396.95) MPa. The reinforcement with nanoclay reduce the impact strength, the lowest impact strength was (0.0055)KJ/m^2 for (blend+5% nanoclay).

Key Words: Mechanical properties, Nano clays, Nanocomposite, Polymer blends.

1. Introduction

Polymer nanocomposites have attracted great interest in recent years because of their remarkable improvement of physical, mechanical and thermal properties at low loadings of fillers. This enhancement is due to the unique characteristics of the nanometer-sized filler particles such as large surface to volume ratio of the filler which highly improves the

properties of the nanocomposites. Nanostructures have gained much attention since the properties of nanocrystals depend not only on their composition but also on their shape, size, structure, phase, and size distribution [1]. Important factors that can affect the physical and mechanical properties of nano-filler-reinforced polymer nanocomposites include nano-filler type, filler loading levels, and processing. Nanoclay, an inexpensive natural mineral, has been reported by many studies as the one of the potential candidates for nanocomposite [2].
Many of the previous instances are nanocomposite materials; this is, they are the combination of traditional materials such as ceramics, metals and polymers with nanomaterials (e.g., nano-silica, nanotubes, nanofibers, etc.) in different phases: continuous, for example nano-silica in a binding matrix (the case of concretes and mortars based on Portland cement) and disperse (with aggregates such as nanofibers, nanotubes and other reinforcements by precipitation) which, when added, would yield nanocomposite materials to improve many of the physical and chemical properties that composite materials normally have[3].
Ahmad Rafiq and Necar Merah studied glass fiber-reinforced epoxy-nanoclay composite plates, with I.30E clay contents ranging between 0 and 5 wt.%, were manufactured by hand layup with hot pressing. Flexural strength of unexposed fiber-reinforced epoxy-nanoclay reached an optimum improvement of 11% for 1.5 wt.%. Scanning electron microscope analysis showed that at this clay loading, better interfacial adhesion of clay with glass fibers was achieved. At higher clay loadings, clay agglomeration and presence of micro-voids led to less strength improvement. The maximum water uptake was found to decrease with increasing clay loading and moisture diffusion at 80°C Exposure to moisture resulted in degradation of fiber-reinforced epoxy-nanoclay flexural properties with about 36% reduction in strength for 80°C and 8% for room temperature[4].
JR Robledo-Ortiz et.al. studied three different types of nanoclays (1.44P, 1.34MN, and Cloisite 15A) which were used to reinforce an injection grade poly lactic acid (PLA). The nanocomposites (NCs) were prepared using three different nanoclay concentration levels (1,3, and 5 wt%) in a twin-screw extruder, Results showed that the three nanoclays significantly increased the tensile and flexural modulus of the injection grade PLA. The 1.34MN NCs also showed improvement in the tensile strength, in addition to improving flexural strength. Additionally, the use of 5 wt% of 1.44P nanoclay allowed an increase in impact strength while using 1.34MN and 15A nanoclays, the impact strength was similar to the one observed for pure PLA. In general, mechanodynamic analysis results showed that storage modulus increased with nanoclay content; while thermogravimetric analysis indicated that none of the nanoclays has a significant effect over the degradation temperature of pure PLA. Differential scanning calorimetry results showed that the crystallinity of PLA is enhanced with nanoclay inclusion. For 1.34MN NCs, X-ray diffraction observations exposed that the mineral clay relative intensity peaks disappeared indicating nanoclay exfoliation, which contributes to the increase in tensile and flexural strength in the NCs.
Nevertheless for 1.44P and 15A nanoclays, an increase in the interlayer distance (intercalation) was detected[5].

2. Experimental work

2.1. materials

2.1.1 Unsaturated Polyester (UP)

A Thermoset polymer that is usually transparent viscous liquid, which solidifies upon the addition of a hardener (methyl ethyl ketone peroxide-MEKP) in a weight ratio 1%. The unsaturated polyester was supplied by Saudi Industrial Resins Limited Company (SIR)™. Table (1) gives some of it's properties.

Table (1). General properties of polyester resin.

Density	1 to 1.3 g\\cm^3 [25°C]
Flash point	30 to 33°C
Physical state	Liquid [clear]
Viscosity	Dynamic:400 to 600 Mpas (400 to 600 CP)
Solubility	Insoluble in cold water

2.1.2 Epoxy Resin

Epoxy resin that used in this work was Sikadur–105, manufactured by Sika™, which is a two component, low viscosity epoxy resin system in the form of transparent liquid (which transforms into solid state after adding the hardener to it in a percentage of 2:1.Table (2) how the properties of epoxy resin .

Flexural properties	14 day 7900 psi(54 MPa)
Viscosity (mixed)	200 cps
Pot life	30 minutes.(60 gm mass)
Tensile properties	14 day 4500 psi(37.2 MPa)
14 day tensile strength	54 MPa
Elongation at break	3.1%
Modulus of elasticity	1400 MPa

Table (2). General properties of epoxy resin.

2.1.3 Nanoclay (Kaolinite)

Kaolinite is known for its abundance and desired properties, leading to its use in a wide range of plastics, cements and catalysts. The nano clay used in this study was provided by Intelligent Materials Pvt. Ltd.™. properties are shown in table (3).

Table (3). General properties of nanoclay.

Molecular weight	258.2 g/mol.
Appearance	White powder
Specific gravity	~2.6
Melting point	>1500 °C
Solubility	Insoluble in cold water

2.1.4 Rockwool:

Sometimes called stone wool, is a type of inorganic fiber. The loose rockwool used in this study was manufactured by Saudi Rockwool™. It is primarily used for thermal and acoustical insulation, typically in buildings, vehicles and other industrial equipments. It has a high tensile strength and modulus, high chemical resistance, high dimensional stability, and has excellent insulation properties. The table (4) shows the general properties of rockwool.

Table (4). General properties of rockwool.

Designated use	Loose filler for heaters e, ovens, cookers, and for brake pads
Desired properties	Reduction of transferred heat, and transmitted noise.
Appearance	Mixed black and yellow
Packing	Loose wool supplied in PE bags

2.2. Method

Hand lay-up method was used to prepare samples with mold size of (250mm*250mm). Unsaturated polyester was mixed with epoxy to form a blend. Five different percentages of epoxy and polyester were mixed, starting from (90:10)%, (80:20)%, (70:30)%, (60:40)%, of epoxy and polyester respectively, and reaching (50:50)% of both. The optimum mixing ratio (OMR) was decided upon the results of the impact test carried out on blends of the above mentioned percentages. The (80:20)% has shown the highest impact strength value, hence was selected as a matrix to the composite, which was reinforced with 2 weight fractions: (5)% and (7.5)% of nano clay, then rockwool was added to prepare two other castings with a volume fraction(10%), hybridized with the previously mentioned weight fractions of nano clay. A single casting was made of solely epoxy/polyester blend for the purpose of comparison.

All castings were left for 24 hours for primary solidification, later they were taken out of the mold and were put in a dryer for curing and stress removal. The castings are shown in figure. (1) below.

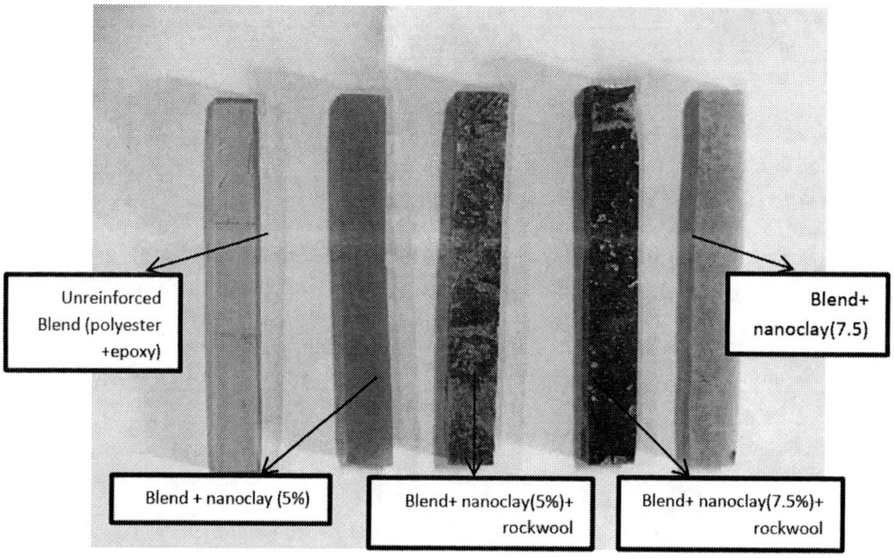

Figure (1). the samples

2.3. The tests

2.3.1 Bending Test (3- Point test)

This test was carried out using the bending set (PHYWE) shown in figure (2), in compliance with the standard specification (ASTM -790). Young's modulus values for the specimens was calculated using the following formula:

$$E = \frac{MgL^3}{48\,IS} \;.....(1\;)$$

Where M: mass (gm), g: Earth's gravitation (9.8 m/s^2), L: distance between the two supports (mm), S: deflection (mm), I: geometrical bending moment (mm^4), and the latter can be calculated using the following equation:

$$I = \frac{bd^3}{12} \;....(\;2)$$

Figure (2). The bending test set

2.3.2 Compression test

Compression test was done using a "Leybold-Harris" hydraulic press, shown in figure (3), in accordance with standard specification (ASTM D-695). The value of compressive strength (UCS) was gained by dividing the maximum load by the cross sectional area of the specimens, and it gives the value of compressive load under which the material fails.

Figure (3). compressive strength test machine

2.3.3 Impact Test Instrument

Charpy test set, manufactured by Amityville Inc.™, had been used to execute the impact strength test for the standard specimens, the set consists of a hammer that is lifted to a highest point, the left to fall freely on the horizontally mounted specimen. As the hammer hits the specimen, it loses its energy and the specimen fractures, the energy of fracture is read off a gauge, and then divided by the cross sectional area to get the impact strength value, according to the following relation

$$I.S = \frac{U_c}{A} \quad \ldots\ldots\ldots\ldots\ldots (3)$$

Where, I.S.: the impact strength (kJ/m^2), Us: the fracture energy (kJ) and A: is the cross sectional area of the sample (m^2). Fig. (4) shows the impact test samples.

.

IOP Conf. Series: Materials Science and Engineering **757** (2020) 012002 doi:10.1088/1757-899X/757/1/012002

Figure (4). Impact test set

3. Results and discussion

3.1 Bending test

From the bending test results, the highest deflection rate was noticed for the unreinforced blend specimen, and the deflection decreased when the blend was reinforced with (5)% and (7.5)%, and the value of Young's modulus reached (746.8)MPa when nanoclay is (7.5%)., Then when hybridized with rockwool (10%) volume fraction, the Young's modulus has further increased significantly to (1887.49) MPa , as show in figure (5).
instead they tended to break apart before the peak,which meant they were brittle. It was because when the nanoclay weight percentage increased, the mixture itself became too viscous, sluggish and more void took place in the samples of high weight fraction. The more the nanoclay added, the more viscous of the clay–resin mixture, leading to a difficulty in even and equal distribution of nanoclay over the whole mixture prior to casting (while still in liquid state). This is the reason for which the higher wt% samples failed. Another reason for failure of higher clay loading is low aspect ratio of clay particles and low contact surface area resulting in weak adhesion between polymer matrix and clay.
furthermore, with high clay loading this behavior was probably attributed to the filler-filler interaction which resulted in agglomerates, induced local stress concentration, and finally reduced mechanical properties of the nanocomposites[6].

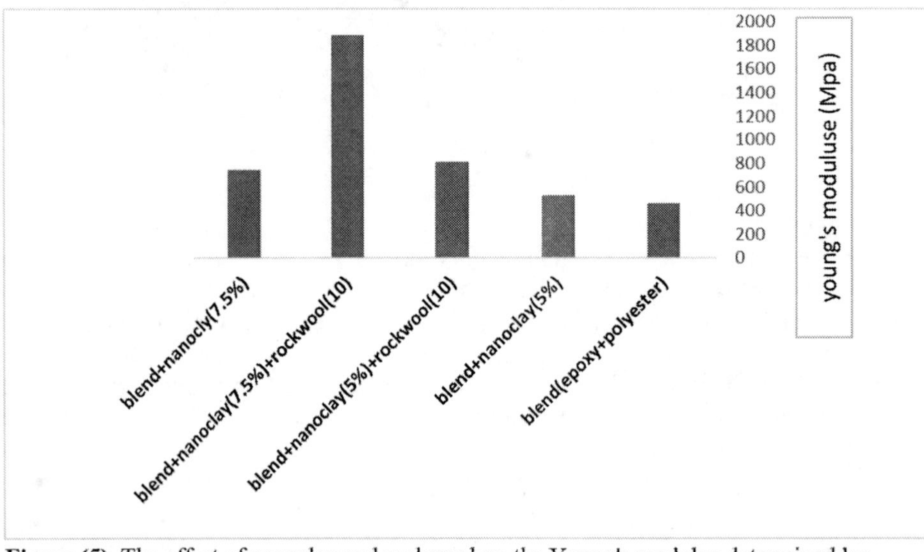

Figure (5). The effect of nanoclay and rockwool on the Young's modulus determined by bending test

3.2 Compressive test

Figure (6) show the effect of (5)% and (7.5)% weight fraction of nanoclay on the compressive strength. It was found that the nanoclay enhanced the compressive modulus of the (epoxy/polyester) blend. However a detrimental effect on the compressive yield strength and failure strain was observed. This shows that the intercalated structure of nanoclay create high localized stresses in the matrix during compression, The results show that the compressive strength for (polyester/epoxy) blend is 210.68 MPa. A slight increase in this value occurred when nanoclay was added with a ratio (5%) weight fraction, The highest value of compressive strength was observed with the hybrid reinforcement of (5%) nanoclay+ (10%) rockwool, to be 396.95 MPa, While reinforcement with (7.5%) weight fraction nanoclay decreased compressive strength of the resulting composite[7]. the compressive strength increased with an increase of the RW(rockwool). The rockwool in the composites acted as a pozzolanic material, giving rise to good mechanical properties, including compressive strength, due to a higher density of the final product. In addition, the specimens containing (10)% rockwool had higher compressive strength that those without rockwool, This might be due to the particle packing effect by the finer RW, which can fill voids [8].

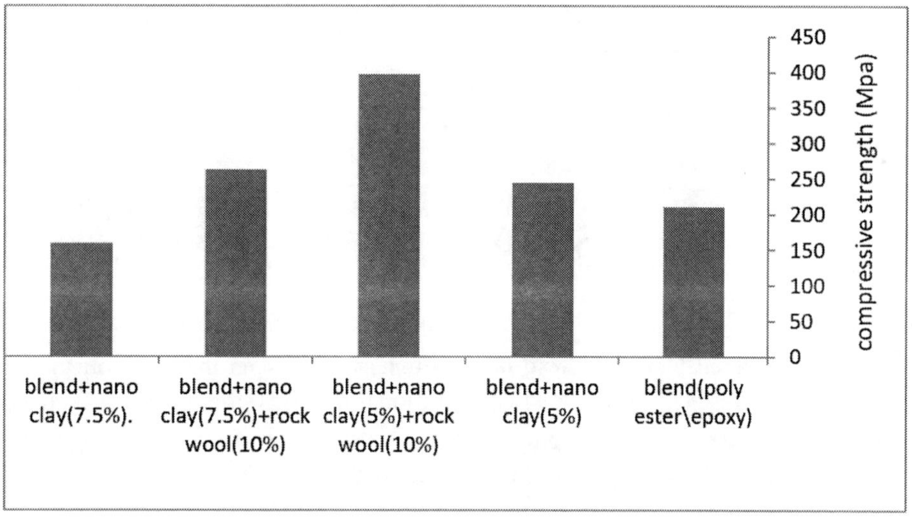

Figure(6). the effect of nanoclay and rockwool on compressive strength

3.3 Impact test

In this test the result show that the highest impact strength was noticed in pure specimen (epoxy/polyester) blend, with a value (0.017) KJ/m^2, the impact strength begin to reduce significantly when reinforce the blend with nanoclay (5%,7.5%) and hybrid with rockwool (10%) volume fraction reached lowest impact strength (0.0055) when the ratio of nanoclay is (5%) weight fraction.

There was a sharp drop of nearly 11.5% in impact strength when (5)% wt.clay was added into the blend, followed by little variation with further increase in the clay content Figure (7).

The failure mechanisms in the impact specimens were somehow different from those of the bending fracture specimens, the specimen with added nanoclay became brittle under the high strain rate loading, leading to nano clay particles acting as stress concentrators in the matrix, which can fracture easily due to the formation of microvoids even before the cracks reached the surrounding matrix material[9]. The fracture in the specimen which contain rockwool fiber is brittle because the rockwool is basically a basalt rocks which brittleness is one of its important characteristics[10].

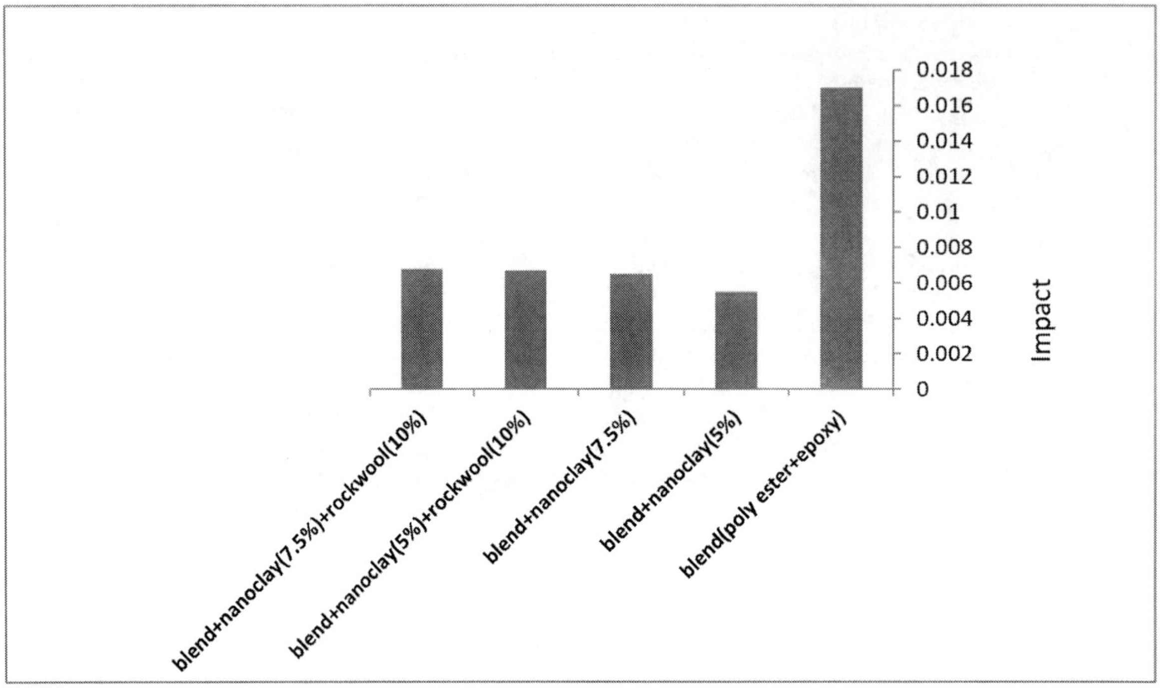

Figure (7). The effect of nano clay and rockwool on the impact strength

3.4 SEM images of the nano composite

SEM technique was used to reveal the microscopic details of the specimens. Figures (8) shows unreinforced blend, with no clear boundaries between the two constituents (epoxy+ unsaturated polyester), even with a magnification down to 50 µm. the image shows a breaking lines along the surface fractography, indicating a brittle fractured surface, along with some voids scattered within the whole volume of the specimen.

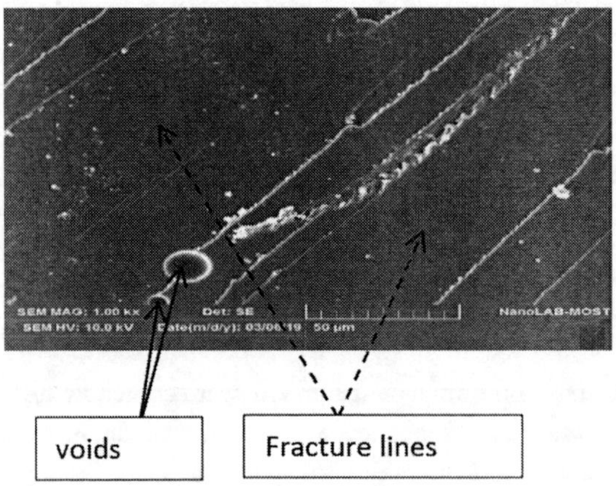

Figures (8). show the unreinforced blend specimen

Figures (9). show the blend reinforced with 5% wt. nano clay (left) and the blend reinforced with 5% wt. nano clay and 10% vol. rockwool (right).

Figures (9) on the left shows the blend reinforced with 5% nano clay as a reinforcement, in this specimen nano clay is agglomerated in small clusters that are scattered throughout the matrix, leaving free space between them (which is free of nano clay). These agglomerates emphasize the nature of nano additions in general, such that even with the use of ultrasonication, the nano clays tend to form clusters, dispersed throughout the matrix. The situation changed a bit after the addition of 10% wt. fraction rockwool to the (5%) nano clay reinforced specimen. Figure (9) on the right, the rockwool fibers are visible even with a relatively low magnification power (100µm), the fibers are used in a random mat shape, which explains the random nature of their dispersion in the matrix (no regular pattern is noticed whether at the macro-, or micro- scale), and the micrographs show small or no plastic deformation to the fractured fibers, since no change in dimensions or necking took place along the fibers.

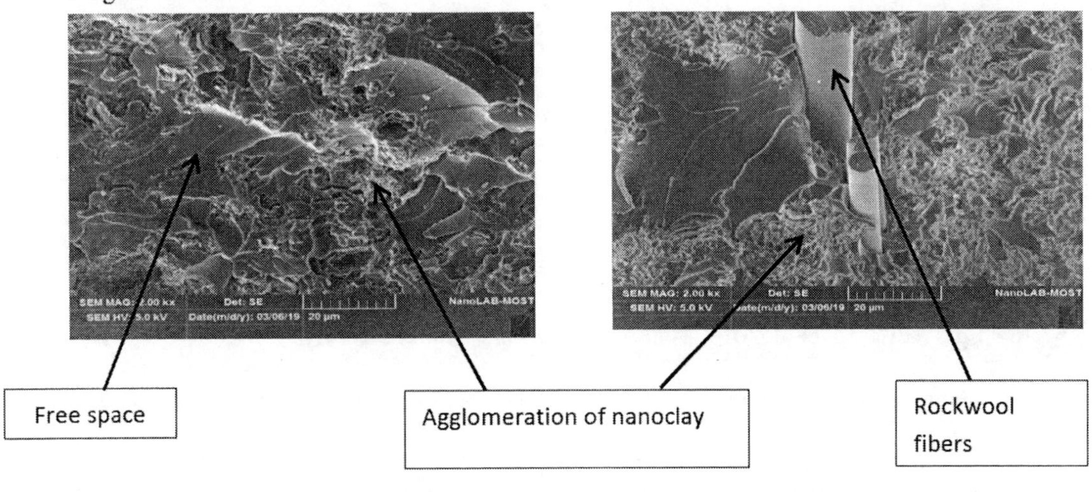

Figures (10). show the blend reinforced with 7.5% wt. nano clay (left) and blend reinforced with 7.5% wt. nanoclay and 10% vol. rockwool (right).

Blend reinforced with 7.5% wt. nanoclay is shown in figures (10) on the left, as with the 5% nano clay reinforced specimen, the 7.5% specimen shows agglomerations of nano clay, co-existing with nano clay-free space in the matrix, with a denser packing of the agglomerates and less free space than that observed in the 5% wt. nano clay specimen. The same figure on the right shows the blend reinforced with 7.5% wt. nano clay and 10% rockwool. Here the surface fractography shows the brittle nature of fracture. Here the fiber cross section is smooth and no necking/ or change in the dimensions of the fibers occurred, beside the flat lines shown on the surface of the nano clay-free volume. Again the nano clays is agglomerated in small gatherings which are scattered though the matrix.

4.Conclusion.

Epoxy and polyester can be mixed in different percentages without visible phase segregation, however, optimum mixing ratio can be decided upon carrying out the impact test, through which the specimen giving the highest value can be chosen to work with. Through the results the highest impact strength was found in the unreinforced blend and the lowest value was for the 5%nanoclay reinforced blend. the result of bending strength proved that the highest deflection was seen in unreinforced blend and the lowest deflection was when reinforced with nanoclay 7.5%wt and rockwool 10% vol fraction. In compressive test the results showed that the addition of nano clay 5%weight fraction, hybridized with 10% vol. fraction rockwool, increased the compressive strength. SEM images showed that there is no interface between the matrix phases so they couldn't be distinguished apart. The surface of fracture was brittle with existence of scattered voids, nano agglomeration and nanoclay-free space.

Reference

[1] Sahraeian, R, Seyed M. D., and Behzad S. H. 2019, *J. Composites Part B: Engineering* **162** 103-111.

[2] Gabr, M H., and Kiyoshi U. 2018, *J. of Thermoplastic Composite Materials* vol.31 **4** 447-464.

[3] M Silverio H., and Sara C. S. de la T. 2017, *Contexto.* **11** 14 63-75.

[4] Rafiq, Ahmad, and Necar M. 2019, *J. of Composite Materials* **53** 2 143-154.

[5] Robledo-Ortíz, J. R., et al. 2019, *J. Polymers and Polymer Composites* **27** 2 43-54.

[6] Chowdary, M. Somaiah, and M. S. R. N. Kumar 2015, *Int. J. Adv. Sci. Techno.* **74** 35-42.

[7] Jumahat, Aidah, et al. 2012, *J. Procedia Engineering* **41** 1607-1613.

[8] Lin, Wei-Ting, et al. 2013, *Materials Characterization* J. **84** 1-9.

[9] Gabr, M H., and Kiyoshi U. 2018, *J. of Thermoplastic Composite Materials* **31** 4 447-464.

[10] Singha, K, 2012, *International J. of Textile Sci.* **1** 4 19-28.

Advanced Techniques In Design Openings Of Smart Facades In Buildings

Abdula Abdulrhman Alsarraf, Maysaa Moffeq Alobaidi

Department of Architecture Engineering, College of Engineering, University of Mosul, Mosul/ Iraq

Abd.88sarraf. @uomosul.edu.iq, maysaa.moffeq @uomosul.edu.iq

Abstract. During the present century, Architecture has concurrence great progress in the field of construction, methods and means of implementation used for the service of architecture, and the use of advanced building techniques is a building with adequate technical and appropriate support to maximize the functions and comfort of occupants with a compromise to reduce operating costs, Where designers used the potential of modern technology to design different parts of the building, including the building's facades as part of the building that achieves environmental balance between the building and the surrounding environment, the research aims to discuss the advanced techniques used by designers in the contemporary interfaces openings in buildings, by building a theoretical framework for global studies in employing the latest technological techniques in architecture and touched on the practical study of computer simulation to handle openings on the facades of building.

Keywords: Advanced techniques, openings, smart facades ,buildings.

1.Introduction

The advanced technology offered enormous solutions and potentials in the development of engineering fields such as architecture. The technology provided advanced solutions that designers and architects have benefited from in finding developed buildings adapted to their environment and creating new forms and ideas for these buildings and thus developing the building covers and their facades. The research discusses the most advanced solutions provided by smart buildings In the design of building openings, as smart buildings one of the most important models of advanced architecture in the twenty-first century.

2. Smart Buildings.

Smart buildings are defined as buildings characterized by their design and information technologies: responsiveness, flexibility, adaptability to changing needs with users and with systems that occupy them throughout their lifetimes, and supports communication between human systems, equipment, built-in communication systems, remote sensing devices and engines..[1] Provides a productive and cost effective environment: maximizing the benefits of its four pillars Basic; Construction, Systems, User Services, Management [2]

3. Characteristics of smart buildings.

The smart building provides efficient productivity, cost effectiveness, internal and natural environment, optimum utilization of structural, management systems and coordination of interrelationships between them, and helps to achieve this set of characteristics: [3]

IOP Conf. Series: Materials Science and Engineering **757** (2020) 012003 doi:10.1088/1757-899X/757/1/012003

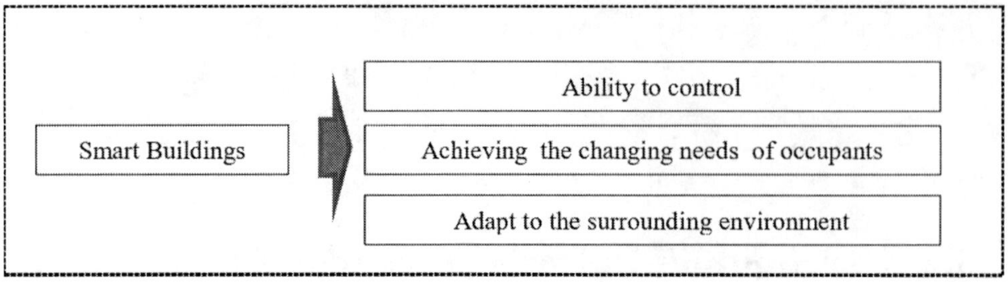

Figure 1. . Smart building (researchers).

1. Automatic control.
2. Achieve the appropriate response to the needs of occupants of the building
3. Adaptation based on changing environmental performance and the needs of occupants of the building [4].

4. Previous studies.

The paper in this paragraph reviews a series of previous studies dealing with smart interfaces and the design of openings in them:

4.1. Study Bugarin and Others,2014

The study discusses a range of technologies that have been used in the construction of smart buildings, which contributed to the reduction in the amount of energy consumed by designing energy efficient glass enclosures[5[

4.2. Study konbr,2016

The study considered that smart systems that can be used in smart buildings are an important entrance to achieve sustainability as they are of environmental and economic importance and achieve environmental balance and provide comfort for the occupants of the building. The study presented several examples of different systems used in buildings[6.[

4.3. Study Omar,2018

The study dealt with intelligent building systems in which the building and its parts can be made intelligent ,Factors related to the design of such systems are eight components of the quality condition. The main factors besides energy, environment and space factors are flexibility, cost effectiveness, customer comfort, work efficiency, safety, culture and technology[7.[

Most of the previous studies examined in intelligent building systems did not examine the design of openings in the smart faced of buildings in detail and the absence of a clear theoretical framework addresses the design of smart openings for buildings

IOP Conf. Series: Materials Science and Engineering **757** (2020) 012003 doi:10.1088/1757-899X/757/1/012003

5. Classification of the techniques used for openings in contemporary facades.

After studying a number of examples, we can classify a number of techniques used in the facades of contemporary buildings such as:

5.1. Dynamic openings/ Facades.

" The building envelope acts like a skin, an intermediate between the interior and the exterior. This skin has evolved over time, with the help of the engineered solutions; the purpose of dynamic facades is to assist in the progress of Sustainable and Responsive architecture. The dynamic facades act as filters between the indoors and the outdoors, facilitating the users providing appropriate shade, sunlight, ventilation and a visual union with the world in motion outside.[11] It is not a new idea, but it is only recently that architects have started utilizing this technology in their structures and design There are a number of types of dynamic facades that we can classify depending on the mechanism of operation and operation and as follows:

5.1.1. User manual control.

In this type, the dynamic building Facades of the building user will be moved manually, where there are no control sensors or the like. It is a simple system, usually a board and a sun louvers that allows the user to control the sun louvers and control the amount of natural light entering the interior of the building space. One example of this system is the Kiefer Technic Showroom in Austria (2007) by Gieselbrecht & Partner figure (2).

Figure 2. the Kiefer Technic Showroom in Austria (2007)

5.1. 2. Light projection by sensors.

In this type, the original facade structure is usually covered with two layers of metal sheets, which are often made of aluminum. The first outer layer of many louvers separates spaces. These louvers are fixed and straight. The second layer is louvers with movable angles. Sensors for natural and artificial l lighting so that their color changes at any time between the night and the day. It also contains artificial light (Variable colors) sensors so that the final product of the person looking at that facade is a multi-wave dance interface. This type of façade is used for aesthetic and special purposes to give illusion. sight For my view of the building, the most prominent example of this type is Multi-use center in Korea by UN-Studio 2010 figure(3).

Figure 3. Multi- use center in Cheonan in Korea (2010)

5.1.3.lens-Light control .

This type integrates the two previous types, although it depends mainly on computer-related sensors, but we can control it manually. This type of lens is similar to the camera lens connected to sensors that control the amount of light input into the space, this type is also a model of environmental control of the building, it also offers a variety gives the building aesthetic of its own, one of the most important examples of this type is Institute du Monde Araba in France, Paris by Jean Nouvel 1987, In this example, we note that the designer borrowed the traditional Arabic decorations and converted them into a number of lenses associated with sensors connected to a group of computers that control the amount of lighting entering the southern facade, adding to the building an aesthetic value as well as an environmental treatment figure(4).

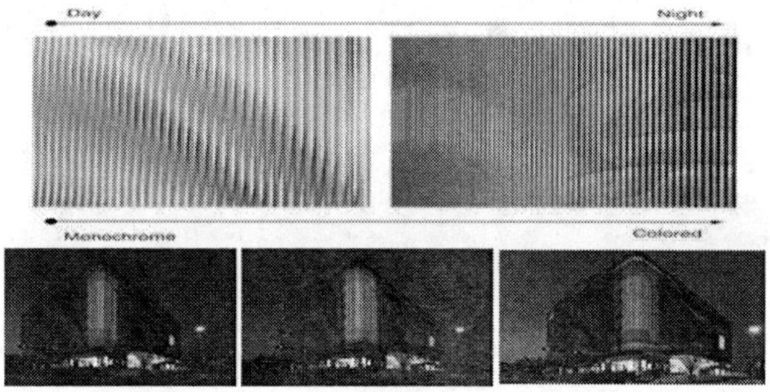

Figure 4. Institute du Monde Araba in France (1987)

5.1.4. shading system of PTFE clad.

It is one of the most modern systems in the packaging of facades. It consists of a layer of PTFE clad, which is connected to a number of temperature sensitive electronic sensors in addition to the amount of carbon emitted from the building outside, which reduces the emission rate of carbon by 40%, was used in many ways Contemporary Towers of the Arabian Gulf Buildings Due to the very sunny weather and high temperatures in this region, one of the most prominent examples of this type is Al Bahar Towers in Abu Dhabi 2011, Where Almstsmon inspired the idea of marshmallows in this building so that the building became more like skin is based on a structure like as Arab Mashrabiya figure(5).

Figure 5. Al Bahar Towers in Abu Dhabi (2011)

5.1.5 Wind sensor system.

The four previous types deal mainly with lighting. In this type, the façade deals with the wind as one of the natural forces that we can invest as a kind of renewable energies. In this type, the facade will be the openings will be small pieces square or rectangular or any shape And a large number to move dynamically with the movement of air in any direction was, these pieces at the same time are sensors that convert this kinetic energy to clean electrical energy used by the building in the process of cheating The most prominent example of this type is Brisbane Terminal Car-park in Australia 2011, The facade of the building consists of 250,000 small square pieces made of aluminum that respond to the movement of the wind. The front of the building is like a piece of wavy moving cloth, giving the integrated (wind) nature between the inside and the outside figure(6).

Figure 6. Brisbane Terminal Car-park in Australia (2011)

5.2. nanotechnologies openings/ Facades.

The term "nanotechnology" is often used as an all-encompassing term for science, engineering, and technology conducted at the Nano scale level- which involves the understanding and control of matter at dimensions between approximately 1 to 100 nanometers (nm). For reference, a single nanometer is one of millionth of millimeter. It would take roughly 50,000 nanometers to span the diameter of an average human hair.[8[

Nanotechnology has been used in many industrial applications, which in turn have led to the construction of buildings. Many nanomaterials have appeared in the installation of buildings such as self-draining glass. There are also nanomaterials that take heat energy from the sun to convert it into electrical energy. Clean The building is provided with the necessary energy used by its occupants. All of these materials can be referred to as smart materials (nanoparticles(. We will discuss in this

IOP Conf. Series: Materials Science and Engineering **757** (2020) 012003 doi:10.1088/1757-899X/757/1/012003

paragraph a number of techniques that are included in the installation of openings for the facades of contemporary buildings:

5.2.1 Geotube system nanotechnology.

This system is one of the most prominent applications on nanotechnology, which is a transparent rod structure works surround the structure of the building to produce a distinctive form of the building's openings, and is made of several layers of nanomaterials, these bars are supplied with saltwater, Of salt-resistant (FRP) nanoparticles (FRP). When salt water passes cold, the salt is deposited on the inner layer. It is negatively ionized by its interaction with the sun that passes through the outer layer containing PV cells and passes inward It aims to give fresh air medicine Inside the building, because the negative air, according to studies, is found in nature when waterfalls and fresh water, which gives the natural air and natural caves, in addition to the introduction of a third layer of PV cells, which provide the building energy through the interaction of sunlight with salt Lying on the inner layer(figure7)[9].

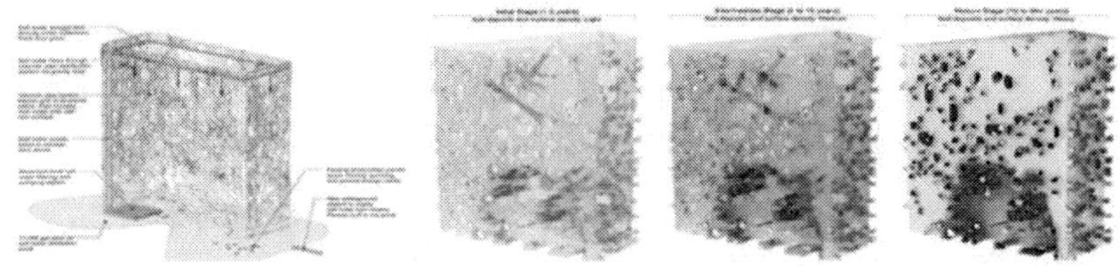

Figure 7. third layer of PV cells

The (GEOTUBE TOWER) by Faulders Studio in Dubai is one of the most prominent examples of this system. The system is used because of the availability of salty Gulf water. This system distinguishes the warm, undesirable air from cool, cool air (figure 8)[10].

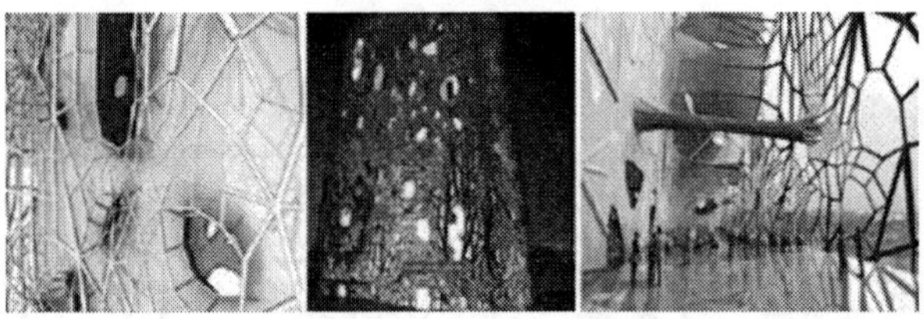

Figure 8. shows the different layers of the structure of Geotube system

5.2.2.Nano_ Ven_Skin (N.V.S) system nanotechnology.

The system is a skin-like plate containing nanostructures that capture sunlight and a nanometer-sized turbine to catch wind. This combined energy is converted into nanomaterials linked to nanowires that transport electrical energy from This interaction to the collective units lies at the end of each plate of these nanoparticles. This nanotube also contains a filter that purifies the air of carbon dioxide so that the air enters purely into the architectural space. Another advantage of this system is the ability to repair itself. If one of the solar cells or one of the solar wind turbines is disrupted, the nanowires will send a signal to the central computer, in turn, to give the matter to the nanomaterials to fix the problem through this wire (figure 9)[11].

Figure9. Shows the composition of the skin used in(N.V.S) system nanotechnology

One of the design companies named (NOS) from Mexico 2008 introduced this system into one of the designs proposed by it. The designer benefited from the beachfront project site (figure 10).

Figure10. Tower proposed by (NOS) in Mexico(2008)

5.2.3 fiber optic rods system nanotechnology.

In this technique, the system consists of a series of long bars, ranging from 3 to 7 meters long and 1-2 meters wide, containing a set of optical fiber nanocompositions, containing the ends of the sensors on the receptor of the light, from outside, these bars are through Sensors at the edges to take the sun and then converted by optical fiber nanoparticles inside to illuminate the interior spaces, at night, the sensors that are located on the inner sides of the space will receive the industrial light of the interior architectural space and transported by the same process of output, but into outer space Which makes the block of the building glow, where the process meets the process of external lighting beautiful night. This technique is usually used in exhibitions, which require an external aesthetic architectural form and the need of interior spaces for natural daylight and industrial outdoor lighting at night. This fiber and along the path of the rods provide photovoltaic sensors that transfer this energy through nanowires to the cooling systems and internal air conditioning of space The most prominent example of this system is Shanghai world expo UK pavilion by Thomas Heatherwick (2010) Where the designer designed this system to meet the needs of the exhibition pavilion of the United Kingdom, using the system of seed cathedrals. In addition, the system in each rod placed a colorful plant seed where light passes through the optical fiber nanoparticles, penetrating the seed and analyzing the colors, which gives a visual diversity of colors a lot (figure 11)[12].

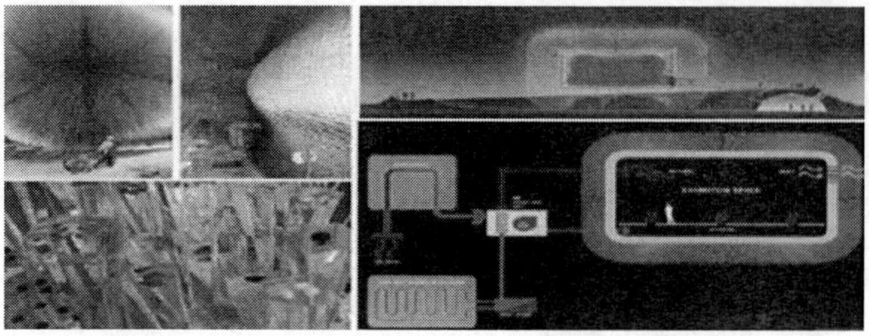

Figure11. world expo UK pavilion in Shangha (2010)

5.2.4.Anti (dust_water) system nanotechnology.

This technique is usually used in towers where it is difficult to access buildings to clean the building openings of dust and plankton, and they are even used in airport control towers where rainwater can obstruct the visibility of these openings. The windows of these openings are usually coated with nanomaterials, for example titanium oxide (TiO2), where they interact with water and dust plankton to react together to form a layer that avoids glass that separates it from it (figure 12)[13].

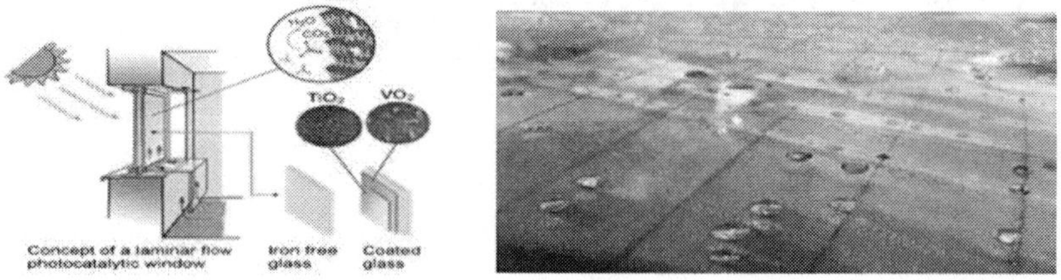

Figure12. Shows the The work of anti (dust_water) system nanotechnology

One of the most prominent examples of this is the proposed airport near the sustainable city of Masdar in Abu Dhabi by Kyosemi Corporation where the sand winds in the region blowing throughout the year, which leads to the difficulty of cleaning the glass in addition to the cost of high-altitude, which led the designer to choose this technology in the openings of the proposed airport, The flight control tower and the proposed roof lighting for the airport space figure (13)[14].

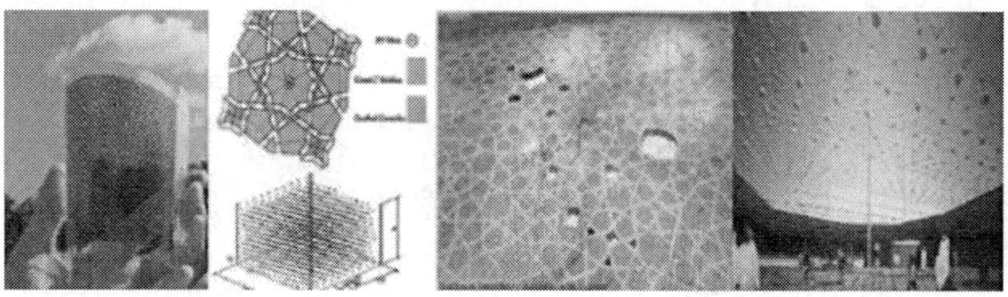

Figure13. airport road near Masdar in Abu Dhabi(2006)

5.2.5. diode (LED) nanopix system nanotechnology

It is a technique based on nano-sized light units spaced 1 μm in background on a reflective coating coated with nano-chemical, which seeks to reduce the area and focus of light produced compared to conventional addition techniques [15].

This technique is used in the facades of buildings and their openings, in order to achieve the aesthetics and the low maintenance costs compared to conventional systems. This technology can be combined with sensors of photoelectric cells so that they light the light of the day and store it and convert it into sustainable electric energy These nanoparticles are one of the most prominent examples of this system being the GreenPix: Zero Energy Media Wall in the Chinese capital of Beijing (2008). The color wall of this technique is like interactive skin. The wall contains photovoltaic cells that store light Confidential and stored to run the screen at night figure(14)[16]

Figure14. GreenPix: Zero Energy Media Wall in Beijing (2008)

6. Theoretical framework.

After classifying the contemporary techniques of smart openings for contemporary buildings and viewing many examples, we can build a theoretical framework described and explain what these openings in terms of the technology used for its work and implementation, as in Table (1):

Table 1. Smart Openings For Contemporary Buildings

Working Mechanism:	Type of technical work	Made materials or compound-made for the purpose of work	techniques used for openings in contemporary facades
physicist	dynamic	Engine louver	User manual control
		Aluminum sheets	Light projection by sensors
		Engine sensor	lens-Light control
			Wind sensor system
	static	Steel louvers fixed	Light projection by sensors
Chemical	Air reaction	salt-resistant (FRP) nanoparticles	Geotube system nanotechnology
		nanotube filter air of carbon dioxide	Nano_ Ven_Skin (N.V.S) system nanotechnology
		skin- nanostructures	Nano_ Ven_Skin (N.V.S) system nanotechnology
		Nanomaterials(titanium oxide (TiO_2))	anti (dust_water) system nanotechnology
	Electrical reaction	nano-sized light	diode (LED) nanopix system nanotechnology
	Interaction between chemicals	optical fiber nano compositions+ sensor Electrical	fiber optic rods system nanotechnology:
Hybrid (chemical + physical)	Electrochemical reaction	solar wind turbines nano structure	Nano_ Ven_Skin (N.V.S) system nanotechnology
		nano-sized light units + reflective coated with nano-chemical	diode (LED) nanopix system nanotechnology
	Aerobic reaction	Chemical sensors	Light projection by sensors
		PTFE clad+ electronic sensors	shading system of PTFE clad

7. Conclusion.

1. The use of modern techniques in the design of contemporary buildings and their various parts, including openings, plays a major role in improving the performance of the building and providing a more interactive and appropriate environment for the occupants of the building, which is one of the important aspects of 21st century architecture.

2. Smart openings for contemporary architectural buildings are those that include one of the smart technologies, whether the physical composition and manufacture, such as dynamic sensors or chemical synthesis and manufacture, such as nanotechnologies.

3. After classifying a number of intelligent technologies for these openings and constructing their theoretical framework, we noticed that most modern intelligent technologies are usually associated with smart materials such as water-based materials such as nanomaterials.

4. The good design of the openings can contribute effectively to increase environmental efficiency, reduce energy consumption, assist designers in the process of architectural design and achieve an aesthetic addition to the facades of their buildings.

5. Through the theoretical framework we observe the technique Nano_ Ven_Skin (N.V.S) system nanotechnology is one of the most classified technologies that fall within the chemical work and agitation that combines chemical and electrical..

IOP Conf. Series: Materials Science and Engineering **757** (2020) 012003 doi:10.1088/1757-899X/757/1/012003

8. References

[1] S Kjeld 2002 *Intelligent Buildings* – a short overview. Alborg University. P . 3.

[2] Seattle Conference. 2001. *Building Smart, Building Green. Intelligent* Buildings Institute (IBI). P3

[3] M Addington & D Schodeck. 2004 *Smart Materials and Technologies for the architecture and design professions*, Architecture Press, an imprint of Elsevier , Linacre House, Jordan Hill, , UK..

[4] I Abiodun,W Wanggen and Chi Z 2011. *Energy Management for Intelligent Buildings, Energy Management Systems*, Dr Giridhar Kini (Ed.).

[5]M Bugarin,S Nizetic and Z Domazet.2014. *Intelligent Glazed Facades Nearly Zero-Energy Buildings,* ttps://www.researchgate.net/publication/305477500.

[6] U konbr , 2016, *Smart Buildings and Sustainability in Egypt - Developing a Concept and Developing a Curriculum.* Journal of engineering sciences 44(4):PP. 472 – 501

[7]O Omar ,2018, *Intelligent building, definitions, factors and evaluation criteria of selection.* Alexandria Engineering Journal (2018) 57, 2903–2910.

[8]I Yasin and D Atiyat, 2017, *The Effect of Nano Technology on Architecture*, Int'l Journal of Advances in Agricultural & Environmental Engg. (IJAAEE) Vol. 4, Issue 1 (2017) ISSN 2349-1523 EISSN 2349-1531.

[9] https://www.faulders-studio.com/

[10] https://www.designboom.com/

[11] P Premier, 2012, *Dynamic Facades And Smart Technologies For Building Envelope Requalification.* ttps://www.researchgate.net/publication/269690115.

[12]https://www.archdaily.com/58591/uk-pavilion-for-shanghai-world-expo-2010-heatherwick-studi

[13]https://www.teknik.uu.se/solid-state-physics+/materials+for+energy-efficiency-and-environmental-applications/#anchor-62983

[14] www.landartgenerator.org

[15]https://www.researchgate.net/publication/224119867_Enhancement_of_light_extraction_in_ultraviolet_light-emitting_diodes_using_nanopixel_contact_design_with_Al_reflector

[16] https://www.archdaily.com/245/greenpix-zero-energy-media-wal

IOP Conf. Series: Materials Science and Engineering **757** (2020) 012004 doi:10.1088/1757-899X/757/1/012004

Study of the Mechanical and Thermal Properties of PMMA/TiO_2 Nano Composite

S. T.Yaseen, A. Sh. Khaleel
Department of Physics Science, College of Sciences, Universityof Baghdad, Baghdad, Iraq.
Corresponding author email: tahasoror@yahoo.com

Abstract:-
This research investigates the effect of adding some reinforcing materials,including (nano and micro) titanium dioxide(TiO_2)particles with different weight fraction(0.1,0.4,0.7,1,3 and5)% to polymethyl methacrylate (PMMA)and studying some mechanical and physical propertiesof composite material. Specimens of this project were prepared using (in-situ) method to form compositematerials. Themechanical tests were performed on these specimens,includingimpact test, thermal conductivity was performed as a physical test.Results of this study showed that the values of impact strength were decreased when adding the (nano-micro) particles in PMMA composite materials, while the values of thermal conductivity weredecreased.
Keywords: PMMA , TiO2, impact test, thermal conductivity.

1.Introduction

Some materials need enhancementin their properties,so this may be achievedby combining two or more materials together to get more developed materials and that's why composite materials were invented[1]. As the world develops,so does the needs of the human body, like a need for replacing an arm , a denture , a tooth and even a valve requires special materials that are compatible and nontoxic to the human skin and human body. Polymer is more preferred in the medical field, because of its properties that are close to a certain degree to the human body. So it became the matrix for most of the medical field materials, with the addition of reinforcements for strength [2]. Each material can havegood properties, which make them suitable for biomedical applications [3].Acrylic resins are the most widely used in dentistry applications and accepted among all denture base materials, and it was estimated that they represent (95%) of the polymer in prosthodontics applications. Poly (methyl methacrylate) is favored for its good properties such as satisfactory mechanical properties, except the impact and fatigue strength, dimensional stability, and it has high Tg temperature, the color whichis similar to the human's internal skin.

PMMA can be fabricated and processed by many methods such as casting, extrusion, injection and thermoforming method [4]. Poly methyl methacrylate (PMMA)is the most widely used of the polymers based on acrylic acid for which one trade name is "Perspex" [5].Polymethylmethacrylateis an amorphous thermoplastic that can be delivered as clear as glass. The material is more known as "Plexiglass". Itis produced by the addition polymerization of methyl methacrylate according to the reaction illustrated in figure (1) [5, 6].The monomer, methylmethacrylate is obtained from acetone cyanhydrin by reacting it with sulphuric acid and methyl alcohol.

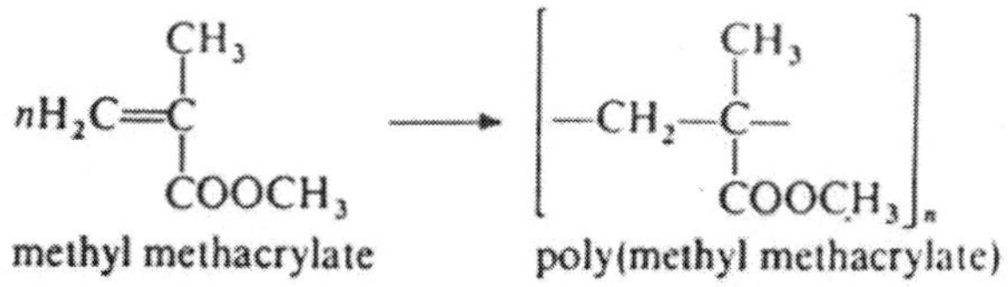

Figure1. Polymethylmethacrylate polymerization [7].

2.Titanium Dioxide(TiO_2)

Titanium dioxide also known as titanium(IV) oxide or Titania, is the naturally occurring oxide or titanium, with chemical formula TiO_2. When used as a pigment, it is called titanium white, Pigment white 6, or CI 77891. It is noteworthy for its wide range of applications, from paint to sunscreen to food colouring when it is given the E number E171. Titanium dioxidis found naturally as the mineral rutile, anatase and brookite. Rutile, anatase and brookite all contain six coordinates titanium. Titanium dioxide is the most widely used white pigment, because of its brightness and very high refractive index (n=7), in which it is surpassed by a few other materials[8].

3. Experimental Work

3.1 Matrix Material

Poly methyl meth acrylate(self cure) has been used as a matrix material in the preparation of composite materials and manufactured by(New Stetic S.A. Cr.53 No. 50-09 Guarne, Antioquia-Colombia)This resin is pink veined color at room temperature and turns into a solid state by adding monomer, which is a transparent, colorless liquid added inthe ratio of 1: 2, ie, 1 g of monomer to 2 g of resin polymethyl methacrylate. This type of materials is distinguished by softer feel, low molecular weight, color stable in the long run, minimized shrinkage, stable polymerization cycle with a perfect end result, the acrylic is long pourable and model able for a long period of time [9].

3.2 Reinforcement Materials

Titanium dioxide nano and microparticlesmanu factured by (Cheng Du Micxy Chemical CO. Lid) were used in this research as additives,with a high purity about (99.9%). The material was tested by an atomic force microscope(AFM) carried out in Baghdad University laboratories. This test is known to give surface shape as well as for the particle size.The particles size of TiO_2 (nano) equal 53.79 nm.

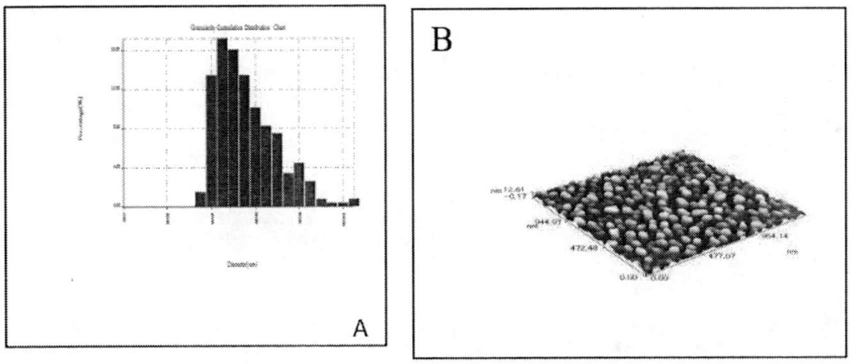

Figure 2.A) AFM analysis image of TiO_2 (nano) (3D),B)Distribution of TiO_2 (nano)

IOP Conf. Series: Materials Science and Engineering **757** (2020) 012004 doi:10.1088/1757-899X/757/1/012004

SPM
Scanning Probe Microscope

Granularity Cumulation Distribution Report

Sample:tio2	**Code:**Sample Code
Line No.:lineno	**Grain No.:**268
Instrument:CSPM	**Date:**2019-01-05

Avg. Diameter:53.79 nm	**<=10% Diameter:35.00 nm**
<=50% Diameter:50.00 nm	**<=90% Diameter:70.00 nm**

Diameter(nm)<	Volume (%)	Cumulation(%)	Diameter(nm)<	Volume (%)	Cumulation(%)	Diameter(nm)<	Volume (%)	Cumulation(%)
35.00	1.49	1.49	60.00	10.07	71.64	85.00	2.61	97.76
40.00	13.43	14.93	65.00	8.21	79.85	90.00	0.75	98.51
45.00	17.16	32.09	70.00	7.46	87.31	95.00	0.37	98.88
50.00	16.04	48.13	75.00	3.36	90.67	100.00	0.37	99.25
55.00	13.43	61.57	80.00	4.48	95.15	105.00	0.75	100.00

3.3 Specimens Preparation

The PMMA denture base materials consistsof polymerpowder and monomer liquid (methyl methacrylate,PMMA).

To prepare a sample of pure resin, a certain quantity of resin (poly methyl meth acrylate) was weighed according to the size of the mold designed at 2: 1 ratio (i.e. 1ml of monomer liquid per 2g ofpowder resin at its temperature). After that, they were mixed for 8 minutes to be homogenized mix and it is preferred to exceed this period of time because it will lead a higher viscosity , which leads to the acceleration of the process of hardening, impeding the process of casting in the mold.The weight of different compound of reinforcement (particles of titanium dioxide(nano and micro) and according to the required weight fraction. To prepare the composite samples, cold cure acrylic resin is moldable for a long period of time, where the mixture was mixed of liquid (MMA)and TiO_2 (micro or nano) particles in the clean and dry container (glass beaker), follow after that by slow addition of dry powder (PMMA) to liquid (MMA) and TiO_2 powder mixture , the final mixture was stirred at room temperature and poured with thin straight line in the center of opening glass moulds ,for curing . The mixture is poured in the mold shown in the figure (1-3)(according to ASTM-D256-87)for impact testand thermal conductivity test samples with diameter (4cm)so that the mixture is distributed to all the areas of the template continuously and add it until the mold is filled with the desired form.The castings of composite specimens after curing process stand at room temperature about (30 min) to complete the cooling and hardening of specimen. After cooling, the specimens was de-mould to remove from the mould carefully and cleaning.

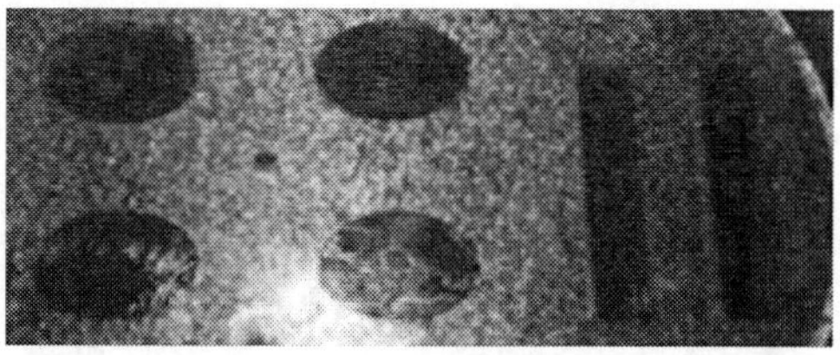

Figure 3. The mold used to prepare test samples

Table 1. Some Mechanical and Physical Properties of Self Cure PMMA resin.

Young's Modulus (GPa)	Impact Strength (KJ/m^2)	Flexural Strength (MPa)	Water Absorption (%)	Density (g/cm^3)
1.63-3	8.3	79	2.5	1.19

4. Mechanical Testing

4.1 Impact strength

In Charpy test, the specimen is supported at each end and struck at the center. A large number of experimental methods have been proposed and adopted to determining the fracture of brittle and plastic fracture [10]. Impact test which is measure the fracture energy of material under the effect of stresses at a high velocity and this is the important parameter to design any structure.A material such as polymer, which has a high toughness, has high fracture energy and it is calculated using this equation:-

$$I.S = \frac{E}{A}$$

Where I.S: impact strength in unit(kg/m^2) .
E:Energy of fracture in unit (kg).
A: Cross section area in unit(m^2).
The impact strength depends on the type of material, stresses system; manufacture condition, and the temperature [11].

5.Thermal Testing

5.1 Thermal conductivity

The thermal conductivity is the quantity of heat flow in unit time through a unit area of substance caused by a unit thermal gradient [12]. Lee's disk were used to calculate the coefficient of thermal conductivity (K) for the prepared samples [13].

6. Results and discussion

This section deals with the results obtained by impact and thermal tests for the prepared samples of PMMA as a matrix and TiO_2(micro and nano) asreinforcements with(0.1,0.4,0.7,1,3and 5) %wt.

IOP Conf. Series: Materials Science and Engineering **757** (2020) 012004 doi:10.1088/1757-899X/757/1/012004

6.1 Impact Test

Figure(1-4)shows the obtained results for impact test, Impact strength valuesfor PMMA composite specimens for prosthetic denture base materials prepared in this research are shown in the table (1), and figure (4 and 5) shows the effect of adding both types of titanium dioxide particles (nano-micro) with different weight fractions on the impact strength of PMMA matrix. This figure how the impact strength values change with increasing the weight fraction for both types of these particles in both groups of PMMA composite materials, because of any increasing in these particles numbers, it will act as points for localized stress concentration regions from which the failure will begin. Furthermore, the nature of these particles is brittle and weak in the ability of resistance to impact loads comparing with PMMA resin.Impact strength of the composite specimen is controlled by two elements: first, the capability of the reinforcing material to stop crack propagation by absorbing energy and the second one, poorbonding between reinforcing and matrix, which cause micro-spaces and result in crack propagation [14].

The energy needed to break the pure polyethyl methacrylate was 4.70 kJ / m^2, but when adding the micro particles and nano particles of titanium dioxide, the energy needed to break the composite samples increased and was higher than the impact of the addition of nanoparticles at the addition of 0.1% and reached 10.915 kg/ m^2 was added to the micro was the highest value of the impact of the addition of the ratio of 0.4%. It reached 11.83 kg / m^2 and when adding other ratios start to decline because of the existence of the interfaces and agglomeration between the matrix material polyethyl methacrylate and the material added titanium dioxide (nano and micro).

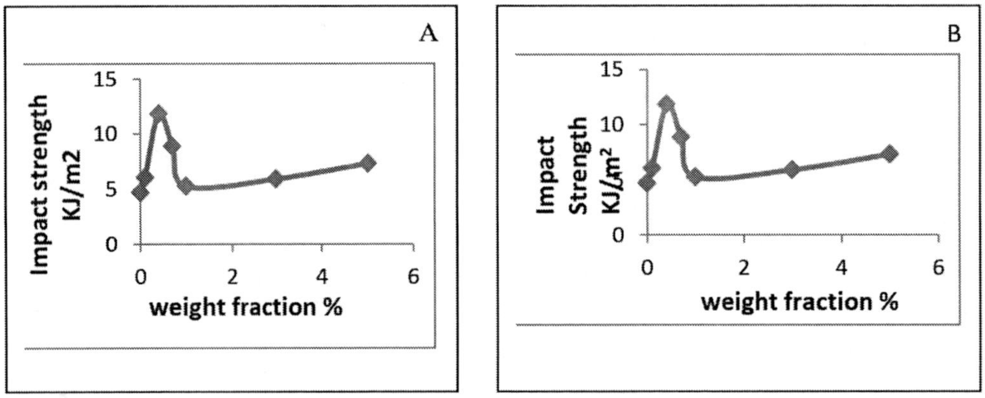

Figure 4. The relationship between impact strength and weight percentage of titanium dioxide(TiO_2) when added:A)nano of TiO_2.B)micro of TiO_2.

IOP Conf. Series: Materials Science and Engineering **757** (2020) 012004 doi:10.1088/1757-899X/757/1/012004

Table 2. Values Impact test

Wt %	Impact Strength KJ/m₂	
	PMMA/MicroTiO$_2$	**PMMA/Nano TiO$_2$**
0	4.70	4.70
0.1	6.05	10.915
0.4	11.83	5.231
0.7	8.860	5.347
1	5.251	7.334
3	5.912	5.889
5	7.324	7.332

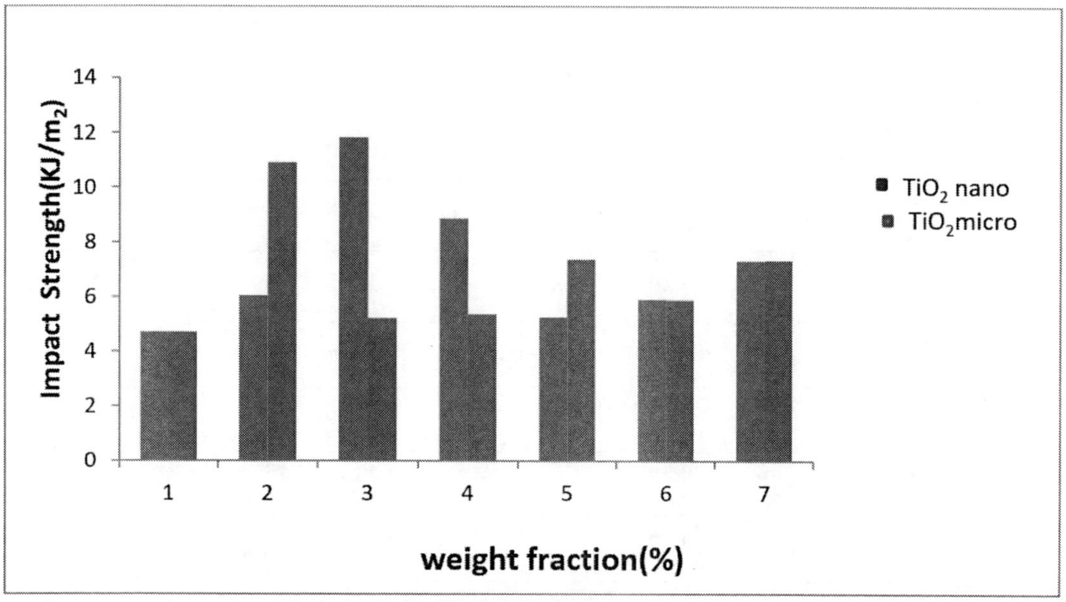

Figure 5. Impact Strength as a function of weight fraction

6.2 Thermal Conductivity

Thermal conductivity is a heat transfer phenomena in which the energy is transferred from one location to another because of the excitation of atoms or particles as a result of the change in the temperature of the medium.

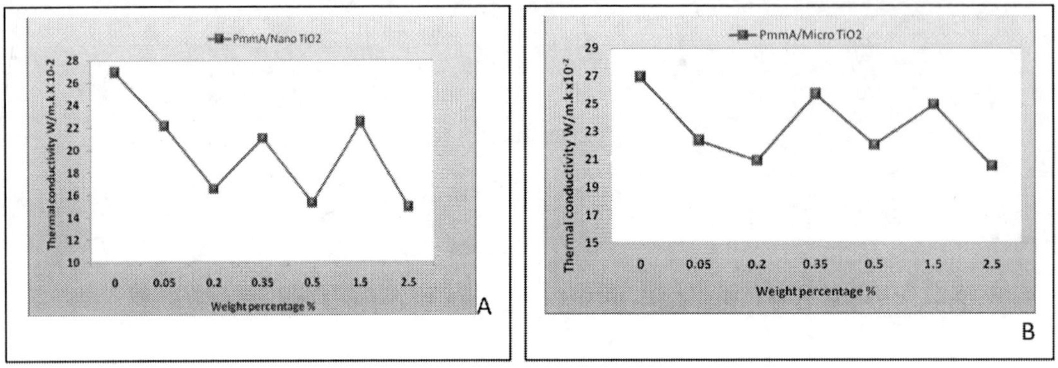

Figure 6. The relationship betweenthermal conductivity and weight percentage of titanium dioxide(TiO_2) when added:A)nano of TiO_2.B)micro of TiO_2.

Figures (1-6), shows the effect of adding (nano-micro) TiO_2particles with different weight fractions on the thermal conductivity of PMMA composite.

In Figure (6 and 7) it can be noticed the thermal conductivity values of PMMAcomposite specimens decrease with increasing the volume fraction of (nano-micro) TiO_2 particles. In addition, the presence of these particles is made to fill or diminish the spaces and voids which were inside the PMMA matrix. Finally, the result leads to facilitate the process of the increase of the thermal conductivity values of the composite specimens [15]. Thus, the thermal conductivity values decrease from (0.2692 W/m.°K) for (PMMA-pure) composite specimen to reach to the higher value of thermal conductivity (0.2248 W/m.°K) (PMMA- 3% nano-TiO_2) and(0.2568W/m.°k)(PMMA- 0.7%micro-TiO_2).

Table 3. Values of thermal conductivity

Wt %	**Thermal Conductivity W/m.k**	
	PMMA/Micro TiO_2	**PMMA/Nano TiO_2**
0	**0.2692**	**0.2692**
0.1	**0.2234**	**0.2210**
0.4	**0.2088**	**0.1653**
0.7	**0.2568**	**0.2104**
1	**0.222**	**0.1530**
3	**0.2489**	**0.2248**
5	**0.2049**	**0.1492**

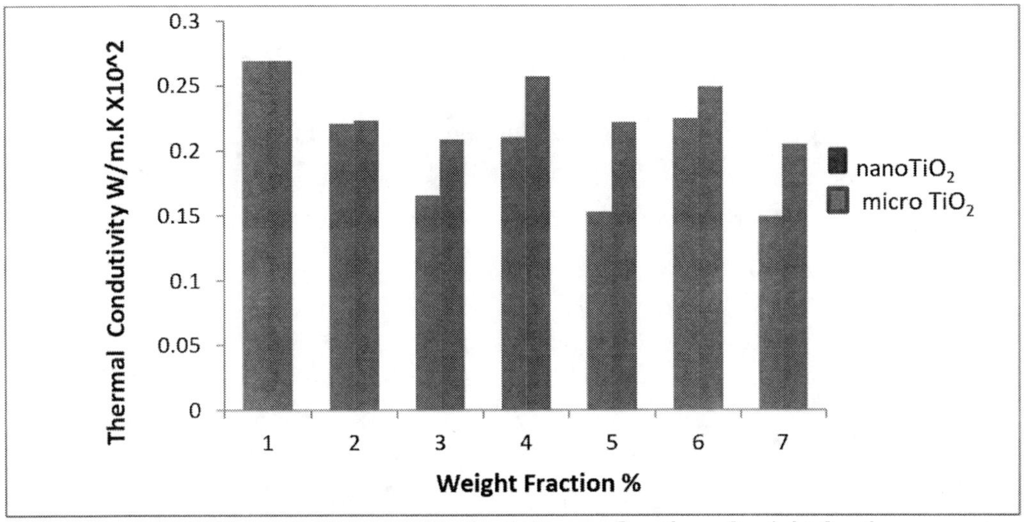

Figure 7. Thermal Conductivity as a function of weight fraction.

7. Conclusions

According to the experimental results of PMMA denture base materials, which prepared in this research, canbe the conclusions the following sentences: The addition of (nano- micro) TiO2 particles has a noticeable effect on most mechanical and physical properties of PMMA denture base materials with increasing the volume fractions of (nano- micro) particles. Some physical properties such as (thermal conductivity) of PMMA composites denture (PMMA - nano TiO2), (PMMA - micro TiO2), were increased with increasing of the volume fractions of (nano-TiO_2 and micro-TiO2) particles. Some mechanical properties such as (impact strength) of PMMA composites denture (PMMA - nano TiO_2), (PMMA - micro TiO_2), decreased with increasing of the volume fraction of (nano-micro) TiO_2 particles.

References

[1] S.L. Kakani 2004 1st ed., Age International Publishers, Ltd., London,.
[2] L.L. Hench 2002 Biomaterials 19:1423-1429.
[3] S. Ramakrishna, J. Mayer, E. Wintermantel, K.W. Leong 2001 Composites Science and Technology **61**:1189- 1224,.
[4] Q.A. Hamad 2017 Eng. & Tech. Journal, Part A **35** 2:124-129.
[5] John, V.B. 1974 The Macmillan Press, Ltd,.

[6] Dug lukKassen, and Annette Midell 2003 Advanced Materials and Structures and their Fabrication Process", 3rd. ed., narvikUniversity College Hin,.

[7] Sheldon, R.P. 1982 Applied Science Pub., London.
[8] F. Rodai, K. Hiroharu, G. Harrison, W. Pamiko 2004 J. Thin Solid Films **454**:162.

[9] KennethShaw, London,.

[10] Grawford, R. J. 1987 2nd, Edition, A. Weaton U. K,.

[11] C.Kittle 2005 Eight Edition, Professor Emeritus, University of California, Berkeley, John Wiley&Sons, Inc,.

[12] Johnson, D. A., Maclen, W. D., and Jacobson, R. 1997 Making & Business from Biomass in Energy, Environment, Chemicals Fiber and Material P.925 www.agroplastics.com/biomass.pdf.

[13] JayatiIaka, A. D. 1979 Applied Science London.

[14] M. Sumaila, I. Amber, and M. Bawa 2013 Asian Journal of Natural & Applied Sciences **2** 1: 39-49,.

[15]S.M. Elie 2007 M.Sc. Thesis, University of Technology, Baghdad, Iraq,.

Influence of Aluminum Fine Powder Content on the Some Mechanical Properties of Epoxy and Polyester Composites

Khadijh S Mohan[1], Noor M Awad[2], Wasan A Hussein[3] and Ahmed J Farhan[4]

[1]University of Al Zahraa for women.
[2]Ministry of Interior.
[3,4]Department of Physics, College of Science, University of Wasit, Wasit. Iraq.
kutpc19@yahoo.com

Abstract: In this study, the effect of aluminum powders on the some mechanical properties of epoxy (EP) and unsaturated polyester (UPE) composites was investigated. Aluminum powders (Al) were added to polymers with (0, 1, 3 and 5) %weight percentages. epoxy /Aluminum and unsaturated polyester /Aluminum composites were fabricated by the cast molding method. Maximum fracture forces, Maximum Extensions, Yield Points, Young Modulus, Tensile Strengths, flexural strength and ultimate compressive strength tests were performed on both composites. The results showed that the tensile strength of epoxy /Aluminum composites increased with increasing of additive percentage of Al powder, and the samples of EP/Al composite gained highest values of tensile strength (22.48 MPa) at ratio percent(5%). While the tensile strength of unsaturated polyester /Aluminum composites decreased with increasing of additive percentage of Al powder. Also the results showed that the samples of EP/Al composite gained highest values of flexural strength (58.63MPa) at weight percentage (1%).While the samples of UPE/Al composite gained highest values of flexural strength (91.88 MPa) at weight Percentage (3%). Finally, the value of compressive strength decreased with increasing of additive percentage of Al powder in both of EP/Al and UPE/Al composites.

Keywords:, tensile strength, flexural strength, ultimate compressive strength and aluminum powder .

1.Introduction

Polymer composites generally consist of two primary ingredients: Organic resin matrix- the organic resin matrix is High molecular weight monomer as epoxy resin, or polyester, which has wide applications due to its properties and Inorganic fillers particles as a reinforcement materials which vary from material to another; each on has its own distinctive characteristics, the particles can be either metallic or nonmetallic. Particulate composites consist of particles of one or more materials

IOP Conf. Series: Materials Science and Engineering **757** (2020) 012005 doi:10.1088/1757-899X/757/1/012005

suspended in a matrix of another material's particles, which can be either metallic or nonmetallic as can the matrix [1]. The physical properties of the resin are affected by the weight percentage of filler in the resin composite, within particle limits. Percentage filler content of a resin composite may be expressed in terms of weight or volume fractions or percentage [2]. The hard particles in a particulate composite place a constraint on the plastic deformation of the matrix material between them, because of their inherent hardness relative to the matrix. The particles also share the load, but a much smaller one, which lies parallel to the direction of load. Thus, these particles are effective in enhancing the stiffness of the composite, but do not offer the potential for much strengthening. Hard particles, placed in brittle matrix, reduce the strength due to stress concentrations [3]. Many particle types of fillers are used to improve the other properties of matrix materials such as mechanical, thermal and electrical conductivity. For example Al2O3 or alumina is one of the most versatile of refractory ceramic oxides and used in a wide range of applications [4, 5]. Epoxy polymers are widely employed as the basis for adhesive compositions, and as the matrix material for composites [6]. Such polymers possess many useful engineering properties, such as a relatively high modulus and strength, good creep resistance and good elevated-temperature properties. However, such epoxies are relatively brittle polymers with poor resistance to crack propagation [7]. Polyesters are one of the most versatile synthetic copolymers. Polyesters are produced in high volume that exceeds 30 billion pounds a year worldwide [8-10]. They are widely used commercially as fibers, plastics, composites and coatings applications, too [11-13].

Alaa A. Abdul-Hamead et al. [14] studied composite material properties of polyester material reinforced with iron weave wires with percentage (5,10,15 and 20)%. And concluded that most physical properties have been improved after reinforcement by metals the value of mechanical properties will increase with increasing percentage of reinforcement. Sihama Issa Salih et al. [15] studied composite material properties of unsaturated polyester with metal powders. The unsaturated polyester resin was reinforced by two types of metal powders Copper (15.598 μm) and aluminum (21.533μm) at selected volume fraction of (0, 5, 10. 15, 20 and 25 %). And concluded that most physical properties include tensile properties (strength at fracture point, elastic modulus and elongation at at break), flexural strength and flexural modulus, impact properties and hardness have been improved and the values increased with increasing percentage of reinforcement.

Srivastava V. and Ansul V. [16] studied the effect of copper and aluminium particles on the mechanical properties of epoxy resin. Copper (Cu) and aluminium (Al) particles were added in the epoxy resin as filler with the variation of weight percentage (1%, 5%, 8% and 10%). The tensile strength, compressive strength, vicker's hardness, friction coefficient and wear properties were evaluated and compared. The fracture behavior was investigated under scanning electron microscope. Experimental data from tensile tests were compared with Bigg's equations. The results indicate that the tensile strength and wear loss gradually reduced with the increase of filler content. The hardness, compressive strength and coefficient of friction increased with an increase of weight percentage of Cu and Al fillers content.

A major aim of the this work was studied the effect of Al micro particle content on the mechanical properties of epoxy and polyester resins includes tensile properties (Maximum fracture forces, Maximum Extensions, Yield Points, Young Modulus and Tensile Strengths), flexural strength, flexural modulus and ultimate compressive strength of epoxy/aluminum and unsaturated polyester/aluminum micro composites.

2. Experimental Procedure

2.1. Materials

Unsaturated polyester resin (UPE) is a liquid with moderate viscosity which can be cured to the solid state by adding (Methyle Ethyle Keton Peroxide, MEKP) as a hardener, while cobalt octoate acts as a

IOP Conf. Series: Materials Science and Engineering **757** (2020) 012005 doi:10.1088/1757-899X/757/1/012005

catalyst to accelerate the solidification process. Epoxy resin (Sikadur-52) was used in this research; which characterized by a low viscosity and free flowing; ideally suited to a wide range of building and civil engineering applications where highly penetrative material is required. Aluminum fine powders (Strem Chemicals, USA) was used in this research.

2.2. Material Preparation

2.2.1. Preparation of Polymer Microcomposites

Epoxy resin was mixed with weight fractions 0, 1, 3, and 5% percentages of aluminum is shown in table 1. then stirred for approximately 1 h, the hardener was added with 2:1 ratio of (epoxy: hardener), then poured into a Teflon mold via cast molding method. The mixture was de-gassed in a vacuum oven for 30 min and cured at 50°C for 3hrs, the unsaturated polyester resin was prepared with weight fractions (0, 1, 3, and 5)% percentages of aluminum. Unsaturated polyester/aluminum microcomposites were prepared by the same manner of preparition of Epoxy /aluminum microcomposites.

Table 1. Materials ratios

Code	Composition	Matrix Polymer(gm)	Reinforcement Al (gm)
A0	EP	100%	0%
A1	EP/Al	99%	1%
A2	EP/Al	97%	3%
A3	EP/Al	95%	5%
B0	UPE	100%	0%
B1	UPE/Al	99%	1%
B2	UPE/Al	97%	3%
B3	UPE/Al	95%	5%

3. Measurements

3.1. Tensile Test

Tensile testing is one of the more basic tests to determine stress – strain relationships. A Tinus Olsen machine model H50KT was used for this test which was carried out a crosshead speed of 5 mm / min. A simple uniaxial test consists of slowly pulling a sample of material in tension until it breaks. Test specimens for tensile testing are generally either circular or rectangular with larger ends to facilitate gripping the sample (ASTM D-638 for plastics). Figure 1. shows the tensile test specimens to carry the test. The typical testing procedure is to deform or "stretch" the material at a constant speed. The required load that must be applied to achieve this displacement will vary as the test proceeds. During testing, the stress in the sample can be calculated at any time by dividing the load over the cross-sectional area $\sigma = P/A$. The displacement in the sample can be measured at any section where the cross-sectional area is constant and the strain calculated by taking this change in length and dividing it by the original or initial length $\varepsilon = \Delta L/L_0$ [17].

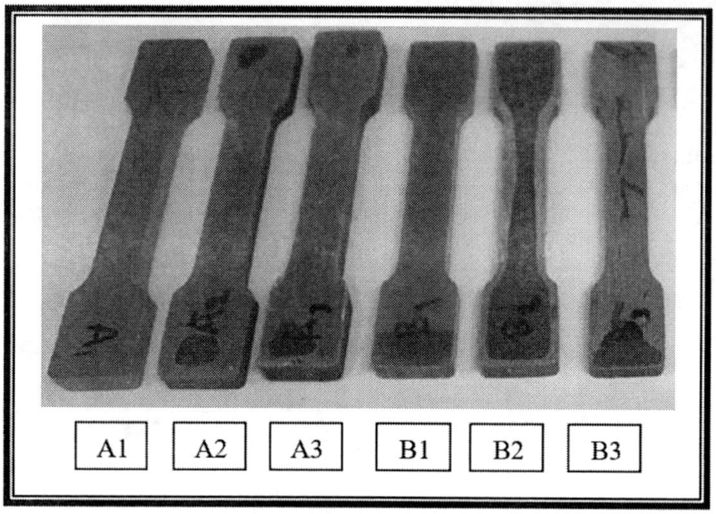

Figure 1. Tensile test specimens.

3.2. Three – Point Bending Test

Three-point bending tests were carried out using a Tinus Olsen machine model H50KT of (5kN) full scale load capacity according to the ASTM D790-10 standard. The test specimen dimensions were (125 mm) long, (96 mm) support span and (10 mm) width. and the test speed was 0.5 mm/min. Flexural strength, flexural modulus, and elongation at break values were obtained and evaluated. Flexural stress was calculated according to Eq.

(1): $\sigma = 3FL / 2\,bd^2$, (1)

where σ is the flexural stress [MPa], F is the load [N], L is the span [mm], b is the specimen width [mm] and d is the specimen thickness [mm].

Flexural modulus was calculated according to Eq.

(2): $E_B = L^3 m / 4bd^3$, (2)

where EB is the flexural modulus [MPa], L is the span [mm], b is the specimen width [mm], d is the specimen thickness [mm], and m is the slope of the linear region of the load-displacement curve.

Bending strain was calculated according to Eq.

(3): $\varepsilon = 6Dd / L^2$, (3)

where ε is the strain [mm/mm], D is the maximum displacement at the central point of the specimen, d is the specimen thickness [mm], and L is the span [mm]. [18]

3.3. Compression Test.

The compression test was done to calculate the compression strength for the samples under test. Hydraulic- piston type Ley Bold Harris No.36110, was used. In this instrument, the sample is fixed on a moving base, then the base is lifted up by a lever found in the instrument till the surface of the sample touches the upper surface of the instrument, then the pointer of the gauge is placed on zero level. The load is applied gradually to the longitudinally fixed sample, and then the reduction in the length of sample is determined via the fixed digital vernier, the increasing of the load continues till sample failure. The value of the maximum load represents the ultimate compression strength of the sample. Figure 2. shows the compression test specimens to carry the test.

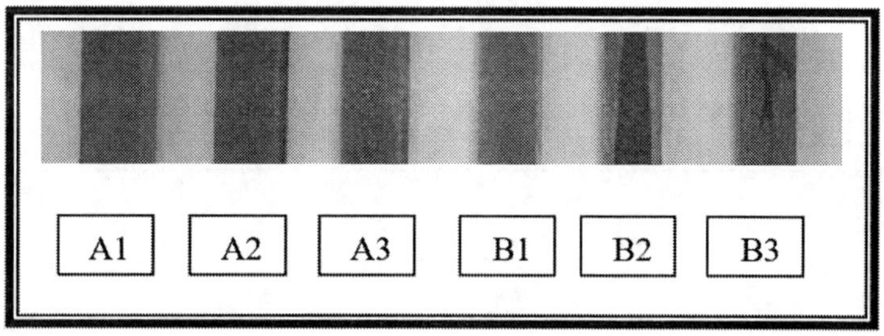

Figure 2. Compression test specimens.

4. Results and discussion

4.1. Tensile Test

The values of tensile strengths, maximum fracture forces, maximum Extensions and yield points for epoxy specimen and (EP/Al) micro composites of all samples are summarized in table 2. The relationship between the tensile strength and aluminum content of (EP/Al) micro composites is shown in Figure 3. It is clear that the tensile strengths of micro aluminum composites increased with increasing Al content ratio. Maximum tensile strength was observed at 5 wt. % of Al (percentage), which is 22.48 MPa. Tensile strength of the polymer composite is also depending upon strength at the interface of matrix and filler material. The enhancement of the tensile strength of the micro composites was due to the homogeneous dispersion of the reinforcement material and the good interaction between the reinforcement and the matrix.

Table 2. Maximum fracture forces, Maximum Extensions, Yield Points, Young Modulus and Tensile Strengths of epoxy and EP/ Al composites.

Sample	Maximum Fracture Force (N)	Maximum Extensions (mm)	Yield Points (MPa)	Young Modulus (MPa)	Tensile Strengths (MPa)
EP	1087.366	7.4	20.28	135.2	20.28
EP / 1% Al	1191	7	21.98	157	21.98
EP / 3% Al	987.5	4.5	20.65	229.4	20.65
EP / 5% Al	1135	4	22.48	281	22.48

Also the Maximum fracture forces, Maximum Extensions, Yield Points, Young Modulus and Tensile Strengths of unsaturated polyester and UPE/ Al composites shown in table 3. It is clear from the results that the tensile strengths of micro-aluminum composites decreased with increasing Al content ratios. This could be that as the increasing Al content caused the weak adhesion between matrix and fillers and led to the decrease of the strength of the unsaturated polyester composite. The tensile test for specimens of EP/Al and UPE/Al micro composites with 1, 3 and 5% of aluminum powder showed that all specimens with Al micro particles suffer brittle fracture during the test course is shown in Figure 4.

Table 3. Maximum fracture forces, Maximum Extensions, Yield Points, Young Modulus and Tensile Strengths of unsaturated polyester and UPE/ Al composites.

Sample	Maximum Fracture Force (N)	Maximum Extensions (mm)	Yield Points (MPa)	Young Modulus (MPa)	Tensile Strengths (MPa)
UPE	2252.991	8.5	33.05	197.06	33.05
UPE / 1% Al	602.5	5	11.82	118.2	11.82
UPE / 3% Al	567.5	2	12.43	310.75	12.43
UPE / 5% Al	542.5	4	12.45	156.75	12.45

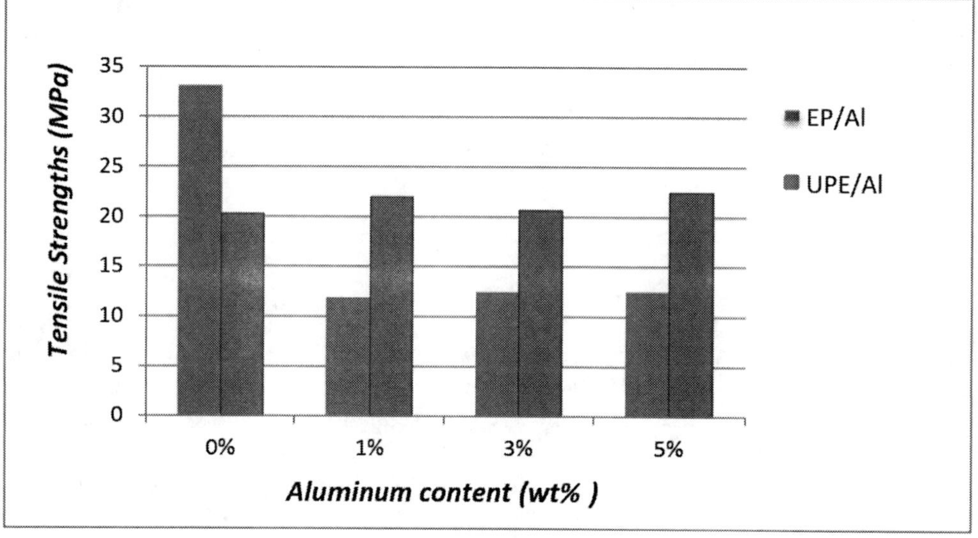

Figure 3. Tensile Strength variation with Al powder content in EP,UPE resins

IOP Conf. Series: Materials Science and Engineering **757** (2020) 012005 doi:10.1088/1757-899X/757/1/012005

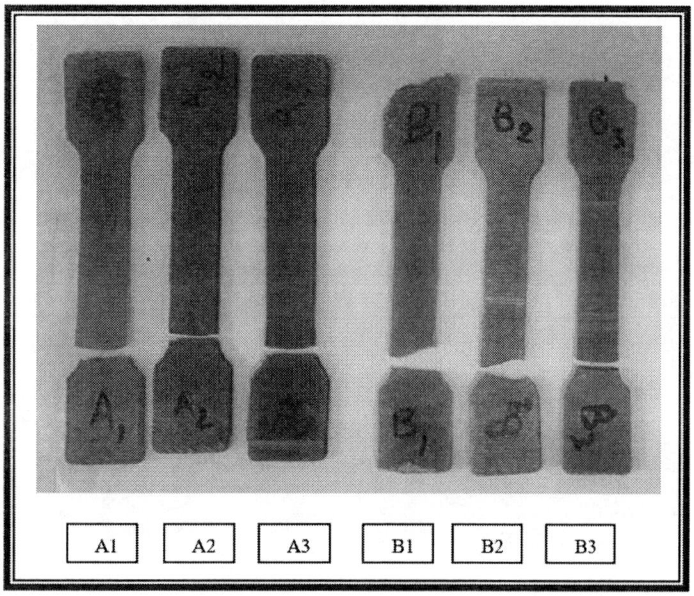

Figure 4. Photographic image that shows the fracture regions of EP/Al and UPE/Al micro composites.

4.2. Three – Point Bending Test.

The flexural strength (F.S) values of (EP/Al) and (UPE/ Al) composites are shown in table 4. Results showed that the flexural strength of epoxy/micro aluminum composites increased to (58.63MPa), when the Aluminum percentage increases from 0 to 1% as shown in figure 5. The increase in flexural strength is due to the dispersion of Aluminum and stronger interfacial bonding between the matrix and filler in the composites An increase in flexural strength of epoxy/aluminium composites due to addition of reinforcing aluminium particles to epoxy. This implies that incorporation of Al micro composites in epoxy matrix enhanced the flexural strength of the developed composites, which implies that both the strength and rigidity of the composites were increased. The increase is linked to a good interfacial adhesion of the particles with the matrix which enabled effective load transfer from the matrix at the bending zones to the reinforcing particles within the entirety of the composites. This explanation is in agreement with literature[19]. When the Aluminum percentage is further increased by 3%, the flexural strength decreases to (47.92MPa), but still higher than the flexural strength value of the neat epoxy samples. When the Aluminum percentage is further increased by 5%, the flexural strength decreases to (45.01 MPa) lower than that of the neat EP/ Al (47.46MPa). Reduction is due to the agglomeration of Aluminum fillers in the composites.

The result also show that the strength of unsaturated polyester/micro aluminum composites decreased to (71.87MPa), when the Aluminum percentage increases from 0 to 1%. When the Aluminum percentage is further increased by 3%, the flexural strength increases to (91.88 MPa). The increase is linked to a good interfacial adhesion of the particles with the matrix which enabled effective load transfer from the matrix at the bending zones to the reinforcing particles within the entirety of the composites. When the Aluminum percentage is further increased by 5%, the flexural strength drop to (64.65MPa) lower than that of the neat UPE/ Al (74.71 MPa) due to the debonding between the reinforcement and the matrix.

Table 4. The effect of Al content (wt %) on the Flexural Strength, (F.S) of (EP/Al) and (UPE/ Al) composites.

Composition	Flexural Strength, F.S (MPa)			
	Aluminum (Al) content (wt) %			
	0%	1%	3%	5%
EP/Al	47.46	58.63	47.92	45.01
UPE/Al	74.71	71.87	91.88	64.65

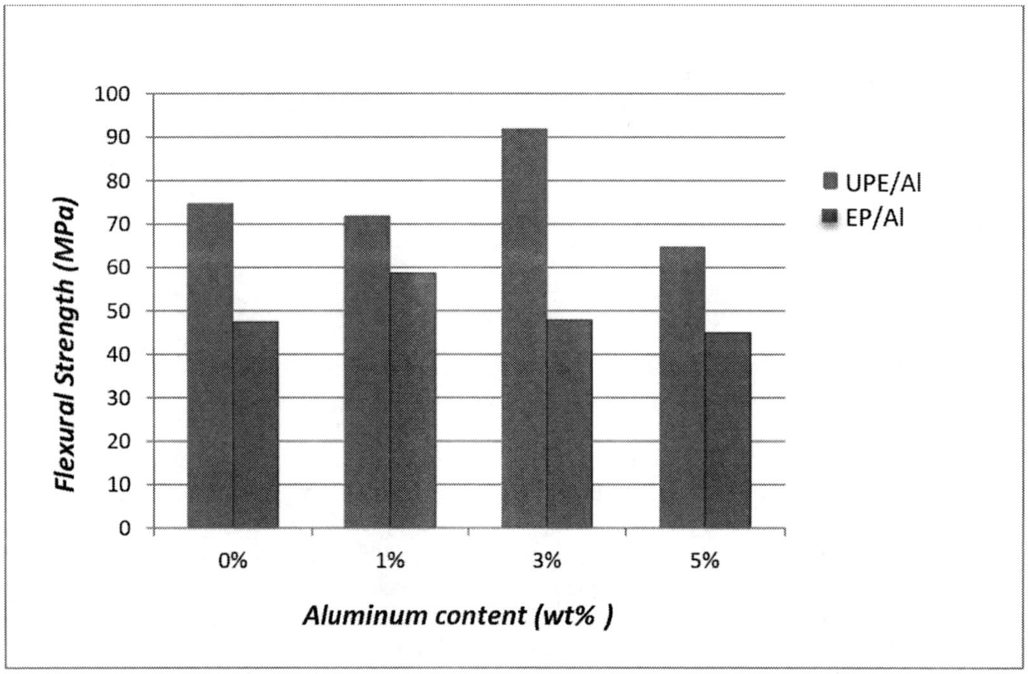

Figure 5. Flexural Strength variation with Al powder content in EP,UPE resins

4.3. Compression Test.

The Compressive test is used to calculate the ultimate compressive strength of the composites that have (0, 1, 3, and 5) % of aluminum powder (Al) added to EP and UPE resins. The ultimate compressive strength (UCS) values of (EP/Al) and (UPE/ Al) composites are shown in Table 5. Compression strength depends on the matrix properties of the material, reinforcement material, strength of binding across interfaces, volume fraction of reinforcement material. Figure 6. shows ultimate compressive strength variation with Al powder content in EP and UPE resins. In this figure, the ultimate compressive strength of the EP/Al composite samples decreased rapidly at the weight

IOP Conf. Series: Materials Science and Engineering **757** (2020) 012005 doi:10.1088/1757-899X/757/1/012005

percentage of Al was in the range of 0% to 1%. But the ultimate compressive strength increased slightly to a range of 1% to 3%. Then the ultimate compressive strength decreased slightly in a weight percentage range of 3% to 5%. The results of ultimate compressive strength still more much lower than the neat epoxy. The results also show the ultimate compressive strength of the UPE/Al composite samples decreased rapidly at the weight percentage of Al was in the range of 0% to 1%. Beyond this ratio, there is some rising in the strength of the composite until reach to the ratio 5% of Al micro particles. But results of ultimate compressive strength still more much lower than the neat unsaturated polyester. It is clear from the results that the ultimate compressive strengths of micro-aluminum composites decreased after the addition of Al content ratios in both of EP and UPE resins. This could be explained as the increasing Al content caused the weak adhesion between matrix and Al powder and led to the decrease of the strength of the epoxy and unsaturated polyester composites.

Table 5. The effect of Al content (wt %) on the Ultimate Compressive Strength of (EP/Al) and (UPE/Al) composites.

Composition	Ultimate Compressive Strength, (UCS) MPa.			
	Aluminum (Al) content (wt) %			
	0%	1%	3%	5%
EP/Al	36.90	28.57	30.04	24.49
UPE/Al	67.92	24.17	29.74	38.14

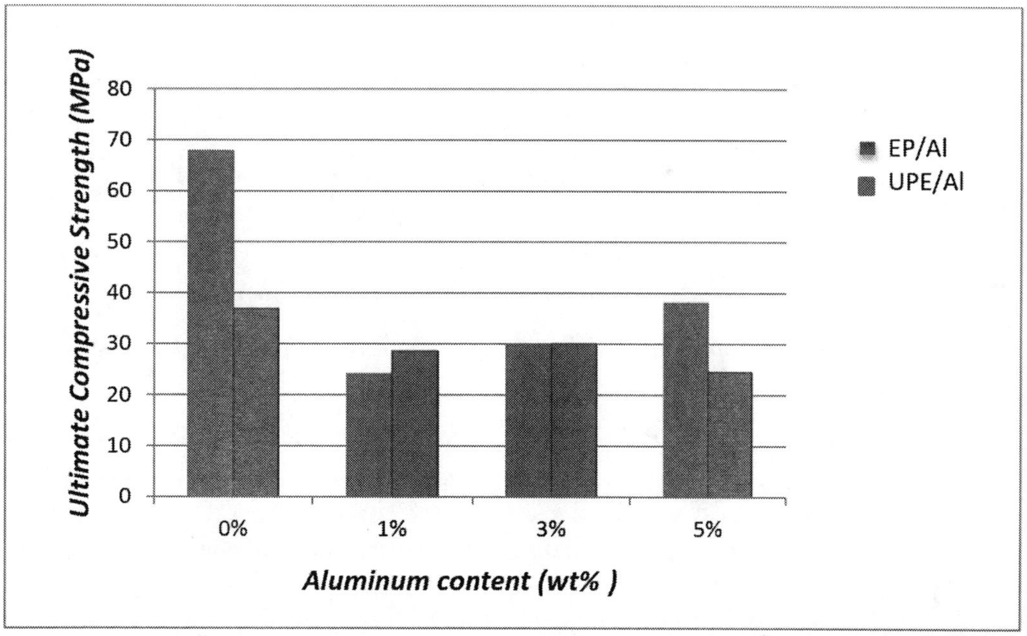

Figure 6. Ultimate Compressive Strength variation with Al powder content in EP,UPE resins.

5. Conclusions

The primary objective of this experimental study was to prepare the composite from aluminum power. The result shows that a Epoxy and polyester resins are a good adhesive material which can use as a matrix with Al fine powder. These composites found to have good mechanical properties. There is enhancement in the flexural strength due to addition small amount of Al in epoxy composite. The highest flexural strength was observed at 3% of UPE/Al composite. Similarly highest flexural strength was observed at 1% Al of EP/Al composite. It is observed that addition of aluminum powder reduce maximum fracture forces, maximum extensions, yield points, and tensile strengths of UPE/ Al composite possibly due to agglomeration. Generally, compressive strength will decrease with increasing aluminum weight percentage wt% to both of epoxy and polyester resins.

6. Reference

[1] Rosenthal D & Asimow R 1971 *Introduction to Properties of Materials* 2nd Ed, Van-Nostrand ,Reinhold Company.
[2] John W, Dean M, and Karyn L. 1987 *Engineers Guide to Composite Materials*, (New York: Joun Willy and Sons, Inc).
[3] Irving S 1957 *Hand Book of Adhesives* (London: Champman and Hall LTD).

[4] Bhatnagar M 2004 *A Text Book of Polymer* **vol. 11** S.C Hand.
[5] Henry L and Kris N 1970 *Hand Book of Epoxy of Resins* (Nem York: McGRAW.HAL Book Company).

[6] Mikell G 1996 *Fundamentals of Modern Manufacturing* (New Jersey: Prentice-Hall, Upper Saddle River).

[7] Kaneko M and Nakamura H1985, *J. Chem. Soc. Comm.*, 441.

[8] Bjorksten Research Laboratories1956 *Polyesters and Their Applications*, (New

York: Reinhold).

[9] Karshak V and Vinogradova S 1965 *Polyesters* (New York: Pergamon).

[10] Goodman I1988 *Encyclopedia of Polymer Science and Engineering* (New York: Wiley) 2nd ed.; **12**;

[11] Goodman I and Rhys J 1965 *Polyesters Saturated Polymers* (London: Iliffe Books).

[12] Goodman I 1968 2nd ed.; **16**: 159, (New York: Wiley)

[13] Morgan P 1965 *Condensation Polymers By Interfacial and Solution Methods* (New York: Inerscience Publishers).

[14] Alaa A, Thekra K, and Awattiff A 2014 *International Journal of Application or Innovation in Engineering & Management (IJAIEM)* **3** 7.

[15] Sihama I, Safaa N and Alyaa H *Eng. &Tech.Journal, ICNAMA* **33 Part (B)**3-4 .

IOP Conf. Series: Materials Science and Engineering **757** (2020) 012005 doi:10.1088/1757-899X/757/1/012005

[16] Srivastava V and Ansul V, 2015 Mechanical Behaviour of Copper and Aluminium Particles Reinforced Epoxy Resin Composites, *American Journal of Materials Science.* **Vol. 5(4)** pp.84-89.

[17] Norman E 1999 *Mechanical Behavior of Materials 2nd ed.* (New Jersey: Prentice-Hall Inc).

[18] ASTM D790-10 2010.*Standard Test Methods for Flexural Properties of Unreinforced and Reinforced Plastics and Electrical Insulating Materials* (West Conshohocken: American Society for Testing and Materials) DOI:10.1520/D0790-10.

[19] Sefiu A, Johnson O, Jeleel A, Nasirudeen K, Raphael G, Abdul F, and Suleiman B.2018 *ENGINEERING JOURNAL* 22 (4).

Study the improvement in the dielectric strength of (UPR/MgO) nanocomposites

Malik J Zayza[1], Najwa J Jubier[2] and Subhi A Al –Bayaty[3]

[1, 2]Department of Physics, College of Science, Wasit University, Kut, Iraq.

njassim@uowasit.edu.iq

[3]Department of Chemistry, College of Science, Wasit University, Kut, Iraq.

Abstract. This work focuses on the preparation of polymer matrix nanocomposite materials by hand lay-up molding which was prepared from the unsaturated polyester resin as matrix reinforced by nanoparticles of Magnesium Oxide(MgO) with particle size (54.19 nm) and weight fraction (0,1,3,5,7,10 and 15)%. Dielectric strength for the prepared samples was studied; the results of tests have shown the increase in the dielectric strength with increasing the weight fraction of additive (MgO) and also with the increases in the rate of voltage. The effect of the cycle's number on the dielectric strength has noticed clearly after the first cycle.The images taken for the electric breakdown points by using optical microscopy has shown the sample coaled in the electric breakdown area as a result of polymeric chains of unsaturated polyester (UPR) destroyed, and there is an appearance of microcracks as a result of electric breakdown extends directly from the point of electric breakdown.

Keywords: polymer nanocomposites, unsaturated polyester resin, MgO nanoparticles, dielectric strength.

1. Introduction

Polymer nanocompositesare materials that have polymer as a matrix material and nanoadditives are used as reinforcement material. The additives can be one-dimensional (nanotubes and fibers), two-dimensional (layered materials like clay) or three-dimensional (spherical particles). Polymer nanocomposites have been gaining considerable attention both in academia and in industries, due to their outstanding mechanical properties such as high elastic stiffness and strength with a smallconcentration of nanoadditives,the other excellent properties of polymer nanocomposites are barrier resistance, flame retardancy, wear resistance, magnetic, electrical and optical properties [1].Many materials such as polymers and ceramics have valence electrons strongly associated with the nuclei of their atoms, so, and without the presence of free electrons, they have a very weak conductivity "According to" the theory of energy bands for solids, the energy gap is large between thevalance and conductionbands and thevalenceband is completely filled with electrons at absolute zerotemperature [2].

The depth (energy difference) of this gap may be (10eV) and above, where even if it applied an electric field, the electrons do not move a large number in one direction, each electron is moving towards a certain to reverse invert another electron is moving in the opposite direction of movement because the band is completely filled, such as these materials are called (Dielectrics) [3, 4].Several reports have been carried out about the investigated such properties as in (2012) Ram Avatar Sharma, et al. have studied the effect of nano and micro silica on the electrical property of Unsaturated Polyester Resin (UPR) composites. The value of the dielectric strength is slightly higher in nano-silica (UPR) composites when compared to micro silica-based composites, dielectric constant keep decreasing. The dissipation factor decreases up to the nano-silica loading of 1.5 percentages [5].as well as in (2016)Najwa.J.Jubier studied some mechanical and physical properties of Polyester/Granite composites by hand lay-up method with different weight fraction, it was found that the values of thermal conductivity and dielectric strength increase with increasing granite content, but it decreased only at weight percentage20% [6]. Also In (2017),Duraid has studied the effect of addition Magnesium Monoxide (MgO) on the thermal and electrical properties of Iraqi Porcelain with different weight percentages (0, 5, 10, 17, 23, and 30) %. The best results for specific heat capacity and thermalconductivity arefor mixture with(30%MgO), thermo - compressive strength for mixture with (23%MgO) and dielectric strength for mixture (10%MgO) [7]. In this research, theUPR/MgO) nanocompositeswere prepared with differentweight fraction (0, 1, 3, 5,7,10 and 15) % of nanoMgO. The dielectric strength of these samples was investigated.

2. Dielectric Breakdown Strength

Dielectric breakdown strength is defined as the highest voltage which samples can stand before they fail electrically, divided by sample thickness, or the magnitude of the electric field required to causing a dielectric breakdown. When applying a strong electric field on the insulator higher than the value of the specific critical, a large number of electrons may suddenly be excited to energies within the conduction band. As a result, the current through the dielectric by the motion of these electrons increases dramatically, sometimes localized melting, burning, or vaporization produces irreversible degradation and perhaps even failure of the material, so that the insulation properties will be lost for the insulator and becomes a conductor. So the voltage that occurs when the breakdown is called (U_{br}), (Breakdown voltage) when divided by the thickness of the samples (h).We obtain (**E_{br}**) dielectric strength in units (kV/mm) or (V/cm) as in Eq. (1) [8].

$$E_{br}= \frac{U_{br}}{h} \qquad .... \; 1$$

A breakdown can be seen in the material by one of the following cases:
Case one, hole in the sample, which occurs when reach the real electrical durability of the insulation.
Case two, burn or melt, which canoccur when the material is heated locally and breakdown [9].There are several factors affecting on the thermal breakdown voltage of the polymers like geometrical structure, size (especially thickness), thermal conductivity, specific heat of polymer, environment heat (ambient), the rate of increase in voltage, the value of the (tanδ) and change its with temperature, and relative dielectric constant [10].The main types of breakdown that occur in insulatorsare intrinsic breakdown andelectrothermal breakdown.

3. Materials and Method

3.1. Experimental

Use unsaturated polyester resin (UPR)made by (Polres) company from Turkey. This resin in the form of a viscous liquid, transparent purple color at room temperature and converts from a liquid to the solid state by adding Hardener Methyl Ethyl Keton Peroxide by rate of 2 %which manufactured by the same company. The Magnesium Oxide(MgO) nanoparticles(NPs) utilized in the present study have been produce from (SkyspringNanomaterials, Inc.USA) (99% purity as per suppliers' data), Atomic Force Microscopy (AFM) was used (SCPMS canning Probe Microscope) to measure the average particles size, surface roughness and Root Mean Square (RMS) of MgO nanoparticles as shown in Table 1, Fig.1 shows image (3D -AFM) of nanoparticles MgO and particles size distribution.Samples of pure unsaturated polyester resin (UPR) and (UPR) with a different weight percentage of MgO nanoparticles (0, 1, 3, 5, 7, 10 and 15) % were prepared by hand lay-up technique and used ultrasonic in the mixing process. All content mixed thoroughly before casting, then the samples were left at room temperature for (24) hours and then for post-curing the samples were left for (1) hour in an oven at temperature 50 °C.

Table 1. Average particles size, surface roughness and root mean square values of nanoparticlesMgO.

Material	Averageparticles size (nm)	Surfaceroughness(nm)	RMS	Peak to peak (nm)
MgO	54.19	2.9	3.36	12.3

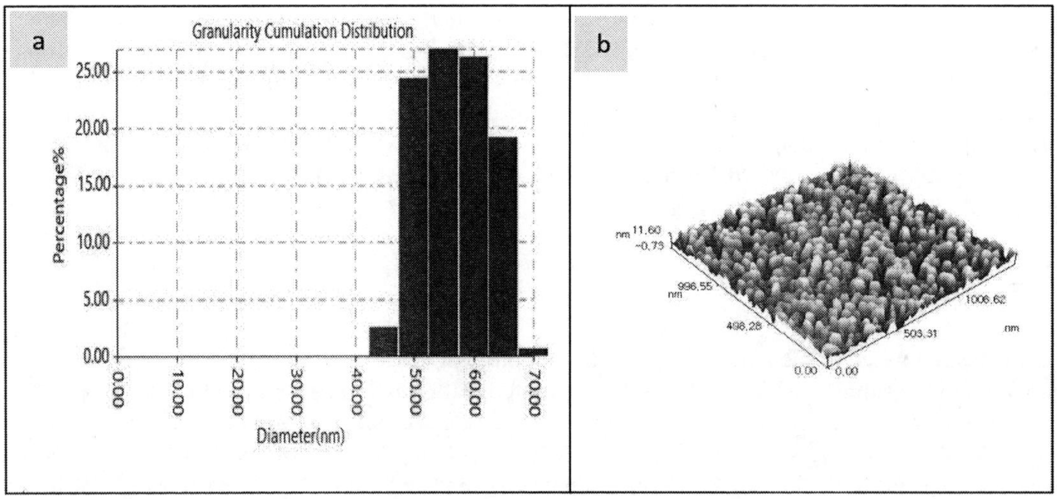

Figure 1. AFM of nanoparticles MgO; (a) Granularity distribution of nanoparticles; (b) 3D image AFM.

The dielectric strength measurement was carried out by (BAUR-PGO-S3) Germany origin with high voltage supplier in the voltage range (0-60 kV) and frequency (50 Hz), The instrument contains liquid with a high dielectric strength (voltage transformer oil (40kV/mm)) to prevent transmittal of the circumstantial spark (Flashover), in addition to rise speed of liquid inflammation, and the oil must change to prevent the ionization of liquid which leads to inaccuracy measurement.It also contains copper poles of good electrical conduction and spherical shape, its diameter about (2mm), as shown in the Fig.2.so the sample put them which are embedded in oil with make sure of touching poles with the sample surface then applying voltage through the sample that occur the breakdown voltage. After knowing the area that breakdown happens which can be distinguishing according to the damage that occurs because of the breakdown. Fig.3 shows the prepared samples.

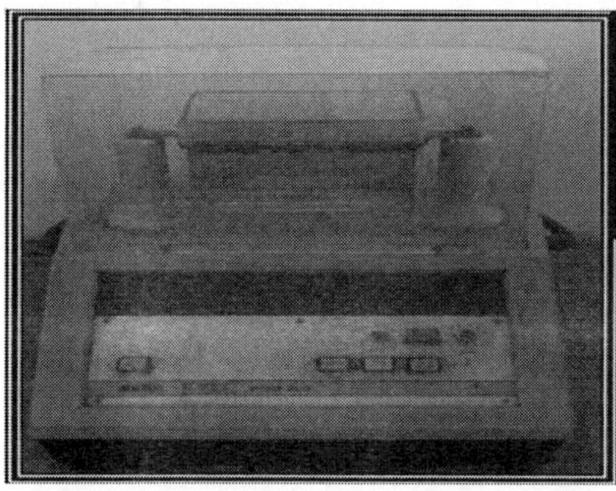

Figure 2. Dielectric strength instrument

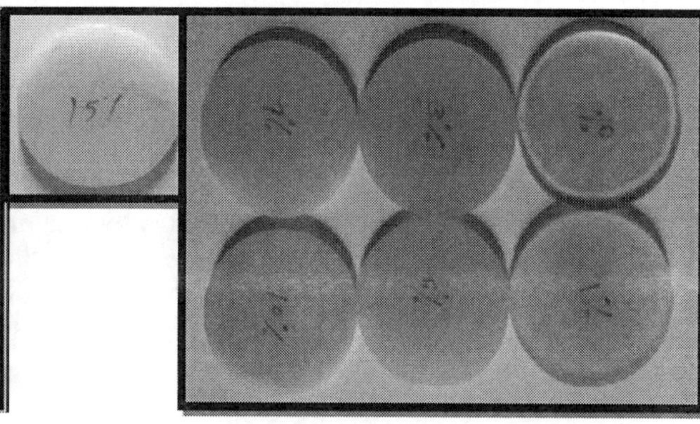

Figure 3. Prepared samples

4. Result and Discussion

4.1. The voltage elevating average

Figure 4. Explains the change in the dielectric strength with the voltage elevating average of polymer nanocomposite (UPR / MgO) at different weight ratio (0, 1, 3, 5, 7, 10 and 15) %. It is observed that the increase of dielectric strength with the increase in the voltage elevating average as the voltage elevating average (0.5kV/sec) leads to increase in heat raised from the leakage currents, where the applying voltage for a long time increases the possibility of occurrence of the electro thermal breakdown. Since the insulating materials decrease electrical resistance due to increased local temperature, so decrease will appear in dielectric strength with decreasing in the voltage elevating average.

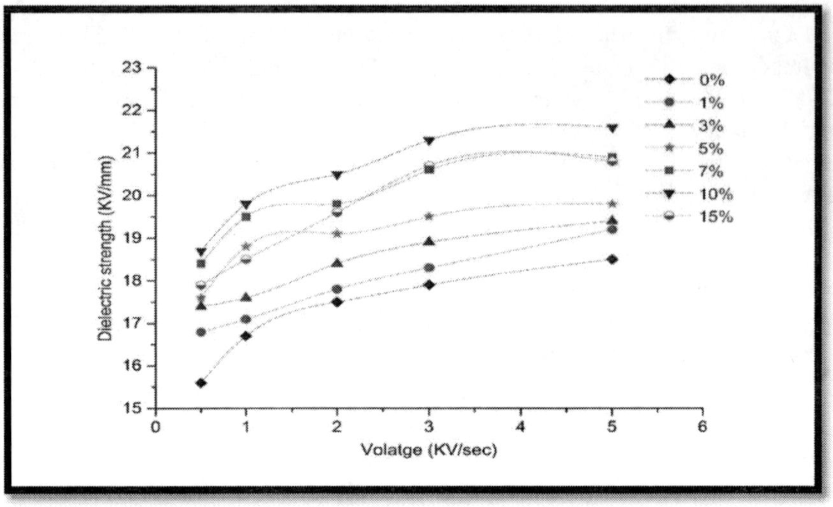

Figure 4. Dielectric strength versus voltage.

The reasons behind the decrease of dielectric strength with a decrease in average time to increase the voltage occur cumulative effects of the collision (chemical and electrochemical) and the corrosion which in turn destroy the materials and speeds up the breakdown by heating. These low values are the same operation values when used insulators in practical applications. We note abnormalities of some points about the general behavior and the reason is attributed defects within the composite material or homogeneous weaknesses of each testing points in the sample. As we note that the electric spark penetrates the insulator material at the time of breakdown which passes through those points that represent points of weakness and this is called the (effect of treeing) in the electric breakdown [10,11,12].

4.2. The effect of the number of periods

The test of dielectric strength changes with the number of periods, which is one of the tests that can know the possibility of using electrical insulators with the electric breakdown and recycle it again in the practical applications that need less voltage.Figures (5 and 6) describe the change in dielectric strength with the number of periods of polymer nanocomposite (UPR/MgO) at the same previous weight ratio at (0.5 and 5)kv/sec. it is noted a decrease in the dielectric strength after the first cycle for all prepared samples, the possible reason is attributed to the large chemical changes that appear with the diffusion of electric spark, because the test leads to burn the material or puncture it at the point of test and therefore the material loses the dielectric property and become conductive.The anomaly tests of some points for the general behavior of the test because the defects have arisen in the prepared samples of nanocomposites, and the effect of thermal and mechanical to the voltage on the sample sometimes may lead to a reduction in the thickness of the sample, or even puncture it, therefore oil enter between the poles which effect on data of examination. There is a significant change in the possibility of using the insulating materials after the first round of electric breakdown, as sometimes a decrease in dielectric strength in the second round to half value of the first round, which causes a significant decrease up to 90% in the value in the first one as in Figure 5.

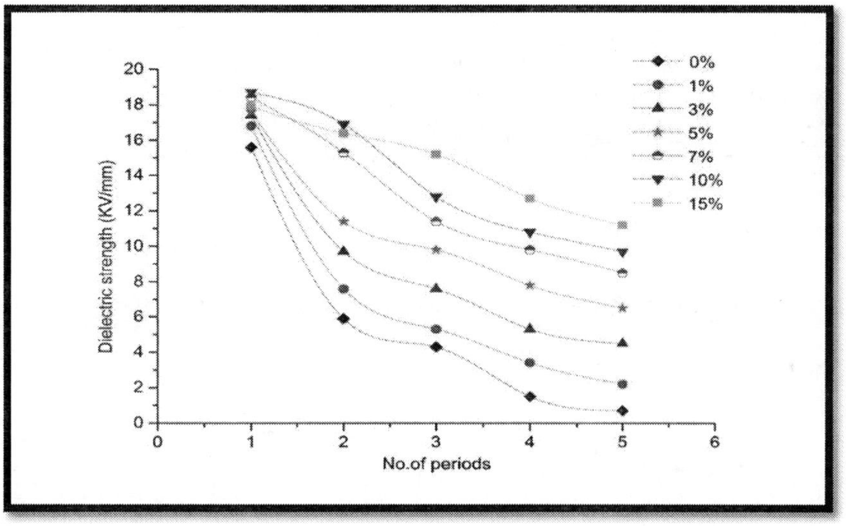

Figure 5. Dielectric strength versus No. of periods at (0.5kv/sec).

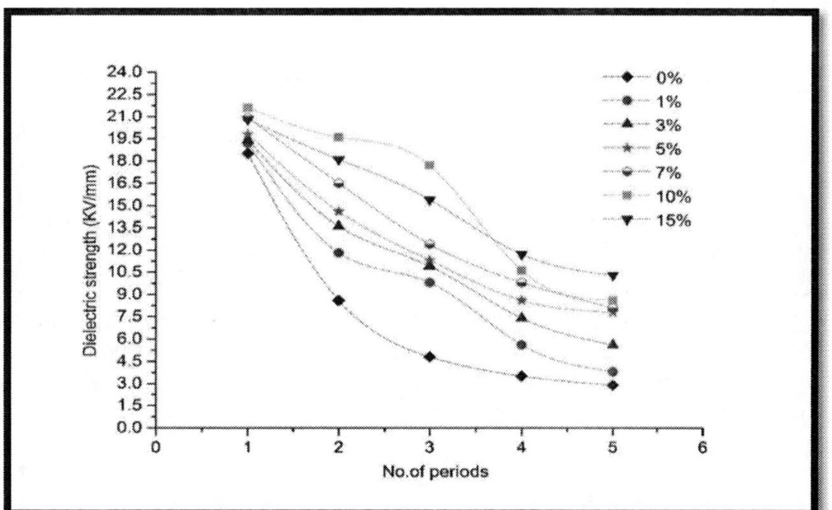

Figure 6. Dielectric strength versus No. of periods at (5kv/sec).

4.3. The effect of additives rates on the dielectric strength

Figure 7.describes the change in dielectric strength with different weight of nanoMgO. It is noted that the dielectric strength of nanocomposites is higher than the polymer and it increases with the increase in the ratio of addition and this result is agreeing with E. Tuncer [13],because the ceramic materials have higher electrical insulation as compared with polymeric materials and the use of nanoparticles lead to the spread within the polymer mixture and thus the high density of widespread nanoparticles, which leads to impeding the passage of current and this result agrees with Saikat[14].

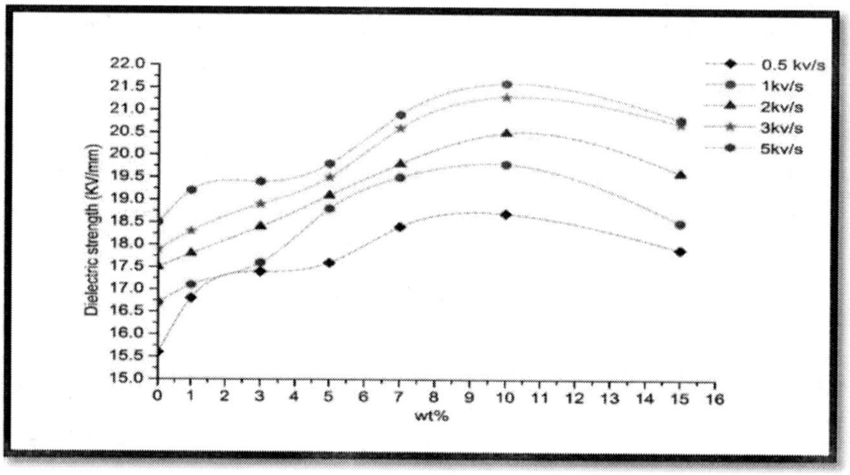

Figure 7. Dielectric strength versus additives rates of nano MgO.

Figure 8.represent images taken for the electric breakdown points by using optical microscopy.The images have shown that the sample coaled in the electric breakdown area as a result of destroying of polymeric chains to unsaturated polyester (UP), and the formation of microcracks as a result of electric breakdown which extends directly to the point of electric breakdown. The optical microscope pictures show that there are micro-cracks were larger and more sinuosity and distortion of the samples, This is because the great convergence of nanoparticles, diffusing, dispersal, high homogeneity, and high stable distribution in all directions, that nanoparticles enjoyed and the most important distinguishing of nano reinforcement.

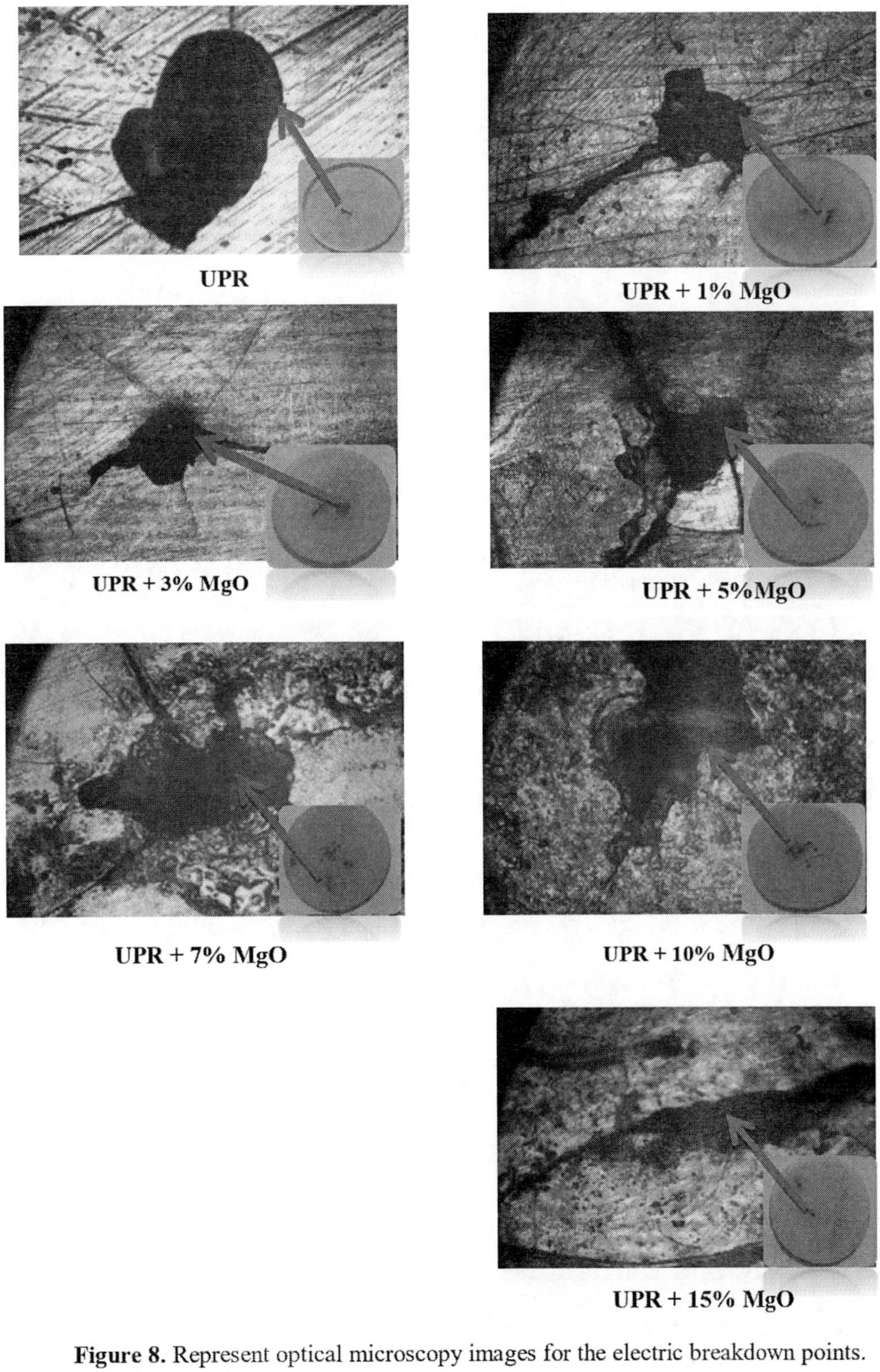

Figure 8. Represent optical microscopy images for the electric breakdown points.

5. Conclusion

The dielectric strength increases with the increase rate of voltage for polymer nanocomposite, while it decreases as the number of the period increased. The dielectric strength also increase when the (MgO) added to the matrix material (UPR), until to reach its heights value at addition rate of (10%) (21.6kV/mm), while at addition (15%), it decreased to (20.8% kV/mm), this is due to the emergence of weak areas in the inter phases, which are areas of low breakdown strength and the presence of aggregates of nanoparticles in addition to the existence of micro bubbles that are difficult to get rid of them.

6. References

[1] Yiu W and Zhong Z 2006 *Polymer nanocomposites* (UK: Cambridge: Wood head Publishing Limited).

[2] Jewetty S 2004 *Physics for scientists and Engineering with Modern Physics* (United state) 6thed p. 1421.

[3] AL-Naimi A, AL-Jawadi A, Ibrahim M and Qassem M 1999 *Modern Applied Physics , Technology University* pp. 179-181.

[4] Harinath A 2011 *National Institute of Technology,* Rourkela, India.

[5] Ram A, Dawid D, Subhendu B, Lokesh C and Sarojini S 2012 *Transactions on electrical and electronic Materials* **Vol. 13**, No. 1, pp. 31-34.

[6] Najwa J 2016 *Wasit Journal for Science &Medicine* **vol. 9(3)**, pp. 99-108.

[7] Duraid F 2017 *Journal of Al-Nahrain University*, **vol. 20 (1)**, pp. 79-88.

[8] Ahmed A, Hashim A and Najaw J 2018 *International Bimonthly*, **vol. 8(49)**, p 14323.

[9] Stanek J 1977 *Electric melting of glass* (Amsterdam: Elsevier Scientific Comp, LTD) pp. 28-31.

[10] Kroschwitz I 1988 *Electrical and Electronic Properties of Polymers* (United State of America A State of Art Compendium) pp. 106-107.

[11] Gingery W, Bowen H, and Uhlmann D 1976 *Introduction to Ceramic* (New York: Wiley) , 2nd.

[12] Razevig D 1978 *High Voltage Engineering*, *Translated from Russian,* (India: *M.P.* Chourasia, Khanna publisher) pp. 96-100.

[13] Hashim A 2008 Study of Some Parameters That Effected on Dielectric Strength for Composites (Ceramic - Glass), *University of Baghdad, Master Thesis.*

[14] Eduard A S, Xiaoli T, Zhiqun L, Nicola B and Michael R K 2010 *polymer* **Vol. 51**, pp 5823-5832.

Fabrication Flexible Solar Cells by Organic Semiconductors

M. A.Abood, [1]*F. I. Mustafa Alattar[2],M. S. Badri[3]
[1, 2, 3]Solar Energy Research Center, Renewable Energy Directorate
[1, 2] Higher Education and Scientific Research Ministry, Baghdad, Iraq
*Corresponding author: morooj_a@yahoo.com

Abstract:
Organic semiconductors require low energy fabrication process coupled with the ease of processing, as well as compatibility with flexible substrates, hence, can be fabricated at low cost as light-weight solar cells. The thin films of Ni-PhthalocyanineTetrasulfonic acid tetrasodium salt and poly (3, 4-ethylenedioxythiophene) poly (styrenesulfonate) on indium tin oxide (ITO glass and plastic) substrates were prepared by drop casting at thickness (100 nm). The optical properties were studied by using UV-Vis absorption spectra. The current–voltage characteristics of solar were measured the efficiency values about 4.17%.

Keywords: Organic, Semiconductor, Phthalocyanines, Flexible Solar Cells

1. Introduction

Flexible solar cells can be used to generate low cost electricity, to afford lightweight, economic solar modules that can be integrated on various surfaces ,inorganic and organic-inorganic solar cells are deposited over flexible substrates by high-through put (often roll-to-roll printing) technologies. Figure1show the structure of the organic semiconductor. The standard conditions of Current conversion efficiencies are in the 3-15 % range. These new photovoltaic technologies are ready to provide cheap, clean electricity to the 2 billion people who lack access to the grid as well as to energy-eager companies and families in the developed world facing the increasing costs of electricity generated using fossil fuel resources [1, 2]. Usually electricity is produce by burning hydrocarbons and, sadly, the coal is much Cheaper, in 2006 nearly half (49%) of the 4.1 trillion kWh of electricity produced in the US used coal as its energy source [3, 4]. China alone needs to add some 25GW of new capacity each year, equivalent to one large coal power station every week. Regrettably, coal contains mercury and along with the production CO2, causing pollution of the oceans and of the food chain.The biggest challenge of our epoch is to abate emissions and stop climate change, is to get electricity directly from the sun [3, 4].

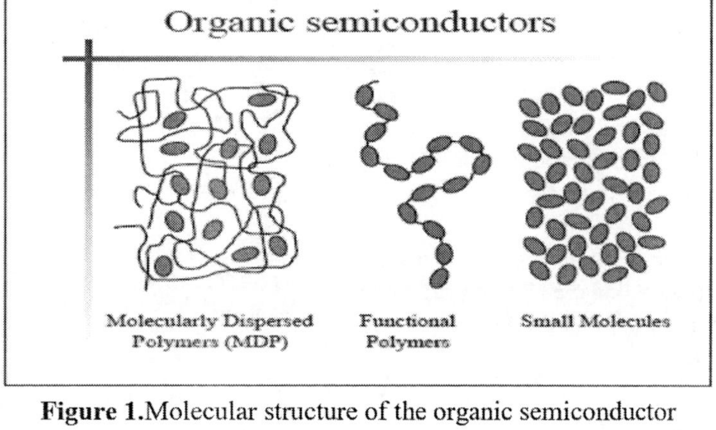

Figure 1. Molecular structure of the organic semiconductor

There are two basic organic solar cell platforms: thin-film heterojunctions and dye-sensitized or photoelectron chemical solar cells as shown in the Figure 1. Thin-film heterojunctions in the thin-film heterojunctions (HJ2) devices, photo-active layers of organic materials (small molecule, polymer or den-drimer) are sandwiched between two electrodes – an anode to collect electrons (a transparent and conducting oxide such as ITO) and a cathode to collect holes (or return the electron to the device from the external circuit – usually a low work function metal such as Al).The drop casting is very simple method. Mix the material in a suitable solvent (the material should not change its form) via ultrasonication and cast by dropping on the target substrate and low-cost, versatile, and potentially scalable casting method is proposed for the fabrication of micro- and nano-thin films, The impingement of a solution drop onto a substrate in a simple process called drop casting, usually results in spreading of the liquid solution and the formation of a non-uniform thin solid film after solvent evaporation [5,6].The aim of this research to manufacturing flexible solar cells from simple method and process.

2. Experimental work

Small molecule organic materials are commonly depositedbydrop casting. It is a surface covering in which a material is deposited on a conductive surface.NiPcTs was dissolved in dyonoyiz water and mix well using magnetic stirrer to be homogeneous. The glass was cleaned by water then acetone and alcohol, ITO glass and plastic wear cleaned by alcohol and Si was etching by HF acid concentration of 0.1 for a ten minutes then was cleaning by acetone and alcohol. The films of NiPcTs:TSS and PEDOT:PPSS thickness of 100 nm, were deposited by Plating over ITO glass and plastic to an area of 1cm^2, A typical device structure is shown in Figure 2 and the device relies upon the formation of a ‘heterojunction’ between two materials of different electron affinities (much in the same way as a silicon pn-junction). One material in the heterojunction is the primary light absorber and electron donor and the second the acceptor. The donor has a lower electron affinity than the acceptor and the formation of an exciton in the donor via photon absorption will enable the transfer of an electron into the acceptor, with the retention of the hole in the donor. The optical measurement include UV-Visible.

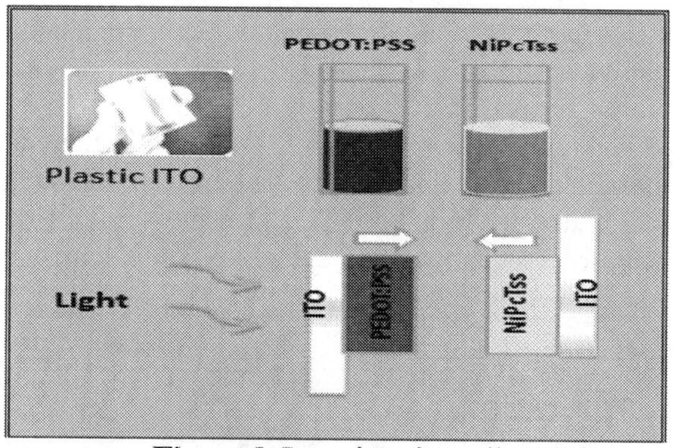

Figure 2. Organic solar cell samples

The characteristicsof current - voltage under dark and light condition were measured by Solar cell efficiency measurement system (C100AAA).Cell efficiency is obtain by:

Isc: Short-circuit current →Current value when V = 0
Voc: Open-circuit voltage→Voltage value when I = 0
P: Power output of the cell P = IV
F.F: Fill factor

$$FF = \frac{I_{mp}V_{mp}}{I_{sc}V_{oc}} , \qquad \eta = \frac{I_{sc}V_{oc}FF}{P_S} \times 100$$

3. Results and Discussion

The absorption and transition spectra of NiPcTsand PEDOT: PSS, and PEDOT: PSSthin films of thickness 100nm deposited with different substrate given in Figures (3 and 4) whichShows Q and B bands. It is also observed that NiPcTs has a higher absorption of PEDOT: PSS at wavelength (600nm) but less transmission,thisexplainsthat use PEDOT: PSSin themanufacture of solar cells.

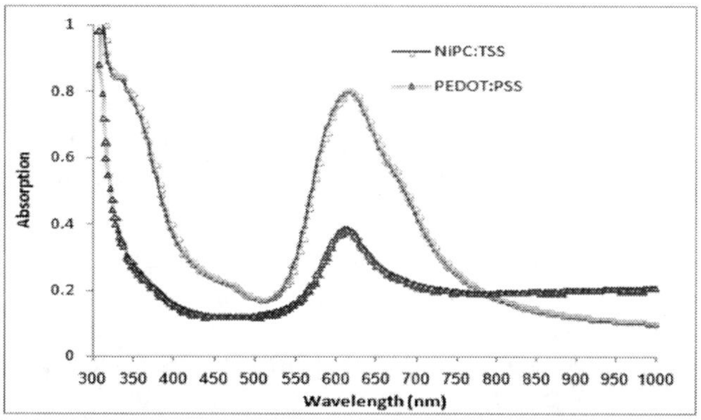

Figure 3. The absorption spectra of NiPcTs and BEDOT:PSS thin film.

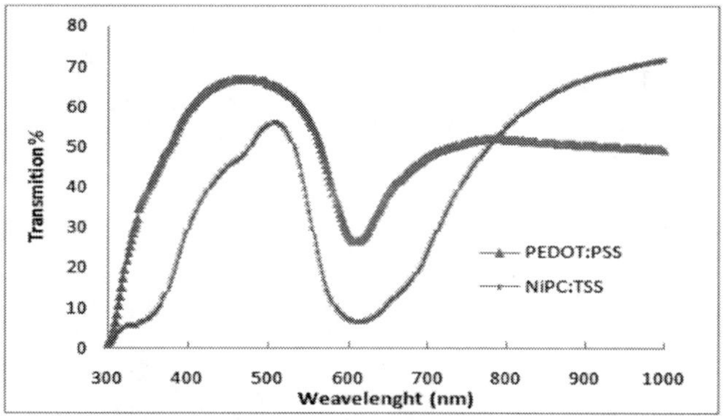

Figure 4. TheTransition spectra of NiPcTs and PEDOT:PSS thin film.

Figure 5 shows that the absorption coefficient (α) of (NiPcTs and PEDOT:PSS) thin films are characterized by a strong absorption in B band in the wavelength region between (280–320) nm also a strong absorption in Q band in the wavelength region between (550–700) nm.In metalophthalocyanines, the central metal atom is in a position of nearest approach to nitrogen atoms in parallel molecules and this increases the orbital overlap between parallel pairs of molecules. Hence, any crystal phase change would affect the energy gap between conduction and valence bands.[7, 8].The higher energy peak of Q-band has been assigned to first p–p* transition on the phthalocyanine macro cycle and The lower energy peak is exciton peak [9,10]. The splitting of Soret band is Davydov splitting. The structural information that can be gathered from analysis of shape of Q and B-bands is that molecules of NiPcTs are in tilted arrangement. NiPc have partially occupied d-bands due to existence of Ni atoms as a central atom in their molecular structures.

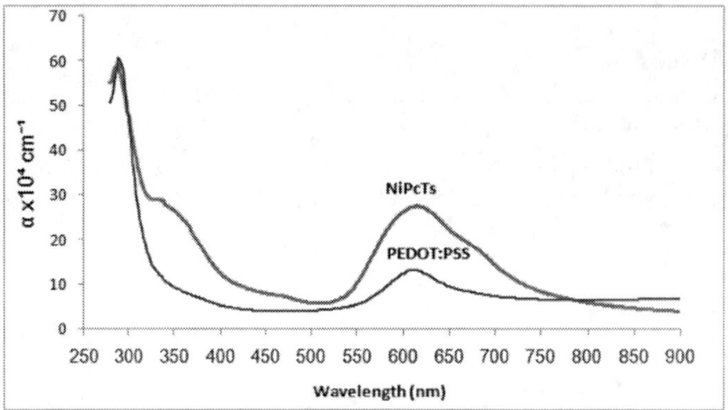

Figure 5. The absorption coefficient of NiPcTs and PEDOT:PSS thin film.

The optical energy gap value (Eg) of NiPcTs and PEDOT:PSS thin film determined by Tauc equation which is used to find the value of energy gap and the type of the optical transition by plotting the relation of (αhv)2 versus photon energy (hv). It is clear that the film has a direct energy gap shows in Figures(6 and 7). The optical energy gap values (Eg) for NiPcTs thin films are(1.75 and 3)eV and for PEDOT:PSSare (1.8 and 3) eV, This result is in agreement with the result have been reported by M.T. Hussein et al .in (2012)[11].

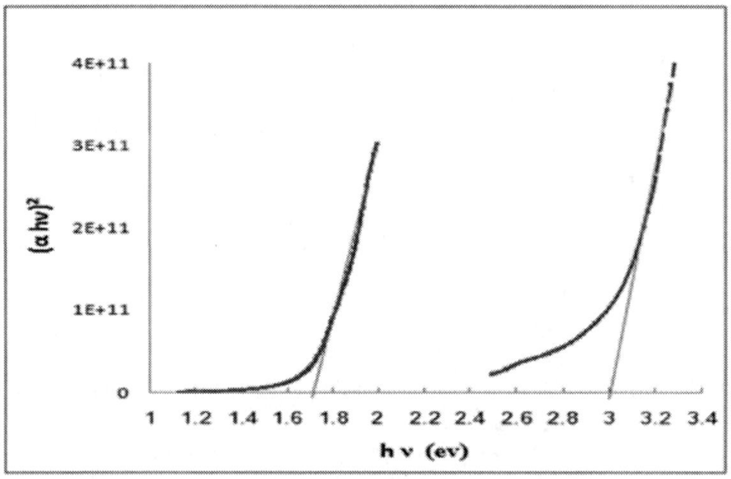

Figure 6. (αhv)2versus photon energy (hv) of incident radiation for NiPcTs thin film.

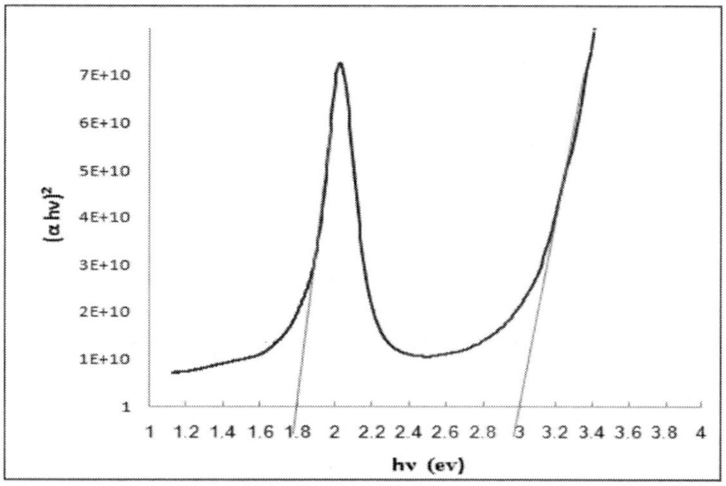

Figure7. $(\alpha h\nu)^2$versus photon energy (hv) of incident radiationfor PEDOT:PSS thin film

3.1 CURRENT–VOLTAGECHARACTERISTICFOR ITO/PWDOT: PSS/NIPCTS

Figures(8 and 9) shows Current –Voltage measurements ofITO/PEDOT:PSS/NiPcTsbulk heterojunctions thinfilms at ITO (glass and plastic), shows that the best efficiency of the models is installed on ITO plastic of (4.1) , but (1.4) on glass substrate, to harmonize the organic conductor cubs with the flexible substrates.The photocurrent is one of the fundamental parameters that play an essential role in solar cells. For the HJ under illumination, which leads to separation of the electron – hole pairs and then increases the photocurrent. The photocurrent is a function of the generation and diffusion of carriers.

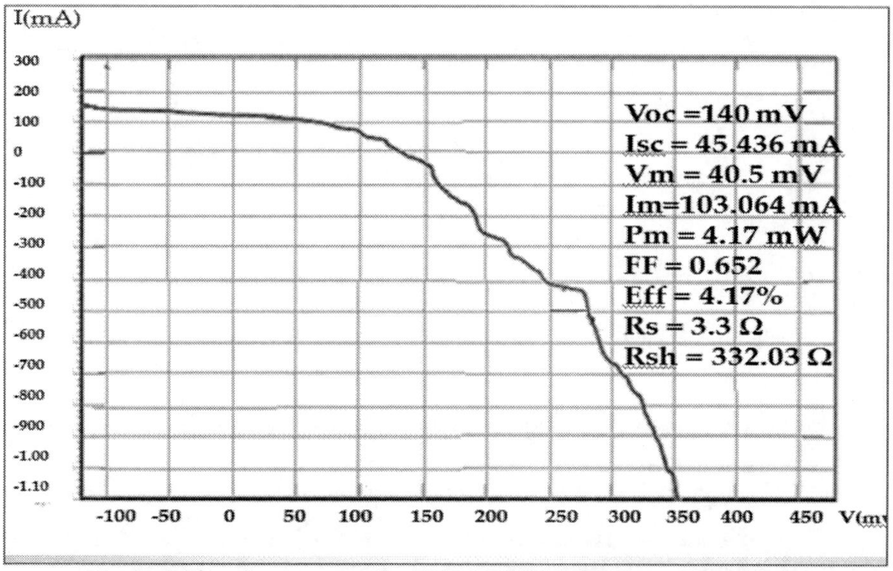

Figure8. The photocurrent for PEDOT:PSS/NiPcTs on ITO plastic.

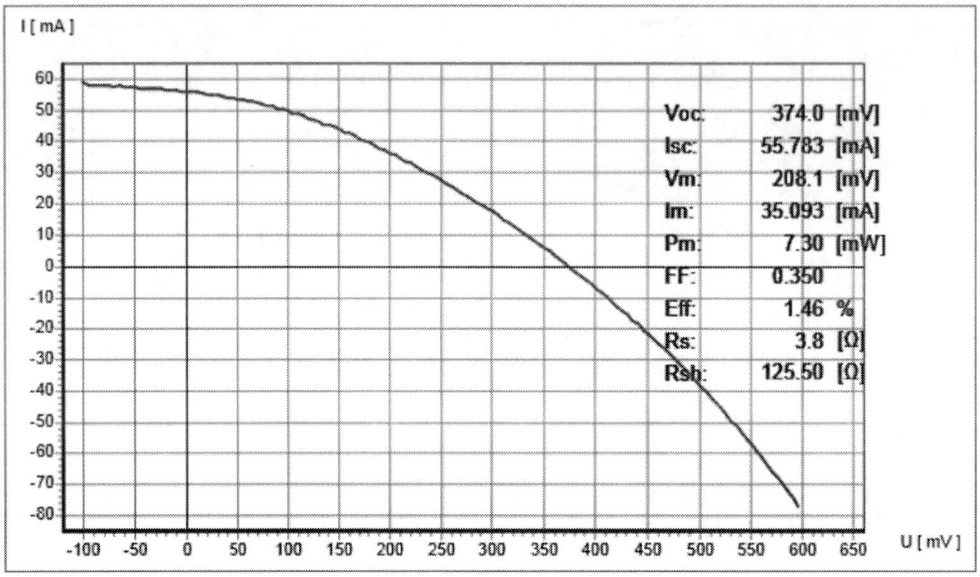

Figure9. The photocurrent for PEDOT:PSS/NiPcTs on ITO glass.

4. Conclusion

The world has turned to manufacturing organic solar cells as an alternative to silicon cells because of its advantages in ease of manufacture, cheap materials and easy deposition on flexible substrates,Where they showed significant potential for organic solar cells compared to conventional silicon cells. Their proficiency in production methods and characteristics seems very promising applications for the future of solar energy.

Acknowledgment

The researcher would like to express the profound sense of appreciation to: Solar Energy Research Center / Renewable Energy Directorate / Ministry of Science and Technology; for their invaluable support and constructive directions.

5. References

[1] Askari. Mohammad Bagher, *2014 Sustainable Energy* "Introduction to Organic Solar Cells", **2** (3) 85-90.

[2] MortezaEslamian* and Fatemeh Zabih *2015Research Letters*.**10** 462

[3] T. Qasuria *2012 Ghulam Ishaq Khan Institute of Engineering Sciences and Technology, Pakistan.*

Gupta 2015Material Sci Eng, 4:6.

[4] B. Lüssem, M. Riede, K. Leo *2013Phys. Status. Solid A* **210** 9.

[5] M. Hiramoto, M. Kubo, Y. Shinmura, N. Ishiyama, T. Kaji, K. Sakai, T. Ohno, M. Izaki *2014 Electronics* **3** 351.

[6] BENNY J. and MENON C. S. 2007 *E-Journal of Chemistry*, **4**(2) 255-264.

[7] L. Kazmerski*1980 "Polycrystalline and Amorphous Thin Films and Devices*", Academic Press.

[8] M.M. El-Nahass a, H.M. Zeyada b, M.S. Aziz b, N.A. El-Ghamaz, *2004 Optical Materials* **27** 491–498.

[9] R.SathyAmoorthy, SA.K.NarayAndass, C.Balasvbrmanian and D.Mangalara *1990 J, Phys. Stt. Sol.*, **117** 495.

[10] M.T. Hussein, E. M. Nasir, A. H. Al-Aarajiy *2012 Int. J. Thin Film Sci. Tec.***1** (2) 71-76 .

Effect Gamma rays effect on Wear rate and Hardness of (Ep-SiC) Nanocomposites for Industrial applications

A K Aobaid and T Y Alqaisy
University of Anbar - College of Education for Pure Sciences - Department of Physics – Iraq.
Ibnalanbar664@gmail.com.

Abstract : The aim of this research is to measure the hardness and wear rate for epoxy resin as a base material, nano ceramic powder (SiC) with particle size (64.85nm) , were used as a reinforcement materials in composites preparation in different weight fractures (2%, 4%, 6%), hand-casting method was used in preparation the research samples, Mechanical properties were measured after irradiation of gamma rays emitted from the CO^{60} with energy rate of 1.33MeV and different irradiation doses (6,8,10) KGy. The results showed an increase in the hardness values after the irradiation compared with the results before irradiation while wear test showed a decrease in the values in normal conditions compare with the result after irradiation.

Keywords: Gamma rays, Hardness , Wear , Epoxy resins , Silicon Carbide.

1. Introduction

In the recent years, modern materials have been needed that will have a lasting impact in modern applications. The need for radiation therapy as an important means of modifying nanoscale structures in the manufacture of high performance polymer compounds has important advantages such as high corrosion resistance, high hardness, and low cost. Due to the extensive use of polymeric materials in coating , packaging , shipbuilding and aircraft wings , as well as their use in various fields of life. These materials are exposed to the sun and the various environmental conditions, and because of this importance in our daily life , it is necessary to study the effect of radiation on composite materials. The use of polymer materials composite of our time a significant shift of the characteristics of these materials of many characteristics led to the use of wide and in various fields. These characteristics include the low cost of fabrication , durability , low density , high thermal and electrical insulation, resistance to chemicals and moisture , corrosion resistance , shapes and sizes , and high surface strength [1,2].

As already mentioned , it is necessary to study the mechanical and physical properties of these materials so that researchers can develop them and use them in their practice fields [3]. Polymeric polymers can be defined simply as those systems resulting from the mixing of two or more substances with specific rules to obtain new materials possessing distinct physical and mechanical properties that differ radically from the properties of their basic material [4,5]. At present the effect of radiation on composite materials , whether ionizing radiation or non-ionizing radiation. Ionizing radiation results in wide variations in the properties of composite materials. Radiation has a significant effect on radiation when it penetrates the radiation , causing a change in the order of atoms and molecules in the composition of matter [6]. The effect of photon radiation in the composite material causes changes in the structure of these materials as a result of the interaction between the photons and their absorbent material. The reaction type depends on the energy of the photons and the atomic number of the target [7]. The evolution of the use of composite materials in the development of advanced computer science and software has motivated scientists to create specialized software in the design and analysis of composite materials and test most of their properties without resorting to traditional methods [8].

This research was carried out to determine the hardness and wear rate property pre-and to post exposure gamma which emitted from the cobalt source (Co^{60}) for the SiC composites and study the effect of these rays in different doses (6 ,8 and 10) KGy on these properties.The effect of these radiations varies depending on the then mixed together of the materials to these rays.

2. Experimental Part

2.1. Matrix material

Epoxy Sikadur-52 , manufactured by Henkel, Australia , is used as a base material in the preparation of composite material. Epoxy resins are characterized by transparent liquids and a density of 1.1-1.4 g / cm^3 The epoxy resins are converted to the solid state after the addition of a metaphenylen Diamine (MPDA) and a plant of the same company by 1: 2.

2.2. Reinforcing material

2.2.1. Silicon Carbide (SiC)

The SiC powder used in this study is the β type .(EV NANO) Technology Co , Ltd. particle size (64.85nm), purity (99%) , Density 3.22g/cm^3.

2.3. Specimens Preparation:

Hand lay-up molding method was used for preparing the samples, as follows :

1-The samples which composed from silicon carbide (SiC) , were prepared as a reinforced material (2%, 4%, 6%), with epoxy resin as a base material mixed with a (1: 2) compound . The mixture is then poured into the mold until the mixture leaks all over the mold and leaves for (48) hours to solidify.
2-After 48 hours , samples were placed in the convection oven at a temperature of (50 ° C) for(6) hours. This reduces the internal stresses produced during shrinkage, as well as obtaining the best interlock and completing the solidification process.
3-The samples were left in the convection oven after it is extinguished until it returns to its normal temperature to obtain a good crystallization of the samples [9].
4- The dimensions of the hardness and wear test samples are thickness 20mm x 10mm x 10mm.

3. Mechanical Tests

3.1. Hardness Test : The surface hardness of the samples had been calculated via used (shore D) method utilizing the tester of (shore D) hardness th210 , an average of three readings were take from each test .

3.2. Wear Test : wear was using performed the slipping wear device used steel disc with a hardness of (269) hb and constant rotational speed of (500) RPM according to the following equation(1) :

$$\text{Wear Rate (W.R)} = \frac{\Delta W}{S_D} \text{ (g/cm)} \quad \ldots\ldots\ldots\ldots. (1)$$

Where; Δw is a mass quantity difference.

S_D is sliding distance [cm].

IOP Conf. Series: Materials Science and Engineering **757** (2020) 012008 doi:10.1088/1757-899X/757/1/012008

$S_D=2\pi\, r\, n\, t$ ....(2)

Where: r-the distance from the sample to the disc center (7cm).

n-number of disc rotations. t-test time.

4. Gamma ray Instrument : . In this research, we used a gamma ray emitted from (Co^{60}) source with an energy rating (1-33-1-17)MeV in order to irradiant the samples .

5. Radiation Effect on The Polymeric Composite Materials:

The radiation (natural or industrial) is classified according to its effect in two types [10,11]

A- Ionizing Radiation: This type of radiation has the ability to ionize the atoms passing through it such as electromagnetic radiation (x-rays, ultraviolet rays, and gamma rays) and particle radiation such as particles of alpha, beta, neutrons and protons. The radiation is in the form of a cosmic or wave emission and its effect in the medium that penetrates depends on energy radiation and its nature where particles and radiation emit different cards and decay over a given time and according to the following equation:

$$A_t = A_\circ e^{-\lambda t} \quad \text{.........}(3)$$

A_t: the radioactivity of the element at time t.
$A_\circ$: Radiation activity at primary time.
λ: the decay constant and given by the relation:

$$\lambda = \frac{0.693}{T_{½}} \quad \text{.......}(4)$$

$T_{½}$: :Is the half-life of the radionuclide.

B- Non-Ionizing Radiation: This type of radiation does not have the ability to ionize the atoms passing through it such as infrared waves (IR) , lasers, visible light , radio and television waves , radar waves and microwave wavelengths .

6. Results and discussion:

6.1. Hardness and wear before irradiation.

The hardness values of all composites prior to irradiation were increase with the increase of the weight fractures(2% ,4% ,6%). The maximum hardness values were at the weight fracture(6%) as shown in figure 1. The addition of silicon carbide nanoparticles, a high hardness ceramic material and resistance to plastic deformation, has led to the penetration of the base material into the core and into the probe and the interstitial spaces formed during the preparation process. This penetration has helped to increase the contact area between overlapping materials and increasing bonding Which leads to the addition of a bonding link in the polymer network, which increases the hardness of the surface [12].

The wear rate values of all composites prior to irradiation were decreased with the increase of the weight fractures(2% ,4% ,6%), The maximum wear rate values were at the weight fracture(6%) as shown in figure 2. This is due to the fact that the minutes of the material play an important role in carrying the large part of the stresses resulting from the load between the surfaces and increase the hardness of the surface of the overlapping random materials distributed randomly as obstacles to distort the base material through pregnancy because of its high strength and lead to increase the severity of the polymer body Its durability and increased thermal expansion coefficient due to the high hardness of (SiC)[13,14].

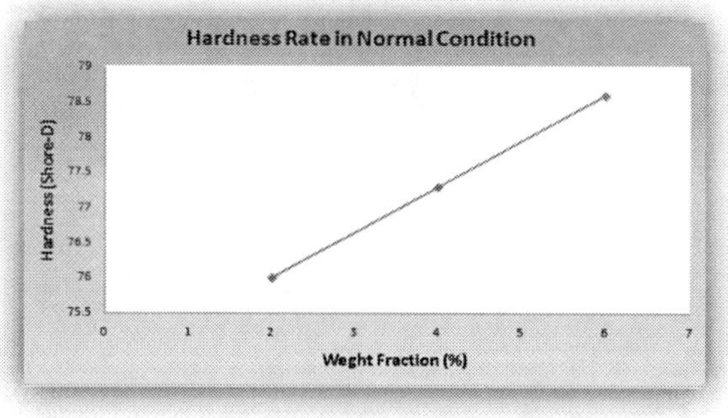

Figure 1. **The relationship between hardness and weight fraction(2,4,6)%**

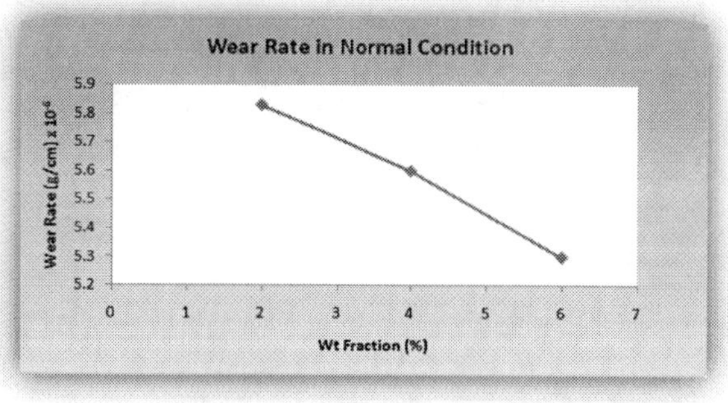

Figure 2. **The relationship between wear and weight fraction(2,4,6)%**

7. Hardness and wear after irradiation :

The exposure of polymeric composite material to gamma radiation with high energy leads to happened excitation and ionization of molecules , this causes broken in chemical bonds and leading to the production of free radicals, and the effect of radiation on the properties of polymeric materials [15].

Hardness : From the figures 3 we observe an increase in hardness values by increasing the rate of radiation doses for SiC composite samples gamma ray irradiation led to increased decomposition and generation of free radicals, which in turn led to an increase in cross-linking between the bonds of the base material and the supporting materials and thus increase the surface hardness of all samples [16].

Wear rate: From the figure 4 we observe a decrease in wear rate values by increasing the rate of radiation doses for SiC composite samples because the gamma absorption of gamma leads to the bonding of the bonds, which means increasing the hardness of the samples and therefore less wear. The greater the irradiation period, the gamma rays absorbed by the sample will increase, which leads to the overlap of the material affects the decomposition of the material and thus increase surface resistance and then wear the sample little [17].

IOP Conf. Series: Materials Science and Engineering **757** (2020) 012008 doi:10.1088/1757-899X/757/1/012008

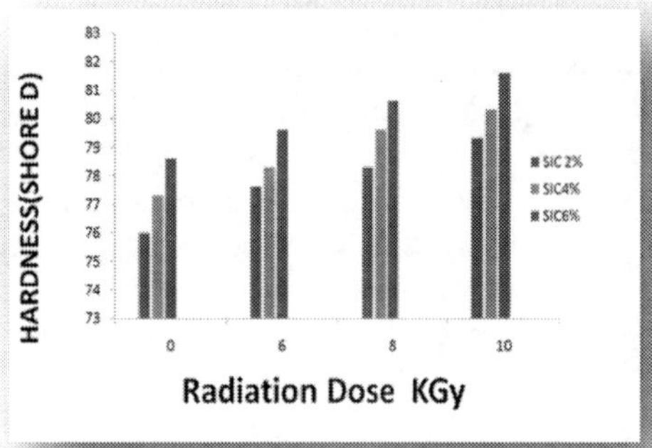

Figure 3. Influence of Ɣ-radiation dose on the hardness of SiC composite for weight fractures (2,4,6)%

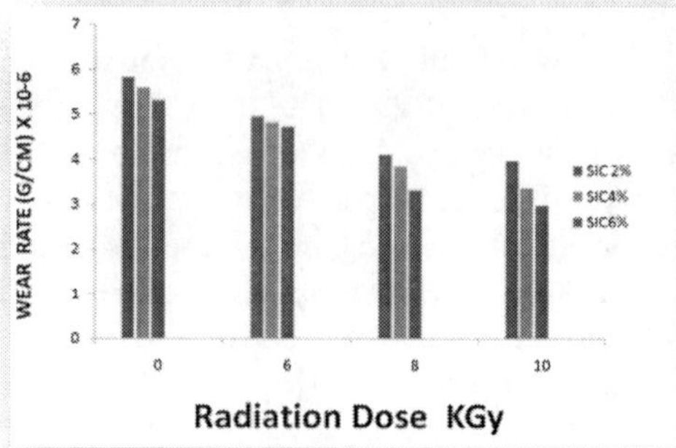

Figure 4. Influence of Ɣ-radiation dose on the wear of SiC composite for weight fractures (2,4,6)%

8. Conclusion:

The hardness of the (EP+SiC) composites increased when radiation doses increase .The hardness of the composite (EP+SiC) increases by increasing the reinforcement ratios, results from others at weight fracture (2%,4%,6%).The wear rate of the (EP+SiC) composites decreased when radiation doses increase .The wear rate of the composite (EP+SiC) decreases with increasing the reinforcement ratios, results from others at weight fracture (2%,4%,6%).

Acknowledgments

I would like to thank all the friends who came with me to complete this research and thank the professors who taught us the science and solve the difficulties to accomplish our work with the best results.

References:

[1] R J Crawford 1987 *Plastics Engineering*, 2nd Edition, Pergamon Press, New York.

[2] D Hull 1981 *An Introduction to Composite Materials*, Cambridge University Press.

[3] Peter A Thornton and Vito J Colangelo 1985 *Fundamentals of Engineering Materials*, Prentice – Hall.

[4] Geoff Eckold 1995 *Design and Manufacture of Composite Structures* Jaico Publishing House.

[5] Hussein Jabbar Hussein 2005 *Study of the Effect of Some Epoxy Properties Supported by Natural Industrial Materials*, Master Thesis, Materials Engineering Department, University of Technology.

IOP Conf. Series: Materials Science and Engineering **757** (2020) 012008 doi:10.1088/1757-899X/757/1/012008

[6] Musa al-Janabi, Atomic Universe 1987 *Atomic Energy Organization* Dar al-Arabiya, Baghdad.

[7] Khaled, O Al-Ahmad 1993 *Introduction to Health Physics* University of Mosul Press.

[8] Reed, K Salman Al-Fahdawi 2002 *Optimal Mechanical Properties of Polymeric Overlay Material Using computer Modeling* Master Thesis, Anbar University Sciences.

[9] Najla Rushdi 2010 *Technical Journal, Vol. 23, No. (1)* Effect and concentration of Particulate Particles on the Mechanical and Thermal Properties of Polymeric Overlay.

[10] Udine and Sanxetre, translated by d. Khalid Younis Al-Yamour 1985 "Principles in Radiological Chemistry" University of Mosul.

[11] Map, Hassan, "Radiation Protection", Arab Atomic Energy Commission, and Syrian Atomic Energy Commission, Damascus, 7/30, (1997).

[12] Huda Jabbar Abdul Hussein, "The Effect of Immersion on the Mechanical and Physical Properties of Polymeric Composites Supported by Nanomaterials of Alumina".

[13] Shehab Ahmed Zidan and Amani Zuhair Mohamed 2017 Ibn al-Haytham Journal of Pure and Applied Sciences, Volume 30, Issue 1, 2017. Study of the Properties of Electromechanical Resistivity of Epoxy Composites Distributed with Silica Powder.

[14] R A Abass 2012 *Iraqi Journal of Science and Technology*, Vol.3, No.3. *Study link Acid Rain through the Evolution of Flexural strain Energy accumulated in the Novolak Composites and Reflection of the Elastic Rebound.*

[15] O'Donell J h 1991 *Chemistry of Radiation Degradation of polymers, in Radiation Effect in polymers* ACHS.

[16] Elsa Reichmans and James , H O Donuell 1990 *The Effect of Radiation on High Technology Polymers* Acs Symposium Series.

[17] B L Oksengendler, N N. Turaeva, I N Ruban, S S Rashidova 2004 *Turk Journal Chemistry* Vol.28, pp 47-51.

IOP Conf. Series: Materials Science and Engineering **757** (2020) 012009 doi:10.1088/1757-899X/757/1/012009

film analysis and quality control from x-rays examinations in some hospitals in Baghdad

M. W. Alhamd , F J Alwan, W O Alobaidi and S S kamoon
Madenat Alelem University college, Baghdad, Iraq

Abstract

There were significant correlation between breathing effects in chest x-ray examination and rejected film.

While the techniques and skills employed contribute to about 50% nearly equally of the rejected films.

The study samples included (300) patients divided males and females. The mean age of subjects was (35±10years) and their mean height was (170±5 cm), and the mean weight was (70±5Kgm)

Introduction

The first phase of the national project in diagnostic radiology concentrate on the film reject rate analysis, quality control for x-ray diagnostic instruments and measurement of patient dose, which was completed within four weeks .

The observed reject rates were found to be nearly independent of the type of x-ray examination, also they reflected an equally 50% by both the nature of technique and skills employed.

Where as the results of measurement of patient dose, which were taken through the first phase of the project, were found to be nearly within the range of the international recommendations given IAEA

Materials and methods

The study was performed with two months, the patients from both sexes were referred to the x-ray sections in (institute of x-ray, specialist surgeries, al-yarmook and Dar - altamreedh hospitals) by various physicians, and they were either inpatients or outpatients

Chest/PA, abdomen/AP, lumber spine /AP, lumber spine /LAT, and pelvis/AP, X-ray examinations were chosen to be studies; they were investigated with general x-ray machine.

Each x-ray room was provided with a daily form for wasted films to be taken .The rejected films were counted according to the x-ray room location and cause of rejection

IOP Conf. Series: Materials Science and Engineering **757** (2020) 012009 doi:10.1088/1757-899X/757/1/012009

Results and Discussion

Rejected film rate analysis:

The overall reject rates by hospitals as a percentage of total films used at institute of x-ray, specialist surgeries, al-yarmook and Dar- altamreedh hospitals, were found to be 6.12%, 4.35%, 4.17% and4.17%, respectively.

Other results in table (2) and (3) indicate that the techniques and skills employed contribute to about 50% nearly equally of the rejected films

Further be observed that the film reject rates are similar across the studies hospitals as shown in figure (1).

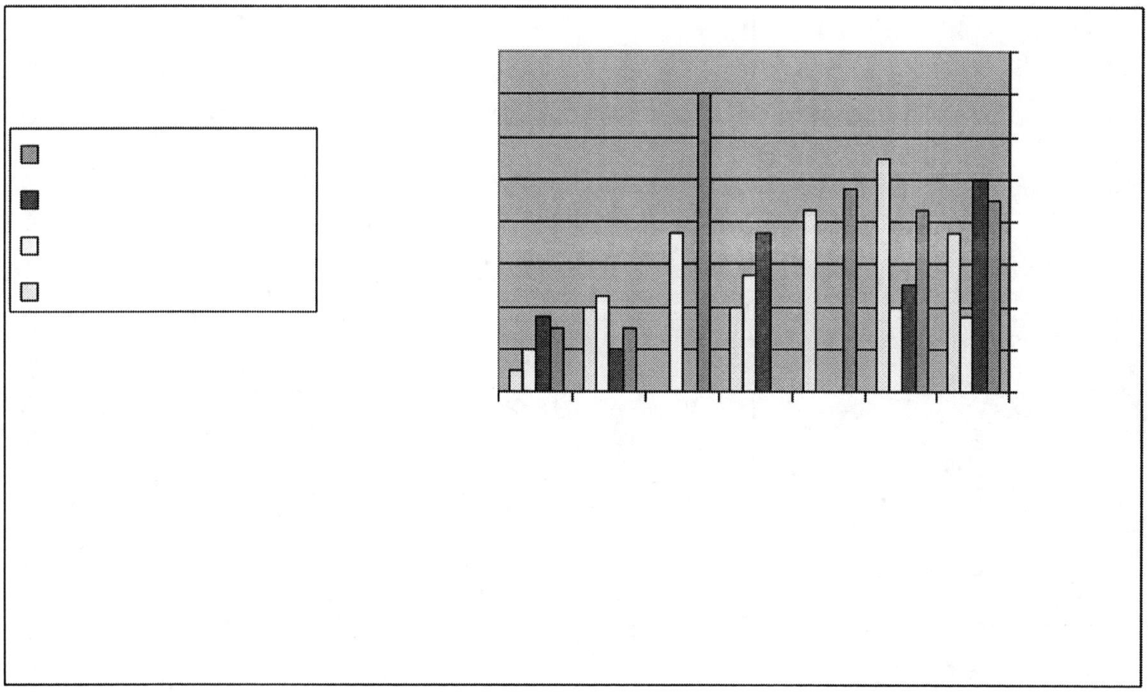

Figure (1): Rejected Rate by Examination For each Hospital as percentage of Film Used in the Examination

IOP Conf. Series: Materials Science and Engineering **757** (2020) 012009 doi:10.1088/1757-899X/757/1/012009
***** Formatting Issue - Best Available Paper/Graphic *****

TABLE (1): rejected rates by reason and room as a percentage of other causes at hospitals.

hospitals	rejected rates by causes%				
	Too dark	Too light	Positioning	Patient movement	other
institute of x-ray Room 1	**55.00**	**25.00**	**5.00**	**5.00**	**10.00**
institute of x-ray Room 2	**33.33**	**33.33**	**0.00**	**33.33**	**0.00**
institute of x-ray Room 3	**12.50**	**37.50**	**25.00**	**12.50**	**12.50**
institute of x-ray Room 4	**0.00**	**55.565**	**22.22**	**22.22**	**0.00.**
specialist surgeries Room 1	**100.00**	**0.00**	**0.00**	**0.00**	**0.00**
specialist surgeries Room 2	**25.00**	**25.00**	**33.33**	**0.00**	**16.67**
al-yarmook Room 1	**53.00**	**27.00**	**5.00**	**5.00**	**10.00**
al-yarmook Room 2	**12.50**	**37.00**	**25.00**	**12.50**	**12.50**
al-yarmook Room 3	**12.50**	**25.00**	**36.00**	**0.00**	**26.00**
Dar – altamreedh Room 1	**4.00**	**4.00**	**11.00**	**25.00**	**20.00**
Dar – altamreedh Room 2	**37.00**	**23.00**	**12.00**	**0.00**	**28.00**
Dar – altamreedh Room 3	**0.00**	**0.00**	**16.67**	**25.00**	**28.00**

TABLE (2): rejected rates by reason and room as a percentage of other causes at hospitals.

hospitals	rejected rates by causes%				
	Too dark	Too light	Positioning	Patient movement	other
institute of x-ray	**26.32**	**14.22**	**13.51**	**9.11**	**36.84**
specialist surgeries	**12.52**	**13.79**	**15.79**	**31.58**	**26.32**
al-yarmook	**20.00**	**30.00**	**30.00**	**5.00**	**15.00**
Dar - altamreedh	**36.93**	**34.48**	**11.34**	**10.34**	**6.90**

The patient reflections on film rejection were observed significantly on chest x-ray examinations and apparently, it is an evidence of breathing effect, as shown in figure (3).

Most of films were used for chest /AP examination (50%) due to the fact that most not all hospitals involved in this project stated to use digital units instead of conventional units like institute of x-ray and Dar – altamreedh as shown in figure (4).

The provision of batter quality films and fresh processing chemicals would improve the results .regular training in related radiation protection matter of the radiology staff would also be useful, for instance, the difficulty in selecting appropriate exposure parameters was observed to be a source of majority avoidably waste films, as shown in figure (5).

***** Formatting Issue - Best Available Paper/Graphic *****

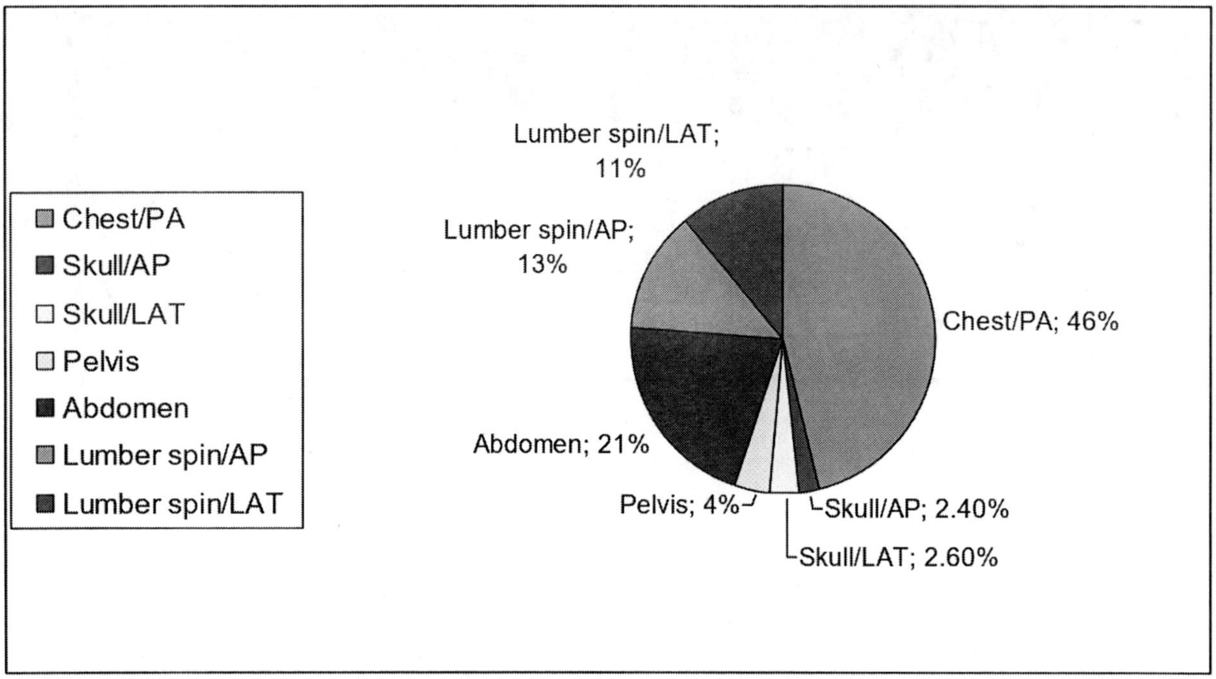

Figure (2): statistical Analysis for Total Rejected Films

Chest/AP	**46%**
Lumber spine / Ap	**13%**
Lumber spine / LAT	**11%**
abdomen	**21%**
pelvis	**4%**
skull/AP	**2.4%**
skull/LAT	**2.6%**

IOP Conf. Series: Materials Science and Engineering **757** (2020) 012009 doi:10.1088/1757-899X/757/1/012009

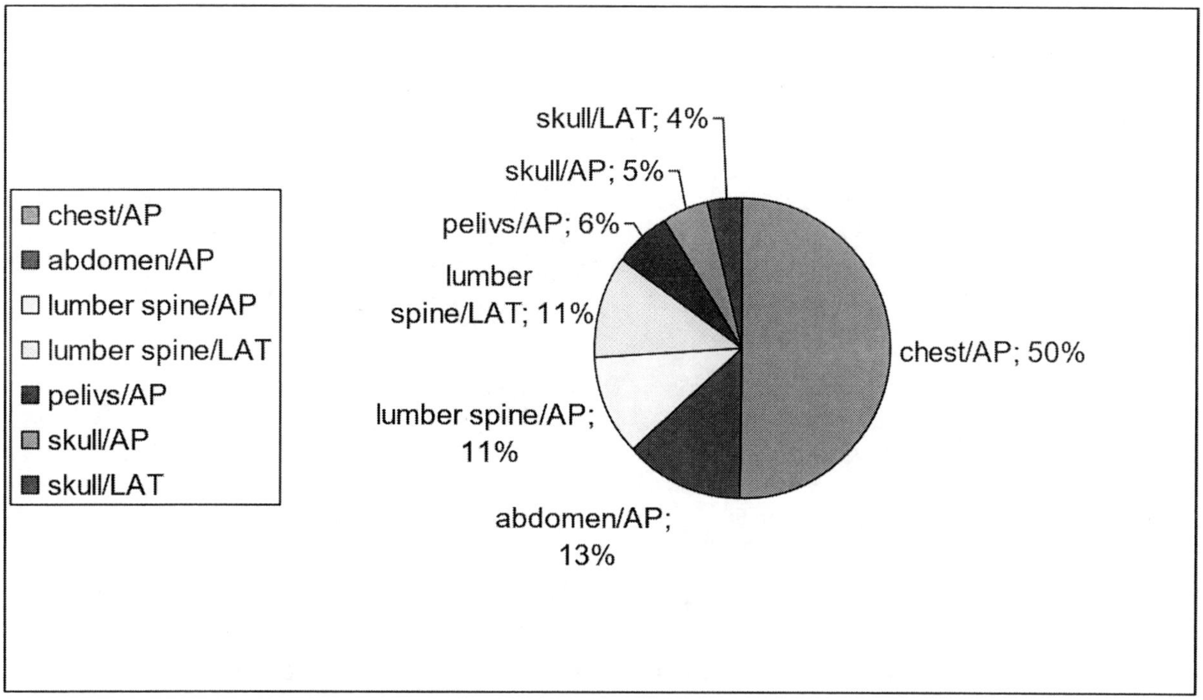

Figure (3): statistical Analysis for Total Films used

chest/AP	**50%**
abdomen/AP	**13%**
lumber spine/AP	**11%**
lumber spine/LAT	**11%**
pelvis/AP	**6%**
skull/AP	**5%**
skull/LAT	**4%**

***** Formatting Issue - Best Available Paper/Graphic *****

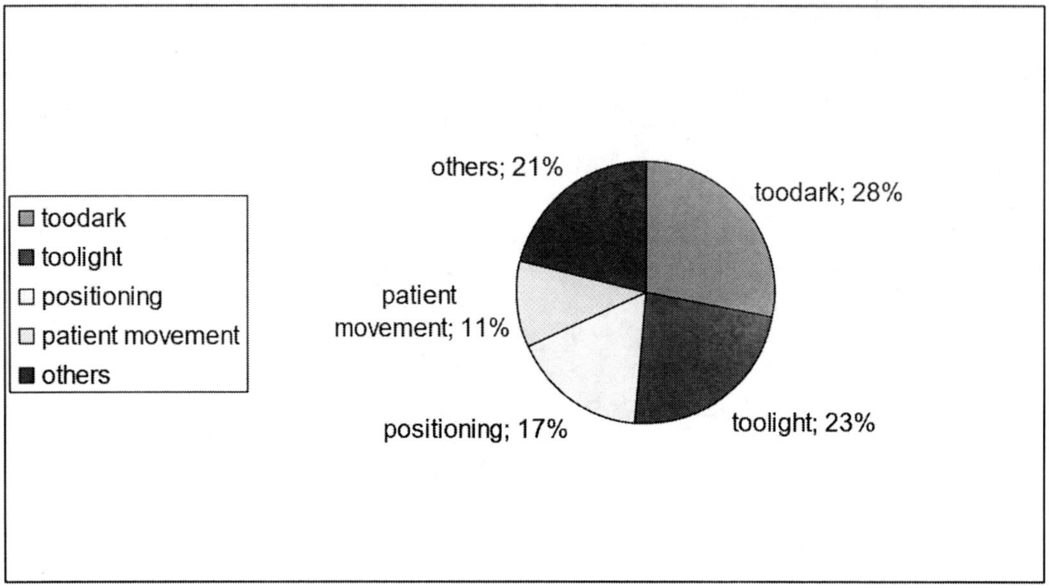

Figure (4): Rejected Rate By cause For All Hospitals As percentage of total Rejected films.

Too dark	**28%**
Too light	**23%**
positioning	**17%**
patient movement	**11%**
others	**21%**

Elevated reject rate in abdomen /AP are evidence of this deficiency as shown in figure (1)

The inadequate training problem manifest it self also in relation the x-ray out put in achieving special diagnostic needs ,for example in x-ray rooms 1,2 and 3 ,we see table (2) recorded significant film reject rate due to characteristic "Too Dark" condition ,and x-ray rooms 4,5,6,7,8 and 9 recorded significant film reject rate due to characteristic "Too Light" conditions. These examples demonstrate the needs to establish regular training for staff in order to reduce the film reject rates and improve related operations.

As we have seen from the previous results, that the majority of waste film were due to lack of the radiation protection culture for radiology staff and absence of in-house quality assurance and programs for immediate corrective action .

***** Formatting Issue - Best Available Paper/Graphic *****

TAEBL (3) : Characteristics of x-ray Equipments.

hospital	x-ray room	x-ray unit					processor	
		Type /SN	Installation date	Generator (pulses)	Total filtration (mm AL)	Focal spot size(mm)	Type	Installation date
institute of x-ray	1	C.GR	1970	Multi pulses	2.5	1.2/1.8	Kodak 5629	1980
	2	Shimadzu	1975	Multi pulses	3	1/2	Kodak10016	2003
	3	Toshiba	1970	Multi pulses	2.5	1.2/1.8	I.M.S A020206	
	4	Philips	1975	Multi pulses	2.5	1/2		1985
specialist surgeries hospital	5	Sedecal	2000	High frequency	2.5	0.6/1.2	Siemens	2004
	6	Sedecal	2000	High frequency	2.5	0.6/1.2	Kodak green/400	
yarmook hospital	7	Shimadzu	1980	Multi pulses	3	1/2	C.G.R	1984
	8	Sedecal	2004	High frequency	2.5	0.6/1.2	C.G.R	1984
	9	Shimadzu	1980	Multi pulses	2.5	1/2	C.G.R	1984
Dar - altamreedh hospital	10	Toshiba	1950	Multi pulses	3	0.6/1	Kodak 9790	1985
	11	Sedecal	2004	High frequency	2.5	1.2/1.8	Sadecal 2682	1999
	12	Sedecal	2004	High frequency	2.5	0.6/1.2		

IOP Conf. Series: Materials Science and Engineering **757** (2020) 012009 doi:10.1088/1757-899X/757/1/012009
***** Formatting Issue - Best Available Paper/Graphic *****

Tables (4): patient information, the mean the corresponding summary of selected techniques at each hospital.

Abdomen/AP first round measurements, results of hospitals and X-ray rooms

No.	Hospitals	Room NO.	Speed class of film screen	Patient weight(Kg)*	Applied potential (kVp)*	Half layer value (mmAL)	Tube current (mAs) *
1	Institute of X-ray	1	200	71.5±1.09	77.00±0.00	3.15	21.72±2.70
		2	200	72.2±1.39	58.80±2.68	3.76	18.64±2.75
		3	400	70.92±1.41	70.35±3.46	3.38	24.33±5.77
2	Specialist Surgeries Hospital	4	400	67.90±1.44	67.10±2.18	3.01	21.65±3.15
		5	200/400	71.5±1.09	77.00±0.00	3.64	19.53±6.67
		6	200/400	71.5±1.09	70.10±4.86	3.01	25.50±9.45
3	Yarmook Hospital	7	200/400	71.6±1.23	65.54±1.71	3.13	23.14±3.42
		8	200/400	72.2±2.19	68.59±3.00	3.4	18.57±2.47
		9	400	68.3±1.45	77.00±0.00	3.31	21.67±2.54
4	Dar-Altamreedh Hospital	10	400	69.4±1.24	75.50±3.63	3.18	28.85±5.14
		11	200	72.2±1.15	70.24±4.86	3.4	23.83±4.13
		12	400	68.3±1.43	73.14±3.26	3.32	24.85±4.89

Chest /AP first round measurements, results of hospitals and X-ray rooms

No.	Hospitals	Room NO.	Speed class of film screen	Patient weight(Kg)*	Applied potential (kVp)*	Half layer value (mmAL)	Tube current (mAs) *
1	Institute of X-ray	1	200	70.3±2.6	117±2.6	3.15	0.72±0.22
		2	200	67.5±1.5	85.9±4.5	3.76	0.62±0.23
		3	400	70.1±1.32	117±00	3.38	0.65±0.12
2	Specialist Surgeries Hospital	4	400	71.6±1.61	117±00	3.01	0.67±0.24
		5	200/400	72.2±2.19	113±3	3.64	0.72±0.15
		6	200/400	72.4±2.12	113±3	3.01	0.82±0.34
3	Yarmook Hospital	7	200/400	68.3±1.44	93±1.5	3.13	0.75±0.10
		8	200/400	69.4±1.36	113±0.9	3.4	0.67±0.14
		9	400	71.5±1.09	94.7±3.6	3.31	0.73±0.13
4	Dar-Altamreedh Hospital	10	400	72.2±1.45	109±00	3.18	0.46±0.21
		11	200	70.92±132	117±00	3.4	0.19±0.24
		12	400	67.90±148	109±00	3.32	0.54±0.20

*** Formatting Issue - Best Available Paper/Graphic ***

Lumber spine /AP first round measurements, results of hospitals and X-ray rooms

No.	Hospitals	Room NO.	Speed class of film screen	Patient weight(Kg)*	Applied potential (kVp)*	Half layer value (mmAL)	Tube current (mAs) *
1	Institute of X-ray	1	200	73.00±2.5	68.60±2.6	3.15	35.00±0.00
		2	200	66.10±4.16	77.00±00	3.76	32.67±2.52
		3	400	76.20±2.3	84.10±1.75	3.38	37.26±3.18
2	Specialist Surgeries Hospital	4	400	79.00±2.00	65.45±2.5	3.01	35.00±0.00
		5	200/400	73.00±2.5	82.00±00	3.64	38.42±2.89
		6	200/400	75.00±2.6	78.00±00	3.01	30.28±1.75
3	Yarmook Hospital	7	200/400	70.75±5.3	77.00±00	3.13	32.67±2.52
		8	200/400	73.00±2.6	67.88±3.4	3.4	36.09±3.05
		9	400	70.8±3.0	80.00±00	3.31	30.28±1.82
4	Dar-Altamreedh Hospital	10	400	66.10±4.2	69.5±3.0	3.18	32.67±2.52
		11	200	72.60±8.0	62.00±1.2	3.4	36.00±00
		12	400	79.00±00	80.00±00	3.32	30.28±0.83

Lumber spine/LAT first round measurements, results of hospitals and X-ray rooms

No.	Hospitals	Room NO.	Speed class of film screen	Patient weight(Kg)*	Applied potential (kVp)*	Half layer value (mmAL)	Tube current (mAs) *
1	Institute of X-ray	1	200	66.00±5.66	95±2.62	3.15	37.08±1.56
		2	200	75.40±2.59	90±3.57	3.76	35.17±2.45
		3	400	71.20±4.69	90±4.72	3.38	38.24±1.71
2	Specialist Surgeries Hospital	4	400	78.67±5.6	93.±2.88	3.01	32.43±5.62
		5	200/400	71.00±8.15	90±3.24	3.64	39.76±7.35
		6	200/400	66.45±3.42	90±2.72	3.01	36.31±4.93
3	Yarmook Hospital	7	200/400	75.40±2.6	94±7.25	3.13	35.25±5.77
		8	200/400	71.30±1.5	92±6.33	3.4	37.34±5.81
		9	400	74.00±3.16	95±2.52	3.31	32.75±3.22
4	Dar-Altamreedh Hospital	10	400	70.75±2.31	93±3.29	3.18	35.16±5.44
		11	200	67.30±1.96	95±2.75	3.4	38.27±3.12
		12	400	73.18±4.5	90±3.42	3.32	33.47±2.63

IOP Conf. Series: Materials Science and Engineering **757** (2020) 012009 doi:10.1088/1757-899X/757/1/012009

Pelvis /AP first round measurements, results of hospitals and X-ray rooms

No.	Hospitals	Room NO.	Speed class of film screen	Patient weight(Kg)*	Applied potential (kVp)*	Half layer value (mmAL)	Tube current (mAs) *
1	Institute of X-ray	1	200	75.40±2.3	85.00±2.5	3.15	28.85±8.43
		2	200	71.20±4.69	83.00±3.7	3.76	32.00±000
		3	400	67.34±2.6	79.14±0.41	3.38	21.72±2.70
2	Specialist Surgeries Hospital	4	400	73.24±3.17	82.75.±1.74	3.01	18.58±2.47
		5	200/400	74.00±3.16	80.00±2.33	3.64	23.83±4.13
		6	200/400	78.67±2.31	80.35±4.15	3.01	24.85±4.98
3	Yarmook Hospital	7	200/400	66.00±5.66	79.85±2.72	3.13	28.85±5.49
		8	200/400	75.00±2.7	82.68±1.6	3.4	32.32±5.81
		9	400	71.34±4.67	85.38±0.25	3.31	21.72±2.70
4	Dar-Altamreedh Hospital	10	400	68.00±3.62	84.43±7.41	3.18	18.68±2.47
		11	200	71.00±8.14	79.75±1.32	3.4	32.00±0.00
		12	400	77.32±5.41	78.72±0.91	3.32	25.46±7.66

Skull/AP first round measurements, results of hospitals and X-ray rooms

No.	Hospitals	Room NO.	Speed class of film screen	Patient weight(Kg)*	Applied potential (kVp)*	Half layer value (mmAL)	Tube current (mAs) *
1	Institute of X-ray	1	200	71.5±1.09	70.10±4.86	3.15	23.14±3.42
		2	200	71.6±1.23	65.54±1.71	3.76	18.57±2.47
		3	400	72.2±2.19	77.00±0.00	3.38	21.67±2.54
2	Specialist Surgeries Hospital	4	400	68.3±1.45	58.80±2.68	3.01	28.85±5.14
		5	200/400	71.5±1.09	70.35±3.46	3.64	23.83±4.13
		6	200/400	72.2±1.39	67.10±2.18	3.01	24.85±4.89
3	Yarmook Hospital	7	200/400	70.92±1.41	77.00±0.00	3.13	21.72±2.70
		8	200/400	67.90±1.44	68.59±3.00	3.4	18.64±2.75
		9	400	71.6±2.18	77.00±0.00	3.31	24.33±5.77
4	Dar-Altamreedh Hospital	10	400	69.4±1.24	75.50±3.63	3.18	21.65±3.15
		11	200	72.2±1.15	70.24±4.86	3.4	19.53±6.67
		12	400	68.3±1.43	73.14±3.26	3.32	25.50±9.45

***** Formatting Issue - Best Available Paper/Graphic *****

Skull/LAT first round measurements, results of hospitals and X-ray rooms

No.	Hospitals	Room NO.	Speed class of film screen	Patient weight(Kg)*	Applied potential (kVp)*	Half layer value (mmAL)	Tube current (mAs) *
1	Institute of X-ray	1	200	71.20±4.69	68.00±00	3.15	21.72±2.70
		2	200	75.00±2.58	60.00±3.00	3.76	19.68±2.47
		3	400	71.00±8.15	77.00±00	3.38	26.32±5.8
2	Specialist Surgeries Hospital	4	400	78.67±3.52	64.80.±5.35	3.01	23.45±5.77
		5	200/400	70.75±5.66	75.00±00	3.64	23.14±3.42
		6	200/400	74.00±3.16	77.00±0.00	3.01	21.67±2.54
3	Yarmook Hospital	7	200/400	68.30±4.42	68.00±0.00	3.13	23.83±4.13
		8	200/400	63.50±4.95	63.00±1.37	3.4	26.60±0.44
		9	400	70.00±00	79.00±1.41	3.31	20.84±0.50
4	Dar-Altamreedh Hospital	10	400	75.00±7.07	64.73±3.00	3.18	24.80±4.98
		11	200	72.50±3.02	76.00±5.00	3.4	21.65±2.45
		12	400	71.70±2.71	73.84±2.52	3.32	24.76±3.96

References

1- International Commission on Radiation Protection, Recommendation of the International Commission on Radiological Commission, ICRP PUBLICATION 60, Pergam on Press Oxford. (1990).
2- H. Geijer, K.-W. Beckman, T. Andersson, J. Persliden ((Radiation dose and image quality with a flat-panel amorphous silicon digital detector European Radiology)) , 11 (2 Suppl. 1) p. S280, (2001) .
3- Zewdeneh, D., Teferi., S and Admassie,((D. X-ray reject analysis in TikurAnbessa and Bethzatha hospitals.Ethiop)). J. Health Dev. 22(1), 63–67 (2008).
4- M.A. Dunn, A.T. Rogers((X-ray film reject analysis as a quality indicator Radiography)) , pp. 29-31 , (2008) .
5- B. Verdonck, R. Nijlunsing, N. Melman, H. Geijer ((Image quality and X-ray dose for translation reconstruction overview imaging of the spine, colon and legs Proceedings of Computed Assisted Radiology and Surgery)) , pp. 500-505, (2012).

IOP Conf. Series: Materials Science and Engineering **757** (2020) 012010 doi:10.1088/1757-899X/757/1/012010

Evaluation of natural radioactivity in selected soil samples collected from northern regions of Oman

A. Al-Harmali[1]

[1] Physics department, Science College, Sultan Qaboos University, Oman, corresponding author: Email: al-amal85@moe.om

Abstract: This study explores the radioactivity concentration in soil in the Northern regions of Oman. The naturally occurring radioactive nuclides of the main series, ^{238}U, ^{232}Th, ^{235}U and ^{40}K and artificial radionuclide ^{137}Cs have been studied by means of low level gamma spectroscopy. Majority of the studied valleys (wadis) are descending from Super Hajar Mountains either towards its east part (coastal area) or the west part (desert). Soil samples were collected from headwaters and estuaries of wadis where residences and agriculture activities exist. In soil samples, the results showed that ^{238}U radionuclide was presented in concentrations of the range (4.5 – 24.0) Bq.kg^{-1}while ^{232}Th was found in the range (2.4-20.6) Bq.kg^{-1}. Ranges of ^{235}U and ^{40}K were reported as (0.6-2.5) Bq.kg^{-1} and (47-296) Bq.kg^{-1} respectively. Artificial radionuclide ^{137}Cs was found in all soil samples in the range (2.8- 11.9) Bq.kg^{-1}. The hazard parameters, dose rate (D), annual effective dose (H_R), equivalent radium activity (Ra_{eq}), internal hazard index (H_{in}), external hazard index (H_{ex}) and gamma representative level index ($I_{\gamma r}$) were all calculated. The values of D, H_R and Ra_{eq} were found in the ranges (6.6-33.5) nGy/h, (0.8-4.1)*10^{-2} mSv/yr and (17.2-88.6) Bq.kg^{-1} respectively. The values of $H_{in,}$ H_{ex} and $I_{\gamma r}$ were all less than 1. These obtained results remained below the optimum admissible values according to UNSCEAR.. The results indicate that the radiation hazard from naturally and artificially radionuclides in soil samples from the area studied in this current work is not significant.

Keywords: Radioactivity- Soil – HPGe Detector- Gamma

1. Introduction

Ionizing radiation arises from radionuclides that produce ionizing α and β particles and γ rays. Two types of radionuclides are recognized, the man made radionuclides and the naturally occurring radionuclides. Higher concentrations of these radionuclides may cause threats to humans in the form of chronic lung diseases, acute, necrosis of the mouth, lung, pancreas, kidney, cancers and leukemia [1].

Under ordinary conditions; artificial radioactivity emitted from the nuclear power plants, industrial plants and research facilities has smaller impact to the overall radiation. Natural environmental radioactivity arises mainly from primordial radionuclides, such as ^{226}Ra, ^{232}Th, ^{40}K and their decay products, which occur at trace levels in all ground formations [2]. Most wadis in Oman are having headwaters descend from the mountains towards the coast or deserts. Majority of the studied valleys are descending from Hajar mountains either towards its east part (coastal area) or the west one (desert). Along these valleys, sub-surface water is available. In result, these regions are agricultural regions and therefore are populous too. Consequently, this study will provide a baseline data of radiation considering 15 locations from northern regions of Oman.

1.1 Gamma (γ) Emission

When a new nucleus is formed after the decay of Alpha or Beta, it is left in an excited state with more energy than usual. This energy is lost by the emission of a gamma ray (photon). For sake of stability, the excited nucleus decays to the ground state by emitting gamma rays with released energy 0.661 MeV (figure 1). This path is followed 94.6% of the time whereas directed β decay carries the rest 5.5% [3].

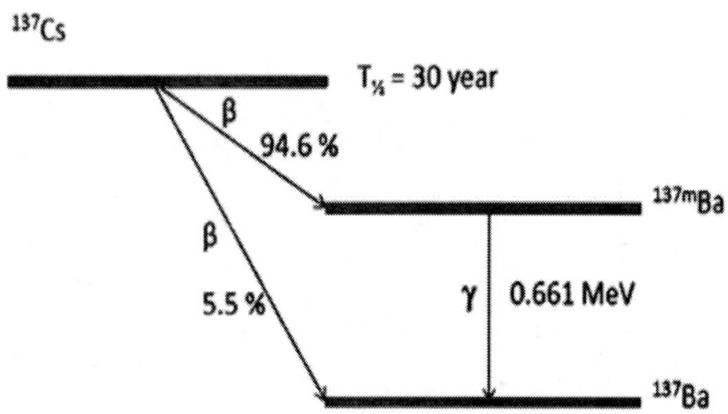

Figure (1): Decay of ^{137m}Ba by Gamma emission.

A radionuclide's activity (A) is calculated by the equation:

$$A = -\frac{dN}{dt} = A_o e^{-\lambda t} \quad .... \quad (1.1)$$

Where A_o is the initial count rate at time (t=0), N is the number of radioactive nuclei, and λ is the decay const. and associated with half-life. Half-life is the time for half of the atoms of a given nuclide to decay. The half-life is a unique characteristic of each nuclide range from millions of a second to millions of years.

Naturally occurring radionuclides are distinguished to three decay chains (^{235}U, ^{238}U and ^{232}Th,). The longest-lived member is ^{232}Th, which has half-life of $1.4*10^{10}$ years [2]. Some radioactive decays are non-series. The most important non-series radionuclide in this study is (^{40}K), which contributes 0.012% of natural potassium. Uranium, thorium and potassium are the major radionuclides contained in the Earth's crust [4]. Table (1) shows radionuclides, their half-lives and major radiation [5].

Table 1: Radionuclides, their half-lives and major radiation

Radionuclide	Half-life	Major Radiation
Thorium-228	1.91 y	α,γ
Thorium-232	1.40 E10 y	α,γ
Uranium-234	2.46 E5 y	α,γ
Uranium-235	7.04 E 8 y	α,γ
Uranium-238	4.47 E9 y	α,γ
Radium-226	1.60 E3 y	α,γ
Radium-228	5.75 y	β,γ
Radon-222	3.82 day	α,γ
Lead-210	22.2 y	α,γ,β
Potassium-40	1.25 E9 y	β,γ
Cesium-137	30.1 y	β,γ

2. Materials and Methods

2.1 Soil samples Collection and Preparation

The sultanate of Oman occupies most of the south east corner of Arabian Peninsula. The distribution of natural radioactivity in the environment depends basically on the geological and geographical properties. In this study, decision was made to collect samples of soil from east and west of western Hajar Mountains. The sampling area considered in this study located at west of north Al Hajar Mountains where the influence of dust storms on the Arabian Peninsula takes place. Heavy rain drifts soil from mountains to the wadis and plain. Disturbed soils cause a decrease in the net surface radiation. Furthermore, the Ophiolites which are igneous rocks constitute the bulk of the geographical configuration of the northern regions of Oman. In terms of natural radioactivity; it is well known that igneous rocks are strongly enriched in thorium and uranium [6]. Samples were collected from 15 different locations. The locations are shown on the map in figure 2).

IOP Conf. Series: Materials Science and Engineering **757** (2020) 012010 doi:10.1088/1757-899X/757/1/012010

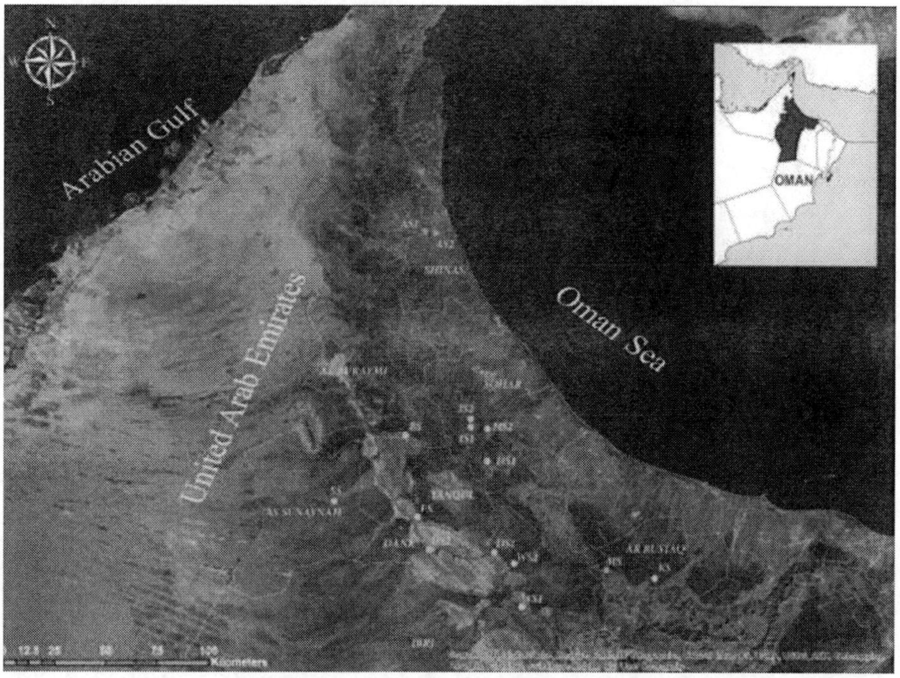

Figure(2): Map shows the locations of collected samples from northern regions of Oman.

In the Lab, soil samples were dried in a slow-airflow at room temperature for several days, and then they were sieved through a 1000 μm mesh sieve in order to homogenize the sample and remove organic materials and stones. Half kilogram of each sample was weighed on a digital balance. Later they were sealed in standard geometry containers (0.5 litter Marinelli beakers). The samples were kept for 4 weeks so that equilibrium would take place between ^{226}Ra and its short lived progenies [3].

3.Results and Discussions

Although thorium has higher activity concentrations in world averages according to UNESCAR [7], the results of the present study showed that the activity concentrations of uranium were higher than thorium in all samples. If the ratio is of $^{232}Th/^{40}K$ in the soil samples is (>0.5) (figure 3), this will ensure the absence of heavy minerals in the study area and vice versa. This indicates the insoluble state of thorium ion Th^{+4}. In contrast, uranium was found in U^{+6} oxidation state which is soluble. This interprets the larger concentrations of uranium paralleling with thorium [8] .

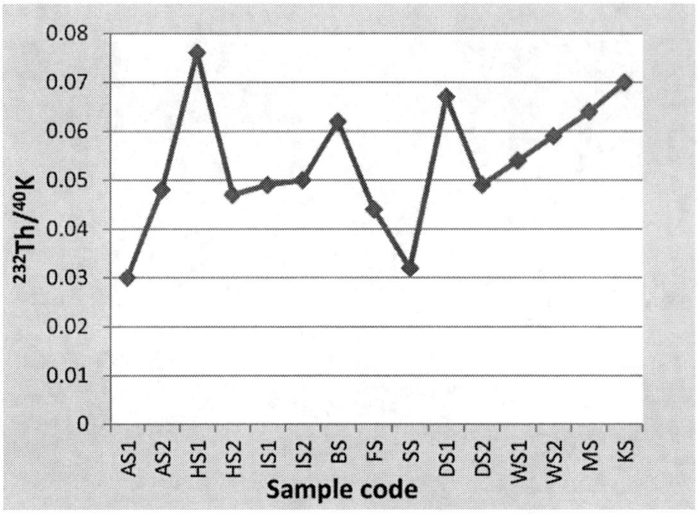

Figure(3): $^{232}Th/^{40}K$ Activity ratios of soil samples

However, the values measured remain below the optimum worldwide average values and appropriately comparable to concentrations reported from different countries (Table 2).

Table (2): A comparison of activity concentrations in soil between this study and different studies reported in other countries.

Country	Activity concentration(Bq.kg^{-1})			References
	^{232}Th	^{40}K	^{226}Ra	
Present study	**2.4 - 20.6**	**47.2 - 224.3**	**10 - 43**	**Present study**
Oman,Musandam	**4.5**	**86.4**	**25.4**	**[9]**
Oman,Musandam	**9.95**	**158.21**	**-**	**[10]**
AlQasseem, Saudi Arabia	**2.5-39**	**212-915**	**1.4-35.3**	**[11]**
Riyadh,Saudi Arabia	**7-25**	**89 - 320**	**11 – 30**	**[12]**
Jedda,Saudi Arabia	**7.4**	**369**	**9.3**	**[13]**
Ad-Dahna, SaudiArabia	**15.8 - 36.7**	**285.3 - 533.2**	**16.2 - 30.6**	**[14]**
Yemen	**18 - 113**	**64 – 1667**	**16.6 – 84.4**	**[1]**
World-wide	**45**	**420**	**25**	**[6]**

3.1 Variation of ^{238}U series activities

The stability in ^{238}U series was checked by plotting a linear fit between ^{238}U and ^{226}Ra (figure 4), which shows a linear regression for their activity concentrations. It showed a statistically correlation with a correlation coefficient of (R= 0.77) and with $P < 0.0004$, where P is the probability of rejecting the truth. However the correlation (R=0.77) is not significant and indicate the result gotten before that uranium is less stable and more soluble (+6 state) in the study area. This explained the two upper shifted points in figure (4).

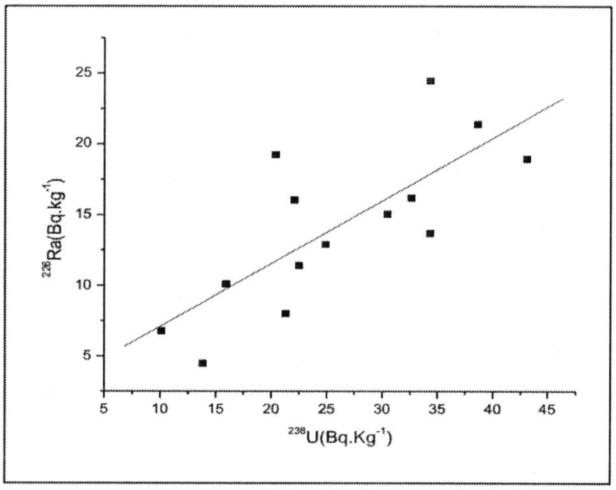

Figure (4): Linear regression of the activity concentrations of ^{238}U versus ^{226}Ra for all soil samples.

3.2 Variation of ^{232}Th series activities

Activity concentrations were calculated in soil for the significant radionuclides recognized in ^{232}Th decay series (see figure 5), which are ^{228}Ac, ^{212}Pb and ^{212}Bi.

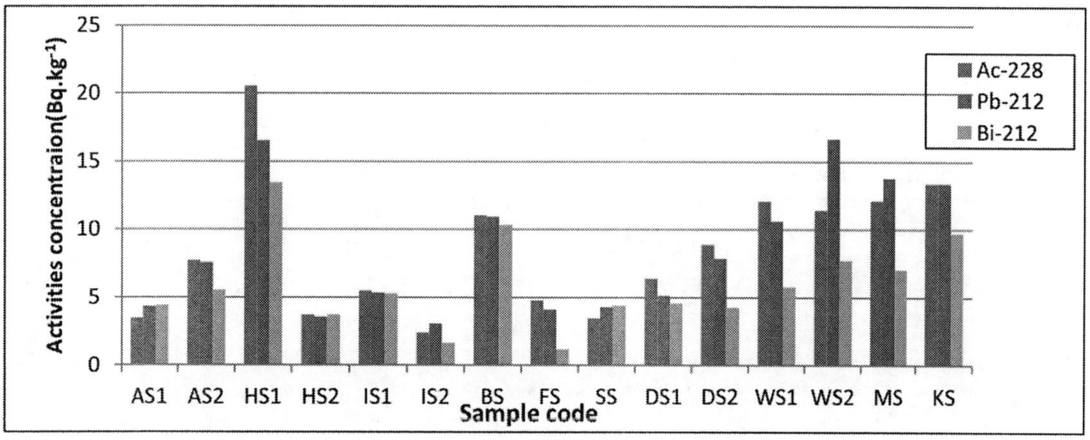

Figure (5): Variation of the concentrations of radionuclides in ^{232}Th series

This study shows that thorium series in R=0.98) is more stable than uranium series (R=0.77) and it is less soluble. The stability between uranium series and thorium series in soil was checked by plotting a linear fit (Figure 6) between the activity concentrations of ^{238}U and ^{232}Th radionuclides. The correlation shows low significance with R=0.83.

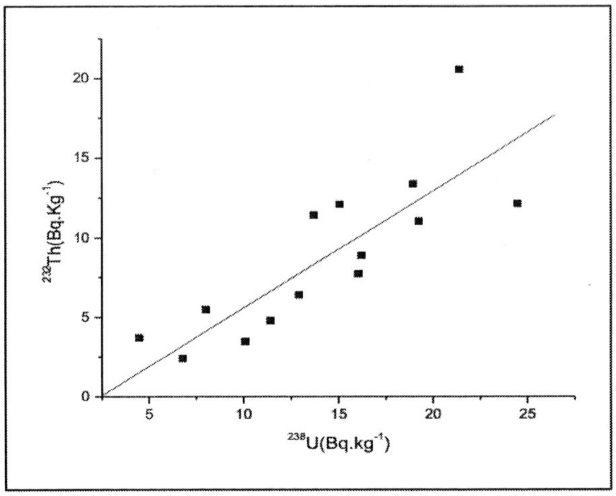

Figure (6): Linear regression between uranium and thorium series for all soil samples.

3.3 Variation of ^{137}Cs activity

The activity concentrations of the artificial nuclide ^{137}Cs from soil samples showed some of the values to be bit higher than the values obtained from other studies conducted in neighboring countries as shown in table (3). The investigated area, Hajar Mountains are very high mountains and peaking at about 3000 meters above sea level. The released cesium in the air blend with clouds accumulated on the top of Hajar Mountains. Later these amounts of ^{137}Cs are deposited on the wadis by rain. However, the influence on humans from such low levels of ^{137}Cs is still insignificant.

Table (3) : Activity concentrations of ^{137}Cs in soil samples in the present study and other studies

Study area	Activity concentrations of ^{137}Cs in soil (Bq.Kg^{-1})	References
Northern Oman	**2.8-11.9**	**Present study**
Oman, Musandam	**1.7**	**[9]**
Kuwait	**4.45**	**[16]**
Iraq	**15.9-70.9**	**[17]**
Jordan	**1.3**	**[18]**
Pakistan	**3.5**	**[19]**
Saudi Arabia	**1.7**	**[20]**
Saudi Arabia	**4.5**	**[21]**

4. Assessment of hazards

Exposure to radiation can be defined in terms of many parameters, such as absorbed dose rate, annual effective dose, radium equivalent activity, Gamma representative level and internal and external hazard indices.
The external absorbed dose rate in air at a height of about 1 m above the ground was calculated by:

$$D\ (\text{nGy}.h^{-1}) = 0.0417A_K + 0.462A_{Ra} + 0.604A_{Th} \quad \text{......... (1)}$$

According the annual effective dose received by a member in mSv.y^{-1} unit, the values obtained for the water samples were calculated using equation:

$$H_R(\mu\text{Sv}.y^{-1}) = 8760\times 0.2 \times 0.7 \times 10^{-6} \times D \quad \text{.....................(2)}$$

The radium equivalent activity (Raeq) was calculated by equation:

$$R_{aeq} = A_{Ra} + 1.43A_{Th} + 0.077A_K \quad \text{.. (3)}$$

The external, internal and gamma representative level hazard indices which are on concern due to their impact on people health were calculated using:

$$H_{ex} = \frac{A_{Ra}}{370} + \frac{A_{Th}}{259} + \frac{A_K}{4810} \leq 1 \quad \text{.. (4)}$$

$$H_{in} = \frac{A_U}{185} + \frac{A_{Th}}{259} + \frac{A_K}{4810} \leq 1 \text{........... (5)}$$

$$I_{yr} = \frac{A_U}{150} + \frac{A_{Th}}{100} + \frac{A_K}{1500} \leq 1 \quad \text{......... (6)}$$

formulas respectively. Where A_{Ra}, A_{Th} and A_K are the activities of ^{226}Ra, ^{232}Th and ^{40}K in Bq.kg^{-1}, respectively [22]. The calculations as shown in table (4) ensured that the values of these indices are less than the unity, the optimum acceptable limit in globe. Hence these areas can be considered as risk free zones from human health perspective.

Table (4): Hazard parameters of soil samples conducted from different studies.

5. Conclusion

Country	D (nGy/h)	H_R (*10^{-2}) (μSv/yr)	R_{aeq} (Bq.kg^{-1})	H_{in}	H_{ex}	I_{yr}	References
Present Study	6.6-33.5	0.8-4.1	17.2-88.6	0.07-0.34	0.04-0.19	0.10-0.53	Present study
Oman, Musandam	-	2.2	38.5	-	-	-	[9]
Oman, Musandam	-	2.5	40.8	-	-	-	[10]
Al-Qasseem, Saudi Arabia	35.2	4	68.1	0.21	0.18	0.54	[11]
Ad-Dahna, Saudi Arabia	51.4	3.2	106	-	0.29	-	[21]
Basrah, Iraq	49.35	-	105.6	0.38	0.29	-	[15]
World-wide	57	7.0	128.7	<1	<1	<1	[6]

IOP Conf. Series: Materials Science and Engineering **757** (2020) 012010 doi:10.1088/1757-899X/757/1/012010

In most locations, soil samples activity concentrations parents' radionuclides and their daughters were lower than the recommended worldwide limits except for six locations activity concentrations of ^{226}Ra were found to be higher than the worldwide average. Artificial radionuclide ^{137}Cs was found in all soil samples as a result of the global fallout contamination from nuclear weapon testing and nuclear accidents around the world.d

References

[1]El-kamel, A. E., El-mageed, A. I. A., Abbady, A. E., Harb, S., & Saleh, I. I. Natural Radioactivity of Environmental Samples and their Impact on the Population at Assalamia-Alhomira Area in Yemen, Geosciences 2(5), 125–132 (2012)..

[2] Simon, S. L., & Ph, D. INTRODUCTION TO RADIATION PHYSICS AND DOSIMETRY.Radiation Epidemiology Branch, National Cancer Institute, U.S.A(2011).

[3]Al-Leswas. Salim, P. Evaluation of Natural Radioactivity in Environmental Samples ., University of Surrey , Master thesis (2010)..

[4] Dash, A., Knapp Jr, F. R., & Pillai, M. R. A. Industrial radionuclide generators: a potential step towards accelerating radiotracer investigations in industry. Rsc Advances journal, 3(35), 14890-1490(2013).

[5]Das, A., & Ferbel, T. Introduction to nuclear and particle physics. (2nd Edition) World Scientific . ISBN: 981-238-744-7 (2003).

[6] United Nations Scientific Committee on the Effects of Atomic Radiation UNSCEAR, B. Sources and Effects Of Ionizing Radiation. New York, USA (2000)..

[7]International-Atomic-Energy-Agency (IAEA). Measurement of Radionuclides in Food and the Environment (2003)..

[8] V. Ramasamy, K. Paramasivam, G. Suresh, M.T. Jose. Spectrochim Acta Part A. 112 340–350 (2014).

[9]Al-kamiyani, S. S. S. Assessment of Gamma-Emitting Radionuclides in Soil and Beach Sediments :Musandam Governorate, Master thesis. Sultan Qaboos University (2011).

[10] Saleh, I. H. Radioactivity of ^{238}U, ^{232}Th,^{40}K, and ^{137}Cs and assessment of depleted uranium in soil of the Musandam Peninsula, Sultanate of Oman. Turkish Journal of Engineering and Environmental Sciences, 36(3), 236-248 (2012).

[11] El-Taher. Radioactivity measurements and radiation dose assessments in soil of Al-Qassim region , Saudi Arabia. Indian journal of pure and applied physics, (52) 147–154 (2014) .

[12] El-Mageed, A. I., El-Kamel, A. H., Abbady, A., Harb, S., Youssef, A. M. M and Saleh, I. I.. 'Assessment of Natural and Anthropogenic Radioactivity Levels in Rocks and Soil in The Environments of Juban Town in Yemen', Radiation Physics And Chemistry journal, Vol. 80, PP. 710-715 (2010).

[13] Poschl, M., & Nollet, L. M. L. Radionuclide Concentrations in Food and the Environment. USA: Taylor & Francis Group, LLC (2007)..

[14] Alaamer, A. S. Measurement of Natural Radioactivity in Sand Samples Collected from Ad-Dahna Desert in Saudi Arabia, World Journal of Nuclear Science and Technology, 187–191 (2012b).

[15]- Jebur, J. H., & Subber, A. R. H. Activity Concentration of 222 Rn Gas , 226 Ra , 232 Th and K in Crops and Soil Taken from Safwan Granges Using Active , Passive and Gamma Spectroscopy Techniques, British Journal of Applied Science & Technology 11(6), 1–12. Article no.BJAST.19983 (2015).

[16] Al-Trabulsy, H. A., Khater, A. E. M., & Habbani, F. I. Radioactivity Levels and Radiological Hazard Indices at the Saudi Coastline of the Gulf of Aqaba. Radiat. Phys. Chem. , 80, 343–348 (2011).

[17] Jebur, J. H., & Subber, A. R. H. Activity Concentration of 222 Rn Gas , 226 Ra , 232 Th and K in Crops and Soil Taken from Safwan Granges Using Active , Passive and Gamma Spectroscopy Techniques, British Journal of Applied Science & Technology 11(6), 1–12. Article no.BJAST.19983 (2015) .

[18] Al-Kheliewi, A. S., Shabana, S. I., & Farouk, M. A. Concentration Of 238Pu, 239+240Pu, 241Am, 90Sr, and 137Cs Radionuclides in Marine Sediments at the Al-Khafji, Mneefa, And Aljobail Coastal Areas of the Arabian Gulf in Saudi Arabia. The Arabian Journal For Science and Engineering, 33(IA), 139-146 (2008).

[19] Akhtar, N., Tufail, M., Ashraf, M., & Iqbal, M. M. Measurement of envzironmental radioactivity for estimation of radiation exposure from saline soil of Lahore, Pakistan. Journal of Radiation Measurements, 39(1), 11-14 (2005).

[20] Alaamer, A. S. Characterization of 137 Cs in Riyadh Saudi Arabia Soil Samples, World Journal of Nuclear Science and Technology 161–164 (2012a).

[21] Al- Mamari, A. S. S, Radiological assessment of selected groundwater in ADhahira Governorate, Sultanate of Oman, Master thesis. Sultan Qaboos University (2013).

[22] Chandrasekaran, A., Ravisankar, R., Rajalakshmi, A., Eswaran, P., Vijayagopal, P., & Venkatraman, B. Spectrochimica Acta Part A : Molecular and Biomolecular Spectroscopy Assessment of natural radioactivity and function of minerals in soils of Yelagiri hills , Tamilnadu , India by Gamma Ray spectroscopic and Fourier Transform Infrared (FTIR) techniques . Spectrochimica Acta Part A: Molecular and Biomolecular Spectroscopy, 136, 1734–1744 (2015).

Assessment of natural radioactivity hazards in selected water samples collected from northern regions of Oman

A Al-Harmali[1]

[1] Physics department, Science college, Sultan Qaboos University, Oman, corresponding author: Email: al-amal85@moe.om

Abstract: This study explores the radioactivity hazards in water samples collected from the Northern regions of Oman. The naturally occurring radioactive nuclides of the main series, ^{238}U, ^{232}Th, ^{235}U and ^{40}k have been studied by means of low level gamma spectroscopy. Majority of the studied valleys (wadis) are descending from Super Hajar Mountains either towards its east part (coastal area) or the west part (desert).Water samples were analyzed by applying the same analysis technique and ranges of concentrations of^{238}U, ^{232}Th, ^{235}U and ^{40}K were found to be (0.35-1.7), (<DL-0.17), (0.01-0.18) and (0.26-0.91) all in Bq.L^{-1} respectively. The radiological hazards, D, H_R and Ra_{eq} were calculated and found at ranges (0.21-7.94) nGy/h, (3.0-9.0)*10^{-4} mSv/yr and (0.02-3.6) Bq.L^{-1} respectively. The $H_{in,}$ H_{ex} and I_{yr} results were less than 1. The assessment of radioactivity in water samples showed values lower than world-wide average according to (WHO. The results indicate that the radiation hazard from naturally and artificially radionuclides in water samples from the area studied in this current work is barely significant.

Keywords: Radioactivity- Water – HPGe Detector- Gamma rays

1.Introduction

Radioactivity is the word used to describe the decay of atomic nucleus. It refers to the property of unstable atoms to move to a more stable. Earlier in 20th century, Radioecology started to be considered an individual natural science [1]. Since the creation of the earth radioactivity exit naturally everywhere, in air, soil, water, plants and food. Depending on the concentration of the radionuclides, the radiation doses delivered by these materials will [2]. The sultanate of Oman occupies most of the south east corner of Arabian Peninsula. The distribution of natural radioactivity in the environment depends basically on the geological and geographical properties. Most wadis in Oman are having headwaters descend from the mountains towards the coast or deserts. Majority of the studied valleys are descending from Hajar mountains either towards its east part (coastal area) or the west one (desert). Along these valleys, sub-surface water is available. In result, these regions are agricultural regions and therefore are populous too. Consequently, this study will provide a baseline data of radiation considering 15 locations from northern regions of Oman.

IOP Conf. Series: Materials Science and Engineering **757** (2020) 012011 doi:10.1088/1757-899X/757/1/012011

2. Materials and Methods

2.1 Water and Grass samples Collection and Preparation

In this study, decision was made to collect samples of water and grass from east and west of western Hajar Mountains. The sampling area considered in this study located at west of north Al Hajar Mountains where the influence of dust storms on the Arabian Peninsula takes place. Heavy rain drifts soil from mountains to the wadis and plain. Disturbed soils cause a decrease in the net surface radiation. Furthermore, the Ophiolites which are igneous rocks constitute the bulk of the geographical configuration of the northern regions of Oman. In terms of natural radioactivity; it is well known that igneous rocks are strongly enriched in thorium and uranium [3]. Some essential aspects were taken into consideration in order to obtain reliable results. Areas of agricultural activities were selected in order to investigate about probable radionuclides that might have eroded into the wadis by fertilizers. For each location, water samples were taken from an area distribution with the same Global Positioning System (GPS) readings to ensure precise comparison with other locations. Samples were collected from 15 different locations as shown in Table (1), Where 1 refers to the headwater of the wadi and 2 to the estuary of the same wadi. In the Lab, water samples were acidified with concentrated nitric acid (HNO_3) to bring the PH to 2 to avoid any adherence of radionuclide to the interior part of the containers and prevent the growth of organisms [4]. Then 1 L of each sample was stored into the Marinelli beakers for 4 weeks.

Table 1: Water samples codes and locations

Sample Code	location	Willayat	LatitudeN (Deg.Min.S)	longitudeE (Deg.Min.S)
AW1	WADI ASWAD1	Shinas	24 52'43.1"	56 17 '03.3"
AW2	WADI ASWAD2	Shinas	24 52'17.0"	56 19'46.8"
HW1	WADI HIBI1	Sohar	23 54'53.4"	56 33'23.5"
HW2	WADI HIBI2	Sohar	24 02'58.1"	56 33'25.8"
IW1	WADI HILTI1	Sohar	24 03'29.9"	56 29'05.3"
IW2	WADI HILTI2	Sohar	24 05'26.9"	56 29'05.3"
BW	BURAIMI	Buraimi	24 01'16.8"	56 11'55.3"
FW	WADI ALFATH	Yanqul	23 40'45.6"	56 15'03.4"
DW1	WADI DHUNK(FEDA)1	Dunk	23 31'48.3"	56 35'15.8"
DW2	WADI DHUNK(FEDA)2	Dunk	23 32'29.0"	56 18'05.9"
SW	WADI ALSAIFA	AlSinaina	23 44'45.6"	55 53'07.1"
WW1	Wadi WAHRA 1	Ibri	23 18'10.8"	56 42'28.6"
WW2	Wadi WAHRA 2	Ibri	23 29'05.0"	56 40'27.3"
MW	Wadi bani Gahafir (MAHBAB)	Rustaq	23 27'28.5"	57 04'59.0"
KW	Wadi bani Gahafir (KHADA)	Rustaq	23 25'20.9"	57 17'56.2"

2.2 Gamma-Ray spectroscopy

The most common detectors in gamma spectroscopy systems are Thallium activated Sodium Iodide (NaI), Lanthanum Bromide (LaBr) and High Purity Germanium (HPGe).The resolution of each detector is different. Resolution of the spectra refers to the sharpness of peaks in a spectrum. The HPGe detector shows narrow and sharp energy lines as shown in figure (1) [5].

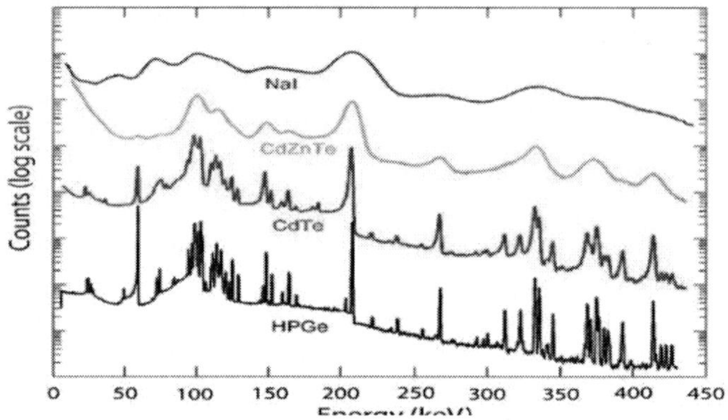

Figure (1): Resolution of most common detectors

A Germanium detector is a semiconductor diode in which the intrinsic region is sensitive to ionizing radiation. The digital signal or spectrum can be analyzed using a personal computer and a Software program called (GENIE 2000). The schematic sketch in figure (2) shows the detection System of γ radiation.

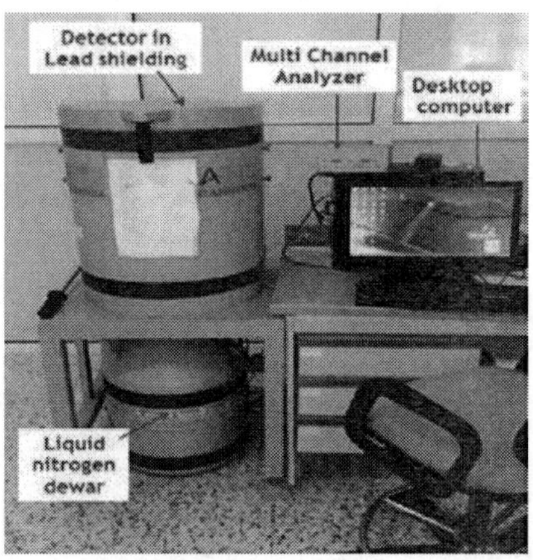

Figure (2): The detection System of γ radiation.

3.Results and Discussions

The maximum activity concentrations of main radionuclides in water ^{238}U , ^{235}U, ^{232}Th and ^{40}K were found all at maximum levels in HW1 sample with values (1.69±0.10 Bq.L^{-1}), (0.18±0.01 Bq.L^{-1}), (0.17±0.02 Bq.L^{-1}) and (0.91±0.01 Bq.L^{-1}) respectively . Whereas the lowest activity concentrations of

^{238}U (0.35±0.06 Bq.L^{-1}) was found in DW2, and the lowest concentrations of ^{235}U, ^{232}Th were found in (HW2) with (0.010±0.001 Bq.L^{-1}), (< *DL*) in 6 locations respectively. The results of the water samples showed that the activity concentrations of uranium were higher than thorium in all water samples figure (3). HW1 is a falaj sample where radionuclides have been presented in the water for long time and contaminated significantly from the soil comparing with sub-surface water which exists temporarily during the heavy rain only and for a short time.

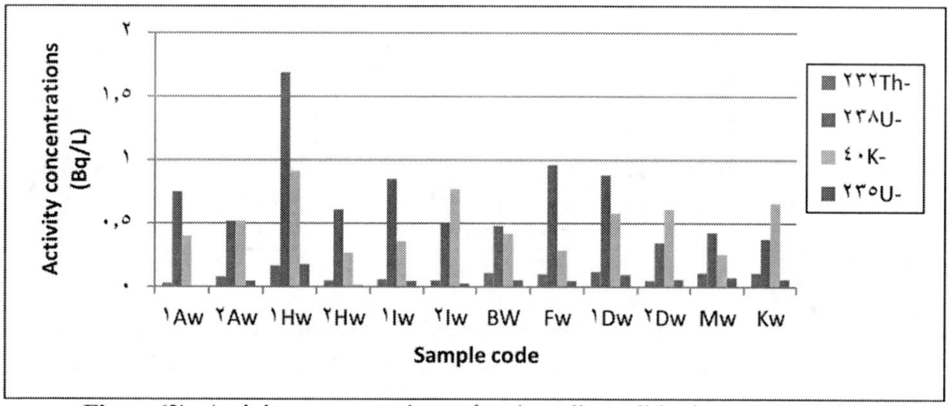

Figure(3): Activity concentrations of main radionuclides in water samples

The values of the concentrations of the present study to be in the range of other studies and is below the recommended values by WHO [6] as shown in Table (2).

Table (2) Comparison of activity concentrations of natural radionuclides in this study with other studies from different countries

Country	Activity concentration(Bq.L^{-1})				References
	^{232}Th	^{40}K	^{226}Ra	^{238}U	
Present study	<DL - 0.17	0.26 - 0.91	<DL - 3.25	0.35 - 1.69	Present study
Oman, A'Dhahira	0.087	0.379	0.73	0.287	[7]
Oman,A 'Dhahira	0.094	0.469	0.775	0.311	[7]
Yemen	1.26	-	3.5	-	[8]
Saudi Arabia Western province	<DL - 3.3	DL - 339.2	-	0.001 - 0.170	[9]
Saudi Arabia, Arar city	9 - 27	63 - 250	11 - 33	-	[10]
Egypt	0.05	-	0.1	-	[11]
Sudan	9.19	-	11.6	-	[12]
Yemen-Ahomria	1.81	<DL	4	-	[8]
Bangladesh	0.250	9	-	0.157	[13]
India	8.0E-04	-	-	-	[14]
Bangladesh, Tarakandi	1.76	11.43	1.59	-	[15]
WHO	0.6	10	1	10	[6]

3.1 Variation of ^{238}U series activities in water

The concentrations of this radionuclide are higher than WHO limit ($1 Bq.L^{-1}$) in four water samples (figure 4). It reflects the geochemical controls on solubility and adsorption, its physical, chemical properties and variation of ^{226}Ra formation in the terrestrial environment [7].

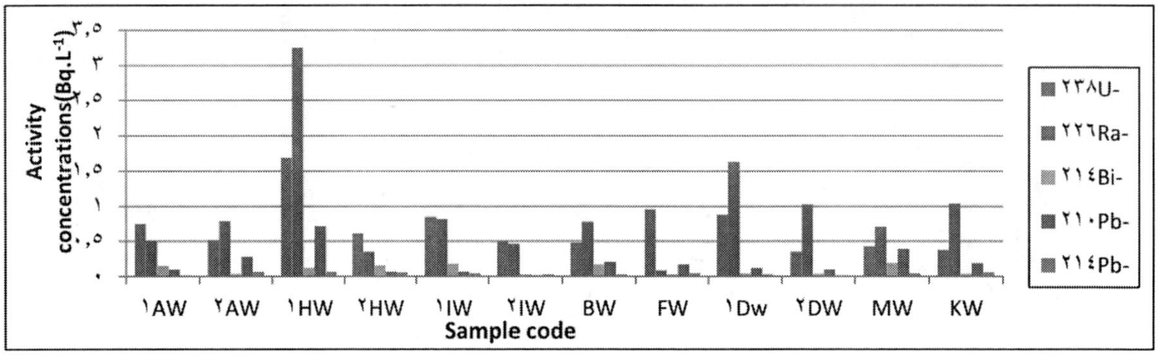

Figure 4: Activity concentrations of radionuclides in ^{238}U series in water samples

The radionuclide ^{210}Pb concentrations were recorded to be less than concentrations of ^{226}Ra in all locations with a range of (DL – 0.72) $Bq.L^{-1}$. This occurred due to the mobility and short half-life of ^{210}Pb compared to ^{226}Ra. The isotopes ^{214}Bi and ^{214}Pb have the shortest half-lives which cause the fastest decay among all radionuclides in the series and this interpreted their lowest concentrations.

3.2 Variation of ^{232}Th series activities

Two significant isotopes from ^{232}Th series were observed in water samples, ^{232}Th (^{228}Ac) and ^{212}Pb.The minimum levels are less than the detection limit (figure5).

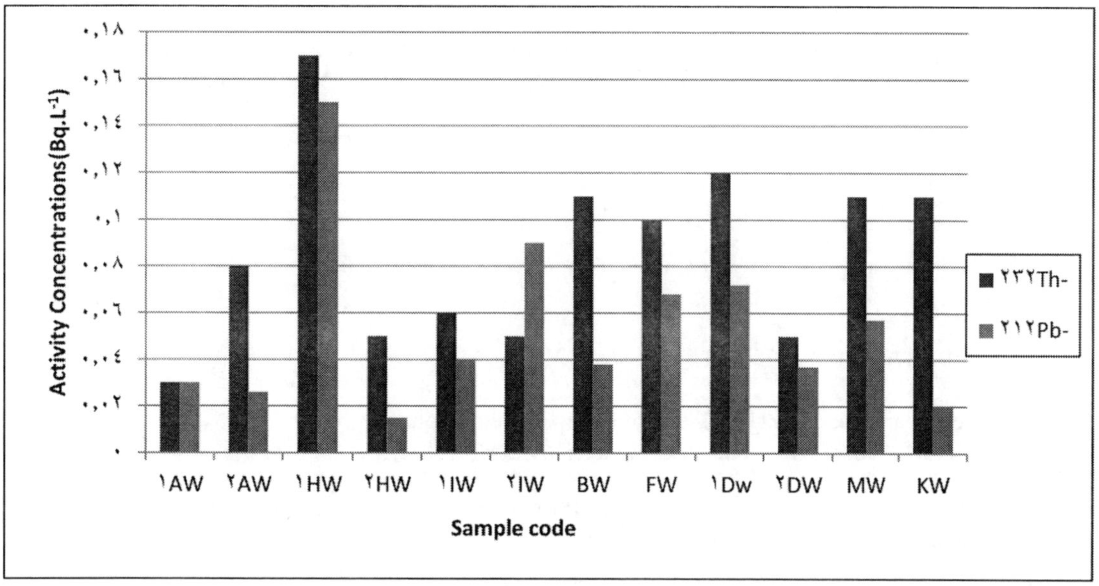

Figure (5): Activity concentrations of radionuclides in ^{232}Th series in water samples

The concentrations of ^{40}K were found to be higher than ^{232}Th (^{228}Ac) concentrations in all water samples. This may be due to the very short half-life of ^{228}Ac (6.13 hours). The activity concentrations of ^{235}U are consistent with that of soil samples. Table (3) compares the activity concentrations of natural radionuclides in this study with other studies from different countries. It showed the values of the concentrations of the present study to be in the range of other studies and is below the recommended values by WHO.

Table (3): A comparison of different activity concentrations of water between different countries.

Country	Activity concentration(Bq.L^{-1}) ^{232}Th	^{40}K	^{226}Ra	^{238}U	References
Present Study	<DL - 0.17	0.26 - 0.91	<DL - 3.25	0.35 - 1.69	Present study
Oman, A'Dhahira	0.087	0.379	0.73	0.287	[7]
Oman, A'Dhahira	0.094	0.469	0.775	0.311	[7]
Yemen	1.26	-	3.5	-	[8]
Saudi Arabia Western province	<DL - 3.3	DL - 339.2	-	0.001 - 0.170	[2]
Saudi Arabia, Arar city	9 - 27	63 - 250	11 - 33	-	[10]
Egypt	0.05	-	0.1	-	[11]
Sudan	9.19	-	11.6	-	[12]
Yemen-Ahomria	1.81	<DL	4	-	[8]
Bangladesh	0.250	9	-	0.157	[13]
India	8.0E-04	-	-	-	[14]
Bangladesh, Tarakandi	1.76	11.43	1.59	-	[17]
WHO	0.6	10	1	10	[6]

3.3 Assessments of hazards in water

samples

Exposure to radiation can be defined in terms of many parameters, such as absorbed dose rate, annual effective dose, radium equivalent activity, Gamma representative level and internal and external hazard indices.
The external absorbed dose rate in air at a height of about 1 m above the ground was calculated by:

$$D\ (nGy.h^{-1}) = 0.0417A_K + 0.462A_{Ra} + 0.604A_{Th} \quad \text{.........} \quad (1)$$

According the annual effective dose received by a member in mSv.y^{-1} unit, the values obtained for the water samples were calculated using equation:

$$H_R(\mu Sv.\,y^{-1}) = 8760 \times 0.2 \times 0.7 \times 10^{-6} \times D \quad \text{.....................}(2)$$

The radium equivalent activity (Ra_{eq}) was calculated by equation:

$$R_{aeq} = A_{Ra} + 1.43A_{Th} + 0.077A_K \quad \text{..} \quad (3)$$

The external, internal and gamma representative level hazard indices which are on concern due to their impact on people health were calculated using:

$$H_{ex} = \frac{A_{Ra}}{370} + \frac{A_{Th}}{259} + \frac{A_K}{4810} \leq 1 \quad \text{..} \quad (4)$$

$$H_{in} = \frac{A_U}{185} + \frac{A_{Th}}{259} + \frac{A_K}{4810} \leq 1 \text{............}(5)$$

$$I_{yr} = \frac{A_U}{150} + \frac{A_{Th}}{100} + \frac{A_K}{1500} \leq 1 \quad \text{............} \quad (6)$$

formulas respectively. Where A_{Ra}, A_{Th} and A_K are the activities of ^{226}Ra, ^{232}Th and ^{40}K in Bq.kg^{-1}, respectively [16].

Table (4): A comparison of different hazard parameters in water between different countries.

Sample Code		D (nGy/h)	H_R (*10^{-4}) (mSv/yr)	R_{aeq} (Bq/L)	H_{in} (*10^{-3})	H_{ex} (*10^{-3})	I_{yr} (*10^{-3})	Category of water
Present study	Aw1	0.36	4.0	0.53	3.0	2.0	2.0	Rain water
	Aw2	0.26	3.0	0.83	4.0	2.0	4.0	Well water
	Hw1	7.94	9.0	3.56	180	460	1140	Falaj water
	Hw2	0.32	4.0	0.44	2.0	2.0	5.0	Rain water
	Iw1	0.44	5.0	0.93	5.0	3.0	7.0	Rain water
	Iw2	0.26	3.0	0.51	3.0	2.0	4.0	Rain water
	BW	0.24	3.0	0.81	4.0	1.0	3.0	Rain water
	Fw	0.46	6.0	0.11	1.0	3.0	6.0	Rain water
	Dw1	0.50	6.0	1.85	9.0	3.0	7.0	Rain water
	Dw2	0.22	3.0	1.14	6.0	1.0	3.0	Rain water
	Mw	0.21	6.0	0.02	5.0	1.0	3.0	Rain water
	Kw	0.27	3.0	1.25	6.0	2.0	4.0	Rain water
Oman, A'Dhahira[7]		0.04	5.1	0.87	-	2.4	-	well water
Oman, A'Dhahira[7]		0.418	6.2	1.06	-	0.10	-	Tap water
Oman, A'Dhahira[7]		0.26	3.2	0.53	-	1.5	-	Falaj water
Saudi Arabia[10]		30	370	-	230	170	-	Well water
Yemen[8]		-	36	6.6	-	-	-	Well water
NorthMalaysia [12]		-	13.7	21.78	-	-	-	Tap water
Bangladesh[14]		-	-	254.14	-	-	-	Tap water
Egypt[13]		-	-	7.13	-	-	-	Well water
Worldwide[6]		18 - 93	0.1	370	<1	<1	<1	WHO

References

[1]Ajayi, O. S., & Adesida, G.. Radioactivity in some sachet drinking water samples produced in Nigeria, Iran. J. Radiat. Res. ; 7(3), 151–158 (2009).

[2] Simon, S. L., & Ph, D. INTRODUCTION TO RADIATION PHYSICS AND DOSIMETRY.Radiation Epidemiology Branch, National Cancer Institute, U.S.A(2011).

[3] United Nations Scientific Committee on the Effects of Atomic Radiation UNSCEAR, B. Sources and Effects Of Ionizing Radiation. New York, USA (2000).

[4] International-Atomic-Energy-Agency (IAEA). Measurement of Radionuclides in Food and the Environment (2003).

[5] Elmenshaz, F. M.. Measurements of Norm in Environmental Samples. Master's thesis, University of Surrey (2009).

[6] World Health Organization . WHO. Nuclear power. The environment. Booklet, 79 (1982)..

[7]Al- Mamari, A. S. S .Radiological assessment of selected groundwater in ADhahira Governorate, Sultanate of Oman, Master thesis. Sultan Qaboos University (2013).

[8]El-kamel, A. E., El-mageed, A. I. A., Abbady, A. E., Harb, S., & Saleh, I. I. Natural Radioactivity of Environmental Samples and their Impact on the Population at Assalamia-Alhomira Area in Yemen,Geosciences 2(5), 125–132 (2012)..

[9]Fakeha, A. A., Hamidalddin, S. Q., & Alamoudy, Z. M. Concentrations of natural radioactivity and their contribution to the absorbed dose from water samples from the Western Province , Saudi Arabia,King Abdulaziz University,JKAU: Sci. 23(2), 17–30 (2011).

[10] Ibrahim, M., Shalabiea, O., & Diab, H. Measurement of some radioactive elements in drinking water in Arar city, Saudi Arabia. *American Journal of Life Sciences*, *2*(1), 24-28 (2014).

[11]Arabi A. M. El, Ahmed N. K. and Salahi Din K. Natural radionuclides and dose estimation in natural water resources from Elba protective area, Egypt. Radiation Protection Dosimetry, 1of 9 (2006).

[12] Almayahi B.A., Tajuddin A.A., JaafarvM.S. Radiation hazard indices of soil and water samples in Northern Malaysian Peninsula. Applied radiation and isotopes journal, 70, 2652-2660 (2012)..

[13]Chakraborty, S.R., Mollah Abdus Staar, Begum Aleya and Ahmad Gias. Radioactivity in drinking water of Bangladesh. Jpn. J. Health Phys., 40(2), 191-201 (2005).

[14] Ahmed, Nour Khalifa. Natural Radioactivity of ground and drinking water in some areas of upper Egypt. Turkish J. Eng. Env.Sci.,28,345-354 (2004)..

[15] Samad M. A., Ali MI, Paul D, Islam SMA. An environmental impact study of Jamuna Urea fertilizer factory at Tarakandi in Jamaplur with radiological indices. J. Environ.Sci. & Natural Resources, 4(2), 27-33 (2011).

[16] Chandrasekaran, A., Ravisankar, R., Rajalakshmi, A., Eswaran, P., Vijayagopal, P., & Venkatraman, B. Spectrochimica Acta Part A : Molecular and Biomolecular Spectroscopy Assessment of natural radioactivity and function of minerals in soils of Yelagiri hills , Tamilnadu , India by Gamma Ray spectroscopic and Fourier Transform Infrared (FTIR) techniques . Spectrochimica Acta Part A: Molecular and Biomolecular Spectroscopy, 136, 1734–1744 (2015).

[17] Saad, R., & Farhan, A. R .Comparative Study of Desert Truffles from Kuwait and Other Countries in the Middle East for Radionuclide Concentration,Humana Press Inc, 309–315. (1999).

Spectral fluctuations in ^{24}Mg nucleus using the framework of the nuclear shell model

Thoraia A Abdul Hussian[1] and Adel K Hamoudi[2]
[1]Radiography techniques department, Al-Turath University, Baghdad, Iraq
[2]Department of Physics, College of Science, University of Baghdad, Baghdad, Iraq
E-mail: thoraiamsc@yahoo.com

Abstract
Random matrix theory is used to study the chaotic properties in nuclear energy spectrum of the ^{24}Mg nucleus. The excitation energies (which are the main object of this study) are obtained via performing shell model calculations using the OXBASH computer code together with an effective interaction of Wildenthal (W) in the isospin formalism. The ^{24}Mg nucleus is assumed to have an inert ^{16}O core with 8 nucleons (4protons and 4neutrons) move in the $1d_{5/2}$, $2s_{1/2}$ and $1d_{3/2}$ orbitals. The spectral fluctuations are studied by two statistical measures: the nearest neighbor level spacing distribution $P(s)$ and the Dyson-Mehta statistics (Δ_3 statistics). For calculations with the full diagonalization of the Hamiltonian, the spectral fluctuations are found to be in agreement with the Gaussian orthogonal ensemble of random matrices.

Keywords: Quantum chaos; Random matrix theory; Spectral fluctuations; Shell model calculations

1. Introduction:
Quantum chaos was investigated intensely through the last three decades [1]. Bohigas et al. [2] suggested a linking between chaos in a classical system and the spectral fluctuations of the analogous quantum system, where an analytical proof of the Bohigas et al. conjecture has been found in [3]. It is now known that quantum analogs of most classically chaotic systems reveal spectral fluctuations that agree with the random matrix theory (RMT) [4,5] while quantum analogs of classically regular systems reveal spectral fluctuations that agree with a Poisson distribution. For time-reversal-invariant systems, the suitable form of RMT is the Gaussian orthogonal ensemble (GOE). RMT was firstly utilized to characterize the statistical fluctuations of neutron resonances in compound nuclei [6]. RMT has become a standard tool for analyzing the universal statistical fluctuations in chaotic systems [7-10].

The mean field approximation can be used to study the chaotic nature of the single particle dynamics in the nucleus. However, the nuclear residual interaction mixes different mean field configurations and affects the statistical fluctuations of the many particle spectrum and wave functions. These fluctuations may be investigated with different nuclear structure models. The statistics of the low-lying collective part of the nuclear spectrum were studied in the framework of the interacting boson model [11, 12], in which the nuclear fermionic space is mapped onto a much smaller space of bosonic degrees of freedom. Because of the relatively small number of degrees of freedom in this model, it was also possible to relate the statistics to the underlying mean field collective dynamic. At higher excitations, additional degrees of freedom (such as broken pair) become important [13], and the effects of interactions on the statistics must be studied in larger model spaces. The interacting shell model offers an attractive framework for such studies. In this model, an effective interactions are available and the basis states are labeled by exact quantum numbers of angular momentum (J), isospin (T) and parity (π) [14].

In the studies [15-19], the distribution of eigenvector components was examined using the framework of the shell model. Brown and Bertsch [17] found that the basis vector amplitudes are consistent with Gaussian distribution (which is the GOE prediction) in regions of high level density but deviated from Gaussian behavior in other regions unless the calculation employs degenerate single particle energies. Later, the study of Ref. [19] also suggested that calculations with degenerate single particle energies are chaotic at lower energies than more realistic calculations.

The electromagnetic transition intensities in a nucleus are observables that are sensitive to the wave functions, and the study of their statistical distributions should complement [11, 12] the more common spectral analysis and serve as another signature of chaos in quantum systems. Hamoudi et al carried out [20] the fp-shell model calculations to study the statistical fluctuations of energy spectrum and electromagnetic transition intensities in A=60 nuclei using the $F5P$ [21] interaction. The calculated results were in agreement with RMT and with the previous finding of a Gaussian distribution for the eigenvector components [15-19]. Hamoudi studied [22] the effect of the one-body Hamiltonian on the fluctuation properties of energy spectrum and electromagnetic transition intensities in ^{136}Xe using an effective interaction for the $N82-$model space defined by $2d_{5/2}$, $1g_{7/2}$ $1h_{11/2}$, $3s_{1/2}$ and $2d_{3/2}$ orbitals. A clear quantum signature of breaking the chaoticity was observed as the values of the single particle energies are increased. Later, Hamoudi et al carried out [23] full $fp-$shell model calculations to investigate the regular to chaos transition of the energy spectrum and electromagnetic transition intensities in ^{44}V using the interaction of FPD6 as an effective interaction in the isospin formalism. The calculated spectral fluctuations and the distribution of electromagnetic transition intensities were found to have a regular dynamic at $\beta=0$, (β is the strength of the off-diagonal residual interaction), a chaotic dynamic at $\beta \geq 0.3$ and intermediate situations at $0<\beta<3$. Recently, Hamoudi and Murad have carried out [24] statistical calculations of nuclear excitation energies in ^{138}Ba nucleus, where this nucleus is described as a core of ^{132}Sn with 6 active protons move in the space of N82. The interaction of N82K has been used to compute the excitation energies required in the study. The obtained outcome has been in agreement with the study [23]. Zelevinsky et al. [14] carried out sd-shell model calculations to examine the spectral fluctuations in ^{24}Mg nucleus using the proton-neutron formalism interaction of Wildenthal (WPN-interaction) [27]. Here the nuclear states are not described by the isospin T but only by the angular momentum (J) and parity (π). The computed spectral fluctuations [14] were intermediate between the Wigner distribution (which characterizes the chaotic dynamic of systems) and Poisson distribution (which characterizes the regular dynamic of systems) due to the absence of mixing and repulsion between levels with different isospin. To explore the effect of hidden integral of motion (i.e., the hidden isospin T) on the spectral fluctuations in ^{24}Mg nucleus, we have (in this study) repeated the calculations of ^{24}Mg as in [14] but this time we have used the isospin formalism interaction of Wildethal (W-interaction) [27]. Here the nuclear states are described by the angular momentum (J), isospin (T) and parity (π).The computed results of the present study is totally different than that of Zelevinsky et al. [14], where our computed spectral fluctuations in ^{24}Mg nucleus are found to be in consistent with the GOE (Wigner distribution) of RMT. This is attributed to the presence of mixing and repulsion between levels with different isospin.

2. Theory

The effective shell-model Hamiltonian, which describes a many-body system, is represented by

$$H = H_0 + H', \qquad (1)$$

where H_0 and H' represent the independent particle (one body) part and the residual two-body interaction of H. The unperturbed Hamiltonian

$$H_0 = \sum_{\lambda} e_{\lambda} a_{\lambda}^{+} a_{\lambda} \qquad (2)$$

describes non-interacting fermions in the mean field of the appropriate spherical core. The single-particle orbitals $|\lambda\rangle$ have quantum numbers $\lambda = (ljm\tau)$ of orbital (l) and total angular momentum (j), projection $j_z = m$ and isospin projection τ. The antisymmetrized two-body interaction H' of the valence particles is written as

$$H' = \frac{1}{4}\sum V_{\lambda\mu;\nu\rho} a_\lambda^+ a_\mu^+ a_\nu a_\rho . \qquad (3)$$

To construct the many-body wave functions with good spin J and isospin T quantum numbers, we use the $m-$scheme determinants which have, for given J and T, the maximum spin and isospin projection,

$$|M = J, T = T_3; m\rangle, \qquad (4)$$

where m span the $m-$scheme subspace of states with $M = J$ and $T_3 = T$.
The matrix of the many-body Hamiltonian

$$H_{kk'}^{JT} = \sum_k \langle JT;k|H|JT;k'\rangle \qquad (5)$$

is eventually diagonalized to obtain the eigenvalues E_α and the eigenvectors

$$|JT;\alpha\rangle = \sum_k C_k^\alpha |JT;k\rangle \qquad (6)$$

Here, the eigenvalues E_α are considered as the main object of the present investigation.

The fluctuation properties of nuclear energy spectrum are obtained via two statistical measures: the nearest-neighbors level spacing distribution $P(s)$ and the Dyson-Mehta or Δ_3 statistics [4, 25]. The staircase function of the nuclear shell model spectrum $N(E)$ is firstly build. Here, $N(E)$ is defined as the number of levels with excitation energies less than or equal to E. In this study, a smooth fit to the staircase function is performed with polynomial fit. The unfolded spectrum is then defined by the mapping [12]

$$\widetilde{E}_i = \widetilde{N}(E_i) . \qquad (7)$$

The real spacings reveal strong fluctuations whereas the unfolded levels $\widetilde{E}_i$ have a constant average spacing.
The level spacing distribution (which exemplifies the fluctuations of the short-range correlations between energy levels) is defined as the probability of two neighboring levels to be a distance s apart. The spacings s_i are determined from the unfolded levels by $s_i = \widetilde{E}_{i+1} - \widetilde{E}_i$. A regular system is forecasted by the Poisson statistics

$$P(s) = \exp(-s) . \qquad (8)$$

If the system is classically chaotic, we foresee to get the Wigner distribution

$$P(s) = (\pi/2)s\exp(-\pi s^2/4), \qquad (9)$$

which is consistent with the GOE statistics.
The Δ_3 statistic (which characterizes the fluctuations of the long-range correlations between energy levels) is utilized to measure the rigidity of the nuclear spectrum and defined by [4]

$$\Delta_3(\alpha, L) = \min_{A,B} \frac{1}{L} \int_{\alpha}^{\alpha+L} \left[N(\widetilde{E}) - (A\widetilde{E} + B)\right]^2 d\widetilde{E}. \tag{10}$$

It measures the deviation of the staircase function (of the unfolded spectrum) from a straight line. A rigid spectrum corresponds to smaller values of Δ_3 whereas a soft spectrum has a larger Δ_3. To get a smoother function $\overline{\Delta}_3(L)$, we average $\Delta_3(L)$ over several n_α intervals $(\alpha, \alpha + L)$

$$\overline{\Delta}_3(L) = \frac{1}{n_\alpha} \sum_\alpha \Delta_3(\alpha, L). \tag{11}$$

The successive intervals are taken to overlap by $L/2$.
In the Poisson limit, $\Delta_3(L) = L/15$. In the GOE limit, $\Delta_3 \approx L/15$ for small L, while $\Delta_3 \approx \pi^{-2} \ln L$ for large L.

3. Results and discussion

Shell model calculations with the computer program OXBASH [26] are performed for ^{24}Mg nucleus (with $T = 0$). This nucleus is assumed to have an inert ^{16}O core with 8 valence nucleons (4 protons and 4 neutrons) move in the sd-shell ($1d_{5/2}$, $2s_{1/2}$ and $1d_{3/2}$ orbitals) model space. The W-interaction [27] is chosen as an effective interaction in the isopspin formalism together with realistic spe's. Many-body basis states $|k\rangle$ were constructed with good total angular momentum *J* (its projection *M*) parity π and isospin *T* (its projection T_3). It is significant to indicate that all computed results are gained with diagonalization of the full Hamiltonian. In the following, we have considered the levels $J^\pi T = 0^+0,\ 2^+0,$ and 4^+0 as test cases in the present analysis.

Table 1 shows the dimensions for the considered $J^\pi T$ states formed with isospin formalism interaction of Wildenthal [27] for the ^{24}Mg nucleus, which is assumed to have 8 valence nucleons (4 protons and 4 neutrons) move in the sd-model space.

Table 1. Dimensions of considered J^π levels in ^{24}Mg generated with isospin formalism interaction of Wildenthal [27].

J^π	Dimensions
0^+	325
2^+	1206
4^+	1311

Figure 1 demonstrates the level density $\rho(E)$ (histograms) of ^{24}Mg nucleus, computed with Wildenthal interaction, for $J^\pi T = 0^+0$, 2^+0, and 4^+0 levels. It is clear from this figure that the distributions of $\rho(E)$ (histograms) have a bell-shaped curve that is symmetric about the mean energy E_0, where these histograms are fitted by the Gaussian shape (the dashed line). The corresponding values of E_0 (the mean energy) and σ (the standard deviation) utilized in the fitting with the Gaussian shape [28] for the considered $J^{\pi}T$ levels in ^{24}Mg nucleus are displayed in Table 2.

Table 2. Values of E_0 (the mean energy) and σ (the standard deviation) for considered J^{π} levels in ^{24}Mg generated with isospin formalism interaction of Wildenthal [27].

J^{π}	^{24}Mg	
	E_0 (MeV)	σ
0^+	41.8153	12.928
2^+	42.3150	12.591
4^+	41.646	11.940

It is obvious from Figure 1 that the level density sharply growths along with excitation energy, arrives its maximum at the middle of the spectrum and then reduces again for the highest energy. This performance of the high energy, and the rough symmetry with respect to the middle of the spectrum, are artificial features of models with limited Hilbert space which is in contrast to real many-body systems. It is noticed that the computed $\rho(E)$ (histograms) of all considered $J^{\pi}T$ levels has a Gaussian shape, which is in agreement with the predicted study [7] for a many body system with two-body residual interaction. It is also noticed that the computed histograms have no dependency on the angular momentum J.

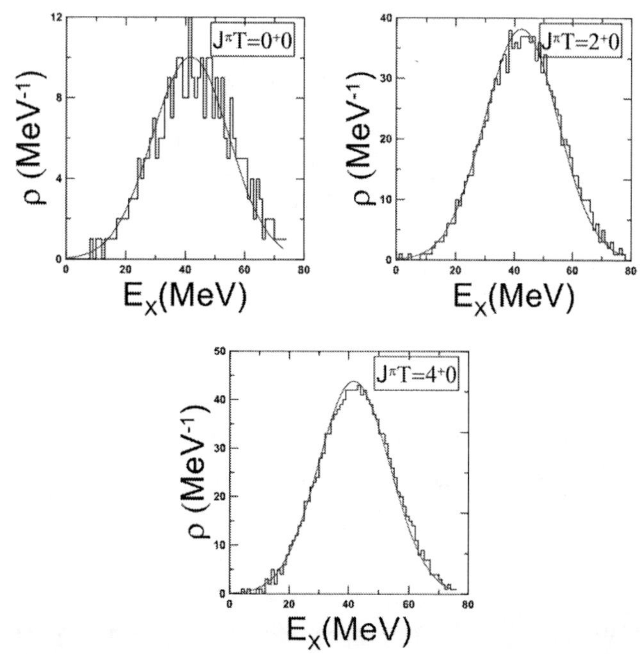

Figure 1. The level density of ^{24}Mg nucleus obtained with full Hamiltonian (histogram) is compared with the Gaussian fit (solid line) for $J^{\pi}T = 0^+0$, 2^+0, and 4^+0 states.

Figure 2 shows the nearest-neighbors level spacing distributions $P(s)$ for the unfolded $J^{\pi}T=0^{+}0,\ 2^{+}0,$ and $4^{+}0$ levels in ^{24}Mg nucleus. The distribution of computed $P(s)$ is displayed by histograms. The GOE distribution (which describes chaotic systems) is displayed by the solid line. The Poisson distribution (which corresponds to a random sequence of levels and describes regular systems) is displayed by the dashed line. The level repulsion at the region of small spacing $0 \leq s \leq 0.2$, which is considered as a distinctive feature of chaotic level statistics, is very well reproduced (not reproduced) by the present study (by the study [14]) due to the presence (absence) of the mixing and repulsion between levels with different isospin. In addition, the occurrence of the level repulsion at the region $0 \leq s \leq 0.2$ is also attributed to the mixing of nuclear energy levels by the off-diagonal Hamiltonian of W-interaction. Moreover, the computed histograms obtained by the present study (by the study [14]) for the unfolded $J^{\pi}T$ levels are in very well agreement with the GOE limit (clearly deviated from the GOE limit, i.e., intermediate between the GOE and Poisson limits). It is so obvious that the computed histograms, displayed in Figure 2, show no dependency on the angular momentum *J*.

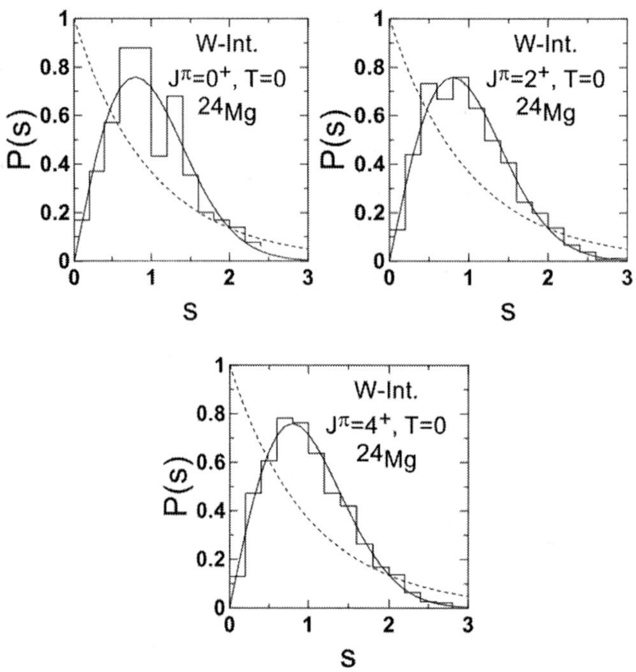

Figure 2. The nearest neighbor level spacing $P(s)$ distributions in ^{24}Mg nucleus for $J^{\pi}T=0^{+}0,\ 2^{+}0,$ and $4^{+}0$ states. The histograms are the calculated $P(s)$ with full Hamiltonian. The solid and dashed lines are the GOE and Poisson distributions, respectively.

Figure 3 illustrates the Δ_3(L) statistics for the unfolded $J^{\pi}T=0^{+}0,\ 2^{+}0,$ and $4^{+}0$ levels in the ^{24}Mg nucleus. The computed distribution of Δ_3(L) statistics is denoted by the symbols of the open circle, the Poisson distribution is denoted by the dashed line while the GOE distribution is denoted by the solid line. Again, the computed Δ_3 statistics of the present study (the study [14]) for the unfolded $J^{\pi}T$ levels are in very good accordance with the GOE distribution (noticeably diverged from the GOE distribution, i.e., in-between the GOE and Poisson distributions). However, the computed distribution of Δ_3 (the symbols of open circle) for $J^{\pi}T = 0^{+}0$ levels demonstrates some slight oscillations around the GOE distribution. These oscillations are attributed to the number of intervals n_{α} (which is related to the dimension of $J^{\pi}T$ states) used in averaging the statistic $\Delta_3(\alpha, L)$, see Eq. (11). A smooth statistic corresponds to a large n_{α} and non-smooth statistic corresponds to a small n_{α}. However, the symbols of the open circle of the $J^{\pi}T = 0^{+}0$ levels are still very close to the GOE limit (the solid line). Once more, the distributions of computed Δ_3 statistics, presented in Figure 3, exhibit also no dependency on the angular momentum *J*.

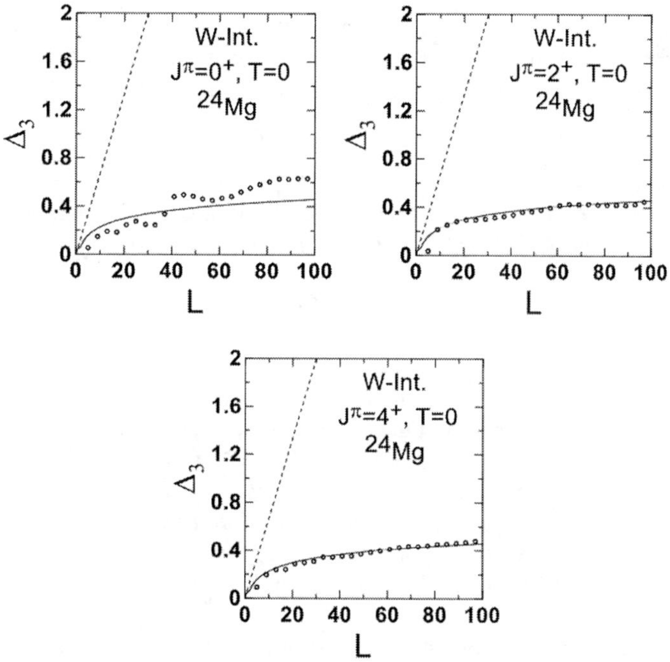

Figure 3. The average Δ_3 statistics in ^{24}Mg nucleus for $J^{\pi}T = 0^{+}0$, $2^{+}0$ and $4^{+}0$ states. The calculated Δ_3 statistics (open circles) correspond to the results obtained with full Hamiltonian. The solid and dashed lines are the GOE and Poisson distributions, respectively.

Additionally, the outcome of Δ_3 statistic, displayed in Figure 3, confirms the outcome that we have obtained from the analysis of the $P(s)$ distributions, displayed in Figs. 2, i.e., the spectral fluctuations of the considered energy spectrum are in accordance with the GOE of RMT.

4. Conclusions

Chaotic properties of the nuclear energy spectrum in the ^{24}Mg nucleus have been analyzed using the framework of RMT. The nuclear energy levels for the states $J^{\pi}T = 0^{+}0,\ 2^{+}0$, and $4^{+}0$ are obtained by carrying out shell model calculations utilizing the OXBASH computer code together with an isospin formalism interaction of Wildethal. It is noticed that the computed level density $\rho(E)$ of all considered $J^{\pi}T$ levels have a Gaussian shape, which is in accordance with the predicted study of Brody et al. [7] for a many body system with two-body residual interaction.

It is seen that the level repulsion, presented in the computed distributions of $P(s)$, at the region of small spacing $0 \le s \le 0.2$ is very well imitated (not imitated) by the present study (by the study [14]) due to the existence (nonexistence) of mixing and repulsion between levels with different isospin. However, the computed distributions of $P(s)$ [Figure 2] and Δ_3 statistics [Fig. 3] for different unfolded $J^{\pi}T$ levels exhibit a very good accordance with the GOE limit (a noticeable discrepancy from the GOE limit, i.e., intermediate between the GOE and Poisson limits). Besides, both $P(s)$ and Δ_3 statistics reveal no dependency on the angular momentum J.

Acknowledgment

The authors would like to express their thanks to Professor B. A. Brown of the National Superconducting Cyclotron Laboratory, Michigan State University, for providing the computer code OXBASH.

References

[1] Haak F 2001, *Quantum signature of chaos*, 2nd enlarged ed. Springer-Verlag, Berlin
[2] Bohigas O, Giannoni M. J and Schmit C 1986, *Phys. Rev. Lett.* **52**, 1
[3] Heusler S, Muller S, Altland A, Braun P, and Haak F 2007, *Phys. Rev. Lett.* **98**, 044103
[4] Mehta M L 2004, *Random Matrices, 3rd ed.* Academic, New York
[5] Papenbrock T, and Weidenmuller H A 2007, *Rev. Mod. Phys.*, Vol. **79**, No. 3, 997
[6] Porter C E 1965, *Statistical Theories of Spectra: Fluctuations* Academic, New York
[7] Brody T A, Flores J, French J B, Mello P A, Pandey A and Wong S S.M 1981, *Rev. Mod. Phys.* **53**, 385
[8] Guhr T, M*ü* ller-Groeling A and Weidenm*ü* ller H A 1998, *Phys. Rep.* **299**, 189
[9] Alhassid Y 2000, Rev. Mod. Phys. **72,** 895
[10] Gutzwiller M C 1990, *Chaos in Classical and Quantum Mechanics* Springer-Verlag, New York
[11] Alhassid Y, Novoselsky A and Whelan N 1990, *Phys. Rev. Lett.* **65**, 2971;
Alhassid Y and Whelan N1991, *Phys. Rev.* C **43**, 2637
[12] Alhassid Y and Novoselsky A 1992, *Phys. Rev.* C **45**, 1677
[13] Alhassid Y and Vretenar D 1992, *Phys. Rev.* C **46**, 1334
[14] Zelevinsky V, Brown B A, frazier N and Horoi M 1996, *Phys. Rep.* **276**, 87
[15] Whitehead R R et al. 1978, *Phys. Lett.* B **76**, 149
[16] Verbaarschot J J M and Brussaard P J 1979, *Phys. Lett.* B **87**, 155
[17] Brown B A and Bertsch G 1984, *Phys. Lett.* B **148**, 5
[18] Dias Het al. 1989, *J. Phys.* G **15**, L79
[19] Zelevinsky V, Horoi M and Brown B A 1995, *Phys. Lett.* B **350**, 141

[20] Hamoudi A, Nazmitdinov R G, Shahaliev E and Alhassid Y 2002, *Phys. Rev.* C **65**, 064311
[21] Glaudemans P W M, Brussaard P J and Wildenthal R H 1976, *Nucl. Phys.* A **102**, 593
[22] Hamoudi A 2011, *Nucl. Phys.* A **849**, 27
[23] Hamoudi A and Abdul Majeed Al-Rahmani A 2012, *Nucl.Phys.* A **892**, 21
[24] Hamoudi A and Murad S 2018, Iraqi journal of Science Vol. **59**, No. 4C, Page 2225-2233
[25] Stephens F S, Deleplanque M A, Lee I Y, Macchiavelli A O, Ward D, Fallon P, Cromaz M, Clark R M, Descovich M, Diamond R M, and Rodriguez-Vieitez E 2005, *Phys. Rev. Lett.* **94**, 042501
[26] Brown A B, Etchegoyen A, Godin N S, Rae W D M, Richter W A, Ormand W E, Warburton E K, Winfield J S, Zhao L and Zimmermam C H, MSU-NSCL Report Number **1289**
[27] Wildenthal B H, Prog 1984. Part. *Nucl. Phys.* **11**, 5
[28] Ribeiro M I 2004, *Instituto Superior Tcnico*, Av. Rovisco Pais, **1**; 1049-001 Lisboa Portugal, Febreuary

Study of Cross Section Areas Ratios for Scattering Interactions of Gamma Photons in He, Fe, Fm, H_2O Materials.

M J R Aldhuhaibat[1], M S Amana[2], A J Farhan[3]* and N J Jubier[4]
[1,2,3,4] Department of Physics, College of Science, University of Wasit, Kut, Wasit, Iraq.
*Corresponding author: kutpc19@yahoo.com

Abstract. In the research, some factors that affect the probability or the so-called interaction cross section were studied. Gamma rays interact with materials through the elastic (coherent) and inelastic (incoherent) interactions represent Rayleigh scattering and Compton scattering respectively. The energies of gamma rays in the range of (0.15 - 100) MeV for this work, and three elements were divided between the light such as Helium, a medium such as Iron, and heavy likes Fermium was selected. Also, the interaction of the H2O molecules was studied as a simulation of live tissue, which water represents more than 70% of its. In addition, the Scattering ratio of Rayleigh to Compton for these energies in chosen materials was studied to illustrate the affect of the energy amount and the material type on the domination of one of the two interactions on the other, also to determine the dependence of Compton scattering interaction on those factors, which represents an important interaction in transfer of energy to the material.The cross sections of the exploration interactions and their proportions are graphically represented to illustrate their dependence on this type of reaction to the energy of the ejected photon and the atomic number or type of target material. Rayleigh scattering to Compton scattering ratios were plotted in diagrams for each material at gamma photon energy (E = 1.25MeV) because the value of energy represents the energy rate of gamma emitted from Cobalt-60 isotope which used in many medical, research, and educational laboratories. In addition, Compton scattering ratios and as well as the Rayleigh scattering ratios in materials were plotted as a function of the sum of these probabilities. Finally, the ratios of probability for both types of scattering and for all materials to the total number of energy exploration potentials (1.25MeV) were plotted to further illustrate the relationship.

Keywords: Cross section, Compton scattering, and Gamma Photon.

1.Introduction

In nuclear physics, the nuclear cross section concept of the atomic nucleus is used to describe the probability of a nucleus entering a nuclear reaction. In physics, this probability is assessed as a "privileged space" around the nucleus where the large area means a high probability of interaction . The nuclear cross section is measured in square centimeter units, and because the nuclear reaction section is too small, physicists have created a small convenient unit that facilitates arithmetic, and that unit is a barn, and 1 barn is equal to $10^{-24}cm^2$. The nuclear section symbolizes σ .The nuclear section can be measured for many nuclear processes that an atomic nucleus can perform, such as elastic collisions with particles, an inelastic collision, or a nuclear interaction with a particle. Thus, we differentiate between an elastic collision section, inelastic collision section, and nuclear interaction section. There is also a neutron nuclear section simply called a neutron section because it is one of the properties of the atomic nucleus. Some atomic nuclei have the ability to absorb particles such as neutrons, and in this Case the nuclear section is called the absorption section._In nuclear physics, it is appropriate to consider the particles falling to atomic nuclei as small-scale with negligible measurements. A nuclear section of a particular process, such as an absorption section or an elastic collision section can be calculated, then a sample of the material in the form of a thin slice, copper, for example, is used, and we throw up the proton flux on it and we measure the number of protons carried out behind the slice. Comparing the proton flux intensity that measured before and after placing the

slice in the way of primary proton flux, we can calculate the interaction of the nucleus of the copper atom with the protons and the density of the copper atoms in the sample is required to complete the calculation. The designated nuclear section is called the total section which can be symbolized by σ_T, and the nuclear radius is approximately about 10^{-12}cm. We may conclude from this fact that the nuclear section is $10^{-24}cm^2$ and barn unit has been selected for this measurement.
In fact, the observed nuclear section is so different from the geometric section, for example, some nuclei absorb heat neutrons (ie, slow) in an interaction, that produces gamma rays and, in some cases, the interaction section is about 1000barn. The nuclear section in other interactions, such as the interaction of the nucleus with gamma ray (as absorption section) only 0.001barn due to different properties of different nuclei of elements.
The cross-section, or so-called probability of interaction, of any nuclear interaction is measured by units of mass area and is called the mass cross-section area and measured either in cm^{-2}/g units or in barns/atom units[1].

2.Previous Studies

Icelli and Erzeneogle [2] (2002) measured the ratio of the Rayleigh scattering cross section to the Compton scattering cross-section of 59.54keV photons energy and at the scattering angles of 55° and 115° for elements of Fe, Ni, Cu, Zn, Nb, Mo, Ag, Sn, Ta, Au and Pb as a target material using Ge(Li) and they found that the proportion of the measured cross section increases by increasing the atomic number and is consistent with the theoretical results calculated by the theoretical model presented by Hubbell et al. (1975) [3] and the relative model provided by Hubbell and Overbro (1979) [4] and the developed model to the relative model provided by Schaupp et al. (1983) [5].
As well as, Singh et al. (2012) [6] measured the ratio of the cross-section of the coherent scattering (Rayleigh) to the incoherent scattering (Compton) of the element with atomic number of $6 > Z > 82$ for the Gamma photon energy of 59.54KV using the high-purity germanium detector (HpGe) and at the angle of 90° to the falling photons beam. The results are shown that the results correspond to previous studies results [3], [4] and [5] and to the practical results of the study by Icelli and Ereneglu [2] for elements with low atomic numbers, but deviate from the theoretical results of the elements with high atomic numbers.
In year of 2016 [7], A. Manjunath and B. R. Kerur Study the Rayleigh to Compton scattering ratio for pharmaceutical ingredients using X-ray, their study shows that one can determination of effective atomic number through the scattering ratio of Rayleigh to Compton that will be the independent of sample density and depends only on the mixture type. Also, Tuba AKKUŞ (2018) [8], measured theoretically the scattering cross-section ratio of coherent to Compton of calcium on the basis of form factors of relativistic and nonrelativistic, and nonrelativistic form factors S-matrix theories, and experimentally using X-ray at 5.9keVenergy and HPGe detector for several polar scattering angles, this study showed that the experimental results were most accordant with the theory of nonrelativistic form factor.

3.Theory and method

Gamma rays are the name given to electromagnetic radiation whose origin is from the nucleus .This radiation has a very short wavelength ranging from (10^{-11}-10^{-15}) cm so its energy is very large ranging from (0.1-10) MeV. The processes that occur when a beam passes through the material are complex processes and some of the properties related to these processes can be understood by the application of classical mechanics (Maxwell's equations), but the correct physical state can only be described using quantum electromagnetism theories. The photon interacts with the material in three main ways: Photoelectric Effect, in which the interactive photons are completely absorbed by the free electrons of the material, Compton Scattering, an inelastic or incoherent scattering, with matter electrons, Pair Production that in which the photon energy is fully absorbed when it interacts with the strong electric field around the nucleus. Another type of interaction of gamma photons with atoms of matter occurs with less probability than the three main reactions above, Rayleigh Scattering, which represents an elastic or coherent of the Gamma photons when interacting with the electrons of matter atoms [9]. In this paper, we will discuss the interactions of the scattering, which include Compton Scattering (CS) and Rayleigh Scattering (RS). CS, or so-called incoherent or Inelastic scattering, occurs between the falling photons of Gamma and electrons of the absorbent material that is often the most common of gamma energies emitted from the radioactive sources or isotopes. This type of interaction is also the most common mechanism in the medial atomic number [9]. A good accuracy formula Compton scattering cross-section, which is offered by Klein-Nishina [10], is:

$$(d\sigma/d\Omega)_{K\text{-}N} = (1/2)r_o^2(\omega_2/\omega_1)^2(\omega_2/\omega_1 + \omega_1/\omega_2 - \sin^2\theta) \quad \ldots\ldots\ldots\ldots\ (1)$$

Where r represents the classical radius of the electron, ω_1 & ω_2 represent the energies of incident & scattered photons. This cross section is equal in magnitude to the differential Thomson cross section per electron ($(d\sigma/d\Omega)_{Th}$), therefore:

$$(d\sigma/d\Omega)_{Th} = (1/2)\, r_o^2(1+\cos^2\theta) \quad \ldots\ldots\ldots\ldots\ldots\ (2)$$

Rayleigh so-called "coherent" or "elastic" scattering is produced when a Gamma photon has been scattered without any change in its energy. This interaction is more common at the low energies of the falling Gamma photons, small angles of scattering, and high atomic numbers for the target material [9]. The mathematical formula that used to calculate the coherent scattering cross section ($\boldsymbol{\sigma_R}$) for any atom can be written as:

$$\sigma_R = \int_{\theta=\pi}^{\theta=\pi} d\sigma_T(\theta)[F(x,Z)]^2 \quad \ldots\ldots\ldots\ldots\ldots\ (3)$$

Where σ_T is the cross section for classical Thomson scattering and from an electron=$8\pi r_e^2/3$=6.652448 $\times 10^{-29}$m^2, and *F(x, Z)* is the relativistic an atomic form Factor, Where the *F(x,Z)* and $\boldsymbol{\sigma_R}$ values can be obtained for a range of energies and for a set of elements from the reference [11].

XCOM (version3. 1) represents an online database through which photonic cross sections of scattering, photoelectric absorption, and pair production can be calculated, as well as the total attenuation coefficients of any element, compound or mixture (Z ≤100) at energies ranging from 1keVolts and 100keV. This work is supported by the National Standards Institute and the Standardized Technology Reference Data Program, and by the US Department of Energy (OHER). The scattering and absorption data of photons (x-rays and gamma rays) are required for many scientific, engineering and medical applications .The number of substances in which photon cross sections are needed is large and increasing .Available tables usually include cross sections of several elements, not all of them, and some tables not possible to meet all the overall requirements across the printed tables. Moreover, cross sections are often needed at photon energies other than those listed [12]. The scattering interactions of Rayleigh and Compton were studied for specific elements that were selected to be a model for light, medium and heavy atoms or elements. In this study, the XCOM database program was used to obtain cross-section areas of required interactions. The elements, that were studied in our current research, are three elements, the first element is the Helium (He) which has an atomic number of Z=26 as an example of the light elements or atoms, the second element is the Iron (Fe), which has an atomic number of Z=26, which represents one of the medium elements, and the element of Fermium (Fm), that has an atomic number of Z=100 to represent a model of heavy elements, was selected, as well as the water H_2O was studied as a simulation of the study of living tissue in living organisms due to its high concentration in those tissues which approximately reach up to 70% [13].

4. Results and Discussions

The results of cross-section calculations of the RS and CS interactions using the XCOM program in unit of cm^2/g were arranged in Table 1. To illustrate the relationship of the cross section area or the so-called probability of the RS and the cross section of the CS of the Gamma ray falling on the target material for each of the above mentioned materials, which were carried out for Gamma energies for the range (0.15-100)MeV which represents a large range of Gamma energies within the Gamma radiation area of the electromagnetic radiations spectrum, the areas of these sections were plotted in Figure 1. as a function of the reactive gamma for the range of energies of (0.15-100)MeV.

Table 1. The cross-section area values of the RS and CS interactions in units of cm^2/g .for different values of gamma ray energies in unit of MeV for He, Fe, Fm and H_2O.

Gamma Energy (MeV)	Scattering Cross Section (cm^2/g)							
	He		Fe		Fm		H_2O	
	RS	CS	RS	CS	RS	CS	RS	CS
1.50E-01	3.41E-04	1.33E-01	1.78E-02	1.20E-01	1.47E-01	9.14E-02	2.44E-03	1.47E-01
2.00E-01	1.92E-04	1.22E-01	1.03E-02	1.11E-01	8.85E-02	8.68E-02	1.39E-03	1.35E-01
3.00E-01	8.54E-05	1.06E-01	4.73E-03	9.79E-02	4.26E-02	7.82E-02	6.22E-04	1.18E-01
4.00E-01	4.81E-05	9.53E-02	2.69E-03	8.81E-02	2.52E-02	7.13E-02	3.51E-04	1.06E-01
5.00E-01	3.08E-05	8.70E-02	1.74E-03	8.07E-02	1.67E-02	6.58E-02	2.25E-04	9.66E-02
6.00E-01	2.14E-05	8.05E-02	1.21E-03	7.47E-02	1.18E-02	6.13E-02	1.56E-04	8.94E-02
8.00E-01	1.20E-05	7.07E-02	6.83E-04	6.58E-02	6.84E-03	5.42E-02	8.79E-05	7.86E-02
1.00E+00	7.69E-06	6.36E-02	4.38E-04	5.92E-02	4.45E-03	4.90E-02	5.63E-05	7.07E-02
1.02E+00	7.36E-06	6.29E-02	4.20E-04	5.85E-02	4.27E-03	4.85E-02	5.39E-05	6.99E-02
1.25E+00	4.92E-06	5.69E-02	2.81E-04	5.29E-02	2.89E-03	4.39E-02	3.60E-05	6.32E-02
1.50E+00	3.42E-06	5.17E-02	1.95E-04	4.81E-02	2.02E-03	4.00E-02	2.50E-05	5.74E-02
2.00E+00	1.92E-06	4.41E-02	1.10E-04	4.11E-02	1.15E-03	3.42E-02	1.41E-05	4.90E-02
2.04E+00	1.84E-06	4.36E-02	1.05E-04	4.06E-02	1.10E-03	3.38E-02	1.35E-05	4.84E-02
3.00E+00	8.55E-07	3.47E-02	4.88E-05	3.23E-02	5.17E-04	2.70E-02	6.26E-06	3.86E-02
4.00E+00	4.81E-07	2.90E-02	2.75E-05	2.70E-02	2.92E-04	2.25E-02	3.52E-06	3.22E-02
5.00E+00	3.08E-07	2.50E-02	1.76E-05	2.33E-02	1.87E-04	1.94E-02	2.25E-06	2.78E-02
6.00E+00	2.14E-07	2.21E-02	1.22E-05	2.06E-02	1.30E-04	1.72E-02	1.56E-06	2.45E-02
7.00E+00	1.57E-07	1.99E-02	8.97E-06	1.85E-02	9.57E-05	1.55E-02	1.15E-06	2.21E-02
8.00E+00	1.20E-07	1.81E-02	6.87E-06	1.68E-02	7.33E-05	1.41E-02	8.80E-07	2.01E-02
9.00E+00	9.50E-08	1.66E-02	5.43E-06	1.55E-02	5.80E-05	1.29E-02	6.95E-07	1.85E-02
1.00E+01	7.69E-08	1.54E-02	4.40E-06	1.43E-02	4.69E-05	1.20E-02	5.63E-07	1.71E-02
1.10E+01	6.36E-08	1.44E-02	3.63E-06	1.34E-02	3.88E-05	1.12E-02	4.65E-07	1.60E-02
1.20E+01	5.34E-08	1.35E-02	3.05E-06	1.25E-02	3.26E-05	1.05E-02	3.91E-07	1.50E-02
1.30E+01	4.55E-08	1.27E-02	2.60E-06	1.18E-02	2.78E-05	9.87E-03	3.33E-07	1.41E-02
1.40E+01	3.92E-08	1.20E-02	2.24E-06	1.12E-02	2.40E-05	9.34E-03	2.87E-07	1.33E-02
1.50E+01	3.42E-08	1.14E-02	1.95E-06	1.06E-02	2.09E-05	8.87E-03	2.50E-07	1.27E-02
1.60E+01	3.01E-08	1.09E-02	1.72E-06	1.01E-02	1.84E-05	8.44E-03	2.20E-07	1.21E-02
1.80E+01	2.37E-08	9.92E-03	1.36E-06	9.24E-03	1.45E-05	7.72E-03	1.74E-07	1.10E-02
2.00E+01	1.92E-08	9.15E-03	1.10E-06	8.52E-03	1.18E-05	7.12E-03	1.41E-07	1.02E-02
2.20E+01	1.59E-08	8.49E-03	9.09E-07	7.91E-03	9.71E-06	6.61E-03	1.16E-07	9.44E-03
2.40E+01	1.34E-08	7.94E-03	7.63E-07	7.40E-03	8.16E-06	6.18E-03	9.78E-08	8.82E-03
2.60E+01	1.14E-08	7.45E-03	6.50E-07	6.94E-03	6.95E-06	5.80E-03	8.33E-08	8.28E-03
2.80E+01	9.81E-09	7.03E-03	5.61E-07	6.55E-03	5.99E-06	5.47E-03	7.18E-08	7.81E-03
3.00E+01	8.55E-09	6.66E-03	4.89E-07	6.20E-03	5.22E-06	5.18E-03	6.26E-08	7.40E-03
4.00E+01	4.81E-09	5.29E-03	2.75E-07	4.93E-03	2.94E-06	4.12E-03	3.52E-08	5.88E-03
5.00E+01	3.08E-09	4.42E-03	1.76E-07	4.11E-03	1.88E-06	3.44E-03	2.25E-08	4.91E-03
6.00E+01	2.14E-09	3.81E-03	1.22E-07	3.54E-03	1.31E-06	2.96E-03	1.56E-08	4.23E-03
8.00E+01	1.20E-09	3.00E-03	6.87E-08	2.80E-03	7.35E-07	2.34E-03	8.80E-09	3.33E-03
1.00E+02	7.69E-10	2.49E-03	4.40E-08	2.32E-03	4.70E-07	1.94E-03	5.63E-09	2.77E-03

The difference between the values of the cross sections areas for both two types of scattering interactions can be seen. Figure 1. shows the high probability or scattering values of the CS from their values for RS interaction and the difference is greater when increasing the energy of reactive gamma. The cross section values of the RS is very close to the cross section values of the CS at the first energies (0.3, 0.2, 0.15)MeV for the Fm element. Thus, we can conclude that at the low energies of gamma radiation interacting with heavy materials or atoms, the cross sections of the scattering interactions are very close and this shows that the interaction of the RS is effective and its effect cannot be neglected in the material with heavy atomic numbers.

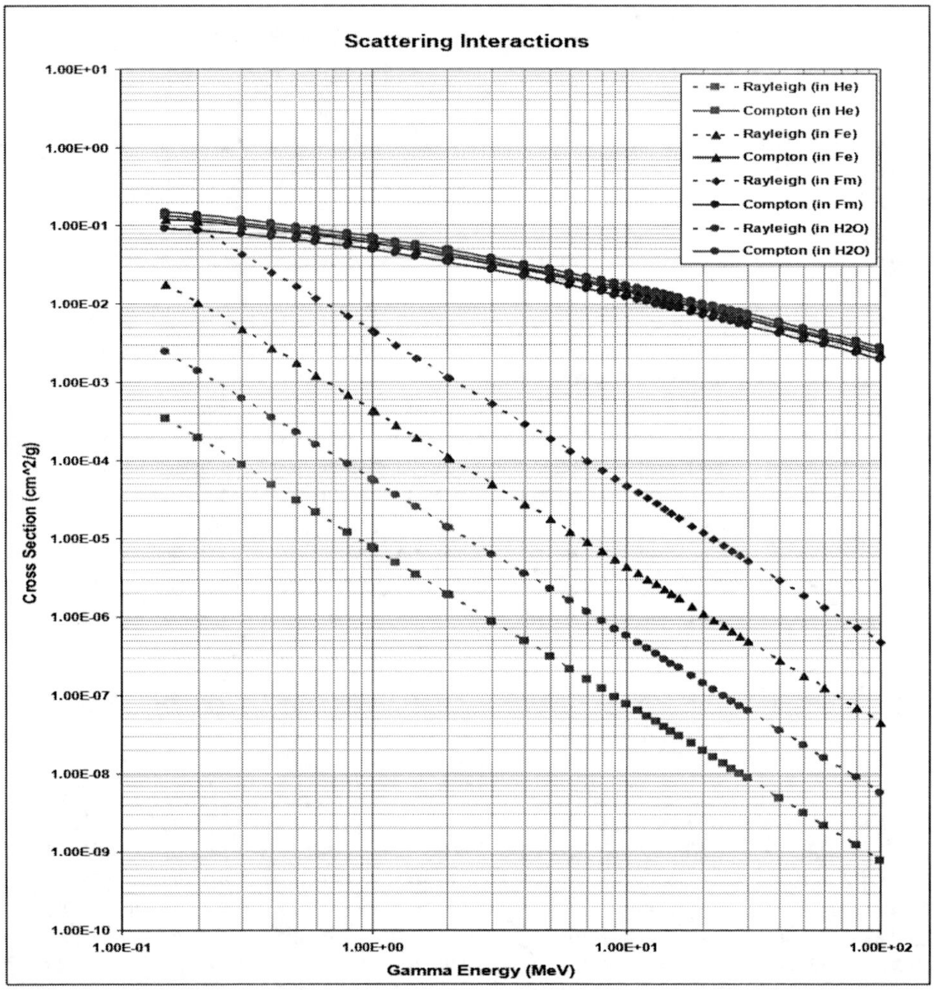

Figure 1. The relationship between the cross-section area of the Gamma radiation interaction represented by Compton, Rayleigh scattering, and the reactive Gamma (logarithmic scale).

From the observation of the figure above, the area of the cross section of the two interactions decreases significantly with the increase in the energy of the interactive radiation and its value is the highest in the case of low energies and exponentially decreases significantly with the increased energy at energies less than 10MeV and this decline decreases after that value of energy. It is also possible to observe that the area of the section for the RS is directly proportional to the atomic number, i.e. the

area of the section increases by increasing the atomic number of the elements used, and for the water, its value is less than that of the iron and the fermium, but the area of the cross-section of the CS is inversely proportional to the atomic number and for all the studied energies, and it has the largest value for the water molecules. As shown in Figure 1, the values of the cross-section of the CS for water molecules are larger than those of all the studied elements. This shows Gamma ray size affection on the water which represents a simulations model for living tissue. Therefore, we can conclude the CS interaction is very effective in living tissue rather than its effect on the elements.
We also studied these probabilities at E=1.25MeV energy, which represents the mean of Gamma photons energies emitted by the cobalt isotope (Co-60) that is commonly used in many researches and studies. To illustrate the ratio of both probabilities or the cross section of the RS interaction and the CS interaction probabilities to the total scattering interaction probabilities. These ratios are represented by the total probability resulting from them, which constitute the total cross section of the scattering in 3-D Pie chart indicating the degree of dependence of these ratios and their variability between one element and another depending on the atomic number of the pure element as well as the water molecule that simulates the living tissue in the living organisms. Figures of (2, 3, 4 and 5) represent the ratio of the cross-section area of the RS and the cross-section area of the CS to the total number of these areas which represent the total scattering probability.
Figure 5. shows that the area of the probability or the CS cross-section represents the largest area, the area of the probability or the RS cross-section starts increasing by increasing the atomic number, and it is very small relative to its value in the CS cross-section for water molecules. It can be concluded from this that the CS has greater effect on the light elements than the medium and heavy elements, and the CS interaction is the most likely or most common in the water molecules of living tissues in this energy and since the CS interaction cause absorption of part of the energy of an interactive photon. Thus, it is possible to say that the effective influence of the target material is the CS because of the transformation of energy from the falling photon to the target material.
The probabilities of the CS interaction were also plotted for all the materials used in figure 6. which shows the value of these two probabilities for the materials relative to the total reactions in the studied materials. The latter figure shows that the ratio is as large as water molecules, and decrease for the three studied elements by increasing the atomic number. The 3-D Pie chart of the total area of the RS interaction of the studied materials was also plotted as in figure 7. which is divided into parts representing the ratio of the area of this interaction to the total scattering of the studied materials.

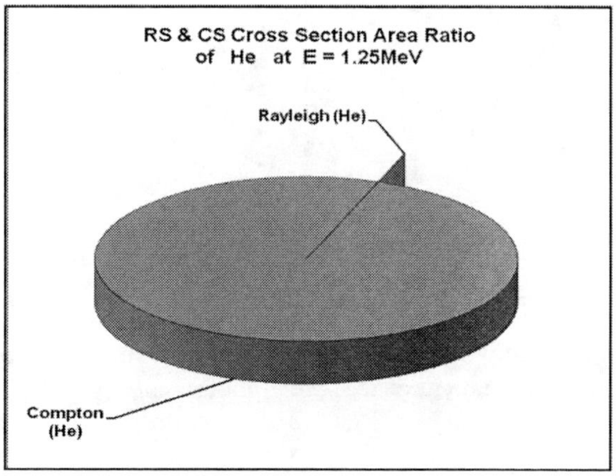

Figure 2. The cross section area ratio of the RS and the CS to the total number of these areas for Helium element at 1.25MeV photon energy.

IOP Conf. Series: Materials Science and Engineering **757** (2020) 012013 doi:10.1088/1757-899X/757/1/012013

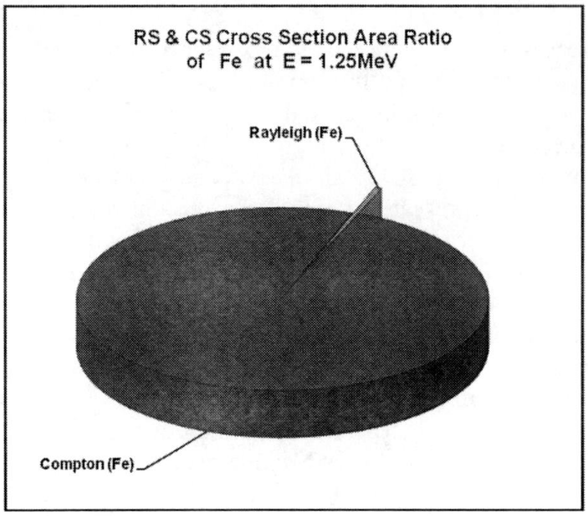

Figure 3. The cross-section area ratio of the RS and the CS to the total number of these areas for Iron element at 1.25MeV photon energy.

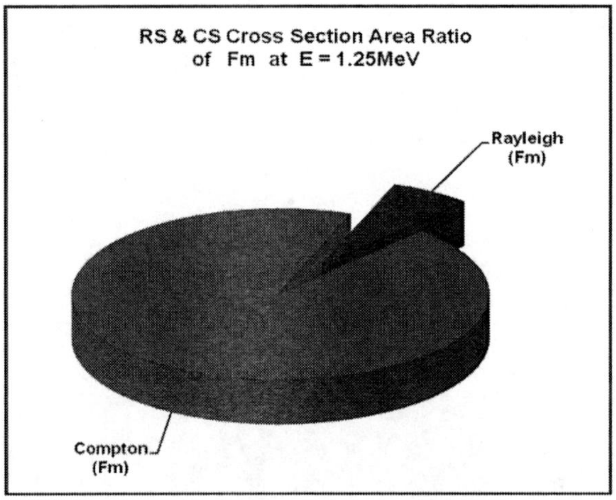

Figure 4. The cross-section area ratio of the RS and the CS to the total number of these areas for Fermium element at 1.25MeV photon energy.

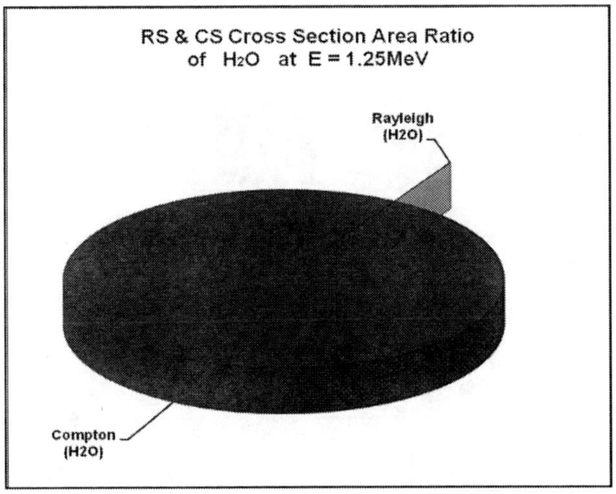

Figure 5 . The cross-section area ratio of the RS and the CS to the total number of these areas for H_2O compound at 1.25MeV photon energy.

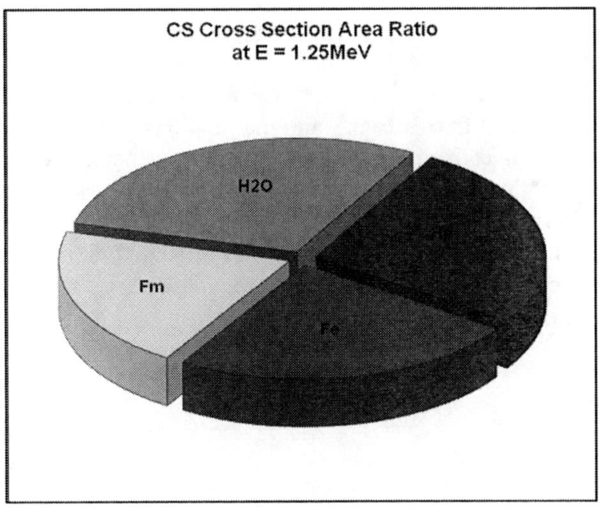

Figure 6. The CS cross-section area ratios He, Fe, Fm and H_2O materials to the total cross-sections at 1.25MeV photon energy.

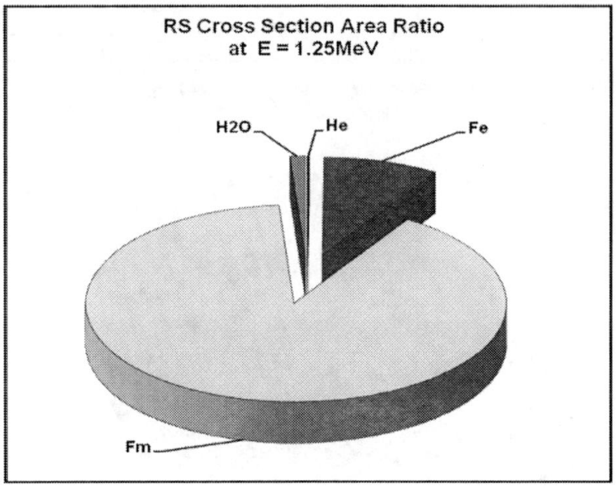

Figure 7. The RS cross-section area ratios He, Fe, Fm and H_2O materials to the total cross-sections at 1.25MeV photon energy.

It is clear from this figure that the area is the largest and with significantly different from the other parts in the Fm element and significantly decrease with decrease in the atomic number of the material. It is worth mentioning that the water molecules have a cross-section area or a probability greater than those measured for the He atoms.
To illustrate the ratios of the probability of interaction of the RS and CS, figure 8. shows the ratio of each of these probabilities for all materials to the total probability of the scattering interactions.
The latter figure gives a clear picture of the area greatness of the cross-section of the CS interaction as compared to that of RS interaction and also the probability greatness of the CS in the living tissue represented by the H_2O particles. In this figure and the previous figures of area or probability ratios at 1.25MeV photon energy emitted from cobalt isotope (Co-60) that the effect of these rays on living tissues has an effect higher than its effect on non-living compounds. Therefore, the danger of these rays that interact with living tissues due to the inelastic or incoherent scattering interactions (CS) because the radiations will lose its energy after several collisions within the living tissue and deposit these energies in it which cause significant damage to this tissues.

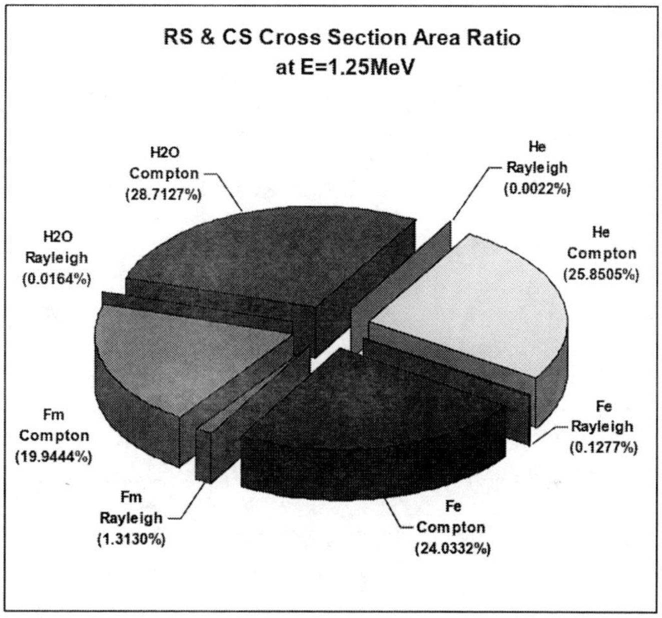

Figure 8. The ratios of the cross-section areas of RS interaction and the CS interaction for materials to the total area of the scattering cross-sections.

5. Conclusions

Several important conclusions were put from this study as follows:

1- The total cross-section area of the RS and CS interaction decreases by increasing the energy of the interactive photons and their values are greater for the CS interaction and for all energy values.
2- The area of the total cross-section of the RS increases by increasing the atomic number of the target material element while the relation is inverse to the total cross-section area of the CS where the area decreases with the atomic number increasing at any value of the Gamma energy.
3- Water molecules (H_2O) showed a behavior as similar as to this for pure materials or elements, but they have cross-section areas greater than the Helium, Iron and Fermium for all Gamma energies for the CS interaction, and the values of these probabilities greater than their values for Helium for all Gamma energies in the case of the RS.
4- RS to the CS ratio values are much greater at low Gamma energies and higher atomic numbers than their values at high Gamma energies and low atomic numbers of the used elements.
5- Water (H_2O), which simulates the living tissue, has the probability ratios of the CS to the total number of these possibilities for all studied elements and materials are greater than those of other used elements.

6. References

[1] Al–Wajidy and Taki A. Musu 1978 Cross–section for the Inactivation of Ribonucleise by Slow Heavy Charged Particles, *Master of Science, Medical Biophysics Department, University of Dundee, Scotland.*

[2] Icelli O. and Erzeneoglu S., 2002, *Spectrochimica Act*, B 57,1317.

[3] Hubbell J. H., Veigele W. J., Briggs E. H., Brown R. T., Cromer D. T. and Howertan R. J. 1975 *J Phys. Chem. Ref. Data*, 4, 471.

[4] Hubbell J. H. and Overbro I., 1979, *J Phys. Chem. Ref. Data,* 8, 69.

[5] Schaupp D, Schumacher M, Smend F, Rullhusen P and Hubbell J H 1983 *J Phys. Chem. Ref. Data*, 12, 467.

[6] Singh M P, Amandeep Sharma, Bhajan Singh and Sandhu B S, 2012, "Experimental measurement of coherent to incoherent cross-section ratio of element in the range 6<Z<82 for 59.45keV gamma photons" *Indian Journal of Pure &Applied Physics* V. 50, pp. 490-493.

[7] A Manjunath and B R Kerur 2016, Study of Rayleigh to Compton scattering ratio for pharmaceutical ingredients at 8 keV to 32 keV X-ray energy, *Materials Today: Proceedings,* Vol. 3, pp. 4134–4139.

[8] Tuba AKKUŞ 2018, An experimental study on the angular dependence of coherent to Compton scattering differential cross-section ratios of calcium, *Turkish Journal of Physics* , Vol. 42, pp. 368 – 377.

[9] S.S.AL-Dargazelli 1987, Detection of Nuclear Radiation.

[10] Klein O and Nishina, Y 1994 "On the Scattering of Radiation by Free Electrons According to Dirac's New Relativistic Quantum Dynamics" *The Oskar Klein Memorial Lectures* Vol 2. Edited by EKSPONG G. Published by World Scientific Publishing Co. Pte. Ltd.. ISBN #9789814335911, pp. 113-129. http://adsabs.harvard.edu/abs/1994okml.book..113K .

[11] J. H. Hubbell and I. Overbo 1979, Relativistic Atomic Form Factor and Photon Coherent Scattering Cross Sections, *J. Phys. Chem. Ref. Data,* Vol. 8, No.1, pp. 69-105.

[12] *XCOM Photon Cross-Sections Database*, http://physics.nist.gov/Phys Ref Data /Xcom /html/ xcom1.html.

[13] E Je´quier and F Constant.2010 "Water as an essential nutrient: the physiological basis of hydration" *European Journal of Clinical Nutrition* . vol.64, 115–123.

Estimation the cancer risk due to ingestion the food spices commonly used in Iraqi kitchen

A H AL-Mashhadani [1], R M Yas[1], W z Majeed[1], A A Saeed[1], N B Naje[1] and W Abdul Hadi[1]

[1] University of Baghdad, College of Science, Physics Department, Baghdad, Iraq

Corresponding author's e-mail: ranamuhi@ yahoo.com

Abstract. Spices are natural substances taken from special plants and have a different taste when added to food and some of them have great benefits for health and body. These plants vary from country to country depending on the type of soil and how they are grown and this affects their quality. In this study, the specific activity of ^{40}K, ^{238}U and ^{232}Th series and ^{137}Cs in some selected natural food spices commonly used in Iraq kitchen were determined using gamma spectrometry and the ingested doses via food consumption were also assessed. The average specific activity of ^{40}K, ^{238}U ,^{232}Th and ^{137}Cs in the samples are 72.00 Bqkg^{-1}, 106.576Bqkg^{-1}, 148.74 and 191.88 Bq.kg^{-1} respectively. The highest activity concentration of ^{40}K, ^{238}U, ^{232}Th and ^{137}Cs was recorded in cumin (401.7) for ^{40}K, laurel paper (260.67) for ^{238}U, briana spices (733.7) for ^{232}Th and Chamomile (833.4) for ^{137}Cs. The maximum value of total annual effective dose received from ^{40}K, ^{238}U, ^{232}Th and ^{137}Cs due to consumption of natural spices by the inhabitants was 169.83 μSvy^{-}. This value is low compared to the average radiation dose of 290 μSvy^{-1} received per caput worldwide due to ingestion the food spices. The excess lifetime cancer risk due to ingestion of radionuclide in food spices was estimated and the values obtained were compared with standards. Based on these results it can be concluded that the foods spices studied in this work do not present any serious radiological effect. Spices are natural substances taken from special plants and have a different taste when added

Keywords: Spice samples, Gamma spectroscopy, Cancer Risk, specific activity.

1. Introduction

Everything on the earth contains a small amount of radioactive material. Heavy metals and gamma radioactive elements are found in fruits, vegetables and plants [1]. Determining the level of radioactivity in the environment and foodstuffs is critical to controlling the amount of radiation to which humans are directly or indirectly exposed [2].

Knowledge of the concentrations of the radionuclides in spices is the important topics as it provides helpful information in the monitoring of environmental radioactivity.

Spices come from the bark, roots, seeds, fruit of plants and trees, they are used as recipes to add flavor to dishes, and they play an important role in the general makeup of culinary art [3]. Foods and spices can be contaminated either through deposition in the atmosphere or transferred by water through the soil. Two types of spices inspired in Iraq the fresh ones and those have been processed into powdered or solid forms. Assessment of contamination of spices consume by the population is very important to know the baseline value, or the level of radiation dose of both natural and artificial radionuclide received by them.

Various workers as in refs [4-7] have reported determination of different radionuclides concentration in food and spices samples.

In this study, the radioactivity levels of ^{40}K, ^{238}U, ^{232}Th and ^{137}Cs in natural food spices were estimated using gamma spectrometry. These spices were purchased from different traders at the mile one market in the Baghdad city, Iraq. This study, therefore, investigated the extent of the exposure of the public due to intake of spices as part of the national effort to establish base line values for the Regulatory Authorities to control the exposure of the public to the natural and artificial radionuclides due to the consumption of spices in Iraq.

2. Materials and Methods

2.1. Sampling and samples preparing

Food spices samples were collected from the Iraqi local market, fifteen samples were collected, every one of these samples weight about 1000 g. The types of samples are listed in Table 1. The homogenized rice samples were sealed in plastic containers and left for at least 4 weeks to reach a secular equilibrium between parent radionuclides and daughters [8].

Table 1. The types of samples

Sample No.	Sample name
1	Carrie
2	Briana spices
3	Cumin
4	Black pepper
5	White paper
6	Laurel paper
7	Black Bead
8	The Maui Roses
9	Cough tea
10	Chamomile
11	Nutmeg
12	Turmeric
13	Carnation
14	Smaug
15	Paprika

2.2. Gamma spectroscopy

In this this study, the technique of spectral analysis of gamma rays is used. The detector and measurement system is shown in Fig. 1, consists of NaI (TI) detector with a size of (3″×3″), the voltage of 691 V, the efficiency of 60%, and the energy resolution (6.5-8.56%) for value energy (661-1332) keV.

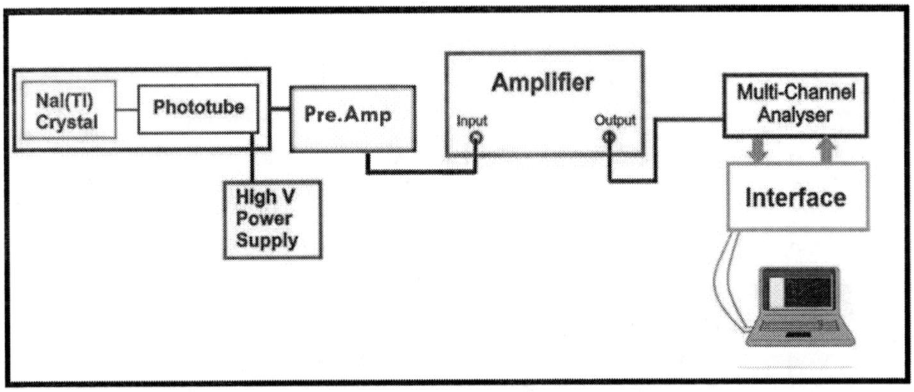

Figure. 1 A scheme for gamma ray spectroscopy with NaI (Tl) detector.

3. Results and discussion

3.1. Specific activity for natural radioactivity in the food spices samples

The specific activity of ^{238}U, ^{232}Th, ^{40}K and ^{137}Cs in samples of food spices commonly consumed by the Iraqi population are presented in Table 2. The activity concentration for ^{40}K in the spices samples ranged from 0.0 $Bqkg^{-1}$ to 401.7$Bqkg^{-1}$ in sample 3, the mean value of ^{40}K in the samples was 72 Bq/kg. The specific activity of ^{238}U in the food spice samples ranged from 0.0 $Bqkg^{-1}$ to 260.67 $Bqkg^{-1}$ in sample 6 and the mean value of ^{238}U in the samples was 106.576 $Bqkg^{-1}$. The activity concentration of ^{232}Th ranged from 0.0 $Bqkg^{-1}$ to 733.7 Bq/kg in sample 2 with a mean value of 148.74 $Bqkg^{-1}$. Fig. 2 represents the activity concentration of ^{238}U, ^{232}Th ^{40}K and ^{137}Cs in various food spices.

Table 2. Specific activity in (Bq/kg) of ^{40}K, ^{232}Th ,^{238}U and ^{137}Cs in food spices samples

Sample No.	Sample name			
	^{238}U	^{232}Th	^{40}K	^{137}Cs
1	0	0	0	0
2	0	733.7	175.3	0
3	122.12	0	401.7	115.5
4	0	0	0	0
5	214.50	0	74.53	629.5
6	260.67	372.6	161.5	323.5
7	71.25	543.47	0	135.4
8	167.48	0	146	0
9	0	289.8	1.187	0
10	13.66	0	0	833.4
11	148.3	0	73.6	633
12	189.5	0	13.24	174.7
13	167.4	253.6	33.25	0
14	180.6	38.04	0	33.25
15	63.17	0	0	0

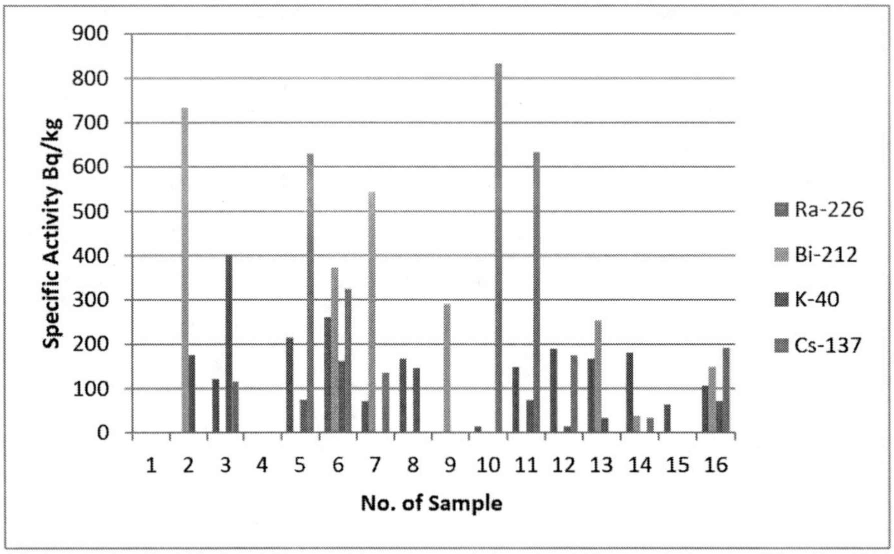

Figure 2. Specific activity of ^{226}Ra, ^{212}Bi, ^{40}K and ^{137}Cs in food spices samples.

3.2. Estimation of Annual Effective Dose and Dose Rate from Ingested Food Spices

The effective dose due to consumption of food spices comes in handy for the possibility of summing up different radionuclides which of course may come from different radioactive sources. These ingested radiation doses can be quantified by measuring the activity concentration (Bqkg^{-1}) of the radionuclide in the food spices and then multiplied by the masses of these food spices consumed over a given timeframe (kg/day or kg/year) and a dose conversion factor (Sv/Bq) given for each radionuclide and can be appropriately applied the equation:

$$E=\Sigma(As \times Is \times DCF) \quad ....(1)$$

where, E = Annual Effective Dose equivalent by ingestion of the radionuclides (Sv/y), As = Activity concentration of radionuclides in the sample (Bq/kg) , Is = Annual intake of the food spices (kg/y), DCF = internal dose conversion factor by ingestion of the radionuclides (Sv/Bq). This is given as 0.28 μSv/Bq for ^{238}U, 0.23 μSv/Bq for ^{232}Th , 0.0062 μSv/Bq for ^{40}K and 0.013μSv/Bqfor ^{137}Cs [9]. In this study the consumption values for locally produced food spices correspond to those of the adult citizen. The scale of the annual intake (1 kg/y) [10] was used. The risk associated with the intake of radionuclides in the body is proportional to the total internal dose delivered by the radionuclides.

Table 3, presents the annual effective dose equivalent of ^{238}U, ^{232}Th,^{40}K and ^{137}Cs radionuclides and the total dose due to the four radionuclides in food spices samples. From this Table, the annual effective ingestion doses due to the intake of ^{238}U varied from 0.0μSvy^{-1} to 72.8 μSvy^{-1} in sample 6. Most samples show insignificant low ingestion doses over the values of 6.3 μSvy^{-1}` reported by UNSCEAR, [11]. The dose received from ^{232}Th due to consumption of food spices varied from 0.00 in samples to 8.75 μSvy^{-1} in sample 14. The values of effective dose from the ingestion of ^{40}K ranged from 0.00 to 2.49 μSvy^{-1} in sample 3. Thus, the contribution to dose from the ingestion of ^{40}K in food spices with its relatively low dose conversion factor will be much less than that for the thorium ^{232}Th but higher than that for ^{238}U. The mean annual effective dose from ^{238}U, ^{232}Th, ^{40}K and ^{137}Cs in food spices were estimated to be 30.5, 38.22, 0.446 and 2.74 μSvy^{-1} respectively. The highest mean annual internal dose was ^{232}Th and all their mean doses are less than annual dose limit of 1 mSvy^{-1} for the general public. The UNSCEAR, [11] report shows that the sub-total ingestion dose of uranium and thorium series is given as 120μSvy^{-1} which is higher than the sub-total dose for the results reported in this study for uranium and thorium. The total effective dose ranged from 0.0 μSvy^{-1}(sample 1) to 152.3 μSvy^{-1} sample (11) with an average value of 64.62 μSvy^{-1} as shown in Fig.3. These values are much less than the world total food dose value of 290 μSvy^{-1} for all foods [11]. The low values of effective dose due to intake of spices are due to low annual intake of only 1 kgy^{-1} when compared to a few hundred kgy^{-1} for the total food intake.

Table 3: Annual effective dose equivalent (AEDE) in (μSvy^{-1}) of ^{40}K, ^{232}Th , ^{238}U and ^{137}Cs in food spices samples and the cancer risk parameters.

Sample No.	Annual effective dose equivalent (AEDE) (μSvy^{-1})				D (μSv/y)	ELCR × 10^{-4}
	Ra-226	Bi-214	K-40	Cs-137		
1	0	0	0	0	0	0
2	0	168.75	1.08	0	169.83	4.2
3	34.19	0	2.49	2.079	38.759	0.97
4	0	0	0	0	33.534	0.84
5	60.06	0	0.462	11.33	71.853	1.8
6	72.8	104.3	1.00	4.2	83.2	208
7	19.95	152.2	0	1.76	21.71	54.27
8	46.9	0	0.9	0	47.8	119.5
9	0	81.1	0.007	0	81.11	202.7
10	3.82	0	0	10.83	45.15	112.8
11	51.6	0	0.46	8.23	152.3	380.7
12	53.06	0	0.082	2.27	55.41	183.5
13	46.87	58.32	0.21	0	47.08	117.7
14	50.57	8.75	0	0.43	51.00	127.5
15	17.69	0	0	0	70.59	176.48
Ave.	30.5	38.22	0.446	2.74	64.62	112.73

3.3. Excess Lifetime Cancer Risk (ELCR)

To estimate cancer risk for an adult person using the following relationship [12]:

$$ELCR = Cd \times RF\ (Sv^{-1}) \quad \ldots\ldots\ldots\ldots\ldots (2)$$

where RF is a risk factor (Sv^{-1}), fatal cancer risk per Sever. For stochastic effects, ICRP 60 uses values of 0.05 for the public [13]. Cd is the lifetime effective dose which is a measure of the total effective dose received over an average lifetime of 50y following ingestion of a radionuclide was calculated using [11]:

$$Cd = 50 \times D \quad \ldots\ldots\ldots\ldots\ldots\ldots\ldots\ldots (3)$$

where D is the total effective dose to an individual.

The calculated cancer risk due to ingestion of food spices was shown in Table 3. The values of cancer risk were ranged from 0.00 to 380.7× 10^{-4} with an average value of 112.73× 10^{-4} as shown in Fig. 4.

The mean value of ELCR is lower than the world average value of 2.9×10^{-4} based on annual dose limit of 1 mSv for the general public [11].

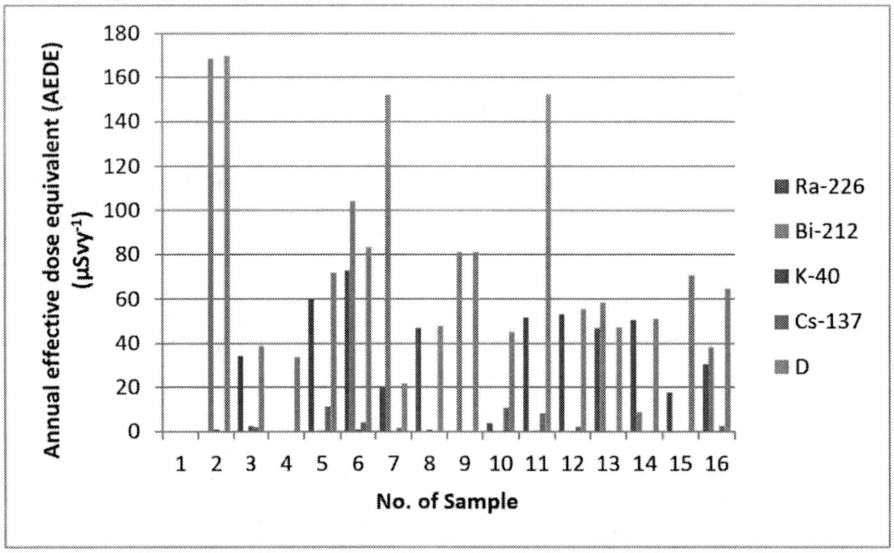

Figure 3. Annual effective dose equivalent (AEDE) (μSvy^{-1}) in food spices

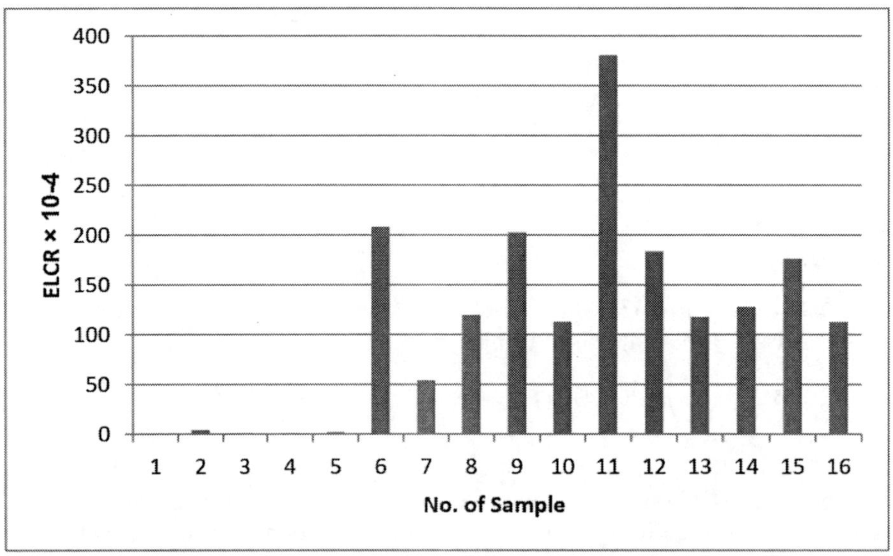

Figure 4. Cancer risk due to the ingestion of food spices.

4. Conclusions

This work was done to estimate the activity concentration of radionuclides ^{226}Ra, ^{232}Th and ^{40}K using gamma ray spectrometry in different spices that are consumed in Iraqi kitchen. The highest activity concentration of ^{40}K, ^{238}U, ^{232}Th and ^{137}Cs was recorded in kimono for ^{40}K, Laurel paper for ^{238}U,

Briana spices for ^{232}Th and Chamomile for ^{137}Cs. The estimated total annual effective dose received from ^{40}K, ^{238}U ,^{232}Th and ^{137}Cs due to consumption of natural spices is far below the average radiation dose of 290 μSvy^{-1} received per caput worldwide due to ingestion of natural radionuclide in food spices. The mean value of cancer risk is lower than the world average value of 2.9×10^{-4} based on annual dose limit of 1 mSv for the general public. From the results, it is clear that the food spices can be used as human diet for the control of activity. Some of these spices should be used in limited amount.

References

[1] Chibowski S 2000 *J. Envir. Study* **9** 249.
[2] Melqiades F L Appoloni C R 2004 *Ciênc Tecnol Aliment Campinas* **24** 501.
[3] Sahu T R 2007 Spices , Economic Botany 470003.
[4] Abdul-Fattah A F and Abdul- Majid S 1995 The fourth Saudi Engineering conference **5** 329.
[5] Al-Ghamdi A H 2014 *J. Am. Sci.* **10** 11.
[6] Kansaana C 2013 *Inter. J. Food Sci., Nutrition and Dietetics* **2** 75.
[7] Ononugbo1 C P, Avwiri G O and Ikhuiwu S O 2017 *J. Sci. Res. and Reports* **16** 1.
[8] Ramli A T 2009 *Malaysia Applied Research* **1** 45.
[9] UNSCEAR 1993 *Sources and effects of ionizing radiation, United Nation Scientific Committee on the effects of atomic Radiation report to the general Assembly, with Scientific Annexes*, New York.
[10] Rogers V C Nielson K K 1991 *Health Physics* **60** 807.
[11] UNSCEAR *2017 Sources, effects and risks of ionizing radiation*, United Nations Scientific Committee on the effects of Atomic Radiation report to the general Assembly, with Scientific Annexes, United Nations, New York.
[12] Taskin H Karavus M Touzogh A, Hindiroglu S and Karaham G 2009 *J. Environm. Radioa.* **10** 49.
[13] ICRP 1991. *1990 Recommendations of the ICRP*. ICRP Publication 60. Ann of the ICRP, 21 (1-3) Pergamon Press, Oxford.

IOP Conf. Series: Materials Science and Engineering **757** (2020) 012015 doi:10.1088/1757-899X/757/1/012015

Radon concentration measurement in a groundwater in Al-Tuz, Salah Al-Din Governorate using nuclear track detector CN-85

Asia H. Al-Mashhadani[1], Huda Saddi Ali[2],Rana M.Yas[1] and KararSameen Ali[2]

[1]Department of Physics, College of Science, University of Baghdad, Baghdad, Iraq

[2]Department of Physics, College of Science, University of Tikrt, Salah Alden, Iraq

E-mail:assia19662006@yahoo.com

Abstract. Thiswork aims to determine the radon concentration in different samples of groundwater in Al-Tuz, Salah Alden. Theconcentrations of radon were studied in 52 samples for different regions using CN-85 nuclear track detector.The results show that the concentration of radon gas in groundwaterwere ranged between the lowest value (0.727Bq/L) in the modern morning well sample(w44) and the highest value (4.001Bq/L) in the west of the district of Tuzwell sample (w47)with an average (1.747).The results were found to be below the allowed limit of radon concentrations ^{222}Rn (11.1 Bq/L). The mean annual dose of drinking water wasranged between the lowest value (2.619μ Sv/y) in sample W44 and the highest value (14.45 μ Sv/y) in the sample W47 withan average (6.292 μ Sv/y).All annual doses of drinking watervalues in this workwere lower than the allowed limit (22.024 μ Sv/y).

Keywords: Radon concentration,water,groundwater, mean annual dose, CN-85 detector.

Introduction

Radiation is generally either natural or man-made for different applications, and human exposure to different levels of radiation [1]. The impact of radiation on the environment in recent years, which can affect the genetic structure of humans and animals, causing genetic defects on the next generations, and that is why the radioactive contamination should have an impact on water and soil can be transmitted through the food chain to humans and animals [2]. The food chain is considered to be the most critical route in the rapid transmission of radioactive contaminants, as the contamination of aggregates in the food chain is polluted and the proportion of radioactive materials such as plants and animals and their products increases. The human body can be contaminated with radioactive materials by eating plants or animal meat that feed on plants that contain radioactive material originating from the main soil [3]. The plant absorbs these materials with other natural materials and is classified in construction as well as drinking water. And fluids where water contains a low dose, air breathing, which is the main source of natural radioactive dose that reaches within the human body and the main source of radon in the Earth's atmosphere resulting from the automatic decomposition of uranium [4]. The aim of this study to measure the concentration of radon (Rn-222), in the groundwater samples in Al-Tuz using the nuclear track detector CN-85.

1.Theoretical concepts

The measurement of radon concentration is based on identifying the constant K, which is basically related to the geometry of the chamber used in irradiation process. The constant K can be related to the number of tracks formed in the detector using the following equation [5].

$$\rho = KCT \qquad (1)$$

where ρ is the track density (Tr/cm^2),K is the diffusion constant measured by length unit (cm),C is the concentration of radon in the air space of the irradiation chamber, in(Bq/cm^3),T is the irradiation time in (s) and D is the track density rate ($Tr.cm^{-2}.s^{-1}$) which could be given by the following equation [6].

$$D = \frac{\rho}{T} = KC \tag{2}$$

for the cylindrical chamber of U- shape, Kcan be given by [5]

$$K = \frac{1}{4} r(2cos\theta_c - r/R_\alpha) \ (3)$$

where r isthe half of the cylindrical tube of 2.4 cm diameter. (θ_c) critical angle of CN-85 detector for particle occurrence. The critical angle of CN-85 is about 25^0. Rα is the range of alpha particles emitted into the air, which is equal to 4.15 cm[5]. Consequently, replace the values above with the parameters in eqn. (3), the constant propagation (K) of the cylindrical tube used to be (0.454 cm) was found using the CN-85 reagent. Consider Eqs. (2) and (3), and the intensity of the path the track density rate ($Tr.cm^{-2} s^{-1}$) is:

$$D_{Rn} = \frac{1}{4} Cr(2cos\theta_c - r/R_\alpha) \tag{4}$$

By which the concentration of radon in the air space of the cylindrical chamber C ($Bq.m^{-3}$) can be estimated. Finally the concentration of radon content within the samples (Cs) can also be calculated using the following equation [6]:

$$C_s = \lambda_{Rn} Ch \frac{t}{L} \tag{5}$$

whereλ_{Rn} is radon decay constant $\frac{\ln 2}{T_{\frac{1}{2}}}$ =0.1814 d^{-1} , h the distance between the sample surface and the exposed surface of the detector = 9.5cm, L sample thickness in the irradiation chamber =1.5 cm and t the exposure time (60 d).

Annual Effective Dose forradon in the water samples(AED_W)[7]

$$AED_W \left(\frac{Sv}{y}\right) = C_S C_{RW} D_{CW} \tag{6}$$

whereC_s(Bq/L) is the radon concentration in a groundwater, $D_{RW}$the annual consumption rate of water is equal(720L/y)[7], and D_{CW} the conversion factor is equal to(5×10^{-9}Sv/Bq).

2.Method and materials

Detector CN-85 of thickness 200μm and dimension 1×1 cm^2 is exposed to groundwater samples for 60 days. The nuclear track detector CN-85, considering long term measurement technique, has been used to investigate the tracks of alpha particles emitted by radon gas released from the sample.

2.1Samplescollection

A total of 52 samples of groundwater were collected for different areas in Al-TuzKhurmatu, Salah –Al-Din Governorate (geographic coordinates between 34°53N 44°38E) and presented in table1and showed inFig.1.Each sample size was 1L and then (5 g) of each samplewas put in a detection process. Place granular samples in U-shaped cylindrical irradiation chambers with an opening diameter of 2.38 cm. The rooms were tightly closed and left for 22 days to obtain an earthly balance of up to 98% between radon and radium decomposed from the contents of the sample. After 22 days, in order to prevent the decomposing radon from the room area, the room covers were quickly replaced. The CN-85 reagents were sealed under tightly closed covers, while the distance between the detectors was maintained at about 9.5 cm to prevent access to the detector. Therefore, the irradiation process was continued for 60 days, and then detectors pieces were raised from the irradiation chambers and chemically etched to reveal the alpha tracks released by radon content within the samples. Then the set of detectors were etched by aqueous NaOH solution of(2.5 N) at 60±1C. The number and density of the tracks were accounted using an optical microscope. The radiation background is measured for the same duration time that of (60 days) using same irradiation chamber with no sample.

Table 1. The most important information about the wells and their locations between 34°53N and 44°38E geographic coordinates.

No.	Well location	Depth of well (m)	History of drilling well	Uses
1	Near the southern control of Tuz	105	2004	The gravel and sand factory
2	Near the village of Yekja	110	2014	Animal Husbandry
3	East of Yenkja village	120	2008	The gravel and sand factory
4	Between the village of Jaragli and the village of Yenkja	120	2006	The gravel and sand factory
5	Between the district of Tuz and the village of Gardagli	100	2015	Animal husbandry and agriculture
6	North of Gardagli village	100	1990	Agriculture
7	South of Gardagli village	72	2009	Agriculture
8	Downtown Gardagli Village Village	70	2009	Drink
9	West of Gardagli village	70	2008	Animal husbandry and agriculture
10	North of Bustamli village	100	2010	Agriculture
11	West of Qarnaz village	40	2013	Agriculture
12	South of Karnaz village	100	1970	Drink
13	North of the village of Albu Hassan	91	2011	Agriculture
14	Between Amri and Qarnaz villages	87	2011	Animal husbandry and agriculture
15	The middle of an old man's hand	80	2011	Agriculture
16	Near the village of Shukr	145	2002	Agriculture
17	Near the village of Wadi	88	1988	Agriculture
18	Near the hand of my husband	120	2007	Animal Husbandry
19	Al - Tayyar District	85	2009	Agriculture
20	Inside Brougli Village	155	1973	Drink
21	North of Braugli village	155	1999	Animal husbandry and agriculture
22	Imam Ahmed Tuz neighborhood	65	2009	Firefighting
23	Imam Ahmed neighborhood	90	2012	Cultivation and drinking
24	The Shrine of the Imam Ahmed A. West of Tuz	95	2000	Wash
25	Between the district of Tuz and the district of Solomon Bek	110	2009	Block Factory
26	Cox Village	71	2004	Drink
27	North of Cox	90	2012	Drink
28	Khadasarli Village	55	2016	Agriculture
29	North Solomon Peck	83	2013	Block Factory
30	Near Solomon Beck	112	2009	Block Factory
31	Central Solomon Beck	105	2013	Block Factory
32	West Solomon Bek	100	2015	Agriculture
33	North of the district of Tuz	144	2012	Agriculture
34	East of the district of Tuz	138	2013	Animal husbandry and agriculture
35	West of the district of Tuz	125	2010	Drink
36	West of the district of Tuz	120	2010	Drink

37	Near Western control of the judiciary	110	2009	Agriculture
38	Aksu-Tuz neighborhood	110	1999	Agriculture
39	Between Tuz and the village of Braugli	40	2009	Drink and cultivate
40	The neighborhood of mud - Tuz	80	2010	Agriculture
41	The village of Albu Sabah	150	2006	Animal husbandry and agriculture
42	North of the village of Albu Sabah	155	2014	Agriculture
43	East of the village of Albu Sabah	140	2005	Animal husbandry and agriculture
44	Modern morning	130	2011	Agriculture
45	The village of Halwa large	140	2014	Animal husbandry and agriculture
46	Hliwa small village	140	2013	Animal husbandry and agriculture
47	West of the district of Tuz	72	2015	Animal husbandry and agriculture
48	West of the district of Tuz	79	2010	Animal husbandry and agriculture
49	Imam Hassan neighborhood - Tuz	130	2003	Agriculture
50	Neighborhood officers - Tuz	60	2013	Agriculture
51	Near the village of Braugli	90	2000	Animal husbandry and agriculture
52	Near the village of Braugli	95	2013	Animal Husbandry

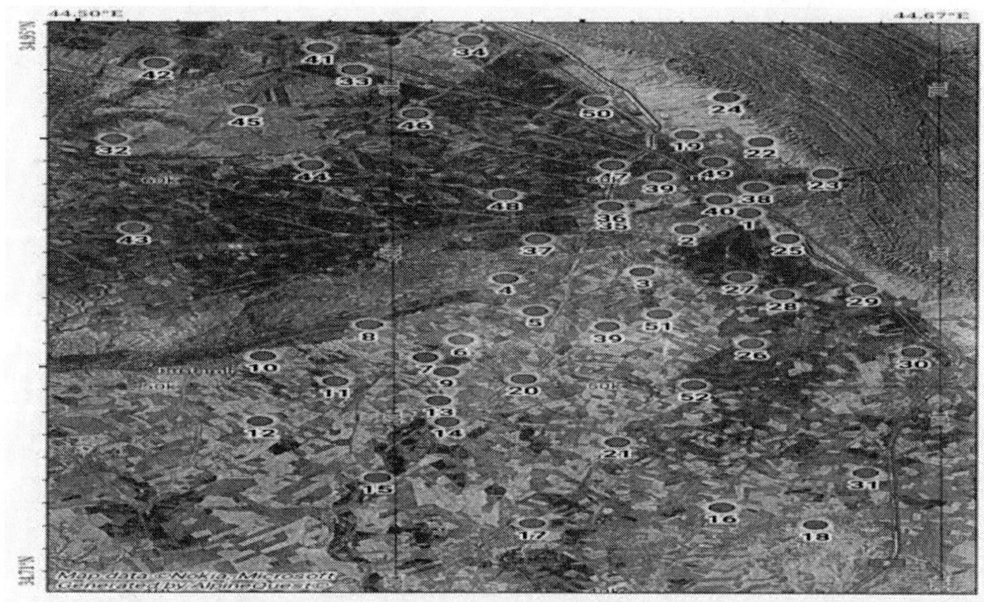

Figure 1: The locations of studied samples

3. Results and discussion

One of the sources that forms large risk on the human being life and it is so important to determine it is the radon gas inhalation from different types of materials into the human body. Table 2 shows the intensity of the tracks (ρ)usingEqu 2in ground water samples and the radon concentration in both the space air of irradiation chamber usingEqu(3) and within the groundwater samples used Equ (4). It is shown that the radon concentration values ranged between the lowest value (0.72 Bq/L) in the sample(w44) and the highest value (4.00 Bq/L) in the sample (w47))with the average (1.74784), its less than the value of the global limit of radon concentrations Rn-222 (11.1 Bq/L)[8]. The mean annual dose of drinking water(AEDw) were calculated using equation (6). Fig.3 shows the lowestAEDw value (2.62μSv/y) in sample (W44) and the highest value (14.41 μSv/y) in the sample (W47)with the average (6.29 μSv/y), resultsof AEDw were lower than the allowedvalue 1mSv/year[9].The spatial variations in radon concentration in wells water could be a function of the geologicalstructure of the area, depth of the water in the well and also differences in the climate, geo-hydrological processes that occur in the area and the age of well.

Table 2: The intensity of the effects and concentration of the radon in the sample and the concentration of radon in the aerobic space as well as the annual effective dose of radon

Sample ID	Intensity of effects (track/mm2)	Radon concentration in the sample(Bq/L)	Concentration of radon in aerobic space (Bq/L)	AEDw (μ Sv/y)
1	10964.29	2.94	0.26	10.58
2	4642.86	1.24	0.11	4.48
3	4428.57	1.19	0.10	4.27
4	3285.71	0.88	0.08	3.17
5	6142.86	1.65	0.14	5.93
6	6000.00	1.61	0.14	5.79
7	5607.14	1.50	0.13	5.41
8	8892.86	2.38	0.21	8.58
9	3142.86	0.84	0.07	3.03
10	5500.00	1.47	0.13	5.31
11	3785.71	1.01	0.09	3.65
12	6071.43	1.63	0.14	5.86
13	4464.29	1.20	0.11	4.31
14	3535.71	0.95	0.08	3.41
15	3107.14	0.83	0.07	3.00
16	4821.43	1.29	0.11	4.65
17	9892.86	2.65	0.23	9.55
18	6035.71	1.62	0.14	5.82
19	5892.86	1.58	0.14	5.69
20	7678.57	2.06	0.18	7.41
21	6642.86	1.78	0.16	6.41
22	6607.14	1.77	0.16	6.38
23	5678.57	1.52	0.13	5.48
24	4571.43	1.23	0.11	4.41
25	3000.00	0.80	0.07	2.89
26	7071.43	1.90	0.17	6.82
27	4607.14	1.23	0.11	4.45
28	5357.14	1.44	0.13	5.17
29	9107.14	2.44	0.21	8.79
30	5285.71	1.42	0.12	5.10
31	5142.86	1.38	0.12	4.96
32	6642.86	1.78	0.16	6.41
33	4678.57	1.25	0.11	4.51
34	3892.86	1.04	0.09	3.76
35	8107.14	2.17	0.19	7.82
36	3821.43	1.02	0.09	3.69
37	12250.00	3.28	0.29	11.82
38	10107.14	2.71	0.24	9.75
39	5214.29	1.40	0.12	5.03
40	4857.14	1.30	0.11	4.69
41	7964.29	2.13	0.19	7.69
42	6857.14	1.84	0.16	6.62
43	6964.29	1.87	0.16	6.72
44	2714.29	0.73	0.06	2.62
45	11571.43	3.10	0.27	11.17
46	7464.29	2.00	0.18	7.20
47	14928.57	4.00	0.35	14.41
48	13000.00	3.48	0.31	12.54
49	9428.57	2.53	0.22	9.10
50	7000.00	1.88	0.17	6.75
51	6928.57	1.86	0.16	6.69
52	7714.29	2.07	0.18	7.44
Average. Value	6520.6	1.74	0.15	6.29

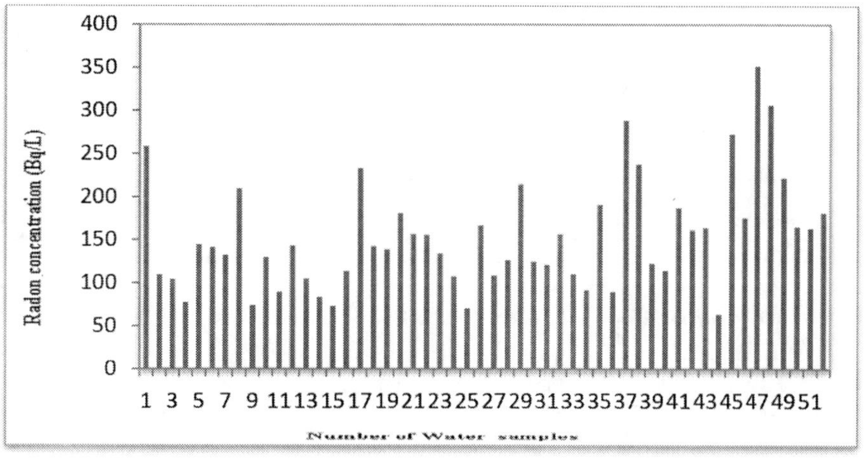

Figure 2: The concentration of radon in the samples

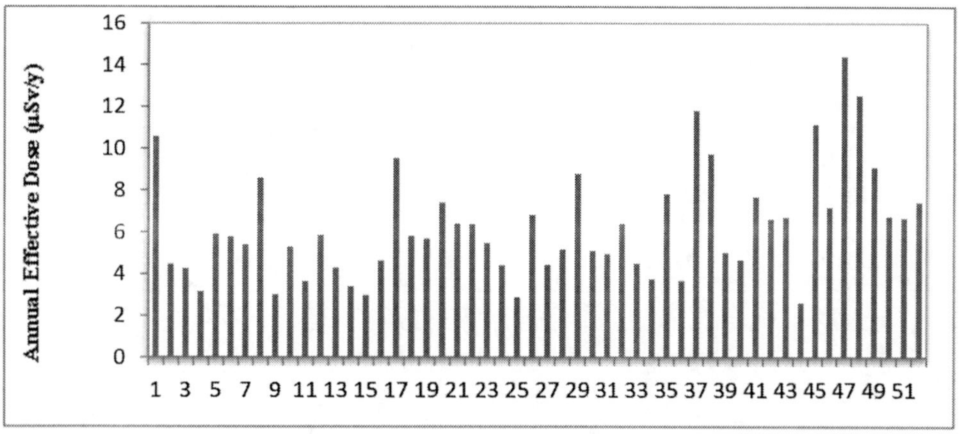

Figure 3: The annual effective dose.

4. Conclusions

One of the most important conclusions of this work is that the concentrations of radon are increasing in the old wells and with increasing depth. The results of the present study show that the average concentration of radon ^{222}Rnin theground waters of the city of TuzKhurmatu cityin Salah Al-Din is lower than the values of the allowed values (11 Bq /L)and the mean annual dose of drinking waterlower than the allowed value (1 mSv/year)[9].

References

[1] Al-Mazuri, N. S. M., 2000 M. Sc. Thesis, Coll. of Engineering, Univ. of Baghdad.

[2] ColmeneroSujo, L.,Montero Cabrera, M. E., 2004*Journal of Environmental Radioactivity* 77 205-219.

[3] Berlin, M. and Rudell, B., 1986 *Uranium. In: Handbook on the toxicology of metals*. 2nd edition.Elsevier Science Publishers, Amsterdam.623-637.

[4] IAEA International Atomic Energy Agency ,*1990 Environment behaviors of radium technical reports* 1 310 192.

[5]Ahmed, M. O., 2009 Proceedings of the First Scientific Conference of College of Science, University of Tikrit, Salahuddin, Iraq.

[6] ICRP, 1993 protection against Rn-222 at home and at work, ICRP Publication 65, Annuals of ICRP 23.

[7] Ridha, A. A.,Karim, M. S. andKadhim, N. F., 2014 *Civil and Environmental Research* 624-30.

[8]Abu-Murad, K.M., Diabat, A.E. and Gharaiba, S.H., 2003Abhath Al-Yarmouk, *Basic Sci. & Eng.***12** 221.

[9] UNSCEAR. 2000 *Sources, effects and risks of ionizing radiation* United Nations, New York, 1-35.

Investigation of the existence of new nuclear magic number in even-even O isotopes using shell model and Hartree–Fock Bogoliubov method

Redhab A Allawi and Ali A Alzubadi

Department of Physics, College of Science, University of Baghdad, Baghdad, Iraq

Abstract : In the present work, the features of some excited states in some even–even $^{14-26}$O isotopes have been investigated. The aim is to predict the evaluated existence of magic numbers in these isotopes using shell model and Hartree-Fock Bogoliubov method based on SLy4, SkC, SkD Skyrme parameterizations. In particular, root mean square radius, binding energies, one and two neutron separation energies, pairing gaps, transition probabilities, excitation energies, energy levels, transition densities and quadrupole deformation parameters have been investigated. The results are compared with the available experimental data.

Keywords: New magic numbers; Hartree–Fock–Bogoliubov method; Skyrme interaction; shell model theory.

1. Introduction

Mayer and Jensen [1,2] identified the magic numbers and their origin in the nuclear shell model and they indicated the existence of closed shells at numbers 8, 20, 50, 82 and 126. One of the most important issues is the change of the shell structure from stable to unstable nuclei which can give rise to the vanishing of known magic numbers and the appearance of new ones. This variation are caused by the monopole effect of the tensor force which was pointed out in the last decades by Otsuka and collaborators [3]. One of the most important methods that widely used to study the nuclear structure of these nuclei and also a wide range of nuclear chart, are the Shell Model (SM) method and Hartree–Fock–Bogoliubov (HFB) method with the Skyrme interactions [4,5], which unifies the self-consistent description of nuclear orbitals, as given by Hartree–Fock (HF) approach and the Bardeen–Cooper–Schrieffer (BCS) pairing theory [6] into a single variational theory. Many experimental and theoretical studies have been performed to investigate the existence of a new magic numbers with nuclear properties of exotic system. The first doubly magic nature of ^{24}O evidences comes from the study of the excited states in $^{20-24}$O through gamma ray spectroscopy [7] using fragmentation reactions of both stable and radioactive beams. Another evidence of ^{24}O doubly magic nature comes from its single-particle structure, which was studied in a knockout experiment [8]. Tshoo et al. (2012) [9], investigated the unbound excited states of the neutron drip-line isotope ^{24}O via proton inelastic scattering in inverse kinematics. In the present work, we concentrate on the SM and on mean field theories by choosing the Skyrme–Hartree–Fock–Bogoliubov method in order to theoretically predict and discuss the signature of new magic numbers in even-even O isotopes located far from stability. As it is known, the magicity in nuclei can be change locally in those which are far away from the stability line. Therefore, the known magic numbers for nuclei located in the stability valley or very close to it can disappear and new one can appear instead. The present work is a part of this current problematic inasmuch as it can concerns the discovery of new magic numbers in the exotic region.

2. THEORY AND METHODOLOGY

In this section, a short description of the methods is given that used for our calculations:

2.1 Shell model method

The many-particles reduced matrix elements of the electric multipole transition operator $\hat{T}^{\lambda}$ for an n-particles model space wave function of multipolarity λ can be expressed as the sum of the product of the elements of the one-body density matrix (*OBDM*) times reduced single-particle matrix elements, and is given by [10]:

$$\langle f \| \hat{T}^{\lambda} \| i \rangle = \langle n\omega_f J_f \| \hat{T}^{\lambda} \| n\omega_i J_i \rangle$$
$$= \sum_{k\alpha k\beta} OBDM(f,i,k_\alpha,k_\beta,\lambda) \langle k_\alpha \| \hat{T}^{\lambda} \| k_\beta \rangle \quad (1)$$

Where k_i and k_f are the single particles states for the initial and final model space states (nw_i J_i) and (nw_f J_f), respectively. The ω indices distinguish the various basis states with the same J value. The OBDM in the proton-neutron formalism is given by [10]:

$$OBDM(f,i,k_{\alpha,t_z},k_{\beta,t_z},\lambda) = \frac{\langle n\omega_f J_f \| [a^{+}_{k_{\alpha,t_z}} \otimes \tilde{a}_{k_{\beta,t_z}}]^{\lambda} \| n\omega_i J_i \rangle}{\sqrt{2\lambda+1}} \quad (2)$$

where $t_z = 1/2$ for neutron and $t_z = -1/2$ for proton. Two different shell model spaces have been used in our work. The first one is the *sd* model space, which consists of the active shells 1*d*5/2, 2*s*1/2, and 1*d*3/2 above the inert ^{16}O nucleus core which remains closed. That model space interactions are USDA, USDB and USDE. While the other one is the *p-sd* model space, which consists of the active shells *1p3/2*, 1*p*1/2, 1*d*5/2, 1*d*3/2, and 2*s*1/2 above the inert ^{4}He nucleus core which remains closed with PSDMK interaction.

2.2 Hartree–Fock Bogoliubov method

In HFB method, a two-body Hamiltonian of a system of fermions can be expressed in terms of a set of annihilation and creation operators (c, c†) [11]:

$$\hat{H} = \sum_{ij} t_{ij} \hat{c}_i^{\dagger} \hat{c}_j + \frac{1}{4} \sum_{ijkl} \bar{\upsilon}_{ijkl} \hat{c}_i^{\dagger} \hat{c}_j^{\dagger} \hat{c}_l \hat{c}_k \quad (3)$$

with the first term corresponding to the kinetic energy $\bar{\upsilon}_{ijkl} = \langle ij | V | kl \rangle$ are anti-symmetrized two-body interaction matrix-elements. The Skyrme interaction for nuclear structure calculations which is the central potential was developed from the idea that the energy functional could be expressed in terms of a zero-range expansion, which lead to a simple derivation of the HF equations that the exchange terms have the same mathematical structure as the direct terms. Thus, when solving the equations, this approximation greatly reduces the number of integrations over single-particle states. The Skyrme effective interaction which leads to a two-body density-dependent interaction that models the strong force in the particle-hole channel and contains central, spin-orbit and tensor contributions in coordinate space and called the standard analytical form, is given by [12,13]:

$$\begin{aligned} V_{Sky}(\vec{r}_1,\vec{r}_2) &= t_0(1+x_0\hat{P}_\sigma)\delta(\vec{r}_1-\vec{r}_2) \\ &+\frac{1}{2}t_1(1+x_1\hat{P}_\sigma)\left[\hat{k}'^2\delta(\vec{r}_1-\vec{r}_2)+\delta(\vec{r}_1-\vec{r}_2)\hat{k}^2\right] \\ &+t_2(1+x_2\hat{P}_\sigma)\hat{k}'.\delta(\vec{r}_1-\vec{r}_2)\hat{k}+\frac{1}{6}t_3(1+x_3\hat{P}_\sigma)\delta(\vec{r}_1-\vec{r}_2)\rho^{\alpha}\left(\frac{\vec{r}_1+\vec{r}_2}{2}\right) \\ &+iW_0(\vec{\sigma}_1+\vec{\sigma}_2)\hat{k}'\delta(\vec{r}_1-\vec{r}_2)\hat{k} \end{aligned} \quad (4)$$

where W_0, *tn* and *xn* are the free parameters describing the strengths of the different interaction terms which are fitted to nuclear structure data. The t_0 term indicates a zero-range central potential, and the t_1 and t_2 terms are non-local, because these depend on the gradient of the densities and have both central and exchange components with the range of the potential associated with $t_1/|t_0|$ [13]. The term consisting of W_0 indicates the spin-orbit part of the nucleon-nucleon interaction and an effective density-dependent three-body interaction is represented by the t_3 term. k, k' are the relative momentum operators with k acting on the right, while k' is the operator acting on the left [13]. In order to estimate the HF equations, we have to estimate the expectation value of the HF Hamiltonian in a Slater determinant $|HF\rangle$ It is given by:

$$E = \langle \phi_{HF} | \hat{H} | \phi_{HF} \rangle = \sum_{i=1}^{A} \langle \phi_i | \hat{T} | \phi_i \rangle + \frac{1}{2} \sum_{ij}^{A} \langle \phi_{ij} | V(i,j) | \phi_{ij} \rangle \tag{5}$$

where T is the kinetic energy operator and *V(i,j)* is the nucleon–nucleon interaction. The full expression for the expectation value of the HF equation with the Skyrme force after substituting the Skyrme interaction terms into the full energy expression, is [12]:

$$\begin{aligned} E = &\frac{\hbar^2}{2m}(\tau_n + \tau_p) + \frac{1}{4} t_0 \left[\rho(\vec{r})^2 (2 + x_0) - (2x_0 + 1)\left(\rho_n(\vec{r})^2 + \rho_p(\vec{r})^2 \right) \right] \\ &+ \frac{1}{24} t_3 \rho^{\eta} \left[(2 + x_3)\rho^2 - (2x_3 + 1)(\rho_p^2 + \rho_p^2) \right] + \frac{1}{8}\left[t_1(2 + x_1) + t_2(2 + x_2) \right] \tau\rho \\ &+ \frac{1}{8}\left[t_2(2x_2 + 1) - t_1(2x_1 + 1) \right](\tau_p \rho_p + \tau_n \rho_n) + \frac{1}{32}\left[3t_1(2 + x_1) - t_2(2 + x_2) \right](\nabla\rho)^2 \\ &- \frac{1}{32}\left[3t_1(2x_1 + 1) - t_2(2x_2 + 1) \right] \times \left[(\nabla\rho_n)^2 + (\nabla\rho_p)^2 \right] + \frac{1}{2} W_0 \left[\vec{J}.\vec{\nabla}\rho + \vec{J}_p.\vec{\nabla}\rho_p + \vec{J}_n.\vec{\nabla}\rho_n \right] \\ &+ \frac{1}{16}\left[(t_1 x_1 + t_2 x_2)\vec{J}^2 - (t_1 - t_2)(\vec{J}_p^2 + \vec{J}_n^2) \right] \end{aligned} \tag{6}$$

The Hamiltonian in Eq. (3) can be expressed in terms of the generalized quasiparticle operators as [14]:

$$\begin{aligned} \hat{H} = &H^0 + \sum_{k_1 k_2} H^{11}_{k_1 k_2} \beta^{\dagger}_{k_1} \beta_{k_2} + \sum_{k_1 k_2} (H^{20}_{k_1 k_2} \beta^{\dagger}_{k_1} \beta_{k_2} + h.c.) \\ &+ \sum_{k_1 k_2 k_3 k_4} (H^{40}_{k_1 k_2 k_3 k_4} \beta^{\dagger}_{k_1} \beta^{\dagger}_{k_2} \beta^{\dagger}_{k_3} \beta^{\dagger}_{k_4} + h.c.) \\ &+ \sum_{k_1 k_2 k_3 k_4} (H^{31}_{k_1 k_2 k_3 k_4} \beta^{\dagger}_{k_1} \beta^{\dagger}_{k_2} \beta^{\dagger}_{k_3} \beta_{k_4} + h.c.) \\ &+ \frac{1}{4} \sum_{k_1 k_2 k_3 k_4} (H^{22}_{k_1 k_2 k_3 k_4} \beta^{\dagger}_{k_1} \beta^{\dagger}_{k_2} \beta_{k_3} \beta_{k_4} + h.c.) \end{aligned} \tag{7}$$

Thus, the HFB equations can be written in matrix form since it is a variational theory that treats in a unified fashion MF and pairing correlations, as:

$$\begin{pmatrix} h - \mu & \Delta \\ -\Delta^* & -h^* + \mu \end{pmatrix} \begin{pmatrix} u_k \\ v_k \end{pmatrix} = E_k \begin{pmatrix} u_k \\ v_k \end{pmatrix} \tag{8}$$

where E_k are the quasiparticle energies, μ is the chemical potential, h and Δ are the HF Hamiltonian and the pairing potential, respectively, and the u_k and v_k are the upper and lower components of the quasiparticle wave functions.

3. RESULTS AND DISCUSSION

In the present work, the OBDM elements are calculated using the SM code NUSHELLX@MSU [15]. It is a set of wrapper codes written by Alex Brown that use data files for model spaces and Hamiltonians to generate input for NUSHELLX [16]. The radial wave functions of the single-particle matrix elements in all our calculations are calculated using a two-body Skyrme interaction potential, from which a one-body potential in Hartrre-Fock theory of types SLy4 [17], SkC [18] and SkD [18] as well as to SM theory of types Skxs25 [19] which can be generated.

3.1 Nuclear Radii

The charge root mean square radius for even- even $^{14-26}$O isotopes plotted as a function of neutron number (N) are shown in Fig.1 obtained from HFB calculations with SLy4, SkC and SkD parameterizations. Inspection of these curves revelled that the charge radii have parabolic trend due to the shell closure of N = 16 which reasonably well reproduced using the SkC parameterization. The shell effect directly influences the deformations of nuclei and thus affects the nuclear rms charge radii.

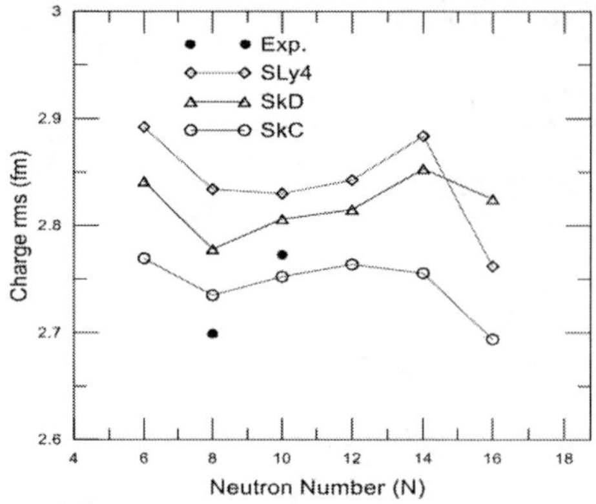

Figure.1. Proton, neutron, charge and mass root mean square radius for even- even $^{14-24}O$ isotopes plotted as a function of neutron number (N) obtained from HFB calculations with SLy4, SkC and SkD parameterizations represented in pink, red and blue lines, respectively. The black circle is the experimental data of charge rms taken from Ref. [20].

3.2 Binding and Separation Energies

The calculated binding energies for the oxygen isotopes using Skyrme interactions are shown in Table 1 in comparison with the experimental data taken from Ref. [21]. The higher value of the binding energy as well as the separation energy is a best sign to indicate the magic nature of nuclei. The behaviour of results is quite smooth and in general have a good agreement with experimental data. The higher value of the binding energy at N=16 indicate the existence of new magic number as compared with the neighbour isotopes. Fig.2a demonstrates the one-neutron separation energies for the oxygen isotopes as a function of neutron number (N) obtained using SM with the Skyrme interactions. Our results are compared with the experimental data [21].In general, the agreement is nicely good. At the neutron rich region, it is evident from the higher value of the separation energy at N=16 as compared with the neighbour isotopes, that the nucleus have a double magic nature. The variation of two-neutron separation energy for oxygen isotopes as a function of neutron number (N) demonstrates in Fig.2b which obtained using SM with Skyrme interaction. The calculated results compared with the available experimental data taken from Ref. [21]. From this figure, the SM predictions with Skyrme interactions give, approximately results in agreement with the experimental data. At N=12,14,16, S_{2n} energy shows a higher value as compared with neighbour isotopes at the neutron drip line. This behaviour corresponds to the appearance of closed shell around N=16.

Table 1: The binding energies in (MeV) obtained with Skyrme and HFB calculations, E(2^+) in (MeV) and B(E2) in (e^2 fm^4) with *sd* and *p-sd* MS, for $^{18-26}O$ isotopes. The experimental data are shown for comparison.

Nucl.	B.E (MeV)					E(2^+) MeV			B(E2) e^2 fm^4		
	Exp.	SHF	HFB Sly4	HFB SkC	HFB SkD	Exp.	*sd* MS	*psd* MS	Exp.	*sd* MS	*psd* MS
^{18}O	139.8	140.2	141.7	139.1	140.44	1.982	1.998	2.228	44.98	23.55	24.23
^{20}O	151.3	151.9	153.4	152.5	153.46	1.674	1.746	2.034	28.221	31.31	31.88
^{22}O	162.1	164.5	160.2	159.6	160.78	3.199	3.158	2.99	20.94	25.99	27.6
^{24}O	168.3	172.4	166.2	168.2	173.82	4.79	5.04	5.754	10.52	10.24	7.33
^{26}O	167.9	171.9	163.5	166.7	170.31	2.0	2.11	1.642	-	12.61	12.59

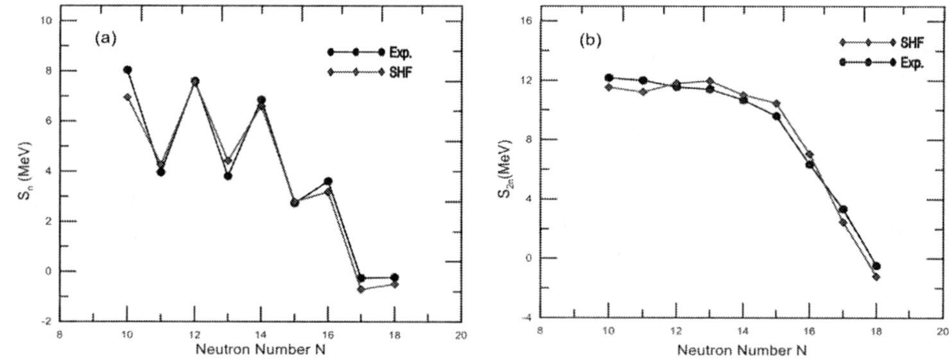

Figure.2. (a) single and (b) two-neutron separation energies as a function of neutron number (N) for the oxygen isotopes. Black circles are the experimental values taken from [21] and red squares obtained using SM with Skyrme interactions.

3.3 Excitation Energy E(2^+_1) and Reduced Electric Transition Probabilities B(E2)

The excitation energies of the 2^+_1 states in the even-even $^{18\text{-}26}$O isotopes is shown in the top panel of Fig.3 while the bottom panel shows the electric transition probabilities B(E2). The calculations are compared with the available experimental data taken from Ref. [21] which includes two model space: *sd* and *p-sd* with USDB and PSDMK interactions, respectively. The calculated E(2^+_1) have a good agreement with the experimental data for both *sd* MS and *p-sd* MS. At N=10 and N=12, the excitation energies have smaller energy values as compared with the neighbour nuclei then it increase rapidly at N=14 up to N=16 then sudden drop at N=18. The calculated *B*(E2) have a well agreement with the experimental data for both *sd* MS and *p-sd* MS. Approximately, the transition probabilities have a reverse behaviour to the excitation energy which mean that N=16, the transition probabilities have the smallest values between neighbour isotopes. It is known from experiment that the excitation energy of the first excited state 2^+1 in even-even nuclei increases at the magic number, while the transition probabilities decreases at that point. According to these two facts, it obvious from the experimental and calculated values the double magic nature at N=16.

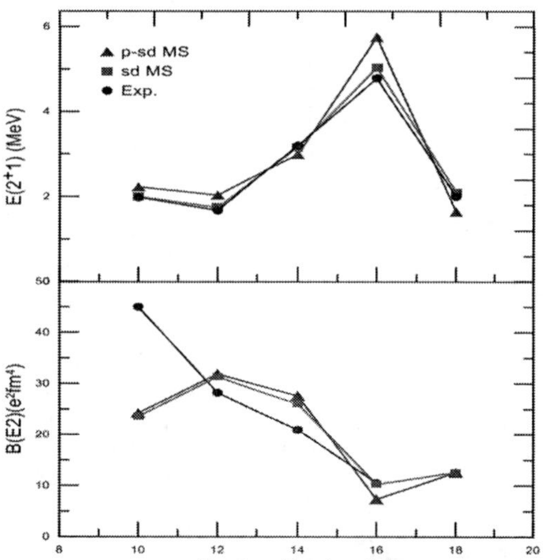

Fig.3 The top panel shows the excitation energies of the 2^+_1 states E(2^+_1) in the even-even $^{18\text{-}26}$O isotopes, while the bottom panel shows the electric transition probabilities B(E2) values for the same nuclei as a function of Neutron number (N). The calculations includes two model space: *sd* as a square shape and *p-sd* as triangle shape. The experimental values (black circles) are taken from [21].

3.4 Structure for the Model Spaces based on Nucleon-Nucleon Interaction

The energy levels of some low laying states in ^{24}O isotopes are calculated in the *sd* and *p-sd* MS with USDA, USDB, USDE and PSDMK interactions, respectively. Fig.4 represents that calculated energy level schemes in comparison with the experimental data taken from Ref. [21]. For the ^{24}O nucleus, the experimental observed state are 2^+ (4.79 MeV) and the presence of large nuclear shell gap gives the onset of the N=16 subshell closure. The calculated 2^+ excited states have a good agreement with the experimental value in all *sd* MS interactions which provide an important implication of evidence for existence the doubly magic nature of ^{24}O isotope.

3.5 Deformation Energy Curves

The results of deformation energy curves plotted as a function of quadrupole deformation parameter β_2 are shown in Fig.5 for (a)^{18}O, (b)^{20}O, (c)^{22}O and (d) ^{24}O which obtained from HFB calculations with SLy4, SkD and SkC parameterizations. In the three parameterizations, it is obvious that these isotopes have a spherical shape since it have an even-even closed shell with (Z=8). Although these isotopes have a spherical shapes, the neutron shell effects evolves the shapes as more neutrons occupy the *sd* shell. This feature is due to a growing collectivity with more neutrons in the *sd* shell, since the mean field and the Skyrme interaction are collective phenomena and the deformation shape is caused by the collective behaviour of the nucleons inside the nucleus. The collective model describes the nuclear motion as a moving mean field, and is based on the breaking symmetry of interaction between the nucleons. Thus, when the interaction strengths are changed, the ground state can manifestation shape transitions between spherical, deformed prolate and deformed oblate. The sharp minima at ^{24}O suggested the magicity at N=16.

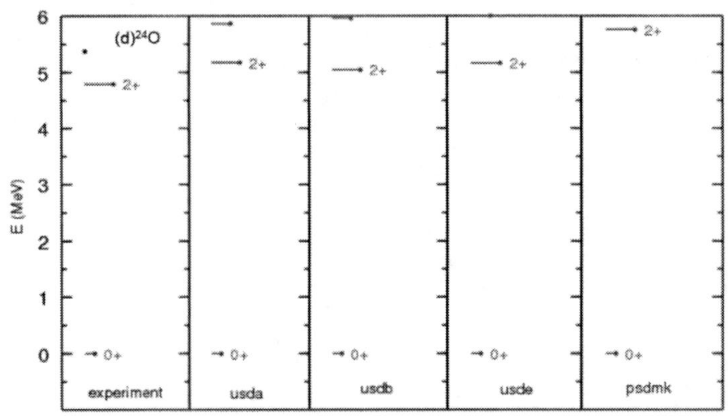

Figure.4 Energy levels ^{24}O. The experimental values are taken from [21] and plotted on the left- hand side with the length of the line proportional to the J value. The red lines end in filled circles for positive parity and the blue lines end in crosses for negative parity. The point on the y-axis indicated the J value that not certain. The calculated values plotted in the middle carried out in the full *sd* MS with USDA, USDB and USDE interactions, while the right- hand side carried out in the full *p-sd* MS with PSDMK interaction.

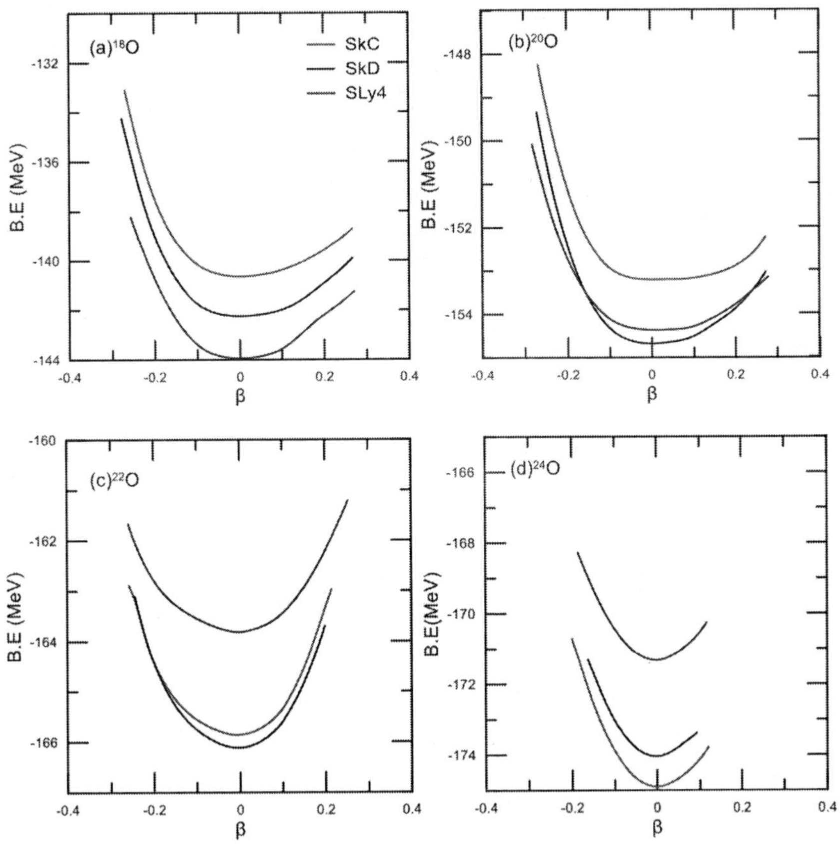

Figure.5. Deformation energy curves plotted as a function of quadrupole deformation parameter β_2 for (a) ^{18}O, (b) ^{20}O, (c) ^{22}O and (d) ^{24}O obtained from HFB calculations with SLy4, SkC and SkD parameterizations .

4. CONCLUSIONS

From the research that has been carried out, it is possible to conclude that the SM with HF method is probably the best method for the investigation magicity of neutron rich nuclei. In particular, in the region far from stability, the HF method is probably the best model for anticipating the total binding energies and single-particle energies of the closed-shell nuclei.

In summary, we have demonstrated the approach for connecting HF and large-basis SM calculations and used it for the system of valence neutrons in the *sd* and *p-sd* shell model spaces and HFB method on the evolution of quadrupole deformation. The calculated results for $^{18-26}O$ isotopes are promising. The obtained results reproduced nicely the available experimental data, which based on the binding energies, one- and two-neutron separation energies, rms radii, low laying excitation energies, pairing gaps, transition probabilities, as well as quadrupole deformation, we reproduced the classic magic numbers and confirmed the existence of N=16 is a new one in the oxygen isotopes. The findings suggest that this approach could also be associate many more excited levels with the experiment, especially in the nuclear island of inversion region and near the proton and neutron driplines.

References

[1] M.Goeppert Mayer, Phys. Rev. **75** (1949) 1969.
[2] O. Haxel, J.H.D. Jensen, H.E. Suess, Phys. Rev. **75** (1949) 1766.
[3] T. Otsuka, R. Fujimoto, Y. Utsuno, B.A. Brown, M. Honma, T. Mizusaki, Phys. Rev. Lett. **87** (2001) 082502.
[4] S. Burrello "Effective interactions and pairing correlations: from nuclei to compact stars" anno accademico (2016).

[5] A. de Shalit and I. Talmi, Nuclear Shell Theory, Dover Publications, (2004).
[6] J. Bardeen, L.N. Cooper, J.R. Schrieffer, Phys. Rev. 108 (1957) 1175.
[7] M. Stanoiu, et al., Phys. Rev. C **69** (2004) 034312.
[8] R. Kanungo, et al., Phys. Rev. Lett. **102** (2009)152501.
[9] K. Tshoo, et al., Phys. Rev. Lett. **109** (2012) 022501.
[10] B. A. Brown, Lecture Notes in Nuclear Structure Physics, National Superconducting Cyclotron Laboratory and Department of Physics and Astronomy, Michigan State University, E. Lansing, MI 48824, 2005.
[11] S. Burrello "Effective interactions and pairing correlations: from nuclei to compact stars" anno accademico (2016).
[12] E. Chabanat, P. Bonche, P. Hansel, J. Meyer, and R. Schaeffer, Nucl. Phys. A **627** (1997) 710.
[13] J. Stone and P. -G. Reinhard, Prog. Part. Nucl. Phys. **58** (2007) 587.
[14] J. Kvasil "Nuclear Structure and Nuclear Processes"(Prague: Institute of Particle and Nuclear Physics, Faculty of Mathematics and Physics, Charles University) (2013).
[15] B. A. Brown and W. D. M. Rae, Nucl. Data. Sheets **120** (2014) 115.
[16] W. D. M. Rae, NuShellX, http://www.garsington.eclipse.co.uk.
[17] E. Chabanat, P. Bonche, P. Haensel, J. Meyer, R. Schaeffer, Nucl. Phys. A**635** (1998) 231.
[18] Y. Xu, H. Guo, Y. Han and Q. Shen, J. Phys. G: Nucl. Part. Phys. **41** (2014) 015101
[19] Brown et al., PRC.**76** (034305) 2007.
[20] I. Angeli, At. Data Nucl. Data Tables **99** (2013) 69.
[21] National Nuclear Data Center (NNDC), http://www.nndc. bnl.gov/.

INVESTIGATION OF THE REACTION CROSS-SECTION FOR PRODUCTION THE RADIOACTIVE ISOTOPES USED IN FABRICATING THE NUCLEAR BATTERIES

Nabeel F. Latooffi [1] **and Ali A. Alzubadi** [2]

[1]Department of Physics, College of Science, University of Anbar, Al-Anbar, Iraq

[2]Department of Physics, College of Science, University of Baghdad, Baghdad, Iraq

Dr.nabeel.fawzi@uoanbar.edu.iq

Abstract. Beta-decay radioactive isotopes have been widely used as a high-energy source in nuclear batteries. The aim of this study is to improve the evaluations of nuclear reactions for the production of some radionuclides with possible use in nuclear batteries. In particular, the 38Ar(n,γ)39Ar, 59Co(n,γ)60Co, 112Cd(n,γ)113mCd, 151Eu(n,γ)152Eu, 153Eu(n,γ) 154Eu, 193Os(n,γ) 194Os and 203Tl(n,γ) 204Tl neutron capture reactions have been studied for an incident neutron energy range from 10-6 eV up to 2 MeV. The evaluations of the capture cross sections are based on the nuclear model calculations with the latest release code EMPIRE-3.2.2 (Malta) and compared with the available experimental ones. The TENDL nuclear data library, which provides the output of TALYS nuclear model code system have been also presented along with the extracted data for some interactions with no experimental support. Simulation of the beta particles transportation and the generated bremsstrahlung intensity through silicon carbide semiconductor has been also investigated.

Keywords: Nuclear batteries; Nuclear cross section; Neutron capture reactions.

1. Introduction

Due to the remote environmental conditions and the growth of long-range space missions and sea voyages and many other different fields such as human-portable energy sources, robotic sensors, developing fields and military missions, the need for alternative energy sources with special specifications, especially longevity ones, has increased. To serve this role Moseley and Harling in 1913[1] have been used the natural radioactive isotope as a substitutional source of energy to the chemical once in what is called the nuclear batteries. These batteries with a good shielding and design can uniquely save much more energy than their chemical counterparts, which can be close to 1,000 watts per hour and the life of it can be in the order of many decades since it is related to the half-life of the radioisotope used. With the availability of low-energy electronics and the fact that radioactive sources are inexpensive, nuclear batteries can be considered as a source of commercial profit as well as a source of high energy [2].

There are many competing types of nuclear batteries: thermoelectric, thermophotoelectric, direct charge collection, thermionic, scintillation intermediate, alphavoltaics, and betavoltaics. Between these types, the most interested ones for the past four decades is the radioisotope thermoelectric generator (RTG) in which the heat produced by the isotopes decay converted to electricity through the Seebeck effect [3, 4]. Selection of radioisotopes must fulfill a good match between the range of radiation and the scale length of the transducer, which in turn depends on the decay type of the radioisotopes. It also depends on the cost of the production and separation of the radioisotopes. The alpha and beta emitters are the most types of radioisotopes used in nuclear batteries. In the present work, the emphasis was

IOP Conf. Series: Materials Science and Engineering **757** (2020) 012017 doi:10.1088/1757-899X/757/1/012017

placed on the nuclear reactions that produce the isotopes with beta particles decay, which have a longer range than the alpha emitters have and can be used in different types of batteries.

The nuclear reaction mechanisms for the nuclear isotopes used in the nuclear batteries can be understood from the information provided by the calculation of the nuclear reaction cross-sections, which is needed also in many other applications such as fusion reactor and nuclear power plants. This study is an attempt to evaluate the cross-sections for some nuclear reactions, which produce radionuclides used in nuclear batteries with the code EMPIRE [5]. Many nuclear models for cross-sections calculations have been included in this complex and flexible system and designed for calculations over a broad range of energies and incident particles.

The reaction of an incident particle with an atomic nucleus can take place in many ways. In studies of light-ion induced nuclear reactions, one distinguishes between three different mechanisms: direct, compound and pre-equilibrium nuclear reactions, which can be subdivided according to time scales or, equivalently, the number of intra nuclear collisions taking place before emission. Furthermore; each mechanism preferentially excites retain parts of the nuclear level spectrum and is characterized by different types of angular distributions.

2. Nuclear reaction cross-section

The nuclear reaction mechanisms of an incident particle with an atomic nucleus can be explained by the three main models or mechanisms mentioned above according to the energy of the projectile and time scales of the interaction. At lower excitation energies (below 10 MeV), the whole system is involved in the interaction and the isotropic distributions of emitted particles in the center of the mass frame can be explained by the compound nucleus mechanism. In this model the theoretical calculations of reaction cross-sections are described by Hauser and Feshbach [6] taken into account the conservation of angular momentum and parity.

$$\sigma_{ab}^{HF}(E) = \pi\lambda_a^2 \sum_{J\pi} \frac{2J+1}{(2i+1)(2I+1)} \frac{T_a T_b}{\sum_c T_c} \tag{1}$$

where J; π are the total angular momentum and parity of the compound state, λ_a is the reduced wavelength in the incident channel (inverse of the wave number k) and $T_a T_b$ are the transmission coefficients of the incident and outgoing channels respectively.

On the other hand, at relatively higher excitation energies, the forward peaked angular distribution of particles indicates the presence of direct reaction mechanism [7, 8] in which only a few nucleons take part in the evolution of the reaction processes. The direct interactions analysis procedure either permanently with the Distorted Wave Born Approximation (DWBA) or by the coupled channels approaches [9].

The theory of nuclear reactions was extended to include a statistical treatment for particles emitted from the target nucleus during the equilibrium in a process called preequilibrium nuclear reaction. Quantum mechanical and phenomenological theories were proposed for pre-equilibrium emission mechanisms such as the classical exciton model including nucleon, cluster and gamma emissions, and Blann's Hybrid model that is limited to nucleon emission for the semiclassical theories. For quantum mechanics, the Feshbach, Kerman, and Koonin (FKK) model [10, 11] and Nishioka, Verbaarshot, Weidenmuller and Yoshida (NVWY) model are the most important models. For the evolution of the projectile-target system from small to large energy losses it is useful to consider separately the states with at least one particle in the continuum which is described in the Multi Step Direct (MSD) theory and the states with all particles are bound, i.e. the Multi Step Compound (MSC) reactions [12, 13].

The multi-step direct cross-sections were defined as the summation of n-step as follows:

$$\frac{d^2\sigma}{d\Omega dE} = \sum_n \frac{d^2\sigma^{(n)}}{d\Omega dE} \tag{2}$$

where the contribution of the n-step cross-sections can be given as a function of probability per energy to find the system in the configuration *c*, *c=npnh,* and the open channel T-matrix:

$$\frac{d^2\sigma}{d\Omega dE} = \sum_{c=[npnh]} P_c(E) \left|T_{c0}^{(n)}\right|^2 \tag{3}$$

with

$$P_c(E) = -\frac{1}{\pi} Im[\int dE'\, g(E - E')(c|G^{intr}(E')|c)]$$

The Lippman-Schwinger formalism gives for the *n*th order (= *n*th step) transition amplitude

$$T_{\gamma^0}^{(n)} = \langle X_E^{(-)}|(\gamma|V^{res}(G^{chan}(E)V^{res})^{n-1}|0|X_0^{+}\rangle$$

where $X_E^{(-)}$, $X_E^{(+)}$ and $G^{chan}(E)$ are the incoming and outgoing scattering waves and the Green's function for the channel γ respectively.

V^{res} represent the residual effective projectile-target interaction which gives the effective Hamiltonian with the averaged energy Hamiltonian of the optical model H^{opt} and intrinsic Hamiltonian H_{intr} of the asymptotically separated nuclei.

Since the multi-step theory is considered in the EMPIRE code for two-step reactions only, then *n* must equal 2, and the cross-section distribution for the second step is:

$$\frac{d^2\sigma^{(2)}}{dEd\Omega} = \sum_{\lambda_1\lambda_2} \int dE_1 S_{\lambda_2}(E.E_1) S_{\lambda_1}(E_1.0) \frac{d\sigma^{(2)}}{d\Omega}(E.E_1)\Big|_{\lambda_1\lambda_2} \tag{4}$$

S_λ the transition strength function and λ is the momentum transfer.

According to NVWY, the average MSC cross-section leading from the incident channel *a* to the exit channel *b* is given by:

$$\frac{d\sigma_{ab}}{dE} = (1 + \delta_{ab}) \sum_{n.m} T_n^q \Pi_{n.m} T_m^b \tag{5}$$

where $\Pi_{n.m}$ is the probability transport matrix.

3. Reaction Modeling

The neutron induced reactions cross-sections on the target nuclei; ^{38}Ar, ^{59}Co, ^{121}Cd, ^{151}Eu, ^{153}Eu, ^{193}Os, and ^{203}Tl targets have been calculated with the latest modular system EMPIRE 3.2.2 Malta [14]. The direct cross-section has been performed with optical model calculations using the ECIS06 code incorporated into the EMPIRE system. The option (DIRECT=1), considering a spherical optical model, were used to calculate the particle transmission coefficients for the emerging channels, while the formalism of the coupled-channel (CC) have been used for transmission coefficients calculations in the incident neutron channel. For non-coupled levels, the direct neutron scattering was considered by a DWBA with dynamic deformations selected to describe available neutron emission spectra. Quantum-mechanical preequilibrium models based on multi-step direct and multi-step compound were taken into account by (MSD and MSC) modules to consider the preequilibrium emission, which are suitable for calculation below 30 MeV of neutron incident energy. The compound nucleus mechanism was described by Hauser-Feshbach statistical model [6].

The global Koning and Delaroche [15] optical potential parameters taken from the Recommended Input Parameter Library (RIPL) have been tested for selected nuclear reactions. This is a dispersive and spherical OMP reproducing all available nucleon scattering data on target nucleus from 0.0 up to 200 MeV. The nuclear level densities can be calculated with the microscopic formalism (Hretree-Fock-Bogoliubov Model HFB) [16]. The parametrization of Gamma Ray Strength Functions (GSF) is particularly important for the study of capture reactions. In our calculations, we used the modified Lorenzian radiative-GSF (MLO) (Plujko MLO RIPL-2) [17] and width fluctuation correction was selected up to 3.00 MeV (for neutron induced) to account the correlation between the incident and exit channels in elastic scattering.

4. Results and Discussion

EMPIRE is a modular system designed for cross section calculations over a broad range of energies starts just above the resonance region of a neutron projectile, this appear clearly in the presence calculations below as a high deviation from the experimental data at very low energy (usually $E<10^{-5}$ MeV). In addition, the cross section calculations of all the capture reactions investigated in this study are not reproducing very well in the resonance region, since the resonance module of the code requires that electronic version of the Atlas be installed. The copy rights for Atlas belong to the publisher (Elsevier), so we will not be able to install Atlas and therefore to use the resonance module of EMPIRE.

The theoretical calculated cross-section for the $^{38}Ar(n,\gamma)^{39}Ar$ reaction were illustrated in Fig. 1(a) along with the available experimental data [18, 19] obtained from the EXFOR library [20]. Inspection of these curves revel that the calculated results are overestimated the data from the TENDL library with the same behavior for all energy range with high deviation at very low energy. The deviation decreases gradually with increasing of the neutrons energies until it reaches approximately the limit of (1%). It can be also seen from this figure that the two points represented the experimental data are coincide with TENDL data. The cross-sections for $^{112}Cd(n,\gamma)^{113m}Cd$ reaction are shown in Fig. 1(b). The calculated results for this reaction have been presented with the only two experimental points available from the EXFOR database.

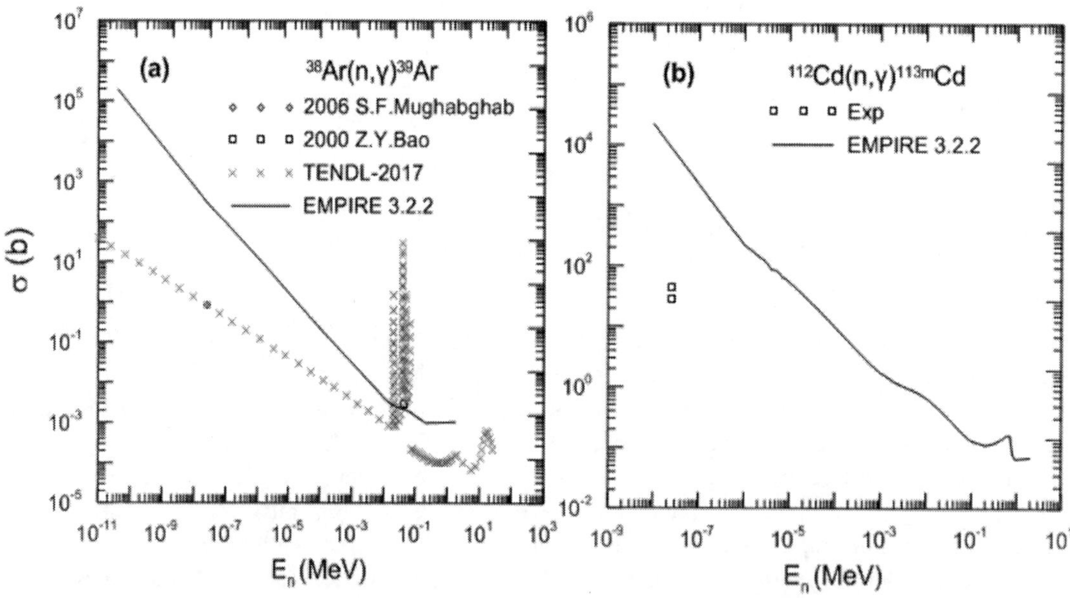

Figure 1. Comparison of available experimental data and calculated neutron-capture cross-sections for the target nuclei (a) ^{38}Ar and (b) ^{112}Cd.

In Fig. 2, $^{59}Co(n,\gamma)^{60}Co$ nuclear reaction cross-sections have been calculated and compared with the available experimental data taken from Refs. [21-25] and with the TENDL data [26]. It is obvious that the calculated results are in good agreement with the experimental data in the neutron energy range 10^{-2}-10^{1} MeV. At low energy range up to 10^{-2} MeV, the calculated data overestimate the experimental data. In this case, the best theoretical estimation is given by TENDL-2017 which coincides with the experimental data for energy range up to 10^{-2} MeV.

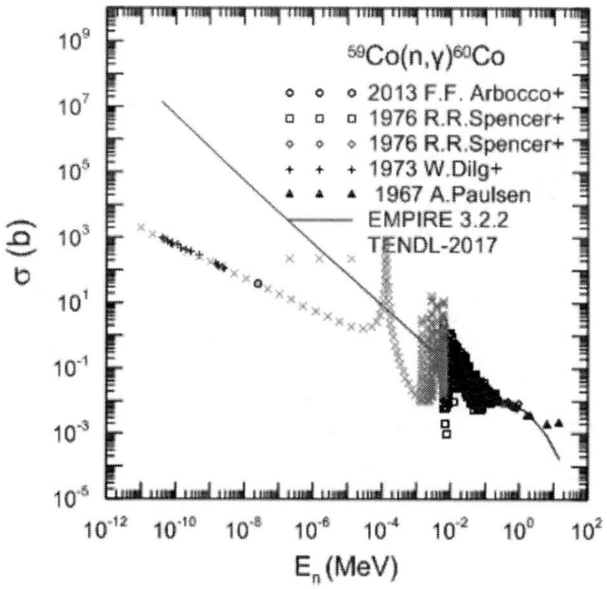

Figure 2. Comparison of available experimental data and calculated neutron-capture cross-sections for the ^{59}Co target nucleus.

The nuclear reactions $^{151}Eu(n,\gamma)^{152}Eu$ and $^{153}Eu(n,\gamma)^{154}Eu$ are examined in Fig. 3. (a) and (b). The experimental cross-section values [27-32] for these interactions have shown a perfect agreement with the calculated results for all neutron energy region of 10^{-6}-10^{0} MeV. The data from the TENDL library have been also shown in these figures for the two reactions mentioned above along with the experimental data and theoretical curves. Regarding to the $^{193}Os(n,\gamma)^{194}Os$ reaction, the calculated results are depicted in Fig. 3 (c) along with the available experimental data [19, 33]. As can be seen from this figure, the calculated results are coinciding with the TENDEL data in the neutron energy range 10^{-4}-10^{1} MeV. Whereas at very high energy from 10 up to 100 MeV, the calculated results underestimate the TENDEL data. In Fig. 3. (d) the cross-section of the $^{203}Tl(n,\gamma)^{204}Tl$ reaction are presented together with the data from the TENDEL library and compared with the experimental data [34-37]. Both results are in reasonable agreement.

In nuclear battery designs, the choice of a suitable radioisotope with a suitable energy of the ionizing radiation is very important. It is depending on the range of the energy emitted from the isotope in the material that the isotope is embedded in. for betavoltaic a p-n junction in a semiconductor is used as a transducer. Beta particles transfer energy to the target material electrons by Coulomb scattering and Bremsstrahlung emission [38]. Since the incident beta particles have a mass equal to the mass of the target electrons, their path of losses energy in a material will be random and undergoes significant scattering. Simulation of the beta particle tracks emitted from isotopes, used in this study, in SiC semiconductor have been investigated and illustrated in Fig. 4 using the PENYLOPE Monte Carlo code system [39]. In these figures we illustrate the YZ section were the beam is coming parallel to the z-axis vertically down on the sample. The energy of the beta particles

IOP Conf. Series: Materials Science and Engineering **757** (2020) 012017 doi:10.1088/1757-899X/757/1/012017

used in this simulation have been chosen with the maximum (E_{max}) energy of the emitted spectrum and listed in Table 1.

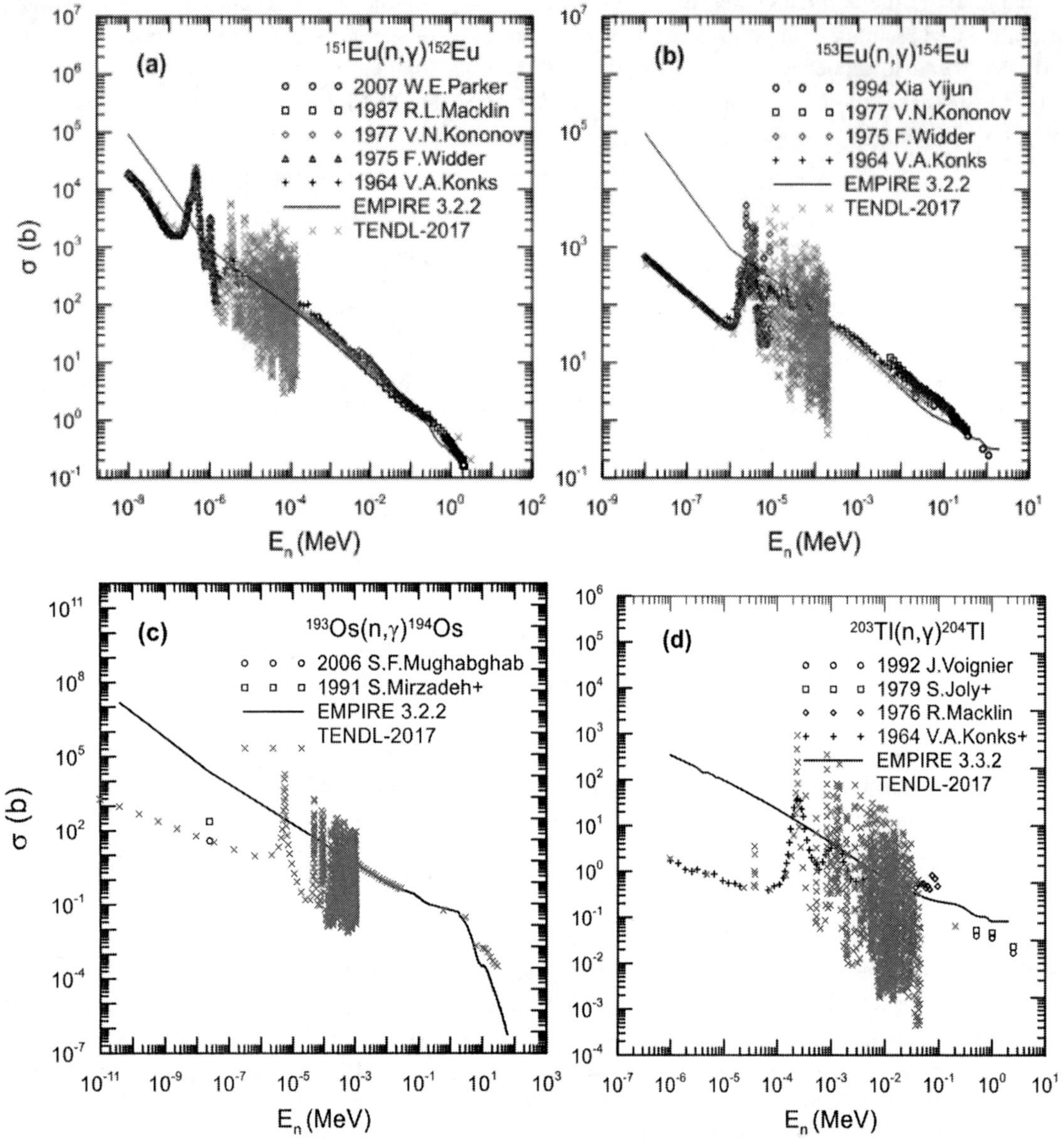

Figure 3. Comparison of available experimental data and calculated neutron-capture cross sections for the target nuclei (a) ^{151}Au, (b) ^{153}Au, (c) ^{193}Os and (d) ^{203}Tl.

Table 1. Characteristics of beta-emitting radioisotopes used in the study [38].

Nuclide	Half-life (Years)	β_{max} (MeV)
Ar-39	269	0.565
Co-60	5.2713	0.318
Cd-113m	14.1	0.58
Eu-152	13.54	1.818
Eu-154	8.592	1.845

Os-194	6	0.0966
Tl-204	3.78	0.763

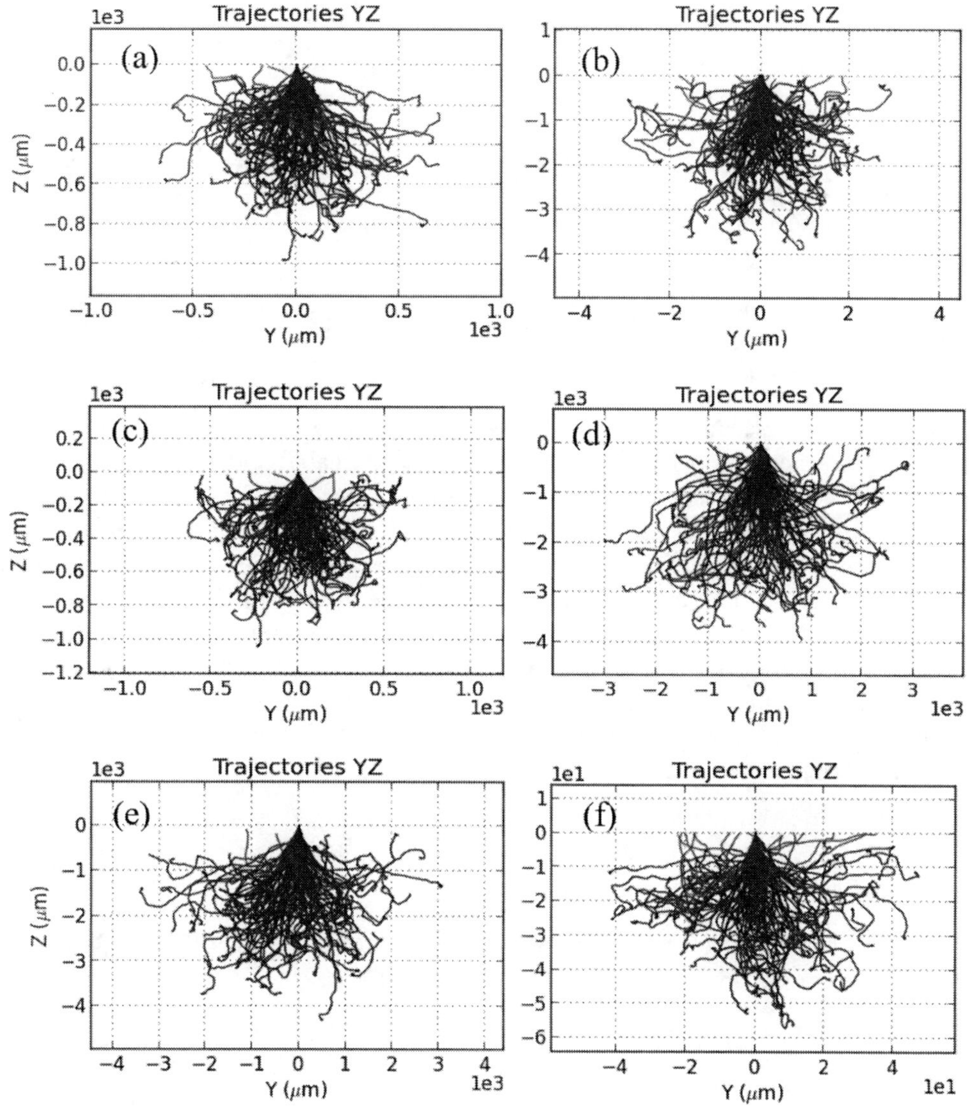

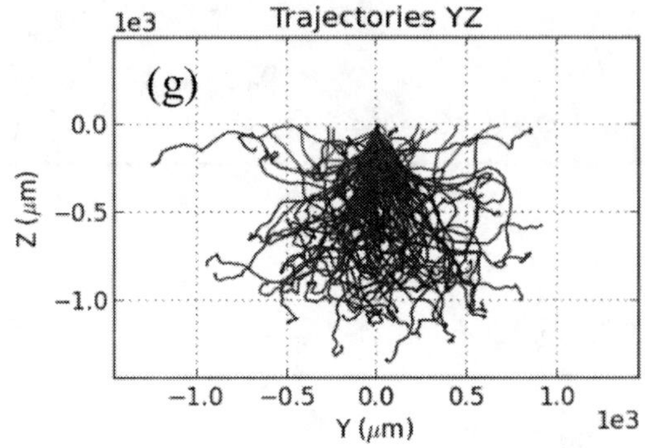

Figure 4. The paths of beta particles decay into silicon carbide with the beam entering a slab. The blue: absorbed within the sample. red: backscattered (beta particle exits from the same surface as it entered); (a) for ^{38}Ar, (b) ^{60}Co, (c) ^{113}Cd, (d) ^{152}Eu, (e) ^{154}Eu, (f) ^{194}Os and (g) ^{204}Tl beta sources.

Due to the difference in the energies of the beta particles one can observe that the beta particles in SiC ranging from (maximum 3.0 mm for ^{154}Eu source) to (minimum 0.04 mm for ^{194}Os source). The backscattering beta particles have been also shown in these figures with a red path.

The generation of secondary gamma photons from bremsstrahlung events, produced from interaction of beta particles with nuclei, require an additional shielding in battery design [40]. Bremsstrahlung energy generation is effective when transducer material has a high atomic number since its scales approximately as Z^2 of the target [41]. In this paper, the silicon carbide semiconductor has been used as a transducer material with relatively low atomic number so that bremsstrahlung spectrum can be neglected. The probability density of the generated bremsstrahlung spectrum has been shown in Fig. 5 for beta particle sources under investigation using the PENELOPE code system [42]. It can be seen from the figures that the probability density is very small and decreases with increasing bremsstrahlung energy and it is approximately equal to zero at high energy photons. In these figures, the absorption energies, elastic scattering parameters and cutoff energies have been used as a default simulation values.

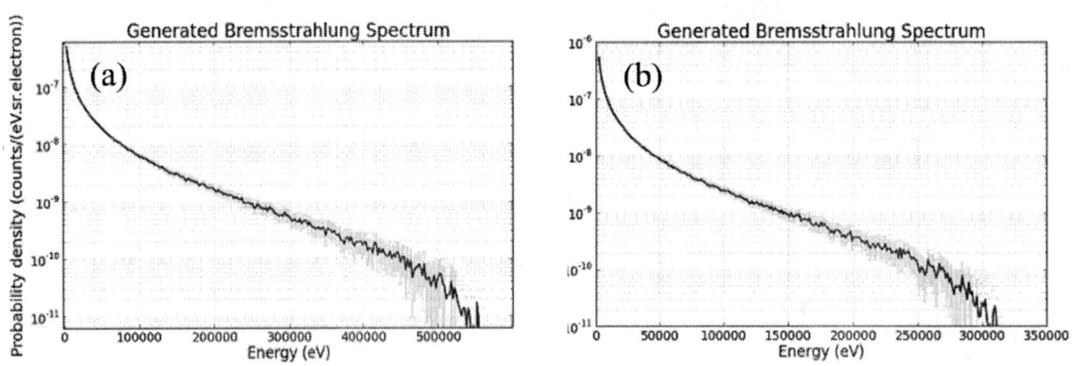

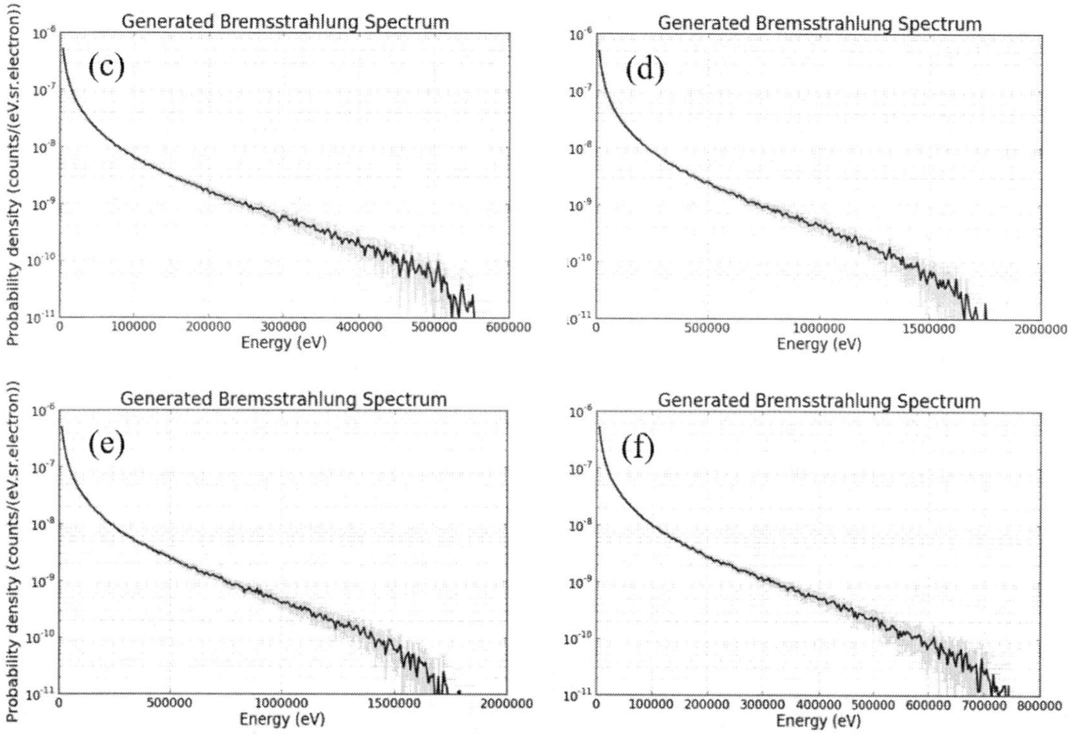

Figure 5. The probability density of the bremsstrahlung spectrum generated by beta particles decay into silicon carbide. (a) ^{39}Ar, (b) ^{60}Co, (c) ^{113m}Cd, (d) ^{152}Eu, (e) ^{154}Eu and (f) for ^{204}Tl source.

5. Conclusions

In the present work, the cross-sections for some reactions produced β^- sources used in nuclear batteries have been calculated at spectrum averaged energies from 10^{-6} eV up to 2 MeV. The obtained data for cross sections provide more responsible estimation values for these reactions and can be considered as additional data for nuclear data evaluation, which can be exercised in the different application fields of nuclear energy as well as in a testing of nuclear reaction models. It shold be mentioned that the resonance mudel shold be activated to representing cross sections of the neutron capture interactions in the low energy region. For nuclear battries design the scale length for the materials contained the isotopes have to be in order of the mean free path of the radiation so that the optimum amount of energy contained in the radiation can be deposited in the transducer wich is in this study is 0.04 mm is for ^{194}Os and gets much larger as the beta energy increases.

References

[1] Moseley H, 1913 Proc. R. Soc. Lond. **A88** 471.
[2] Bertulani C, Campbell C, Glasmacher T 2003 Commun Phys. Comput.**152** 317.
[3] Blanke B, Birden J, Jordan K, Murphy E 1960 Mound Lab., Miamisburg, Ohio.
[4] Board S S, Council N R, Committee R P S "Radioisotope power systems: an imperative for maintaining US leadership in space exploration" (National Academies Press, 2009).
[5] Herman M, Capote R, Carlson B, Obložinský P, Sin M, Trkov A, Wienke H, Zerkin V 2007 Nucl. Data Sheets **108** 2655.
[6] Hauser W and Feshbach H 1952 Phys. Rev. **87** 366.
[7] Lane A 1959 Nucl.Phys. **11** 625.
[8] Lane A and Lynn J Nucl.Phys. **11** 646.

IOP Conf. Series: Materials Science and Engineering **757** (2020) 012017 doi:10.1088/1757-899X/757/1/012017

[9] Austern A "Direct Nuclear Reaction Theories, Interscience Monographs and Texts in Physics And Astronomy" Vol, XXV (RE Marshak, ed., Wiley-interscience) 1970.
[10] Feshbach H, Kerman A and Koonin S 1980 Ann. Phys. **125** 429.
[11] Lemmer R H, Multistep Direct Reactions, Workshop On (World Scientific, 1992).
[12] Agassi D, Weidenmüller H A and Mantzouranis G 1975 Phys. Rep. 22 145.
[13] Tamura T, Udagawa T and Lenske H 1982 Phys. Rev C **26** 379.
[14] Herman M, Capote R, Sin M, Trkov T, Carlson B, Oblozinsky P, Mattoon C, Wienke H and Hoblit S 2013 Brookhaven National Laboratory (BNL) National Nuclear Data Center.
[15] Koning A and Delaroche J 2003 Nucl. Phys. A **713** 231.
[16] Demetriou P and Goriely S 2001 Nucl. Phys. A **695** 95.
[17] Capote R, Herman M, Obložinský P, Young P, Goriely S, Belgya T, Ignatyuk A, Koning A J, Hilaire S and Plujko V A 2009 Nucl. Data Sheets **110** 3107.
[18] Bao Z, Beer H, Käppeler F, Voss F, Wisshak K, Rauscher T 2000 At. Data Nucl. Data Tables **76** 70.
[19] Mughabghab S F "Atlas of Neutron Resonances: Resonance Parameters and Thermal Cross Sections. Z= 1-100 Elsevier, 2006.
[20] N.R.D.C. Network, Experimental Nuclear Reaction Data (EXFOR/CSISRS), http://www. nndc. bnl. gov/exfor/exfor00. htm.
[21] Arbocco F F, Vermaercke P, Smits K, Sneyers L and Strijckmans K 2013 J. Radioanal. Nucl. Chem. **296** 931.
[22] Dilg W, Mannhart W, Steichele E and Arnold P 1973 Z. Phys. **264** 427.
[23] Paulsen A 1967 Z. Phys. **205** 226.
[24] Spencer R and Beer H 1976 Nucl. Sci. Eng. **60** 390.
[25] Spencer R and Macklin R 1967 Nucl. Sci. Eng. **61** 346.
[26] Rochman D, Koning A J, Sublet J C, Fleming M, Bauge E, Hilaire S, Romain P, Morillon B, Duarte H and Goriely S, 2017 EPJ Web of Conferences, EDP Sciences.
[27] Macklin R and Young P 1987 Nucl. Sci. Eng. **95** 189.
[28] Parker W, Sheets S, Agvaanluvsan U, Becker J, Becvar F, Bredeweg F, Clement R, Couture A, Esch E nad Haight R, in: International Conference on Nuclear Data for Science and Technology, (EDP Sciences, 2007).
[29] Konks V A 1964 Inst. Nucl. Res., Dubna, reports **1845** 100.
[30] Kononov V A, Manturov G N, Poletaev e D, Timokhov V M and Shorin V S 1977 Yadernye Konstanty **22** 29.
[31] Widder F "Neutron capture cross section measurements in the energy region from 0.01 to 10 electron volts (EIR, 1975).
[32] YiJun X, Xianguan L, Xiaobing L, Zhihua Y, Mantian L, Fuqing H, Chunhuo W, Xiufeng P, Jingfu Y and Hanlin L 1994 American Nuclear Society, Inc., La Grange Park, IL (United States).
[33] Mirzadeh S, Knapp F and Callahan A "Nuclear Data for Science and Technology" (Springer, 1992).
[34] Joly S, Voignier J, Grenier G, Drake D and Nilsson L 1979 Nucl. Sci. Eng. **70** 53.
[35] Konks V, Shapiro F and Eksperim Z H 1964 Zh. Eksperim. i. Teor. Fiz. **47** 795.
[36] Macklin R and Winters R 1967 Astrophys. J. **208** 812.
[37] Voignier J, Joly S and Grenier G 1986 Nucl. Sci. Eng. **93** 43.
[38] Prelas M, Boraas M, Aguilar F D L T, Seelig J, Tchouaso M T and Wisniewski D "Nuclear batteries and radioisotopes" Springer 2016.
[39] Salvat F, Fernández-Varea J M and Sempau J, Workshop Proceedings 2006.
[40] Prelas M A, Weaver C L, Watermann M L, Lukosi E D, Schott R J and Wisniewski D A 2014 Nucl. Prog. **75** 117.
[41] Knoll G F "Radiation detection and measurement" John Wiley & Sons, New York 2010.
[42] Issy-les-Moulineaux F 2003 PENELOPE–a code system for Monte Carlo simulation of electron and photon transport (Paris: OECD Publications).

IOP Conf. Series: Materials Science and Engineering **757** (2020) 012018 doi:10.1088/1757-899X/757/1/012018

Computational analysis for electrical breakdown in air due to streamer discharge in rod-to-plane arrangement

Maha F Abd Alameer and Thamir H Khalaf *

Department of Physics, College of Science, University of Baghdad, Iraq.

*Corresponding author e-mail: drthamirhameed@gmail.com

Abstract: In addition to the theoretical and experimental, the numerical simulation is the third way to study the electrical discharge phenomenon. In this work, a numerical simulation was executed within a rod-to- plane electrodes arrangement to track the streamer discharge between the two electrodes for different rod dimensions and air gap length distances. The simulation results show the path of the streamer channels, and their branches, from the rod to the plane. Also, the electric potential and field distributions were presented as contour plots. The plots show clearly the sites of the streamer initiation and growth and how the distributions were affected by the streamer growth between the two electrodes. The dc applied voltage was variated, and the minimum value that caused the streamer to bridge the air gap and touch the plane was indicated as the breakdown voltage. It was found that, the breakdown voltage increased with the increasing of the air gap length distance, the electric field threshold value, the rod diameter, and less of that with the increasing of the rod lengths.

Key words: *Streamer discharge, Numerical Simulation, Breakdown voltage, Air gap breakdown.*

1. Introduction

There is a wide attention to study the electric breakdown in air, mainly, in two directions. The first is to prevent the occurrence because the air is a good natural and low cost insulator used in transmission lines and high voltage equipments. The second is to produce plasmas can be used in different applications.

In order to explain the phenomenon of low pressure, Townsend proposed the gas breakdown theory. In low pressure, Townsend discharge theory can be a good method to describe gas breakdown. But when Pd is large (near atmospheric pressure), the growth of breakdown voltage is no longer a linear relationship, and Paschen's law cannot be used in conditions with a large Pd. So that, Meek, Leob and Reather established a theory about streamer discharge development to illustrate high pressure gas discharge, such as air breakdown at atmospheric pressure [1].

A streamer electric discharge may be formed when a sufficiently high voltage is applied to a gas filled electrodes' gap. It is an electrical discharge phenomenon characterized by a narrow and weakly ionized non-thermal plasma channel with a self-propagating head consisting of a thin curved layer of either net positive or negative charge [2, 3]. In air at ground pressure, streamers are formed through electron avalanches in electric fields greater than about 30 kV/cm and propagate with typical speeds of 0.1−10 mm/ns [2].

The streamers can be distinguished by the starting place to positive streamer and negative streamer: when the applied voltage on the discharge gap is sufficiently high, then if electrons leave the cathode and ionize the gas molecules in the gap, the ionized electrons would accelerate by the electric field and go on to ionize other gas molecules. After the original electrons move a distance in the direction of the applied electric field, they generate new electrons with a very rapid process. The new electrons will accumulate to build a single electron avalanche. As the electron avalanche moves forward, the ionization process on its

IOP Conf. Series: Materials Science and Engineering **757** (2020) 012018 doi:10.1088/1757-899X/757/1/012018

head will be more intense. When the electron avalanche is through the whole space, the space charge density in the head part has been so great that the electric field of the tail is greatly strengthened to radiate a lot of photons around, photons cause space photoionization and the newly formed photoelectrons are attracted by the positive space charge of the head part of the main electron avalanche. The photoelectrons create a new electronic collapse in the electrical field strengthened by distortion, which is called the secondary electron avalanche. The secondary electron avalanche moves to the main electronic avalanche and the electrons of the secondary electron avalanche head part move into the positive space charge region of the main electron avalanche head part (the electrons of the main electron avalanche have gone into the anode mostly). Due to the smaller field strength here, mostly electrons form as negative ions. A large number of positive and negative charged particles constitute a plasma stream which is called a positive streamer. Positive flow starts from the anode to the cathode after the initial avalanche head arrives at the anode. When the streamer reaches the cathode, the entire gap was run through by the plasma channel with good conductivity. So the gap breakdown is completed. Described above is a situation where the electron avalanche needs to go through the entire gap to form the streamer when the voltage is low. This voltage is the breakdown voltage [1].

Streamer discharge has many industrial applications because of its ability to enhance the activity of many chemical reactions without using a heat. Examples include production of ozone, water cleaning and removal of pollutants such as SO_2, NO_x and fly ash [4, 5, 6]. Streamers also appear in nature in the form of so-called sprites which forms high above thunderclouds at an altitude of about 80 km. they are known to be the first self-propagating stage of a lightning discharge [7].

In last decades, numerical simulation of streamer discharges has got a wide interest, due to fast numerical algorithms and more powerful computers [8]. Numerical simulation makes it possible to obtain a much deeper understanding of streamer dynamics than what can be obtained only through experiments [2]. In addition, an accurate performing of experiments requires a cost measuring equipments and particularly designed set-ups. So that, numerical simulations offer a low-cost, valuable and multilateral option increasingly being desirable in industry.

The details of microphysics knowledge for streamer channels were reached, such as models of electron energy [5], and of transport coefficients and cross-sections of the most reactions, at least the main for air and other gases. On the other hand, we barely understand most macroscopic processes in a fully developed corona or streamer tree involving hundreds or thousands of mutually interacting plasma filaments [9]. Studies for the growth of electrical discharge trees are mostly based on the dielectric breakdown model that was proposed by Niemeyer *et al* [10] to explain the fractal properties of some electrical discharges. In their model, a discharge tree expands in discrete time-steps by the stochastic addition of new segments with a probability that depends on the local electric field. The original dielectric breakdown model assumed that the channels in the tree are perfectly conducting, but there is strong experimental evidence that the electric potential decreases a long a discharge channel. A. Luque *et al* [11] modeling for streamer discharges assuming the streamer channels as imperfect conductors.

In this paper, we assume the streamer channel as weakly ionized plasma channel or a conductor has high resistance. i.e. there is a voltage drop cross it. And simulate the streamer discharge to indicate the breakdown voltage cross air gaps.

2. Modeling

Many studies have proposed the streamer discharge channels as good conductors and the breakdown occurs when the streamer crosses the electrodes gap. The main parameter that affects the streamer discharge is the electrostatic external or applied field, in addition to some stochastic probabilities. The model, here, based on stochastic models, [12,13], and follows the following considerations:

1- The streamer initiates at a site in air when the local electric field value at that site $E_{thr} \geq 26$ kV/cm [14]. The local electric field can be evaluated at each streamer step using Laplace's equation as follows:

$$\varepsilon \nabla^2 V = 0 \dots\dots\dots\dots\dots (1)$$

V is the potential in each point between the two electrodes at the surrounding domain in Volt (V), and ε is the dielectric permittivity (in air, $\varepsilon \cong \varepsilon_o$). The value of the local electric field intensity, can be given as;

$$\boldsymbol{E} = -\nabla V \dots\dots\dots\dots\dots (2)$$

Or, in 2-D the electric field intensity is calculated in terms of the variation of the electrical potential as:

$$\boldsymbol{E}_{loc} = -\left(\frac{\partial V}{\partial x}\hat{\imath} + \frac{\partial V}{\partial y}\hat{\jmath}\right) \dots\dots\dots (3)$$

2- The evolution of the streamer from step to the other step spending a stochastic time τ given as [14]:

$$\tau = -\frac{\ln(\delta)}{r(E)} \dots\dots\dots\dots\dots (4)$$

Where r is a field depended growth rate function and given as:

$$r(E) = A\left[\frac{E_{loc}}{V/d}\right]^n \dots\dots\dots\dots\dots (5)$$

The parameter A is a constant and during the simulations took the value 3.7×10^5 sec^{-1} [15].n is a number that controls the variation of the growth rate with the local electric fieldE_{loc}, V is the applied potential at the anode and d is the gap distance between the electrodes.

3- The streamer channels were assumed as conductors have high resistance, and the voltage drops, $E_s$along in magnitude of(4-5) kV/cm [9,15].

4- The main channel of the streamer will branch into only two branches at certain condition given as [17]:

$$MxQ \geq 2CrQ \dots\dots\dots\dots\dots (6)$$

Where

$$MxQ = max\int_0^x (\alpha - \eta)dx \dots\dots\dots (7)$$

And the parameterCrQ, is defined as the natural logarithm of the chargeN_c at which the avalanche is able to convert itself to a streamer, and given as:

$$N_c = exp\left[\int_0^x (\alpha - \eta)dx\right] \dots\dots\dots (8)$$

Whereα is the Townsend's primary ionization coefficient andη is the coefficient of attachment.

5- Branches of the streamer will be followed for only one step, because they usually will decay and only the main will bridge the gap between electrodes.

3. Implementation of the Model

3.1 Configurations

The model was implemented within Rod-to-plane electrodes arrangement, figure 1, with different dimensions of the plane and the rod, with air gaps distances between electrodes start from 1 cm to 3 cm in a 0.5 cm step increment are used for measurement of air breakdown voltages. The rod electrode is a cylinder long enough (ℓ), ended with hemisphere tip of radius r. The plane electrode is a disk plane, with diameter *D*. One electrode (rod electrode) is stressed by Positive voltage is applied and the other (plane electrode) is grounded. This electrodes arrangement is producing a non-uniform field because the surfaces of both the electrodes are not similar.

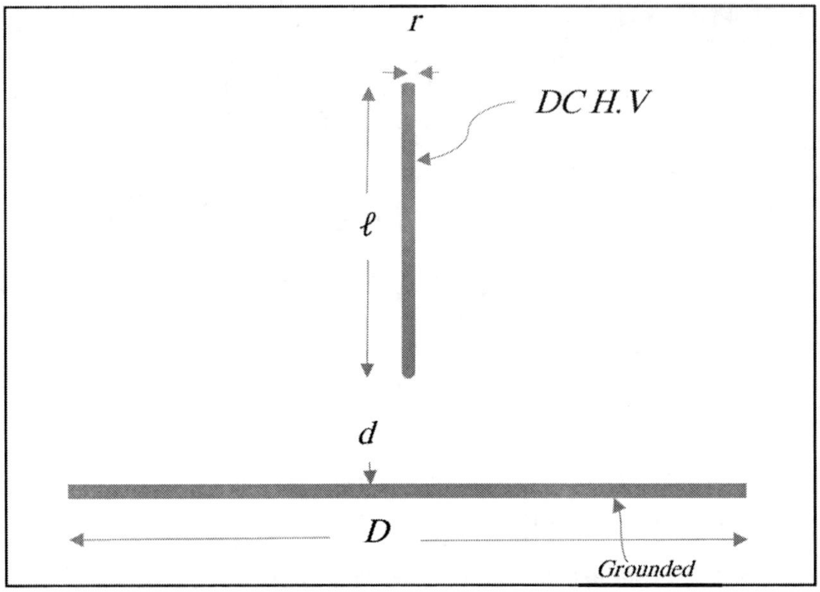

Figure 1. Rod-to-plan electrodes configuration.

3.2 Numerical Solution

As mention before, the electric field is main parameter that affects the streamer discharge behavior. To determine the distribution of the electric field within the electrodes configuration, firstly the potential distribution must be determined. In other words, Laplace's equation must be solved. finite element method is a numerical procedure for solving this type of problems. This is because the method basically relies on solving a large set of algebraic equations and entails considerable manipulation. The fundamental concept of the finite element method is that any region is made up of elements; therefore, the general behavior of a system can be determined by considering the behavior of its component (subsystem). The finite element analysis of any problem involves basically four steps:

- Discretizing the solution region into a number of sub regions or elements.
- Deriving the governing equations for typical elements.
- Assembling of all elements in the solution region.
- Solving the system of all equations that were obtained.

3.3. Simulation Procedure

The implementation of the model within the configuration in figure 1, or in other words, the simulation of the streamer discharge phenomenon was done as in following procedure

1- Mesh generation: In 2D geometries the domains are discretized into triangular elements. When curves are included in the geometry, especially on the boundaries, there is a risk that the surrounding area will not be properly divided and the resulted solution, even if it will be acceptable, it will not be the optimal. A fix to this is to try to use mesh structures that smoothly change in size and resolution near these boundaries, where the solution is exposed to more steep changes and certainly is an area of higher interest [14]. AUTOMESH 2D Multi-Domain V1.5 package was used to generate the suitable meshes for the cases of 1cm, 1.5cm, 2cm, 2.5cm, and 3cm air gaps for the different diameters. The meshes were designed to have high density elements around the tip of the rod and low density far away because of the expected high variation of the voltage and the electric field around this region. An example of a generated meshes for the geometry of rod-plane electrodes is shown in figure 2. The division of the domains seems quite uniform, with a small change of the distribution of the elements in the vicinity of the electrode

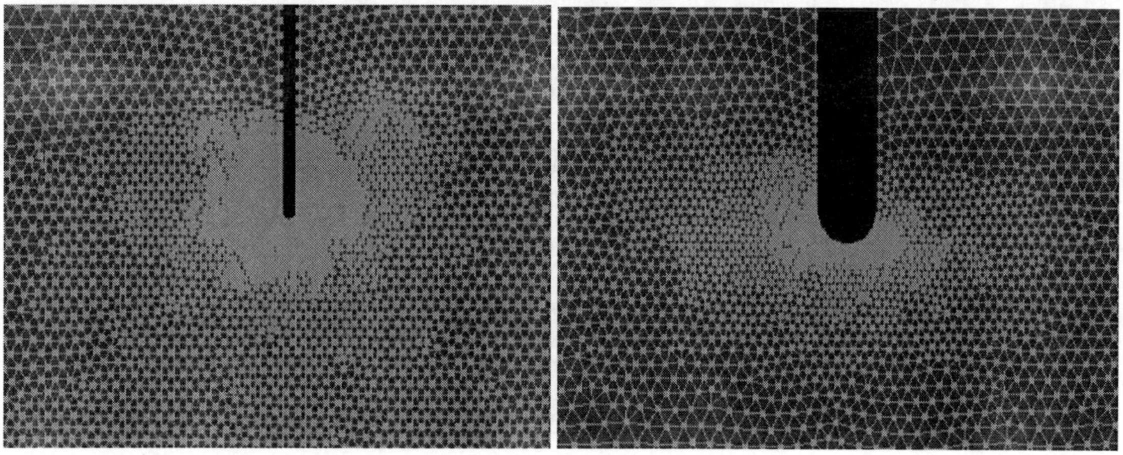

Figure 2. The discretized mesh as examples for two cases.

2- Solving of Laplace's Equation: Based on the meshes that was generated as described in (1), the simple_2d solver [17] was used to solve Laplace's equation numerically in the two dimensions' area of the longitudinal cross section of the electrodes arrangement. The electric potential of all nods of the mesh that belong to the dielectric (air) is calculated by solving the Laplace equation with the boundary conditions on the electrodes and the discharge pattern.
3- Tracking the first step of the streamer between the electrodes: it was done as follows,
   a) Using the potential at the nodes of each element, the electric field at the center of each element was calculated as E_{loc} for each element.
   b) The values of$E_{loc}$for all elements was compared with the threshed value $E_{thr}$and the elements that have values equal or greater thanE_{thr} were determined.
   c) The stochastic time τ was calculated for all elements that were determined in (b).
   d) The site of the streamer initiation was selected at the element which have the smallest τ.
   e) The branching condition in equation (6) was checked to indicate if there is a branch or not.
   f) The voltage drop cross the segment, from the selected element in (d) to the tip of the rode, was calculated and subtracted from the applied voltage at the rod electrode. The result voltage was given for the nodes of the selected element as the site of the streamer.
4- Tracking the other steps of the streamer between the two electrodes: that requires updating the boundary conditions by add the nodes of the selected elements to the nodes of the rod electrode and repeating the procedure from 2 to 4 until the streamer reach the plane electrode.
5- The minimum applied voltage, which caused the streamer to reach the plane, was indicated as the breakdown voltage of a specified air gap within the electrodes arrangement.

The above procedure was executed by a program written in Fortran langue.

4. Results and Discussions

The results, that were presented here, are outputs of the execution of the Fortran program. It is simulated the streamer discharge Phenomenon within air gap between rode and plane electrodes, figure 1, in different dimensions. The air is considered as dray air at atmospheric pressure and room temperature. The simulation was tracked the time evolution of streamer steps from the rod to the plane and indicated the minimum applied voltage can have caused the breakdown.

4.1. Streamer Tracking

The streamer initiation and growth was tracked within the solution region between the two electrodes. The streamer initiated at the elements that have electric field values greater thanE_{thr}. In the rode-plane configuration, the highest values, as expected, was found around the tip of the rod. The model was implemented different rod diameters.

Figure 3 shows the streamer path between the two electrodes when the rod length of 2.0 cm and diameter of 0.1cm, 0.8cm, and 1.0 cm. the drop voltage was averaged as 4.5kV/cm, electric field thresholdE_{thr} of 26kV, and the air gap distance length of 1.5 cm. the plane electrode is a disk has diameter of 7.5 cm.

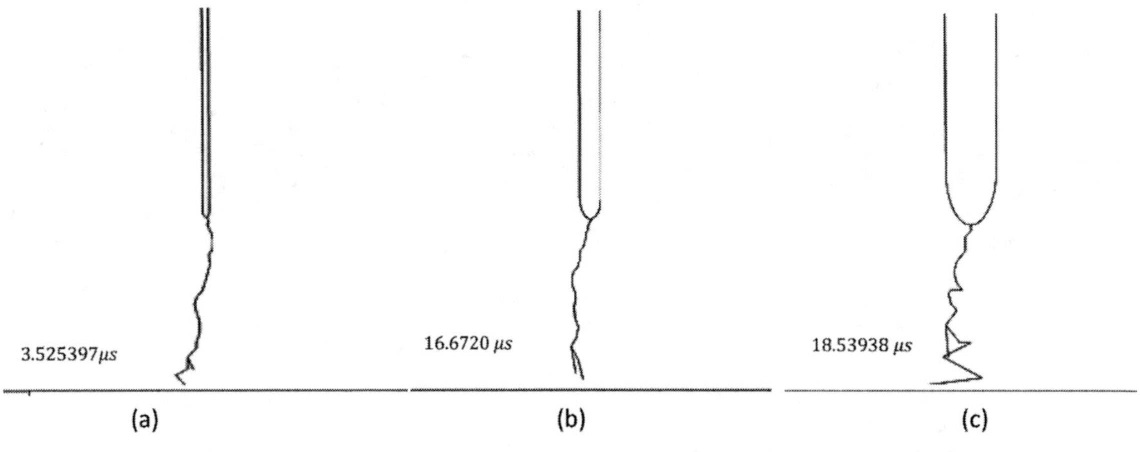

Figure 3. Enlargement of the streamer initiation and growth from the rod to the plane electrodes when the rode diameter of a) 0.1 cm, b) 0.8 cm, and c) 1.0 cm.

From the figure above, one can note the effect of the increasing of the rod diameter on the behavior of the streamer in path, branching and the reaching time. The reaching time, which was indicated below each subfigure, increased with the increasing of the diameter which means the streamer velocity decreases with the increasing of the diameter. That can be explained by the effect of distributions of electric potential and field within the simulation region in the section below.

4.2. Electric Potential and Field within the Simulation Region

Figures 4 and 5, show a contour plotting for the effect of the streamer growth on the distributions of the voltage and electric field magnitudes of the rod-to-plane configuration for different diameters of the rod, at minimum breakdown voltage in the air gap of1.5cm length distance.

The sites that have high voltages and the high electric fields between the electrodes are the weakest insulation. In this case, the weak sites that was prepared for streamer initiation and growth is identified to be the sites where the magnitude of the electric field is the highest. According to figures 4 and 5, one can observe that the streamer moves according to the high voltage/E field sites from the rod down to the plane electrode.

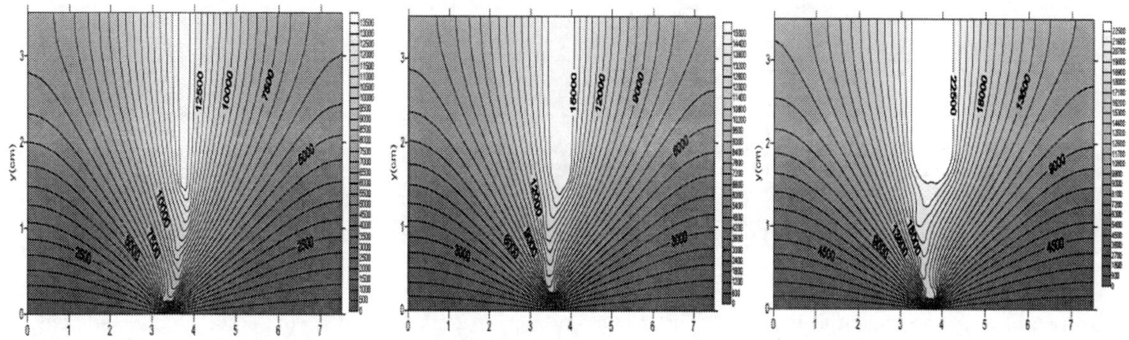

(b) (b) (c)

Figure 4. The potential distribution within the electrode arrangement according to the streamer growth from the rod to the plane when the rode diameter of a) 0.1 cm, b) 0.8 cm, and c) 1.0 cm.

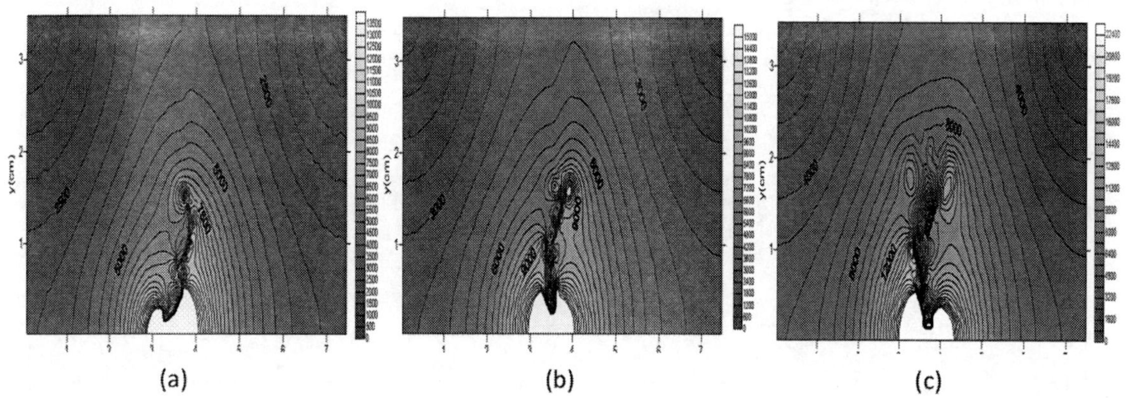

(a) (b) (c)

Figure 5. The electric field distribution within the electrode arrangement according to the streamer growth from the rod to the plane when the rode diameter of a) 0.1 cm, b) 0.8 cm, and c) 1.0 cm.

4.3. Effects of the Rod Dimensions

The rod has the important effect on the streamer discharge and then the values of breakdown voltage for any gap. That is because it is the site of the applied high voltage. Therefore, the simulation was expanded to show the effect of the rod diameter and length on the breakdown voltage values for different air gap lengths and different values of the electric field threshold that found in literatures.

The same producer above was followed and the breakdown voltage was indicated as the minimum applied voltage that caused the streamer to bridge the gap and reach the plane. The results for different diameters (0.1cm, 0.4cm, and 1.0) cm were presented in table 1 below:

Table 1. The breakdown voltage for different diameters of the rode electrode.

Gap length (cm)	Breakdown Voltage (kV)														
	Rod Diameter (cm)														
	0.1					0.4					1.0				
	Thr (kV/cm)					*Thr (kV/cm)*					*Thr (kV/cm)*				
	21.3	25	26	30	31	21.3	25	26	30	31	21.3	25	26	30	31
1	8.6	9.8	10.2	10.8	11.3	10.1	11.8	12.3	14.2	14.7	14.6	17.1	17.8	20.6	21.2
1.5	12.0	13.1	13.8	15.0	16.1	12.7	14.9	15.5	17.8	18.4	18.5	21.7	22.5	26.0	27.0
2	14.9	16.7	17.4	18.8	19.6	15.1	17.7	18.4	21.3	22.0	22.6	26.5	27.6	31.8	32,9
2.5	18.7	21.1	22.0	21.9	24.7	17.2	20.2	20.8	24.3	25.1	27.6	32.4	33.7	38.8	40.0
3	21.8	24.5	25.2	25.1	28.9	19.0	22.4	23.2	26.8	27.7	30.2	35.4	36.8	42.5	43.9

From table 1., one can note that the breakdown voltage, in general and as expected, increased with the increasing of gap length and threshold electric field. Also, it is clear that, the breakdown voltage increased with the increasing of the diameter of the rod electrode. That can be explained as the increasing of the diameter of the rod or the diameter of the hemisphere means decreasing in the electric field in the shortest distance between the electrodes which require higher applied voltage to causes the streamer to bridge the gap and the breakdown occurrence.

Also, the length of the rod was examined for the different diameters. The results show no considerable effect on the breakdown voltage values, as presented in table 2, because the little effect on the electric field distribution between the electrodes.

Table 2. The breakdown voltage for different diameters of the rode electrode.

Rode diameter (cm)	Breakdown Voltage		
	Rode length (cm)		
	1.0	1.5	2.0
0.1	17.6	17.7	17.9
0.2	18.2	18.7	18.8
0.4	18.8	19.0	19.5
1.0	25.8	26.3	26.4
1.2	27.1	28.1	28.4

5. Conclusions

From the simulation results, one can indicate the following conclusions

1. The physical shape of the electrodes gives a variation in the breakdown voltages.
2. The breakdown voltage of the rod-to-plane gaps with the positive dc voltage linearly increases with the gap spacing.
3. The distribution of the electric field in a rod-plane gap is strongly affected by the geometry (shape and dimensions) of the electrodes and the arrangements of the air gap.
4. The threshold electric field valueE_{thr} exerts a strong influence on the value of the minimum breakdown voltage.
5. The values of the breakdown voltage influenced by the dimensions of the rod especially the rod diameter.

References

[1] Dengming X 2016 *Gas discharge and gas Insulation* (Shanghai and Springer-Verlag Berlin Heidelberg: Jiao Tong University Press/Springer).
[2] Dubinova A A 2016 *Modeling of streamer discharges near dielectrics* PhD thesis, Eindhoven University of Technology.
[3] Singh S. 2015 *Computational framework for studying charge transport in high - voltage gas – insulated Systems* PhD thesis, Chalmers University of Technology.
[4] Ebert U Montijn C, Briels T M P, Hundsdorfer W, Meulenbroek B, Rocco A and van Veldhuizen E M 2006 *Sources Science and Technology* **15** 118–29.
[5] Bourdon A, Pasko V P, Liu N Y, C´elestin S, S´egur P, and Marode E 2007 *Plasma Sources Science and Technology* **16** 656–78.
[6] Kogelschatz U 2003 *Plasma Chemistry and Plasma Processing* **23** 1–46.
[7] McHarg M G, Stenbaek-Nielsen H C and Kammae T 2007 *Geophysical Research Letters* **34** 1–5.
[8] Georghiou G E, Papadakis A P, Morrow R, and Metaxas A C 2005 *J Phys D App Phys* **38** 303–28.
[9] Luque A, Ebert U 2014 *New J. of Phys.***16** (2004) 013039.
[10] Niemeyer L, Pietronero L and Wiesmann H 1984 *J. Phys. Rev. Lett.* **52** 1033-36.
[11] Luque A, Gonz´alez M, Gordillo F J 2017 *Plasma Sources Sci. Technol.* **26** (2017) 125006.
[12] Yamamoto O, Hara T and Hayashi M 1991 *Electr Eng Japan.* **111** 576-83.
[13] Allen N L and Mikropoulos P N 1999 *Electrostatics* **163**, 49-52.
[14] Charalambakos V P, Stamatelatos C P, Agoris D P and Pyrgioti E C 2005 Proceedings of the 5th WSEAS Int. Conf. on Power Systems and Electromagnetic Compatibility Corfu Greece 29-33.
[15] Allen N L and Boutlendj M 1991 *IEE Proceedings-A* **138**.
[16] Akyuz M, Larsson A, Cooray V and Strandberg G 2003 *J Electrostatics* **59**, 115-41.
[17] Silvester P P and Ferrari R L 1996 *Finite elements for engineers* (Cambridge: Cambridge University Press, 1996.

IOP Conf. Series: Materials Science and Engineering **757** (2020) 012019 doi:10.1088/1757-899X/757/1/012019

Investigation of The Thermal Conductivity of Binary Polymer Blends

Ahmed J Farhan[1] and Harith I Jaffer[2]

[1]Department of Physics, College of Science, University of Wasit, Kut, Wasit /Iraq

kutpc19@yahoo.com

[2]Department of Physics, College of Science, University of Baghdad, Baghdad /Iraq

harithibrahem@yahoo.com

Abstract In this research, binary polymer blends are formed from epoxy, unsaturated polyester and polyurethane. The concentration were chosen with varying percentages of components, resulting in three groups; polyurethane / epoxy (PU/EP), epoxy/unsaturated polyester (EP/UPE) and polyurethane /unsaturated polyester (PU/UPE) blends. Hand lay-up technique was used to prepare the samples with 6 mm thickness prepared from simultaneous mixing of the two constituents in the required percentages.

Thermal conductivity K for these blends was carried out by using Lee's disk method at room temperature. The experimental results showed that an increment and /or decrement of thermal conductivity depend on type and blend concentration. Also the experimental results indicated that the thermal conductivity has the highest values of (K) (0.78 $J/m^2.k$) at ratios (PU 10% / EP 90 %), (0.70 $J/m^2.k$) at ratios (EP 50% / UPE 50%) and (0.50 $J/m^2.k$) at ratios (PU 40% / UPE 60%) respectively. This increment of (K) implies that the penetration of molecules of both polymer components in the same direct then a good transfer of energy by the vibration and rotation of the chain molecules.

Also the experimental results indicated that the thermal conductivity has the lowest value of (K) (0.52 $J/m^2.k$) at ratios (PU 40% / EP 60%), (0. 0.53 $J/m^2.k$) has obtained at ratios (EP 30% / UPE 70%) and (0. 0.31 $J/m^2.k$) has obtained at ratios (PU 60% / UPE 40%) respectively.

E-mail address of the corresponding author: kutpc19@yahoo.com

IOP Conf. Series: Materials Science and Engineering **757** (2020) 012019 doi:10.1088/1757-899X/757/1/012019

1. Introduction

In recent years there has been rapid growth in the development of polymer blends and their end user applications. These applications associated with lower cost with improved properties have spurred more research activities on polyblends. Polyblends of the type thermoset – thermoset offer property advantages such as enhanced processability, thermal properties, strength and adhesion properties, chemical resistancy, optical properties, flame retardancy, etc. [1 - 3].

Polyurethane (PU) has been used in several technical applications due to high tensile strength, chemical resistance, process ability, and mechanical properties [4, 5]. In polyurethane, hydrogen bonding is sufficient to produce a physical link between polymer chains to enhance overall improved properties. Phase separation in PU has been found to influence by hard and soft segment structures, molecular weight, polydispersity, and crosslinking ability. Epoxy polymers are tough and flexible with good corrosion and chemical resistance. The epoxy reactions are highly temperature dependent, epoxy oligomer/monomer dependent, and energetic [6, 7].Polyester resins have a number of advantages, including versatility in structure and properties, overall low cost, and ease of application. However, they suffer from some decisive drawbacks, such as low mechanical properties, alkali resistance, and low hardness. So, to improve these drawbacks of polyester resins, blending with other suitable resins, such as epoxy resin, amino resin, silicone resin, and ketonic resin, can be performed, as polyester resins have good compatibility with a wide variety of other resins. The better compatibility comes from the relatively low viscosity of the resin and from the structure of the resin, which contains a relatively polar and aromatic backbone and aliphatic side chains with low polarity. The blending technique may be used effectively to improve the inferior properties of both the components. Miscible polymer blends produce a new improved material from less superior individual components, where as well established miscible polymer blends are very rare to obtain. However, semi-miscible blends with a uniform distribution of components also improve properties to an acceptable range .[8-11]

There are numerous numbers of previous studies related to polymer blends and their industrial applications, for examples: Awham et al. [12] prepared binary and ternary polymer blends by hand lay-up method with different weight ratios. The binary blends are epoxy/unsaturated polyester (EP/UPE) and epoxy/polysulfide rubber (EP/PSR) while the ternary blend is epoxy/unsaturated polyester/polysulfide rubber (EP/UPE/PSR). Impact, tensile and flexural characteristics for these blends were investigated at room temperature. The morphology of impact fracture surfaces for the prepared blends was examined by using the optical microscopy. The results revealed that the blending process of epoxy resin with UPE, PSR or both together within certain weight percentages forms compatible blends with improved toughness. The brittleness phenomenon of epoxy resin disappears with increasing the content of added polymer

Kalid [13] prepared binary polymer blends by hand lay-up method with different weight ratios from epoxy, unsaturated polyester, polyurethane, polymethyle methacrylate and polystyrene in 10 groups; thermoset-thermoset, termoset-thermoplastic, and thermoplastic- thermoplastic blends. Impact test, hardness test, differential scanning calorimetry test, thermogravometry test, and morphology by optical microscopy and scanning electron microscopy were carried out to investigate whether these blend are classified as compatible blends. The main aim of the present study was to characterization and comparison among three groups of binary polymer blends; polyurethane / epoxy (PU/EP), epoxy/unsaturated polyester (EP/UPE) and polyurethane /unsaturated polyester (PU/UPE) blends. In this study the mixing ratios effect on the thermal conductivity have been studied.

IOP Conf. Series: Materials Science and Engineering **757** (2020) 012019 doi:10.1088/1757-899X/757/1/012019

2. Experimental Procedure

2.1. Materials

Epoxy resin (Sikadur-52) was used in this research; it is a low viscosity, free flowing and fast curing injection resin and primer/coating based on a two component solvent free epoxy resin; ideally suited to a wide range of building and civil engineering applications where highly penetrative material is required. Unsaturated polyester resin (UPE) is a liquid with moderate viscosity which can be cured to the solid state by adding (Methyle Ethyle Keton Peroxide,MEKP) as a hardener, while cobalt octoate acts as a catalyst to accelerate the solidification process. Platine clear polyurethane gloss-057ME is a highly durable two-component polyurethane varnish with high gloss, excellent weathering and UV resistant properties. Theoretical spreading rate: 16.7 m^2/litre.

2.2. Material Preparation

2.2.1. Epoxy Resin preparation

A clean disposable aluminum container was put on sensitive electronic balance, and the required epoxy resin was then poured in. For100 gm of EP resin a sufficient amount of hardener (Sikadur-52-B) was added. The ratio of hardener of epoxy was approximately 1:2 by weight. The content was thoroughly mixed to be ready for further applications; blend specimens casting.

2.2.2 Unsaturated polyester Resin preparation

With the same above procedure the required unsaturated polyester resin was weighed. 100 gm of UPE resin were mixed with 0.5 gm of cobalt naphthenate as accelerator and 2 gm methyl ethyl keton Peroxide (MEKP) as hardener. The resin was mixed thoroughly with accelerator and hardener until a homogeneous state of the mixture evolved, and ready for blend specimens casting.

2.2.3. polyurethane preparation

With the same above procedures the required polyurethane resin was weighed. For 100 gm of PU resin (polyol) a sufficient amount of the hardener (isocyanate) was added. The ratio of hardener to resin was approximately 1:4 by weight. Again both were thoroughly mixed to be ready for blend specimens casting.

2.3. Polymer Blends Preparation

Epoxy/ Polyurethane, Epoxy/Unsaturated Polyester and Unsaturated Polyester / Polyurethane were added together and thoroughly prepared mixed to obtain the polymer blend required table(1), shows the percentages from which we have obtained the controlled and suitable blends, however the unsuccessful blends that were ignored due to their complete immiscibility between the two polymers. The mixture was stirred for 30 sec, then cast in a mold and finally left overnight at room temperature for complete curing .

Table 1. The blend percentages of three group.

Group No.	Blend No.	Material Concentration		Description
		PU%	EP%	
1 PU/EP	1	0	100	success
	2	10	90	success
	3	20	80	success
	4	30	70	success
	5	40	60	success
	6	50	50	success
	7	60	40	success
	8	70	30	success
	9	80	20	success
	10	90	10	success
	11	100	0	success
		EP%	UPE%	
2 EP/UPE	12	0	100	success
	13	10	90	Failure
	14	20	80	Failure
	15	30	70	success
	16	40	60	success
	17	50	50	success
	18	60	40	success
	19	70	30	success
	20	80	20	success
	21	90	10	success
	22	100	0	success
		PU%	UPE%	
3 PU/UPE	23	0	100	success
	24	10	90	success
	25	20	80	success
	26	30	70	success
	27	40	60	success
	28	50	50	success
	29	60	40	success
	30	70	30	success
	31	80	20	success
	32	90	10	success
	33	100	0	success

3 Measurements

3.1. Thermal Conductivity Test Instrument

Lee's disc instrument showed in figure 1, manufactured by the Griffen and George Company, was used to calculate the thermal conductivity of the samples under test. The figure below shows this instrument which consists of three discs of brass (40 mm diameter by 12.25 mm thickness) and a heater. The sample (S) is placed between the discs A and B, while the heater is placed between B and C. Heater was supplied with voltage (6 volt) and the current value through the apparatus was about (0.25A). The heat transfers from the heater to the near two discs then to the third disc across the sample. The temperature of the three discs (T_A, T_B, and T_C) is measured by using a thermometers placed inside them. After reaching thermal equilibrium, the temperatures were recorded[14].

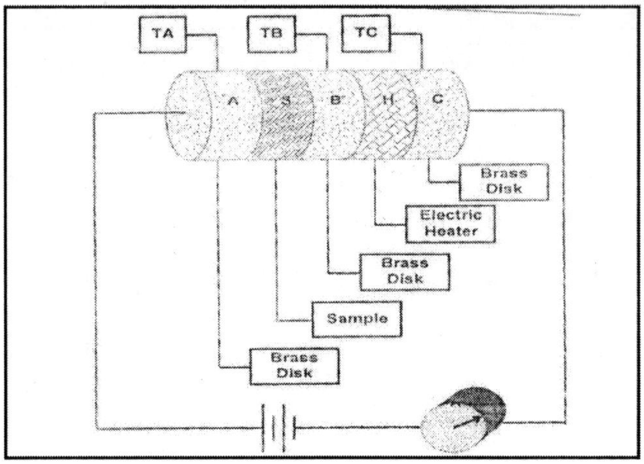

Figure 1. Thermal Conductivity Test Instrument.

4. Results and Discussion

Thermal conductivity (K) is the ability of a material to conduct heat. This quantity represents the rate of heat flow per unit time in a homogenous material under steady conditions, per unit area, per unit temperature gradient in a direction perpendicular to area [15]. Polymers are often utilized as thermal insulators because of their low thermal conductivities. For these materials, energy transfer is accomplished by the vibration and rotation of the chain molecules [16,17].

Figure 2 shows the relation between the thermal conductivity (K) for blending ratio of (PU/EP) blends. It can be seen that the highest value of (K) (0.78 $J/m^2.k$) has obtained at ratios (PU 10% / EP 90%) respectively. This increment of (K) implies that the penetration of molecules of both (PU) and (EP) in the same direct then a good transfer of energy by the vibration and rotation of the chain molecules. Also it was noticed the thermal conductivity (K) reach to minimum value (0.52 $J/m^2.k$) at ratios (PU 40% / EP 60%), but the results of thermal conductivity still higher than that of neat polyurethane.

Figure 3 shows the relation between the thermal conductivity (K) with blending ratio of (EP/UPE) blends. It can be noticed that the highest value of (K) (0.70 $J/m^2.k$) has obtained at ratios (EP 50% / UPE 50%), but results of thermal conductivity still lower than that of neat unsaturated polyester. Also it can be noticed that the lowest value of (K) (0. 0.53 $J/m^2.k$) has obtained at ratios (EP 30% / UPE 70 %), but results of thermal conductivity still lower than that of both neat epoxy and unsaturated polyester. In general due to nature of cross link properties some blends are unsuccessfully prepared at ratios (EP 10% / UPE 90%) and (EP 20% / UPE 80%) respectively.

Figure 4 shows the relation between the thermal conductivity (K) with blending ratio of (PU/UPE) blends. It can be noticed that the highest value of (K) (0.49 $J/m^2.k$) has obtained at ratios (PU 40% / UPE 60%), but results of thermal conductivity still only lower than that of neat unsaturated polyester. This figure also shows that the lowest value of (K) (0. 0.31 $J/m^2.k$) has obtained at ratios (PU 60% /

UPE 40%), but results of thermal conductivity still lower than that of both neat polyurethane and unsaturated polyester.

Finally the experimental results indicated that the thermal conductivity (K) has the highest value at ratios (PU 10% / EP 90%). Also it has the lowest value at ratios (PU 70% / UPE 30%) as shown in Figure 5.

Table 2. thermal conductivity test results

Group No.	Blend No.	Material Concentration		Thermal conductivity
		PU%	**EP%**	(K)
	1	0	100	0.588891108
	2	10	90	0.780901142
	3	20	80	0.748959999
	4	30	70	0.662203083
1	5	40	60	0.522667544
PU/EP	6	50	50	0.591965052
	7	60	40	0.614968254
	8	70	30	0.60131201
	9	80	20	0.641581322
	10	90	10	0.576047154
	11	100	0	0.458610635
		EP%	UPE%	
	12	0	100	0.674592014
	13	10	90	0
	14	20	80	0
	15	30	70	0.53751102
2	16	40	60	0.605051827
EP/UPE	17	50	50	0.700881057
	18	60	40	0.690017215
	19	70	30	0.644594821
	20	80	20	0.589127944
	21	90	10	0.579927478
	22	100	0	0.723933312
		PU%	**UPE%**	
	23	0	100	0.674592014
	24	10	90	0.397359579
	25	20	80	0.447877533
3	26	30	70	0.43544589
PU/UPE	27	40	60	0.495948859
	28	50	50	0.446673486
	29	60	40	0.315848843
	30	70	30	0.360776919
	31	80	20	0.447030908
	32	90	10	0.408571945
	33	100	0	0.458610635

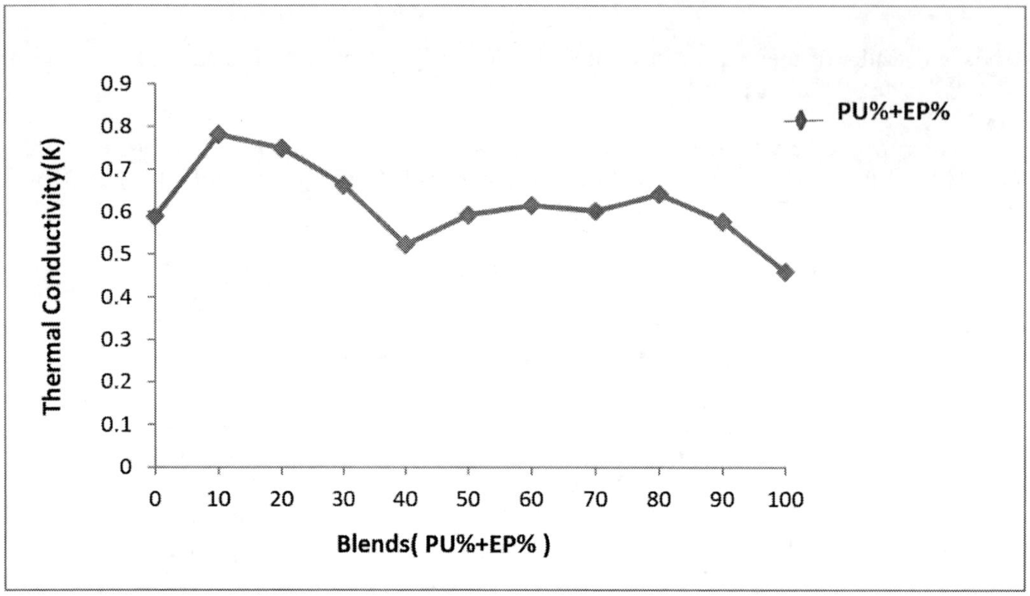

Figure 2. Thermal conductivity (K) values as a faction with blending ratio of (PU/EP) concentrations.

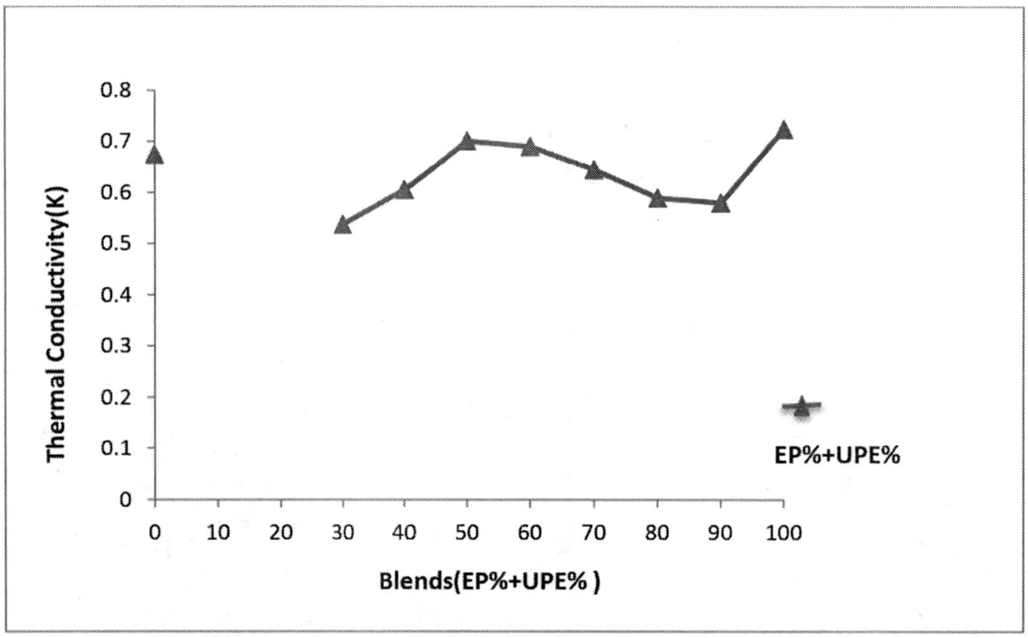

Figure 3. Thermal conductivity (K) values as a faction with blending ratio of (EP/UPE) concentrations.

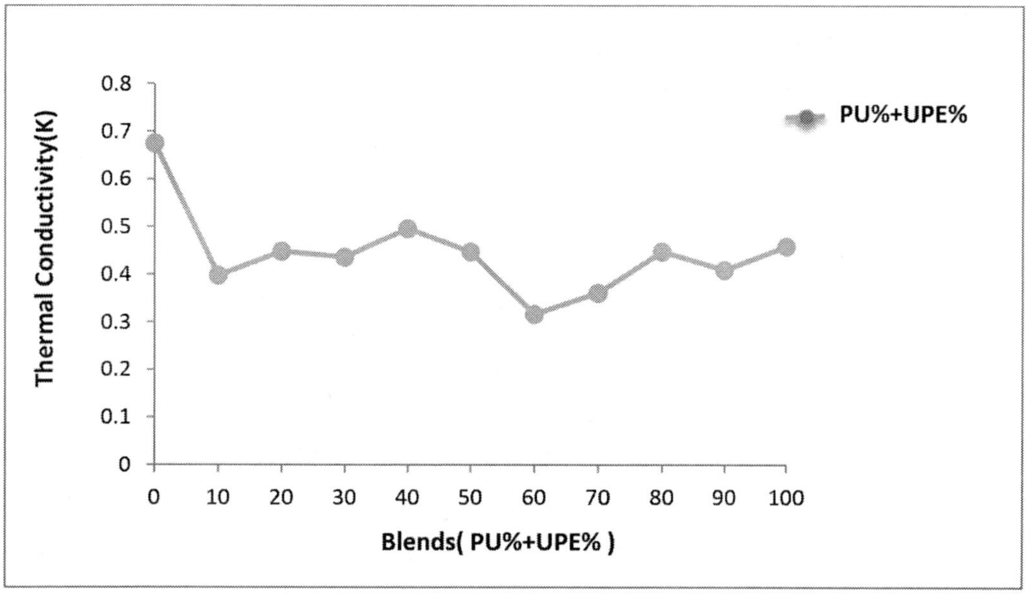

Figure 4. Thermal conductivity (K) values as a faction with blending ratio of (PU/UPE) concentrations.

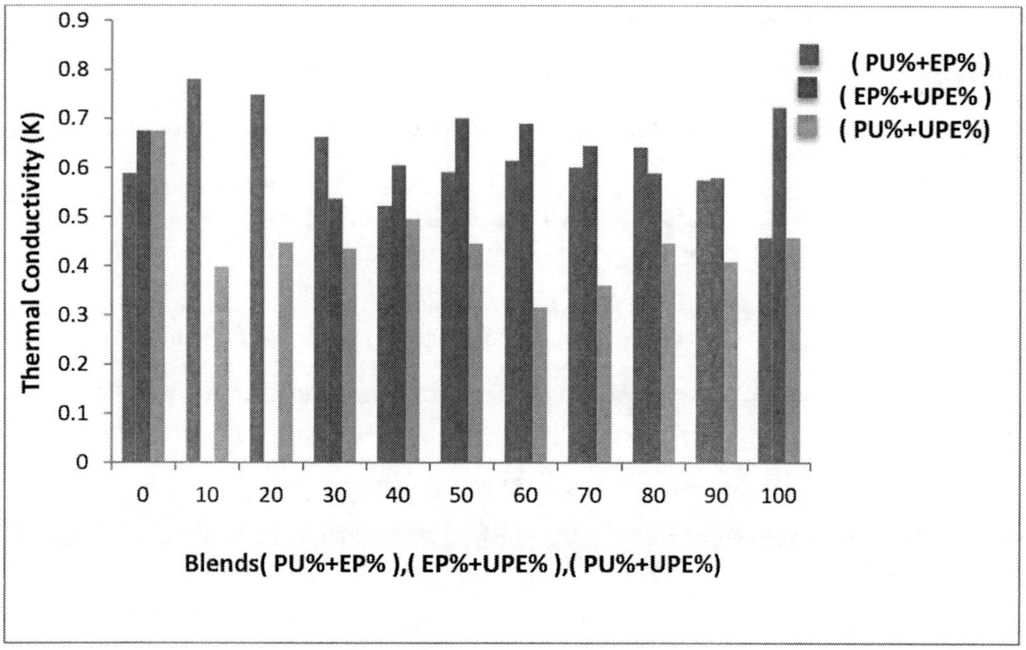

Figure 5. Thermal conductivity (K) values as a faction with blending ratios of (PU/EP), (EP/UPE) and (PU/UPE) concentrations respectively

IOP Conf. Series: Materials Science and Engineering **757** (2020) 012019 doi:10.1088/1757-899X/757/1/012019

5. Conclusions

Thermal conductivity decreases when increasing the ratios of PU to EP concentrations. From the experimental results noted that thermal conductivity (K) has the highest value at ratios (PU 10% / EP 90%). Also it has the lowest value at ratios (PU 70% / UPE 30%), that still lower than that of all neat epoxy, unsaturated polyester and polyurethane.

6. References

[1] Kumaraswamy G N and Ranganathaiah C 2006 *Polym. Engg. Sci.* 1231.

[2] Rajkumar P R and Nanthini R 2011 *Rasayan J. Chem.* **4**, 567.

[3] Mishra S P and Deopure, L 1985 *Polymer Bull.* **26**, 5.

[4] Usman A, Zia K M, Zuber M, Tabasum S, Rehman S, Zia F. Chitin and chitosan 2016 *International Journal of Biological Macromolecules.* **86** 630- 645.

[5] Mahmood N, Yuan Z, Schmidt J, Xu CC 2016 *Renewable and Sustainable Energy Reviews.* **60**: 317- 329.

[6] Kausar A. 2017 *American Journal of Polymer Science & Engineering.* **5**: 1- 12.

[7] Qu T, Yang N, Hou J, Li G, Yao Y, Zhang Q, He L, Wu D, Qu X. 2017 *RSC Advances.* **7**: 6140- 6151.

[8] Li, J.; Jian Guo, Z. *Polym. Plast. Tech.Eng.*, **50** (10), 996–999.

[9] Chinnakkannu K C, Muthukaruppan A, Rajkumar J S and Periyannan G 2007 *J. Polym. Res.* **14**, 319–328.

[10] Uday K, Gautam D, Niranjan K and Mesua ferrea L 2011 J. *Appl. Polym. Sci.* **121**, 1076–1085.

[11] Guo, J, Han-Xiong H and Zhao-Ke C 2011 *Tech. Eng.* **50** (10), 1035–1039.

[12] Awham M H and Ryam E H 2015 School of Applied Sciences, University of Technology, Res., vol. (**11**), no. (2), April-June pp. 19-24.

[13] Kalid.R.AL-Raui 1998 "Ph. D Thesis "Physics Dep. College of Science, University of Baghdad.

[14] Al-Janabi R H 2004 "Studying The Effect of Weathering Conditions on Some Properties of Epoxy Composites", M.Sc these, *the School of Applied Science*, University of Technology.

[15] Oronzio M, Biagio M and Giuseppe R 2005 Dipartimento *Ingegneria Aerospaziale Meccanica, ttp://event.ua.pt/*ds/manca3.pdf. Italy.

[16] Harper C A 1975" Handbook of Plastics & Elastomers", *McGrew-Hill ,Inc.,* NewYrok.

[17] William D C 1994 " Materials Science & Engineering ", third edition, *John Wiely & Sons, Inc.*, NewYork.

IOP Conf. Series: Materials Science and Engineering **757** (2020) 012020 doi:10.1088/1757-899X/757/1/012020

Study of effect reinforcement by coconut fiber on some mechanical and physical properties of thermoset polymer

S. M Hashim*, W. B Salih.
College of Education for Pure Scince, University of Anbar, Anbar, Iraq
*Corresponding author email: saadmalaniph@yahoo.com

Abstract
In this paper, a polymeric composite consisting of unsaturated polyester resin as a base material reinforced with coconut fiber was produced by volumetric fractions percentage (0%, 0.5%, 1%, 1.5%, 2%, 2.5%, 3%) of coconut fiber and at room temperature. The properties of compressive strength, hardness and thermal conductivity of the composite material were studied. The results showed an improvement in the mechanical properties (compressive strength and hardness) while the thermal conductivity values decreased with the increase of the volumetric fraction of the fibers. The heat insulation of the composite material is increased with the increase of the volumetric fracture of the fiber..

Keywords: Composite Material , Coconut Fiber , Unsaturated Polyester , Compressive Strength , Hardness , Thermal Conductivity.

1. Introduction

For need to manufacture materials for various industrial uses led to the emergence of alternatives with a good characteristics in terms of light weight , cost and durability , which can be used in various areas such as radars , cars , ships , aircraft and others , so produced composite materials that its material resulting from the merger of two materials has different properties to create new properties that cannot be obtained from the original materials [1,2]. Composite materials which have polymer - based are materials used in technological and engineering applications. The main requirements for their use are high performance , good durability , resistance to internal and external stresses and resistance to ambient conditions of pressure and temperature [1]. The composite materials of fiber - reinforced resin polymers were used in the mid - thirties of this century and at the beginning of the Second World War. Glass - reinforced polyester was used in the manufacture of radar domes and planes such as epoxy - Kevlar and epoxy - carbon In the marine industry , building industry and buildings , as well as in the medical and chemical industries [3]. The choice of ceramic , plastic , mineral or carbon additive is according to the type of use if it is industrial or structural. It is used in electrical and thermal fields and it is in different forms either in the form of continuous or non - continuous fiber or wool or Flakes or Particles or Fillers. The composite materials are used in various applications that depend not only on external properties and mechanical and physical properties but on their ability to change their properties according to the materials used [1]. The high cost of traditional reinforcement and construction materials, such as synthetic fiber , is the most important factor hindering researchers' work. Due to increased economic and environmental requirements , alternative natural materials such as cannabis , kaizaran , coconuts , pineapples , flax and others were needed to strengthen polymeric materials , so efforts have been directed towards the use of waste materials such as the use of their fibers and outer shell , The second stage is the exploitation of these wastes to support these materials. Therefore, the subject of recycling and reuse are the most important pillars of the industrial world today [4]. Environmental waste is a byproduct of human activity. Some natural fibers are one such waste, which is defined as a type of renewable energy source and a new

IOP Conf. Series: Materials Science and Engineering **757** (2020) 012020 doi:10.1088/1757-899X/757/1/012020

generation of support of polymers. The development of composite materials with natural fibers or environmentally friendly compounds rather than traditional industrial materials has been an important issue recently due to increased environmental awareness [5]. All of the above has led us to this research, which included the use of composite materials polymers with a basis of unsaturated polyester , which is supported by the fibers of the coconut fiber, which is part of environmental waste in nature and study some of the mechanical and physical properties of them. The main objective of this research is to invest the environmental waste (Coconut Fiber) to reinforcement polymeric material (Unsaturated Polyester Resin) (UPE). The research also aims to manufacture and study a thermoplastic polymeric compound consisting of unsaturated polyester as a base material supported by nut fiber India, also aims to study the effect of changing the volumetric fracture of these fibers on some mechanical and physical properties, which included the examination of the Compressive Strength, hardness and thermal conductivity of polymeric composite material that was prepared.

2. Experimental work

2.1. Matrix Material

Unsaturated Polyester Resin was used as a base material in the preparation of the polymeric composite material which manufactured by SIR Saudi Arabia company. It is in the form of a transparent pink liquid at room temperature. It is a type of thermosetting polymer, This resin turns from liquid state to solid state by the addition hardener (Methyl Ethyl Ketone Peroxide) which manufactured by the same company SIR , which is in the form of a transparent liquid added to the unsaturated polystyrene resin at (2 g) per (100 g) of resin at room temperature. Its density (1.2 g/cm^3).

2.2. Reinforcement Material

Coconut fiber used in this research as a reinforced material with a density of (1.2 g/cm^3) is a high quality fiber and widely used applications in the form of cut-off fibers, Its color is brown.

2.3. Preparation of Samples

In this study, samples were prepared from the composite substrate of unsaturated polyester, reinforced by of coconut fiber at a length of (5 mm). The hand lay up method was adopted by casting and the compression, hardness and thermal conductivity test was carried out for the purpose of evaluating the final models. A mold made of aluminum has been used with the required sample dimensions. All mixtures prepared at room temperature are mixed and the mixing phase continues for (8-10 minutes) so that the mixture is homogenized to distribute the reinforcement material homogenously within the base material in order to avoid bubbles or aggregations during the mixing process then we pour the liquid mixture into the mold regularly until the mold is filled and according to the required level. The sample is left for (48 hours) in the mold to become completely solid at room temperature. The samples are then placed in an electric oven for two hours at (50 °C) for Heat treatment process according to the company's instructions. The process of preparation of the samples is done in several steps after the coconut fiber is sliced to reach short lengths at a rate of (5 mm) length and as follows :

1- By Apply volume fracture law V_f of fiber added (0, 0.5%, 1%, 1.5%, 2%, 2.5%, 3%), get weigh the fiber and find its size and then after that find the volume of unsaturated polyester and its hardener according to mixing rates.

2- Add the hardener with (2g) per (100g) of unsaturated polyester and mix well using the electric mixer for (2-3 minutes). Add the fiber with the mixing process until a homogeneous mixture is obtained and then pour the mixture into the mold for casting.

3- Leave the sample for 48 hours to complete Curing.

4- For post cureall samples were put in an oven for 2 hour at 50 C.

5- The samples are then removed from the convection oven and tested.

2.4. Mechanical Tests

Compressive Strength Test : Compression strength test on the samples was performed at room temperature by using (LARYEE Your Tasting Solution). In this test, compressive strength for all prepared samples are examined. When the sample is subjected to a compressive load until the failure, the compression test samples were tested and measured according to American standards (ASTM D695) [6]. Figure (1) represents a schematic diagram of the sample dimensions used in this test. This test is performed by applying a load at (5mm / min) at laboratory temperature, using the instrument diagram, the results are obtained directly. Figure (2) shows the samples of the compression test.

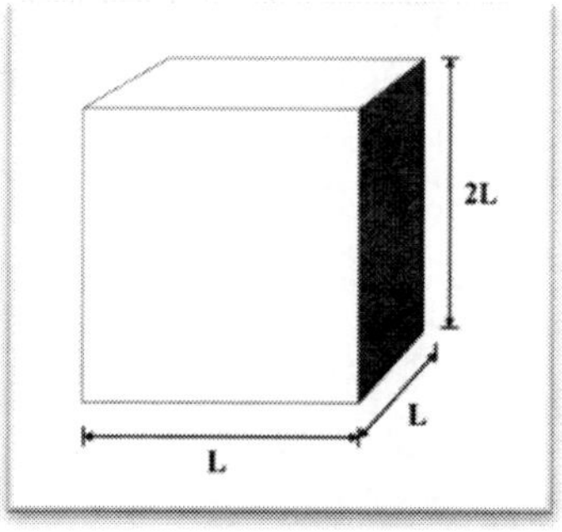

Figure (1). Dimensions of the compression strength test samples.

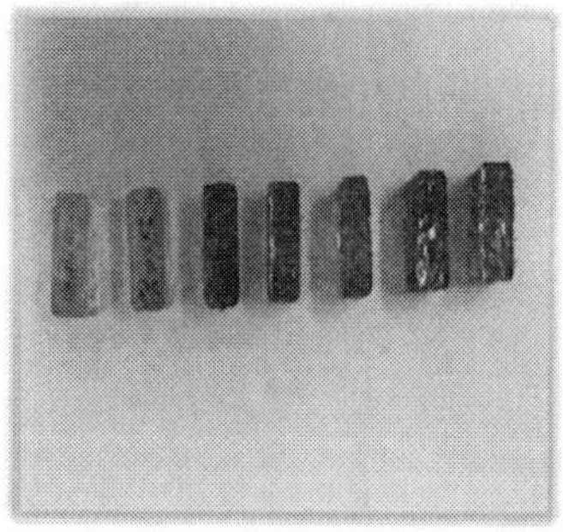

Figure (2). Samples used in compressive strength test.

Hardness Test : The hardness of the samples was measured in (Shore D) method at laboratory temperature. This method is suitable for thermally thermoplastic materials. The sample used in this test should have a smooth surface and a thickness of not less than (3 mm) and diameter greater than (30 mm) and not have been subjected to mechanical vibrations previously. Therefore, the samples used in this test are thickness (5 mm) and diameter (40 mm). These dimensions are taken according to ASTM D2240 [7]. As shown in figure (3). Figure (4) shows samples used in this test.

The device used is a similar figure to the compass containing a needle in the middle. The method of the test involves placing the device vertically on the sample to measure the hardness so that must touch the surface of the sample to measure its strength to stitch the needle in the surface of the material is waiting for (5s) for all samples, The hardness values were taken in three different locations for each sample and the values take according to the reading rate.

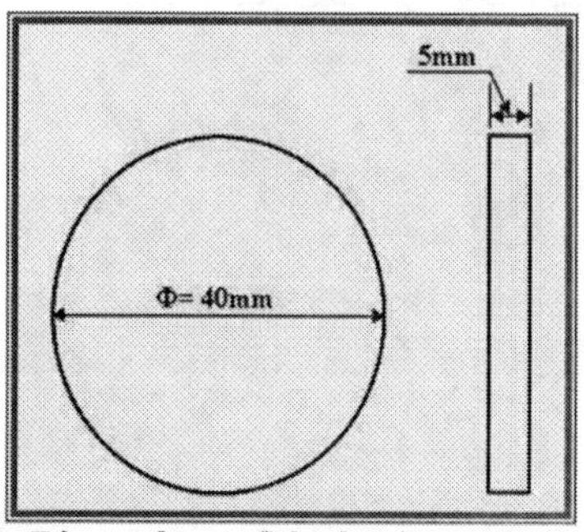

Figure (3). Dimensions of the hardness test samples.

Figure (4). Samples used in hardness test.

2.5. Phisical Tests

Thermal Conductivity test : For the purpose of conducting the thermal conductivity test one can use a Lee's Disk instrument , circular samples with diameter (40 mm) and thickness (5 mm) were cut according to international standards. Lee's Disk instrument factory by Griffin and George Figure (5) . For the purpose of this test, the test sample (S) is placed between the two disks (A, B) , and the heater

(H) placed between the disc (C,B). Heat transferred to the next disc until reaching the final disc. When thermal stability is achieved, the temperature of the three disks (T_A, T_B, T_C) is determined by using the placed syringes. In applying these two relationships, thermal conductivity can be calculated [8] :

$$K\left(\frac{T_B-T_A}{T_S}\right) = e\left[T_A + \frac{2}{r}\left(d_A + \frac{1}{4}d_s\right)T_A + \frac{1}{2r}d_S d_B\right] \quad ..... (1)$$

$$H = IV = \pi r^2 e(T_A + T_B) + 2\pi re\left[d_A T_A + d_s \cdot \frac{1}{2}(T_A + T_B) + d_B T_B + d_C T_C\right] \quad .... (2)$$

Where e : represents the thermal energy passing through the unit of space in the time unit $\left(\frac{W}{m^2.c^{\circ}}\right)$. H : The time-rate of energy delivered in the device. T_A, T_B, T_C : represents the temperature of the disks respectively (°C). d_A,d_B,d_C : Representing the thickness of copper disks (A,B,C) (mm). d_s : Sampling thickness (mm). I : Current in the circuit (Amp). V : Processed voltages for circuit (Volt). R : The radius of the copper disk (mm).

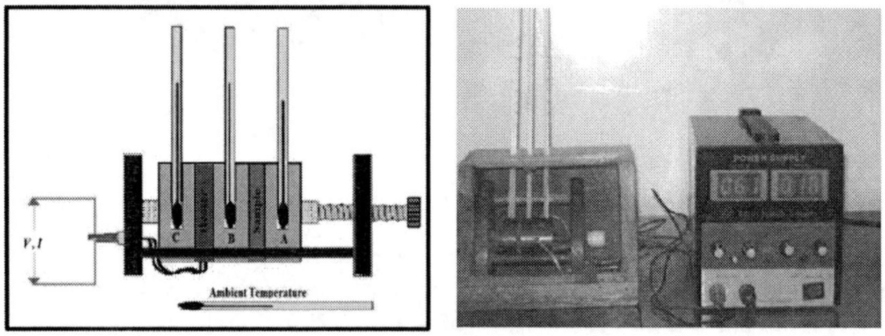

Figure (5). Thermal conductivity test device.

It is important to ensure that the surfaces of the disks are clean and the disks are well aligned to obtain the best heat transfer across them.

Figure (6) represents a diagram of the dimensions of the thermal conductivity test samples. While figure (7) shows samples of thermal conductivity test.

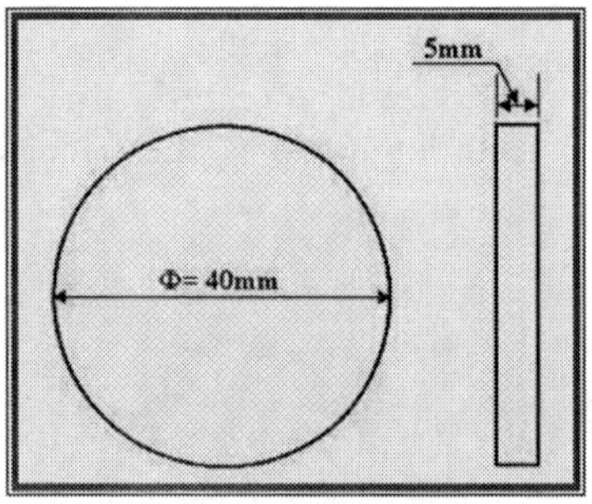

Figure (6). Dimensions of the thermal conductivity test samples.

Figure (7). Samples used in thermal conductivity test.

3. Results and discussion

3.1. Results and Discussion of Compression Strength Test

From the diagram of the device, the results of compressive strength were obtained directly. Figure (8) shows the relationship between the volume fraction of the coconut fiber and the compressive strength of the prepared samples. It is noted from this figure that the compressive strength is increased by increasing the volume fracture of fibers in the polymeric composite material. This is due to the role of these fibers, And these fibers greatly affect the transfer of stress as well as the shear strength of the composite materials. The presence of these fibers will also impede the movement of the polymer chains within the composite and thus increase their strength. Resin penetration in coconut fiber will reduce the possibility of separation or withdrawal of fiber from the base material during pregnancy , this result is agree with [9].

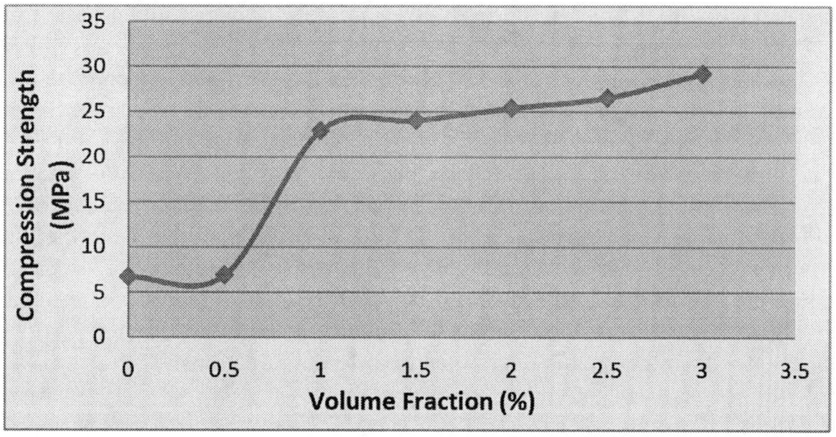

Figure (8). The relationship between volume fracture of coconut fiber and compressive strength of prepared samples.

3.2. Results and Discussion of Hardness Test

Hardness tests depend on the material's resistance to penetrations at its outer surface. There are different methods that represent the hardness guide , In this study, the Shore D hardness test was performed. Figure (9) represents the relationship between the volume fraction of the coconut fiber and the hardness values of the prepared samples. It is noted that the hardness values of the samples are increased with the increase of the volume fraction of the coconut fiber in the polymeric composite material because these fibers have high hardness , in other word the presence of fiber increases the resistance of the material to plastic deformation. By increasing the strength and bonding of the atoms or molecules, the material stiffens and thus increases its resistance to scratching , this is due to the distribution of the load on the fibers, which reduces the penetration rate of the surface of the composite material and raises the values of its hardness , this result agree with [10,11].

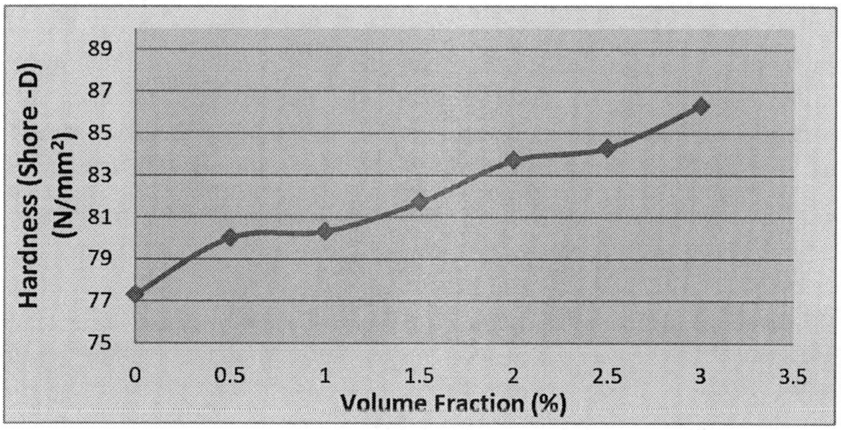

Figure (9) The relationship between the volume fraction of the coconut fiber and the hardness values of the prepared samples.

3.3. Results and Discussion of Thermal Conductivity Test

Figure (10) illustrates the relationship between the volume fraction of coconut fiber and the values of thermal conductivity , one can noted a decrease in the heat coupling coefficient with increasing volume fracture of coconut fiber in the polymeric composite material , this means that the prepared samples become more heat-neutral , this is due to the fact that coconut fiber is considered a heat insulation material due to the nature of this fiber because it does not contain electronic abundance or ionic decomposition to facilitate the transfer of phonons, which is the only means of heat transfer in polymer composites and insulating materials. The ability of insulation here depends on the presence of fine fiber filaments, which leads to the transfer of thermal energy in two ways (conduction and pregnancy). The elastic waves (phonons) pass through the base material and the solid part of this fiber by the vibrational movement of the atoms and by the covalent effect. When the phonons reach the capillary part of the fiber, the phonons are blocked in their movement because of the different structural structure of this medium (because it has atoms and bonds different from the former), which will lead to lower values of thermal conductivity. Other reasons attributed to the low thermal conductivity values are due to the presence of thousands of branched and interconnected compounds and the irregularity of regularity in the composite structure , this impedes the flow of heat in one direction and moves it to the other, but is distributed in different directions.

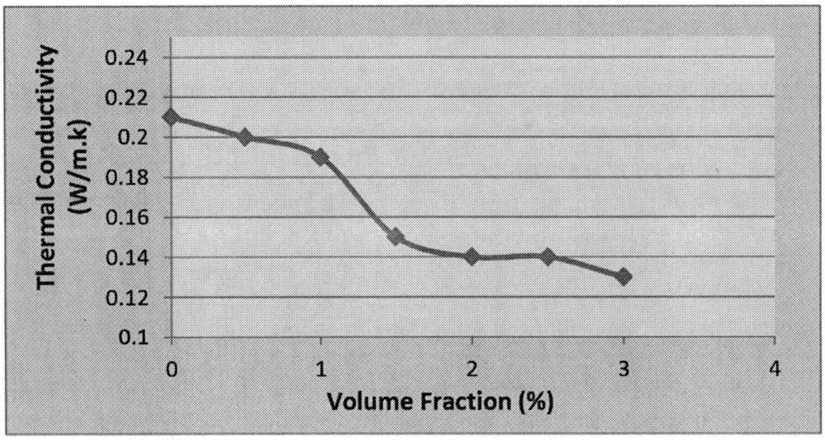

Figure (10). The relationship between the volume fraction of the coconut fiber and the thermal conductivity values of the prepared samples.

4. Conclusions

The addition of coconut fiber has resulted in the improvement of the mechanical and physical properties of the composite material prepared. The values of mechanical properties (compressive strength and hardness) are increased with the increase of the volume fraction of the added fibers. The thermal conductivity values shall be reduced with the increase of the volume fraction of the added fiber, which means an improvement in the thermal insulation properties of the material.

Reference

[1] Noor S, 2017, Materials, Effect of the Change of Volume Fraction and Curing Time on Some Electrical and Mechanical Properties for Polymeric Composites. Master Thesis (Msc.). Iraq-Anbar : University of Anbar ; 2017.

[2] Sihama I. Saleh , Waleed B. Saleh , Salam O. Abdulghani 2017 Diyala Journal For Pure Sciences **13(1)** : 60-71.

[3] Tahsin Sh. Bdioui. Study of the Effect of Aluminum or Copper Powder on the Flexible Fracture Properties of a Polyester - Based Substance. Master Thesis (Msc.) , Iraq-Baghdad : University of Technology ; (2008).

[4] Hind Basil Ali 2013 Engineering and Technology Journal **31(17)** : 296 – 308.

[5] M. R. Sanjay , G. R. Arpitha, L. Laxmana Naik, K. Gopalakrishna and B. Yogesha Applications of Natural Fibers and Its Composites. Scientific Research Publishing Inc. 201 7 : 108-114.

[6] Astm, D. 695 Standard Test Method for Compressive Properties of Rigid Plastics. ASTM materials standards, (2002).

[7] Annual Book of ASTM Standard. "Standard Test Method for Plastics Properties-Durometer Hardness", D (2240-03), PP.(1-12),(2003).

[8] R. C. Hibbert, " Material ", 3rd Edition Russell, C. H. published partic Hall, Inc, (1997).

[9] Obasi H.C, Iheaturu N.C , Onuoha F.N, Chike-Onyegbula C.O, Akanbi M.N and Ezeh V.O 2014 American Journal of Engineering Research (AJER) 03 (02):117-123.

[10] Idrus M.A., Hamdan S., Rahman M.R. and Islam M.S 2011 Journal of Biomaterials and Nanotechnology 2:435 – 444.

[11] Rola Abdul-Khader Abbas 2012 Journal of Engineering and Technology 30(12):284-301.

IOP Conf. Series: Materials Science and Engineering **757** (2020) 012021 doi:10.1088/1757-899X/757/1/012021

Sustainable manufacturing process applied to produce magnesium oxide from sea water

A K Jassim[1], S A S Almtori[2]and J A Jassam[3]
[1]The State Company for Iron and Steel, Research and Development Department, Iraq, Basra, ahmadkj1966@yahoo.com
[2]University of Basra, Materials Engineering Department, Iraq, Basra, dr.safaaalmtori@yahoo.com
[3]The State Company for Iron and Steel, Planning and Studying Department, Iraq, Basra

Abstract. Magnesium oxide is one of the most important materials that used to produce a basic refractory material which is applied for liningsteel and iron furnaces as well as cement furnaces. In this research, magnesium oxide was produced from sea water that content effective percentage of magnesium hydroxide instead of magnesium carbonate rock. The process has been done by adding additive materials which include high alumina, iron oxide, sodium phosphate and sodium silicate to reduce the sintering temperature to 1400 °C instead of 1750 °C. Moreover, kiln furnace was used instead of rotary furnace to do the sintering process. The results show that there is a possibility to produce magnesium oxide with purity of 56%. It is stable at temperature less than 1600 °C and resists the humidity and water. The process of producing magnesium oxide could be considered as a sustainable manufacturing process due to the applied lower sintering temperature which leads to reduce cost, energy and pollution.

Keywords: Magnesium, sea water, refractory, sintering, sustainability.

1. Introduction

Refractory material is defined as a heat-resistant material that designed to resist the decomposition by heat and pressure, withstand the chemical attack and high temperature. It is commonly consists of ceramic and mineral materials [1]. Moreover, they are defined according to ASTM C71 as non-metallic materials have chemical and physical properties that make them applicable for structures, or as components of systems that exposed to environments above 538 °C [2]. Refractory materials can be classified into acidic refractories such as alumina, silica, and fireclay, basic refractories materials like magnesite and dolomite and neutral refractories materials such as carbon, silicon carbide, chromite and zirconia. Refractory materials are used for lining hot surfaces for many industrial processes like furnaces, kilns and reactors etc.,.Today, 70% of all refractory material have been produced for iron and steel industry application and metal casting sectors. It must be chemically and physically stable at high temperature and resistant to thermal shock [3-5].
Magnesium oxide is the most important basic refractory materials which can be considered as industrial magnesium compound that used for steel and iron industry applications. Moreover, it is largely used in many other industrial sectors like food and animal feed industries. Magnesium oxide can be produced from both natural magnesium carbonate and synthetic magnesia as magnesium chloride from seawater and brines. The production process involves decomposition of the $MgCO_3$ to MgO and CO_2 as gas. However, sea water megnesite made by a chemical process of $Mg(OH)_2$. The whole refractory magnesia used in UK, USA and Japan comes from the sea water magnesite plants [6-7].

Refractory material is usually sintered in lamp form with different temperatures in the range of 600-2000 °C. The first type of magnesium oxide produced at temperatures in the range of 1500 – 2000 °C as dead burned magnesium oxide that extensively used in steel production serving as protective and replaceable linings for equipment used to handle molten steel. The second type of magnesium oxide produced at temperatures ranging of 1000 - 1500°C as hard-burned which is typically used in applications where slow degradation or chemical reactivity is required with animal feeds and fertilizers. The third grade of magnesium oxide is produced by calcination at temperatures ranging from 600 – 1000 °C as light-burn or caustic magnesia which is used in plastics, rubber, paper and pulp processing, steel boiler additives, adhesives, and acid neutralization. Different types of magnesium oxide are produced by using the dry process route, such as dead burned magnesia, caustic calcined magnesia and fused magnesia. Manufacture of magnesium oxide is high energy consuming because it is manufactured at very high temperatures which consumed (6 and 12) GJ/t of magnesium oxide [6-11].

Pilarska et al., had studied the effect of type and concentration of modifiers on the physicochemical and functional properties of magnesium hydroxide which is obtained by precipitation from magnesium sulphate and sodium dioxide that subjected to thermal decomposition at 450 °C with calcination magnesium oxide. The results show that the presence of a modifier did not change the crystalline structure of the products [12].

J. Chen and et al, had studied the effect of using preparation spherical magnesium hydroxide with a cured like could be successfully by magnesium sulfate, ammonia water and sodium hydroxide solution. The results show that the products obtained with smooth surface and the percentage of SO4 will decrease [13].

Z. Xing and et al, investigated the chemical kinetics model of the hydration reaction with different active magnesium oxide under an unforced and ultrasonic condition. The result indicated that the rate order of hydration was different in the hydration process under an unforced condition [14].

2. Experimental work

2.1 Materials

Magnesium hydroxide that prepared from sea water has been used to produce magnesium oxide. It was mixed with sodium phosphate, iron oxide, sodium silicate, and aluminium oxide as an additive material to reduce the sintering temperature. Tables 1-6 show the main chemical and physical specifications of magnesium hydroxide and their additives.

Table 1. Specifications of magnesium hydroxide

Chemical Symbol	Purity %	Cao %	Colour	Density (gm/cm^3)	Shape
$Mg(OH)_2$	85	3	White	0.9	Fine powder

Table 2. Specifications of Tri Sodium Phosphate

Chemical Symbol	Purity %	Colour	Water soluble %	PH %	Shape
$Na_3PO_4.H_2O$	92	White	42	11.8	Fine powder

Table 3. Specifications of Iron Oxide

Chemical Symbol	Purity %	Fe %	Colour	Shape
Fe_2O_3	100	68	Red	Fine powder

Table 4. Specifications of Hexa Meta Phosphate

Chemical Symbol	P_2O_5 %	Fe %	Solubility %	Colour	PH	Density (gm/cm^3)	Shape
$Na_3(PO_4)_6$	68	68	5	Red	7.5	1.4	Fine powder

Table 5. Specifications of Sodium Silicate

Chemical Symbol	SiO_2 %	Na_2O %	H_2O%	Density (gm/cm^3)	Shape
Na_2SiO_3	37.5	12.5	50	1.5	Liquid

Table 6. Specifications of Aluminium Oxide

Chemical Symbol	Purity %	SiO_2 %	(CaO+Mgo) %	Fe_2O_3 %	Shape
Al_2O_3	57	37	2	2.5	Fine powder

2.2 Production procedure

The procedure was started by mixing magnesium hydroxide with additive materials by using mechanical stirrer to get a good homogeneous mixture. The water was added to make good mixture for all materials. Table 7 present the main experimental work that done in the laboratory to prepare cylindrical specimens with diameter of 40 mm and length of 10 mm. The specimens were prepared by pressing the mixing materials in hydraulic press with pressure of 40 tons. Then, the specimens burnt in furnace with a temperature in the range of 1300 – 1500 °C.After that, the best mixing percentages were determined as shown in table 8 which are used to produce high quantity of magnesium oxide in the refractory plant. Moreover, the quantity of mixing materials that used in refractory plant shown in Table 9.

The blending materials that get after mixing with quantity of 1562 kg was burnt by kiln furnace with temperature up to 600 °C as first burnt. Then press by hydraulic press with pressure of 50 bars to form green block for easy handling and transportation. After that, the press blocks were burnt in the same kiln furnace with temperature up to 1400 °C for 72 hours and cooling gradually in the same furnace with decreasing rate of 25 °C per hour for 48 hours. Therefore, the total period of time that need for first burnt is 120 hours. The result quantity of this burnt is 893 kg with loss of ignition equal to 43%. The burnt quantity that get from the first burnt was crashing into fine powder and mixing again with 5% liquid sodium silicate, 5% of sodium phosphate and 5% of water to form as green block by pressing with pressure of 150 bars. The result blocks dried and burnt for 120 hours as in the first burnt procedure to get final products of magnesium oxide.

Table 7. The main experimental works in the laboratory to prepare magnesium oxide

Experiment No.	$Mg(OH)_2$ %	$Na_3(PO_4)_6$ or $Na_3PO_4.H_2O$ %	Al_2O_3 %	Fe_2O_3 %	Na_2SiO_3 %	H_2O %	Temp. °C	Burnt Time (hr)
1	90	10	-	-	-	-	1250	18
2	70	30	-	-	-	-	1250	18
3	50	50	-	-	-	-	1250	18
4	30	70	-	-	-	-	1250	18
5	85	15	-	-	-	-	1250	18
6	85	15	-	-	-	-	1250	18
7	77	13	5	5	-	-	1250	18
8	75	15	-	10	-	-	1250	18
9	75	15	-	10	-	-	1250	18
10	70	15	5	5	5	5	1250	18
11	70	15	10	5	-	10	1250	240

Table 8. The best mixing percentagesin the laboratory to prepare magnesium oxide

Experiment No.	$Mg(OH)_2$ %	$Na_3(PO_4)_6$ or $Na_3PO_4.H_2O$ %	Al_2O_3 %	Fe_2O_3 %	Na_2SiO_3 %	H_2O %	Temp. °C	Burnt Time (hr)
1	70	15	5	5	5	5	1250	18
2	70	15	10	5	-	10	1250	240

Table 9. The main experimental works in the laboratory to prepare magnesium oxide

Materials	$Mg(OH)_2$ Kg	$Na_3PO_4.H_2O$ Kg	Al_2O_3 Kg	Fe_2O_3 Kg	Na_2SiO_3 Kg	H_2O %
Quantity	1130	163	95	50	50	5

3. Results and discussion

3.1 Chemical composition
The chemical analysis of the final magnesium oxide produced in this work shown in table 10. It has been found that the percentage of magnesium oxide increased with increasing the number of burns. The percentage of magnesium oxide after the first burnt with temperature of 1400 °C for 120 hours is 43% and 56-58% after the second burnt at 1400 °C for 240 hours while calcium oxide is found to be less than 13%. Aluminium oxide, sodium phosphate and sodium silicate were used as bindery materials to get good adhesive between powders.

Table 10. Chemical composition of magnesium oxide produced in this research work

No. of burnt	Burnt time (hr.)	MgO %	R_2O_3 %	SiO_2 %	CaO %	SO_3%	Sum %
1	120	43	19	2.56	9.53	1.54	99.42
2	240	56	23	4.45	12.79	1.83	99.091
3	240	59	27	1.31	10.65	1.95	99.37

3.2 Loss of ignition
Magnesium oxides that produced at laboratory and factory were evaluated to determine their loss of ignition. The results show that, the loss of ignition for laboratory samples is equal to 38% with shrinkage of 13.5% in dimensions, while for factory samples are equal to 24% for the first burnt and 0.12% for the final burnt as shown in Table 11.

Table 11. Loss of ignition for magnesium oxide produced in this research work

No. of burnt	Burnt time (hr.)	L. O. I%
1	120	24
2	240	0.37
3	240	0.12

3.3 Sintering temperature
Additive materials were used to reduce the sintering temperature of magnesium hydroxide to be 1400 °Cinstead of 1750 °C by using kiln furnace instead of rotary furnace to reduce the waste of materials.

4. Conclusions

The main conclusions of this research work include the possibility to produce magnesium oxide with purity of 56% from seawater by using kiln furnace instead of rotary furnace in order to reduce the waste of materials and temperature of 1400 °C instead of 1750 °C to reduce energy and pollution. Moreover, the maximum time that required for producing dead burnt magnesite is 240 hours using two stages of burnt. In addition, additive materials help to reduce the sintering temperature to be 1400 °C instead of 1750 °C. Finally, magnesium oxide that produced in this work can be withstanding the high temperature up to 1600 °C and high humidity.

References

[1] Yildirim M.and Akarsu H. 2010 *Journal. of Physicochemical Problem Mineral Processing*, 44,257-272.

[2] Hideto Minami, Keigo Kinoshita, Tomoki Tsuji, and Hiroshi Yanagimoto, Preparation of Highly Crystalline Magnesium Oxide and Polystyrene/Magnesium Hydroxide Composite Particles by Sol–Gel Processes in an Ionic Liquid, J. Phys. Chem. *C*, 116 (27), PP. 14568–14574,2012.

[3] I. Ganesha, S. Bhattacharjeea, B. P. Sahaa, R. Johnsona, K. Rajeshwarib, R. Senguptab, M. V. Ramana Raob and Y. R. Mahajana, An efficient $MgAl_2O_4$ spinel additive for improved slag erosion and penetration resistance of high-Al2O3 and MgO–C refractories, Journal of Ceramics International, Vol. 28, Issue 3, PP. 245-253, 2002.

[4] Toru Shimizua, Kazuhiro Matsuurab, Harumi Furuea and Kunio Matsuzaka, Thermal conductivity of high porosity alumina refractory bricks made by a slurry gelation and foaming method, Journal of the European Ceramic Society, 33, Issues 15–16, 3429-3435, 2013.

[5] F. Raupp -Pereiraa, D. Hotzab, A. M. Segadãesa and J. A. Labrinchaa, Ceramic formulations prepared with industrial wastes and natural sub-products, Journal of Ceramics International, 32, 173-179, 2006

[6] Addis ababa and Ethiopia, magnesium oxide powder manufacturing industry establishment, www.confindustria.vicenza.it/notiziario/...nsf/.../executive%20summary-%20final.pdf, 2015

[7] J. H. Cheslers, refractories production and properties book, the iron and steel institute, London, 118-177, 1973

[8] H. Dong, Z. Du, Y. Zhao and D. Zhou, Preparation of surface modified nano-$Mg(OH)_2$ via precipitation method, Elsevier, Powder Technology, Vol. 198, issue 3, PP. 325–329, 2010

[9] J.L. Booster, A. van Sandwijk, M.A. Reuter, Conversion of magnesium fluoride to magnesium hydroxide, Minerals Engineering, Vol.1, PP. 273–281, 2003

[10] J. Kang and P. Schwendeman, Comparison of the effects of Mg(OH)2 and sucrose on the stability of bovine serum albumin encapsulated in injectable poly (D, L-lactide-co-glycolide) implants, Biomaterials, Vol. 23, PP. 239–245, 2002.

[11] R. Richards, W.F. Li, S. Decker, C. Koper, V. Zaikovski, A. Volodin, T. Rieker and K.J. Klabunde, Consolidation of metal oxide nanocrystals. Reactive pellets with controllable pore structure that represent a new family of porous, inorganic materials, Journal of the American Chemical Society, Vol. 122, PP. 4921–4925, 2000

[12] Agnieszka A. Pilarska, Marcin Wysokowski a, Ewa Markiewicz b and Teofil Jesionowski, Synthesis of magnesium hydroxide and its calcinates by a precipitation method with the use of magnesium sulfate and polyethylene glycols, journal of powder Technology, PP 148 -157, 2012

[13] J. Chen, W. Zhang and Y. Song, Synthesis of spherical magnesia and application in degradation of phenolic wastewater, Journal of Functional Materials, 2015.

IOP Conf. Series: Materials Science and Engineering **757** (2020) 012021 doi:10.1088/1757-899X/757/1/012021

[14] Zhibo Xing, Limei Bai, Yuxin Ma, Dong Wang, and Meng Li, Mechanism of Magnesium Oxide Hydration Based on the Multi-Rate Model, Journal of Materials, Vol. 11(10), 2018.

Numerical Analysis of CW Raman Amplifier in Silicon-on- Insulator Nano-Waveguides

Z S Khaleefa[1]*, Sh S Mahdi[1] and S Kh Yaseen[1]

[1] College of Science for Women, Physics Department, University of Baghdad, IRAQ,
*corresponding author: E-mail: zainabsalam6480@gmail.com

Abstract: Numerical analysis of continuous wave (CW) on-off optical Raman gain in Silicon-On-insulator (SOI) nano-waveguides has been investigated. Two types of waveguide structures rib and strip waveguides have been incorporated. The waveguide structures have designed to be 220 nm in height. Three different widths of (350, 450, 1000) nm were studied. At telecommunication wavelength of 1550 nm, Raman amplification was calculated to be about 65 dB in rib SOI waveguide. The obtained Raman amplification is the heist reported value at relatively low pump power. High Raman gain amplification in SOI nano-waveguides presents an important step towards integrated on-chip optoelectronic devices.

Keywords: Silicon photonics, Raman Amplifier, Optical Waveguides, Photonic Integrated circuits.

1. Introduction

Silicon photonics technology has made rapid progress and especially based on SOI (silicon on insulator) attracted great deal attention in recent years. It is seen as the Key enabling technology for integration of highly complex optical circuit [1]. CMOS (complementary-metal-oxide-semiconductor) technology is sufficiently advanced to fabricate virtually all silicon photonics components [2]. This offers an efficient technique to produce such devices with low cost, in addition to, the possibility to integrate Si photonics devices and electronic control functions together.Which opens a wide range of applications from telecommunication to chip-to-chip interconnects on chip communication, also possible applications in emerging areas such as biomedical and sensing devices [3]. Silicon with bandgap of 1.12 eV is transparent at wavelengths used typically for optical communication transmission rang from 1270 nm to 1625nm. It has many of important properties which make it a good material, such as wafers of silicon are incredibly pure and low defect density, also has high thermal conductivity, which is Useful for an active device substrate. Furthermore. Therefore it presents as an ideal platform for electronics and integrated optics focused on achieving active functionality, mostly light amplification, modulation and wavelength conversion in silicon waveguides.

The approach that has been investigated for light generation and amplification in our work is the Raman Effect. Raman amplification process used to amplify signal wave via Raman interaction with the pump power. The Raman Scattering process is inelastic scattering. When light beam propagates inside an optical silicon waveguide, the photons are scattered to new frequencies causing change in the energy of scattered photons. In this interaction, the energy may be losing leading to Stokes shift or gains energy resulting anti-Stokes shift. See Figure. 1A. Stimulated Raman scattering SRS process in silicon-on-insulator waveguides (SOI) is an attractive way of amplifying optical signal based on Raman Effect as shown in Figure (1).

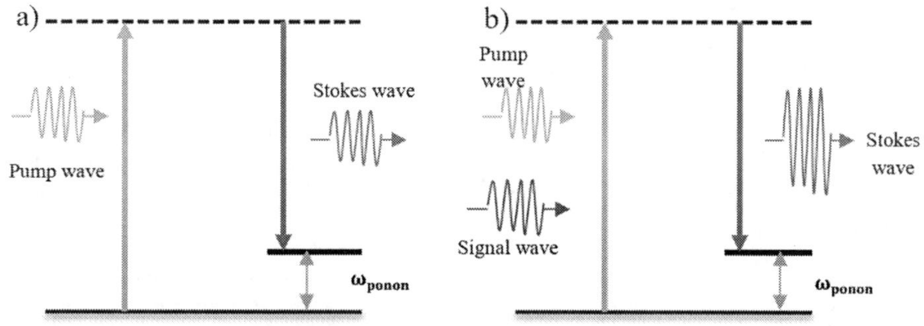

Figure 1. a) Schematic of Spontaneous Raman scattering phenomenon b) Stimulated Raman scattering phenomenon [4].

Small cross sections with low loss waveguides are required for the realization of high performance devices. Surface roughness produces strong scattering and high propagation loss Because of the high index contrast between the waveguide core and the cladding [5],[6]. For example, the losses of silicon waveguides tend to increase with reduction in dimension of cross-section. SOI waveguide fabrication processes by CMOS technology promise submicron cross section with low loss waveguides [5]. Raman net gain of 13 dB has been achieved by using a pulsed pump power at peak pulse power of 39 W with a pulse width of 3 picosecond and a repetition rate of 25 MHz in SOI waveguides with 2 μm^2 modal areas and a length of 25 mm [7]. A micrometer-sized laser was recently satisfied on a continuous-wave (CW) Raman Si laser based on a photonic-crystal high-quality (*Q*)-factor nanocavity, with a resonator size of 10 μm and an ultralow threshold of 1 μW, it generated from the most widely used silicon material in the electronics industry [8]. A Raman silicon laser has been achieved by using a high-quality (high-*Q)* nanocavity design at optically-pumped sub microwatt threshold [9].

In this work, Raman amplification in SOI Nano-waveguides has been investigated by numerically solved coupled equations, which describe the evaluation of the pump and Stokes wave via commercial software. Different dimensions of waveguide structures were relied of (3, 6, 8, 10, 12, 16, 20, 25, 30, 35, 40, 45, 50, 55, 60, 65, 70, 75, 80, 85) mm in length, widths of (350, 450, 1000) nm, and compatible height with CMOS process of 220 nm. Two types of waveguide structures rib and strip waveguides have been incorporated. The inner core (in the order of hundreds of nanometers) is designed with a high refractive-index silicon material surrounded by a low-refractive-index of silica cladding offering high light confinement.

2 Simulation Part

Mode profiles were calculated for all of our rib and strip waveguide structures by using FullWave commercial software. We suppose that our waveguide structures are covered with a SiO_2 layer to reduce the contamination effect. Refractive indices of 3.47 and 1.45 for Silicon

IOP Conf. Series: Materials Science and Engineering **757** (2020) 012022 doi:10.1088/1757-899X/757/1/012022

and SiO_2 were adopted in the mode profile calculations. The calculations show the possibility of coupling or exciting only the fundamental mode into waveguide structure of 350 nm in width, and coupling more than the fundamental mode in the waveguide widths of 450, 1000 nm. The results of our simulation have been shown in figures (2, 3), where only the fundamental mode has been presented. The effective refractive indices are shown in the figures calculated by the software. The right side shows the computed mode profile for both figures. The left side shows the rib and strip waveguides. The effective area of the core waveguide is defined as [10]:

$$A_{eff}\frac{[\iint I(x,y)dxdy]^2}{\iint I^2(x,y)dxdy} \tag{1}$$

Where I(x, y) is the transverse intensity of the fundamental mode (TE mode) of a waveguide (the integration is over the transversal plane). Considering only the calculation of the transverse mode profile for the fundamental mode, the effective core area could be estimated for each waveguide.

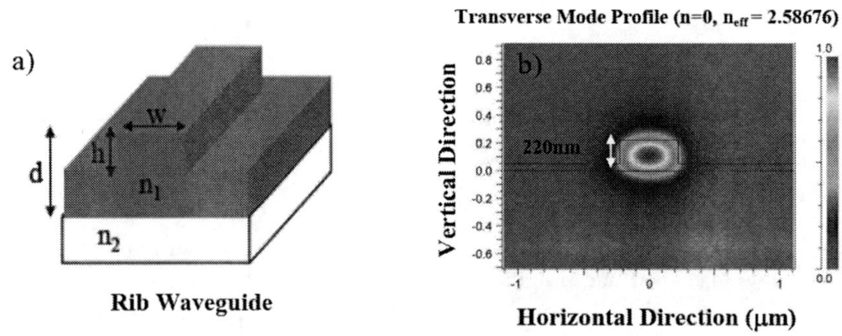

Figure 2. a) Basic structure of rib waveguide. b) computed transverse Mode profile of Stokes radiation in the SOI waveguide (always z is the propagation direction)

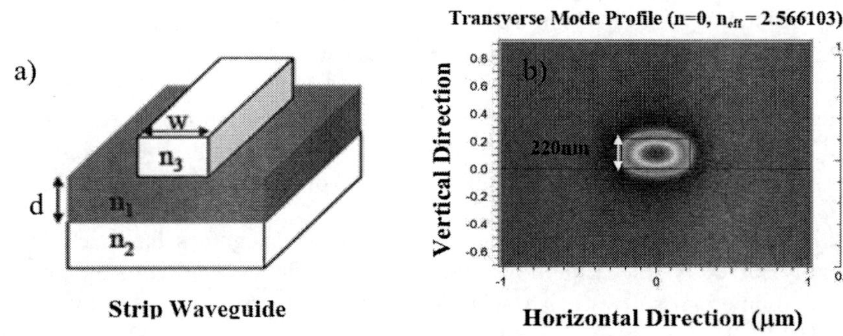

Figure 3. a) Basic structure of strip waveguide. b) Computed transverse Mode profile of Stokes radiation in the SOI waveguide.

IOP Conf. Series: Materials Science and Engineering **757** (2020) 012022 doi:10.1088/1757-899X/757/1/012022

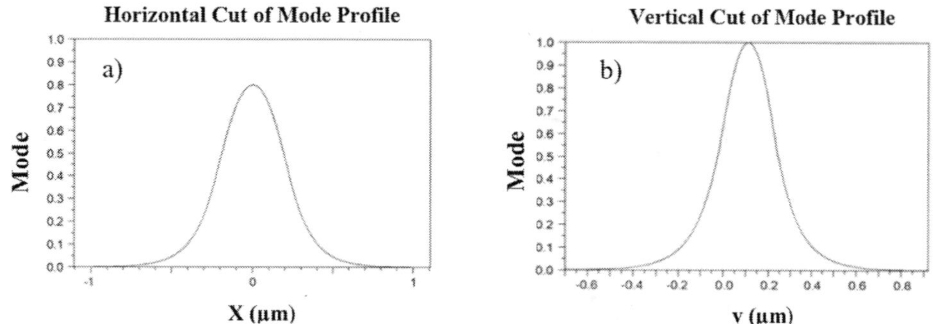

Figure 4. The computed transverse fundamental mode profile of Stokes radiation SiO_2-covered for rib waveguide width (W) of 450nm, a) horizontal cut and b) Vertical cut.

The effective core areas could be estimated for each waveguide, as shown in Figure – 4- and table -1- . Assuming that only the fundamental mode coupled out of the waveguide at stokes radiation, the value of the full width at half maximum (FWHM) of the outcoupled mode profile can be determined from the vertical and horizontal cut transverse mode profile, as shown in Figure - 4 - .

Table 1: The calculated effective core area (A_{eff}), effective refractive indices (n_{eff}), group index (n_{gr}) of waveguides with 350, 450 and 1000 nm in width, 220 nm in height for rib and strip waveguide.

Rib/Strip – width Slab – height	N_{eff} at 1550nm	A_{eff} (μm^2) at 1550nm	N_{gr} at 1550nm
350nm, 0nm	2.102149	0.113977	4.45
450nm, 0nm	2.386479	0.126323	4.31
1000nm, 0nm	2.746367	0.244736	3.7059
350nm, 50nm	2.276466	0.138204	4.01
450nm, 50nm	2.464671	0.112799	4.08
1000nm,50nm	2.7524445	0.252889	3.6896

3 Coupled Equation of Raman Propagation and Simulation Results

3.1 Mathematical Model.

A weak signal and strong pump are launched into the optical waveguide, leading to amplify the signal due to the Stimulated Raman Scattering process. The difference in the energy between the pump and signal is called vibration state ($\Omega = \upsilon_p - \upsilon_s$), which determines the frequency shift of the pump and Stokes ($\upsilon_{p,s}$) and shape of the Raman gain curve[4].

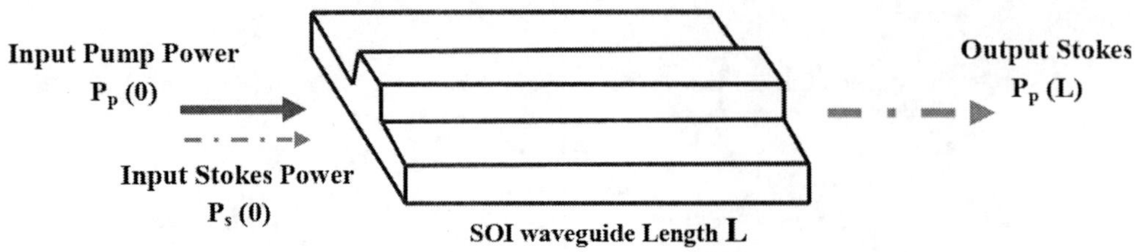

Figure 5.Schematic diagram of a basic silicon Raman amplifier [10].

Figure 5 shows the schematic of the SOI Raman amplifier setup we analyze in this paper. It consists of the SOI waveguide of length (L), into the left-hand side of which pump and Stokes power light source and right-hand side of which Stokes output.

We use a simple model for the SOI Raman amplifier; we describe the amplifier in terms of the longitudinally varying optical powers. By taking into account pump depletion due to stimulated Raman scattering [11]. We generalize the differential equations are describing the longitudinal distributions of the pump power (P_p) and the signal power (P_s) inside the SOI waveguide structure are given in [12],[13], [14].

$$\frac{1}{P_p}\frac{dP_p}{dL} = -[\alpha_p + \varphi\, \lambda_p^2\, N_{eff}[z] - \left(\frac{\beta}{A_{eff}}\right) P_p \tag{2}$$

$$\frac{1}{P_s}\frac{dP_s}{dL} = -[\alpha_s + \varphi\, \lambda_s^2\, N_{eff}[z] - \left(\frac{g_s - 2\beta}{A_{eff}}\right) P_s \tag{3}$$

Where ($\alpha_{p,\,s}$) is the linear absorption coefficient for pump and signal wavelengths, (β) is the two photon absorption coefficient, z is the longitudinal coordinate system, and α represents the linear Waveguide losses, (g_s) is the Raman gain coefficient. (A_{eff}) is the effective area of the waveguide, calculated to be 0.11, 0.12, and 0.24 μm^2 for strip waveguides and 0.1, 0.11, and 0.25 μm^2 for rib waveguides, respectively. Where (ΔN) is the density of electron hole pairs given by[15]:

$$\Delta N = \beta * (\frac{P}{A_{eff}})^2 * \frac{(\tau_{eff})}{2hv} \tag{4}$$

Where v_p is the frequency of pump beam, (h) is Planck's constant and (hv_p) is the photon energy, and (τ_{eff}) is an effective recombination lifetime for the free carries.

From Equation (2) the on-off (pump-on-pump-off) Raman gain is given in[16]:

$$G_{on\text{-}off} = \frac{Ps(L=Z)}{Ps(0)} \tag{5}$$

Raman net gain for considered as in [17]:

$$G_{net} = \frac{Ps[L=Z]}{Ps[0]Exp(-L\, \alpha_S)} \tag{6}$$

Where P_s [L] is the signal output power, which is obtained in equation (3), L is the waveguide length.

3.2 Simulation parameters

We base the simulations in this section on the parameters given as: Raman-gain of g = 70 cm/GW and TPA coefficients $\beta_{p,s}$ = 0.5 cm/GW [18], pump and Stokes wavelengths of λ_p = 1450nm and λ_p = 1550 nm, respectively, and effective areas taken from Table 1.The amplifier has a various lengths and widths.

Table2: linear loss ($\alpha_{p,\,s}$) in the Silicon waveguide structure.

linear losses of $\alpha_{p,\,s}$ (dB/cm)	Reference
0.5	[19]
1	[10], [13]
2.8	[20]
3.6	[21], [22]
4	Estimated
5	Estimated

3.3 Results and discussion

The most important factor in determining the Raman gain in SOI waveguides is the effective carrier lifetime (τ_{eff}), which plays an important role reaching maximum transmittance CW power into the silicon waveguide structures. It is clearly shown in figure -6- that maximum transmitted CW power is limited by nonlinear losses, where shorter carrier lifetime results smaller carrier density leading to increase Raman gain. The effective carrier lifetime in figure (6) are reported by various research groups (see Table (3)). As shown in the calculation results, the output power starts dropping at input pump power exceeding 1 W. Due to the fast free carrier recombination mechanisms and smaller diffusion time at sub-microns silicon waveguides size, the effective carrier lifetime (τ_{eff}) becomes short, which is promising to have high Raman gain.

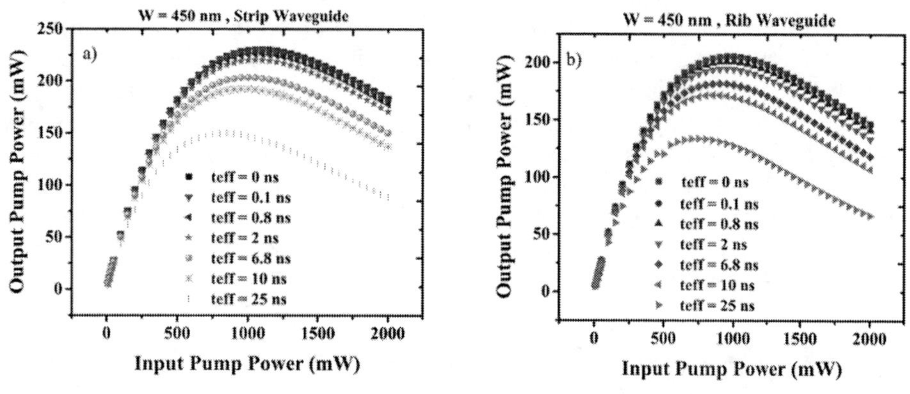

Figure 6. Output pump power characteristics for silicon Raman amplifiers with various effective carrier lifetimes for a) Strip, b) Rib waveguides.

Table 3: Effective (Carrier) lifetime in Si Waveguide.

Carrier Lifetime (ns)	Reference

IOP Conf. Series: Materials Science and Engineering **757** (2020) 012022 doi:10.1088/1757-899X/757/1/012022

0.1	[10]
0.8	[15]
2	[15]
6.8	[6]
10	[23]
25	[12]

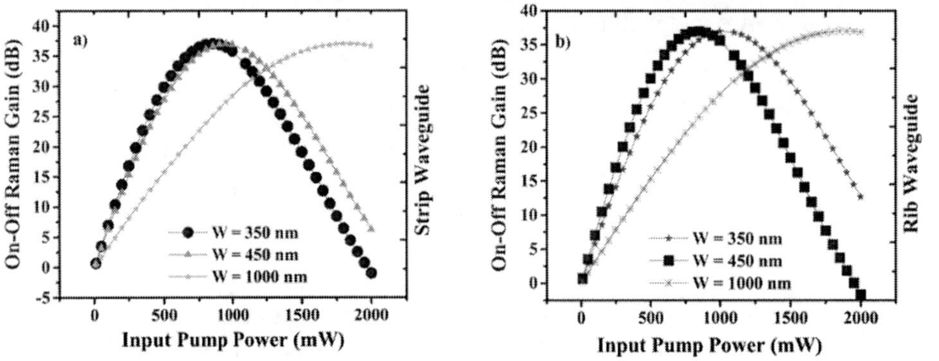

Figure 7.on-off CW Raman gain versus input pump power of the SOI waveguides with various widths of (350, 450, and 1000) nm and length of 3 mm, a) Strip waveguide ,b) Rib waveguide

Figure -7- shows the on-off Raman gain as a function of the input pump power of a passive strip and rib waveguides at various widths of (350, 450 and 1000) nm with a fixed length of 3 mm. At these dimensions, maximum pump-on-pump-off Raman gain of about 36.9 dB was obtained for all our SOI structures but at different input pump powers. For the waveguides of (W = 350 and 450) nm, maximum gain was of about 0.85 W input pump power. While for the waveguide of W = 1000 nm, input pump power of about 1.8 W was required to achieve maximum gain. It found that there is a tolerance of about 1 W input power between the waveguides of (W= 350, 450) nm and the waveguide of W=1000 nm starting Raman gain reduction. Also figure 7 shows that the Raman gain increases sharply with increasing the pump power for the waveguides of (350 and 450) nm in widths, because of the relatively small waveguides effective area. Then drops dramatically at pump power of about 0.9 W, nonlinear losses become more significant with increasing the coupled pump power. While Raman gain for the waveguide of 1000 nm is raising linearly with increasing the pump power reaching maximum Raman gain, then drops at pump power of about 2 W. These results improve the high dependence of waveguide dimensions on the nonlinear losses due to two photon absorption induced free carrier absorption in silicon waveguides as well as Raman amplification. The equivalent in the values of Raman gain between the strip and rib waveguides at the same length is due to the agreement of the effective area, where the mode distributions are nearly the same (see figures 2 and 3).

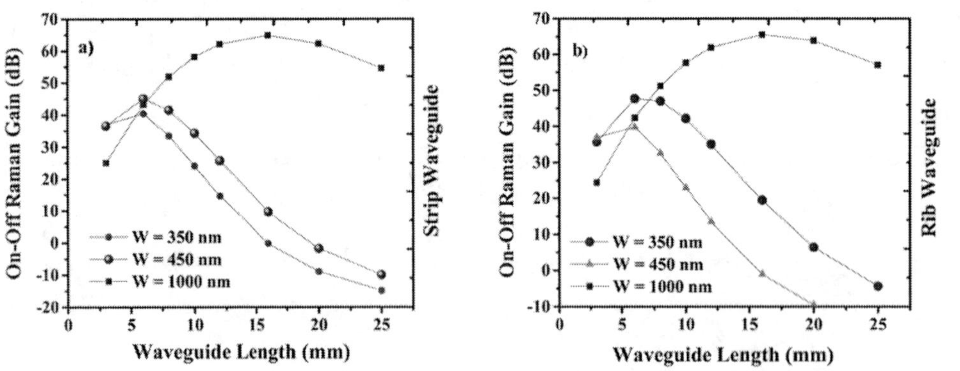

Figure 8. on-off Raman gain versus various waveguide lengths at 850 mW pump power and different waveguide widths of 350, 450 and 1000 nm., a) Strip waveguides, b) Rib waveguides,

Figure -8- shows the calculated on-off Raman gain at various waveguide length, which computed by solving the mathematical mode equations for strip and rib waveguides of (350, 450, 1000 nm) in widths. The overall on-off Raman gain increases to a maximum value with the optimum lengths and then drops. It is obvious in figure 8 that short lengths in the range of several millimeters of (W= 350 and 450) nm SOI waveguides are enough to satisfy the maximum value of the on-off Raman gain. Maximum on-off Raman gain at pump power of 850 mW and optimum waveguide length of 6 mm was calculated to be (39.8, 40.4) dB in a strip waveguides of 350 and 450 nm widths respectively. While these values were found to (45, 47.7) dB for the rib waveguides of 350 and 450 nm widths, respectively see (figure 8 a, b). 12 to 16 mm lengths are the optimum waveguide lengths (L) to obtained maximum Raman gain for a waveguide 1000 nm in widths at 850 mW pump power. Maximum on-off Raman gains of strip and rib waveguides with (W= 1000 nm and L=16 were calculated to be (65.5 and 64.9), respectively. These results show the tradeoff between the waveguide effective area and waveguide length. The pump power depletes by lengthening the relatively small waveguides due to the linear loss because of the sidewall roughness. In addition to linear loss due to generated free carriers.

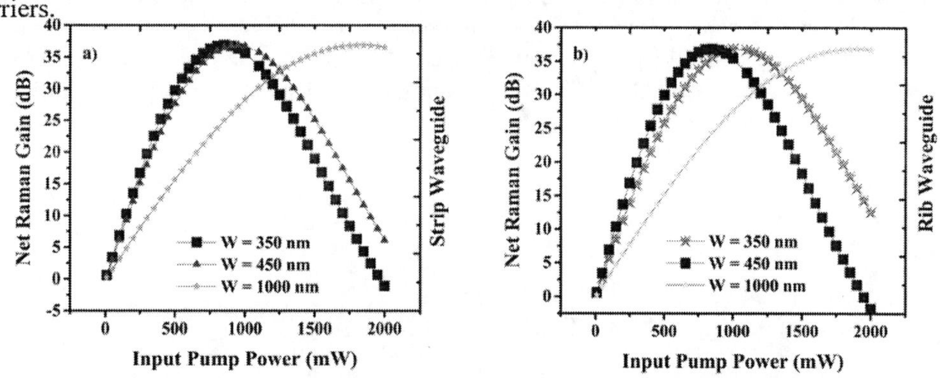

Figure 9. Raman net gain as a function of the input pump power at various waveguides of (W = 350, 450 and 1000) nm, a) Strip waveguides. b) Rib waveguides.

Since the Raman net gain is calculated without taking in considering the linear losses. Figure 9 shows the calculated Raman net gain for our previously considered SOI waveguides. Laser action could be achieved by introducing optical feedback, in addition to high Raman net gain [10].

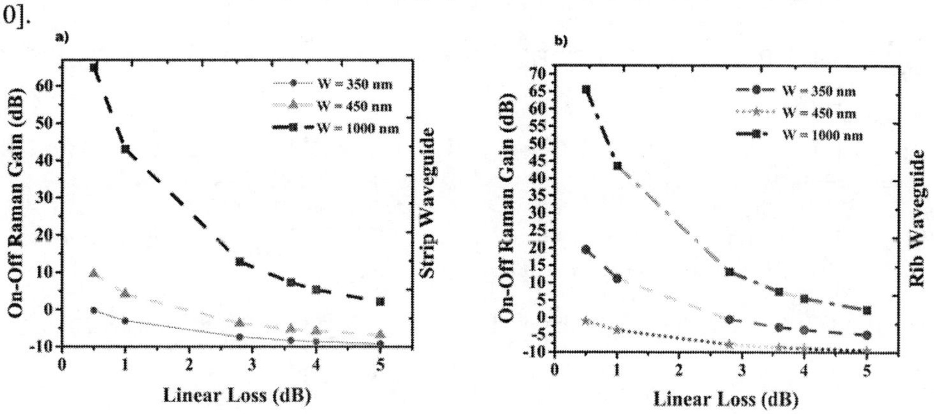

Figure 10.on-off Raman gain as a function to the linear loss at a) Strip. b) Rib waveguide widths W = 350, 450 and 1000 nm.

Our results in figure (10), shows that the Raman gain decreases by increasing the linear losses ($\alpha_{p,\ s}$). From hand, the maximum possible gain can be increased equally well by decreasing the linear losses (for example, by waveguide sidewall smoothing to reduce surface scattering [24]).

In the other hand, scaling down the waveguide dimensions has a significant effect to reduce the nonlinear absorption in SOI waveguides. The previous results show that there is a tradeoff between the linear losses and nonlinear loss into SOI waveguides leading to an increase in Raman gain.

4.Conclusion

In summary, Numerical analysis using beam propagation method was investigated to calculate CW Raman amplification in sub-micron SOI nano-waveguides at telecommunication wavelength. Two different waveguide structures (Strip and Rib) were simulated. Raman amplification was calculated to be 65.5 dB in rib SOI waveguide of 1000 nm in width and 16 mm in length. The obtained Raman amplification is the heist reported value at relatively low pump power 0.85 W. Silicon photonics technology is an important step towards integrated on-chip optoelectronic devices. Farther development, silicon photonics components and circuits on SOI platform for example, ring resonator, photonic crystal as well as arrayed waveguide gratings.

References

[1] Chen L and Lipson M 2009 *Opt. Express* **17** 7901–6

[2] Liu L, Kumar R, Huybrechts K, Spuesens T, Roelkens G, Geluk E-J, De Vries T, Regreny P, Van Thourhout D and Baets R 2010 *Nat. Photonics* **4** 182

[3] Meister S, Al-Saadi A, Franke B A, Mahdi S, Kuhlow B, Voigt K, Tillack B, Richter H H, Zimmermann L and Ksianzou V 2010 *Silicon Photonics V* vol 7606 (International Society for Optics and Photonics) p 760616

[4] Kumbhare V R 2015 *Int. J. Comput. Appl.* **975** 8887

[5] Jalali B, Raghunathan V, Dimitropoulos D and Boyraz O 2006 *IEEE J. Sel. Top. Quantum Electron.* **12** 412–21

[6] Liu A, Rong H, Jones R, Cohen O, Hak D and Paniccia M 2006 *J. Light. Technol.* **24** 1440

[7] Raghunathan V, Boyraz O and Jalali B 2005 *Conference on Lasers and Electro-Optics* (Optical Society of America) p CMU1

[8] Takahashi Y, Inui Y, Chihara M, Asano T, Terawaki R and Noda S 2013 *Nature* **498** 470

[9] Yamashita D, Takahashi Y, Kurihara J, Asano T and Noda S 2018 *Phys. Rev. Appl.* **10** 24039

[10] Krause M, Renner H and Brinkmeyer E 2004 *Opt. Express* **12** 5703–10

[11] AuYeung J and Yariv A 1979 *JOSA* **69** 803–7

[12] Liu A, Rong H, Paniccia M, Cohen O and Hak D 2004 *Opt. Express* **12** 4261–8

[13] Claps R, Raghunathan V, Dimitropoulos D and Jalali B 2004 *Opt. Express* **12** 2774–80

[14] Cutolo A, Iodice M, Spirito P and Zeni L 1997 *J. Light. Technol.* **15** 505–18

[15] Liang T-K, Nunes L-R, Tsang H-K and Tsuchiya M 2005 *IEICE Electron. Express* **2** 440–5

[16] Rukhlenko I D, Dissanayake C, Premaratne M and Agrawal G P 2009 *Opt. Express* **17** 5807–14

[17] Herziger G, Lindner H and Weber H 1964 *Zeitschrift für Angew. Phys.* **17** 67–8

[18] Rong H, Liu A, Nicolaescu R, Paniccia M, Cohen O and Hak D 2004 *Appl. Phys. Lett.* **85** 2196–8

[19] Rickman A G and Reed G T 1994 *IEE Proceedings-Optoelectronics* **141** 391–3

[20] Claps R, Dimitropoulos D, Han Y and Jalali B 2002 *Opt. Express* **10** 1305–13

[21] Espinola R L, Dadap J I, Osgood R M, McNab S J and Vlasov Y A 2004 *Opt. Express* **12** 3713–8

[22] Dadap J I, Espinola R L, Osgood R M, McNab S J and Vlasov Y A 2004 *Opt. Lett.* **29** 2755–7

[23] Rong H, Liu A, Jones R, Cohen O, Hak D, Nicolaescu R, Fang A and Paniccia M 2005 *Nature* **433** 292

[24] Sparacin D K, Spector S J and Kimerling L C 2005 *J. Light. Technol.* **23** 2455

Diagnosis of copper plasma by laser induced breakdown spectroscopy

*N.M .Saadoon[1], N.M.Hadi [2] ,S. H.Sabeeh [1]

[1]Applied science department, University of technology, Iraq

[2] laser optoelectronic center, Ministry of sciences and technology, Iraq

*Corresponding author email: noor_musawi@yahoo.com

Abstract .In this paper influence of laser energy on the emission spectrum of laser copper-interaction and plasma parameter for plasma Copper by used a first harmonic generated wavelength of Nd:YAG exaction source. The metal plasma that had detection by HR4000 detection system (ocean optics) was use to different excited. plasma temperature via the Boltzmann plot diagram with measuring the electron density number by McWhirter CRITERION. Values of measuring (N_e), (T_e) found between (61178776 K°) and (N_e) is ($1.77*10^{15}$, $2.01*10^{15}$ cm^{-3}) at 1064nm.
Keywords: LIBS; Copper plasma; plasma parameter, electron number density, self-absorption

1. Introduction

Laser induced breakdown spectroscopy (LIBS) is a technique, which was first reported in early 1960s and then reviewed by several researchers. This sensitive technique is based on an optical detection for certain atomic and molecular species by monitoring their emission signals from the laser induced plasma. It provides a useful tool to determine the chemical and physical properties of a wide range of materials including metals, liquids, aerosols, plastics, minerals and biological tissues, etc [1-3]. It is a simple technique as compared with many other types of elemental analysis because of its straightforward experimental set-up. In this technique a pulsed laser is required to generate microplasma on the target surface. Elemental analysis is accomplished by study the emission from plasma plume. The characteristics of laser induced plasma depend upon several parameters characterizing the features of the target, properties of the ambient medium, laser wavelength and pulse duration etc. The first direct spectral analysis made by LIBS can be attributed to RUNG et al [4-5]. In 1980s, this technique was increasingly applied as an analytical tool in research [6].Copper is the most widely used metal for various applications of plasmonic materials due to its high electrical conductivity and optical properties.[7,8] The emission characteristics and plasma parameters of Copper are of great importance not only in the laboratory plasma diagnostics but also for a number of technological applications, such as for nano-particles, cluster production, contacts of low-voltage switching devices, and as metal films.[9–10] Shuaibov etal.[11] investigated the time-averaged spectra (200–600 nm) of Ag produced by the 1064 nm laser irradiation in vacuum (3–5 torr) at a distances of (1-7mm) from the target surface. Chuchmann and Shuaibov[12] reported the temporal dependence of the populations of excited atomic states of silver and copper at the (1-7mm) distance from the target surface. Roberts et al.[13] reported the femtosecond laser (130 fs) ablation of silver foil using the single and double pulse configurations. Rashid et al.[14] reported the comparative study on the effect of the inter-pulse delay and the ratio of the laser pulse energies on the silver emission line intensities, enhancement for collinear and orthogonal pre-ablation dual pulse configurations using the (1064nm) and (532 nm) of Nd:YAG laser. The objective of the present work was to characterize the laser induced Copper plasma in term of spectral line intensities and plasma parameter (n_e and T_e) under different laser power energy using LIBS technique.

2. Experimental and method.

1. Nd:YAG (Quantel Brilliant) which use as laser source, with pulse duration of 6 ns , 10 Hz repeated rate and laser energy can deliver until 850 mJ per pulsed at 1064 wavelength.
2. The exaction source energy has changing used flash lamp Q-switch delay through laser controller and measuring in intensity joule meter (Nova-Quantel P/niz01507).
3. The laser pulsed to the point above the surface of copper by convex Lens of 10 cm focal length. The copper foil has moving on 3-D target holder and moved to remove the no uniform pitting of simple surface.
4. The distance between the focusing lens and the sample was kept less than the focal lengths of the lens to forbidding any discharging environments before the target surface. The spectrum was taking in average 10 point of malty pulsed under setup parameters.
5. The spectrum generated from plum have been collect in a fiber optics (High –OH, core diameter: 600_{μ}m) own a collimating lens (45° field of view) put in the 45◦ angle in direction of the source pulsed. That fiber was joined with (ocean optic spectroscopy HR4000).
6. The emission sign was deal with it by tracing the black single detector through OOH software the data obtained by spectrometer were allowed saved on PC through the HR4000 software for diagnosis method.

3. Result of Emission Spectrum

Plasma produce by a interaction power high with pulsed expand normally on target surface according to the shocks wave [14]. We have produced copper plasma by used Nd: YAG laser with the fundamental wavelength 1.064nm with different energy of laser pulsed between (20-30mj) and 6ns pulsed duration has focusing on the copper foil surface. The spectrum of the plum produce around the Cu surface is recording at fixed distance (5mm) and different laser energy. In this experiment we used a reflected mirror as a focused with angle 45◦. The plasma plum emission was recorded as a definition of laser intensity to the Cu surface metal. The ground state configuration of copper is $3d4P^2P_{5/2}$ his yield many state Fig (1) show the spectrum emission cover the window in 300-400 nm, it noted that most strong line belong to the natural copper the line at 324.75 and 327.39nm are the strong line recognized $4S3d4S^2S_{1/2to}3d4P^2P_{3/2}$ and $3d4S^2S_{1/2}$ to $3d4P^2P_{1/2}$, consequently.

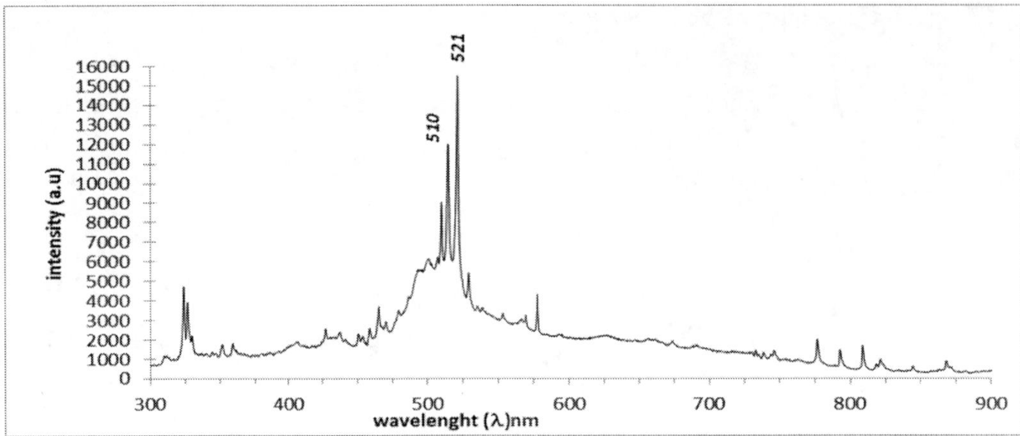

Figure 1: Spectrum of Cu foil from (300-900nm) generated at 1064nm and 20mj laser enenrgy

The Figure (2), show the strong line of emission area in 400-600nm. natural copper emissions dominate line belong to the cuI. The strong line at 521.28nm as $3d4d^2D_{5/2}$ to$4p^2p_{3/2}$ and 515.32nm identified as $3d4d^2D_{3/2}$ to $4p^2p_{1/2}$.

IOP Conf. Series: Materials Science and Engineering **757** (2020) 012023 doi:10.1088/1757-899X/757/1/012023

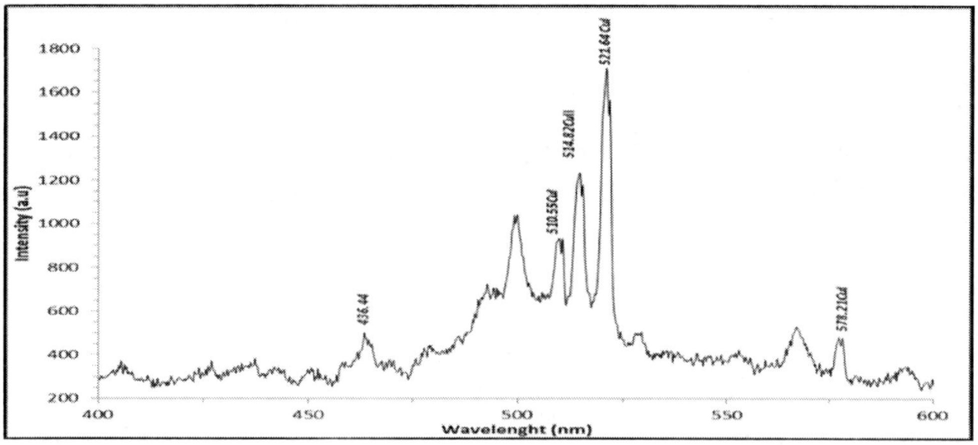

Figure 2:Spectrum of Cu of the region (400-600nm) generated by 1064nmlaser at 20 mj laser energy

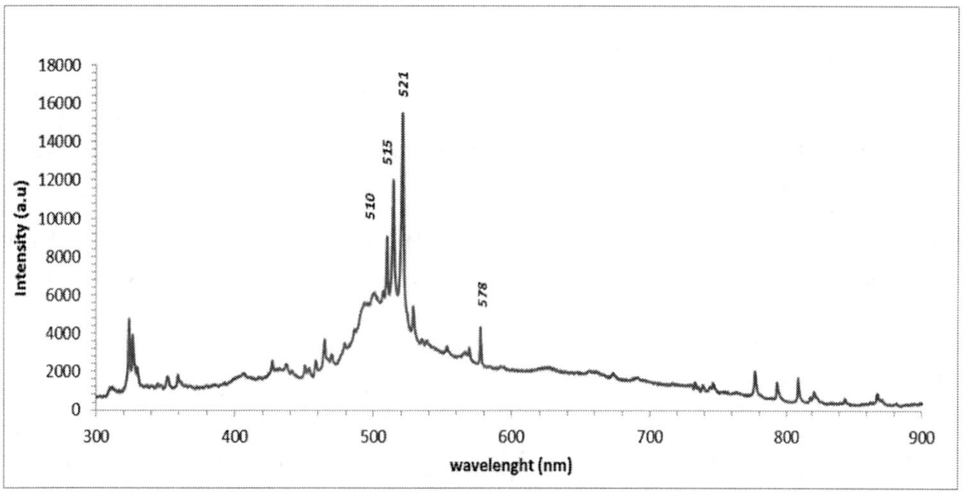

Figure ٣: Spectrum of Cu foil from (300-900nm) generated at 1064nm and 30mj laser enenrgy

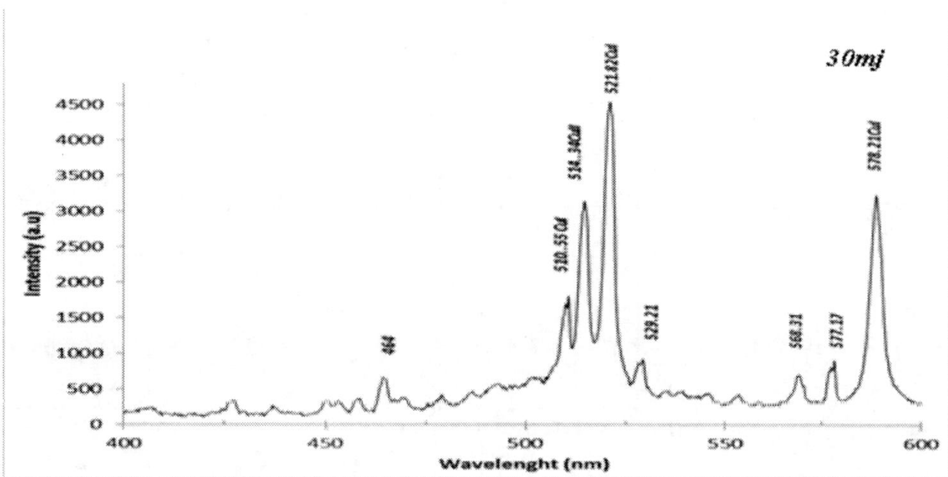

Figure 4: plum emission of target cover the window (400-600nm) generated by 1064nmlaser at 30 mj laser energy

4. Result of Determine the plasma parameter

4.1.Plasma temperature

The plot Boltzmann equation with the Thermodynamic Local Equilibrium (LTE) assumption was used for the calculation the Te and Ne, the Boltzmann method use to the determine of that figurative is give in [13,14]:

$$\ln kkk\ ^{\lambda\lambda}gg^{\underline{kkkk}}\ kk = \iota\iota\iota^{NN}\_\_UU(^{(}TT^{TT})) - KKTT^{EE\ \underline{kk}} \ldots\ldots\ldots\ldots\ldots.1$$

the m, n, I_{mn}, g_m, and A_{mn} are an upper state, lower state , the line intensity, level degeneracy, with the transition probability, consequently n(t), u(t), E_m, K, with T are the number of plum, the partition function, the upper energy level, the Boltzmann constant, and the electrons temperature, consequently. The parameters of selection spectrum line in this work had been used from NIST database [15, 16] they listed in table (1). the electrons temperatures can be calculated by using the slope of line which have relation to the selection energy lines driven above the plum species. The spectroscopic parameters like g_m, A_{mn}, and E_m can be retrieved from the available database.

Typically Boltzmann plot at 1064 nm wavelength and 20mj laser energy is dhow in fig 3 in, a linear fitting with the data are summarized. from the diagram Te is 6113K. The profile of spectral line from LIBS technique has various broadening effects like Doppler broadening, Resonance broadening, and Stark broadening [16]. The Stark broadening was the observed broadening in this methods .

Table 1. Spectroscopy parameter of the singly / second ionized and Cu line data take from reference [15].

Atom/ ion	λ(nm)	g_j	Aji($^{s-1}$)	Ej (ev)	Transition
Cu I	510.55	4	$2.0*10^6$	3.82	$4S^2\ ^2D4P^2P$
Cu I	515.32	4	$6.6*10^7$	6.191	$4P^2P4d^2D$
Cu I	521.82	6	$7.5*10^7$	6.192	$4P^2P4d^2D$
CuI	578.21	4	$1.65*10^6$	3.76	$4S^22D4P^2P$
Cul	324.31	4	$1.37*10^8$	33.78	3d4p2p3/2

Where E_k=Upper level energy, A = Transition probability

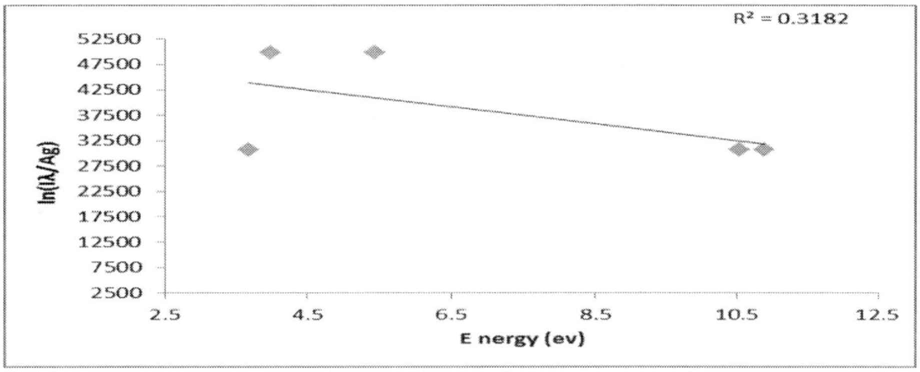

Figure5: Boltzmann plot for 5 neutral spectral line in 20mj by using 1064nm for Nd:YAG laser

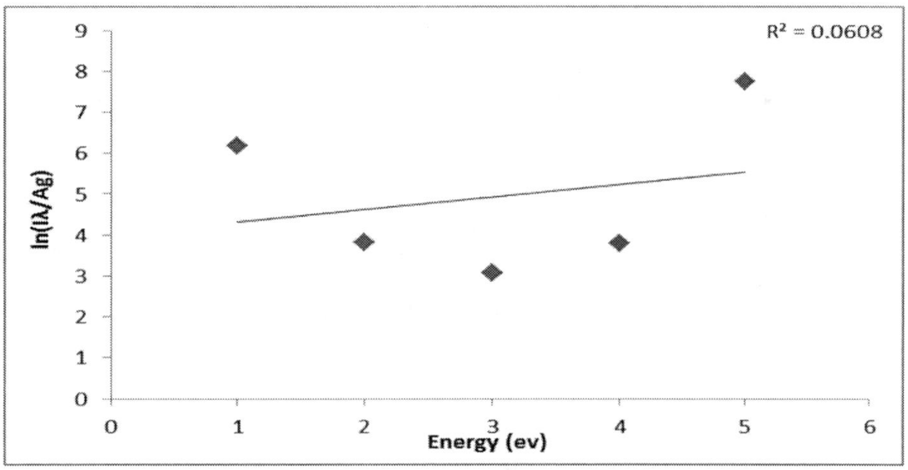

Figure 6: Boltzmann plot for 5 neutral spectral line in 30mj by using 1064nm for Nd:YAG laser

The plasma temperature was determined for verging value energy laser by Nd:YAG laser with 1064 nm as wavelength. It is noticed with increasing the energy laser the intensity and width of spectrum line increased. The electron temperature increased between 6113 to 7776.4 k also the electron number density was calculated for different energy laser. As show in result note that END varying between $1.77*10^{15}$to$2.01*10^{15}cm^{-3}$.It was observed increasing in electron temperatures and END with increasing the E_i it is came from absorption ,reflection of laser photon by plum. With increase the exaction flounce, more species, ion and electron are generated the laser pulsed interacts with species. resulting in further heat and ionisation and respectively increasing consumption of incoming laser pulsed [17,14].

4.2. Local thermodynamic equilibrium

Emission spectroscopy used to measurement of the Te and Ne necessitates optically thin spectrum line .the self-absorption depend on the oscillator strength, level energy degenerate, broader enviormante and also on the plum parameter. The target plum is dominate to by optical thin in case of the self – absorption a strongest lines appears to a dipping at the fixed frequently [18].Here we didn't find any penetration at the central frequency of plume line. The condition that the atomic state should be populated and depopulated predominantly by electron collision, rather than by radiation required an electron density is given by McWhirter equation , which is the condition for reached to smallest Ne to ensure the warrant of the LTE [19,20].

$$N_e \geq 1.6*10^{12} T^{1/2} (\Delta E)^3 \quad (2)$$

When t (K) as plum temperature and ΔE (eV) it energy different between levels must been with local thermodynamic equilibrium. In our experiment the electron number density range satisfy the local thermodynamic equilibrium .when the laser focused on the target the ablated Cu surface take a placed with according to the density gradient plasma fasting expansion[11].

5. Conclusion

Nd:YAG lasers in the 1064nm wavelength to diagnosis the copper plasma. ET and END was measured on the axis direction of plum. It observed was noted that the plasma temperature and END increased with increasing laser energy

6. References

[1] A. W. Miziolek, V. Palleschi, and I. Schechter 2006 Laser-Induced Breakdown Spectroscopy (LIBS), Fundamentals and Applications (Cambridge University Press, Cambridge).

[2] R. Noll, Laser-Induced Breakdown Spectroscopy, Fundamentals and Applications 2012 Springer, New York.

[3] G. Cristoforetti, A. D. Giacomo, M. Dell'Aglio, S. Legnaioli, E. Tognoni, V. Palleschi, *et al* 2010 Local thermodynamic equilibrium in laser-induced breakdown spectroscopy: Beyond the Mc Whirter criterion. Spectrochim. Acta B 65 86–95

[4] D. A. Cremers, and L. J. Radziemski, Handbook of Laser-Induced Breakdown Spectroscopy 2006 John Wiley & Sons, Haboken.

[5] F.A,A.C ,M.L.Higuera,"Laser- Induced Breakdown Spectroscopy : Fundamental ,Application , and Challenges 2012 ,Int scholarly research network 19

[6] R. Brennetot, J. L. Lacour, E. Vors, P. Fichet, D. Vailhen, S. Maurice and A. Rivoallan, 2002 Laser induced plasma Spectroscopy and Applications, OSA Technical Digest , Trends in Optics and Photonics 81- 9

[7] V.K. Unnikrishnan, K. Alti, V.B. Kartha, C. Santhosh, G.P. Gupta and B.M. Suri, Pramana2010 J. Phys.74 983.

[8] C.Y. Ho, Y.H. Tsai, C.S. Chen and M.Y. Wen, Curr.2011 Appl. Phys. 11 S301 .

[9] P. Stavropoulos, C. Palagas, G.N. Angelopoulos, D.N. Papamantellos and S. Couris 2004 , Spectrochemica. Acta B, 59 1885 .

[10] J.L.H. Chau, C. Y. Chen, M. C. Yang, K. L. Lin, S. Sato, T. Nakamura,C. C. Yang and C.-W. Cheng 2011 Mater. Lett 65 804.

[11] J. P. Singh, and S. N. Thakur 2007 Laser Induced Breakdown Spectroscopy (Elsevier, Amsterdam.

[12] M. Hanif, M. Salik,M. A. Baig 2012 Diagnostic Study of Nickel Plasma Produced by Fundamental (1064 nm) and Second Harmonics (532 nm) of an Nd: YAG Laser. J. of Modern Physics 3 1663-1669

[13] A. Ciucci, M. Corsi, V. Palleschi, S. Rastelli, A. Salvetti, and E. Tognoni 1999 New Procedure for quantitative element analysis by laser-induced plasma spectroscopy, Appl. Spectrosc. **53** 960–964

[14] M. Hanif, M. Salik and M. A. Baig2011 Quantitative Studies of Copper Plasma Using Laser Induced Breakdown Spectroscopy, J. of Optics and Lasers in Engineering, 49. 12 14561461.

[15] NIST Atomic Spectra Database, Kurucz Output Atomic Spectra Line Database from R.L.

[16] Kurucz's Kurucz atomic spectral line database, http://www.pmp.unihannover.de/cgibin/ssi/test/kurucz/sekur/sekur.html.

[17] H. R. Griem 1974 Spectral Line Broadening by Plasmas. New York and London: A Subsidiary of Harcourt Brace Jovanovich ,Academic Press, New York.

[18] C. Aragon, J. Bengoechea and J. A. Aguilera 2001Influence of the Optical Depth on Spectral Line Emission from Laser-Induced Plasmas," Spectrochemica Acta Part B 56 2 619628.

[19] A.M.E. Sherbini, H. Hegazy, T.M.E. Sherbini 2006 Measurement of electron density utilizing the Hα-line from laser produced plasma in air", Spectrochim Acta B 61 532–539 .

[20] R. W. P. McWhirter 1965 Plasma Diagnostic Techniques, Academic Press, New York.

Structural and Optical properties of CdTe:CuThin Films by Pulsed Laser Deposition technique

N Naeema[1], A Kudher[2] and Ghuson H Mohammed[3]
[1,2]Department of physics, College of Science, University of Wasit,Iraq.
[3]Department of physics, College of Science, University of Baghdad, Iraq.
kutpc19@yahoo.com

Abstract. Polycrystalline cadimumtelluride : copper composite thin films were deposited on a glass substrate by pulsed laser deposition technique(PLD).Using Nd:YAG with 1064 nm wavelengthconstant energy 600 mJ , repetition rate (6) Hz and No. of pulse (150) for each sample.The pure of bulk (CdTe) and Mixture of bulk (CdTe:Cu) percentages (x=0.0, 0.3, 0.5and 0.7)wt% were prepared . The structural, optical and morphological properties were inspected for the produced thin films.The energy band gap decreasing from 2.4 to 1.7eV with increase of Cu concentration. XRD patterns showed the domination of polycrystalline nature of cubic structure but in adding 0.5 , and 0.7% Cu the nature of thin films became amorphous. AFM results demonstrated the increase of particle size from 73.66 nm to 76.67 nm at percentage of 0.5% Cu. The optical properties has been including the study of measuring the spectrum of transmittance, absorbance, Absorption coefficient (α), Optical Energy Gap (Eg), Reflectance (R), Refractive index, Extinction Coefficient, Dielectric Constant and Optical Conductivity (σ) of the wavelength range (320-1100) nm. The results have been show that the films have good transmittance for all films around 60-94% in the visible region and NIR of the spectrum that we can use them as optical windows in solar cells.

Keywords: CdTe thin films, Cadmium Telluride- Copper, PLD, nanostructure.

1. Introduction

Inrecent years the exceptional physical and chemical properties possessed by II–VI semiconductors have paying attention of great deal of interest. Because of high absorption coefficient greater than 5 $x10^5$ /cmand direct energy gap (~1.5 eV) that are well-suited to the solar spectrum. Therefore, cadmium telluride (CdTe) is considered to be one of the ideal materials for solar cells [1,2]. TherforeCdTewas sutible material for several applications such as: radiation detectors [3], photovoltaic cells[4], high act electro-optic modulators and infrared detection[5]. CdTe shows both n and p types of conductivity, and it can exhibit a semi- insulating state as well, which makes diode technology and field effect transistors probable,. Device grade CdTethin films can be formed in both homojunction and heterojunction configuration [6,7].CdTe in the shape of thin film can be polycrystalline or nanocrystalline,which can contribute towards the change of its characterstics[8]CdTe NPs can be synthesized by many techniques, like thermal evaporation, chemical deposition, sol-gel,sputtering, and pulsed laser deposition (PLD)[9].To characterize obtained films crystalline By X-ray diffraction by using Bragg's law it can be deduce the d_{hkl} spacing between crystals planes.

$n\lambda = 2d_{hkl}\, sin(\theta)$--------------------(1)

where θ is diffraction angle and λ is the X-ray wavelengthfor kα. transition from Cu target (1.5406 Å). To calculate crystallite size one can used Scherrer equation formula [10].

$G.S = \frac{0.9\lambda}{\text{FWHM}.\cos(\theta)}$ --------------------(2)

FWHM is full width at half maximum.A (double beam UV-Vis-NIR Metertech) spectrophotometer was used to measure the absorbance of the thin films synthesized by a pulsed laser deposition of the targets at various Cu concentration of (x=0.0 , 0.3, 0.5, and 0.7) wt.% in the range (190 – 1100 nm). The background correction was taken for each scan.

The equation 3.was used to calculate the refractive index [11]:

$n = [(4R/(R-1)^2)-k^2]-(R+1)/(R-1)$ .... (3)

where n is the refractive index, R is the reflectance.The extinction coefficient (k) is related to the exponent decay of the wave as it passes through the medium and is defined to be :

$k=\alpha\lambda/4\pi$.. (4)

where λ is the wavelength of the incident radiation.αOptical absorption coefficient In its medium it is possible to measure the material's potential to absorb the radiation at a given wavelength and in an inverted measure of distance(cm^{-1}).

An absorbing medium characterized by a complex dielectric constant:

$\varepsilon = \varepsilon_r - i\varepsilon_i = 1+4\pi\varphi$ .... (5)

where ε_r is the real part of dielectric constant, ε_i is the imaginary part of dielectric constant and φ is the sum of polarizabilities resulting from free carriers, so that[12]:

$\varepsilon_r= n^2-k^2$ .... (6)

$\varepsilon_i= 2nk$ .... (7)

The morphological characteristics of the CdTe, and CdTe:Cu thin films wereone using atomic force microscope (AFM) type Angstrom advanced incorp. SPM-AA3000 with contact mode to observe the ultrafine structure (nanostructure) of the thin film with high resolution accurately 0.26nm lateral and 0.1nm vertical. AFM examines the minerals surface, determines morphology and measures the size of nanoparticles. Morphology of the CdTe, and CdTe:Cu thin film.AFM image used in PLD experiment to measure the size of nanoparticles of the CdTe, and CdTe:Cu thin films. with magnification of 1000000X. In this study, we have prepared pure CdTe and its composite, which mixed with Cu via PLD technique to produce Cu mixed to CdTenanocomposite thin films by PLD.Special attention was paid to the influence of the processing parameters, such as mixing concentration on the structural and optical properties of the films during the deposition.It has been needed to study the optical, structural, morphological [13].

2. Experimental

Cadmium telluride , with a purity of 99.99% and copper with purity of 99.99%, were mixed together at different Cu concentration of (0.0, 0.3, 0.5, and 0.7) using agate mortar. Then the mixture was pressedbyhydraulic manually type (SPECAC), under pressure of 6 tons for 10 minutes to formed a pellets of 2.8 cm in diameter and 0.3 cmthick.. The CdTe: Cu films deposited on glass substrate of size (1.5x1.5)cm^2. The substrate was washed with distilled water then with alcohol using ultrasonic process for 15 minutesfor each oneto deposit the films at room temperature. PLD technique was used to deposit the films under vacuum of (2.8×10^{-2} mbar) using Nd:YAG with (λ= 1064 nm) SHG Q switching laser beam at 600 mJ, repetition frequency (6Hz) for 150 laser pulse is incident on the target surface making an angle of 45° with it. The distance between the target and the substrate was (2.5 cm) at room temperature.Figure 1. Shows a schematic diagram of PLD experiment with λ=(1064nm). In this work, X-ray diffractometer type RigakuMiniFlex II, USA, a power diffraction system with Cu-Kα X-ray tube at 1.54 Å with voltage 30.0 kV and current 15.0 mA and speed scan 6 deg./minute was used. The scans of X-ray are approximately perform between 2θ values of 20° and 90°. The crystal structure of the prepared pure CdTe and CdTe mixed with Cu thin film were investigated. Optical properties and energy band gap were calculated at room temperature for CdTe and CdTe/Cu thin films.

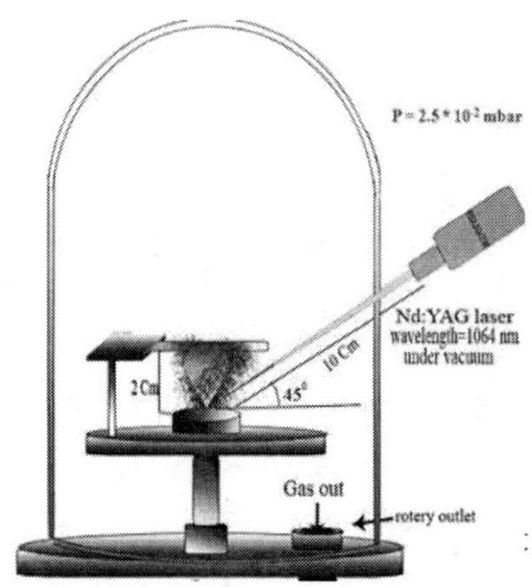

Figure 1. : with λ=(1064nm).

3. Results and Discussion

3.1. *Structural properties*

XRD pattern of CdTe, and CdTe: Cu powders with different concentrations at RT are shown in figure 2 with (JCPDS Card No: 96-900-8841CdTe, and 96-901-3015 Cu) at room temperature.The FWHM , phase and G.S shows in table 1 for undoped an Cu- doped CdTepoweders. XRD pattern of CdTe and CdTe: Cu thin films with different concentrations at RT deposited on glass substrate atroom temperature using (150) pulses of laser energy equal to (600 mJ) at vacuum are shown in figure 3. All the diffraction peaks of pure CdTe thin films, at 2 θ = 23.67^0, 39.23^0, $46.49^0$and 56.80^0, corresponding to the (111), (220), (311) and (400) identical with (JCPDS Card No: 96-900-8841) for rutile As shown in figure 3.For Cu- doped CdTe thin films the diffraction peaks just at 0.3 Cu at 2 θ= 23.78^0 and 43.28^0 corresponding to the(111),(111) identical with (JCPDS Card No: 96-900-8841 for CdTe , and 96-901-3015 for Cu). All parameters and phase were mentions in table 2.

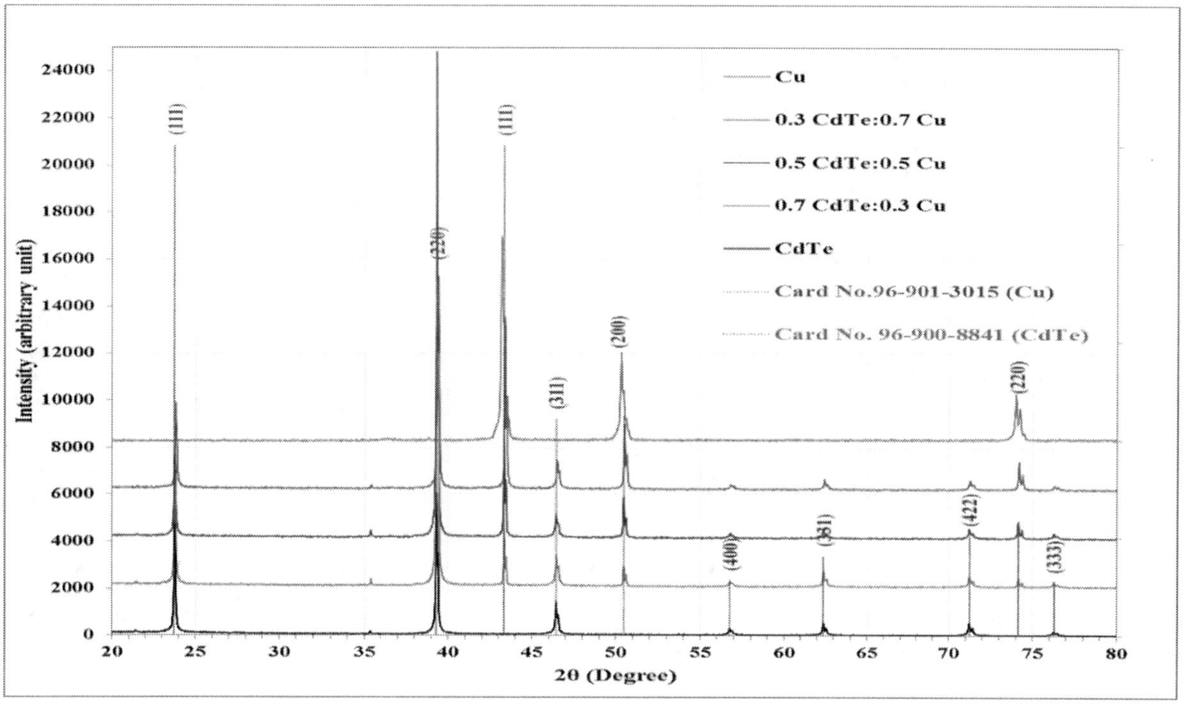

Figure2. Show XRD pattern of CdTe, and CdTe: Cu powders with different concentrations at RT

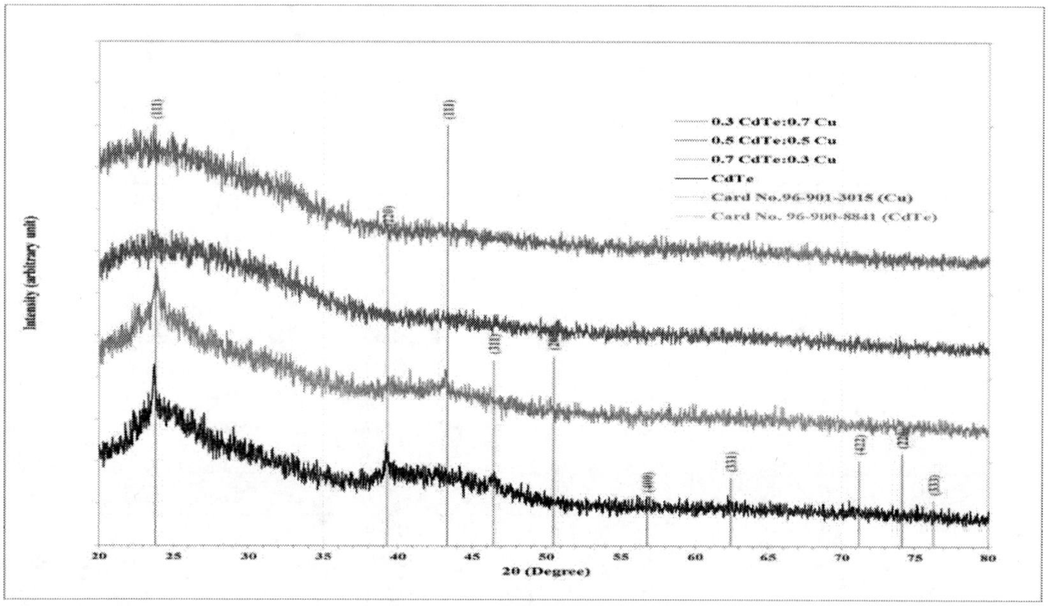

Figure 3.shows XRD pattern of CdTe and CdTe: Cu thin films withdifferent concentrations at RT.

IOP Conf. Series: Materials Science and Engineering **757** (2020) 012024 doi:10.1088/1757-899X/757/1/012024

In figure 3. for Cu doped CdTe thin films . All films (exept 0.3 Cu) possess amorphous nature . This is because, during the film deposition at room temperature , the low thermal activation energy leads to the formation of some unsaturated bounds along with the saturated bonds due to the insufficient number of atoms and hence the atoms may not be able to sit in its lattice position, which leads to distortion .[14,15]

Table 1. illustrate the diffraction angle (2θ), interplaner distance (d), the full width at half maximum (FWHM), a grain size (G.S.) and the phases which are identified along with (hkl) CdTe planes and (CdTe)1-x(Cu)x nanocompositepowders at room temperatures .

x	2θ (Deg.)	FWHM (Deg.)	d_{hkl} Exp.(Å)	G.S (nm)	d_{hkl} Std.(Å)	Phase	hkl	card No.
	23.7847	0.2083	3.7380	39.0	3.7412	Cub. CdTe	(1 11)	96-900-8841
	39.2708	0.1736	2.2923	48.6	2.291	Cub. CdTe	(220)	96-900-8841
	46.4236	0.2430	1.9544	35.6	1.9538	Cub. CdTe	(311)	96-900-8841
0.0	56.7708	0.2778	1.6203	32.5	1.6200	Cub. CdTe	(400)	96-900-8841
	62.3958	0.2778	1.4871	33.4	1.4866	Cub. CdTe	(331)	96-900-8841
	71.2153	0.3472	1.3230	28.1	1.3227	Cub. CdTe	(422)	96-900-8841
	76.2847	0.3473	1.2472	29.1	1.2471	Cub. CdTe	(333)	96-900-8841
	23.7500	0.1736	3.7434	46.8	3.7412	Cub. CdTe	(1 11)	96-900-8841
	39.3403	0.2083	2.2884	40.5	2.291	Cub. CdTe	(220)	96-900-8841
	43.3333	0.1736	2.0864	49.2	2.0860	Cub.Cu	(111)	96-901-3015
0.3	46.4583	0.2083	1.9530	41.5	1.9538	Cub. CdTe	(311)	96-900-8841
	50.4514	0.2083	1.8074	42.2	1.8065	Cub.Cu	(200)	96-901-3015
	56.8056	0.2778	1.6194	32.5	1.6200	Cub. CdTe	(400)	96-900-8841
	62.3958	0.2083	1.4871	44.6	1.4866	Cub. CdTe	(331)	96-900-8841
	71.2153	0.2778	1.3230	35.2	1.3227	Cub. CdTe	(422)	96-900-8841
	74.1319	0.1389	1.2780	71.7	1.2774	Cub.Cu	(220)	96-901-3015
	76.2847	0.2083	1.2472	48.5	1.2471	Cub. CdTe	(333)	96-900-8841
	23.7847	0.1389	3.7380	58.4	3.7412	Cub. CdTe	(1 11)	96-900-8841
	39.3403	0.2430	2.2884	34.7	2.291	Cub. CdTe	(220)	96-900-8841
	43.3333	0.1389	2.0864	61.5	2.0860	Cub.Cu	(111)	96-901-3015
0.5	46.4236	0.2430	1.9544	35.6	1.9538	Cub. CdTe	(311)	96-900-8841
	50.4514	0.1736	1.8074	50.6	1.8065	Cub.Cu	(200)	96-901-3015
	56.7708	0.2430	1.6203	37.2	1.6200	Cub. CdTe	(400)	96-900-8841
	62.3958	0.2083	1.4871	44.6	1.4866	Cub. CdTe	(331)	96-900-8841
	71.2153	0.2431	1.3230	40.2	1.3227	Cub. CdTe	(422)	96-900-8841
	74.1319	0.1736	1.2780	57.4	1.2774	Cub.Cu	(220)	96-901-3015
	76.2847	0.1736	1.2472	58.2	1.2471	Cub. CdTe	(333)	96-900-8841
	23.8194	0.1389	3.7326	58.5	3.7412	Cub. CdTe	(1 11)	96-900-8841
	39.3403	0.2084	2.2884	40.5	2.291	Cub. CdTe	(220)	96-900-8841
	43.3681	0.1736	2.0848	49.2	2.0860	Cub.Cu	(111)	96-901-3015
0.7	46.4931	0.2083	1.9517	41.5	1.9538	Cub. CdTe	(311)	96-900-8841
	50.5208	0.1736	1.8051	50.6	1.8065	Cub.Cu	(200)	96-901-3015
	56.8403	0.3472	1.6185	26.0	1.6200	Cub. CdTe	(400)	96-900-8841
	62.4653	0.1736	1.4856	53.5	1.4866	Cub. CdTe	(331)	96-900-8841
	71.2847	0.2083	1.3219	46.9	1.3227	Cub. CdTe	(422)	96-900-8841
	74.1667	0.1389	1.2775	71.7	1.2774	Cub.Cu	(220)	96-901-3015
	76.3542	0.1736	1.2463	58.2	1.2471	Cub. CdTe	(333)	96-900-8841
	43.1944	0.2778	2.0928	30.8	2.0860	Cub.Cu	(111)	96-901-3015
1.0	50.3125	0.2777	1.8121	31.6	1.8065	Cub.Cu	(200)	96-901-3015
	73.9931	0.3472	1.2801	28.6	1.2774	Cub.Cu	(220)	96-901-3015

Table 2. . illustrate the diffraction angle (2θ), interplaner distance (d), the full width at half maximum (FWHM), a grain size (G.S.) and the phases which are identified along with (hkl) CdTe planes and (CdTe)1-x(Cu)x nanocompositethin films at room temperatures .

x	2θ (Deg.)	FWHM (Deg.)	dhkl Exp.(Å)	G.S (nm)	dhkl Std.(Å)	Phase	hkl	card No.
	23.6794	0.6307	3.7544	12.9	3.7412	Cub. CdTe	(1 1 1)	96-900-8841
0.0	39.2378	0.5781	2.2942	14.6	2.291	Cub. CdTe	(220)	96-900-8841
	46.4915	0.6833	1.9517	12.7	1.9538	Cub. CdTe	(311)	96-900-8841
0.3	23.7845	0.7885	3.7380	10.3	3.7412	Cub. CdTe	(1 1 1)	96-900-8841
	43.2852	0.6308	2.0886	13.5	2.0860	Cub.Cu	(111)	96-901-3015
0.5				amorphous				
0.7				amorphous				

3.2. Morphological properties.

All sample have a smoth morphology with many particles on the surface , thes particles have sizes ranging from 48.66nm-76.67nm.The AFM results displayed an increase in nanoparticles grain size at the 0.5 ratio of Cu. Table 3 shows that the higher the diameter of the granule the lower the roughness of the surface and note this at the rate of 0.5 of copper where the roughness of the surface is less than for the rest of the membranes, which is 0.237 so the surface is more soft. The XRD and AFM analysis indicate that the grains are composed of small crystallities. It is noted that increasing the proportion of copper in the thin film does not make the increase is correct for the radius and also for the roughness of the surface, that is not a direct relationship and this corresponds to the source [16].

Table 3. The results of AFM examination of CdTe, and CdTe: Cu thin films of different concentrations at room temperatures.

Sample at RT (K)	Avg. Diameter (nm)	R.M.S (nm)	Roughness Average (nm)
CdTe	73.66	2.03	1.69
0.7 CdTe : 0.3 Cu	48.66	3.58	3.1
0.5 CdTe : 0.5 Cu	76.67	0.275	0.237
0.3 CdTe : 0.7 Cu	68.62	1.3	1.13

3.3. Optical properties

Optical properties of CdTe,andCdTe:Cu thin films have been deduced from transmission (T) and reflection (R) spectra of the films deposited on glass substrate using UV-visible spectrophotometer fig. 5. Displays the transmission spectra of (CdTeandCdTe_{1-x}:Cu_x) nanoparticles thin film at different X (0. 0.3, 0.5, and 0.7) at room temperature.The optical characteristics involve:The absorption coefficient (α), the optical energy gap (Eg), and the optical constants (i.e. refractive index (n), extinction coefficient (k), real dielectric constant (εr) and imaginary dielectric constant (εi))which are studied within the range (190-1100) nm.

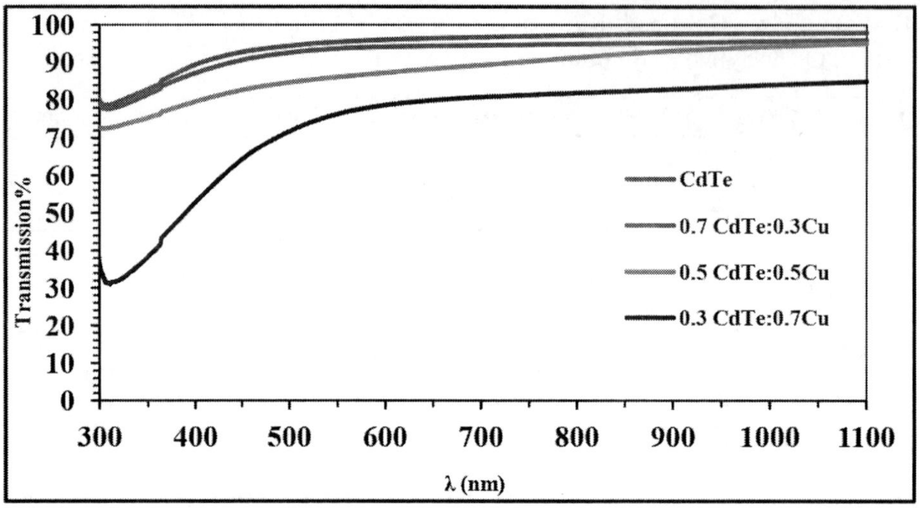

Figure 4. Transmission spectra of (CdTeandCdTe$_{1-x}$:Cu$_x$) nanoparticles thin film at different X (0. 0.3, 0.5, and 0.7) at room temperature

As seen from figure 4. the average transmission was found 70%-95%, which is good for opto-electronic device, especially for solar cell window layers.The higher transmittance can be attributed to less scattering effects,structural homogeneity and better crystallinity[17].Furthermore, sharp fall at band edge is an indication of good crystallinity of CdTe in the films .

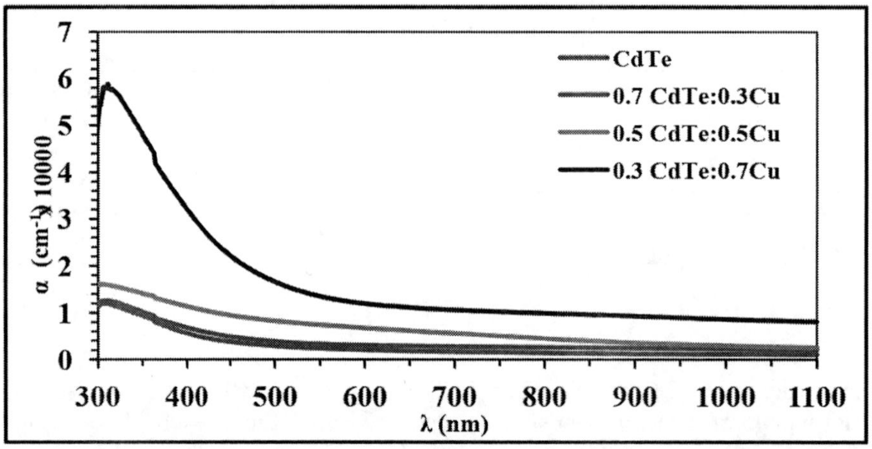

Figure 5. Variation of absorption coefficient and wavelength of pure and Cu doped cdTe films at room temperature

The absorption coeficint determines how the light of particular wavelength penetrates into the material. The variation of absorption coefficient and wavelength of pure and Cu doped cdTe films at room temperature is shown in Figure 5. It is seen that the material exhibits sharp edge in their absorption coefficient infers that the light which has energy lower than the band gap cannot excite electrons from the valence band to the conduction band. Consequently, the light of those wavelengths is not absorbed. The absorption coefficient of pure CdTe thin film is higher than the doped ones, which may be due to the existence of trap levels and the scattering loss as a consequence of Cu (metal) insertion into the film matrix [18].

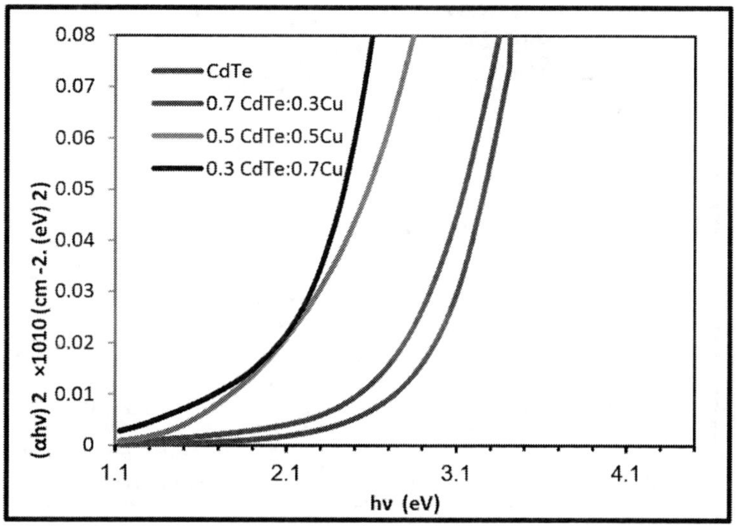

Figure 6. Tauc plot of pure and Cu doped CdTe thin films deposed at room temperature

It is Known that the efficiency of Photovoltaic devices depends on the amount of photons absorbed by the material, which in turn related with the energy of the photon and the band gap of the material. In this aspect, we have evaluated the optical energy band gap of pure and CdTe:Cu films using the Tauc relation shown in eguation (8) .

$$(\propto h\vartheta) = B\left(h\vartheta - E_g\right)^n \quad \ldots\ldots\ldots\ldots\ldots(8)$$

Where B is a constant which arises from the FermisGolden rule of fundamental electronic transition within the frame work of parabolic approximation for the dispersion relation and **n**determins the type of transition. CdTe is a direct band gap material and it is verified from the slope gives the value of 0.5 (graph is not shown here)[19]. Hence, the Tauc plot was drawn between ($\propto h\vartheta$) vs $h\vartheta$ (for direct band gap system) ,and Tauc plot of pure and Cu doped CdTe thin films deposed at room temperature is shown in figure 6. The extrapolated linear region at $\propto$ =0 gives the energy band gap value (E_g) of the films. The energy band gap value of pure CdTe film is 2 eV that was agoodCompatible with and it decreases to 1.1 eV when we icreasing Cu content. The decrease in optical energy band gap(as shown in table 3. with increasing Cu content may be due to the occupation of Cu impurity in the host lattice, which acts as acceptor level, becomes degenerate, and the energy level shifts thereby reducing the energy band gap.[20] It can be mentiond that Abbas Shah et al[21] reported that the energy band gap of Cu doped CdTe thin films is found to decrease with the increase of Cu concentrations, which is consistent with our observation. The reduction in energy band gap may also be due to the extention of localized states in amorphous structure[18]. As observed form the XRD pattern, the unsaturated bond in the amorphous structure due to the insufficient number of atoms causes defects that are responsible for the formation of localized states.

Table 4. The energy gap values for direct allowed transmission for CdTeand and CdTe:Cu thin films

X(gm)	Energy gap (eV)
CdTe	2
0.7 CdTe: 0.3 Cu	1.8
0.5 CdTe: 0.5 Cu	1.3
0.3 CdTe: 0.7 Cu	1.1

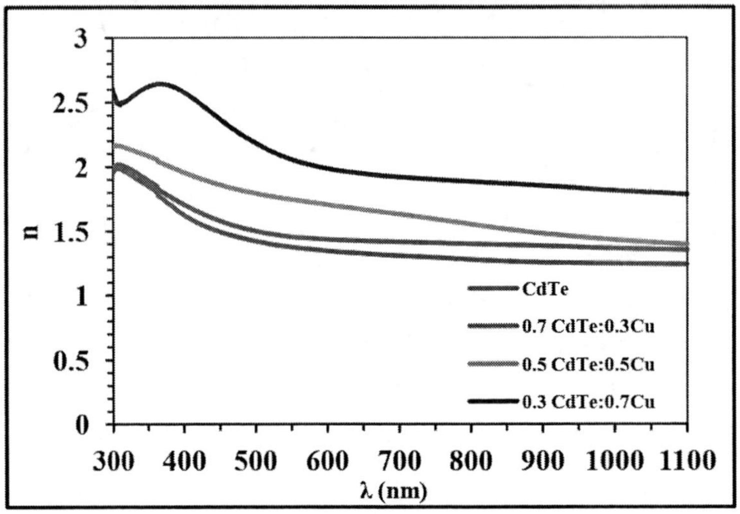

Figure 7. Relationship between refractive index (n) and wavelength (λ) of the nanoparticles thin membrane at room temperature.

The Relationship between refractive index (n) and wavelength (λ) of the nanoparticles thin membrane at room temperature are shown in fig.7. The refractive index is found to increase with Cu content. This increase in refractive index infers the enhancement in interactions between photons and electrons in the Cu doped CdTe films.

4. Conclusion

CdTe, and $CdTe_{1-x}$: Cu_x (x= 0.0,0.3,0.5,and 0.7)%wt films were grown by pulsed laser deposition technique at room temperature. Structural and optical properties of these films were investigated by using characterization techniques. Low angle XRD analysis showed that CdTe film are cubic polycrystalline and has preferred orientation in (111) direction.For Cu doped CdTe thin films . All films (exept 0.3 Cu) possess amorphous nature. The UV-Visible spectroscopy analysis showd that the optical band gap decreases from 2 eV- 1.1 eV when we increasing copper concentration.Further,we have discussed the dielectric constant and refrectiveindex.Webelive that these results would be useful in developing CdTe based absorber layer with controlled level of Cu doping for solar cell applications.

5. References

[1] Britt J and FerekidesC, 1993 *Appl. Phys*. **Lett. 62** 2851.

[2] Younghun Jung, Gwangseok Y, Seungju C, Donghwan K, and Jihyun K 2013 Applied Physics Letters **103**, 231910.

[3] Csaba S 2004 *Physica Status Solidi (b)* **241**(3) 783-790.

[4] Osborne M 2012 NREL Confirms Latest CdTe Module Efficiency Record from First Solar *Solar Media Limited*, 17 January 2012.

[5] Rogalski A 2011 *Infrared Physics & Technology* **54**(3) 136-154.

[6] Triboulet R, 2003 *Cryst. Res. Technol*, **38**(3-5) 215-224.

[7] Bengul Z, Murat B and Sertap K, 2006 *Journal of Arts and Sciences Sayt*, **1**(5)103-116.

[8] Dharmadasa I, Bingham P, Echendu O, Salim H, Druffel T, Dharmadasa R, Sumanasekera G, Dharmasena R., Dergacheva M, Mit K, Urazov L, Bowen K, Walls M and Abbas A 2014 *Coatings* **4**(3)380-415.

[9] Riyam A, Ahmed K, A. K. Abbas, and Abdulhussein K, 2014,*International Journal of Application or Innovation in Engineering & Management (IJAIEM)*,**3**(7) 1-7.

[10] Ramírez-Velázquez L and Olvera M2015, *Superf. y Vacío*, **28**,(1)25–29.

[11] Peidong Y 2003 *The Chemistry of Nano Structured Materials* (Printed in Singapore.: World Scientific Publishing) Co. Pte. Ltd., p. 396.

[12] Anabitarte F and Adolfo C, 2012 *International Scholarly Research NetworkISRN Spectroscopy,* **2012** (285240) 12.

[13] Harilal S, Beau O and Mark S 2005, *Journal of Applied Physics* 98(3) 036102-036102-3.

[14] Sabastine C, 2012 Synthesis and characterization of copper nanoparticles and copper-polymer nanocomposites for plasmonic photovoltaic applications *MSc thesis, The University of Western Ontario.*

[15] Shaaban E, Yahia I, and El-Metwally E 2012 *Acta Phys. Pol. A 121, 628.*

[16] Punitha K, Sivakumar R, Sanjeeviraja C, Vasant S, and Ganesan V 2014 Journa*l of Applied Physics 116, 213502.*

[17] Li J , Yang L , JianJ , ZouH and Sun Y 2008 , *Journal of Semiconductors*, **30**(11).

[18] Kulkarni R, Rondiya S, Pawbake A, Waykar R., Jadhavar A, Jadkar V, Bhorde A, Date A, Pathan H, and Jadkar S 2017 *Energy Procedia 110* pp. 188-195.

[19] Jata Dzhafarov, Serco Yesilkaya, Yilmaz C, and Murat C 2005 *Energy* Mater. Sol. Cells 85(3)371-383.

[20] Garadkar K, Pawar S, Hankare P, and Patil R 2010 *J.Alloys compd , 491,77 Jan* 2010.

[21] De Moure-Flores F, Quiñones-Galván J, Guillén-Cervantes A 2012 *J. Appl. Phys.* 112(11).

Study of the Spectroscopic Performance of Laser Produced CdTe, and CdTe:Ag Plasma.

Nadia Naeema[1], Ahmed Kudher[2] and Ghuson H Mohammed[3]

[1,2]Departmentofphysics , College of Science, Wasit University.

[3]Department of physics, College of Science, Baghdad University.

kutpc19@yahoo.com

Abstract. This article, plasma parameters [((Te) electron temperature, (n_e) electron density, (fp) Plasma frequency, (λD) Debye length , and (ND) Debye number], were studied by using a spectrometer that collects the laser spectrum produces cadmium telluride Plasma in different energies. As well as for CdTe: Ag for different card ratios. The electron temperature results for the CdTe range (0.699-0.738) ev also has an electron density of (2.867-4.430)* 10^{18}Cm $^{-3}$ was measured under vacuum up to 2.5 x 10^{-2} mbar. As well as for CdTe:AgTe range (0.738-0.606) and ne range(4.430-4.691))* 10^{18} Cm $^{-3}$.The properties of CdTe and CdTe: Ag were determined throughoptical transmission method using the ultraviolet spectral spectrum inside range 190-1100 nm.

Keywords. Optical Emission Spectroscopic (OES), Cadmium telluride (CdTe), Laser-Induced Plasma Spectroscopic (LIPS).

1.Introduction

The characteristics of the plasma that introduced from interactions between Laser with solid targets depends upon numerous parameters describing the solid target has features like: pulse duration, laser wavelength, properties of the ambient medium, etc.[1].That plasmas are a promising subject for much major and connected research. There are three stages of ablation method using lengthy pulse period lasers (greater than one ns). The first step, ray of the laser interrelates with the solid resultant speedy target surface ionization into a plasma on time scale short associated with the pulse duration. |The second step, ray of laser competently absorbance thru the plasma was grows iso-thermally, and after ending of laser pulse, are resulting plasma cloud grows quasiadiabaticall on media, which can contain vacuum or gas, within or without applied fields that is the third stage [2]. A plasma is formed, Whenthe sample materialheats, ablates, atomizes, and ionizes by the energy from the laser pulse. A method based on the analysis of the radiationspectral gives out by plasma cause by focusing an intense laser pulse on the sample surface its known as Laser-induced breakdown spectroscopy (LIBS) , also named laser-induced plasma spectroscopy (LIPS) or laser ablation spectroscopy (LAS) . The intense laser pulse concenters on the sample surface causes evaporation, atomization and ionization of the material and produces a plasma, which expands and cools very rapidly . The ablation process in typical LIBS conditions , is stoichiometric: analyze the atomic and ionic lines give out by the plasma provides the ID of the species current in the taster and their quantification .[3]

The composite semiconductor cadmium telluride [CdTe]is a II-VI and originate most right candidate for the production of thin-film solar cells proper to its optimal energy band gap (1.44eV) at room temperature and in the visible range(>105 cm^{-1}) high absorption coefficient . [4]

A main task is to determine of Plasma parameters in order to achieve the best understanding as a way to improve their application. Its main parameters such as the temperature, the electron density and the number densities of the different type current in the plasma. Spectroscopic methods are normally determine electronic temperature as the most chief parameter . In this kind of plasmas at atmospheric pressure the strong electron density allows ,in general the assumption of the local thermodynamic equilibrium (LTE) and then use of the Boltzmann . Surface hardness versus normalized ionic to atomic lines intensities. The calculation for the plasma temperature by Saha–Boltzmann equations. The plasma electronic temperatures T_e was conclude by using Boltzmann equation (1):

$$\mathrm{Ln}(I\lambda/gA) = -E_k/kT_e - \mathrm{Ln}(4\pi Z/ hcN_0) \qquad (1)$$

where the line intensity is I , the line wave length is λ, E_k is the energy of the upper state, Z is generally taken as the statistical weight of the ground state it is the partition function , h is the Planck constant, c is thespeed of light, and N_0 is the population of the ground state . When the left hand of Eq. (1) is plotted versus E_k, from the slope of straight line we can deduce the plasma temperature, which is equal to $-1/kTe$.[5]

Stark broadening of spectral lines is under probe since the orgination of the effect in 1913. With the variegation of the available plasma sources and the increasing attentio for plasma diagnostic tools, the theoretical and experimental studies devotedto Stark broadening became favored in the 1960s. Since that time, several analysis papers have been published to summarize the results get by a big number of research groups all over the world. count on the moderate worth of Doppler and Stark widths, the line shapes are telling of by Gaussian, Lorentzian or Voigt profiles. The Doppler width is measured as stated by to plasma temperature and atomic mass of the emitting kind. The width Stark is get by using .

$$\Delta\lambda_{Stark} = w(n_e^{ref} /ne)^m \qquad (2)$$

where w is the Stark width at the reference electron density n_e^{ref}.[6]

The electric field that causes Stark effect in laser-induced plasmas results primarily from collisions with electrons, with small contributions due to collisions with ions Therefore, so can be simplified the equation .:

$$n_e = \left[\frac{\Delta\lambda}{2\omega_s}\right] N_r \qquad (3)$$

$\omega_s$is the theoretical line full-width Stark broadening parameter, calculated at the same reference electron density $N_r \approx 10^{17} cm^{-3}$

Debye shielding gives the plasma its quasi-neutrality characteristic.The shielding caused by reaction of charged particles to lower the effect of local electric fields is called Debye shielding. a distance λ_D, called the Debye length and defined as [7]:

IOP Conf. Series: Materials Science and Engineering **757** (2020) 012025 doi:10.1088/1757-899X/757/1/012025

$$\lambda_D = \left(\frac{\varepsilon_o k T_e}{n_e e^2} \right)^{1/2} = 743 * (T_e / n_e)^{½} \qquad \text{\_\_\_\_\_\_\_\_(4)}$$

where :- λ_D : Debye extent (cm) ,L: dimension of system(cm) ,n_e : is the density of the electron $\left(m^{-3}\right)$, T_e : is the electron Temperature (K) ,e : is the electron charge (C).

2. Experimental setup

In this work a lens has a focal length of 10 cm has been used. A small beam waist can be produce by a shorter focal length lens, and for that reason, tougher breakdown, but it besides has afocus of smaller deepness, a graphic diagram for the LIBS setup shown infigure1.

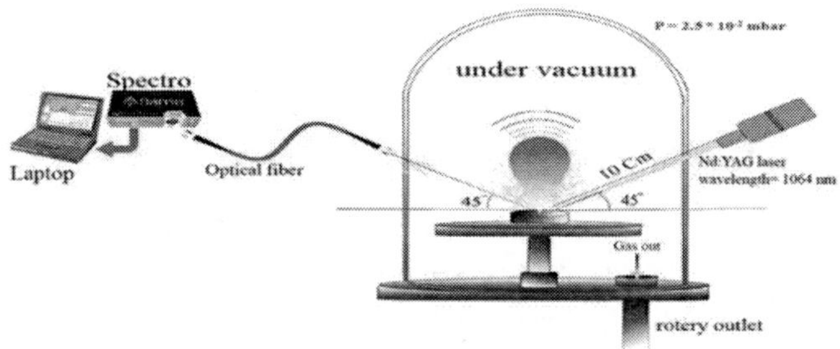

Figure 1. Schematic of the experimental setup for the emission spectroscopy

of the laser-induced plasma.

By changing the distance between the target and the laser lens can be changed the diameter of laser spot. (9 ns) pulse duration using(6 Hz)duplication rate frequency. an exactly distance pending the measurements for precision. By using the emitted light from sample bombed by pulse laser the spectrometer was analyzed. To analyze emitted light it has been used the short response time spectrometer from Ocean Optics (HR 4000 CG-UV-NIR) in the setup . The optical fiber has been collected by the light produced by the ablated plasma . The optical fiber has been set at angle of about 45° degree to axes of the laser beam to avoid splashing and then guided to the entrance slit of the spectrometer . The spectrometer has a high accuracy depending on grating used in it, react wavelength from(200to900) nm with 3648 pixels. At 1064 nm NEODYMIUM: YAG laser is tightly fixated on the target to produce plasma Plume . In order to insure exposing a fresh surface after every train of shots the target surface was rotated manually . The spectrum of plasma with different value of energies varied from 500 mJ to 1000 mJ collect the Spectra of CdTe, and 1000 mJ collect the Spectra of CdTe:Ag each spectrum was gotten over a 300-800 nm wavelength range. In conclusion the results were studied and compared with Nationality Institute of Standards and Technology data (NIST) [8]. Then calculated parameters of plasma.

IOP Conf. Series: Materials Science and Engineering **757** (2020) 012025 doi:10.1088/1757-899X/757/1/012025

3. Results and discussions

Figure 2. show the CdTe target plasma induced by emission spectra of laser , which(CdTe plasma) confined in air in the spectral range (300-800) nm with different pulsed laser energy E=(500,600,700, 800, 900 and 1000) mJ. The optical emission spectra of CdTe: Ag target Plasma which confined in air were noted down using (OES) technique with pulse laser energy E=(1000)mJ with different concentrations (0,0.3,0.5,and 0.7).

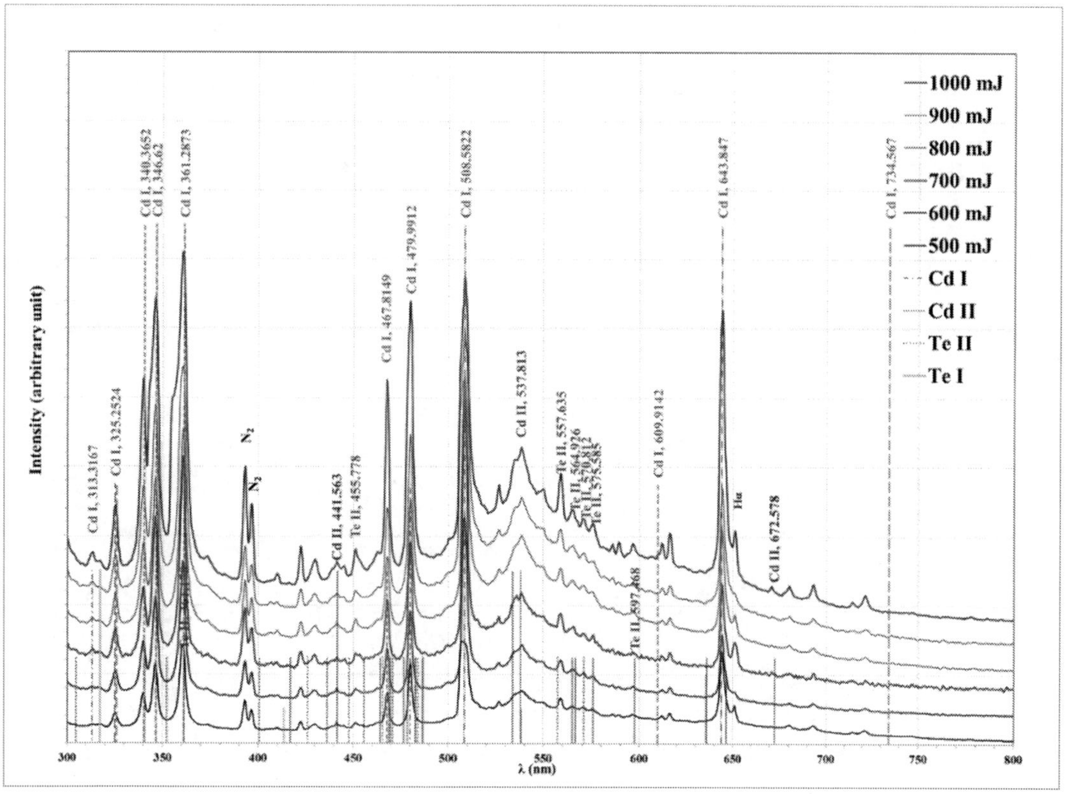

Figure 2. Emission spectra of laser-induced CdTe target in the air with different laser energies.

Table 1. display the calculated electron temperature (T_e), fall width half maximum (FWHM) , electron density (n_e), plasma frequency (f_p) and Debye length (λ_D) for CdTe target at different laser pulse energies. Measured all plasma parameters (λ_D, fp and n_e) were contented the criteria for the plasma . It shows that fp decrease with decrease laser energy because it is proportional with n_e, while λ_D increasing .

Table 1.parameters of plasma for CdTe inair with laser of different energy.

E(mJ)	Te (eV)	FWHM (nm)	n_e*10^{18} (cm^{-3})	f_p (Hz) *10^{12}	λ_D *10^{-7}(cm)
500	0.699	1.100	2.867	15.204	3.669
600	0.716	1.200	3.127	15.880	3.555
700	0.721	1.400	3.648	17.152	3.302
800	0.732	1.500	3.909	17.754	3.216
900	0.734	1.700	4.430	18.901	3.025
1000	0.738	1.700	4.430	18.901	3.034

The calculated values of the electron temperatures (Te) by using Boltzmann plot equatio1. show that the electron density and electron temperature are increased thru increasing the laser pulse energy, in the ambient air as shown in figure 3 . At higher laser peak energy, n_e becomes almost stable, because the plasma becomes opaque to the laser beam which shields the target. Plasma shielding occurs when the plasma itself reduces the transmission of the laser peak power along the beam path.

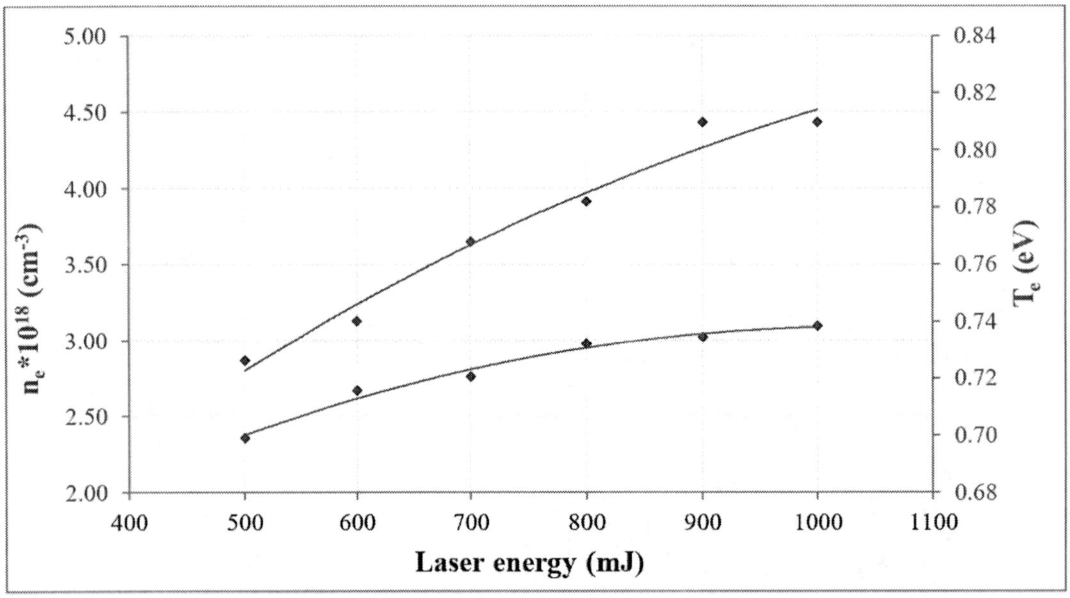

Figure 3. The variation of the laser energy versus (Te) and (ne) for CdTe in air

Boltzmann plot needs peaks that originate from the same ionization stage and the same atomic species, (peak is used Cd II species at 441.5) nm for CdTe powder (pellet) in the ambient air as shown in figure (4) . For each element it has been used the National Institute of Standard Technology database (NIST) for the experimental plots .The energies of higher levels, statistical weights, and transition probabilities have been obtained .As well, the electron temperature equals to the invert of slope of the fitting line (the slope of the fitted line equals to -1/ T). R2 is a statistical coefficient representingbest of the line-fitting which takes a value between (0, 1). The equation of fitting and the R2 were shown in the figure for all fitting lines. The better one has R2 value closer to 1.

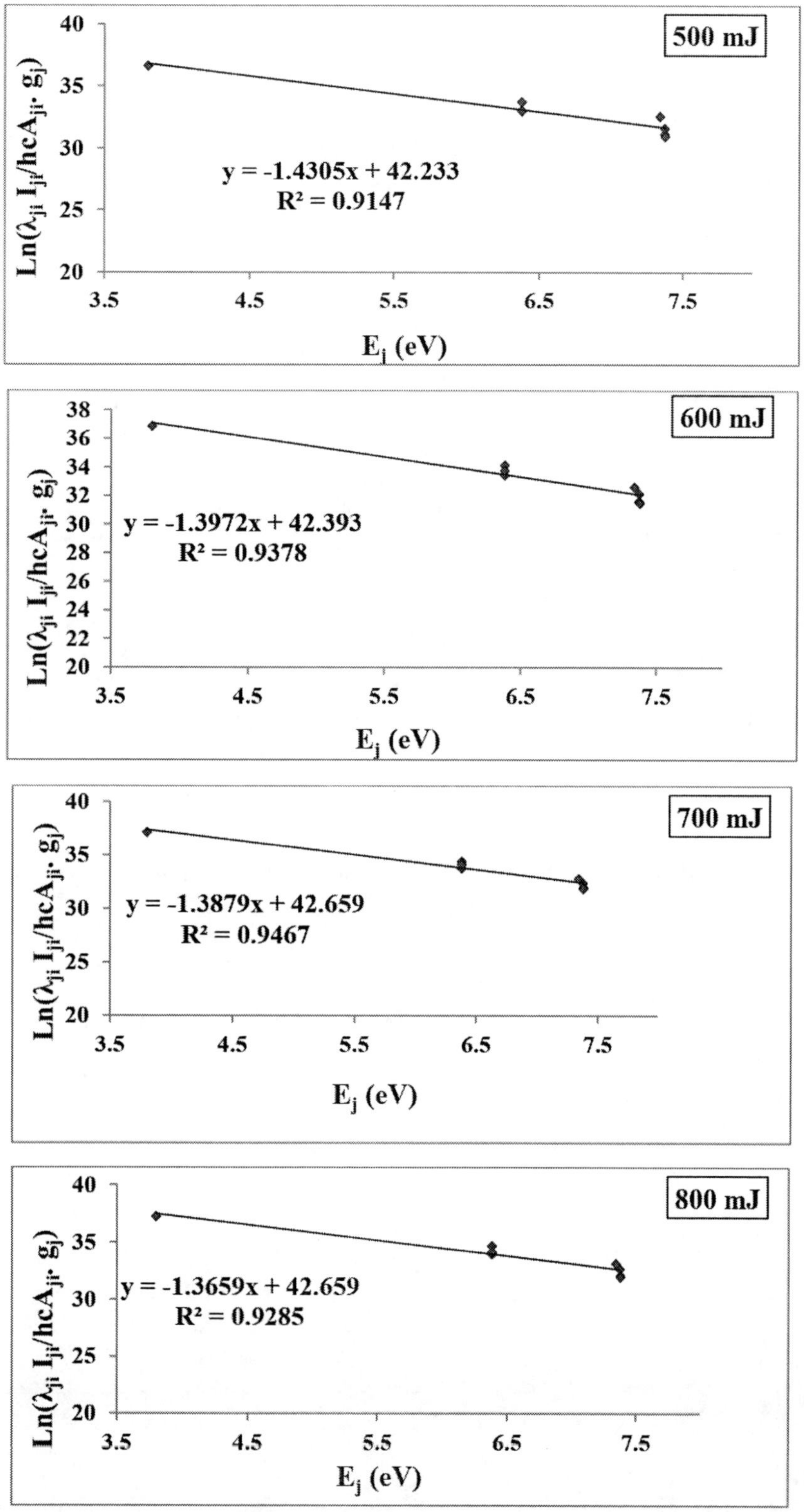

Figure 4. Boltzmann plots of CdTe target with different laser energies in the ambient air

The electron densities are calculated by using stark broadening equation (3) as shown in figures 5. The results collisions of charged species in the broadening of the line and the shift of the top wavelength results or form the Stark broadening of spectral lines in plasmas.

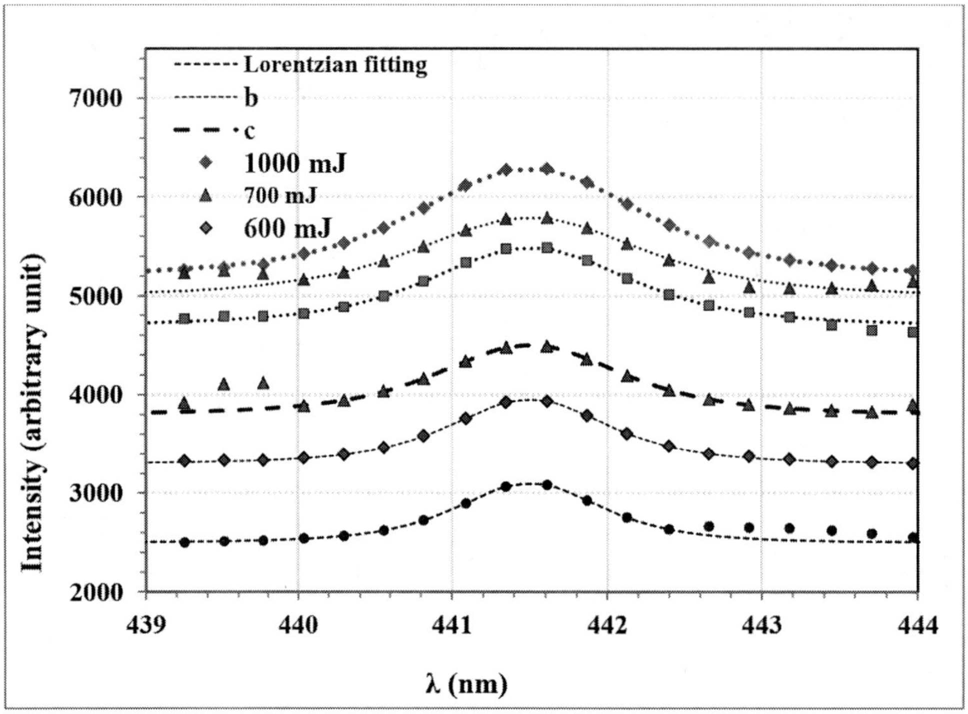

Figure 5.Variantof width&intensity of the Cd(II) positions at (441.5) nm with changed values of 1064 nm laser wavelength for CdTe in the ambient air.

The optical emission spectra of CdTe: Ag target plasma is shown on the figure (6)

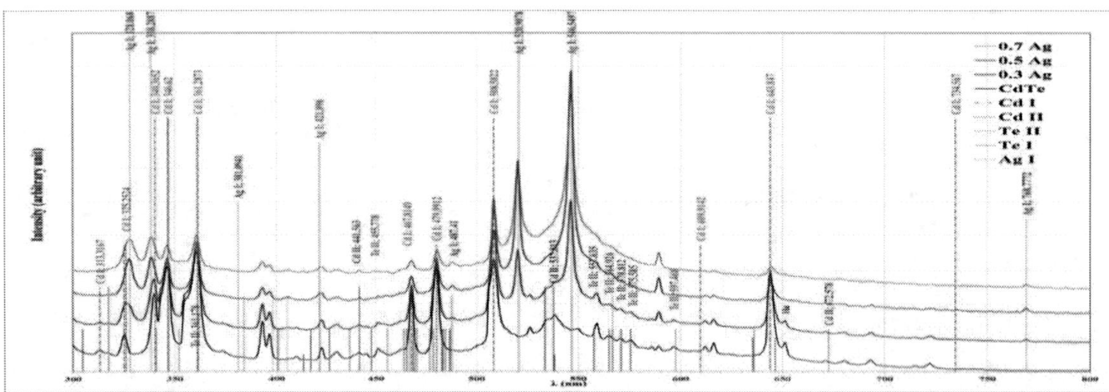

Figure 6. Emission spectra of laser-induced CdTe:Ag target in air with different Ag concentration.

Table 2. shows that fp decrease with decrease of Ag ratio for the reason that the frequency of plasma proportional with n_e, however, λ_D growing and increasing.

Table 2. plasma parameters for CdTe Ag in air with different Ag ratio

Ag ratio	Te (eV)	FWHM (nm)	n_e*10^{18} (cm^{-3})	f_p (Hz) *10^{12}	λ_D *10^{-7}(cm)
0.0	0.738	1.700	4.430	18.901	3.034
0.3	0.700	1.770	4.613	19.286	2.894
0.5	0.624	1.800	4.691	19.449	2.709
0.7	0.606	1.800	4.691	19.449	2.670

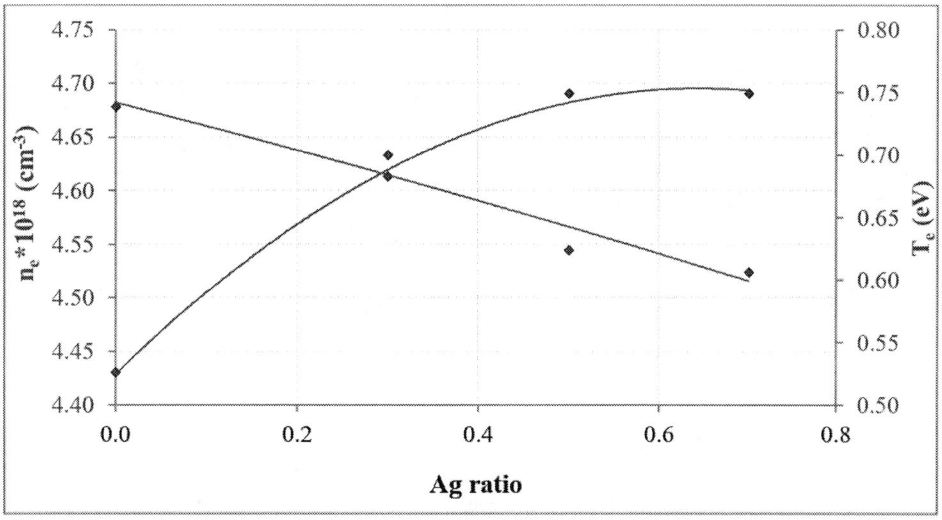

Figure 7. (Te) and (ne) variation vs. different Ag concentration used for CdTe :Ag in ambient air.

The calculated values of the electron temperatures (Te) by using Boltzmann plot equation(1) show that the electron temperature is decreasing with increasing of Ag ratio but the electron density is increased with increasing the Ag ratio, in the ambient air as shown in figures (7).

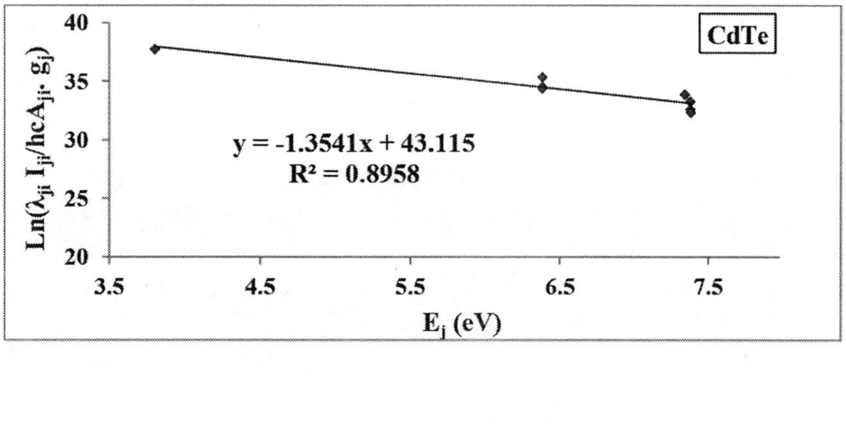

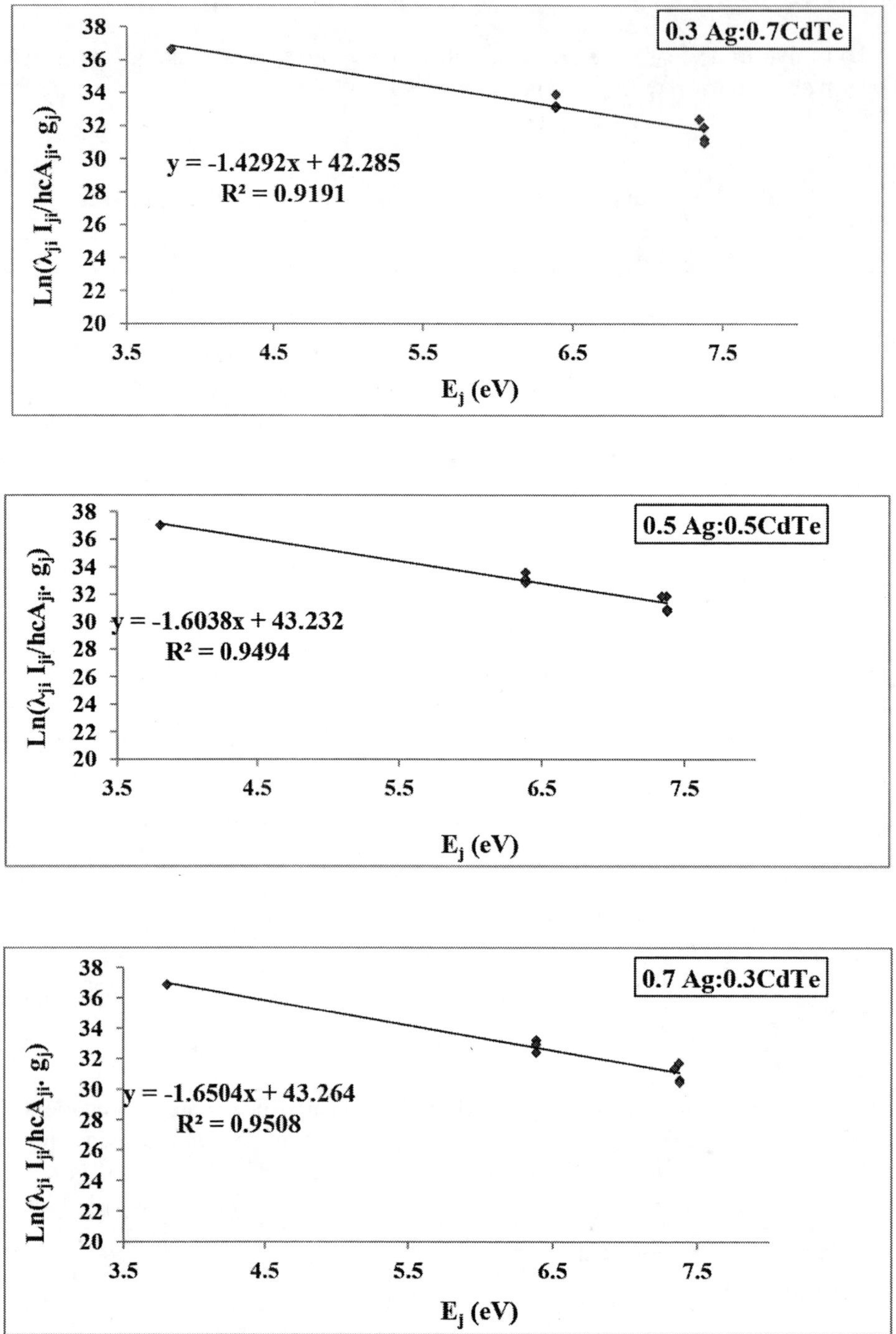

Figure 8. Boltzmann plots of CdTe:Ag target with different Ag ratio in the ambient air.

The electron densities are calculated by using stark broadening for peak 441.5 nm and it be fitting by using lorentzain fitting.

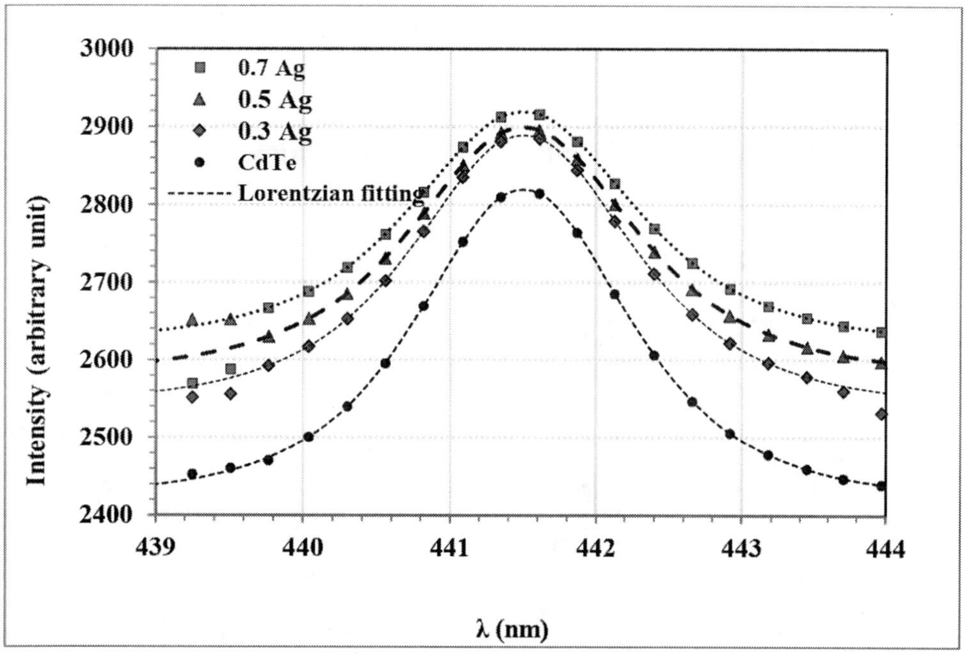

Figure 9. intensity and width Variant for Cd(II) positions at (441.5) nm with different values Ag ratio for CdTe : Ag in the ambient air.

5. Conclusions

For studying the plasma CdTe, and CdTe:Ag produced by the laser , a Nd: YAG laser with a fundamental-wavelength Q-switched (1064 nm) laser was used. The plasma emission spectrum represents the transfer of neutral atoms and ionized Cd ions. The spectral line from the plasma laser emission intensity shows a Strong adoption on environmental circumstances. It has been discovered that the intensity of the power of different laser peaks increases as the peak laser power rises for CdTe Plasma, and its increases as the Ag ratio rises for CdTe:Ag Plasma. It has been found that the electron temperature, electron density, Debye length(plasma parameters), and plasma frequency are very effective with laser energy and with Ag concentration rises. The results show that the change in an electron density , and electron temperature with laser energy displays that both increases per laser energy for CdTe Plasma but For CdTe: Ag Plasma, electron density rises and electron temperature decreases with increasing of Ag Ratio. An electron temperature of 1064 nm was calculated.

5. References

[1] Khalil A, 2013 *J.Laser Phys.* **23** no. 1.
[2] Nek M, Tao Y, Burdt R, Yuspeh S, Amin N,and Tillack M, *J. Phys.Conf. Ser.***244** no. 4, 2–5.
[3] Maria M and et al. 2017, *Trends in food science & technology* **65** 80-93.
[4] Chander S and Dhaka M, 2015 *Low-dimensional Systems and Nanostructures* **73** 35-39.
[5] Aberkane S and et al. 2014 *J. Applied Surface Science* **301** 225-229.
[6] Burger M and Jörg H 2016 *J. Spectrochimica Acta Part B: Atomic Spectroscopy* **122** 118-126.
[7] Akhtar M and et al. 2018 *Spectrochimica Acta Part B: Atomic Spectroscopy* **148** 143-151.
[8] Kramida, A, Ralchenko, Yu, Reader J and NIST ASD Team 2017 *National Institute of Standards and Technology (NIST) Atomic spectra database* **5**.

IOP Conf. Series: Materials Science and Engineering **757** (2020) 012026 doi:10.1088/1757-899X/757/1/012026

Influence of titanium oxide additions with co_2 laser treatment on the microstructural properties of glazes layer-dental zirconia ceramics

Z A Salman[1], Sh A Zaidan[2] and K A Hubeatir[1]

[1] Department of Laser Engineering and Electronic Optics, University of Technology, Baghdad, Iraq
[2] Department of Applied Sciences, University of Technology, Baghdad, Iraq

Abstract: This paper presents the effect of CO2 laser on the glazes layer-dental zirconia ceramics after adding TiO2 nanoparticles to the glazes and study the its improvement. The specimens has been prepared by using CAD/CAM dental machine and sintered at 1530°C. Then the surface was glazed by VITA Glaze with (5% and 10%) of nano-TiO2 additions. The glaze layer were irradiated by continuous CO2 laser of 15Watt as indicated power. It was found that the main phases of the ceramic substrate are tetragonal zirconia with a percentage of alumina in the corundum phase. The glazing of the ceramic substrate led to a change in the X-ray diffraction pattern by the emergence of new phases, thus changing the crystallite size and percentage of lattice strain. The change in grain size measured by AFM technique, ranging from 88.46 nm to nm 62.18 in addition to changing surface roughness due to the emergence of centers of nuclei and grain growth. Moreover, the addition of TiO2 and laser irradiation changed the residual stresses on the surface, and this was reflected in the increasing hardness values from 575 kg/mm2 to 1215 kg/mm2 after added 5% TiO2 with laser treatment. Generally, addition of 5% TiO2 is better than 10% in terms the structure of the glaze layer surface and hardness, as the additions of 10% need a higher glazing temperature.

Keywords: CAD/CAM; Dental ceramics; CO_2 laser; Zirconia; Glazing; Residual stress, Grain size.

1. Introduction

The Stone Age and the Bronze Age were named according to the materials which dominated these historical periods. And therefore Modern era, facing an increasing variety of ceramic materials for industrial or biomedical use, therefore this marked the era of ceramics. Ceramic materials used for medical and dental purposes are called bio-ceramics. Over the years that have passed, Zirconia technology has been encouraged The rapid development of metal-free dental minerals, which provided a high proportion of Dynamic compatibility, improved aesthetic teeth and increased strength [1].The addition of "Stabilizing" oxides to pure zirconia allows to generate multiphase materials known as Partially Stabilized Zirconia (PSZ) whose microstructure at room temperature generally consists of cubic zirconia as the major phase, with monoclinic and tetragonal zirconia precipitates as the minor phase [2]. Various oxides, such as yttrium oxide (Y_2O_3), calcium oxide (CaO) or magnesium oxide (MgO), can be added to zirconia for stabilization, allowing for a quadrilateral shape at room temperature after flocculation [3]. In the early stages of the development, several solid solutions (ZrO_2-MgO, ZrO_2 -CaO, ZrO_2-Y_2O_3) were tested for biomedical applications. But in the following years the research efforts appeared to be more focused on zirconia-yttria ceramics, characterized by fine grained microstructures known as Tetragonal Zirconia Polycrystalline (TZP) [4]. In order to produce glass ceramics, we use controlled crystallization method. We apply heat on the glass material, thus crystals of the structure get nucleated, and they become larger and embedded in the glass matrix [5]. In fact, surface compression layer was found to occur on a wide range of ceramic materials following different treatment processes that acts to strengthen ceramic material. It can be achieved by thermal tempering, machining and polishing and the application of a glazing layer with a lower coefficient of thermal expansion than the adjacent ceramic material. The CO_2 laser is well suited for the treatment of glazes because its emission wavelength is almost totally absorbed by glass. It was suggested that laser treatment of glass-ceramic surfaces inhibits the formation of microcracks, leading to greater mechanical resistance of the glazes [6]. Therefore, the aim of this study was to evaluate the surface hardness, surface morphology and structure of zirconia dental surface glazing layer. The originality in the subject is to convert part of the amorphous glaze layer into crystallized glass ceramic by adding Nano titanium oxide and laser treatment for better specification.

2. Material and methods

2.1 *Specimens preparation*

The used Zircon material is from (Easy co) company - Origin of high purity (99.9%) Zircon. The specimens were cut to small cylinders by CAD/CAM dental machine with diameter (0.5 cm) and height (0.5 cm). The specimens were sintered at temperature 1530°C for 120 min as hold time. Glazing process of the surface uses mixture of the following; VITA AKZENT Plus Glaze LT (Low Temperature) of weight (0.060 g) with 3 drops of (VITA AKZENT PLUS, POWDER FLUID) and with two different percentages of TiO_2 (5% and 10%) nanoparticles (< 26 nm), purity ≥ 99.8% supplied by Cheng Du Micxy Chemical Co Ltd – China was added to glaze mixture to produces two different glazing layers. A special low fusing glaze material is now available for ceramics with low firing temperatures. After the completion of glazing process on specimens surfaces, the specimens was dried and firing up to 800°C.

2.2 *Irradiation parameters*

The experimental procedure of laser engraving is done using a CNC CO2 laser machine with 10.6 μm wavelength, the beam spot size was used in treatment is 145 μm and output power range from 1-70Watt. Because of the high potential temperature of irradiated porcelain with this wavelength and safety of ceramics, the value of energy output power used is 15Watt, with Overlapping space 0.01 mm. The irradiation was done offline After 10 mm, the size of the beam spot is aimed at the entire specimen surface with speed 500 mm/min and time 4.14 min. The laser hand piece was keep perpendicular to the irradiated surfaces.

2.3 *X-ray diffraction*

The technique give us information about the measurement phase contrast, crystal size, the crystallization degree and composition material. X-ray diffraction (Shimadzu 6000-XRD) scan range is (20-80 deg.). The mode of scan is a continual scan, scan velocity is (10 deg. /min), and XRD device has X-ray tube Cu(Filter) (1.5406 Å) , This examination helps us to provide the data necessary to identify the crystal structure of output and phase information.
The crystallite size of the structure is estimated by using Scherer's formula:

$$D_c = \frac{K\lambda}{\beta \cos\theta} \quad (1)$$

Where K=0.94 and λ is the wave length of X-Ray used, which is Cu K_α radiation (λ=1.54 A°) and β is the full width at half maximum (FWHM) of the diffraction peak corresponding to a particular crystal plane. The lattice strain (ϵ) was calculated using the formula [7, 8, 9].

$$\epsilon = \frac{\beta}{4\tan\theta} \quad (2)$$

2.4 *Atomic Force Microscopy*

The Atomic Force Microscopy device (AFM) type SPM-AA 3000 Angstrom (USA) is used in surface science laboratories to obtain grain size and distribution, as well the images with atomic resolutions of 10^{-10} m or one tenth of nanometer. This type of microscopy can be effectively applied in the field of specimen to study the surface characteristics of a specimen. The instrument is based on the principle that when a tip, integrated to the end of a spring cantilever, is brought within the interatomic separation between the tip and sample, interatomic potentials are developed between the atoms of the tip and the atoms of the surface.

2.5 *Hardness test*

Before you perform a hardness test, the surfaces of all specimens were meticulously inspected by a magnifying glass to exclude Samples with no surface defect. The micro hardness was measured by Vickers micro hardness apparatus. for all specimens after surface roughness measurement to ensure that the specimens have the same hardness =330.7 HV. The measurements were made with load of 30 kg for 15 second as a hold time. Three measurements were taken at the center of laser spot and averaged to one value.

3. Results and Discussion

'Figure 1' shows the results of crystallite size calculated from X-ray diffraction data. We note from the figure that the crystallite size was 38.08 nm in the case of glaze with laser treatment of the zirconia surface without any addition. That value decreased to 21.34 nm when 5% of nano TiO_2 powder was added and the change caused by TiO_2 in terms of the appearance of other crystalline phases and decreased surface transparency turned white.

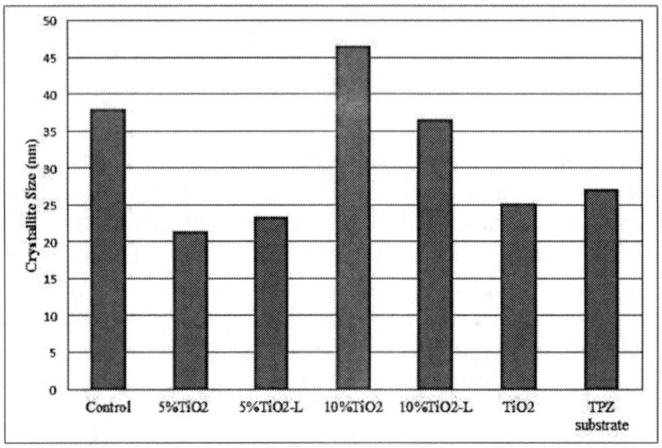

Figure 1. Crystallite size of specimens: (control: TZP with glazes and with laser treatment).

The decrease in crystalline size is due to the increase in the full width at half maximum (FWHM) of the diffraction peak corresponding to a particular crystal plane from 0.23 to 0.4 in degrees when adding 5% TiO_2 compared to control specimen, as shown in 'Figure 2'. M. Vashista & S. Paul [12] found that there is a close relationship between the FWHM and the amount of residual stresses between the glazes coating layer and the surface of the ceramic substrate, where increasing the FWHM means increasing the residual stresses. In contrast, it was observed that laser treatment did not significantly affect the crystallite size and FWHM, therefore expect that the laser will not affect the residual stresses between glaze coating layer and ceramics.

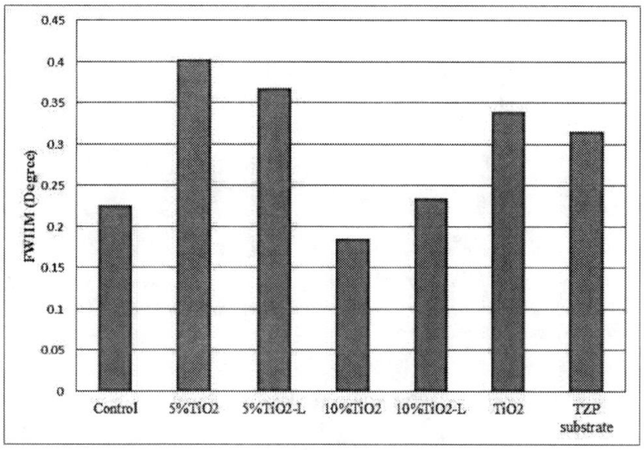

Figure 2. Variation the full width at half maximum (FWHM) with TiO_2 addition and laser treatment.

The increase in TiO_2 percentage up to 10% in the glaze mixture was led to a clear change in crystallite size and increased up to 46.56 nm. This can be attributed to the increased crystalline phases due to increased TiO_2 interaction with alumina on the ceramic surface. The doubling of the TiO_2 percentage has reduced the FWHM to 0.19 in degrees, which means that the residual stresses will decrease. The removal of the residual stresses significantly may be detrimental to the success of the glazing process, in which a certain amount of residual stresses, especially the compressive stresses should be present between glaze layer and ceramic substrate. This undesirable decline of the residual stresses in the glazing of the ceramic surfaces has been processed and improved by laser treatment. This was observed by increasing the FWHM to 0.23 in degrees with decreasing crystallite size up to 36.7 nm.

In the case of a stress-free material, the inter planner spacing d, for a particular reflection (hkl), is constant from one crystallite to another, when it is deformed elastically, the lattice spacing of the crystallites change from their stress free values, and cause a shift in the Bragg angle. The strain calculated from this shift is termed the lattice strain. The lattice strain will depend upon the orientation of the reflecting group of the crystallites with respect to the direction of stress [13].

Percentage of lattice strain behavior is similar to FWHM behavior and the effect of laser treatment is evident on the strain values as shown in 'Figure 3'.

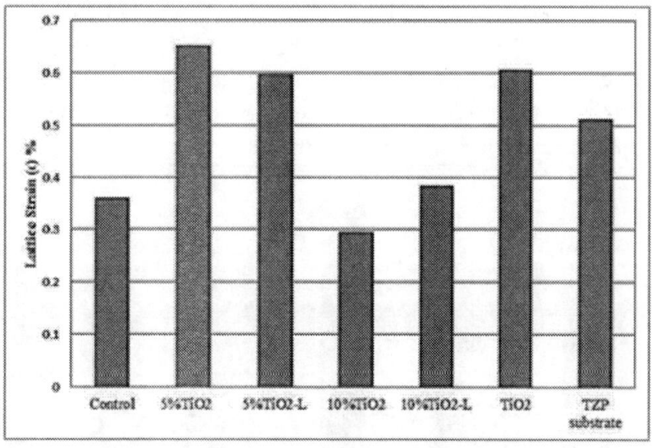

Figure 3. Variation the lattice strain with TiO2 addition and laser treatment.

The average of grain size was determined using the Atomic force microscopy (AFM) device as well as the grain size distribution. 'Figure 4' (a) shows the statistical distribution of grain size to the (TZP-Glazes) control specimen, the grain size ranged between 60 and 120.

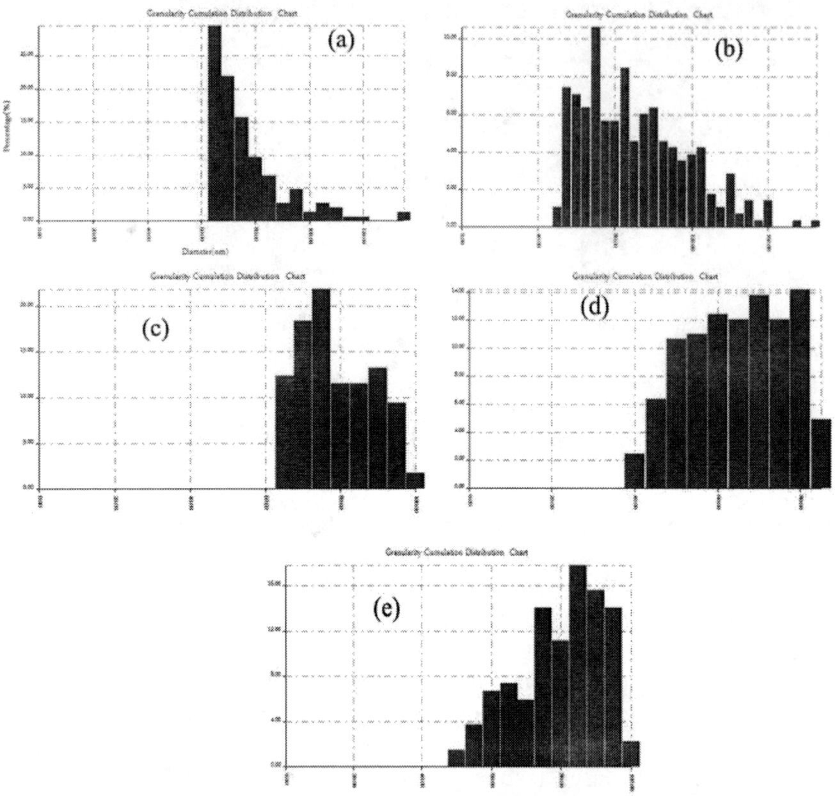

Figure 4. Grain size distribution for: (a) control specimen (TZP + glazes with laser), (b) TPZ+glazes+5% TiO2, (c) TPZ+glazes+5% TiO2 with laser, (d) TPZ+ glazes + 10% TiO2, (e) TPZ+glazes+10% TiO2 with laser.

The statistical distribution diagram of glazes without addition of TiO_2 is more skewed to the right and the average grain size of this specimen is 74.26 nm as shown in 'Figure 5'. The addition 5% of Nano TiO_2 for specimen (TZP-Glazes+5% TiO_2) led to a change in the grain size distribution of the glaze layer and led to decreasing the amount of the Nano grains. The effect of the addition seems clear in the distribution of granular sizes to extend from 50 nm to greater than 140 nm and the shape of statistical distribution is like to bimodal and symmetrical due to the wide range of granular sizes, as shown in 'Figure 4'(b). The average of grain size increased up to 88 due to the emergence of crystallized phases of TiO_2 reaction with glazing mixture components as shown in 'Figure 5' [14].

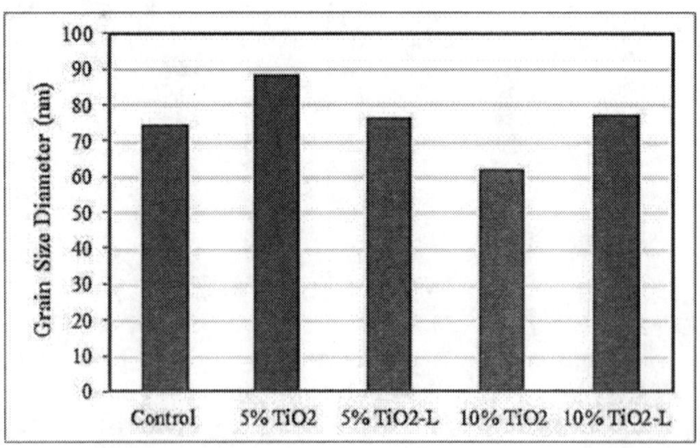

Figure 5 Variation of the average grain size diameter for different specimens.

After the treatment by laser of the specimen (TZP-Glazes+5% TiO_2 with laser), there is a clear change in the grain distribution of the glaze surface. The statistical distribution is narrow to a specific range within nanoscale, which is the range from 65 nm to 100 nm as shown in 'Figure 4'(c). While the average of grain size was decreased to 76.4 nm due to the formation of nuclei formed between the glazes and TiO_2 components with alumina separated on the surface of the ceramic substrate as shown in 'Figure 5'. The statistical distribution of grain size was repeated by adding 10% of TiO2, and this distribution was corresponded to the bimodal and symmetrical with in max Nano size granular range less than 100 nm, as shown in 'Figure 4'(d). The symmetry distribution means more homogeneity in grain size and also obtaining uniform surface in granular distribution, and inhibits the continued growth of granular of glazes mixture. The average grain size was at its lowest value of 62.18 nm, which is due to increased crystallization due to the increase in the amount of TiO_2, which is an important factor in the nucleation and glass-ceramic formation. The laser effect on the 10% TiO_2 enhanced glaze was clear, and the grain size distribution of the specimen in this case is called negatively or left skewed as shown in 'Figure 4'(e). So that the extent of the distribution of grain sizes is wider to range from 50 nm to 100 nm and the average of grain size increased up to 77.47 nm, due to the effect of the laser on the structure of the glaze layer through the fusion of crystalline nuclei together.

The topography images of glazed specimens taken by using AFM were illustrated in the 'Figure 6', the measurement range for all specimens is (2×2 μm). This figure shows 3D topographies profiles under different TiO_2 additives, where the max high (peak-peak) of the grain size was increased with 5% TiO_2 compared to the TiO_2-free coating layer. The value has become 45.32 nm for glazes layer without TiO_2 and decreased to 8.85 nm when 5% TiO_2 was added due to the emergence of crystallization nuclei. The laser effect was evident in removing the high peaks of the grains to reach the value of 2.61nm. The height of the peaks does not change much when added 10%TiO_2 to glaze mixture and reach to 2.39 nm. The maximum value of the peak-peak of the grain size was obtained when the 10% TiO_2 surface was treated with laser to reach 58.26 nm. This means that there was a large variation in the grain size distribution due to the grain growth caused by the laser treatment.

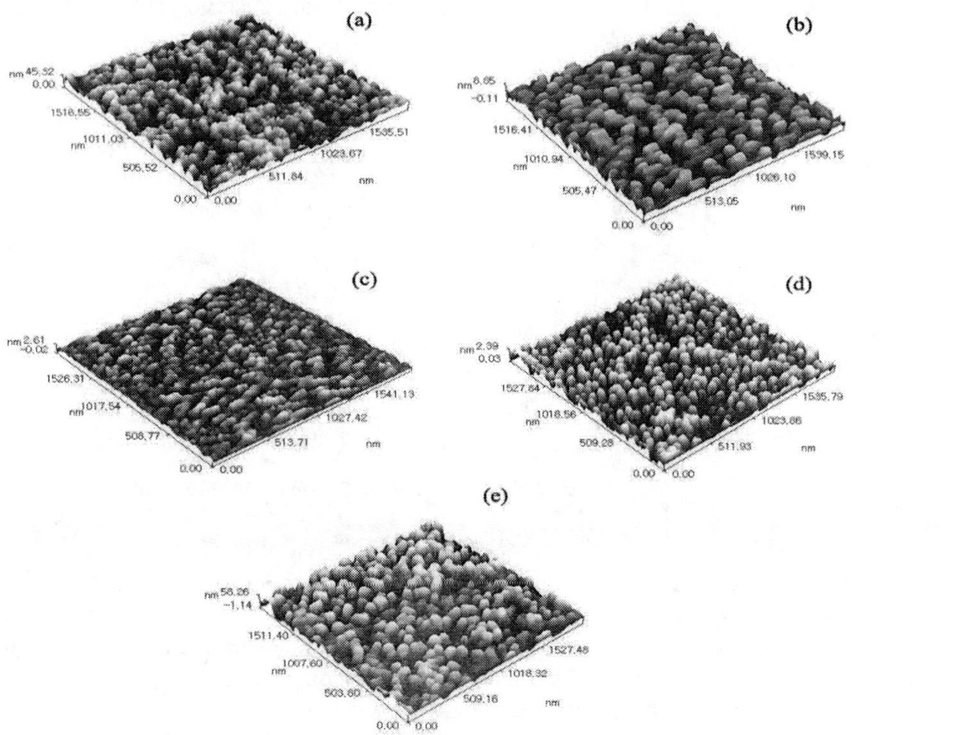

Figure 6. 3D AFM scan data for: (a) control specimen (TZP + glazes with laser), (b) TPZ+glazes+5% TiO_2, (c) TPZ+glazes+5% TiO_2 with laser, (d) TPZ+ glazes + 10% TiO_2, (e) TPZ+glazes+10% TiO_2 with laser.

The roughness was clearly depend on the TiO_2 additions and laser treatment as shown in 'Figure 8' The interpretation of the variation in the roughness values corresponds to the change in grain size peak-peak with grain size average and statistical distribution.

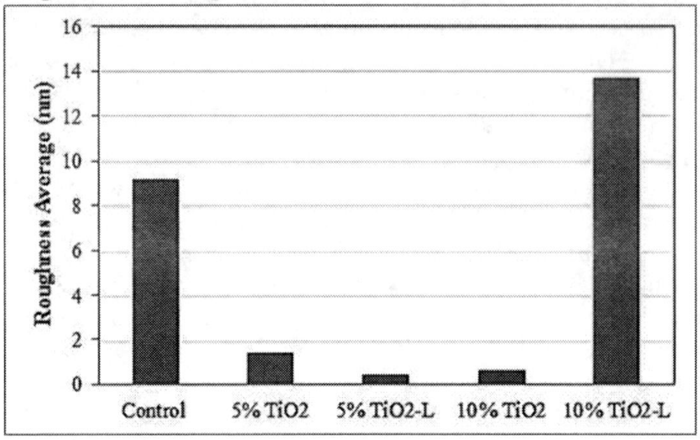

Figure 7. Variation of roughness average for different specimens.

Hardness is one of the important mechanical properties of the material surface. It is a measure of the surface toughness of the scratch and wear resistance. The increase in hardness as shown in 'Figure 8' is due to increased crystalline phases for glazes added 5% TiO_2 and irradiated by laser, but the hardness dropped after adding 10% of TiO_2 and resumed rising after laser treatment. Increasing of hardness not necessarily to increase the toughness of glazed surface, because the toughness is heavily depends on the grain sizes and distribution. In glazed surfaces, increased hardness can be estimated as an increase in residual surface stresses arising from reaction between the glaze components and ceramic substrate. This is useful if the remaining stresses are in a compressive state and this is due to the difference between the coefficient of thermal expansion of the ceramic substrate and glaze layer [15].

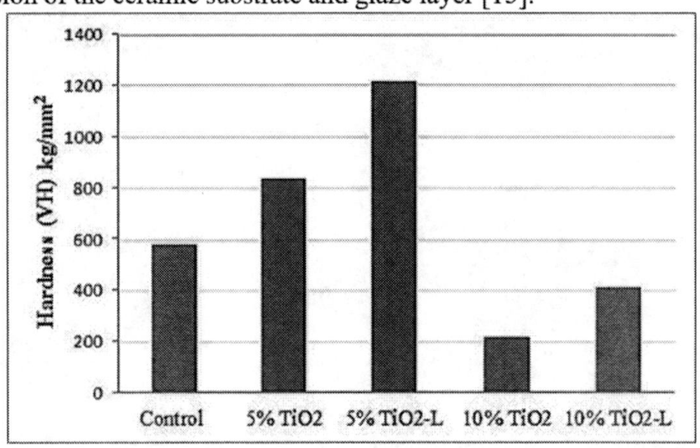

Figure 8. Variation of Hardness for different specimens.

'Figure 9' represents the x-ray diffraction pattern for all specimens sintered at 1530°C and glazed with different percentages of frit glass supported with nano titanium dioxide powder. Scheme TZP shows the main structures of the substrate ceramic material without glazing, and it is clear that the components are tetragonal zirconia (t-ZrO_2) and alumina (α-Al_2O_3), which represent the components of toughed zirconia. Alumina increases the toughness when added to zirconia at limited percentage. [10] We also notice that the percentage of cubic phase is low and the monoclinic phase is absent. After glazing without any addition of TiO and with laser surface treatment, we observe the appearance of amorphous phases between the crystalline phases represented by the peaks of the TZP as in the scheme (TZP-Glazes). These amorphous phases led to atrophy of the α-alumina phase (corundum) [11]. Adding 5% of TiO_2 to the glazes and without laser treatment, we notice a slight change in the components of the glass layer as in the scheme (TZP-Glazes+5% TiO_2), and thus the little (5%) ratio of the rutile enhanced the appearance of the ceramic glass. While after the treatment of the surface by laser, this process led to the emergence of crystalline phases clearly as in the scheme (TZP-Glazes+5% TiO_2 with laser), and this indicates the increase in the proportion of glass-ceramic composite.

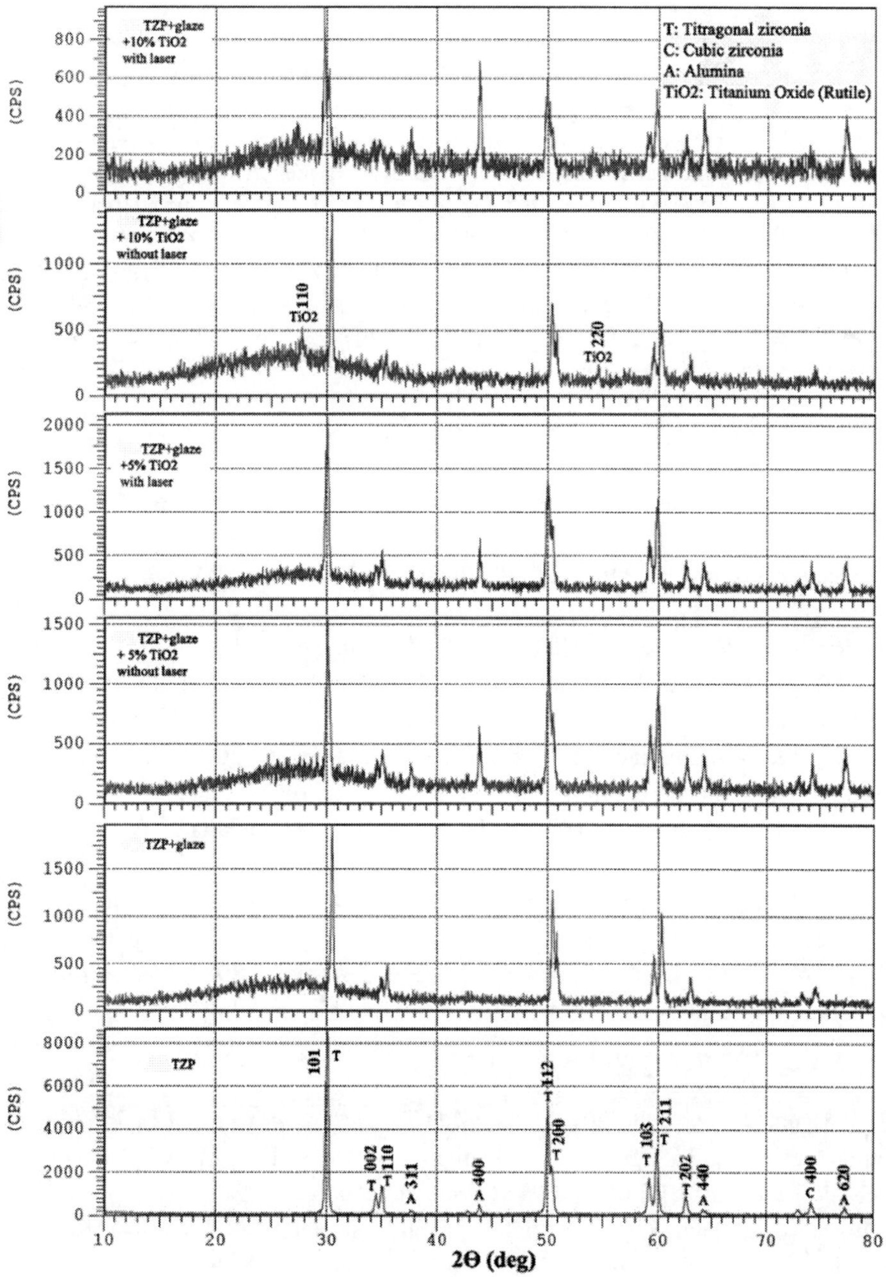

Figure 9. X-ray diffraction patterns of specimens.

The increase of the TiO to 10% without laser treatment led to the emergence of the rutile phase with the increased appearance of crystalline phases in the glass mixture, or the transformation of part of the amorphous glass compounds into crystalline phases, as shown in scheme (TZP-Glazes+10% TiO_2) . Thus, new characteristics may emerge that may help to improve the properties associated with synthetics such as mechanical properties. After laser treatment, all crystalline phases have become clear, and new phases can be introduced if the surface is treated with higher temperatures. This is evident from sche me (TZP-Glazes+10% TiO2 with laser).

4. Conclusions

The addition of titanium oxide is an important factor in the appearance of crystallized and nucleating agent of ceramic phases due to the interaction with the components of glaze mixture and with alumina in the zirconia components type TZP. Surface treatment by CO_2 laser, enhanced the appearance of crystalline phases and the appearance of overlay of glass-ceramic. In addition to the control of grain size, especially in the large additions of titanium oxide (10% TiO_2) , where the laser increases the grain growth of the glaze layer and thus grain size average was increases. One of the most important roles played by laser irradiation is to control the residual stresses. If their values are high due to grain growth or the difference in thermal expansion coefficients, the laser reduces these stresses. If the residual stresses are low, laser irradiation works to increase them so that the glaze layer remains in permanent compressive stress, which is the most important factor in the success of the glazing process. CO_2 laser reduces roughness on the surface of 5% TiO_2 specimen, while it cannot do by 15 W when doubling the amount of TiO_2 to 10% due to the centers of glass-ceramic nuclei increases and that need higher laser energies or longer irradiation times, and because the glazing coted layer needs high hardnees, so the optimum conditions to achieve this is by adding 5% nano Tio_2 ceramics and the power of laser irradiation is 15 Watt.

References

[1] Vagkopoulou, T., Koutayas, S. O., Koidis, P. and Strub, J. R. (2009). *European Journal of Esthetic Dentistry*, *4*(2).

[2] Mahalaxmi Sekar, V Sujatha, Renjith Babu and Ajit George Mohan; Article, *ISSUE* 1, January 2014.

[3] Thamaraiselvi, T. and Rajeswari, S. (2004). *Carbon*, *24*(31), 172.

[4] Minh, N.Q., Badwal, S.P.S., Bannister, M.J. and Hannink, R.H.J., 1993. Technomic, Lancaster, pp.652-687.

[5] Rezvani, M., Marghussian, V. K. and Eftekhari Yekta, B. (2011). *International Journal of Applied Ceramic Technology*, *8*(1), 152-162.

[6] Abdallah, R.M., Hammouda, I.M., Mohammed, M.K., Abouelatta, O.B. and El Fallal, A.A.,

[7] Rajesh Kumar, B. and Hymavathi, B. (2017). *Journal of Asian Ceramic Societies*, *5*(2), 94-103.

[8] Aly, K. A., Khalil, N. M., Algamal, Y. and Saleem, Q. M. (2016). *Journal of Alloys and Compounds*, *676*, 606-612.

[9] Aly, K. A., Khalil, N. M., Algamal, Y. and Saleem, Q. M. (2017). *Materials Chemistry and Physics*, *193*, 182-188.

[10] Maji, A. and Choubey, G. (2018). *Materials Today: Proceedings*, *5*(2), 7457-7465.

[11] Yamamoto, L. T., Rodrigues, V. A., Dornelles, L. S., Bottino, M. A., Valandro, L. F. and Melo, R. M. D. (2016). *Brazilian dental journal*, *27*(5), 543-547.

[12] Vashista, M. and Paul, S. (2012). Philosophical Magazine, 92(33), 4194-4204.

[13] Balasingh, C. and Singh, A. K. (2000). *Metals materials and Processes*, *12*(2 amp;), 269-280.

[14] Chavoutier, M., Caurant, D., Majérus, O., Boulesteix, R., Loiseau, P., Jousseaume, C. and Lecomte, E. (2014). *Journal of Non-crystalline solids*, *384*, 15-24.
[15] Wang, Y., Darut, G., Luo, X. T., Poirier, T., Stella, J., Liao, H. and Planche, M. P. (2017). *Ceramics International*, *43*(5), 4606-4611.

Structural and optical properties of Ag_2S nanotubes prepared by laser ablation in liquid

Hanan A .Rawdhan , Duha S. Ahmed ,Raid A.Ismail

Department of Applied Science, University of Technology, Baghdad, Iraq

Corresponding Author email: duhasaadi2015@gmail.com

Abstract: Synthesis of Ag_2S nanotubes (NTs) by laser ablation of silver target in Thiourea (Tu) aqueous solution without using catalyst was demonstrated. The structure, morphology, size and elemental composition of Ag_2S NTs were investigated. X-ray diffraction (XRD) results show that the Ag_2S NTs are crystalline with monoclinic phase. The optical energy gap of Ag_2S NTs is around 1.6 eV at fluence 4.7 Jcm^{-2}. Transmission electron microscope TEM investigation reveals the formation of NTs Ag_2S morphologies with the diameter of 26-30 nm and lengths of 200-400 nm. Raman spectra of Ag_2S nanorods show the presence of active Raman peaks indexed to A_g modes and the stretching bending Ag-S bonds and 2LO mode of Ag_2S.

Keywords: Silver sulfide nanotubes; Laser ablation; Thiourea; Laser fluence

1. Introduction:

In recent years, the synthesis of nanometer-sized semiconductors has attracted great interest due to their shape- and size-dependent physical and chemical properties [1-4].
As one of the most important semiconductors, Ag_2S is a favorable prospective photoelectric and thermoelectric material with potential application in optical and electronic devices such as photoconductive cells, infrared detectors, photoconductors and magnetic field sensors, etc. [5-6] Various techniques for the synthesized of Ag_2S nanostructures and their formation mechanism have been studied [7-11].
For example, Zhao et al. prepared rod-like Ag_2S nanocrystals by using sodium thiosulfate $Na_2S_2O_3$ as a sulfur source via gamma-ray irradiation of aqueous solutions at room temperature [12]. Recently, the research of one-dimensional (1D) nanomaterials, including nanorods, nanobelts, nanotubes and nanowires, is the most active area due to their potential application in nanodevices. A number of methods have been developed to fabricate and assemble 1D nanostructure such as organometallic precursor, sacrificial template, gamma-ray irradiation, solvo-thermal, sonochemistry and microwave irradiation method [13-14] .This

paper focuses on the preparation and characterization Ag_2S nanotubes by using laser ablation in pure thiourea (Tu) aqueous solution without catalyst.

2. Experimental work

Colloidal Ag_2S nanotubes were prepared by laser ablation of high purity Ag target (99.9%) in 0.3M thiourea (Tu) aqueous solution. The Tu represents a source of sulfur. The laser used was for this purpose was second harmonic Q-switched Nd:YAG laser operating at λ=532 nm,7ns pulse width and 2Hz repetition frequency. High purity silver pellet with thickness of 10mm and diameter of 20mm was located in the bottom of glass vessel filed with Tu aqueous solution and the height of solution was 2mm above target. The laser was focused on the Ag pellet by using positive lens of 10cm focal length. The laser fluence for ablation was 4.7 Jcm^{-2}/pulse and the ablation time was set 15min for each sample. Figure 1 shows the experimental schematic diagram of pulsed laser ablation in liquid PLAL system used in this work.

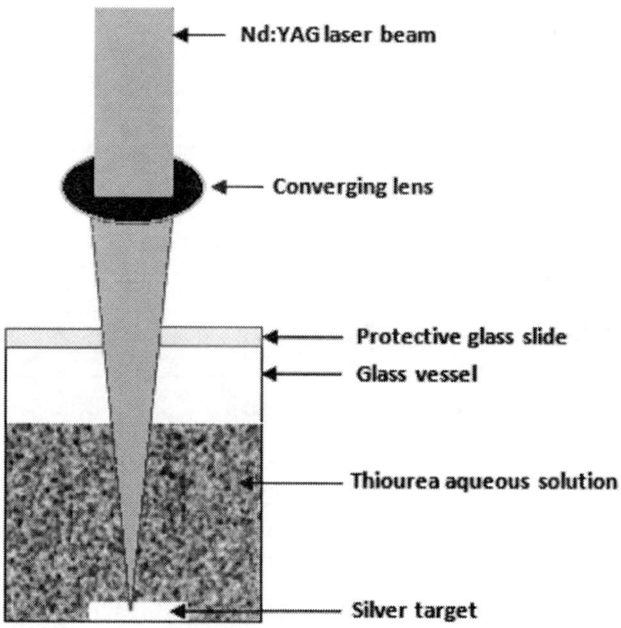

Fig 1. Experimental set-up of PLAL system

2.1 Characterization

The optical absorption of colloidal Ag_2S nanotubes was measured by using UV–Vis double beam spectrophotometer (Lambda 750, Perkin Elmer). To study the structural properties of Ag2S nanotubes, x-ray diffractometer (XRD-6000, Shimadzu) was employed. Raman spectroscopy of the Ag_2S was performed using Raman spectrometer (Bruker Optics, Germany). The morphology and size of nanotubes were transmission electron microscope TEM (EM208, Philips) with EDX analysis to determine the chemical composition of the product.

3. Results and discussion

The XRD patterns of Ag_2S nanotubes prepared at laser fluence of 4.7J cm^{-2} is shown in Figure 2.

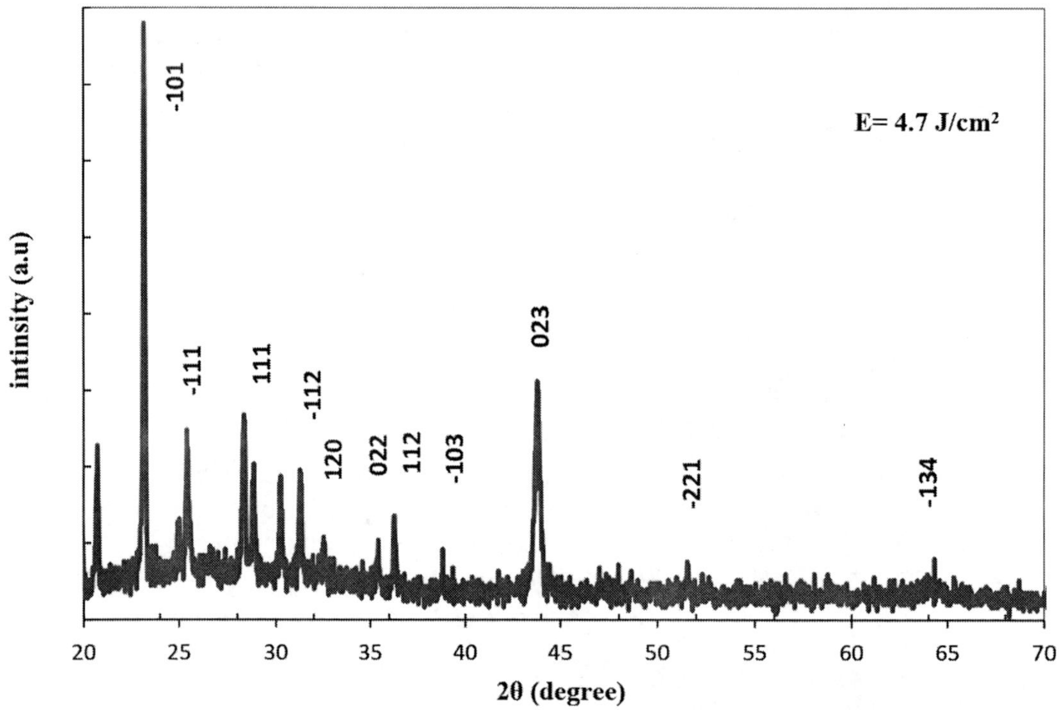

Figure 2. XRD patterns of Ag_2S NTs

From the Figure 2, we found there are several XRD peaks at 2θ =23.16°, 25.4°, 28.34°, 31.3°, 32.54° , 35.42°, 36.26, 38.82°,43.8°, 51.54° and 64.32° corresponding to (-101),(-111),(111),(-112),(120),(022),(121),(-103),(023),(-221) and (-134) planes. All these peaks are belonged to crystalline Ag_2S nanotubes with cubic phase according to (JCPDS Card # 652356) [15]. These XRD peaks indicate formation Ag_2S NTs with highly crystalline and pure Ag_2S. . The average crystallite size D along of Ag_2S nanotubes along (-101) plane was calculated from XRD peak broadening using Debye–Scherrere quation:

$$D_{ave} = \frac{0.9\lambda}{\beta\, cos\theta} \quad \ldots\ldots\ldots (1)$$

IOP Conf. Series: Materials Science and Engineering **757** (2020) 012027 doi:10.1088/1757-899X/757/1/012027

Where the β is the full width at the half maximumi nunits of radians and λ is the X-Ray wavelength of CuKα source (0.1549nm). According the XRD analysis of Ag_2S nanotubes, the average diameter of Ag_2S nanotubes was 43 nm.

Figure 3 (a) shows the TEM image of Ag_2S nanotubes which confirms the formation of monodisperse nanotube adhered with some of nanoparticles. The diameter of nanotubes was around 75nm and about 2μm in length. The nanoparticles adhered to nanotubes are spherical with average diameter of 85nm.As shown; the nanotubes are distributed over the glass substrate with high density. EDX data shows the presence of two peaks related to S and Ag with small-off stoichiometry Ag_2S NTs as shown in Figure 4[16-17].

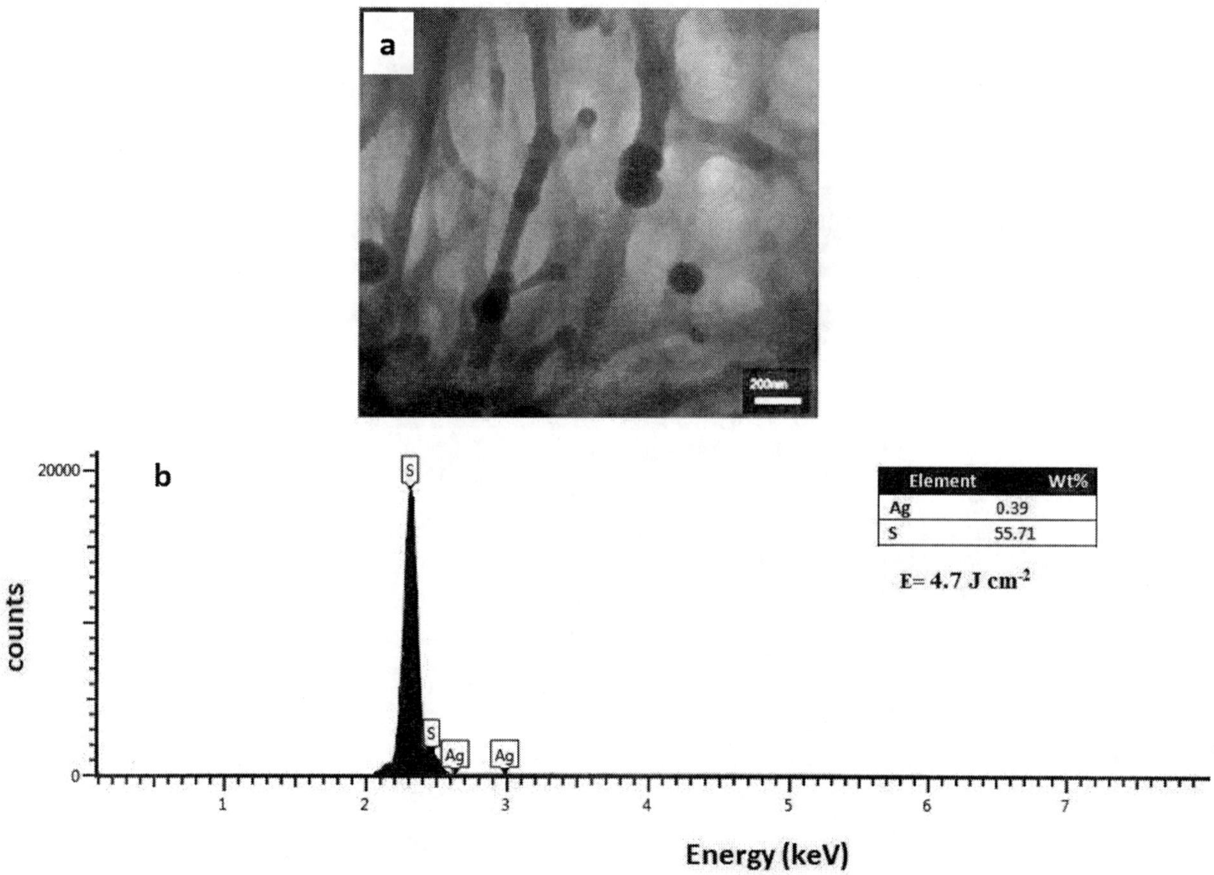

Figure 3.a) the TEM images with b) EDX of Ag_2S nanotubes at laser fluence 4.7 Jcm^{-2}

According to EDX analysis as shown in Figure (3 b), the content of Ag and sulfur in Ag_2S NTs formation is about 0.39 and 55.71 wt.%, respectively. These results reveal that the content of sulfur is higher than Ag due to the stoichiometry of Ag_2S at high laser laser flunce [18].

Figure (4 a), shows the UV-Vis absorption of colloidal Ag_2S NTs prepared at laser fluence 4.7 Jcm^{-2}. It is obviously reveal that the optical absorption peak appear at about 315 nm due to the absorption edge of the Ag_2S. Besides, the optical energy gap of Ag_2S NRs was calculated using Tauc plot and the extrapolation of the linear part of the curve as shown in Figure (4 b) is about 1.6 eV. The energy gap of Ag_2S NTs was larger than bulk Ag_2S (1.5eV) at room temperature due to quantum size effect.

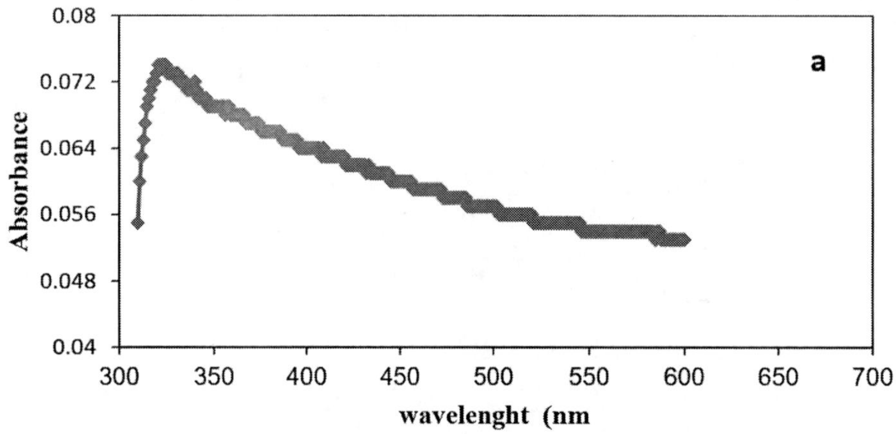

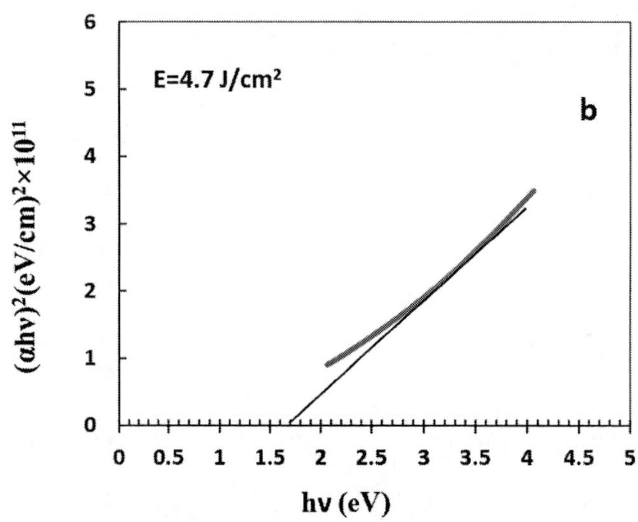

Figure 4: a) Optical absorption of Ag_2S NTs prepared at laser fluence 4.7 Jcm^{-2} and b) $(\alpha h\nu)^2$ versus photo energy plot

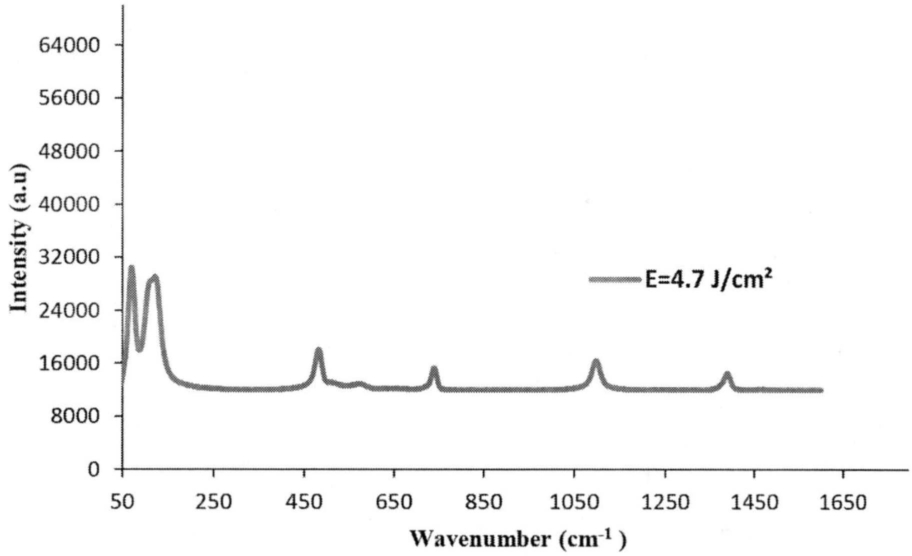

Figure 5: Raman spectra of Ag_2S NTs prepared in Tu solution laser fluence 4.7 Jcm^{-2}

Figure 5 displays Raman spectrum of Ag_2S NTs. Raman peaks are noticed at 67 cm^{-1} and 100 cm^{-1} which can assigned to A_g modes and the stretching bending Ag-S bonds. The peak at 475 cm^{-1} is related to second order overtone 2LO mode of Ag_2S NRs. The peaks positions at 738, 1093 and 1385 cm^{-1} are related to the optical transitions connected with Ag vacancies state [19].

4. Conclusion

In this study, we report on the preparation and characterization of Ag_2S nanotubes by laser ablation in liquid. XRD reveals the formation of crystalline cubic Ag_2S. SEM investigation confirms the formation of nanotubes with some of spherical nanoparticles. The optical energy gap of Ag2S NTs was 1.6eV at room temperature. Raman spectrum confirms the presence of Ag modes and the stretching bending Ag-S bonds and 2LO mode of Ag_2S.

5. References

[1]Selinsky R S, Ding Q, Faber MS, Wright JC, Jin S 2013 Chem Soc Rev **42**:2963–2985

[2] PanL LT, Liu X, Lu T, Zhu G, Sun Z, Sun CQ 2012, Catal Sci Technol **2**:754–758

[3] Brelle MC, Zhang JZ 1998, Chem J Phys **108**:3119

[4]Bruhwiler D, Leigener C, Glaus S, Calzaferri G 2002. J Phys Chem B **106**:3770

[5] F. Gao, Q. Lu, D. Zhao 2003, Nano. Lett. **3**: 85–88.
[6]X. Liu, Y. Zhu, Q. Zhang, Z. Li, B. Yang 2012 ,Mater. Res. Bull. **47** :4263–4265.

[7] Buda C, Chen X B, Narayanan R, Sayed EI 2005, Chem. Rev. **105**(4) 1025-1102.
[8]Zhang W, Xu L, Tang K B, Li F, Qian W T 2005, Eur J Inorg Chem **4**:653–656

[9]Liu Q, Guo F, Komarneni S 2004, J Am Chem Soc **126**:54–55

[10] Kuang D, Xu A, Fang Y, Liu H, Frommen C, Fenske D 2003, Adv Mater **15**:1747–1750

[11] Chen X, Wang X, Wang Z, Yang X, Qian Y 2005. Cryst Growth Des **5**:347–350

[12]Zhao, Y., Zhang, D.W., Shi, W.F. 2007 Mater. Lett. **61**(14–15): 3232–3234

[13] Du N., Zhang H., Sun H., Yang D., 2007 Mater. Lett. **61** 235–238.

[14] Yan S., Wang H., Zhang Y., Li S., Xiao Z. 2008, J. Non-Cryst. Solids **354** 5559–5562.

[15]Yan-Li Z, Xin-Yuan X, Ming L, Shu-Ming X, Jie-Mei C and Wen-Jie Z. 2016 Nanotechnology **27**(22) (8pp)

[16] Chen A, Wang H, Li X, 2005 Chem Commun **14**:1863–1864

[17] Zhai H, Wang H, (2008) Mater Res Bull **43**:2354–2360

[18]S.I. Sadovnikov a, A.I. Gusev , A.A. Rempel 2015 Superlattices and Microstructures **83** (2015) 35–47.

[19] Alekberov O., Jahangirli Z., Paucar R., Huseynova S., Abdulzade N., Nakhmedov A., Wakita K., Mamedov N., 2015 Phys. Status Solidi **C 12**, 672

Functionalization, characterization, and antibacterial activity of single wall and multi wall carbon nanotubes

M. K A Mohammed, M. R Mohammad, M. S. Jabir and D. S Ahmed*
Department of Applied Science, University of Technology, Baghdad, Iraq
*Corresponding author email: duhasaadi2015@yahoo.com

Abstract

The raw-SWNTs and raw-MWNTs were chemically oxidized with a mixture of sulfuric acid and nitric acid under ultrasonic vibration. The functionalized-SWNTs (F-SWNTs) and functionalized-MWNTs (F-MWNTs) were characterized by using ultra violet-visible spectrophotometer, X-ray diffraction, and field emission scanning electron microscopy (FESEM). The antibacterial efficiency of SWNTs and MWNTs towards *Escherichia coli* (*E. coli*) as gram-negative bacteria was evaluated via viable count method and fluorescence microscopy. The results of viable count method showed that the SWNTs and MWNTs have higher inhibitory effect after being treated with H_2SO_4/HNO_3.The *E. coli* images under fluorescence microscopy exhibited that almost red color for dead cells, which confirms the efficient lethal ability of F-SWNTs and F-MWNTs.

Keywords: Carbon nanotubes, functionalization process, The antibacterial activity of CNTs, fluorescence microscope,pathogenic cells.

1. Introduction

Carbon nanotubes (CNTs) is one of the most useful substance as antibacterial agent due to its phenomenal properties such as high surface area, high microbial adsorption ability compared with other adsorbent materials, and high bactericidal ability comes from its tubular shape and presence of metals used as catalyzed in production procedure of CNTs [1,2]. In previous studies, CNTs have been applied by researchers on broad range of bacteria, both Gram negative and Gram positive [3,4]. But, one crucial parameter extensively influences on antibacterial activity of CNTs, which is the poor dispersion stability in many solvents [5]. The functionalization process of CNTs based on chemical treatment following by ultrasonic vibration can be utilized to increase dispersion stability of CNTs suspension in the solvent [6]. Using of aggressive acids leading to introduce oxygen groups with negative charge, which separate tubes aggregation from each other due to reducing the Van der Waals force between tubes [7]. In the present work, single wall nanotubes and multi wall nanotubes were functionalized with the mixture of H_2SO_4/NHO_3 and characterized by UV-vis spectroscopy, XRD, and FESEM. The antibacterial activity of CNTs against *E. coli* was assessed and discussed.

2. Experimental work

2.1. Chemicals

SWNTs (purity ~99 wt%; diameter~1-6 nm; length~0.5-3 μm) and MWNTs (purity>95 wt%; diameter~7–16 nm; length~12-50 μm) were provided from (cheaptubes.com, Grafton, USA). Other chemicals includes anhydrous ethanol (C_2H_5OH, 99.9%), sulfuric acid (H_2SO_4, 95%), and nitric acid (HNO_3, 65%) were purchased from Sigma-Aldrich.

2.2. Functionalization of SWNTs and MWNTs

One gram of raw-SWNTs was oxidized with 100 ml in a 3:1 mixture of 65% H_2SO_4/HNO_3 to generate OH- and/or COOH functionalized-SWNTs (F-SWNTs). Then,thesolution was placed in an ultrasonic device at 40 °C for 30 min, and diluted with deionized water, filtered through a 0.22 μm cellulose nitrate membrane, and dried at 120 °C for 14 h. The same step was repeated for raw-MWNTs to prepare functionalized-MWNTs (F-MWNTs).

2.3. Viable count method

The viable cell wasassessed with the traditional count method [8]. The *E. coli*were seeded on MuellerHintonmedia at 37 °Cforone day. The samples (raw-SWNTs, raw-MWNTs, F-SWNTs, and F0-MWNTs) with concentration 1mg/ml were dispersion in9 ml of saline and sterilized in an autoclave at 121 °C for 40 min [9,10]. Bacterial solution was adjusted to obtain cell samples with initial concentration about 10^7 CFU/ml and fit with the McFarland turbidity solution. After that, 1ml of bacterial solution was added into the samples and kept in a shaker incubator at 37 °C(160 rpm) for 40 min. The solution of 1ml of samples-cell was serially diluted with saline (1:1000) [11,12]. The viable cell in each plate was assessed by spreading of 100 μl of the dilution onto Nutrient media plate for incubation at 37 °C. After 24 h, colonies on each plate were measured and viable cell number was evaluated as colony forming-units per milliliter (CFU/ml) [13,14].

2.4. Fluorescence imaging

The fluorescence microscope wasutilized to image the dead and live cells after treating with CNTs.This done with acridine orange/ethidium bromide (AO/BO) dual staining[15,16]. At first,40μl of cells-samples solution was added with 40 μl of the (AO/BO) fluorescent dye at concentration 20 μg/ml for 10 min. Afterwards, a drop of the solution was spread on a glass slide, and fluorescent images were taken with the fluorescence microscopy [17].

3. Results and discussion

3.1. Optical properties

The optical properties of F-SWNTs and F-MWNTsdissolved in ethanol were measured as a function of wavelength in the range of 200 to 600 nmasdepicted in figure 1. The samples show a feature absorption band locatedathighest intensity of spectra, which are assigned with π–π* transition of CNTs structure. The functionalization of CNTs causes red shift compared with raw-CNTs due to the electronic interaction of molecular orbitals of oxygen groups withCNTs, making a new molecular orbital and decreasing the band difference [18].

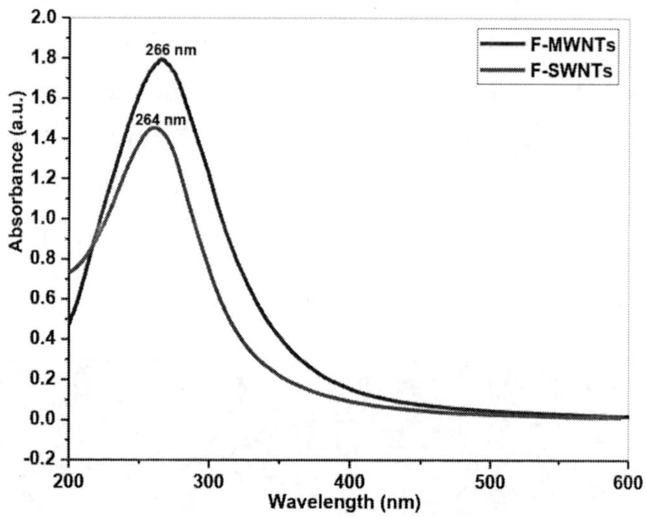

Figure 1.UV-Vis spectra for F-SWNTs and F-MWNTs

3.2. Structural properties

Figure 2 exhibits XRD patterns of F-SWNTs and F-MWNTs. The high dominate peaks located at 2θ = 25.68° for F-SWNTs and 2θ = 25.78° for F-MWNTs, can be ascribed to the (002) diffraction plane, which are match well with the graphite structure (JCPDS no. 01-0646). The weak intense peaks centered at 2θ = 43.96° for F-SWNTs and 2θ = 43.54° for F-MWNTs are characteristic peaks for the (100) diffractionplane. Since, the structure of SWNTs and MWNTs is protected even after undergoing the acid treatment .Besides, the XRD pattern of Raw-SWNTs shows broad diffraction peak at 2θ = 25.68° while narrow broadening peak at 2θ = 25.78° due to the small interlayer distance of 3.5˚Ain SWNTs as compared to MWNTs [19].

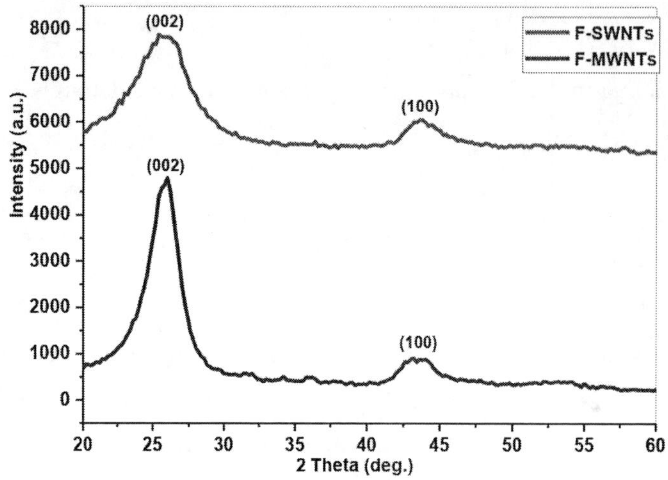

Figure 2.XRD pattern for F-SWNTs and F-MWNTs.

3.3. Morphological properties

The morphological properties of F-SWNTs and F-MWNTs were characterized via FESEM analyses. Figures 3 (a) and (b) shows FESEM micrograph of raw MWNTs and F-MWNTs, respectively. While, Figures 3 (c) and (d) display FESEM micrograph of raw-SWNTs and F-SWNTs, respectively. It can be easily found that raw-MWNTs and raw-SWNTs images show aggregatednanotubes like bundles, due to Van der Waals force exists betweenthem.The F-MWNTs and F-SWNTs show less aggregates and tangled clusters with shorter tubes and more defective sites were found on the surface compared with than R-MWNTs and R-SWNTs.

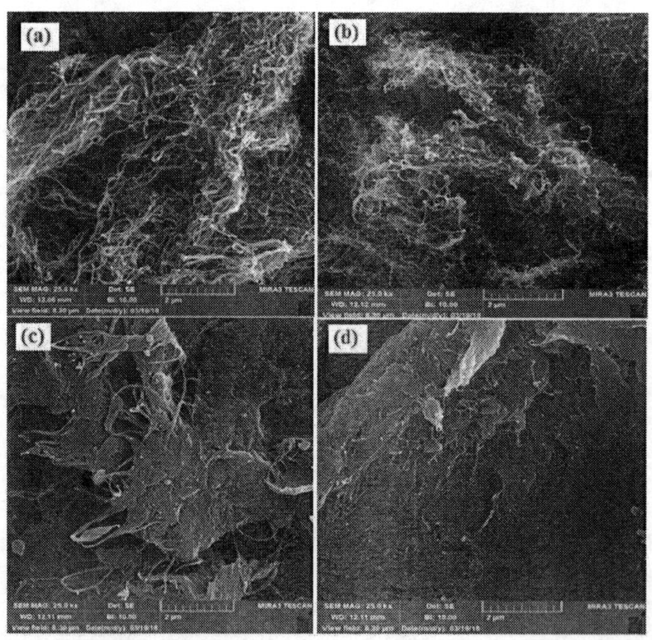

Figure 3.FESEM images for a) R-MWNTs, b) F-MWNTs, c) R-SWNTs, and d) F-SWNTs

3.4. Antibacterial activity

The Antibacterial activity was done using the viable count method. At first, the antibacterial activity of the R-SWNTs and R-MWNTs alone was tested against *E. coli* as shown in Figures 4 (a) and (b). Both R-SWNTs and R-MWNTs were inefficient in killing the *E. coli* at 1 mg/mL after incubation.On the other hand, the number of *E. coli* colonies of decreases significantly after being treated with F-SWNTs and F-MWNTs as demonstrated in Figure 4 (a) and (b). This is causing by effecting of oxygen functional groups. The mechanism of CNTs bactericidal ability is mostlyaffectedby length, residual catalyst, electronic structure, surface functional group, and surface chemistry [20,21,22]. Essentially, the length of CNTs is important during interactions with the bacteria membrane. Longer tubes have shown higher antibacterial activity than shorter ones. The short length of tubes are more likely to self-aggregate without including a large number of bacterial cells, while longer tube agglomerates are more bio-effective as they affect a larger number of cells in the aggregates as shown in Figure 5.

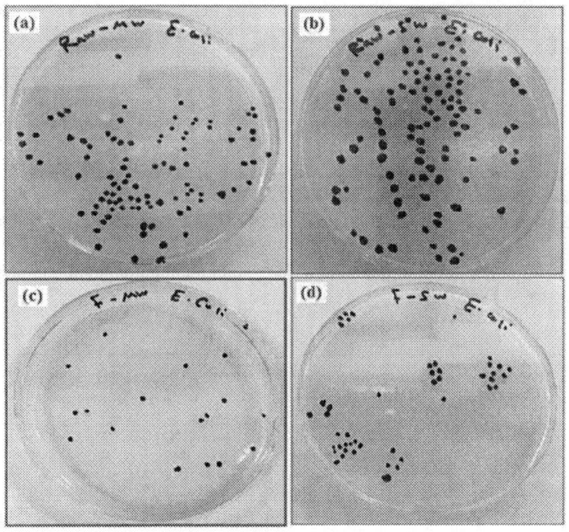

Figure 4. Survival *E. coli* cells incubated with a) R-MWNTs, b) R-SWNTs, c) F-MWNTs, and d) F-SWNTs.

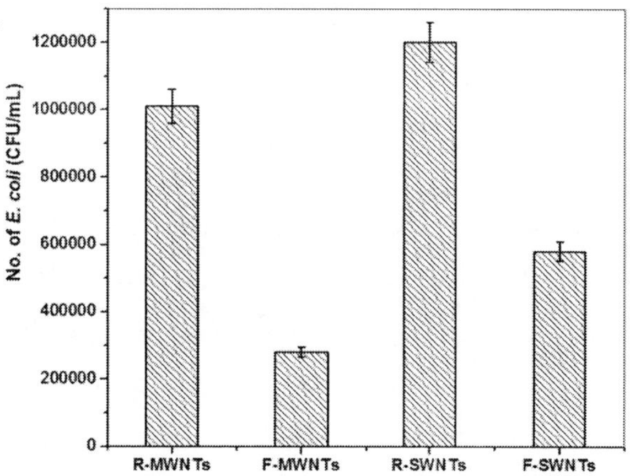

Figure 5.*E. coli* viable number after incubated with R-MWNTs, R-SWNTs, F-MWNTs, and F-SWNTs.

Besides, in Figure (5) revealed that the R-SWNTs (before acid treatment) and F-SWNTs (after acid treatment) possessed higher antibacterial activity towards *E. colis* bacterial strains compared withF-MWNTs as well as R-MWNTs. These results were attributed to the physical interaction with the cell membrane, the large formation of cell-samples aggregates, and disruption of membrane function, hence resulting in cell death.

Figure 6 shows fluorescence microscope images of the *E. coli* cells after being treated with F-SWNTs and F-MWNTs. It can be seen that fluorescence microscopy image of the distinctive bactericidal action was displayed with almost all the dead cells with red color [23,24,25], which indicated F-SWNTs and F-MWNTs have a potential application as an antibacterial agent for inactivating pathogenic cells.

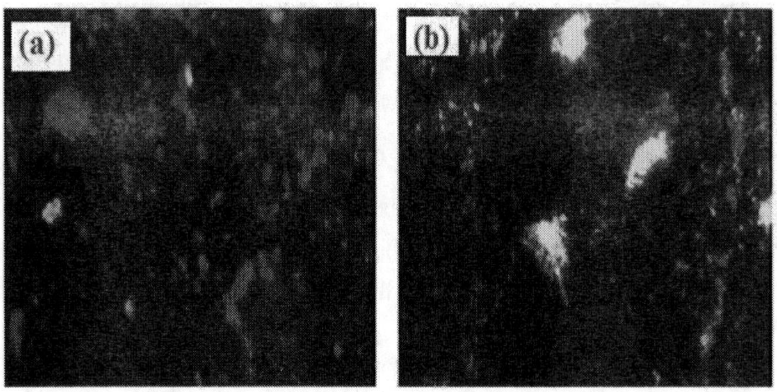

Figure 6. Fluorescence images of *E. coli* incubated with a) F-MWNTs and d) F-SWNTs.

4. Conclusions

The raw-SWNTs and raw-MWNTs have been successfully functionalized by the mixture of sulfuric acid and nitric acid. The UV-Vis spectroscopy showed that the characteristic absorption band for SWNTs and MWNTs due to π–π* transition. The SWNTs and MWNTs hexagonal structure was confirmed by XRD pattern. The tube-like structure of SWNTs and MWNTs was successfully verified by FESEM images.Finally, the study demonstrated that the F-SWNTs and F-MWNTs have higher antibacterial performance in comparison with R-SWNTs and R-MWNTs.

Reference

[1] Salisu N, Mohd Z, Zulkarnain Z, NorA.2018,*Materials***11** 1-24.

[2] DizajS M, Mennati A, Jafari S,Khezri K , Adibkia K2015 , *Adv. Pharm. Bull.***5** 19–23

[3] Liu H,Zhang L,Yan M,Yu J, 2017*Mater J., Chem. B***5** 6437–6450

[4] Botta L, Bizzarri B M, Crucianelli M, Saladino R ,2017*J. Mater. Chem. B***5** 6490-9510

[5] Jia G, Wang H,Yan L, Wang X , Pei R , Yan T, Zhao Y, GuoX, 2005*Environ. Sci. Technol.***39** 1378–1383

[6] Yang K, Li Y, Tan X,Peng R, Liu Z, 2013 *Small* **9** 1492-1503

[7] Taqi Z. J.,Hamad M. A., Jabir M.S. 2019. *Research Journal of Biotechnology* **14** 156-159.

[8] Alsaedi I. I, Taqi Z. J., Abdul Hussien A. M., Sulaiman G. M., Jabir M. S., 2019 *Materials Research Express* **6** 095413.

[9] Ibraheem S. A., Haitham A K, Sarmed A H, Majid S J, Raluca G, Marcela P, Irina G, Mihailescu D F, 2019, *Romanian Biotechnological Letters* **24** 286-293.

[10] Kadhim W. K ., Nayef U. M, Jabir M. S., 2019 *Surface Review and Letters* 1950079.

[11] Khashan K. S., Jabir M. S., Abdulameer F. A., 2018 *Materials Research Express* **5** 035003.

[12] Haitham A K, Sumayah A I, Majid S J, Afraa A K, 2019 *Nano Biomed. Eng* **11** 35-43.

[13] Khashan K. S., Jabir M. S., Abdulameer F.A ., 2019 *Surface Review and Letters* 1950078.

[14] Majid S J, Uday M N, Abdul K, Waleed K, 2019 *Nano Biomedicine and Engineering* **11** 56-63.

[15] Ali I. H., Jabir M. S., Al-Shmgani H., Sulaiman G.M., Sadoon A.H., 2018 *Journal of Physics: Conference Series* **1003** 012009.

[16] Kadhem H. A., Ibraheem S. A., Jabir M. S., Afraa A. K., 2019 *Nano Biomed. Eng* **11** 35-43.

[17] Ali S. H., Sulaiman G.M., Al-Halbosiy M. F., Jabir M.S., Hameed A. H., 2019 *Artificial cells, nanomedicine, and biotechnology* **47** 378-394 .

[18] Duha S A, Adawiya J H, Mohammad M R,2013 *Energy Procedia***36** 1111-1118

[19] Adawiya J H, Amin D T, Duha S A, Mohammad M R 2016 ,*AIP Conf. Proc.* **1758** 030003

[20] Radhi M. M., Abdullah H. N., Jabir M. S., Emad A. J., 2017 *Nano Biomed. Eng* **9**103-106.

[21] Jabir M. S., Nayef U. M. , Jawad K. H., Taqi Z. J., Ahmed N. R,2018 IOP Conference Series: Materials Science and Engineering **454** 012077.

[22] Younus A. H, Ahmer S. , Jabir M. S., 2019 *Research Journal of Biotechnology* **14** 131-133.

[23] Abdulkadhim W. K., Jabir M., Nayef U. 2019 *Research Journal of Biotechnology* **14** 89-93.

[24] Jabir M. S., Nayef U. M., Abdul Kadhim W. K., 2019 *Nano Biomedicine and Engineering* **11** 56-63.

[25] Ali Z. U., Jabir M. S. and Al- Shammari A.M., 2019 *Research Journal of Biotechnology* **Vol** 79-82.

Adopting Texture Features to Detect and Recognize Brain Tumors in Magnetic Resonance Images

Aya S Derea[1], Heba Kh Abbas[1], Anwar H Al-Saleh[2], and Haidar J Mohamad[3]

[1]Department of Physics, College of Science for Women, University of Baghdad, Baghdad, Iraq
[2]Department of Computer Science, College of Science, Mustansiriyah University, Baghdad, Iraq
[3]Department of Physics, College of Science, Mustansiriyah University, Baghdad, Iraq

Abstract. The detection and recognition of tumor have a very important role in many applications of medical imaging. The use of computer in this process required high accuracy as it relates to human life especially that the error rate should be as low as possible. In this paper, Magnetic Resonance Imaging (MRI) images are tested using image processing (segmentation-based threshold) and supervised classification (minimum distance) to detect the region of interest. The location of the tumor is located using histogram algorithm with run length features. The geometric measurements calculated to identify tumors dimensions, location, volume and area. The results were highly efficient and highly accurate to determine the location and dimensions of the tumor.
***Keywords* — Brain Tumor, Morphological Operators, Segmentation, Gray Level Co-occurrence Matrix.**

Introduction

Medical imaging techniques produce images, including a lot of information about the anatomical construction and valuable for making correct diagnoses [1]. Magnetic resonance imaging (MRI) worked to align nuclear magnetism using the magnetic field. The scanner detects the change in magnetic alignment generated by radio frequencies [2]. This signal processed to obtain the internal structure of brain details. Manual image analysis is very difficult due to large amounts of data in it. Image segmentation has become mandatory to detect a brain tumor in clinical diagnosis [3]. The word tumor is a synonym for a word neoplasm, which formed by an abnormal growth of cells. Tumor is something very different from cancer; malignant tumor is a term that typically used for the description of cancer. Malignant is a medical term that describes a severe progressing disease [4,5]. It greatly improves human diagnosis, especially that the error rate should be as low as possible, with good software to help doctors to detect and diagnose the tumor better [6]. Many studies have addressed this domain like Siva Sankari S. *et.al.* (2014) were used MRI images, Gabor feature extraction algorithm with help of k-means Clustering Segmentation. Results extracting and compared with these standard metrics [7]. In 2017[8], Kapoor *et.al.* attempted to detect and classify tumor in benign stage. Their proposed method consists on feature extraction related to MRI images using Gray Level Co-occurrence Matrix (GLCM) based methods and classification images using K-nearest neighbor (K - NN) classifier. The extracted features are used to recognize the class of the tumor as normal and abnormal. In 2017[9], Moitra used median filter to remove noise from the image. Moreover, GLCM used to extract the needed features from the image so that they can find whether the tumor is present or not. Fuzzy c-means used for image segmentation and artificial neuro fuzzy technology. Artificial Neural Network Fuzzy Inference System (ANFIS) used for classification of type of brain tumor. In 2017[10], Bahadure *et.al.* study the texture features for several image. The performance of the features pull out from the various texture methods such as histogram, Gray Level Co-occurrence Matrix (GLCM), Gray-Level Run-Length Matrix (GRLM). The Support Vector Machine (SVM) is used to classify the extracted features into benign or malignant. The performance of texture methods are assessed using a variety of statistical performance measurements such as sensitivity, specificity and accuracy the experiment results show that there is considerable performance variability among the various texture methods. The histogram features and GLRLM features performances are considerably poor. In 2018, S. Sri Devi [11] suggested brain segmentation based on fuzzy k-means with Otsu Thresholding method to segmentation the brain tumor from magnetic resonance images and extract texture features using the gray level co-occurrence matrix. In this study, a designed program suggested to detect and recognize region of interest in MRI. Co-occurrence features adopted to use the number of features like contrast, energy, coarseness and angular second moment. As well as the adoption of the histogram in determine the location of the tumor. The flow chart of the adopted features to calculate the geometric characteristics like area, size, dimensions of the tumor and the image of the presence and absence of tumor (complementary image) shown in figure 1.

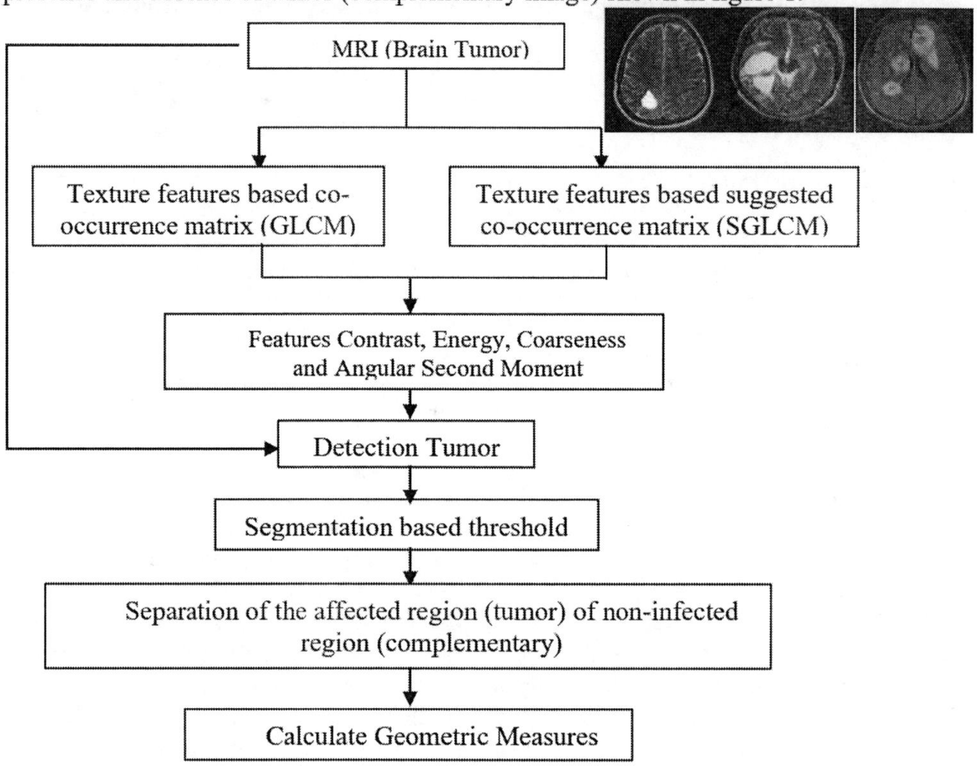

Figure 1. Flow chart of the suggested algorithm.

Experimental part

Tumor Detection

Image Segmentation divides an image into its constituent regions or objects. Segmentation of non-trivial images is one of the difficult tasks in image processing. Segmentation accuracy determines the eventual success or failure of the computerized analysis procedure [12]. Segmentation algorithms based on one of two basic properties of intensity values discontinuity and similarity. First category is to partition an image based on abrupt changes in intensity, such as edges in an image [13]. Second category based on partitioning an image into regions that are similar according to a predefined criterion. Histogram thresholding approach falls under this category. Histograms constructed by splitting the range of the data into equal-sized bins (called classes). The gray-level histogram corresponds to an image, f(x, y), composed of dark objects in a light background, in such a way that object and background pixels have gray levels grouped into two dominant modes. One obvious way to extract the objects from the background is to select a threshold 'T' that separates these modes [14]. Then any point (x, y) for which f(x, y) > T is called an object point, otherwise, the point is called a background point [15]. In this study threshold is chosen manually, by trying a range of values of *T* and check which one works best at identifying the objects of interest which includes two morphological techniques called cleaning through reconstruction and filling through reconstruction [16], as shown in figure 2.

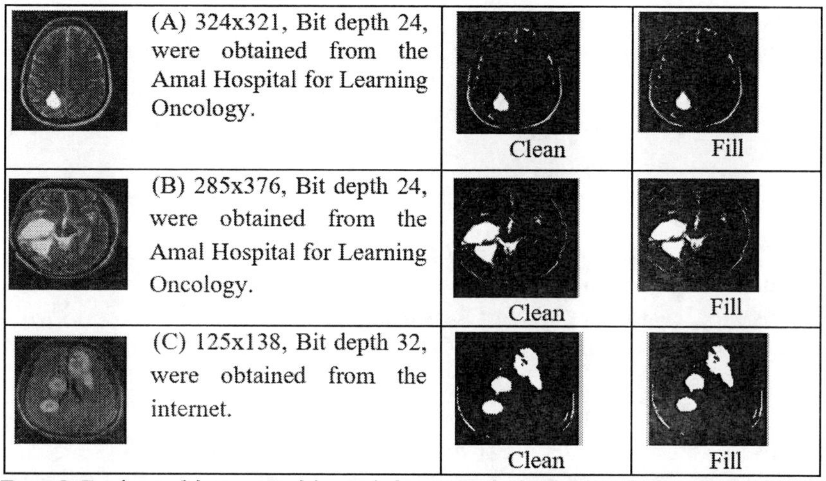

Figure 2. A, B and C adopted images with applying morphological techniques (Filling and cleaning)

Feature Extraction

From each co-occurrence matrix (GLCM) and suggested co-occurrence matrix (SGLCM) method, a set of four-features extracted in different orientations for the training of the segmentation based threshold model. The GLCM is a feature to identify texture in an image by modelling texture as a 2-Dimensional array gray level variation and consider a spatial relationship of pixels. Its features calculated in four directions 0°, 45°, 90° and 145° and four distances (1, 2, 3 and 4). Four properties namely contrast, energy, coarseness and angular second moment computed using the number of gray levels in an image that determines the size of GLCM. The matrix element F (m, n | Δx, Δy) is the relative frequency with two pixels separated by pixel distance (Δx, Δy), which occurs within a given neighborhood, one with intensity M and other with intensity N. A gray level co-occurrence matrix (GLCM) contains information about the positions of pixels having similar gray level values. A GLCM F [m, n] is defined by first specifying a displacement vector d= (dx, dy) and counting all pairs of pixels separated by *d* having gray level m and *n*.

The suggested co-occurrence matrix SGLCM method is the same mechanism as the traditional co-occurrence matrix, but SGLCM based on the diagonal diameter for different distances and angles, then calculate the same statistical features of the proposed method and conventional. These features explained by equations (1, 2, 3 and 4). These equations are; contrast (Con) which is a measure of intensity of a pixel and its neighbour over the image and it defined as[17]:

$$Con = \sum_{x=0}^{m-1}\sum_{y=0}^{n-1}(x-y)^2 F(x,y) \tag{1}$$

Energy (E) is the quantifiable amount of the extent of pixel pair repetitions and represents measure the similarity of an image, and it defined as [16]:

$$E = \sqrt{\sum_{x=0}^{m-1}\sum_{y=0}^{n-1} F^2(x,y)} \tag{2}$$

Coarseness (C_{ness}) is a measure of roughness in the textural analysis of an image, represented by[18]:

$$C_{ness} = \frac{1}{2^{m+n}}\sum_{x=0}^{m-1}\sum_{y=0}^{n-1} F(x,y) \tag{3}$$

IOP Conf. Series: Materials Science and Engineering **757** (2020) 012029 doi:10.1088/1757-899X/757/1/012029

Angular Second Moment (ASM) measures the image homogeneity, known as Uniformity or Energy. It is the sum of squares of entries in the GLCM. ASM is high when image has very good homogeneity or when pixels are very similar [19], represented by:

$$\textbf{ASM} = \sum_{x=0}^{m-1} \sum_{y=0}^{n-1} [\, F\,(x,y)\,]^2 \tag{4}$$

The Results and Discussion

This work includes several parts. The first part dealing with texture features were it extracted by using the GLCM function using MATLAB software for the adopted images (A, B and C) as shown in figure 2. A filter window (7×7) passed on image for different gray level (g), angle (θ) and distance (d) by function offset [θ d]. Grayscale image (imgy) determined by its clarity (imgF), as shown in figure 3.

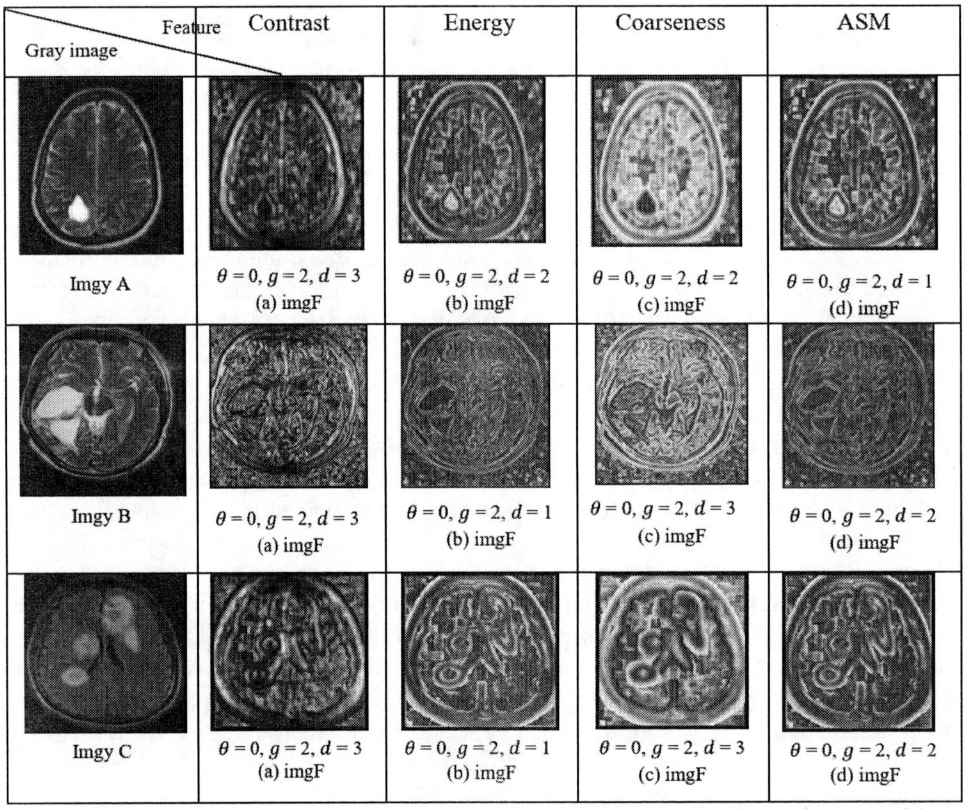

Figure 3. The gray image (imgy) and texture features GLCM (imgF) for each adopted image.

The second part dealing with texture features were it extracted by using the SGLCM for Adopted images (A, B and C) as shown in figure 2. A filter (7×7) passed on image for different gray level (g), angle (θ) and distance (d) by function offset [θ d] for each grayscale image (imgy) which is determined by its clarity and take diagonal coordinate only from resulting image GLCM (imgFS). Then texture features approved, as shown in figure 4.

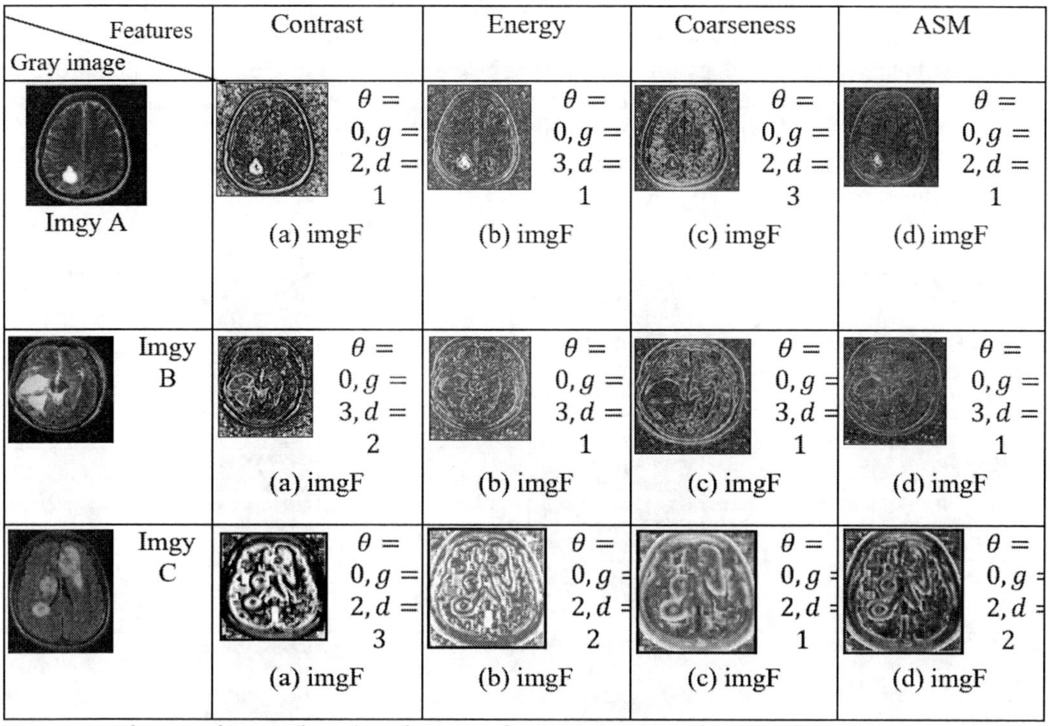

Figure 4. The gray image (imgy) and texture features SGLCM (imgFS) for each adopted image.

Texture analysis adopted in this study GLCM and SGLCM compared by measuring the values of the texture features of the images in the figures (3 and 4) as shown in tables 1 and 2, respectively.

Table 1. Texture feature extraction of GLCM value for each adopted image.

	Contrast	Energy	Coarseness	ASM
Image A	0.9750	0.4746	0.7745	244.7092
Image B	3	0.5510	1.2031	486
Image C	0.8182	0.7146	1.3393	215.7619

Table 2. Texture feature extraction of SGLCM value for each adopted image.

	Contrast	Energy	Coarseness	ASM
Image A	0.6071	0.2898	0.9394	94.6301
Image B	0	0.7551	0.4167	154.1664
Image C	0.2143	0.5510	0.4702	113.0238

IOP Conf. Series: Materials Science and Engineering **757** (2020) 012029 doi:10.1088/1757-899X/757/1/012029

The third part consists of tumors detection; from figures 3 and 4, and tables (1 and 2) for SGLCM method of using the segmentation-based threshold, each image has a specific threshold (Th). The detection process included recalling the original image (imgy-A) and feature image (imgFS) as in figure 4. The number of objects determine in the image using morphological processes (fill and clean) and find the location (imx) and area of the largest object (max A) in the image. The tumor detected in the original image (TO) and feature image adopted (TF). Then the task is find the complementary image (without the tumor) for the original image (imC) and the feature image adopted (imCF). A plot histogram of the original image (I), feature image (IF), the tumor (IT) and the complement (IC and ICF), as in figure 5a.

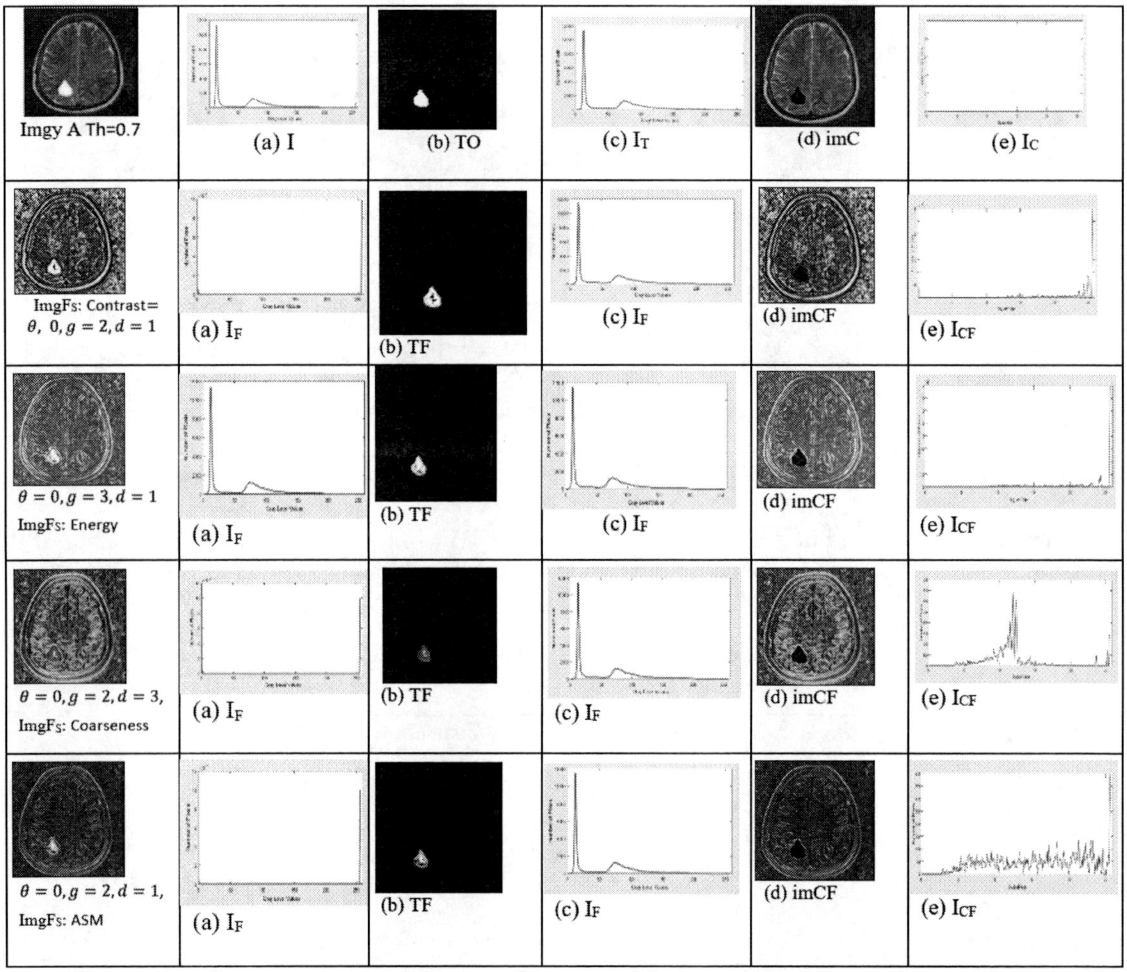

Figure 5-a. Demonstrates the detection of tumors of the imgy-A and feature image adopted

The detection process included recalling the original image (imgy-B) and feature image (imgFS) in figure 4, then determine the number of objects in the image using morphological processes (fill and clean) and find the location (imx1 and imx2) and the area of the largest object (maxA1 and maxA2) in the image. The tumor detected in the original image (TO) and feature image adopted (TF). After that is finding the complementary image (without the tumor) for the original image (imC) and the feature image adopted (imCF). The final step is plot histogram of the original image (I), feature image (IF), the tumor (IT) and the complement (IC and ICF), as shown in figure 5-b.

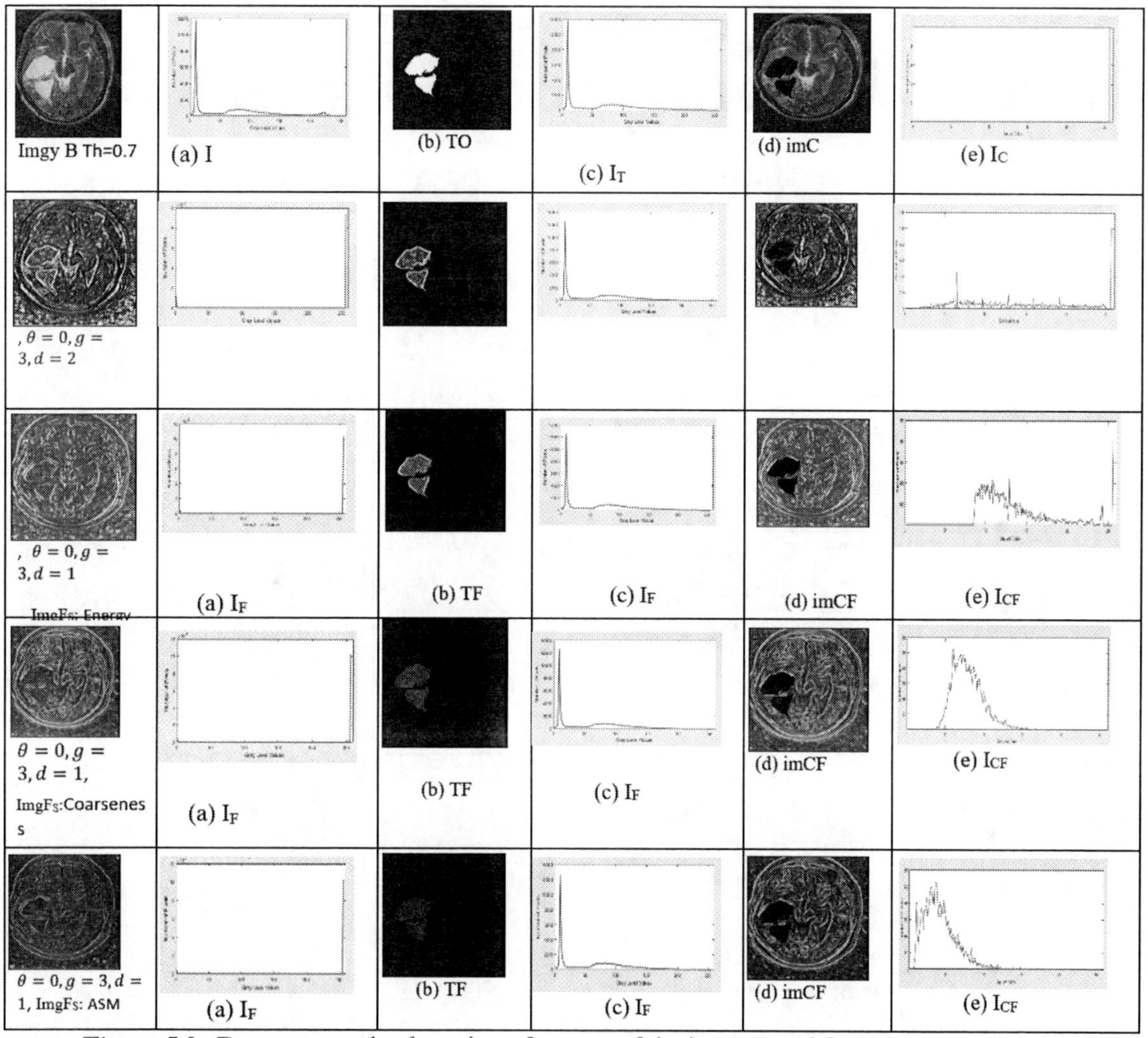

Figure 5-b. Demonstrate the detection of tumors of the imgy-B and feature image adopted

The detection process included recalling the original image (imgy-C) and feature image (imgFS) in figure 4 then determine the number of objects in the image using morphological processes (fill and clean) and find the location (imx1 , imx2 and imx3) and the area of the largest object (maxA1, maxA2 and maxA3) in the image. The tumor detected in the original image (TO) and feature image adopted (TF). Then, it is necessary to find the complement image for the original image (imC) and the feature image adopted (imCF). Finally, histogram plotted of the original image (I), feature image (IF), the tumor (IT) and the complement (IC and ICF), as shown in figure 5-c.

IOP Conf. Series: Materials Science and Engineering **757** (2020) 012029 doi:10.1088/1757-899X/757/1/012029

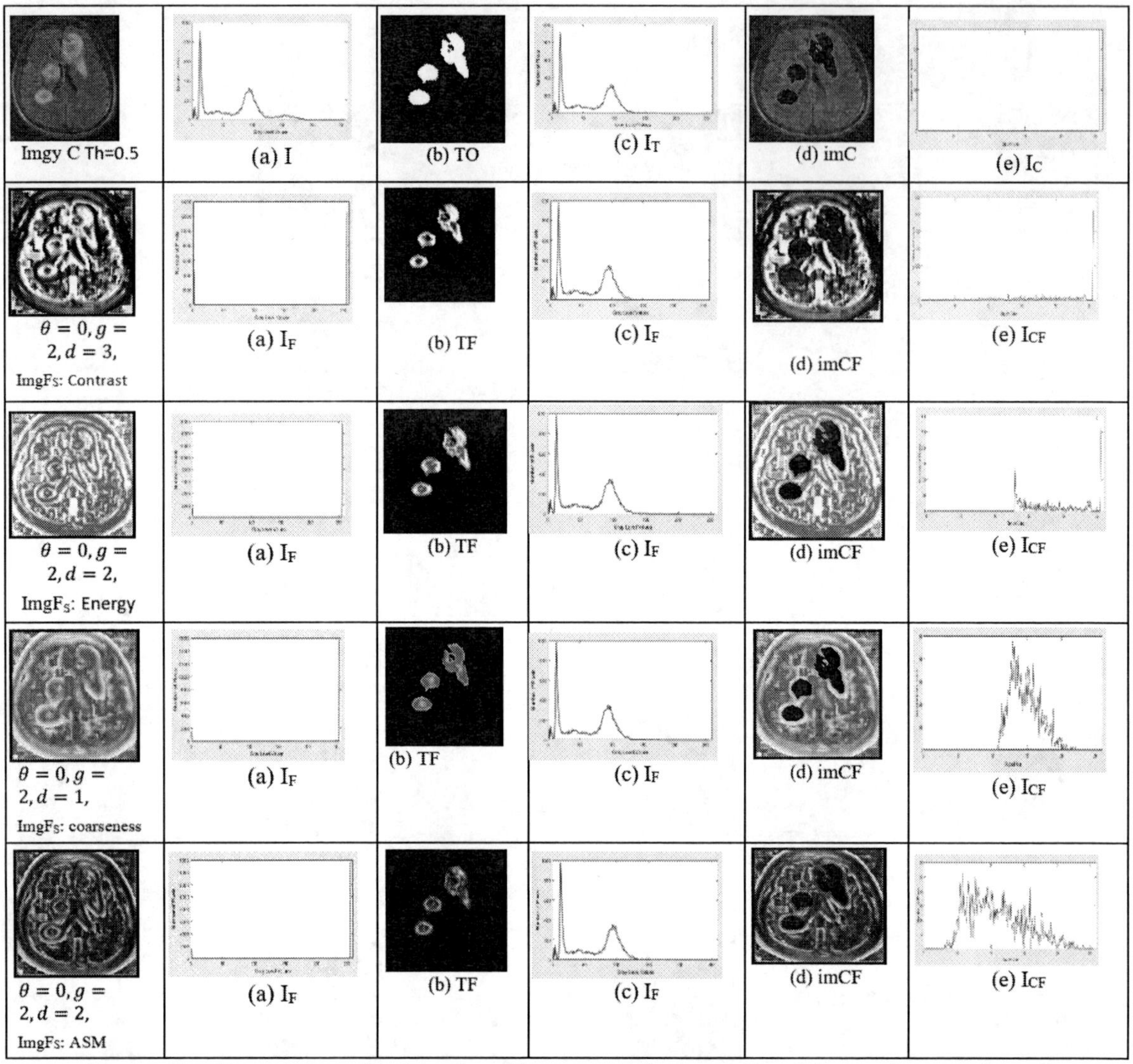

Figure 5-c. Demonstrate the detection of tumors of the imgy-C and feature image adopted

From figures 5-a,b and c, it is observed from the histogram of the original images (I) and feature images (IF) that the (IF) is more accurate when there is shift of values. Moreover, a high increase in the number of pixels than original images (I) and high accuracy in the separation of the affected region (Tumor) from the healthy region (complement). The histogram of the complement images (ICF) represents the behaviour of each supported feature.

The fourth part calculates the complement image area (imC) which represent (B), then determine the number of objects in the image using morphological processes (fill and clean).
Calculate the geometric characteristics of the tumor image (TO) and convert it to a binary image with a threshold of 0.2. Then identify the affected region with a block (length and width) using *regionprop* function to find image area for (TO) which represent (A), radius (R), dimensions of the tumor (T1 and T2) where some conditions set as follows:

- If T2> T1 then rate =T1/T2 else T2/T1
- If rate > 0.6 compute size of image (TO) as in the equation 5

$$\text{Size V(TO)} = \frac{4}{3}\pi R^3 \quad (5)$$

where $R = \left(\frac{A}{\pi}\right)^{1/2}$

Then calculate the area (Ablock), size (Vblock) and dimension (TT) of the infected area (tumor only) specified in the block as shown table (3) as equation 6, where:

$$Ablock = T1 \times T2 \times \pi$$
$$TT = T1 + T2/2 \qquad (6)$$
$$Vblock = \frac{4}{3}\pi \times TT^3$$

Table 3. Geometric properties of approved images, tumors and complementary images

Complement image	Image with Tumor			Tumor only				
B (pixels)	A (pixels)	R	V (pixels)	T1	T2	TT	A block (pixels)	V block (pixels)
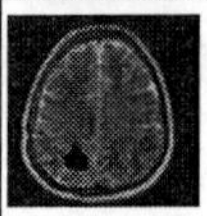		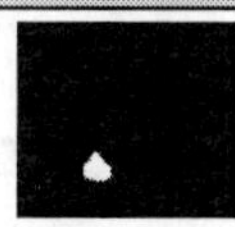				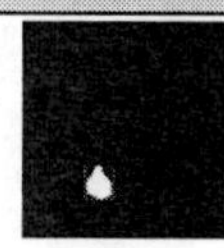		
25751	331	10.2645	$4.5301x10^3$	21	24	22.5	$1.5834x10^3$	$4.7713x10^4$
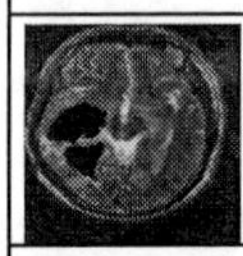		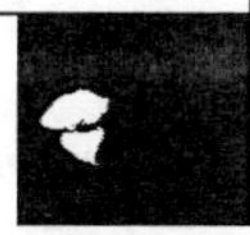						
1011250	5910	43.3729	$3.4178x10^5$	72 53	82 73	77 63	$1.8548x10^4$ $1.2155x10^4$	$1.9123x10^6$ $1.0474x10^6$
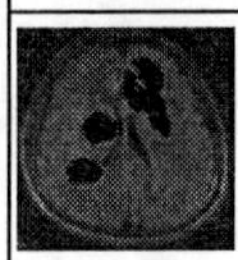								
15886	1364	20.8369	$3.7895x10^4$	17 22 22	21 20 20	19 21 21	$1.1215x10^3$ $1.3823x10^3$ $1.3823x10^3$	$2.8731x10^4$ $3.8792x10^4$ $3.8792x10^4$

Conclusions

This paper presents the design of programs capable to detect tumors within MRI image that have high complexity. An effective method proposed for the segmentation of MRI brain images, which contains a natural, benign and malignant tumor. Where the method of segmentation based on the threshold is very efficient in the process of detection of tumors at a very high accuracy. The suggested method (SGLCM) was of high quality in extracting the texture features comparing with the traditional method GLCM. The results show that the properties succeeded in separating the infected region (Tumor) from the healthy region (complement). The geometric characteristics of the images were well calculated by the presence of the tumor, the absence of the tumor, and the determination of the measurements of the tumor. The program was able to overcome these difficulties by enabling the doctor to better diagnose the tumor in medical images and clear, which improves the performance of the automated diagnosis and reliability of the comprehensive.

References

[1] V. Y. Borole;, S. S. Nimbhore;, and S. S. Kawthekar, International Journal of Emerging Trends & Technology in Computer Science **4**, 28 (2015).
[2] G. Kaur and A. Oberoi, International Journal of Computer Sciences and Engineering **6**, 404 (2018).
[3] A. Mustaqeem, A. Javed, and T. Fatima, International Journal of Image, Graphics and Signal Processing **4** (2012).
[4] D. J. Withey and Z. J. Koles, Int J Bioelectromag **9** (2007).
[5] J. Prince and J. Links, *Medical Imaging Signals and System*(2006).
[6] R. Popilock, K. Sandrasagaren, L. Harris, and K. Kaser, Journal of nuclear medicine technology **36**, 79 (2008).
[7] M. Al-Tamimi and G. Sulong, *Tumor Brain Detection Through MR Images: A Review of Literature* 2015).
[8] L. Kapoor and S. Thakur, in *2017 7th International Conference on Cloud Computing, Data Science & Engineering - Confluence*2017), pp. 582.
[9] D. Moitra, INTERNATIONAL JOURNAL OF COMPUTER SCIENCES AND ENGINEERING **5**, 3 (2017).
[10] N. Bahadure, A. Ray, and H. P. Thethi, International Journal of Biomedical Imaging **2017**, 1 (2017).
[11] S. S, International Journal for Research in Applied Science and Engineering Technology **6**, 1911 (2018).
[12] H. Abdalla and M. Esmail, *Brain Tumor Detection by using Artificial Neural Network* 2018).
[13] M.Karuna, International Journal of Research in Engineering and Technology **02**, 586 (2013).
[14] H. K. Abass, Phd Thesis. Al-Mustansiriyah University, 2013.
[15] F. Mahmood and W. A. Abbas, Iraqi Journal of Science **57**, 279 (2016).
[16] P. Mohanaiah, P. Sathyanarayana, and L. Gurukumar, International Journal of Scientific and Research Publications **3**, 1 (2013).
[17] A. Lakshmi and T. Arivoli, Journal of Theoretical and Applied Information Technology **64**, 561 (2014).
[18] H. K. Abbas, S. F. Ahmed, and A. H. J. I. J. o. S. Mahdy, **58**, 1546 (2017).
[19] H. Abbas, S. J. A. i. N. Ahmad, and A. Sciences, **11**, 268 (2017).

IOP Conf. Series: Materials Science and Engineering **757** (2020) 012030 doi:10.1088/1757-899X/757/1/012030

Using spatial analysis techniques to Preparing maps for distribution of pollutant concentrations in Shatt al-Arab waters in November 2015.

Ebtesam F. Khanger[1*], Ban A. Al Razaq[1], Rafah Rasheed Ismail[1] and Zena F. Rasheed[2]

[1] Department of Astronomy and Space, College of Science, University of Baghdad, Baghdad, Iraq.

[2] College of Engineering, University of Baghdad, Baghdad, Iraq.

*E-mail: dr.efk33@yahoo.com

Abstract

Water has a great self-generating capacity that can neutralize the polluting interventions carried out by humans. However, if human activities continue this uncontrolled and unsustainable exploitation of this resource, this regenerating capacity shall fail and it will be jeopardized definitively. Shatt Al-Arab River in South of Iraq. It has an active role in providing water for irrigation, industry, domestic use and a commercial gateway to Iraq. in the last five years Shatt Al-Arab suffered from a rise in pollutants due to the severe decline in sewage networks, irregular networks and pesticide products, as well as the outputs of factories and companies that find their way to water sources and lead to a widespread collapse of water quality. In present work , by using Data observation with the integration between remote sensing and GIS techniques to prepare maps of the distribution of concentration materials in Shatt al-Arab River south of the province of Basra in January 2015 to determine the level of pollution in the river. These include pH, dissolved oxygen (DO2), phosphates (PO4), nitrates (NO3), calcium, magnesium, potassium, Total soluble solids (TDS), electrical conductivity (EC) as well as alkaline salts (ALK.) The quality of polluted water has been observed at the sites of the study due to the increase in wastewater flowing into the river, especially river branches and the illegal discharges of industrial waste and sewage. In addition to the severe shortage of water levels in the last five years.

Key word: Shatt Al-Al Arab, pollution, Water Quality, remote sensing, GIS.

1. Introduction

Water is considered polluted if some substances or condition is present to such a degree that the water cannot be used for a specific purpose. Water pollution is the presence of excessive amounts of a hazard (pollutants) in water in such a way that it is no long suitable for drinking, bathing, cooking or other uses. Pollution is the introduction of a contamination into the environment. It is created by industrial and commercial waster, agricultural practices, everyday human activities and most notably, models of transportation. No matter where you go and what you do, there are remnants earths environmental and its inhabitants in many ways. [1]

1.1. Water Quality Assessment

Water quality study is the process of determining the chemical, physical and biological characteristics of water bodies and identifying the possible contamination sources that degrade the quality of water [2]. Degradation of the quality of water resources may result from waste discharges, pesticides, heavy metals, nutrients, microorganisms, and sediments. Different water quality standards have been developed to aid in checking the extent of water pollution, and consequently to maintain these quality standards.

1.2. Remote sensing and Geographic information system (GIS)

Remote sensing is the science of obtaining and interpreting information from a distance, using sensors that are not in physical contact with the object being observed. The science of remote sensing in its broadest sense includes aerial, satellite, and spacecraft observations of the surfaces and atmospheres of the planets in our solar system, though the Earth is obviously the most frequent target of study.[3]

IOP Conf. Series: Materials Science and Engineering **757** (2020) 012030 doi:10.1088/1757-899X/757/1/012030

GIS is a manual or automated system, which can store, retrieve, manipulate, and display environmental data in a spatial format. It has the capabilities to use different set of operation for working with spatially referenced geo-data. It uses several manual data elements like maps, aerial and ground photograph, statistical report etc. transportation planning, physical planning, land use planning or zoning, non-point source pollutants, monitoring hazardous waste sites etc.[4]

2. Interpolation Methods

Spatial interpolation is tool in GIS used to find the values of unknown points. It can be defined as a procedure of estimating the values of properties at un-sampled locations based on the set of observed values at known locations.

2.1. Inverse Distance Weighted (IDW)

This is one of the simplest and most readily available methods. It is based on an assumption that the value at an unsampled point can be approximated as a weighted average of values at points within a certain cut-off distance, or from a given number m of the closest points (typically 10 to 30) [5].

$$z(s_\circ) = \sum_{i=1}^{N} \lambda_i \,.\, z(s_i) \text{ --------- } 1$$

Where:

$z(s_\circ)$ Represents a portended value at the unsampled site s0.

N represents a measured sample point's number within the neighborhood determined for s0.

λi represents associated distance-dependent weights with each sample point.

$z(s_i)$ Represents the observed value at site si.

The obtained by weights are using:

$$\lambda_i = \frac{d_{io}^{-p}}{\sum_{i=1}^{N} d_{io}^{-p}} \text{ ----------- } 2$$

$$\sum_{i=1}^{N} \lambda_i = 1 \text{ ---------- } 3$$

Where:

d_{io} is the prediction site s0 and the measured site location si distance.

P represents the power parameter that determines the reduction rate of the weights as distance increases.

IDW is forced to be an exact interpolator to avoid the division by zero that occurs when di0 = 0 at the sampled points

3. Study area Shatt Al-Arab

Shatt al-Arab is a river of some 200 km (120 mi) in length, formed by the confluence of the Euphrates and the Tigris in the town of al-Qurnah in the Basra Governorate of southern Iraq. The southern end of the river constitutes the border between Iraq and Iran down to the mouth of the river as it discharges into the Persian Gulf. It varies in width from about 232 meters (761 ft.) at Basra to 800 meters (2,600 ft.) at its mouth. It is thought that the waterway formed relatively recently in geologic time, with the Tigris and Euphrates originally emptying into the Persian Gulf via a channel further to the west.[6]

4. Geography

The general climate of the Shatt al-Arab region is subtropical, hot, and arid. At the northern end of the Persian Gulf is a vast floodplain formed by the Euphrates, Tigris, and Karun Rivers, featuring huge permanent lakes, marshes, and forest. The aquatic vegetation includes reeds, rushes, and papyrus, which support numerous species. The marshy land is home to water birds, some stopping here while migrating, and some spending the winter by feeding on lizards, snakes, frogs, and fish. Other animals found in these marshes are domestic water buffalo, two endemic rodent species, antelopes, gazelles, small animals such as the jerboa, and several other mammals.[7]

Figure 1. Scene of the Shatt Al- Arab

5. Material and Methods

Five monitoring stations were selected on the Shatt al-Arab, For the purpose of monitoring pollution in the study area to prepare pollution maps, as shown in Table (1).

Table 1 - Description of monitoring stations on Shatt al-Arab

The province	Station Code	Location	Station coordinates	
			Longitude (E)	Latitude (N)
AL-Basra	SH1	Before the Karmat Ali River / water project (25) million and the unified Basra	47° 45' 56"	30° 39' 01"
	SH2	Almeql / near Khalid Bridge	47° 46' 70"	30° 34' 45"
	SH2B	Abo Al - Khasib - Al snaker / desalination plant water in Al snaker	47° 56' 67"	30° 27' 95"
	SH3	Al Saybah / near Seihan water project / Seihan border post and raw water pumps for Seihan water project	48° 11' 66"	30° 19' 60"
	SH4	FAO / Mooring Boats	48° 29' 04"	29° 58' 51"

Data on the results of laboratory tests of the five monitoring stations in Shatt al-Arab water were used during the month of November on 2015 as shown in table (2) , these tested includes pH, dissolved oxygen (DO2), phosphates (PO4), nitrates (NO3), calcium, magnesium, potassium, Total soluble solids (TDS), electrical conductivity (EC) as well as alkaline salts (ALK.) .

Table 2 - Results of laboratory tests of the five monitoring stations in Shatt al-Arab water

The province		AL-Basra				
Station Code		SH1	SH2	SH2B	SH3	SH4
date for first monitoring		03/11/2015	11/11/2015	04/11/2015	15/11/2015	15/11/2015
date for second monitoring		24/11/2015	25/11/2015	16/11/2015	28/11/2015	28/11/2015
Results of tests	PH	7.95	7.95	7.7	7.6	7.5
	DO2	7.85	6.82	7.55	8.1	6.835
	PO4	0.3	0.335	0.255	0.345	0.295
	NO3	3.285	3.52	3.59	3.48	4.005
	Ca	204	240	176	156	212
	Mg	175.5	208.5	134.5	65	230
	TH	1240	1470	1000	660	1490

	K	29.55	35	16.05	6.15	31.35
	Na	1450	1870	1050	505	3235
	CL	1699	2015.5	1121.5	568.5	3423
	TDS	5009	6081	3541	2095.5	8148
	EC	7740	9015	5220	3035	10950
	ALK.	208	217.5	200	177	177

By using the pollution data in Table (2) and using remote sensing and Geographic information system (GIS), we prepared pollutions maps for Shatt al-Arab water. These maps shown in figures (2-14).

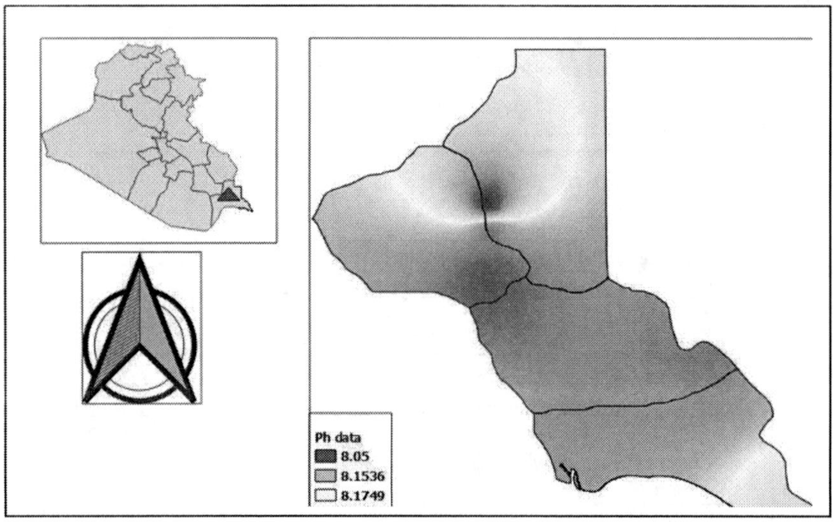

Figure 2. PH Concentration map in the Shatt al - Arab

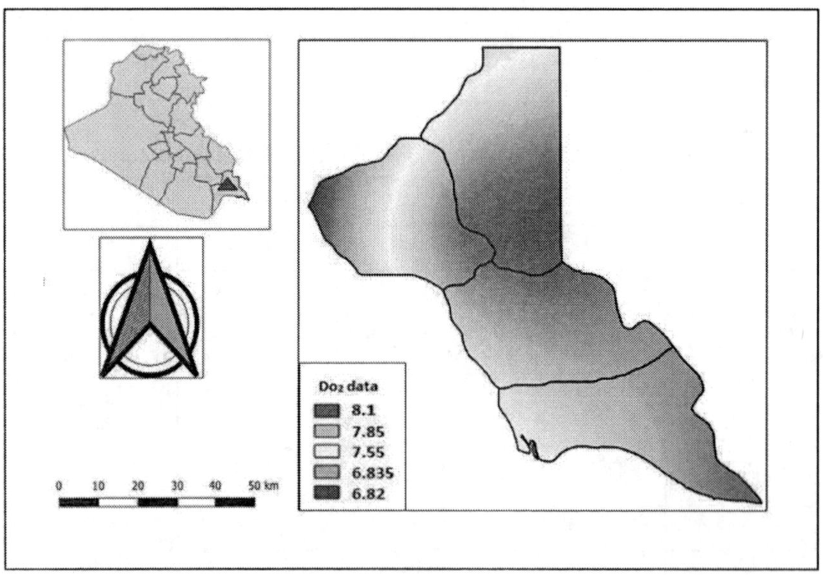

Figure 3. DO_2 Concentration map in the Shatt al - Arab

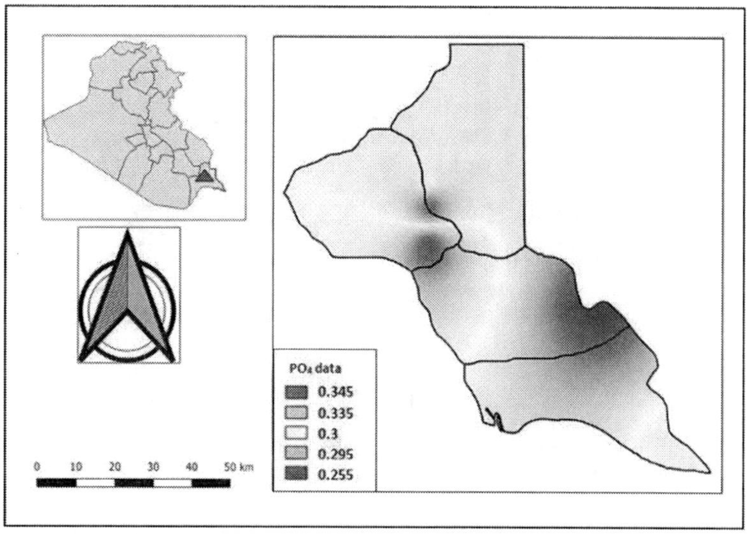

Figure 4 . PO_4 Concentration map in the Shatt al - Arab

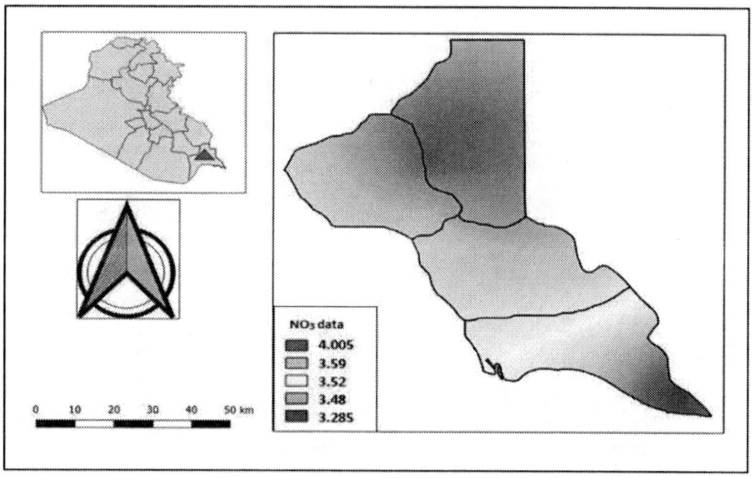

Figure 5 . NO_3 Concentration map in the Shatt al - Arab

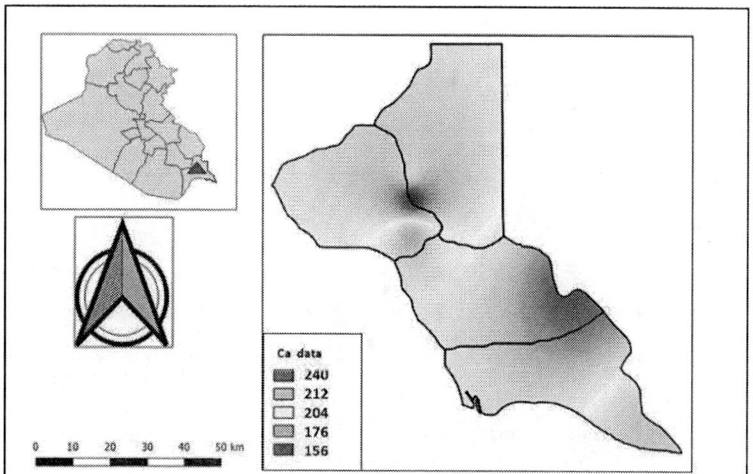

Figure 6 . Ca Concentration map in the Shatt al - Arab

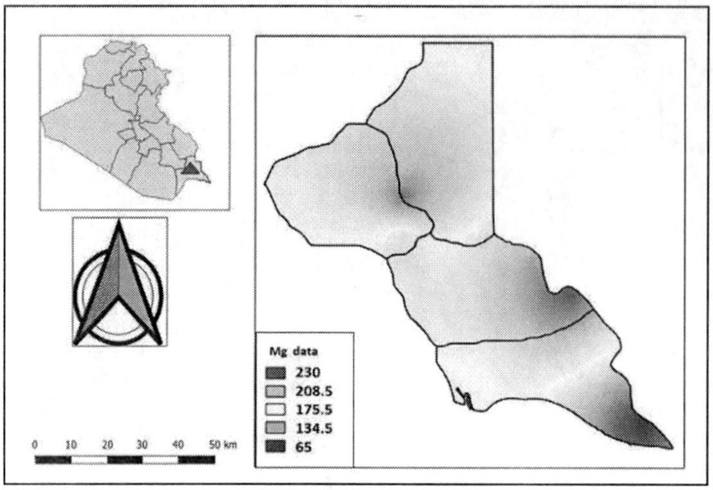

Figure 7. Mg Concentration map in the Shatt al – Arab

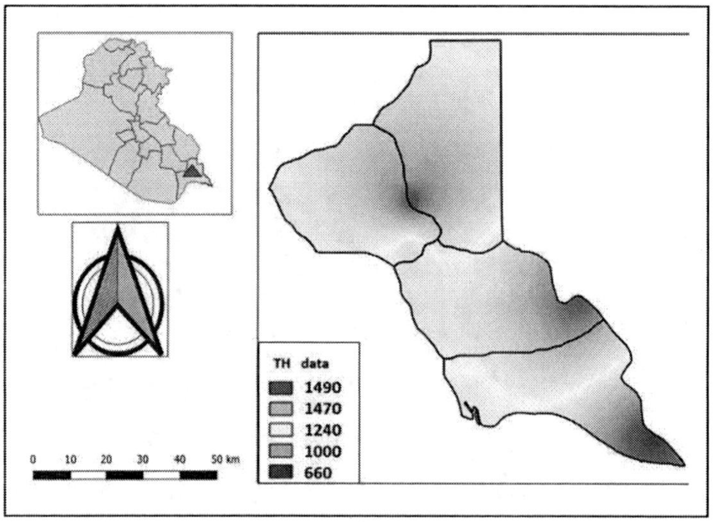

Figure 8. TH Concentration map in the Shatt al - Arab

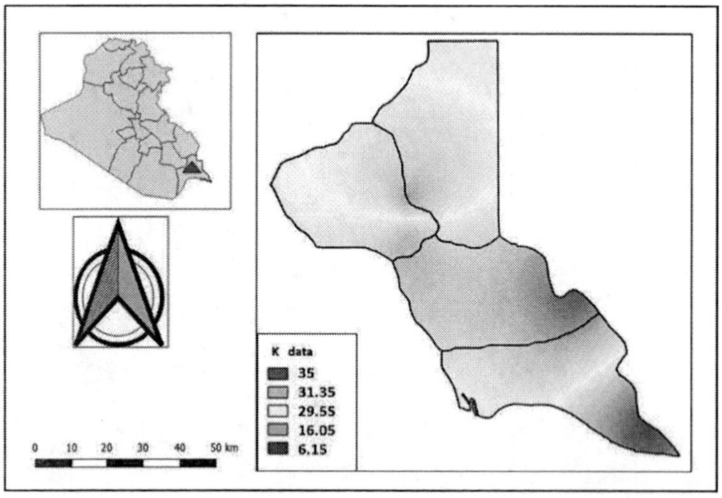

Figure 9. K Concentration map in the Shatt al - Arab

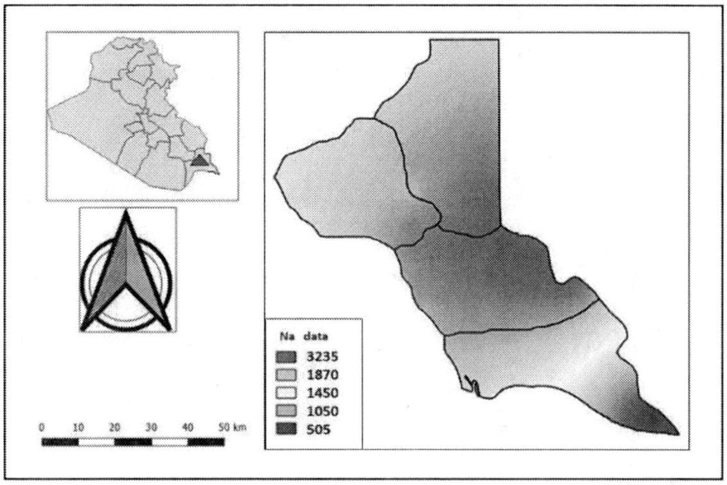

Figure 10 .Na Concentration map in the Shatt al - Arab

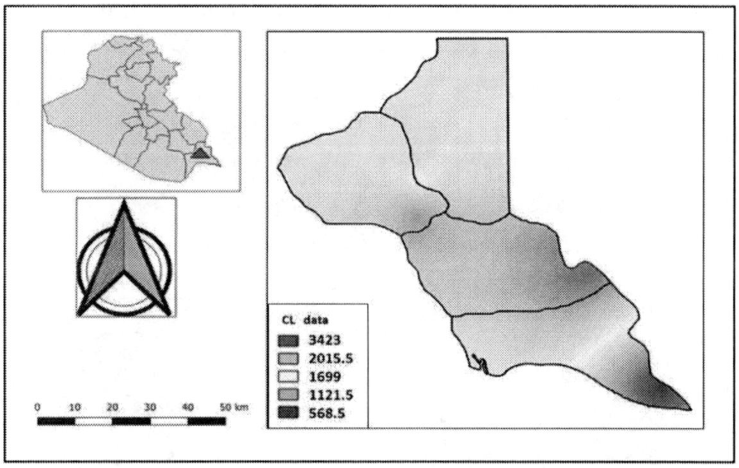

Figure 11. Cl Concentration map in the Shatt al - Arab

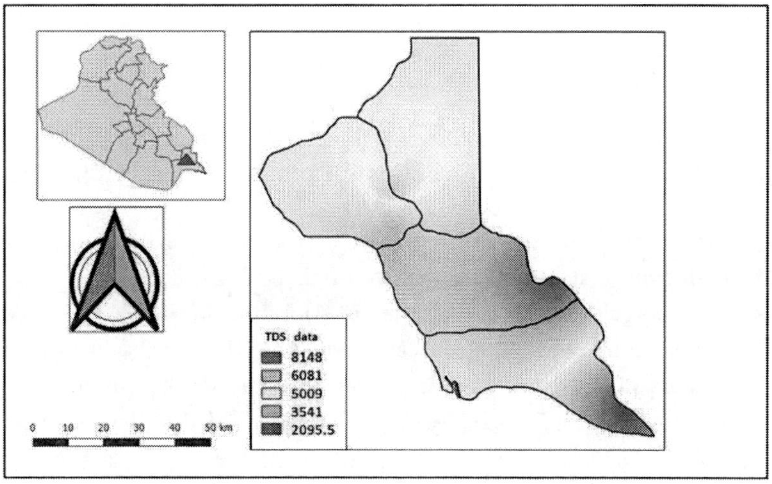

Figure 12. TDS Concentration map in the Shatt al - Arab

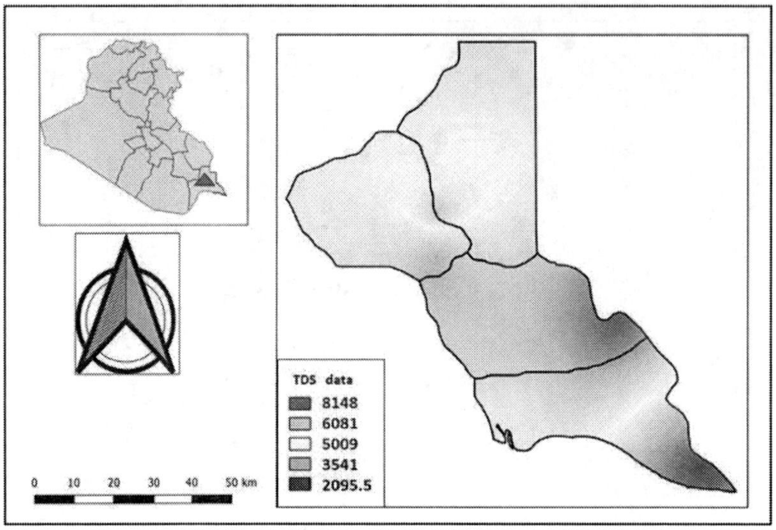

Figure 13 . EC Concentration map in the Shatt al - Arab

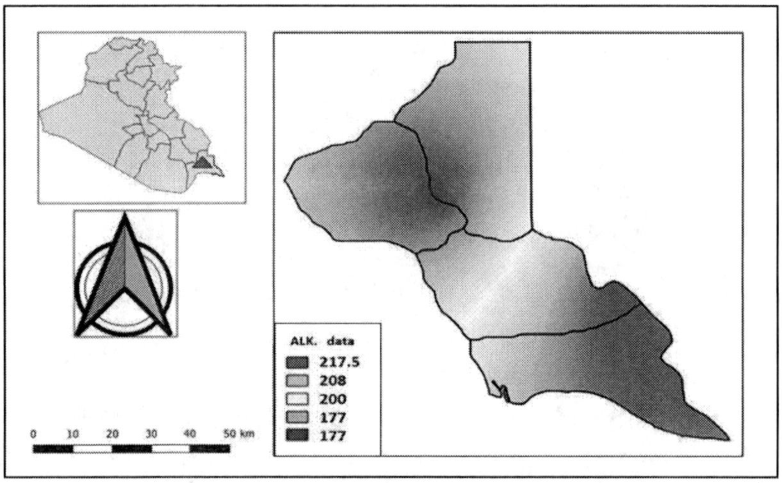

Figure 14 .ALK. Concentration map in the Shatt al – Arab

6. Conclusions

In this study, maps were prepared for the distribution of different pollutant concentrations in water of Shatt Al-Arab by using the integration of remote sensing techniques and geography information systems.

Five stations were selected in the waters of the Shatt al-Arab south of the province of Basra in March 2015 to identify pollutants that affect the quality of the river water throughout its course. These include pH, dissolved oxygen (DO2), phosphates (PO4), nitrates (NO3), SO4, calcium, magnesium, potassium, Total soluble solids (TDS), electrical conductivity (EC) as well as alkaline salts (ALK.) The quality of polluted water has been observed at the sites of the study due to the increase in wastewater flowing into the river, especially river branches and the illegal discharges of industrial waste and sewage. In addition to the severe shortage of water levels in the last five years.

7. References

[1] Abolude, D. S., Davies, O. A., & Chia, A. M. 2009. *Research Journal of Envi-ronmental Sciences*, **1**(2), 39-44.

[2] Taurai, B. 2011. The Diatom assemblages as indicator of field and lab conditions inlotic systems: conservation and water quality management in Sao Carlos SP catchh-ment, Brazil. pHD Dessertation Submitted to the Department of Biological Science,Federal University of Sao Carlos, Brazil.

[3] Ebtesam F.Khanjer,Mohammed A.Yosif MathemA.Sultan 2015. Iraqi Journal of Science, Vol **56** (1) C, pp: 839-845

[4] Ebtesam F. Khanjer & Nidaa ch. Jaheed . 2017. *International journal of advanced biological Research.* IJABR, **7** (4) 2017: Galley Proof

[5] C. Giardino, M. Bresciani, P. Villa, A. Martinelli, 2010 Water Resource Management. **24**(14),pp. 3885-3899.

[6] Oke, A. O., Sangodoyin, A. Y., Ogedengbe, K., & Omodele, T. , 2013. Acta Geographica Debrecina. Landscape & Environment Series, **7**(2) 48.

[7] Al-Malikey, R. N. , 2011 Al-Mustansiriyah Journal of Science, **22**(6), 1-12.

IOP Conf. Series: Materials Science and Engineering **757** (2020) 012031 doi:10.1088/1757-899X/757/1/012031

Prepare Maps For Greenhouse Gases With Some Weather Elements For Baghdad City Using Data Observation And Arc-GIS Techniques.

Ebtesam F. Khanjer[1] , Lina Awad[1], Wassan A. Hassan[2]. and Ban S. Ismael[1]
1 Department of Astronomy and Space, College of Science ,University of Baghdad, Iraq.
2 Forensic DNA research and training center University of Al Nahrain, Iraq

Abstract
Air pollution refers to the release of pollutants into the air that are detrimental to human health and the planet as a whole.In this research, the air pollutants concentration measurements such as Total Suspended Particles(TSP), Carbon Monoxides(CO),Carbon Dioxide (CO2) and meteorological parameters including temperature (T), relative humidity (RH) and wind speed & direction were conducted in Baghdad city by several stations measuring numbered (22) stations located in different regions, and were classified into (industrial, commercial and residential) stations. Using Arc-GIS program (spatial Analyses) , different maps have been prepared for the distribution of different pollutant concentrations of the city of Baghdad. These maps indicate where the highest and lowest concentrations are based on the color gradient of the map.

Key word : green house gases, CIS, Baghdad city, Spatial analysis .

1. Introduction

The air that a human breathe is a mixture of gases, small solid and liquid particles. Some substances come from natural sources while others are caused by human activities such as our use of motor vehicles, domestic activities, power generation and industry. Air pollution occurs when the air contains substances in quantities that could harm the comfort or health of humans and animals, or could damage plants and materials. These substances are called air pollutants and can be either particles, liquids or gaseous [1].

The air pollutants can be classified as primary or secondary pollutants. The primary air pollutants are harmful chemicals which directly enter the air due to natural events and human activities. Asecondary air pollutant is a harmful chemical produced in the air due to chemical reaction between two or more components. That is a primary pollutant combines with some component of the atmosphere to produce a secondary pollutant [2].

With the rapid industrialization and urbanization after the industrial revolution, human activities accentuated the degradation of air quality for the last several centuries. Severe air pollutant events drew public attention to the influences of air pollutants on human health, and ideological studies proved the effects on mortality and diseases [3].

1.1 Sources of Air pollution

The sources of air pollution can be either natural or artificial. The natural sources of air pollutants include forest fires, wind erosion of soil, volcanic eruption, evaporation of volatile organic matter and bacterial decomposition. Most of the potential air pollutants artificially added to the atmosphere due to human activities including the burning of fossil fuels in power plants and industries and in motor vehicles [4].

1.2 Global Warming

Global warming is an environmental phenomenon that guarantees human life and living organisms on earth. Global warming protects the heat energy from the sun from being lost. Otherwise, temperatures would be less than zero by 18 degrees Celsius, making it difficult for organisms to live on Earth. How heat is prevented by blocking thermal energy through a range of gases within moderate proportions, thus contributing to the provision of climatic conditions suitable for living organisms on Earth. Over time, global warming, which protects the planet, has become a serious problem affecting the environment and living organisms. This is caused by negative human activities that have increased the proportion of gases, such as water vapor, methane, chlorine fluoride and carbon dioxide, In the balance of this phenomenon, and cause great danger to the life of living organisms [5,6].

1.3 Spatial Analysis in GIS

The need for spatial interpolation models in the monitoring environment has grown in the past few years. The EPA is using these models to review choices on monitoring network design and to calculate the efficacy of release control programs. Due to the limited number of monitoring sites across the country for ambient concentrations of ozone and fine particles, there is anecessity to use spatial interpolation to predict ambient concentrations in unmonitored locations.Support for these methods has occurred from researchers and state/local/ EPA agencies in recent workshops[7]. . The general consent is that it is now probable to model the spatial requirement of air pollution data to reliably predict concentrations in unmonitored locations along with associated uncertainties for use in developing regulatory policy [8].

2. Interpolation Methods

Interpolation is the process of using points with known values or sample points to estimate values at anther unknown points. It can be used to predict unknown values for any geographic point data, such as elevation, rainfall, chemical concentrations, noise levels, and soon. The available interpolation methods are listed below[9].Assessment the air quality of Baghdad city through determining the concentration of air pollutants in the ground level. By using the Arc_GIS program Version 10.3 to prepare maps for distribution of concentration many gases in Baghdad.

3. Results and Discussion

This section includes the results of the measurements and analysis of the pollutants in air at many selected stations of Baghdad city that have been obtained from the sampling operations. study of air quality by using the ground measurements of Baghdad city in order to recognize the concentrations of pollutants in 22 station , the stations are classified in to industrial, commercial and residential, also they are distributed on regular
basis to cover most area of Baghdad city.

3.1 Study area

Baghdad city is located in central of Iraq, within the sector of flat sedimentary plains. The border of the municipality of Baghdad encompass fourteen administrative units, eight in Rusafa (east of Tigris river) and six in Karkh (west of Tigris river).It lies on latitude 43 east and longitude 34 north .Baghdad is suited in a plain area of elevation between 31-39 m above sea level. The climate of Baghdad region (which is part of the a plain area at the central of Iraq and has same climatic characteristics) is define a semi-arid subtropical and continental, dry, hot, long summer, cool winter and short spring and autumn

Figure 1. show the satellite image of Baghdad city.

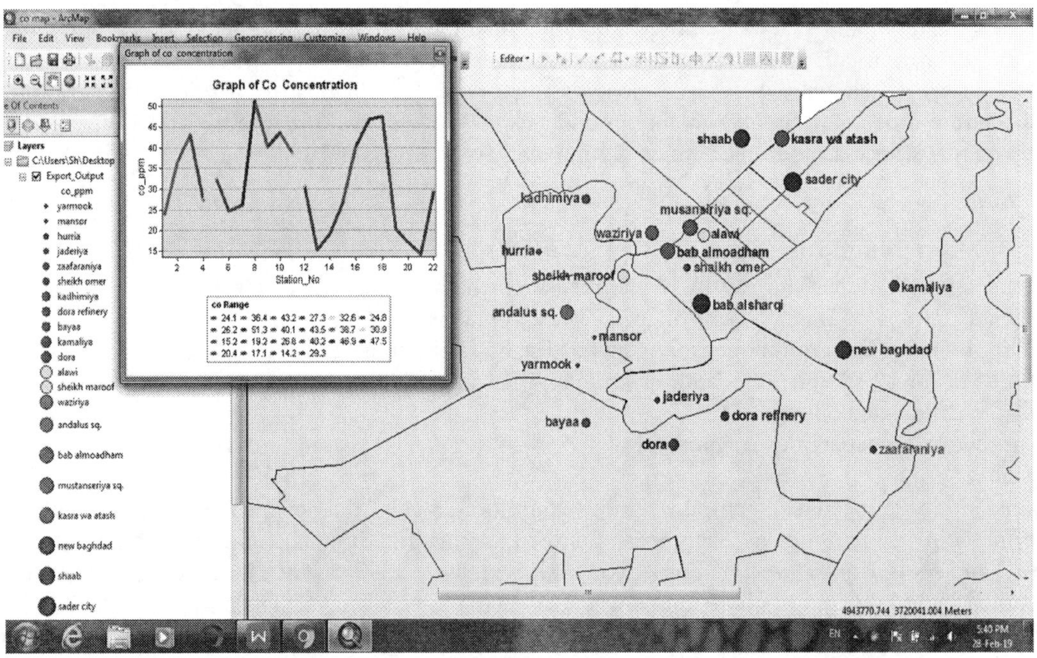

Figure 2. Concentration of carbon monoxide in Baghdad areas.

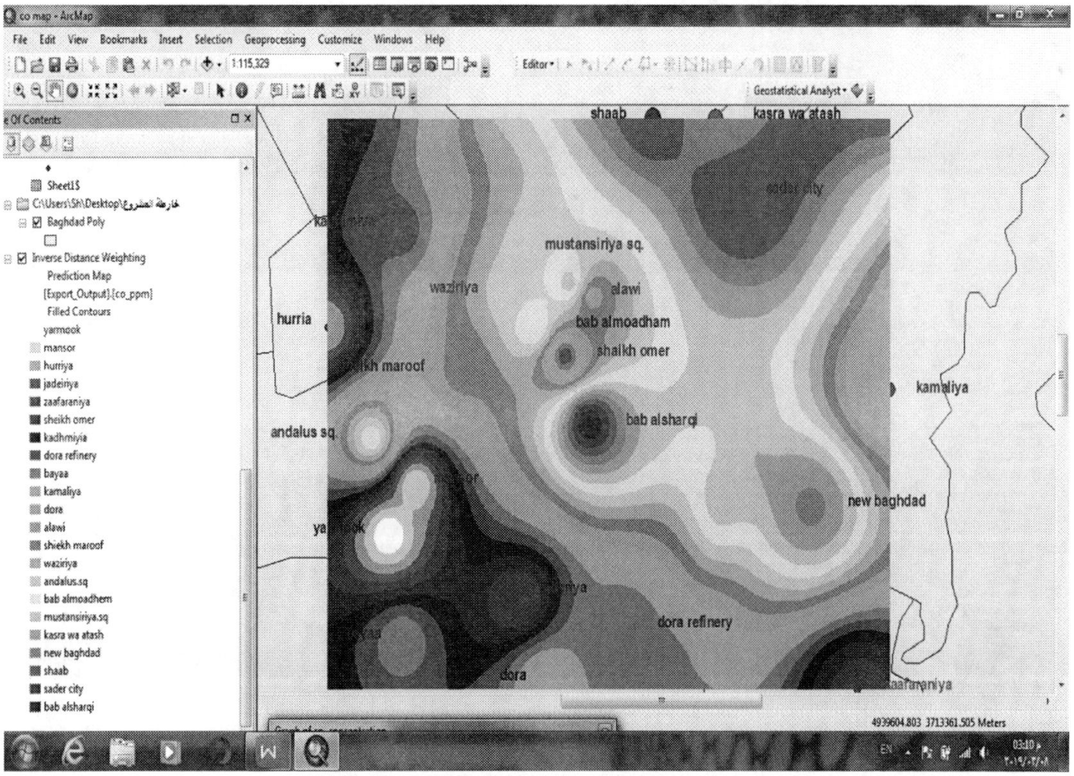

Figure 3. Geostatistical Analyst of CO.

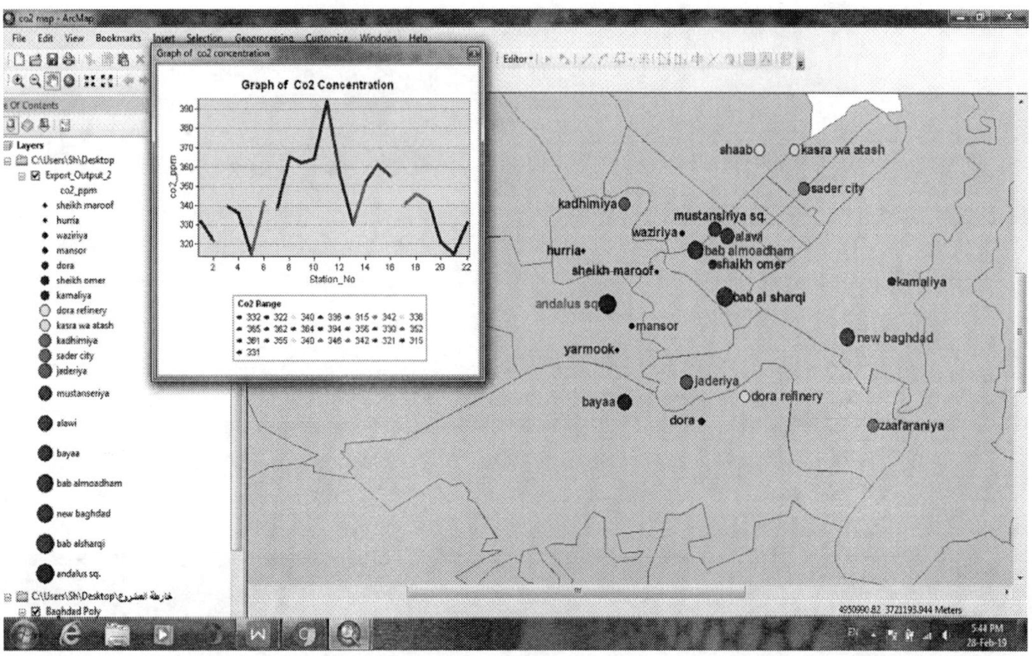

Figure 4. Concentration of carbon dioxide in the areas of Baghdad.

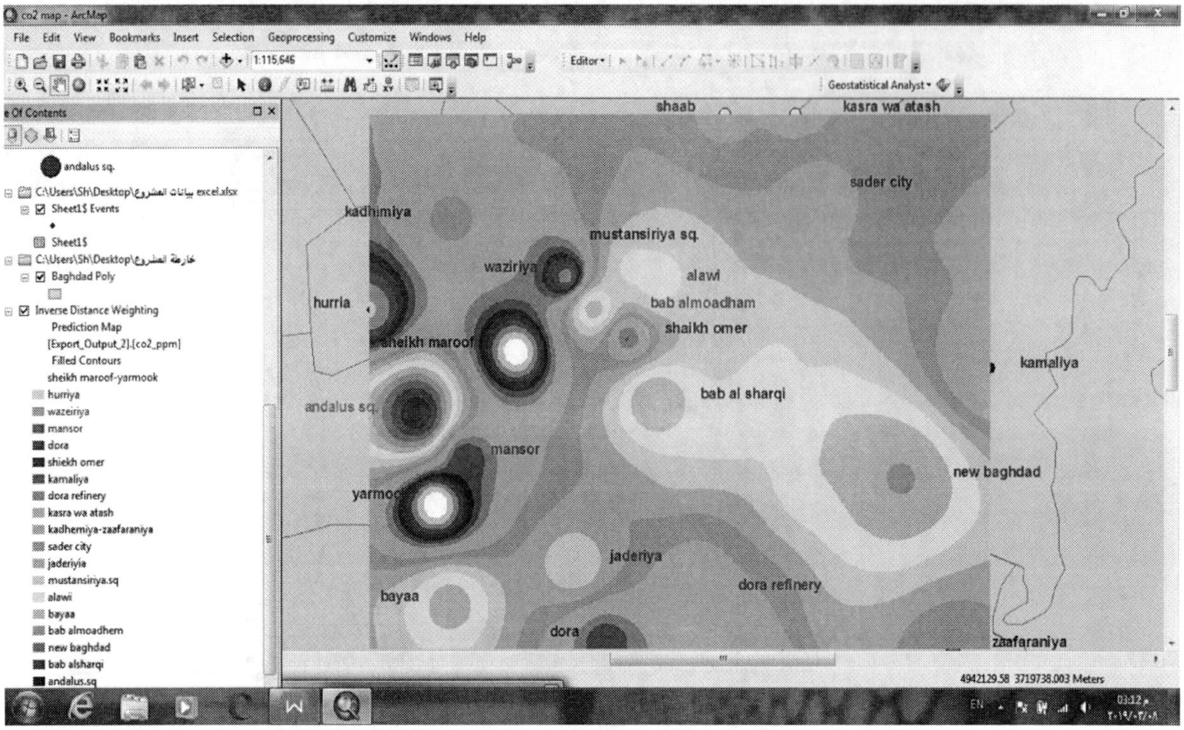

Figure5. Geostatistical Analyst of CO2.

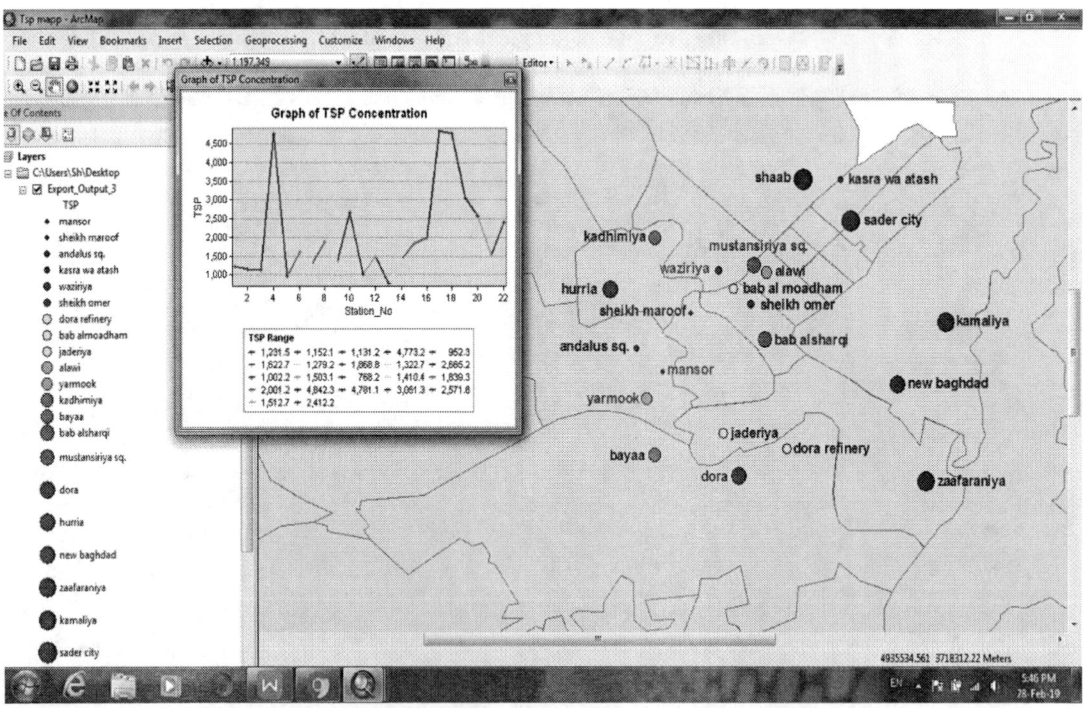

Figure 6. Ratio of TSP in Baghdad areas.

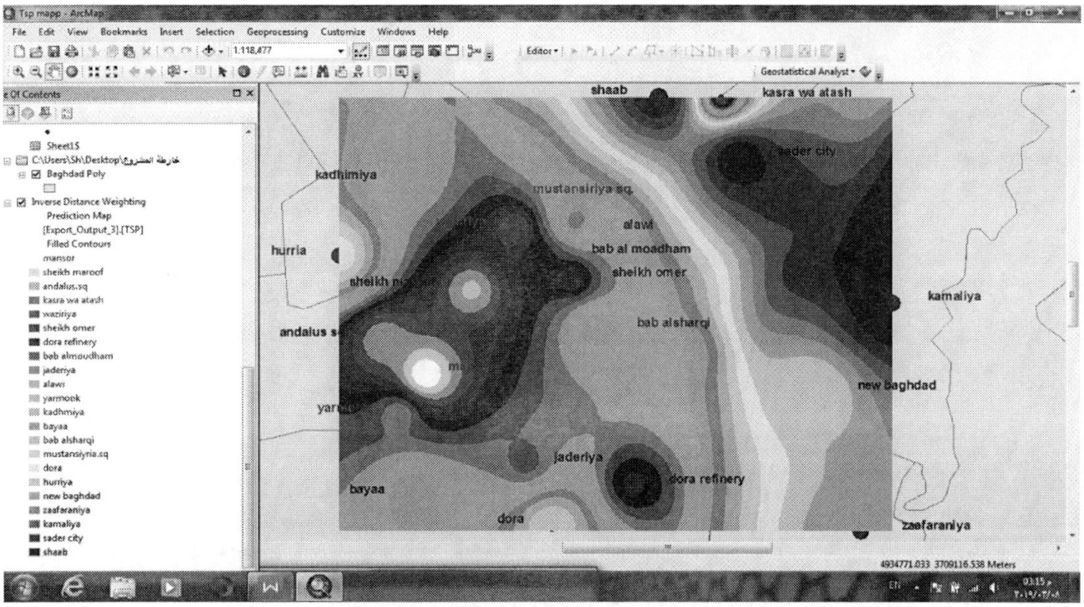

Figure7. Geostatistical Analyst of TSP.

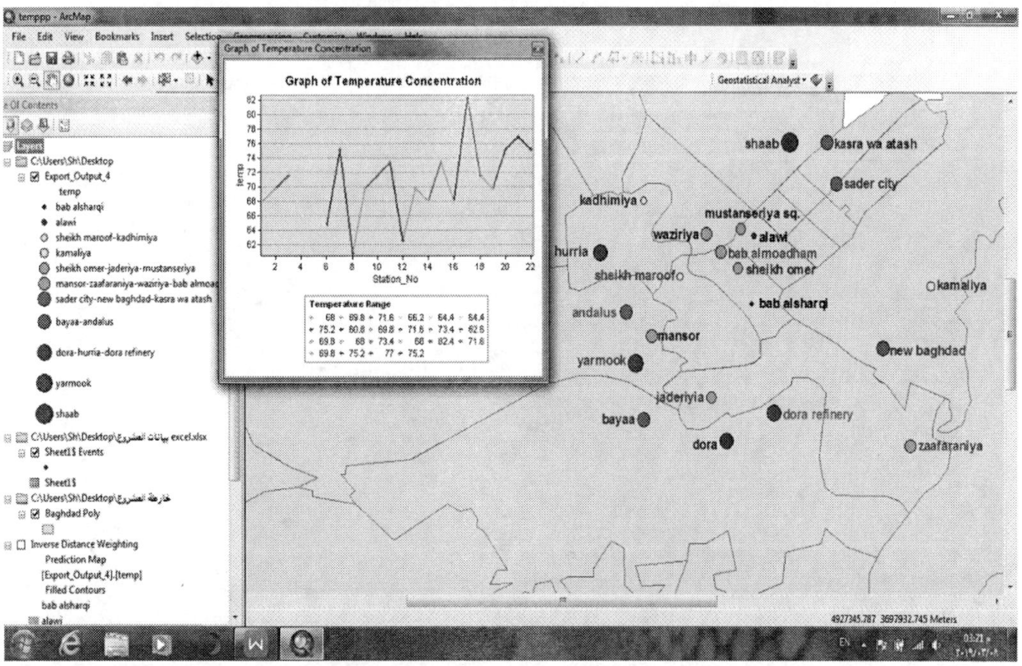

Figure 8. Temperature of Baghdad areas.

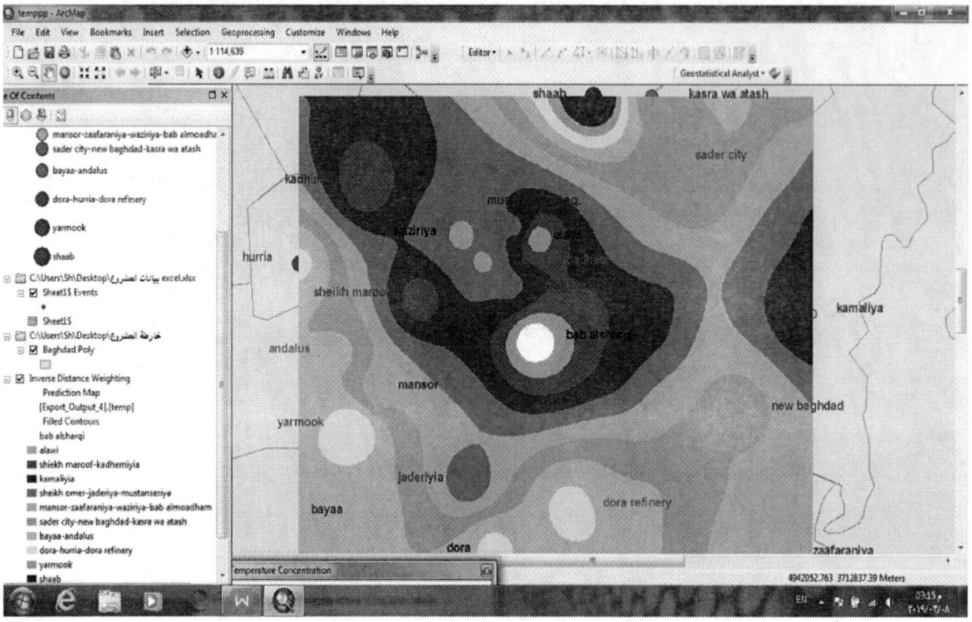

Figure 9. Geostatistical Analyst of Temperature.

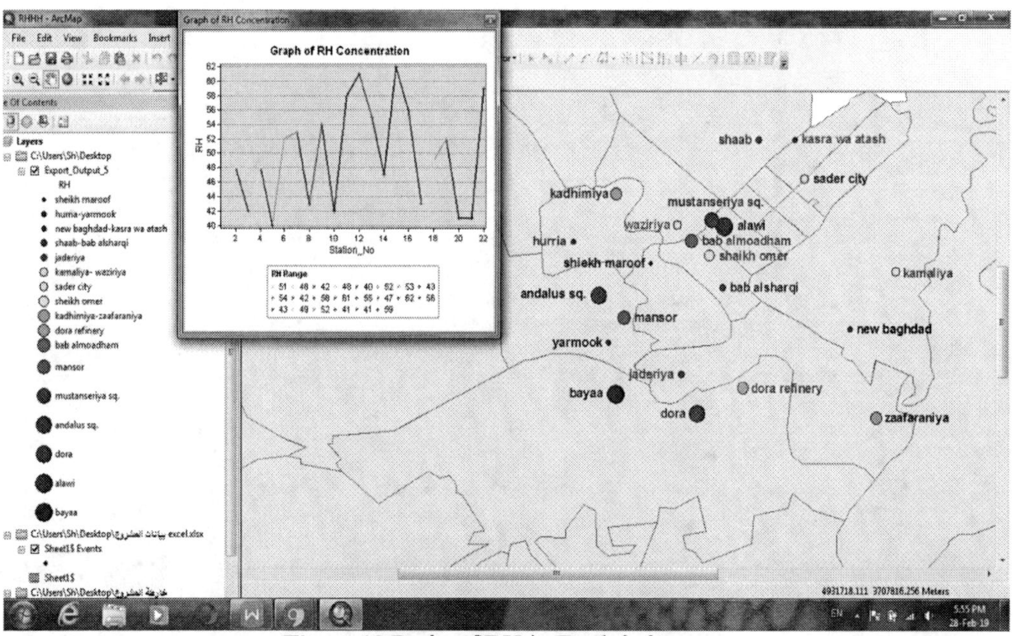

Figure10. Ratio of RH in Baghdad areas.

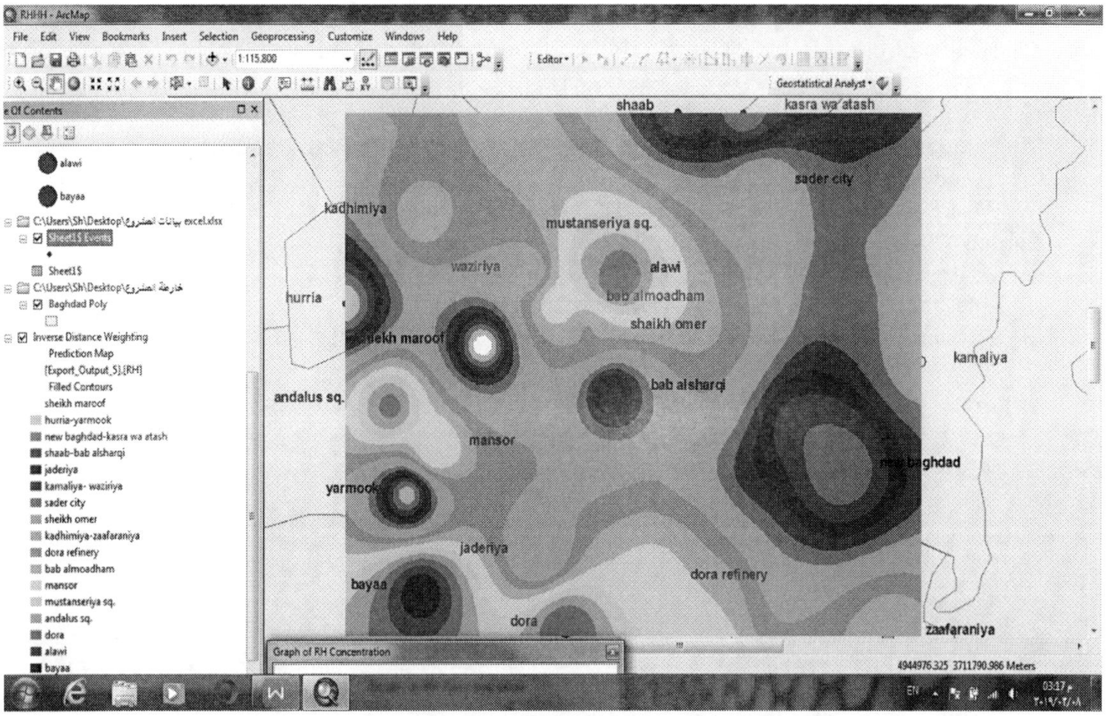

Figure 11. Geostatistical Analyst of RH.

3.2 Conclusion and Recommendations

1. The results indicate that the air pollution levels were severe over the Baghdad city with the concentrations of CO, CO and TSP in the ground measurements, the highest concentrations of TSP, CO, and CO2are recorded in many stations (4, 6, 8, 10, 16, 17, 18), which are characterized by high population and increase in the number of cars as well as poor fuel, the age of the vehicles and industrial activities inside the city

2. This measurement was characterizing the local variation of air pollution in Baghdad city.

3.The increased concentrations of CO and CO2 in the air of Baghdad are due to increase in anthropogenic emissions, mainly due to the increase in fossil fuel use for heating,the burning of the fuel used in various types of vehicles, as well as the low quality of fuel and the age of the vehicles and the use of generators ,the industrial activity, high population density and traffic volumehave direct effect on the increasing criteria air pollutants concentration.

4. Conclusions

As a part of departure the current study has shown some suggestions to limit the Environmental pollution through the following:

1. Increasing the environmental consciousness in the citizens by specified programs that share to limit pollution.
2. Reduce the traffic volume at the main roads especially near the highly polluted locations;reduce the amount of air pollutants emission from the industrial factories and the development of electric train system.
3. Using clean technology in product operations (or processes).
4. Planting trees in the city to delete its environmental effect.

5. References

[1] Geneva 1997a, Who, Health and Environment in sustainable Development Five years after Earth summit ,.

[2] Elsom D. 1992,"Atmospheric pollution :AGlobal problem ", second Edition oxford :Brackwell publishers,.
[3] Wallace, John M. and Peter V. Hobbs,2006.Atmospheric Science: An Introductory Survey. Elsevier. Second Edition. ISBN 978-0-12-732951-2. Chapter 1

[4] Jump up to: "The height of the tropopause". Das.uwyo.edu. Retrieved 2012-04-18.
[5] "Annex II Glossary". Intergovernmental Panel on Climate Change. Retrieved 15 October 2010.
[6] Who Result from the WHO project "systematic Review of Health aspect of Air pollution in Europe "2004.
[7] M. Azpurua and K. Dos Ramos 2010 Research M **14**:135 –145,.
[8] Ebtesam F. Khanjer & Nidaa ch. Jaheed 2017 IJABR **7** 4.
[9] Ebtesam F.Khanger, Nawal.K Ghazal, Ban Abd.Al Razak 2017 I.J.S.N. **8** 2 : 202-208.

IOP Conf. Series: Materials Science and Engineering **757** (2020) 012032 doi:10.1088/1757-899X/757/1/012032

Study of Some Factors That Affect ECG Device

Haitham T. Tafash[2], Ahmed H. Ali[1,3], Maath mohammed[2], Mohammed D. Salman[1], Abdul Aziz Ali[2], Asmaa Maher[1], Teba Yaseen[2], Safa saade[2], Manar sabar[2], Abdullah Mohammed[2], Hend Jasim[2], Sarah Huthaifa[2]

[1]Department of Medical Physics, College of Applied Science, University of Fallujah, Iraq.

[2]Collage of medicine, University of Fallujah, Iraq

[3]Biotechnology and Environmental center, University of Fallujah, Iraq

[1]Corresponding author: dr.ahmedphysics@uofallujah.edu.iq

Abstract Changes are studied that appear on the electrocardiogram due to the effect of electromagnetic waves emitting by mobile devices. The research was carried out at Fallujah Teaching Hospital by using a MAC1600-Morta ECG device. The work included a study on the effect of the presence of mobile devices during the measurement of electrocardiograms and the discussion of all cases, as well as the impact of the ordinary watch near the electrodes. The results showed a disturbance in the electric waves and return wave will return to its normal shape after the demise of the effect.

1. Introduction:

The process of producing an electrocardiogram (ECG or EKG) is Electrocardiography , and it is recording a graph from voltage versus time of the electrical activity of the heart by using electrodes placed on the skin. Reveal these electrodes a small electrical changes that are a effect of cardiac muscle depolarization followed up via repolarization during each cardiac cycle (heartbeat) [1] .

The first attempts to record the electrical activity of the muscle by a number of the scientists began in 1666 by (Franciscuride), and in 1773 he managed (Walsh) to recognize the formation of some fish muscles for electricity, and in 1792 was completed published a report to the Italian world (Luigi Giuliani), which applied a preliminary registration process of the muscle. It followed six decades and until 1849, when the world showed that (Doubeez, Raymond) also recorded the activity of the involuntary muscle. In addition, he managed (MORI) recording the electrical activity of the working muscle. Use the worlds (Jasser, 1922) to display the electrical signals caused by constriction muscular.

The methods of recording such electrical signals quickly and extensively evolved 1930 research then used advanced electrodes to capture signals broad when studying muscles.

The history of ECG returns to 1903 when Einthoven invented the electrocardiogram, which was very complex, and need five people to operate it [2]. Fig.1 shows the old machine of the EGC in 1901 [3].

IOP Conf. Series: Materials Science and Engineering **757** (2020) 012032 doi:10.1088/1757-899X/757/1/012032

Figure 1. Old ECG machine [3].

2. Components of Cardiac Planning:

The devices of Heart-shaping share in the same principle but only different slightly in the terms of ingredient. Generally this device consists from the following parts[4]:

1. Calibration: This part of device effectively adjusts a device and calibrates it correctly prior beginning to the process of cardiac planning, if a square wave is made one millivolt ,exhibit that the device is in the state perfect.[4]
2. Sensitivity point: This part is extremely important in maintaining the sensitivity of this device, as in its normal state (1mV) via utilizing the sensitivity point, the wave can be amplified or reduced according to the patient's case during which we relate body of human and the device.
3 .Ground: As usually, excessive charge deposition, and the electrical shock protection
4. Poles: The device Consisting of five electrodes put in specified places in the body of human used.
5. Speed limitation: The ECG has two of speeds (25-50 mm/s) employ each speed according to the existent case and specified via the doctor back to the heart if the patient is old age has a slightly weak pulse, so a low (25 mm/s). If the young is a fast pulse, utilize the high speed unto we get keeping pace with the planning of the patient's case.
6. The Screen: When the doctor is absent of the paper or not necessary for reading continuous to the heart.
7. The division: Of the protection circuits in the device, utilizing a circuit of protection from hight currents and voltages it is indeed a succeeded method in all the organs.
8. The candidate: A currency is limited in the winding up of exterior effectes that can effect on the cardiac planning, because side effects such as neurons and other devices in the same room testing has a big role in acquiring to the wrong planning [4].

The heart Anatomy:

The heart's wall is consist of three layers [4]:

The epicardium, the myocardium, and the endocardium

- The Epicardium is the external layer
- The myocardium, is the midst layer, that consists the largest part from wall of the heart . This layer of muscle tissue contracts with each heartbeat.
- The endocardium, the deepest layer in heart.

2.1. Champers of the heart:

The heart consist of four chambers two atria and two ventricles. The right and left atria serve as large tank for the blood being send to the ventricles.

- The right atrium receiving deoxygenated blood return from the human body.
- The left atrium receives the oxygenated blood from the two lungs during the four pulmonary veins [4].

right and left ventricles serve as the pumping chambers for the heart.

- The right ventricle receives blood from right atrium and pumping it during the pulmonary arteries to lungs, where it picks the oxygen and drops off the carbon dioxide.

- left ventricle receives the oxygenated blood from left atrium and pumping it during the aorta and thereafter out to the rest of the body [4].

2.2. The electricity of the heart:

:[2] The conductive system of the human heart consists from five specialized tissue
1- Sinoatrial node "SA" node .
2 - Atrioventricular node "AV" node
3- His bundle is a group of the human heart muscle cells specializing
4- Left bundle branch "LBB" and right bundle branch "RBB"
5- Purkinje fibers.

As impulses arise in SA node and traverse through atria, they cause depolarization of the atria, From the atria impulses reach AV node, where there is some delay, this delay will allow the atria to contract and pumping blood for the ventricles.

This impulse is later spread along branch of His, left and right branch branch and lastly, during Purkinje fibers causing ventricular depolarization. The predominant pacemaker is "SA node". Atrial cells, " AV node", bundle of His, bundle branch, Purkinje fibers and myocardial cells are the other pacemaker sites. When SA node fails, they can initiate impulse at a slow rate [2]. This entire process take a less than 200 parts from one second to form one beat.

2.3. How ECG Works:

In a classic ECG that component of 12 lead , ten electrodes are placed on the patient's limbs and on the surface of the chest for the patient. The total volume for the heart's electrical potential is then measured from the twelve various angles "leads" and is registered along a interval of the time (commonly ten seconds). In this way, the total volume and direction of heart's electrical depolarization is captivated at each moment during the cardiac cycle [5].

In ECG the lead registration the heart's electrical activity utilizing a chain of electrodes tacked on the patient's extremities and chest wall. The 12 leads contain to three dipolar limb leads "I, II, and III" up, down, and across scanning up, down, and across the heart, every lead transferred data about a various region. The waveforms acquired from every lead change counting on the place of the lead link to the wave of depolarization, or electrical stimulus, by passing throughout myocardium [4].

recording electrical activity for the six-limb leads in the heart's frontal plane. This plane is a view during the midst of the heart from the top to the bottom. Electrical activity is registered from the frontal to the rear axis [3]. The six precordial leads supply information about electrical activity in heart's horizontal plane, a transversal view during the midst of the heart, dividing it into upper and lower parts. Electrical activity is on record from either a top or an inferior approach [4]. Every lead consist of metal strip that connect the electricity from heart.

2.4. The 12 leads:

To recording the bipolar limb leads (I, II, III) the unipolar limb leads (aVR, aVL, and aVF) placed electrodes on both of the patient's arms and on his left leg. The leg right also receives an electrode, however that electrode acts as a ground and does not participate into the wave form. Three limb leads (aVR, aVL, and aVF), and six unipolar precordial, or chest, leads (V1, V2, V3, V4, V5, and V6). These leads providing 12 various views of the heart's electrical activity [4].

The placed of electrodes (leads) [4]:
Placement of limb leads:
"RA" on the right arm
"LA" on the left arm
"RL" on the right leg
"LL" on the left leg
Placement of Chest Leads:
"V1" fourth intercostal space at right sternal boundary
"V2" fourth intercostal space at left sternal boundary
"V4" fifth intercostal space at left midst clavicular line
"V3" midway between V2 and V4
"V5" at the itself horizontal level as V4 in the anterior axillary line
"V6" at the same horizontal level as V4 in the midst axillary line.

2.5. Monitoring limb leads:
Lead "aVR": This lead links right arm (positive magnate) with the heart (negative magnate).
Lead "aVL": This lead links left arm (positive magnate) with the heart (negative magnate).
Lead "aVF": This links left leg (positive magnate) with the heart (negative magnate).
Lead I: This lead links right arm (negative magnate) with the left arm (positive magnate).
Lead "II": This lead links right arm (negative magnate) with the left leg (positive magnate).
Lead "III": This lead links left arm (negative magnate) with the left leg (positive magnate) [4].
Criterion limb leads (dipolar)
(I): Side wall
(II): Inferior wall
(III): Inferior wall
Augmented limb leads (unipolar)
"AvR": No specified view
"aVL": Side wall
"aVF": Inferior wall
Precordial, or chest, leads (unipolar)
(V1): Septal wall
(V2): Septal wall
(V3): Frontal wall
V4:Frontal wall
(V5): Side wall
(V6): Side wall

2.6. The heart rate

The number of times the heart beats per minute (bpm) is the heart rate, and it varies according to physical necessity such as the need for absorb oxygen and remove carbon dioxide, usually the heart rate is equal or closer to the measuring pulse in any endpoint.

- Activities that can affect the heart rate are exercises, sleeping, worry, stress, disease, and ingestion of drugs like B-blockers and atropine [4].

The tracing on an ECG , bpm is usually calculated as the number of QRS(wave) complexes. The rate is measured of the R-R entr'acte, the distance between one R wave and the next. The atrial rate (the number of P waves) and the ventricular rate the number of QRS complexes wave the analysis may show them as different rates, one atrial and one ventricular. The way selection to calculate HR(the heart rate) varies according to rate and regularity on ECG Where the normal rate of heart rate into an adult at the rest time is 60-90 bpm tracing[4]. and recognized the tachycardia when heart rate override 100 bpm at the rest time, and heart rate goes down the 60 bpm also known bradycardia at the rest time.

2.6.1. Heart rate calculation:
The Heart rate has many features as a clinical parameter: It 's quick and inexpensive to measurement and is easy understand. In spite of the passable boundaries from the heart rate this convenience was based on the scale of a squares on are among 60 and 100 beats per min, the electrocar diogram paper; The better definition of natural sinus heart rate may become among 50 and 90 beats per min[6].
The standard textbooks of physiology and the medicine mention that the heart rate "HR" is easily calculated of ECG as following:
(HR = 1,500/RR) period in mm, (HR = 60/RR) period in seconds, or (HR = 300/number of large squares between consecutive R waves). At every case, the writer are really indicate to the instantaneous (HR), which is the number of times the heart would beat if consecutive (RR) periods were constant. However, because of the top formula refer is nearly always aforementioned, students limitations (HR) this method without looking at ECG any further.
2.6.2. Paper of ECG graph:
Waveforms generated via the heart's electrical current are registrated on graphed the ECG paper via a stylus. The paper of ECG consists of horizontal and vertical lines formation a grid. An ECG strip or tracing is a piece of ECG paper [2]. A horizontal axis of the ECG strip is represents the time, every little bulk equals (0.04 sec), and five little bulks form a big bulk, which equal to (0.2 sec). This time increase is specific via multiply (0.04 sec) (for one small bulk) via 5, the number of little bulks that form a large bulk. Five number of large bulks equal to (1 sec) (5 X 0.2). At measure or calculate a heart rate of patient's[4].
The strip of ECG cephalic axis measures the amplitude in millimetres or electrical voltage in millivolts . Every little bulk representing (1 mm) or (0.1 mV); Any large bulk (5 mm) or (0.5 mV). For determine the amplitude of the wave, segment, or interval, counting the number of little bulks from the base line to a high or low point of the wave, segment, or interval [4].
Paper of ECG graph
For better understanding the ECG graph as shown Fig.2 [2]:
1 (mm) = 0.04 (sec)
2 (mm) = 0.08 (sec)
3 (mm) = 0.12 (sec)
4 (mm) = 0.16 (sec)
5 (mm) = 0.20 (sec)
10 (mm) = 0.40 (sec)
15(mm) = 0.60 (sec)
20 (mm) = 0.80 (sec)
25(mm)= 1.00 (sec)

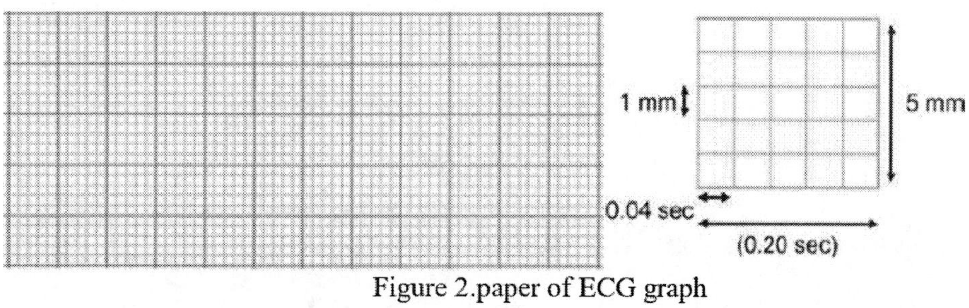

Figure 2.paper of ECG graph

2.7. A normal wave of ECG graph:
The wave of the ECG result consisting of five of waves ("P", "Q", "R", "S", and "T"). Each wave of these represents the electrical events as shown below:

• Wave P symbolize atrial de-polarization conduction of an electrical impulses during the atria, a first wave seen very small, up-right (positive wave indicating atrial constriction) and it's amplitude is (2 to 3 mm) high and its period (0.06-0.12 sec)[7].
• (PR interval): The distance among the first of (P wave) and first of (QRS complex) Measurments time through which a depolarization wave trips from atria to ventricles and it's period (0.12 -0.20 sec)[7].
• (QRS complex): Three of deviations after (P wave) mentions ventricular de-polarization and constriction (Q Wave): First negative deviations, (R Wave): First positive deviations, (S wave): First negative deviations following (R wave) and it's amplitude is (5 to 30 mm) high and it's period (0.06-0.10 sec).
• (ST-segment): The distance among (S wave) and the beginning of (T wave) measurment of time among ventricular de-polarization and the beginning of re-polarization and its period minimal than (0.10 sec).
• (T wave): Rounded up-right (the positive) wave following (ORS) represents ventricular re-polarization and it's amplitude (0.5 mm) in leads (I, II, III) and up to (10 mm) in the precordial leads.
• (QT interval):The distance inter beginnings of (QRS) to end of (T wave) representing overall ventricular activity and it's period varies; generally continues from (0.36 to 0.44 sec) as illustrate in Figure 3.
• (U waves): The small rounded, up-right wave that following (T wave) most readily seen with a slew heart rate represents re-polarization of Purkinje fibers [7].

Indicates the height of wave to the strength of voltage exerted and the width of wave indicates to the time wanted, the distance among the waves symbolizes to the time needed for transmission the electricity from one point to another point[4].

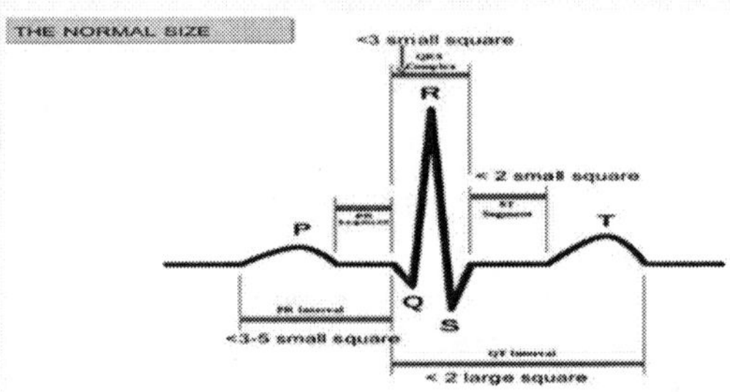

Figure 3. ECG of the heart in natural(Heart's wave) sinus rhythm [6].

2.7.1. Normal ECG pattern:

Fig.4 shows the normal waves in each lead

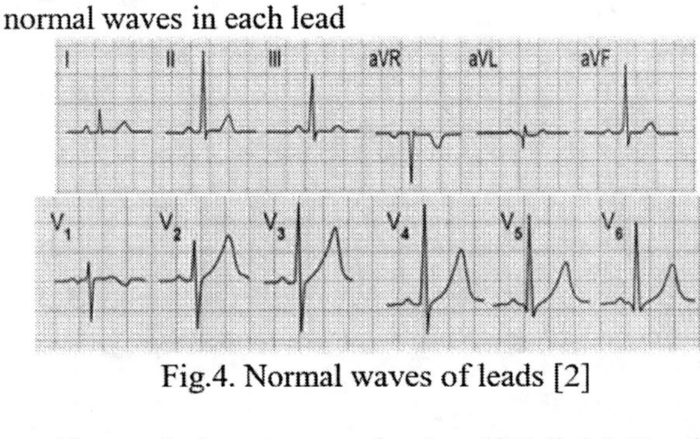

Fig.4. Normal waves of leads [2]

The heart of a normal human being was examined at Al Fallujah Teaching Hospital by using a MAC1600-Morta ECG device as shown in Fig. 5.

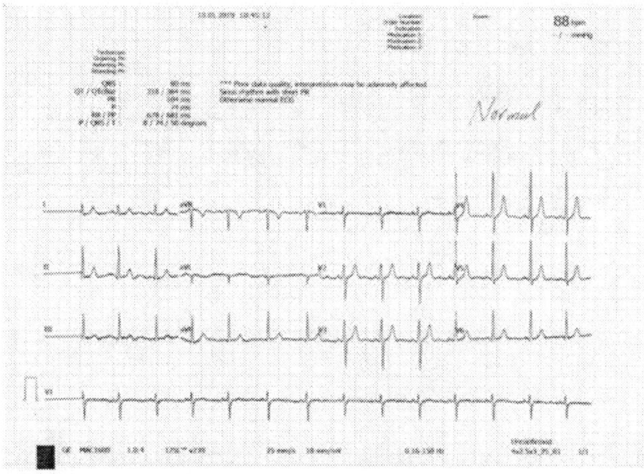

Figure 5. Normal record for the ECG of the patient in Fallujah Teaching Hospital.

3. Result and discussion

3.1. Effect of the phone:

In last years, there has been an increasing in the utilize of telecommunication devices, which has become an simple means of communication [8]. During the last decade, the utilize of mobiles has become more Pronounced. So, it has result to the structure of transmission turret in large numbers, both in urban, as well as in rural regions inclusively other sparsely populated regions. Many attempts are made to investigate the effect of the mobile phones (MP) on the health of human [8].

A form of energy appear wave like behaviour as it travels during space is electromagnetic radiation, the electromagnetic radiation has each electric and magnetic field components, which oscillate in phase vertical form to each other and vertical form to the direction of energy propagation,Though the amount of electromagnetic energy due to the cell phones is very small in compare to other radio frequency sources, the increased utilize of wireless mobile phones world wide (about 3.8 billion mobile users) has focused solicitude on it's probable side affects, and the health potential impacts [8]. Mobile phones can effect on the heart rate, TP segment and time of T wave. So, it appears that long time utilize can effect the heart of human. Based on several reports on the influence of these waves on biological processes, precautionary measures should be taken about employ mobile phones [9].

The cell phone is not a safety device, so, we must use it as little as possible, and in necessary situations. Keeping in mind, don't put it behind heart or ear for long receiving call time, since its effects on human body temperature, as a result, the temperature has been increased so the mobile effect on human rhythm, heart rate...etc. Mobile phone cells send electromagnetic waves and these waves interfere with the electricity wave of the human sent by the heart and therefore affected the signal received by the ECG device [10]. It has been illustrate that through ECG recordings shouldbe turned off the mobile phone or must be (7.5cm) from every pole of the ECG electrodes [10].

It must be notice that even when mobile phone is not utilized for any voice or information communication it is periodical sending low-frequency bursts during the phone ring, the cell phone sent high electromagnetic waves that cause interference to the signal received by the ECG [10] therefore inaccurate planning when the ECG is recording and the mobile phone ringing. As shown in this Fig.6 and Fig.7.

IOP Conf. Series: Materials Science and Engineering **757** (2020) 012032 doi:10.1088/1757-899X/757/1/012032

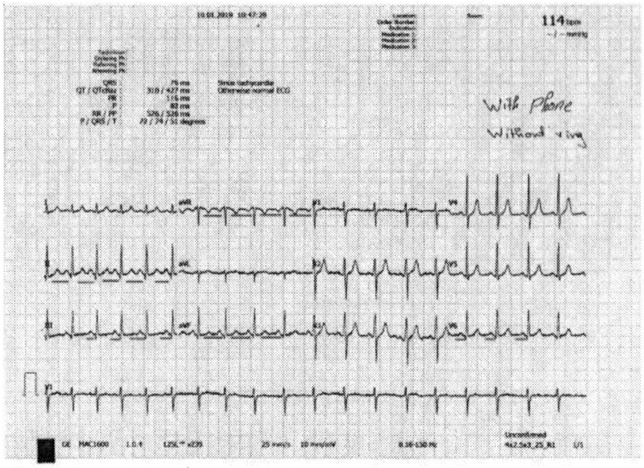

Figure 6. ECG record with effect of phone without ring

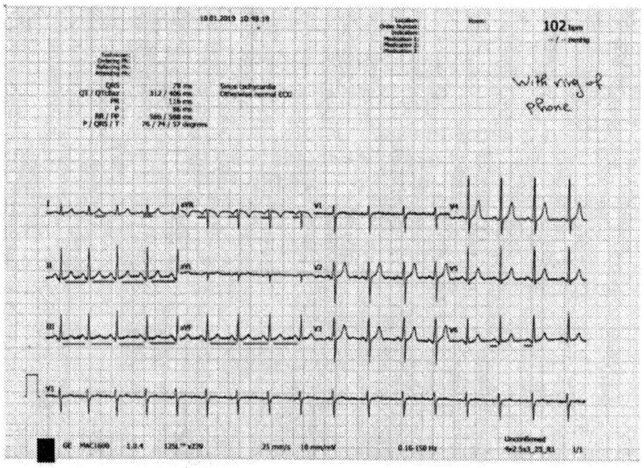

Figure 7.Effect of the ring of phone on ECG

3.2. Effect of the watch:

The ECG readable patient must get rid of all metals including the ordinary or digital clock because they effect there adding of the ECG due to the interference of the wave. In the present work not effect from the watch on ECG as shown in Fig.8.

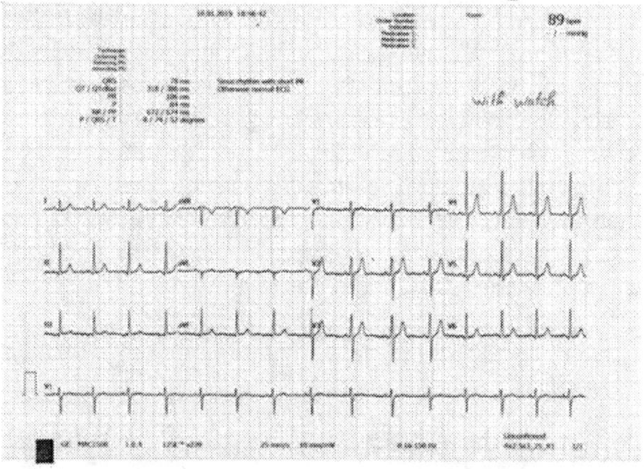

Figure 8. ECG record show little effect of ordinary watch

3.3. Effect of the hair:

Before beginning the ECG recording, the hair on the chest should be shaved off it. Because the presence of hair causes an improperly electric field, it results because the electrodes do not stick well on the skin. Hair causes a resistance between the skin and the leads so all of this referred above occur.

3.4. ECG Gel:

The gel of ECG is a water depending conduction (transport) gel with high conductivity characteristics. The ECG gel generation has been formulate with high viscosity and is specific destined to decrease resistance among the skin and electrodes in a most cardiac electrical operations. It is likewise destined to transfer weak electrical signs in a high precision method to improve the precision testing of ECG . Hight conductivity stay constant even through long procedures [11]. The gel of ECG includes non-toxic components and non-toxic preservatives tested in terms of toxicity by an independent third party[11].

The gel put on the skin to decrease the resistance among the skin and the electrodes.

3.5. Apple watch:

Recently Apple company detected watch measures the rhythm of the heart and heart rate that allow the patient to observe his/her heart [12].

Principle work of apple watch:

The sensor of optical used to measure the pulse of the heart in Apple watch utilizes what a recognized as scaling. This technicality is depended on an extremely easy fact: the blood is red because of reflect red and absorb green. Apple Watch utilizes green lights of LED as well as light sensitive light diodes to expose the amount of blood flowing during the wrist at any given moment. When the heartbeats, the blood streams in the wrist and the green absorption is maximal. While lower among each pulse and others. Apple Watch calculates number of heartbeat times per min. via flashing of the LED lights hundreds of times every sec., thus calculating the pulse rate of your heart. The light sensor for pulse measurement supports a range of 30 to 210 pulses per min. In addition, this light of sensor is designed to recompense for low sign levels via increasing the LED of light rate and the sampling rate [12].

4. Conclusions:

There are numerous interventions that influence the quality of the ECG recording process and will indicate to each other in summary the most important of them is the electrical interference. All mobile devices effect on the ECG recording process because overlapping the waves that are sent from the device that leads to addling on the ECG wave recorded so the patient should get rid of mobile devices. Also all metals including the ordinary or digital clock because they effect there adding wave on the ECG due to the interference of the waves.

References

[1] Lilly, Leonard S, ed.. Pathophysiology of Heart Disease: A Collaborative Project of Medical Students and Faculty (sixth ed.). Lippincott Williams & Wilkins. 74. 2016.

[2] Sajjan M,"Learn ECG in a day systematic" ,Jaypee,2013

[3] Einthoven W. Un nouveau galvanometer. Arch Neerl Sc Ex Nat; 6: 62533. 1901.

[4] Chris Burghardt. ECG interpretation made incredibly easy fifth editions", Lippincott Williams & Wilkins. 2011.

[5] Aswini Kumar MD. "ECG- simplified". LifeHugger. Archived from the originalon. 2. 2017.

[6] Spodick, DH "Operational definition of normal sinus heart rate". Am J Cardiol. 69 (14): 1245. 1992.

[7] Shirley A. Jones, MS Ed, MHA, EMT-P ,"ECG notes interpretation and management guide second edition",F.A.Davis company .philadelphia, 2010. ertsch, Marc. 2004.

[8] Dawser Hussain, Alyaa H. Ali, Sabah N. Mazhir, Aya Juma .Study the Effect of Mobile (Cell Phone) on the Heart Electricity. Int. J. of Appl. Inf. Sys. (IJAIS), FCS, New York, USA, 2014.

[9] Gholamreza Komeili ,Sima Nabizadeh Sarabandi, Studying the Effects of Mobile phone waves on Electro Cardiogram Parameters of Students in Zahedan University of Medical. Int J High Risk Behav Addict., 1(2): 75, 2012.
[10]T. Buczkowski, D. Janusek, H. Zavala-Fernandez, M. Skrok, M. Kania, A. Liebert,"Influence of Mobile Phones on the Quality of ECG Signal Acquired by Medical Devices", measurement science review, 13, 5, 2013
[11] Telepaper magazine in website: www.telepaper.com/ar/products/medical-paper/gel-liquid/ecg-gel/
[12] Apple's company website: www.support.apple.com/ar-sa/HT204666.

Characterization of nickel oxide nanocatalyst electrodes for an alkaline fuel cell

N. .A. Ali[1] ,H. J. Mohammed,[2]
[1] University of Baghdad, department of physics, Iraq
[2] Ministry of Science and Technology. Baghdad, Iraq
[2] alhamdaniya2003@yahoo.com

Abstract. In this paper had been studied the characterization of the nanocatalyst (NiO) Mesh electrodes. For fuel cell. The catalyst is prepared and also the electrodes The structural were studied through the analysis of X-ray diffraction of the prepared nanocatalyst for determining the yielding phase and atomic force microscope to identify the roughness of prepared catalyst surface, Use has been nanocatalyst led to optimization of cell voltage, current densities & power for a fuel cell.

Keywords: (NiO) nanocatalyst , electrolysis cell, an alkaline fuel cell, X-ray diffraction, atomic force microscope.

1. Introduction

Since the first decade of the twenty-first century. There has been a tremendous revolution of nanotechnology and many researchers have talked about it. The introduction of nanotechnology, that contributes considerably to sustainable economic growth, has broad applications in advanced technologies because of its characteristic properties [1], One amongst the foremost promising trends in nanotechnology is that it's utilized in electrolysis and fuel cells. The nanomaterial's obtained with this method are incredibly promising as light selective coatings, greatly increasing corrosion resistance and protection capability. [2] These nanomaterial's are characterized by reasonable electrical, [1, 3], To develop these cells and obtain low-cost nanotechnology alkaline fuel cell, by developing the catalyst efficiency. [4]. The researcher Yasuhiro Mizukoshi and Katsumi Yamada evaluated the electrochemical properties of NiO/Au. High-velocity electrochemical changes were obtained within one second by improvements in penis length and film thickness for the NiO film [5],The researcher Tomokazu Sakamoto and his group NIO/ Nb2O5 /C, catalysts were manufactured for electro-hydrazine electro-oxidation by evaporation. thermodynamic nanomaterials have been extensively tested in recent decades because of their high surface area, which severely affects their physical properties. And on reported nanostructure shapes [6], The NiO-CNT catalysts were supported Pd , By the researcher S. S. Hossain et al, that the NiO-CNT / Pd electrical catalysts show the high electrochemical activity and also the long-run stability of chemical oxidization were synthesized , characterized, and assessed of formic acid for fuel cell [7].

IOP Conf. Series: Materials Science and Engineering **757** (2020) 012033 doi:10.1088/1757-899X/757/1/012033

2. Experimental work

2.1. The mesh stainless-steel

The prepared mesh stainless-steel of type 316 was cleaned with is up and water, Before immersion in HCl solution (8-10%) for 5 minutes, then washed with acetone, ethanol and at last in ultrapure water via an ultrasonic treatment for 10 min then dried.

2.2. The prepared nano (NiO) catalyst electrode

The prepared mesh stainless-steel was cut into 3 cm long. One ends of stainless-steel of type 316 were connected to a cathode. The plating resolution has consisted of nickel sulfate, sodium acetate, and sodium sulfate mixture of temperature. Nickel sheet was used as an anode. The nanoporous nickel oxide film was deposited on the surface of the stainless. The deposited film is very porous, As shown in figure (1). Once deposited, the deposited stainless-steel was rinsed many times with deionized water, then dried.

Figure1. nano (NiO) catalyst electrode

2.3. Design of electrolyser cell

consists outer wall of organic glass (14.5x14.5) cm2 to prevent leakage of gases from the cell, 0.1 cm thick of electrode. as well as consists of two electrodes isolated from each other, to isolate each gas separately (hydrogen and oxygen).

Figure2. The electrolysis method set up

2.4. An alkaline fuel cell

consists of nano (NiO) catalyst anode and cathode, with These two electrodes are separated by an electrolyte (KOH). An oxidant is fed to the cathode to supply hydrogen while fuel is fed into the anode to supply hydrogen. The outer wall of the cell consists of organic glass sheets.

3. Results and discussion

3.1 Structural properties

The most famous methods for studying the determination of the crystalline structure of any material or skinny layer are X-ray diffraction. The diffraction pattern is analyzed on the spacing and spacing of the material within the examined material [8], In figure. (3) and table (1) , Analysis of nano nickel oxide, Showed a series of diffraction peaks at 2 θ of 37.39 °, 42.089 ° , 43.092 °, 63.700 °, 75.15 °, and 78.93° can be assigned to (111), (200), (220), (311) (211) and (222) planes, The diffraction peaks show good crystalline nanoparticles and match very well with ideal lattice constants, In this paper, This result shows that the physical phases of the NiO nanoparticles have higher purity prepared as agreeing with [9,10,11].

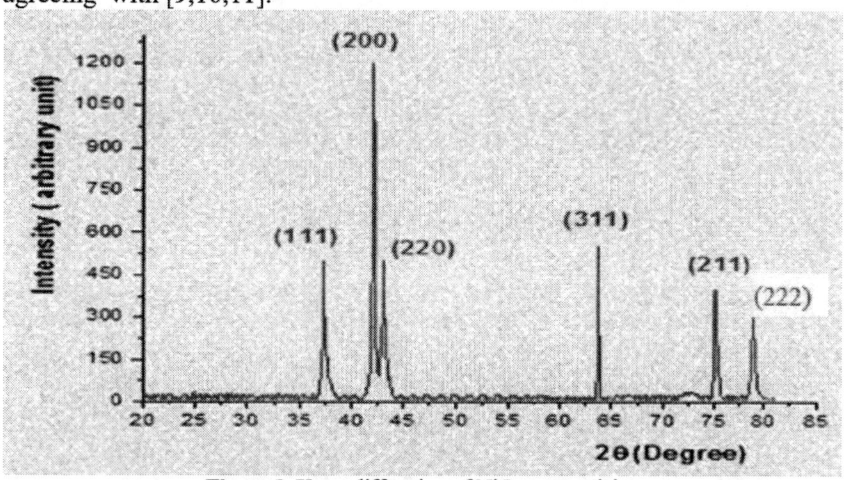

Figure 3. X-ray diffraction of NiO nanoparticles

Table 1. Peak data used for crystallite size estimation using the Scherrer formula for nano NiO

hkl	2theta [deg]	FWHM sample	Crystallite size [nm]	Average Crystallite size [nm]
(111)	37.39	0.42314	14.91670825	14.68331975
(200)	42.089	0.19908	29.61348227	
(220)	43.092	0.62618	9.264579907	
(311)	63.700	0.13679	25.73184801	
(211)	75.15	0.38451	5.293644395	
(222)	78.93	0.46512	3.279655669	

3.2 Morphological properties

The average grain size and surface morphology can be obtained by (AFM) analysis. Surface morphology of (NiO) was obtained very highly spaced, Randomly oriented and highly correlated of pores. It has found that the average particle size has 42.15 nm, As shown figure (4) a,b, As agree with [12,13].

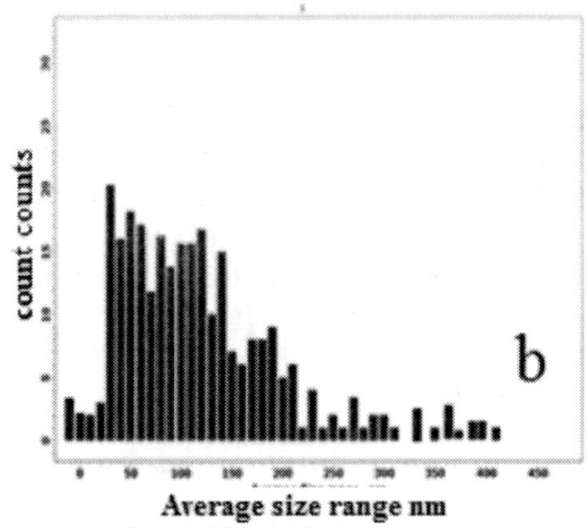

Figure. 4(a,b) .Atomic force microscope of nano NiO catalyst

3.3 Study the parameters for an alkaline fuel cell

Electrical conductivity increases with an increase in the current density, As shown in Figure 5. and table (2). The voltage of fuel cell decreased with the current, As shown in Figure 6 a,b. The energy produced by the alkaline cell depends on the purity and quantity of hydrogen supplied to it.

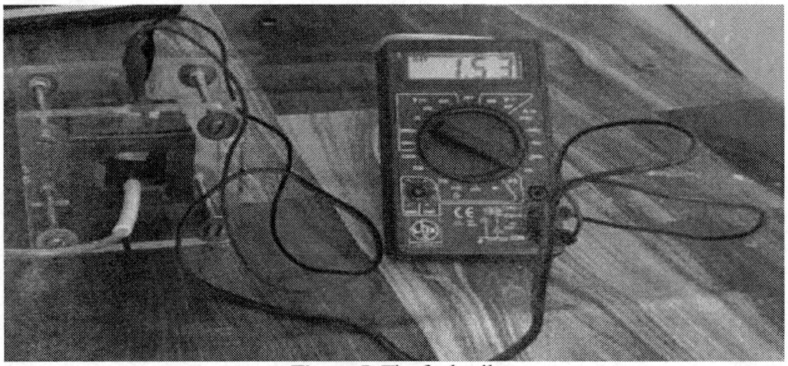

Figure 5 .The fuel cell

Table.2. Show the relationship between the current and voltage with electrical conductivity

Electrical conductivity (S\cm)	Current (A)	Voltage (V)	Power (Watt)	Time (min)
0	1.8	1.53	2.75	5
0.9	2	1.5	3	10
1.2	2.4	1.47	3.53	15
2	2.7	1.44	3.89	20
2.5	3.1	1.42	4.26	25
3.7	3.4	1.39	4.73	30

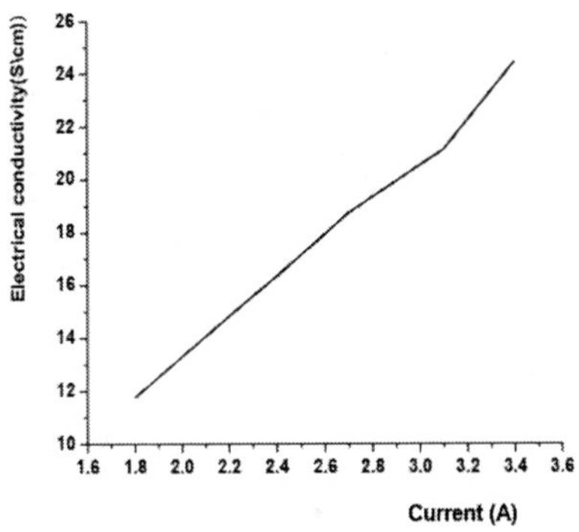

Figure 6a The relationship between current -Electrical conductivity

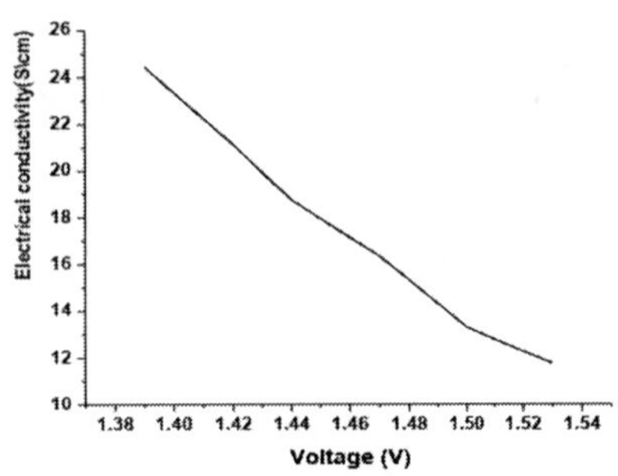

Figure 6b. The relationship between voltage - Electrical conductivity

4 Conclusions

In this research, nano nickel oxide catalysts were synthesized successfully for membrane for fuel cell application. Manufacture of electrode for fuel cell of mesh stainless steel coated with a layer of nanocatalyst is considered inexpensive compared to other methods using precious and expensive metals.

It was found that the voltage of the fuel cell decreased with increasing the current for the same flow rate. While electrical conductivity increases with increasing the current

Reference

[1] N. Cirovic, P. Spasojevic, L. Ribic-Zelenovic, P. Maškovic, M. Spasojevic 2015 Science of Sintering, 47 ,347-365.

[2] Zh. I. Bespalova, A.V. Khramenkova 2016 nanosystems physics, chemistry mathmatics, 7 (3), P. 433–450.

IOP Conf. Series: Materials Science and Engineering **757** (2020) 012033 doi:10.1088/1757-899X/757/1/012033

[3] Josiane Dantas Costa, Mikarla Baía de Sousa, José Jaílson Nicácio Alves, Bianca de Oliveira Evaristo, Raíssa Alves Queiroga, Aureliano Xavier dos Santos, Theophilo Moura Macie 2018 Int. J. Electrochem. Sci., 13 ,2969 – 2985.

[4] Ksenia Novikova, Alexandra Kuriganova, Igor Leontyev, Ekaterina Gerasimova, Olga Maslova, Aydar Rakhmatullin, Nina Smirnova,Yuri Dobrovolsky 2017 Springer Science,Business Media, LLC .

[5] Eileen Hao Yu, Xu Wang,a Ulrike Krewer,bc Lei Lid and Keith Scotta 2012 Energy Environmental Science Cite this: Energy Environ. Sci., 5, 5668 .

[6] Yasuhiro Mizukoshi and Katsumi Y amada 2014 Bull. Soc. Photogr. Imag. Japan. Vol. 24 No. 1: 12–17.

[7] S. S. Hossain, J. Saleem, A. Al. Ahmed, M.M. Hossain, M. N. Shaikh, S. U. Rahman, G. Mc Kay 2016 Int. J. Electrochem. Sci., 11 ,2686 – 2708.

[8] J. Solgom 2007 Vol. I, Structure and Dynamics, Springer Berlin Heidelberg New York.

[9] Andreas Leineweber, Herbert Jacobs, Steve Hull, 2001 Inorganic Chemistry 40, 5818-5822 .

[10] D. S. Bai, R. P. Suvarna, and C. B. M. Krishna 2016 Indian J. Adv. Chem. Sci., 4(1) 98-101.

[11] Azhagu Raj , Mohamad S. Al Salhi , and Sandhanasamy Devanesan 2017 Materials, 10, 460.

[12] K. Anandan and V. Rajendran 2012 Nanoscience and Nanotechnology: An International Journal Universal Research Publications. All rights reserve ISSN: 2278 -1374.

[13] K .C. Liu, M.A. Anderson 1996 J. Electrochem. Soc. 143 ,124–130.

Heterojunction Solar Cell (NiPcTs/CdS) Organic /Inorganic

A S Hassan[1], M A Aboode*[2]and I H Shallal[3]

1,3 Department of Physics, College of Education for Pure Science /Ibn Al-Haitham, Baghdad University, Baghdad, Iraq

[2]Solar Energy Research Center, Renewable Energy Directorate, Higher Education and Scientific Research Ministry, Baghdad, Iraq

*Corresponding author: morooj_a@yahoo.com

Abstract

In the present study, thin films of organic semiconductors Nickel PhthalocyanineTetrasulfonic Acid Tetrasodium Salt (NiPcTs) and inorganic semiconductor (CdS) prepared from the mixing of liquids for these two materials with different size ratios by the spin coating method on prepatterned (Fluorine-doped Tin Oxide) FTO coated glass substrates and then the manufacture of solar cells. The properties of solar cells through the study of the optical properties (absorption spectra, absorption coefficient, energy gap) and electrical characteristics included (DC. Electrical onductivity, Hall Effect and solar cell efficiency measurements). The results obtained That the efficiency of a multiple solar cell ranging from (0.16-13.2) %

Keywords: Organic ,Inorganic, NiPcTs/CdS, Solar cell

1. Introduction

Over the past years decades, organic photovoltaic devices have gained a broad interest in the last few years due to their potential for large-area low-cost From the first reports on molecular thin film devices more than 30 years ago, their power conversion efficiencies is (0.001%) in 1975 to (1%) in 1986 [1] and have increased considerably more recently to (5.5%) in(2006). This was seen in Organic compound Nickel (II) phthalocyanine-tetra sulfonic acid tetrasodium salt (NiPcTs) is a conjugated molecule. The molecular formula for NiPcTs is ($C_{32}H_{12}N_8Na_4NiO_{12}S_4$) and has (979.4 g/mol) molecular weight, the structure of NiPcTs in where that polar (SO3Na) is attached to the angles of benzene rings and make this compound water-soluble [2]. The molecular structure of NiPcTs which used as active material in this work is shown in Fig 1. Organic photovoltaic devices including organic solar cells, gas sensors, field-effect transistors, Organic thin film deposition techniques like roll-to-roll processing, ink-jet printing or spin casting are widely used, due to their cost-effectiveness,[3]. CdS is an II-VI compound semiconductor and has an energy band gap of 2.42eV. CdS thin films belonging to the Chalcogenide family are used as a window material for CdS/CdTe solar cells [4]. They are also useful as buffer materials for solar cells and continue to be a subject of intense research due to their potential applications in high-efficiency solar devices. Also, CdS thin film is one of the important materials for application in electro-optic devices such as photo-conducting cells, photo-sensors, transducers, laser materials, optical wave-guides, and non-linear integrated optical devices[5]. There are many methods of fabricating CdS thin films, these included spray pyrolysis, chemical bath deposition [6], molecular beam epitaxy, close space sublimation, successive ionic layer adsorption and reaction, screen printing, physical vapor deposition, and spin coating method which was used in this study. Hybrid solar cells, which combine the advantages offered[7] by both organic and inorganic materials, are also highly versatile in terms of the device structure they can adopt depending on how the photoactive layer is constructed. Two basic solar cell device architectures can be obtained, i.e. bulk heterojunction (BHJ) and bilayer heterojunction as in the (fig. 2 a).Bilayer solar cells have the merit of a more direct charge transport path way at the expense of

poorer exciton dissociation as compared to their BHJ counterpart as in the(fig.2 b) [8]. Furthermore the bilayer solar cell with its simple architecture essentially consisting of a stack of two photoactive layers, provides an inexpensive and straight forward platform for device fabrication without the necessity for extensive morphological optimization[9]. In This work synthesis and electrical properties of thin films organic (NiPcTs), inorganic (CdS) and blend BHJ (NiPcTs-CdS) and manufacturing solar cell by spin coating method (FTO/PEDOT: PSS/blend/AL) and by bilayer (FTO/NiPcTs/CdS/Al) and Sandwich (FTO/NiPcTs/CdS/FTO).

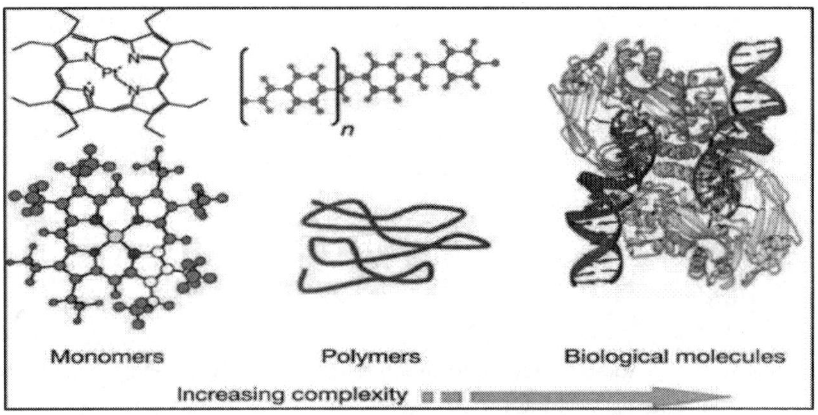

Figure 1. Types of organic materials in order of increasing complexity

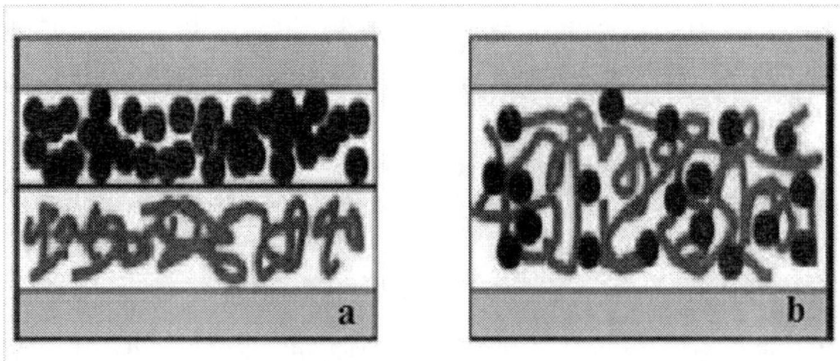

Figure 2. (a) bilayer heterojunction (b) BHJ

2. Experimental Work

2.1. *Thin films preparation*

Thin films of Nickel PhthalocyanineTetrasulfonic Acid Tetrasodium Salt(NiPcTs) were synthesized on pre-patterned (Fluorine-doped Tin Oxide) FTO coated glass substrates of thickness (3.2 mm) and (10 × 10 cm) dimensions, that were cleaned ultrasonically in acetone, ethanol, and deionized water for 10 min., respectively. After drying in 70◦C, the substrates were cut into four equal pieces with dimensions (2.5x2. 5 cm) Thin films of NiPcTs prepared by sol-gel spin coating methods, with thickness of (200±7) nm deposited by dissolving (0.65gm) organic powder of (NiPcTs), which was provided by in (10 ml) deionized water and used without any purification, then using magnetic starrier to mix the solutions over 30 minutes and with using a special syringe to deposited the solution on FTO glass substrate, then leave samples for 60 minutes in spin coater type [Laurell-WS-650Mz-23NPP]with speed of (1500) rpm, then drying process of the sample in (70◦C) temperature. Two (Al) electrodes of 200 nm thickness were deposited through a shadow mask by using the thermal evaporation method onto the glass substrate with 40 ± 5 μm gaps between them. The evaporation rate of (0.3 nm/s) was choosing to deposit electrode films, and the vacuum chamber maintained at a pressure of (5× 10−5) mbar throughout the evaporating process. CdS thin films prepared by two methods:

1. CdS thin films were deposited on glass substrates by a spin coating method. Two Solutions for CdS were prepared from dissolving (1.83gm) cadmium chloride (CdCl2), placed in a glass baker with a piece of magnet bar than using magnetic starrier to mix the solutions over 60 minutes and then using a special syringe to deposited the materials on FTO glass substrate and leave it for 10 minutes for an hour. Turning it into a transparent liquid. Then, (0.76gm) of Thiourea was dissolved in(25ml) of 2- ethanol and added drop wisely to the mixture under stirring which was continued for 5 hours. The solution was dropped onto the glass substrate about 5 and 10 cycles at speeds of (1400) rpm for 30 sec. for each cycle. After deposition, the films were dried in air at 150 °C for an hour.
2. CdS thin film preparation: Two Solutions for CdS were prepared first from dissolving (2gm) polyethylene glycol in (20 ml) of ethanol in glass baker for one hour on the magnetic mixer and inside it a small magnetic piece to mix the materials well. Then adding (1.83gm) cadmium chloride (CdCl2).The second method by adding (0.65gm) cadmium acetate powder as a source of cadmium element dissolving in (2gm) polyethylene glycol in (20 ml) of ethanol in glass baker for one hour until completely dissolved. In the process, we dissolve (0.25gm) thiourea powder as a source of sulfur in (5ml) ethanol for about an hour and then mix the two solutions for 6 hours in a row and filtered then get a transparent liquid, with using a syringe especially, we draw
(0.5ml) of the fluid and deposit it on the prepared substrate glass and leave it for 5 min at room temperature and then place it in the spin coating device at a speed of

(1500 rpm) for 30 seconds, then at a temperature of (200 ◦C) for one hour and thus get with thickness of (130 nm).

3. Preparation of thin film for (Heterojunction structure) is done by mixing the amounts of liquids of (NiPcTs) with the (CdS) to the preparation by preparing the cadmium chloride solution once and cadmium acetate once in different volume quantities and placed on the magnetic stirrer for two consecutive hours after that a certain amount is withdrawn by a special syringe and deposition on substrate glass and (FTO) proportion and leave it for 5 minutes at the temperature of the room and then put them in spin coating and the speed (1200) for (30 Second) to ensure the distribution of liquid on all substrate and then put them at (150 ◦C) a temperature.

2.2. *Prepare of a solar cells :*

1. *We lay a thin film (CdS) on the substrate (FTO), and deposition of a thin film (NiPcTs) and then we match the two substrates as in Figure3.*

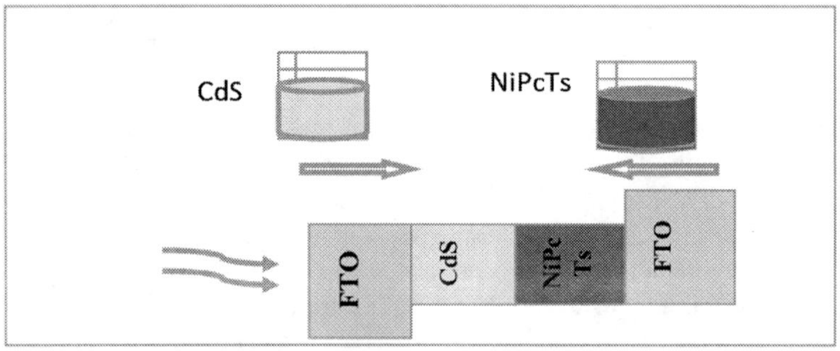

Figure3. Sandwch solar cell FTO/ CdS/NiPcTs/FTO

2. Layer by Layer:By depositing a thin film (CdS) on the Substrate (FTO) and thickness (130nm) and after cool deposition deposited a thin film (NiPcTs) and thickness (150nm) and after the formation of the poles as in the Figure 4

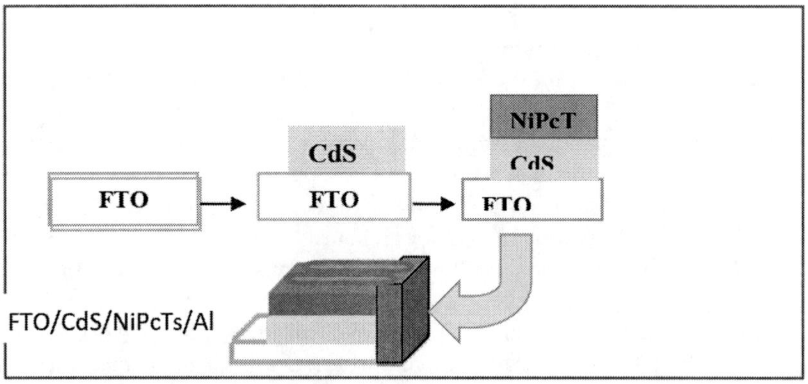

Figure4. .Layer by layer solar cell FTO/ CdS/NiPcTs/Al

3. By depositing Thin film (PEDOT: PSS) organic connector on the substrate (FTO) and after preparation deposition Thin film (Heterojunction blend) and put it at the temperature (70 ◦C) and then the electrodes as in Figure 5.

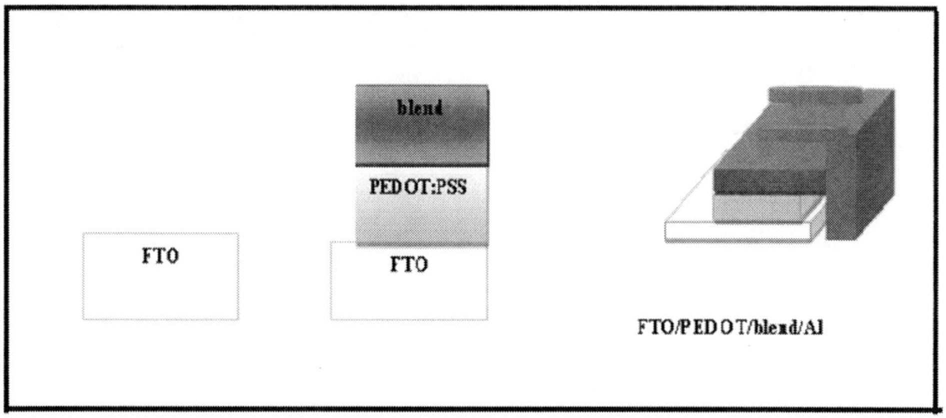

Figure 5. blend Heterojunction cell FTO/ PEDOT:PSS /Al

4. Results and discussion

3.1. Optical properties

The optical properties (Absorption, absorption coefficient, Transmittance spectrum, and energy gap) of the prepared (NiPcTs) and (CdS) prepared in cadmium chloride, cadmium acetate and Heterojunction Blend (NiPcTs/ CdS) were studied. The permeability has an accompanying behavior to the concept of absorptive, it is usually the opposite of the absorption behavior, and depends on the properties of the material and thickness of thin films by Spin Coating. The thickness of the thin film was (200-150 nm) on the (FTO)substrate. Figure 6(a) represents the permeability spectrum of cadmium sulfide packaged from cadmium chloride and cadmium acetate, the highest permeability of cadmium chloride solution was equal (77%).Figure6b, which represents the Transmittanceof of the thin film (NiPcTs), the precipitation on the floor of the glass and the thickness (200 nm). It is noted from the figure

that the transmittance spectrum has two growing regions, the first two areas of permeability at wavelength(525 nm)Where it starts with a gradual increase at wavelength (300 nm) to wavelength (525 nm) where the transmittance is 50% and then starts gradually increasing until the transmitting (65%) of the wavelengths(640-1100nm) and the gradual reduction to the lowest spectrum at wavelengths (530-640nm). However, the permeability spectrum of the first junction is higher than that of the second junction (10%)figure 6 (c).

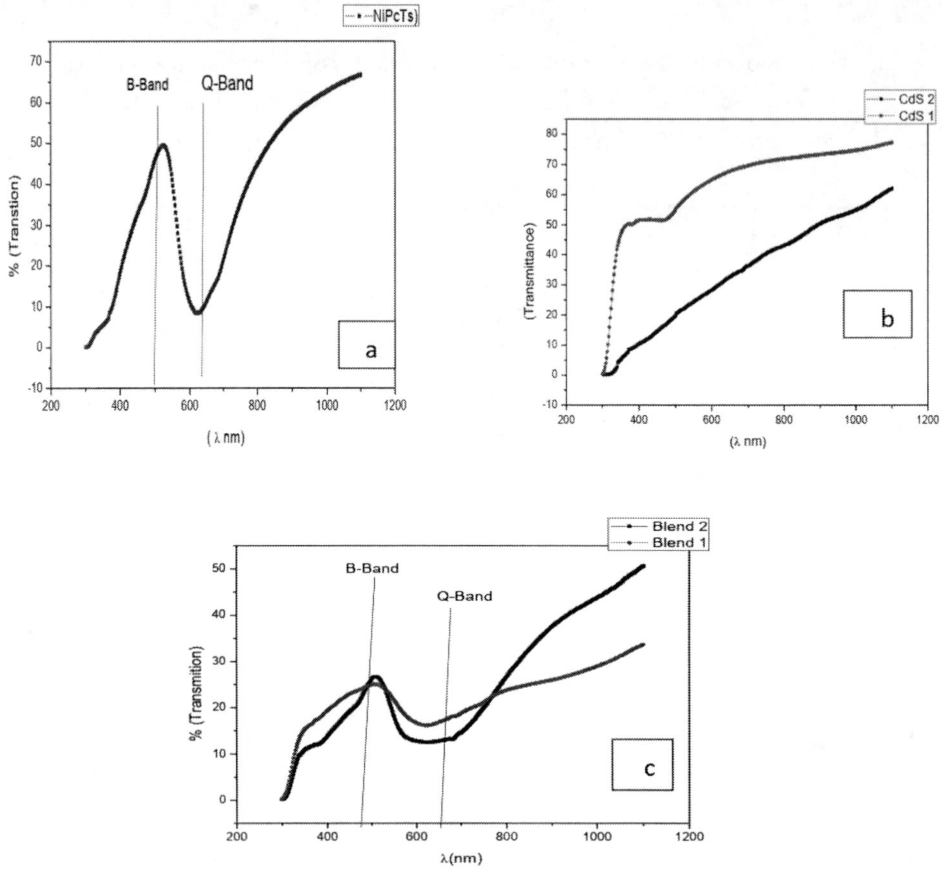

Figure 6.Spectrum of transmition for (a)NiPcTs thin film (b)CdS (C) BHJ cell FTO/ PEDOT:PSS /Al

Absorption depends on the energy of the photons that fall on the semiconductor material, the surface roughness, the type of material and the nature of its crystalline structure, The spectrum of thin films (NiPcTs, CdS, HJB (CdS / NiPcTs)) was calculated for wavelengths (300-1100 nm) within the visible region (UV-Vis), Figure 7 shows the Film absorbance spectra (NiPcTs)Prepare by spin coating method and thickness (200nm) as a function of wavelength As shown by the absorptive form it increases sharply and rapidly until its peak (550 nm). The appearance of two spectrum peaks at wavelengths (325, 625 nm)Which range back to the range (Q, B)respectively and the same behavior as for the cadmium sulfide samples as seen in Figure (6b), where the absorbance spectra, are decreasing significantly by increasing wavelengths, particularly from cadmium chloride solution and cadmium cells for wavelengths (800-1100nm), This is due to the density of the local levels formed by the atoms of the material in the similar material between the parity and conductivity packages.

The absorption coefficient (α)for all films prepared from the thin film adsorption spectrum (NiPcTs, CdS, Heterojunction Blend) was calculated according to the equation:

$$\alpha = 1/d \ln (T)........(1) \quad [8]$$

Where is the film thickness. All thin films have a coefficient of absorption ranging from 4-(6x10^5)and thickness (200,130,200nm), respectively, This means that it has a direct energy gap and the value of (α) for CdS prepared by cadmium acetate solution (4.5 x10 cm) and Blend sample (CdS) (35 x 10^4 cm^{-1}). The blend prepared by the cadmium acetate solution has an absorption coefficient(55 x 10^5 cm^{-1}) and the sample (NiPcTs) has an absorption coefficient (57 x 10^5 cm^{-1}). All calculated were within the Visible region. The relationship between the absorption coefficient (α) and the wavelength of the falling photon (λ) of the prepared membranes was plotted as shown in Figure 8.

The energy gap for NiPcTs, CdS, Blend was calculated by plotting the relation between $(\alpha hv)^2$ as a function of the photon and the(hv) observed in Figuer9. The photon energies (the range of the visible ray area) are linear, indicating that the energy gap is a direct energy gap and that the nature of transitions is transient Direct. In order to determine the optical energy gap (Eg) from the optical absorption spectra, measured around the absorption edge, we have usedTauc equation [9]

$$(\alpha hv)^{1/n} = B (hv-Eg)........(2)$$

where B is a parameter depending on the transition probability and n is a number caracterizing the transition process.

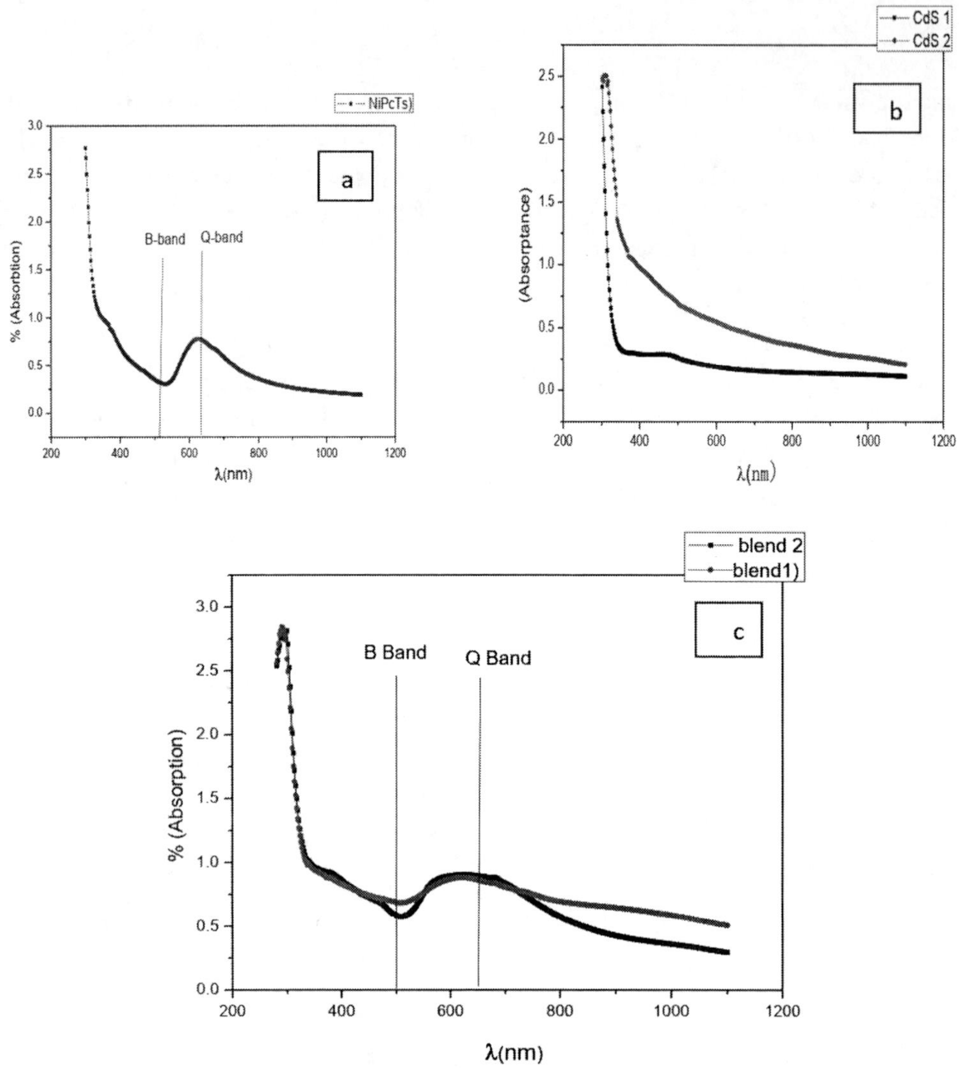

Figure 7. Absorption (a) NiPcTs (b)CdS (C) HJB (CdS-NiPcTs)

IOP Conf. Series: Materials Science and Engineering **757** (2020) 012034 doi:10.1088/1757-899X/757/1/012034

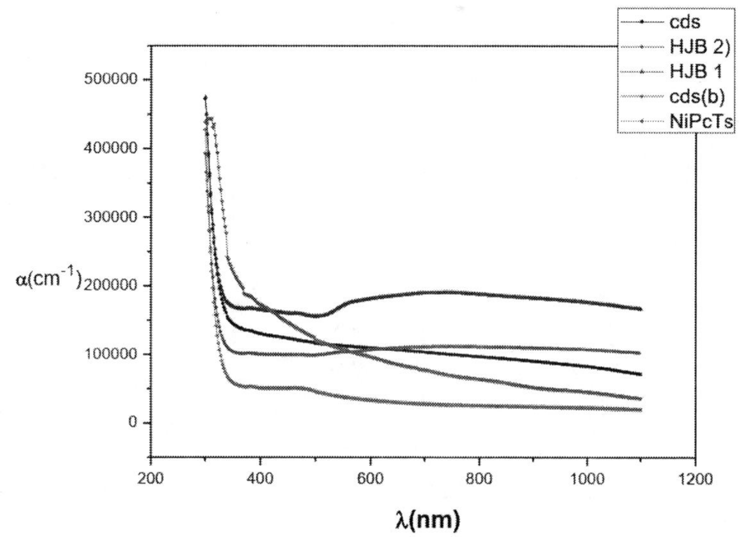

Figure 8. Absorption coefficient for all prepared films

The energy gap for NiPcTs, CdS, Blend was calculated by plotting the relation between (αhv)2 as a function of the photon and the(hv) observed in Figuer9. The photon energies (the range of the visible ray area) are linear, indicating that the energy gap is direct and that the nature of transitions is transient Direct. To determine the optical energy gap(Eg) from the optical absorption spectra, measured around the absorption edge, we have usedTauc equation [9]in Figuer9]:

(αhv)1/n = B (hv-Eg).........(3)

where B is a parameter depending on the transition probability and n is a number characterizing the transition process.

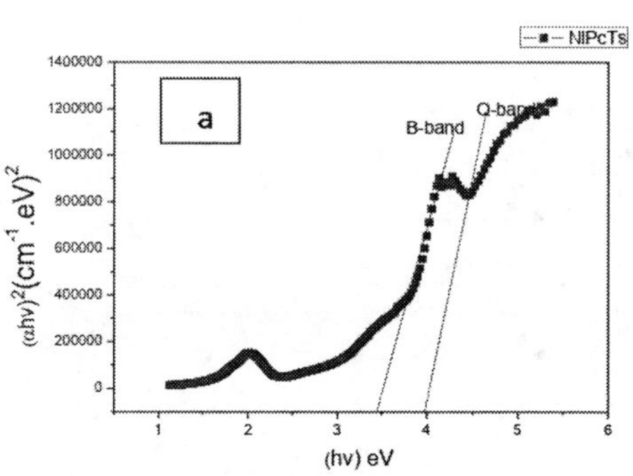

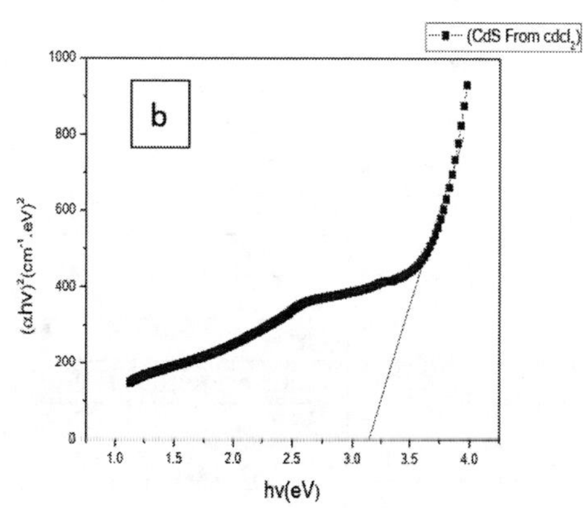

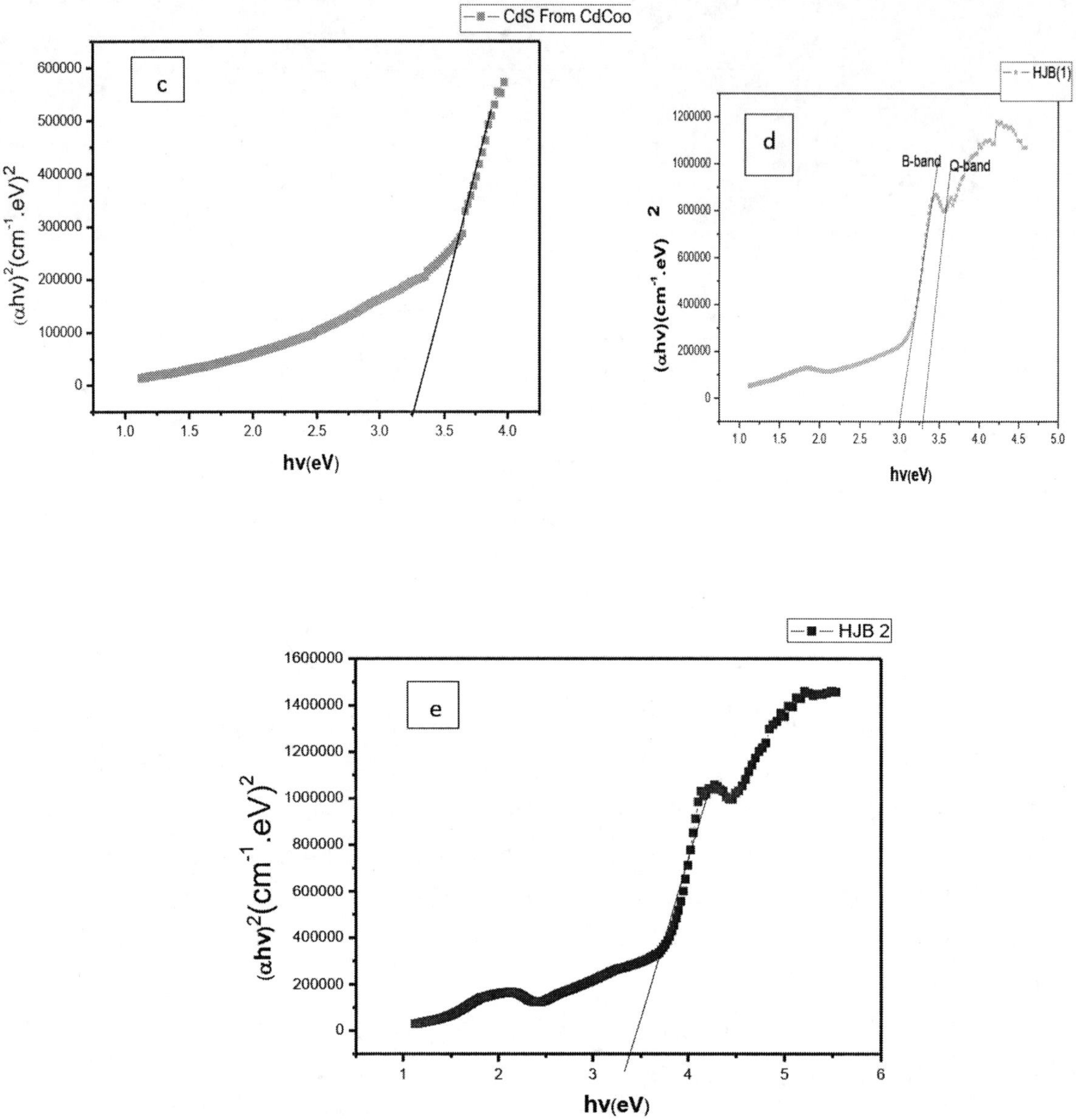

Figure 9.energy gap (a)NiPcTs (b) CdS From CdCl2.sol (c) CdS From CdcooH3.sol (d) HJB 1(e) HJB 2

Table 1.Energy gaps for preoared thin films and BHJ.

Samples of thin films	Energy gap (eV)
NiPcTs	3.1
CdS (1)	3.4
CdS(2)	3.2
Heterojunction (Blend)1	3
Heterojunction (Blend)2	3.2

3.2 Electrical Properties:

The conductivity of all thin film prepared by spin coating (NiPcTs, CdS, HJB (CdS / NiPcTs) was calculated By measuring electrical resistance (R) as a function of temperature change within the thermal range (30-150◦C) and from the creation of resistivity values (ρ)Then find the electrical conductivity of those inverted values, Drawing the natural logarithm of the latter as a function of the inverse absolute temperature (1000 / T) figure 10. The resulting graph after being hit by Boltzmann's constant, The values of activation energies are obtained from the electrical conductivity of thin films (NiPcTs), CdS, and hybrid (CdS / NiPcTs) (HJB). The presence of two bending regions in the curved decree for each sample which indicates two energizing mechanisms, one at low temperatures (30-90 C)) and the other at high temperatures (90-150 C) and for all samples prepared as shown in table 2.

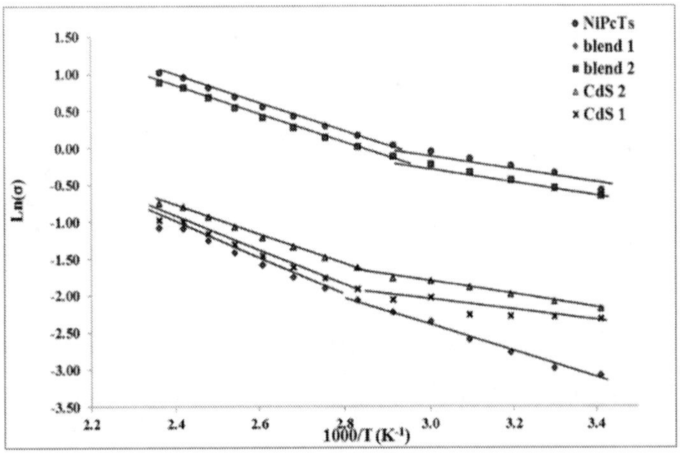

Figure 10 . Continuous conductivity to thin film

Table 2 Activation energies and DC.conductivity

Samples	E_{a1}(eV)	Tem.Range (K)	E_{a2}(eV)	Temp. Ranage(K)
CdS 1	0.051	293-343	0.180	243-423
CdS 2	0.075	293-343	0.166	243-423
Blend Heterojunction 2	0.091	293-343	0.162	243-423
Blend Heterojunction 1	0.155	293-343	0.191	243-423
NiPcTs	0.100	293-343	0.158	243-423

Hall effect measurements were made to determine the concentration of the major load carriers, the electrical conductivity and the electrical conductivity values of the thin films prepared (NiPcTs, CdS, HJB) on a glass substrate.Table 3 shows that the concentration of charge carriers of the organic thin film (NiPcTs) is lower than that of the inorganic semiconductor thin film As for the concentration of the HJB charge carriers and the cadmium sulfide sample of both solvents (cadmium chloride and cadmium coke), we note a difference in the concentration of the carriers of both types due to the different solutions of the crystalline structure in both cases As for the dynamics, we observe that the organic values of the organic substances are represented in a sample(NiPcTs) of the same rank (102) but slightly higher than the inorganic sample, HJB and recorded from the cadmium sulfide membrane produced from cadmium chloride solution are higher than those of the recorded hybrid junction of the cadmium sulfide sample produced from the cadmium acetate solution Due to the difference in the crystalline structure of the two decomposers. The low values of the motility are obtained from the process of clarification by the granules.

Table 3. Measurments of Hall Effect

Samples	$n*10^{13}$ cm^3	μ_H*10^2 ($m^2/V.s$)	$\rho*10^2$ ($\Omega.cm$)	$\sigma*10^{-3}$ ($\Omega.cm^{-1}$)	R_H*10^6 (cm^2/c)
NiPcTs	**0.19**	**4.545**	**0.722**	**0.1383**	**3.285**
CdS from $CdCl_2$ Sol.	**-1.448**	**2.696**	**0.1599**	**0.6253**	**-0.4311**
CdS from $CdCOOH_3$ Sol.	**-8.827**	**2.029**	**0.3485**	**0.2869**	**-0.7707**
BHJ CdS/ NiPcTs $CdCl_2$ Sol.	**-2.425**	**9.640**	**2.671**	**3.744**	**-0.2575**
BHJ CdS/ NiPcTs CdCOOH3 Sol.	**-2.393**	**2.699**	**9.670**	**1.034**	**-20.08**

3.3 Current-Voltage measurements of solar cells

Figures11,12 shows Current-Voltage measurements of solar cells were prepared from the cadmium chloride solution and cadmium acetate solution follows:

1 – FTO/PEDOT: PSS/HJB/AL 2-FTO/CdS/NiPcTs/AL 3-FTO/CdS/NiPcTs/FTO

Shows that The best efficiency of the models is FTO/PEDOT: PSS/Blend1/Al equal to (13.24%) as shown in Figure11. Table (4) represents the efficiency values of the prepared solar cells. It is clear from the table that the highest efficiency value was for the hybrid junction (FTO / PEDOT: PSS / blend1 / Al of 13.2%). The high value of this value is due to the presence of the polymeric material (PEDOT: PSS) and the high conductivity of the blend of CdS / NiPcTs and the formation of a group of excitons as a result, which becomes an electron-gap as an active substance. The lowest efficiency was for hybrid (FTO / CdS / NiPcTs / FTO (0.16%) and prepared by the cadmium sulfide sample produced by the cadmium acetate solution. This is due to the method of manufacturing the hybrid divider

IOP Conf. Series: Materials Science and Engineering **757** (2020) 012034 doi:10.1088/1757-899X/757/1/012034

where the two-layer method (Sandwich) was used in the preparation method. The characteristics of current-voltage under dark and light conditions were measured by the solar cell efficiency measurement system (C100AAA). Cell efficiency is obtained by:

$$F.F = \frac{J_m V_m}{V_{OC} J_{sc}} \text{……..(4)}$$

$$\eta = \frac{P_m}{P_{in}} = \frac{V_{oc} I_{sc} F.F}{P_{in} A} \text{……….(5)}$$

I_{SC}: Short-circuit current: Current value when V = 0, V_{OC}: Open-circuit voltage :Voltage value when I = 0, P: Power output of the cell P = IV, F.F: Fill factor. Table (4) represents the efficiency values of the prepared solar cells.

Table (4) represents the efficiency values of the prepared solar cells

Samples	**F.F**	**I_m μA**	**V_m mV**	**P_m μW**	**I_{sc} μA**	**V_{oc} mV**	**η %**
FTO/BEDOT:PSS Blend1(CdCl2 Sol.)	0.401	292.85	452.1	132.39	557.97	591	13.2
FTO /CdS/NiPcTs CdS (CdCl2 Sol.)	0.349	216.02	442.1	95.51	463.3	591	9.5
FTO /CdS/NiPcTs/ FTO CdS (CdCl2 Sol.)	0.24	51.464	297.5	15.31	139.20	515.0	1.53
FTO / BEDOT:PSS Blend2 from ($CdCOOH_3$ Sol.)	0.4	252.28	424.9	107.18	448.8	594.7	10.7
FTO/CdS/NiPcTs/Al CdS from($CdCOOH_3$ Sol.)	0.3	245.92	352.7	86.74	421.38	591.1	8.67
FTO /CdS/NiPcTs/ FTO CdS from($CdCOOH_3$ Sol.)	0.45	11.05	205.8	2.28	13.44	377	0.16

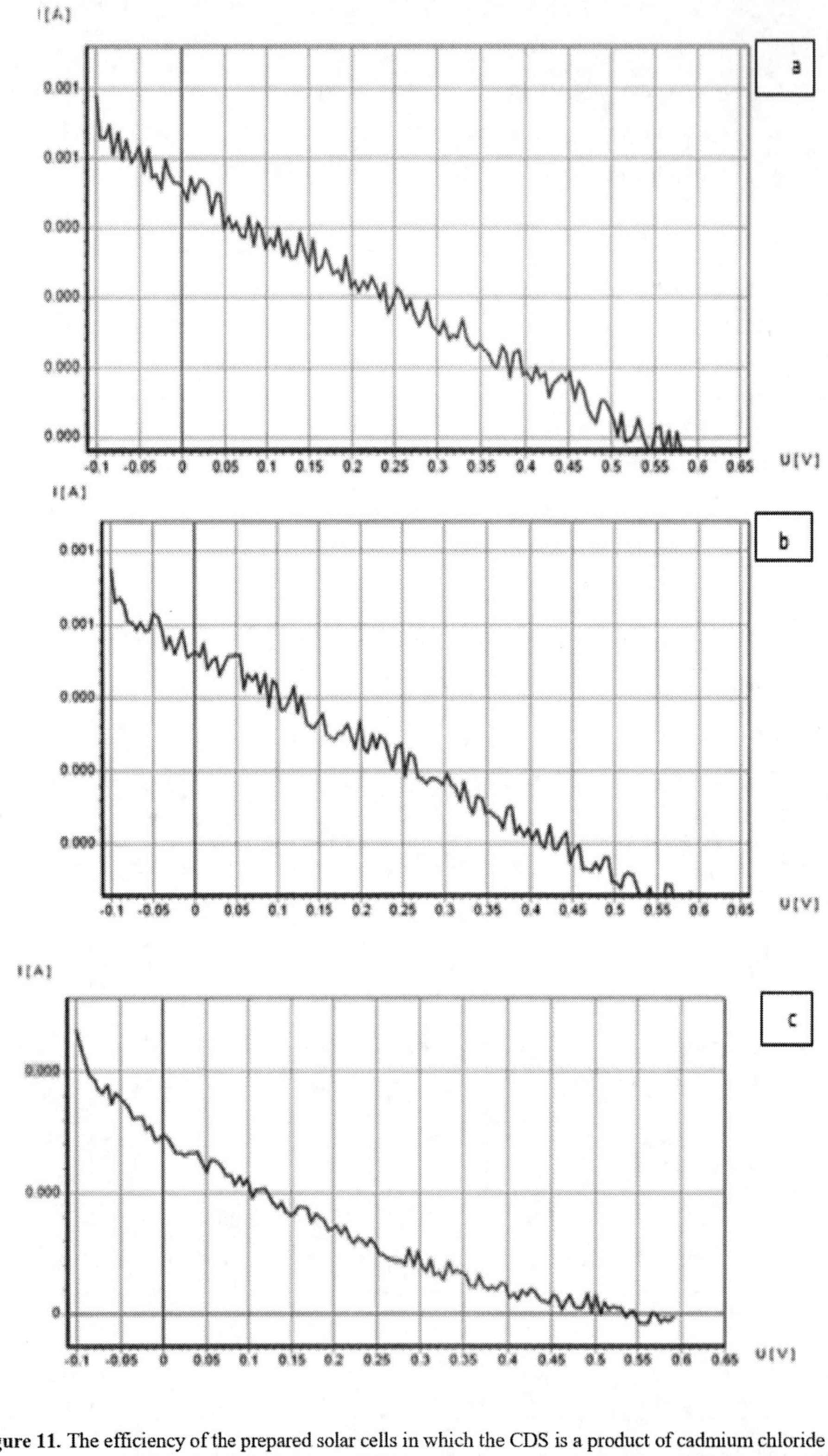

Figure 11. The efficiency of the prepared solar cells in which the CDS is a product of cadmium chloride solution (a) FTO / PEDOT: PSS / Blend / AL (b) FTO / CdS / NiPcTs / Al (c) FTO / CdS / NiPcTs / FTO

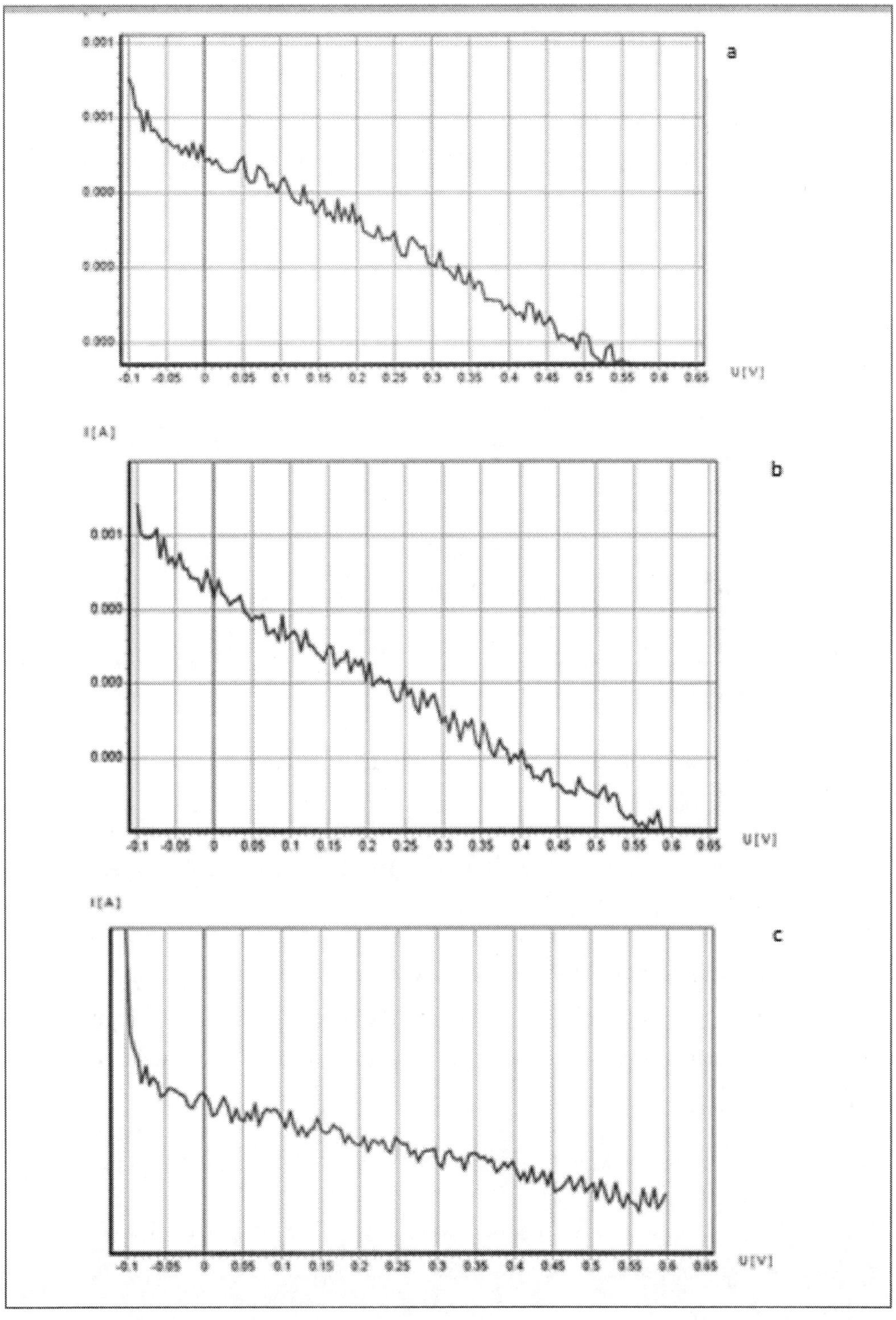

Figure 12. The efficiency of the prepared solar cells in which the CdS is a product of cadmium acetate solution (a) FTO / PEDOT: PSS / Blend / Al (b) FTO / CdS / NiPcTs / Al (c) FTO / CdS / NiPcTs / FTO

IOP Conf. Series: Materials Science and Engineering **757** (2020) 012034 doi:10.1088/1757-899X/757/1/012034

4. Conclusions

Organic NiPcTs and inorganic CdS semiconductors have been mixed because of their good price, low conductivity, and easy deposition by a very simple Spin Coating technology to manufacture highly efficient solar cells andThe best efficiency was for hetrojunction (FTO/PEDOT:PSS/HJB).

References

[1]Ameer F. Abdulameer, 2017 Fabrication and characterization of NiPcTs organic semi conductors based surface type capacitive–resistive humidity sensors .

[2]J. Fraden, 2016 Handbook of Modern Sensors: Physics, Designs, and Applications *Springer, Cham*

[3] P. Kumar, L. Morawska, C. Martani, G. Biskos, M. Neophytou, S. DiSabatino, M. Bell, L. Norford, R. Britter, 2015 *Environ. Int.* **75** 199–205()

[4] Q. Guo, G. M. Ford, R. Agrawal and H. W. Hillhouse, 2013 *Progressin Photovoltaics: Research and Applications* **21** 64–71

[5] T. K. Todorov, J. Tang, S. Bag, O. Gunawan, T. GokmenY. Zhu and D. B. Mitzi, 2013*Adv. Energy Mater* **3** 34–38

[6]The origin of high efficiency in low-temperaturesolution-process able bilayer organ metal halidehybrid solar cells

[7]Atyaf Hussian Kadhum Alkanani 2017 Synthesis and Characterization of Organic Semiconductor Heterojunction Blend NiPcTs/Alq3 for Humidity Sensor ApplicationSubmitted to the Council of the College of Science for WomenUniversity of Baghdad

[8] J. Wu, P. Li, S. Hao, H. Yang and Z. Lan, 2007 *ElectrochimicaActa* **52** 5334.

[9]R.K. Hazra, M. Ghosh and S.P. Bhattacharyya, 2007 *Chemical Physics* **333** 18.

IOP Conf. Series: Materials Science and Engineering **757** (2020) 012035 doi:10.1088/1757-899X/757/1/012035

Morphological and Electrical Properties of Capped CdSe Nanoparticles

Kahtan A Mohammed *, Alaa S AL-Kabbi ,Kareema M Ziadan ,

Department of Physics, College of Science, Basrah University, Basrah, Iraq

*Corresponding authors:, E-mail address: kahtan444@gmail.com

E-mail address:kmziadan16@gmail.com
E-mail address: as.alkabbi@yahoo.com

Abstract In this study, cadmium selenide (CdSe) were synthesized by a chemical route process using complex agent (tri sodium citrate) as a ligand. cadmium acetate as a Cd source and Na2SeO3 as a Se precursor. The synthesized nanoparticles were characterized for their compositional, morphological and electrical properties using FESEM, EDAX and I-V characterization.

1. Introduction

Due to their unique structural, optical and electronic transport characteristics resulting from the quantum containment of electrons, nanostructured semiconductors have drawn extensive attention. They are structurally, morphologically, optically and electrically dependent on size[1]

High quality, stability and reliability of semiconductors remains a challenge, posing severe barriers for their use for practical need. High quality II-VI semiconductors with controlled defects and impurities are needed in electronic and optoelectronic applications[2]

Because of their prospective uses in many applications, much attention has been paid to II – VI materials.. Binary metal chalcogenides synthesis of group II – VI semiconductors in the form of nanoparticles was a quickly increasing area of studies due to their significant nonlinear optical characteristics, luminescent properties, quantum-size effects and other significant physical and chemical characteristics[3]. All these properties are size dependent

Nanoparticles ' size control remains a key method for successfully applying nanotechnologies..Understanding of the role of the capping agent, because of surfactant adsorption on the surface of the particulate, is essential for the growth of regulated synthesis of nanoparticles, preven ting nanoparticles from being aggregated.. A adequate knowledge of the role of surfactants can allow the control by various surfactants of particle size[4,5].

NCs consist of an inorganic core stabilized in ligands with a layer of organic molecles and the resulting inorganic hybrid provides versatility and flexibility in the processing and surface engineering of solutions.. The characteristics can therefore be modified by playing with the organic surfactant layer to manipulation of surface chemistry[6].

The nanocrystal size is regulated by the adsorption of longchain surfactants, capsule materials to the surface of nanocrystals, forming inverse structures of micelles[7]

Aqueous synthesis do not require any special equipment or high temperature. Bare synthesized NPs are very unstable, highly reactive which results in photochemical degradation. Therefore a stabilizer or surfactant is required to synthesize stable and prevent aggregation of NPs and can be further used for conjugation or various applications . The stabilizer also helps in controlling the size of NPs[8]

Cadmium selenide is a classic II-VI semiconductor with a direct Eg of 1.74 eV[9]. It is photosensitive in visible wavelength regions due to its suitable Eg. It is widely used in solar cells[10] as well as photoelectrochemical cells[11]

in this paper we study the electrical and morphological properties of CdSe capped with TSC nanoparticles.

2. Experimental

2.1 *CdSe Nanoparticles Preparation*

The procedure of CdSe-TSC NPs. preparation , optical and structural properties had been discussed in our paper[12].

3. Results and discussions

The surface morphology of TSC capped CdSe thin films was analyzed by using a Scanning Electron Microscope (SEM). SEM is a suitable and useful technique to study and analyze the surface morphology of thin films.

The surface morphology of CdSe-TSC is shown in Figure 1 shows that the surface is smooth and has spherically shaped grains. the surface is covered with uniformly size grains. No cracks or holes are observed in the thin films .The mean grain sizes of films were about 30 nm .

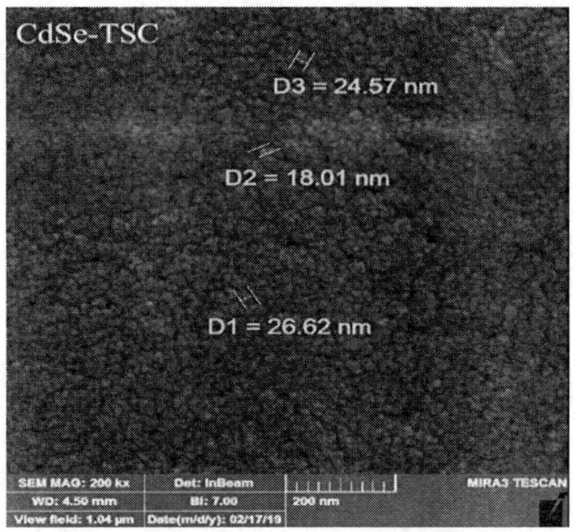

Figure 1. FESEM images of CdSe-TSC nanoparticles

EDX measurements were performed on the sample to study the chemical composition. Figure 2 shows an EDX spectrum for the illustrative CdSe-TSC NPs. The EDX form confirms the existence of cadmium and selenium in the compound. It also displays that the ratio of Cd and Se is 54.18 and 21.67, respectively.

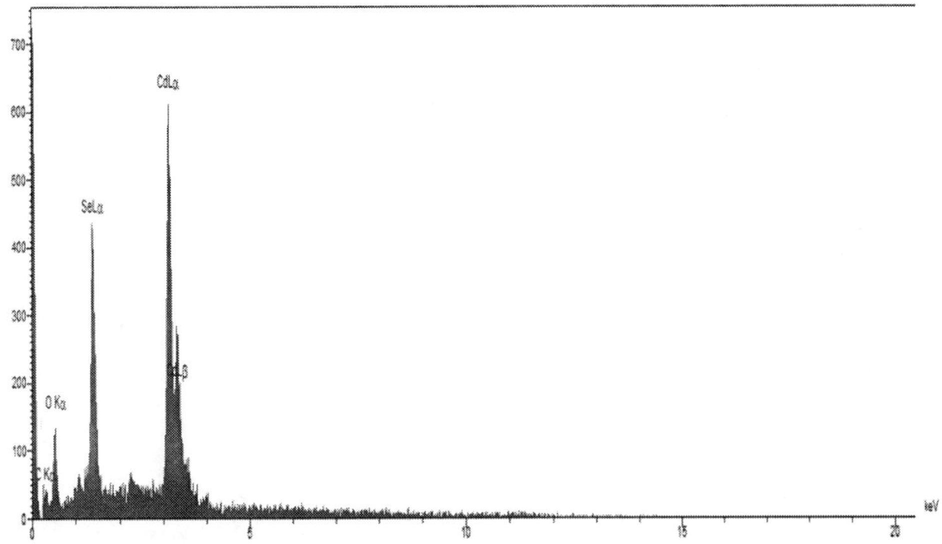

Figure 2. EDX analysis of CdSe-TSC nanoparticles.

Figure 3 gives the FTIR spectrum in the frequency range (500–4000 cm-1) of TSC and CdSe-TSC. The observed band positions in the spectra of TSC capped CdSe match well with the IR bands of pure TSC, except wavenumber shift in certain bands[13], [14]. In the spectrum of pure TSC, we observe band at 1582 cm^{-1} which can be assigned to the asymmetric and symmetric stretching vibration respectively, of COO– group. The large downshift of the former band to 1554 cm^{-1} in the spectra of CdSe-TSC has been appeared. the band at 3446 cm^{-1} is can be assigned to OH stretching of sodium citrate in the TSC spectra is downshift to 3345 cm^{-1} in the CdSe-TSC spectra.

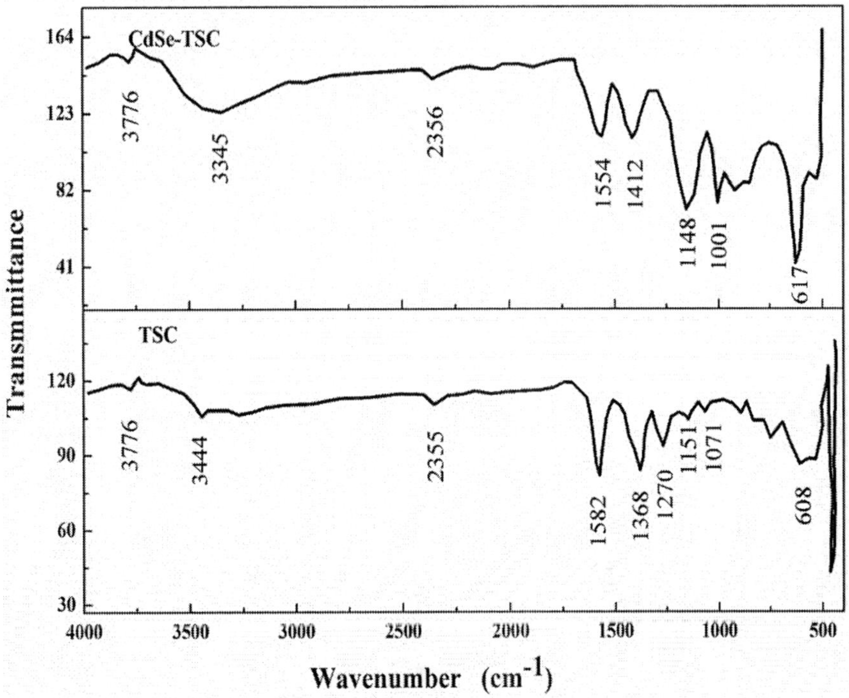

Figure 3. FTIR spectra of trisodium citrate and CdSe capped with trisodium citrate

Interdigitated electrodes (IDE) assembled from two individually addressable interdigitated comb-like electrode structures have frequently been used to study the electrical properties of CdSe NPs.

Figure 4 shows electrode that contains of interdigitatted Aluminium lines on an glass substrate . It can be achieved using minterdigitated electrodes to measure the surface conductivity of the samples from the following relationship[15].

$$\sigma\ s= [I/V]\ [\ L/Wt\ell] \qquad (1)$$

where, t is thickness of polymer, W is the distance fingers (10mm),ℓ is number of fingers is to be (10), and L is the space between electrodes (100μm). So that;

$$\sigma_s= (I/V)*10^{-3}(S.\ m^{-1}) \qquad (2)$$

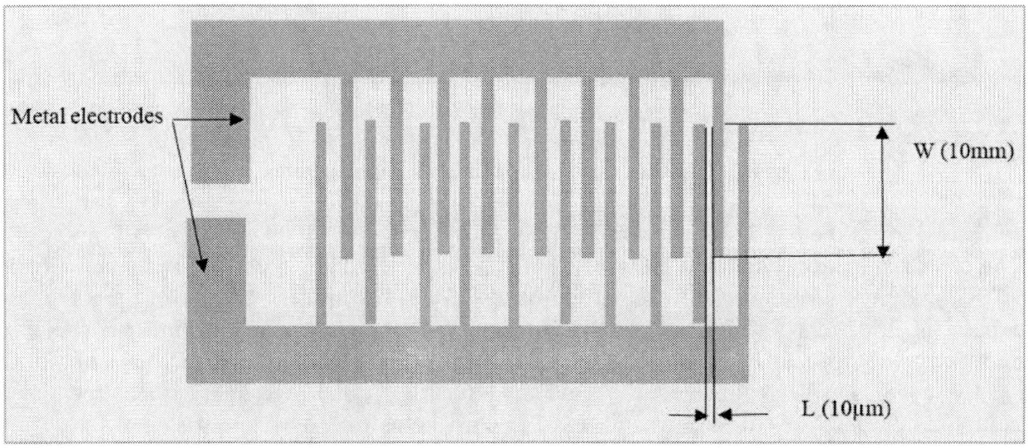

Figure 4. shows electrode that contains of interdigitatted Aluminium lines on an glass substrate

To study the electrical performance and the mechanisms of conduction in CdSe NPs., current-voltage (I–V) characteristics in dark was perfomed by two-probe method in voltage range (1-40) V for CdSe capped with TSC , naoparticles . I–V behavior of CdSe shown in Figure 5

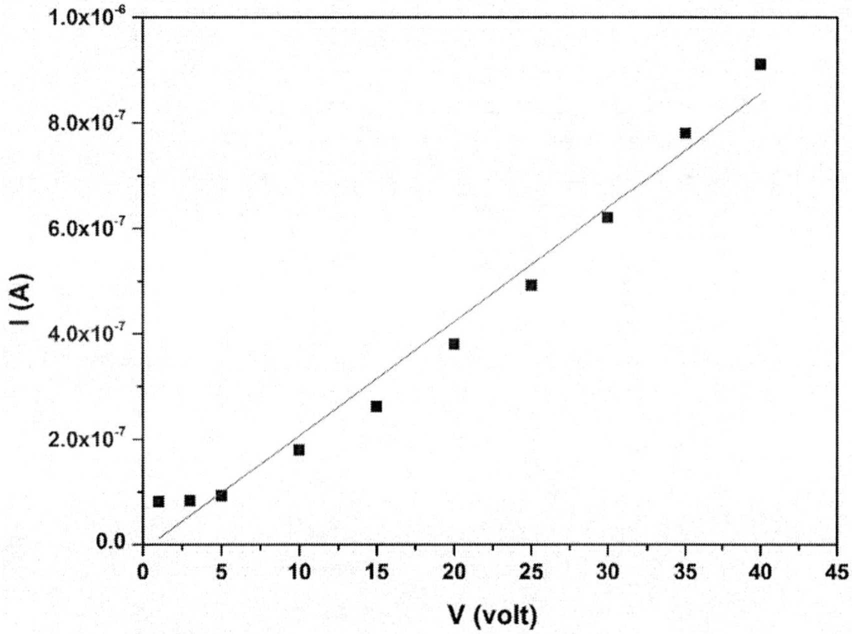

Figure 5. Plot showing variation of I current voltage

DC electrical conductivity is a very important property of semiconducting materials for seeking their applications in photosensitive/photovoltaic devices[16]. The electrical conductivity (DC) of CdSe NPs. in dark was performed using twoprobe method in the temperature range 293–373 K. DC conductivity at room tempreture was equl to $6.3*10^{-7}$

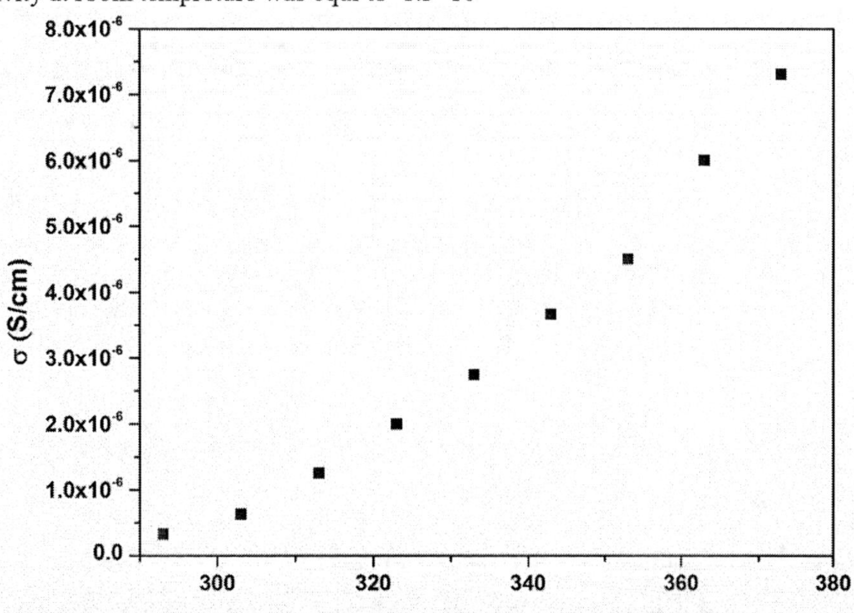

Figure 6. Variation of conductivity with temprature

The temperature dependences of electrical conductivities for the as-prepared NPs are shown in figure 6. All display a semimetallic conductivity and thermal response, the conductivity as a function of temperature can be well expressed by the relation[17]

$$\sigma = \sigma_o \exp(-\Delta Ea/KT) \quad (3)$$

where ΔEa is activation energy, k the Boltzmann's constant, T temperature and σo is pre-exponential factor. Arrhenius plots of conductivity of CdSe NPs. are shown in figure 7. The exponential increase in DC conductivity with temperature. The value of the activation energy (ΔEa) is estimated with the help of the slope of ln σ versus 1000/T plot. The graph shows straight lines, which indicate that the conduction in the synthesized samples is through an activated process with activation energy equal 0.3427 ev.

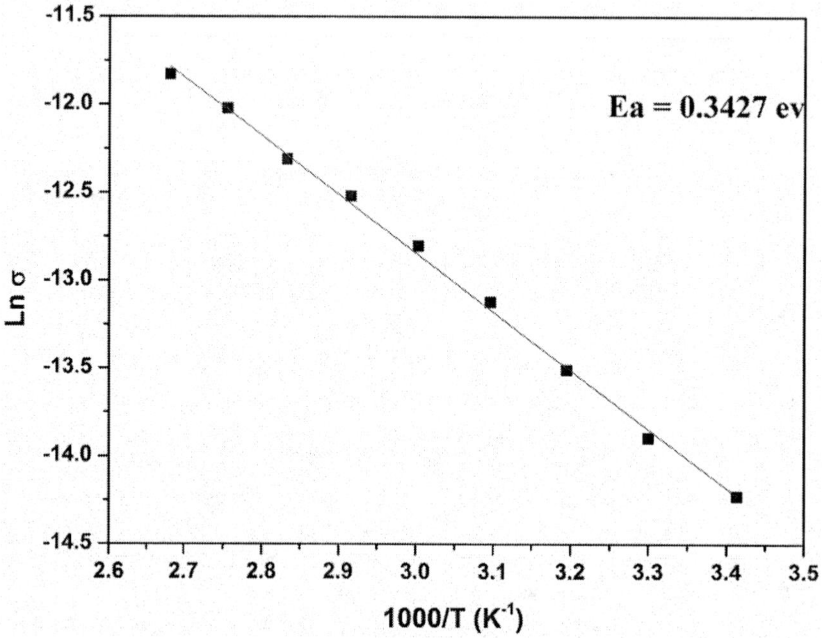

Figure 7. Variation of Ln σ verses 1000/T (K^{-1})

4. Conclusions

CdSe in the presence of TSC as a capping agent was synthesized by chemical method. Charge transport properties like I-V and electrical conductivity as a function of temperature have been thoroughly investigated. FESEM images demonstrated to be of spherical shape morphology and nearly well distributed with particle size of about 30 nm. The EDX results showed that the percentage of Cd to Se is 2:1. TSC capping CdSe was confirmed by FTIR a spectroscopy.

5. References

[1] A. S. Al-Kabbi, K. Sharma, G. S. S. Saini, and S. K. Tripathi, "Effect of doping on transport properties of nanocrystalline CdSe thin film," *Thin Solid Films*, vol. 586, pp. 1–7, 2015.

[2] S. Mathuri, K. Ramamurthi, and R. Ramesh Babu, "Effect of Sb incorporation on the structural, optical, morphological and electrical properties of CdSe thin films deposited by electron beam evaporation technique," *Thin Solid Films*, vol. 660, no. 2017, pp. 23–30, 2018.

[3] A. S. Al-Kabbi, K. Sharma, G. S. S. Saini, and S. K. Tripathi, "Determination of the transport parameters of nanocrystalline CdSe:Cu thin films," *Physica Scripta*, vol. 87, no. 2, 2013.

[4] A. L. Suherman, G. Zampardi, H. M. A. Amin, N. P. Young, and R. G. Compton, "Tannic acid capped gold nanoparticles: Capping agent chemistry controls the redox activity," *Physical Chemistry Chemical Physics*, vol. 21, no. 8, pp. 4444–4451, 2019.

[5] C. M. Phan and H. M. Nguyen, "Role of Capping Agent in Wet Synthesis of Nanoparticles," *Journal of Physical Chemistry A*, vol. 121, no. 17, pp. 3213–3219, 2017.

[6] S. Yadav, A. Singh, L. Thulasidharan, and S. Sapra, "Surface Decides the Photoluminescence of Colloidal CdSe Nanoplatelets Based Core/Shell Heterostructures," *Journal of Physical Chemistry C*, vol. 122, no. 1, pp. 820–829, 2018.

[7] J. K. Lorenz and A. B. Ellis, "Surfactant-semiconductor interfaces: Perturbation of the photoluminescence of bulk cadmium selenide by adsorption of tri-n- octylphosphine oxide as a probe of solution aggregation with relevance to nanocrystal stabilization," *Journal of the American Chemical Society*, vol. 120, no. 42, pp. 10970–10975, 1998.

[8] S. Bharti, S. Singh, S. Jain, G. Kaur, and S. Gupta, "Interaction of TGA with CdSe nanoparticles," AIP Conference Proceedings **1953**, 030157 (2018); doi: 10.1063/1.5032492.

[9] C. Li *et al.*, "Characterization of sputtered CdSe thin films as the window layer for CdTe solar cells," *Materials Science in Semiconductor Processing*, vol. 83, no. March, pp. 89–95, 2018.

[10] M. Sohail *et al.*, "Hole transfer from CdSe nanoparticles to TQ1 polymer in hybrid solar cell device," *Journal of Molecular Structure*, 2018.

[11] S. Thanikaikarasan, "Effect of annealing on microstructural, compositional, optical, and photoelectrochemical properties of CdSe thin films," *Ionics*, no. 001, pp. 1–8, 2017.

[12] Kahtan A. Mohammed, Alaa S. AL-Kabbi ,Kareema M. Ziadan , "Synthesis and Characterization of CdSe Nanoparticles " *AIP Conference Proceedings* 2144, 030009 (2019); https://doi.org/10.1063/1.5123079.

[13] A. K. Thottoli, A. Kaliani, and A. Unni, "Effect of trisodium citrate concentration on the particle growth of ZnS nanoparticles," *Journal of Nanostructure in Chemistry* vol. 3, no. 1, p. 1, 2013.

[14] S. H. Frayne, S. N. Barnaby, N. Nakatsuka, and I. A. Banerjee, "Growth and Properties of CdSe Nanoparticles on Ellagic Acid Biotemplates for Photodegradation Applications," *Materials Express*, vol. 2, no. 4, pp. 335–343, 2012.

[15] K. M. Ziadan, H. F. Hussein, and K. I. Ajeel, "Study of the electrical characteristics of poly (o-toluidine) and application in solar cell," *Energy Procedia*, vol. 18, pp. 157–164, 2012.

[16] V. Kumar, D. K. Sharma, K. Sharma, and D. K. Dwivedi, "Structural, optical and electrical characterization of vacuum-evaporated nanocrystalline CdSe thin films for photosensor applications," *Applied Physics A: Materials Science and Processing*, vol. 122, no. 11, pp. 1–6, 2016.

[17] S. Islam, S. Ahmad, and H. Khan, "Enhancement in the dielectrics of poly (o -toluidine)/ single wall carbon nanotubes (POT / SWCNTS) polymer nanocomposites for electrical energy storage," *Journal of Materials Science: Materials in Electronics*, vol. 0, no. 0, p. 0, 2018.

Electrochemical evaluation of polyaniline/multi-walled carbon nanotube composite synthesized by microwave plasma polymerization as a supercapacitor electrode

Nada I Ibrahim[1] and Ahmed S Wasfi[2*]

[1]Department of Physics, College of Science, University of Diyala, Iraq
[2]Department of Physics, College of Science, University of Baghdad, Al Jadriah, Iraq

[*]corresponding author: e-mail: a.wasfi@scbaghdad.edu.iq

Abstract. PANI/MWCNT nanocomposite supercapacitor electrodes were synthesized by in-situ microwave plasma polymerization of aniline in the presence of MWCNT-COOH. The SEM image of the composite illustrates the formation of the core-shell MWCNT-PANI ordered structure that has a strong effect on the electrical conductivity and porosity, in addition to the appearance of agglomerated polymerized islands. The electrochemical performance of PANI/MWCNT composite was evaluated with respects to the obtained specific capacitance, energy and power density, and cyclic stability, using cyclic voltammetry, galvanostatic charge/discharge at a constant current and electrochemical impedance spectroscopy. The highest obtained specific capacitance was 112 F/g at a charging current density of 1.6 A/g, provides the highest energy density of 34.8 Wh/Kg and a power density of 1670 W/Kg. While the highest achieved power density was 8381 W/Kg with an energy density of 12.8 Wh/Kg. The delivered energy density crossed the threshold energy density limit of the battery, which meets the scientific motivation to use supercapacitor as an independent energy source.
Keywords: Plasma polymerization, Carbon nanotubes, Polyaniline, Nanostructure composite, Core-shell structure, Supercapacitor.

1. Introduction

Supercapacitors (also known as electrochemical capacitors) are attracting significant attention due to their high capacitance and potential applications in electronic devices. They can be categorized into two classes according to their charge storage mechanism; electrochemical double-layer capacitors (EDLCs) and pseudocapacitors [1-6].

EDLC can be accomplished using materials that have a high accessible surface area as electrodes, like porous carbon materials (activated carbons, carbon nanotubes, carbon nanofibers and reduced graphene oxide etc.) [7-10], while the electrode materials in the case of pseudocapacitors are transition-metals oxides (RuO_2, SnO_2, NiO, MnO_2, etc.) or conductive polymers, such as polypyrrole (PPy) and polyaniline (PANI) [1-5, 11-17]. Amongst the electrically conducting polymers, PANI has gained further attention as a potential pseudocapacitive candidate material owing to its benefits over the others; for instance, low cost, uncomplicated synthesis, high environment constancy, and facile doping and de-doping which offer a reversible control of the electrical properties [18, 19]. On the other hand, it undergoes poor cycling stability due to repetitive swelling and shrinkage of its backbone structure during the charge-discharge process. As well as, low achievable doping besides mass transfer drawback within massive polymer layers [11-13]. In the direction of resolving these difficulties, PANI should be highly crosslinked [20], nanostructured, or can be grown on several nanomaterials as composites [21], and the polymerization process can control all these problems. Thus, the choice of the polymerization scheme has always been regarded as a substantial filed of interest. Conductive polymers are usually formed employing electrochemical and chemical polymerization; however, in recent decades, plasma polymerization is recognized as a considerable single-step low-cost approach to producing films of conductive polymers [22-25]. The conductive polymers that are plasma polymerized possess chemical structures that are rather unlike the polymers produced by electrochemical and chemical polymerization and are reliant on the plasma polymerization parameters [23, 24]. The produced plasma polymerized films are considerably more uniform, smoother, greatly adherent, and pinhole-free. The polymer chains are randomly branched and shorter, with a high degree of crosslinking. However, a higher conductivity can be achieved with electrochemically and chemically synthesized polymers due to the in situ doping through these methods of polymerization, whereas in the plasma polymerization, the synthesized polymer is an insulator and needs to be doped to get a conductive polymer film [23-27]. Consequently, the in situ plasma polymerization of aniline monomer in the presence of CNTs is a standout technique amongst the most proficient approaches to combine PANI nanocomposite due to the proper interfacial interaction between the CNTs and the PANI matrix. The main idea in the use of CNTs in the field of the polymer-based composite is to transfer their electrical properties to the insulating polymer matrices to obtain conducting composites, such composites have shown a high potential as supercapacitor electrodes due to their conductivity, high surface area, and capability to store energy by means of two charge storage mechanisms [14-16, 28-30].

In the present work, a microwave plasma polymerization technique was used to synthesize the PANI/CNT composite that was tested as a supercapacitor electrode. The aniline monomer with dispersed multiwalled carbon nanotubes (MWCNTs) was directly injected into the plasma induced by the microwave discharge. The morphological and electrochemical characterizations using cyclic voltammetry (CV), galvanostatic charge-discharge (CD), and electrochemical impedance spectroscopy (EIS) revealed the detailed features of these composites and their performance as a supercapacitor electrode.

2. Experimental

2.1 Materials and characterization instruments:

The aniline of purity 99.99% was provided from Hopkin and Williams Ltd., England, and distilled before used. Nanostructured and Amorphous Materials Inc. USA, provided the functionalized MWCNTs-COOH, with a purity of 95%, 10-30 nm diameter, and 10-30 μm length. A Titanium of 0.2 mm thickness was used as a substrate for the deposition of the composite. The morphological investigation was carried out using a scanning electron microscope (SEM), Philips XL 30, and KYKY-EM3200. While the electrochemical performance of the composite electrodes was examined using IVIUMSTAT.XR (Electrochemical Interface & Impedance Analyzer). A conventional three electrodes cell with a piece of Ti sheet that coated by active materials as working electrode, Platinum wire, and an Ag/AgCl (Argental, 3M KCl) were used as counter and a reference electrode, respectively. The EIS test was conducted in a wide frequency range of 10 mHz -100 kHz with a perturbation amplitude of 5mV.

2.2. Synthesis of PANI/MWCNT composite electrode

A microwave plasma CVD apparatus that was built in our laboratory was used in the plasma polymerization process and was described in detail in earlier publications [31, 32]. The functionalized MWCNTs, of different loading percentages, were dispersed ultrasonically for 30 min in the aniline monomer to form a colloidal solution. Then the colloidal solution was injected into the plasma vessel as pulses by a car fuel injector. The deposited film thickness depends on the injected colloidal solution amount in the plasma vessel as pulses of controlled period, and the MWCNTs-COOH loading percentage in the monomer. However, the period control of the pulse and their repetition rate are essential to maintain the vessel pressure within 0.1 - 1.0 mbar, which is the plasma working pressure range. The solution minute droplets of the spraying pulse will directly evaporate when they entered the evacuated plasma reactor and turned into high concentration radicals nearby the substrate as a result of the surrounding plasma activity in the vessel. Two simultaneous processes will take place within the formation of the PANI/MWCNT-COOH composite; the plasma surface functionalization of the MWCNTs, and the aniline plasma polymerization. Whereas the walls of the MWCNTs-COOH adsorb the aniline due to their high surface area and chemical activity. Therefore, the polymerization process begins on the MWCNTs walls in their pathway to the substrate through the plasma, and the produced composite film will enclose MWCNTs that were implanted and chemically bonded with the PANI matrix. A Titanium substrate of 1 cm^2 was polished with emery paper and used after a suitable cleaning process for the composite deposition; it was fixed in the vessel at a location that is properly exposed to the plasma. The maximum achievable plasma density and electron temperature were10^{17} cm^{-3} and 0.8 eV, respectively. They were measured using optical emission spectroscopy [33].

3. Results and discussion

The characteristics of PANI, MWCNTs, and their nanocomposite films were investigated using many techniques to evaluate its electrochemical performance.

3.1. Scanning Electron Microscopy Analysis

The SEM images of pure PANI that was plasma polymerized revealed a spherical granular morphology of massive spheres, up to many micrometers in dimension, uniformly distributed on the Titanium substrate. Each large sphere composed of aggregations of small nanospheres, which relatively offer a high surface area to the structure, as it can be shown in figure (1- a, and b). While figure (1-c) illustrates the SEM image of the functionalized MWCNTs film, it can be seen that the MWCNTs are entangled and interconnected, creating 3D network, with an arrangement of macroporous and mesoporous structures, facilitating excellent electrical conductivity. Whereas, the SEM images of PANI/MWCNT nanocomposite of 1.5wt. % MWCNTs loading are shown in figure (1- d, and e), which represent the top and the cross-sectional view, respectively. The SEM top-view image illustrates a high-density MWCNTs which are well dispersed, implanted, and completely enclosed with PANI, creating a continuous matrix. Moreover, a core-shell MWCNT-PANI structure was formed, which enhanced the composite porosity and could offer a high surface area for excellent electrode/electrolyte interfacial contact. This kind of open structure can provide better accessibility to the electrolyte [21, 30]. Also, it can be seen the appearance of big agglomerated polymerized islands that is unfavorable growth tends to block the pores structure of the electrode and reduce the surface area. The SEM cross-sectional view image of the composite indicates the formation of a thick composite layer of thickness about 35 μm, which may have a negative effect on increasing the electrode material resistance.

3.2. Electrochemical Investigations

The electrochemical behavior of pure PANI, MWCNTS, and PANI/MWCNT composite films was examined by cyclic voltammetry (CV), galvanostatic charge-discharge (CD) and electrochemical impedance spectroscopy (EIS) measurements, to evaluate their performance as a supercapacitor electrode. The experiments were done in a conventional three-electrode glass cell of a platinum counter electrode, Ag/AgCl reference electrode, and a Titanium with the deposited materials as a working electrode, in 1M H_2SO_4 aqueous electrolyte. Figure (2-a) shows the current response profile of the CV voltammograms of pure PANI, functionalized MWCNTs-COOH and PANI/MWCNT-COOH composite electrodes at a scan rate of 50 mV/s and a potential range of -0.5 to 1 V. The CV voltammograms of the MWCNTs show a nearly rectangular-like symmetrical shape which is an electric double-layer capacitive behavior [1, 7]. The same behavior is recognized with PANI CV voltammograms as a result of its insulating state (leucoemeraldine form) due to its preparation method by the plasma polymerization [31, 46], and its EDLC behavior coming from its nanostructure surface morphology. However, the CV voltammograms of PANI/MWCNT-COOH composites of 0.2 and 0.5 wt.% MWCNTs loading show the same previous trend with a bit increase in the peak current due to the enhancement of the nanopores structure of the composite, but they are still suffering from the poor

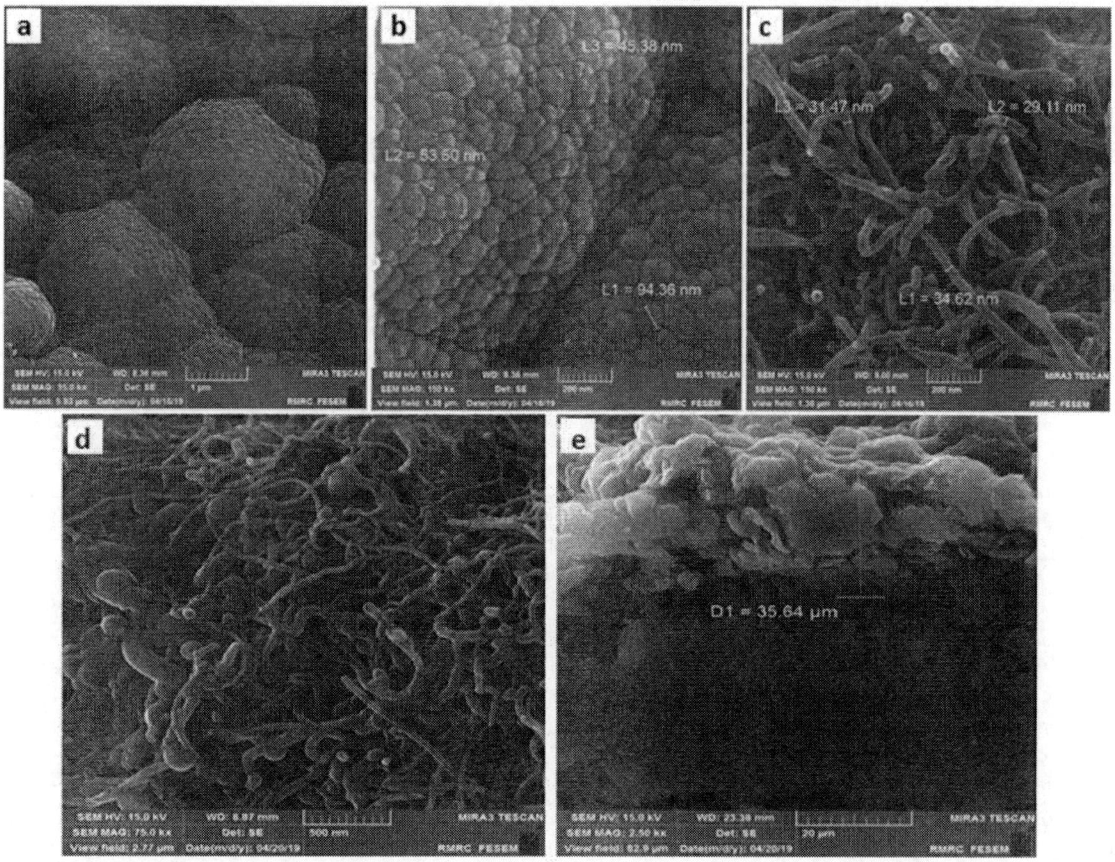

Figure 1. SEM images of (a, b) Pure PANI, (c) MWCNTs and (d, e) PANI/MWCNT composite films deposited under similar circumstances.

electronic conductivity. Whereas, the CV voltammograms of the 1.5 wt.% MWCNTs loading composite shows a good rectangular shape with a jump in the peak current representing an increase in the capacitance. Furthermore, two pairs of redox peaks were appeared, caused by the insertion/removal of counter ions that lead to the reversible redox transition. That is a pseudocapacitance behavior due to the synergistic effect between PANI and the CNTs [13-16]. It appears that the doping of PANI with 1.5 wt. % of the functionalized MWCNTs were adequate to transfer PANI to its conductive electronic state (polaronic emeraldine form), which increase the electrical conductivity of the nanocomposite [16, 31], and simplifies the electron transport through the CV scan process. In addition to the significant enhancement of the nanopores structure of the composite, which offers a large surface area that permits better accessibility to the transport of electrolyte ions and providing lower internal resistance. It is well known that the dynamic interaction among the π-bonds in the aromatic ring of PANI and the graphitic structure of the CNTs should powerfully enable the charge-transfer reaction between the two components [34-37]. The CV voltammograms of PANI/MWCNT-COOH composite of 1.5 wt. % MWCNTs at different scan rates extending from 10 to 300 mV/s in the given potential range are shown in figure (2-b). Through the increase of the scan rate, the redox current increased, which revealed a good rate ability. The symmetric and nearly rectangular shape of the curves and their fast current response at the voltage reversal points even at high scan rates specify the rapid diffusion of the electrolyte in the film [5, 9].

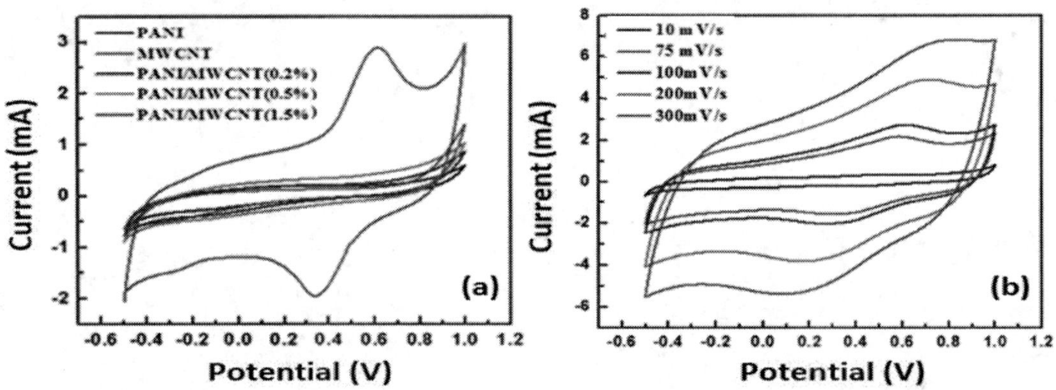

Figure 2. (a) Cyclic voltammograms of pure PANI, MWCNTs, and PANI/MWCNT composites at a scan rate of 50 mV/s in1M H_2SO_4. (b) Cyclic voltammograms of 1.5 wt. % MWCNTs composite at different scan rates in1 M H_2SO_4.

The rate capability and the electrode capacitance are best determined from the galvanostatic charge/discharge cycling curves at a constant current. Figure (3-a) shows the charge/discharge behavior of PANI, MWCNT-COOH, and PANI/MWCNT-COOH electrodes in the potential range of -0.5 to 1.0 V at a current density of 1.6 A/g in 1M H_2SO_4 aqueous electrolyte. The charge/ discharge curves through this potential range have nearly triangular shapes with almost mirror images that indicating an excellent capacitive behavior of the PANI/MWCNT-COOH nanocomposite electrodes [17]. The initial part of the discharge curves exhibit an IR voltage drop owing to the internal resistance of the electrodes [17, 38], and the rest of the curve is linear for the non-faradic materials like the nanostructure PANI in its insulating state. While the pseudocapacitive and hybrid system exhibits a deviation in linearity created by varying the capacitance with the voltage as it can be seen clearly in the nanocomposite electrode of 1.5 wt.% MWCNT loading [17]. The specific capacitance (Cs) of the electrodes in the cell was calculated from the discharge portion of the curve using equation (1) [16, 17, 38, 39]:

$$Cs = \frac{I}{\left(\frac{\Delta V}{\Delta t}\right) m} \tag{1}$$

where I is the applied constant current, ΔV is the change in potential, Δt is the discharge time, ΔV/Δt is the slope of the discharge curve after the IR voltage drop, and m is the mass of the active material on the electrode. The calculated Cs of pure PANI was 16 F/g, and for MWCNTs was 65 F/g, while Cs of

PANI/MWCNT-COOH nanocomposite films were 26, 36, and 112 F/g for MWCNTs loadings of 0.2, 0.5, and 1.5 wt.%, respectively. The low capacitance of pure PANI may be attributed to two reasons, the first is due to the insulating conductive state of the produced PANI by the plasma polymerization method, and this is clear in the absence of any redox peaks in its CV voltammograms. The second reason is due to the structural nature of the deposited PANI that is of the form of big aggregations of macro sizes, in spite of the nano granular morphology of its surface, causing a great lack in the PANI porosity, which is a decisive factor affecting the capacitance greatly. Figure (3-b) presents the charge/discharge curves of the nanocomposite film of 1.5 wt. % MWCNTs loading at various current densities. From this figure, it can be realized that the curves have an almost triangular shape with the discharge curves approximately symmetric with their corresponding charge counterparts. This trend suggests that the electrode has good charge propagation across the composite that means good charge/discharge performance and electrochemical reversibility [16,17]. Moreover, the slight asymmetry, which is clear in the charge/discharge curve at a low current density of 1.6 A/g is due to the strong influence of the pseudocapacitance since a redox reaction usually spends more time than pure EDL formation [17]. The IR drops are seen at the high charge current densities, decreased with lowering the current density, as marked in figure (3-b). The specific capacitances can be determined from the discharge curves using equation 1, where Cs is 112, 78, 68, 62, and 59 F/g at a constant charging current density of 1.6, 3.3, 6.6,10, and

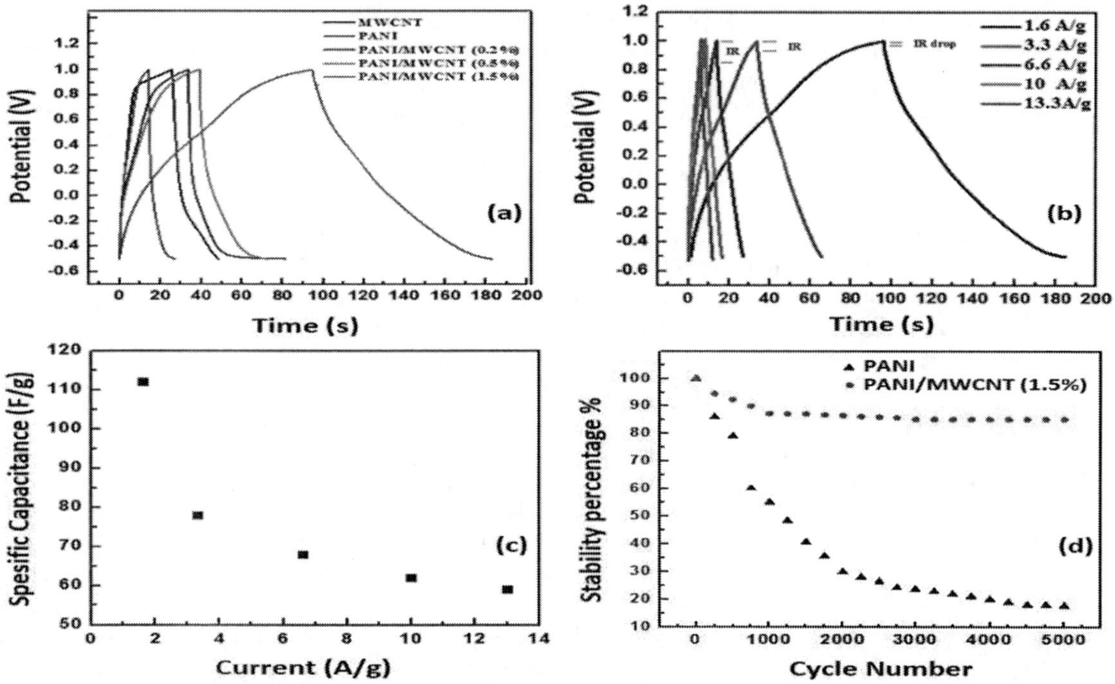

Figure 3. (a) Charge/discharge curves of PANI, MWCNTs, PANI/MWCNT composites electrodes at a current density of 1.6 A/g, in 1 M H_2SO_4 . (b) Charge/discharge curves of 1.5 wt. % MWCNTs composite at different charging current densities in 1 M H_2SO_4. (c) Variation of the specific capacitance of 1.5 wt. % MWCNTs composite with different charging current densities in 1 M H_2SO_4. (d) Cyclic performance of PANI and 1.5 wt.% MWCNTs composite electrodes at charging current density of 13.3 A/g in a potential range of -0.5 to 1 V.

13.3 A/g, respectively. Figure (3-c) illustrates the relation between Cs and the charging current density. At high charge current densities, the specific capacitance decreases because only the outer pores of the active materials at the electrode/electrolyte interface are accessible to the electrolyte ions since they will not have enough time to reach the deeper ones to enable the doping/dedoping process [40]. Cyclic stability is a crucial requirement for supercapacitor applications.

The stability test of 5000 charge/discharge cycle was done for pure PANI and PANI/MWCNT electrodes at a constant current density of 13.3 A/g in a potential window ranging from -0.5 to 1.0 V.

Figure (3-d) illustrates the stability of the two electrodes in the relation of losing their capacitance as a stability percentage. The electrode stability can be calculated from equation (2):

$$\text{Stability} = \frac{Cn}{C1} \times 100 \tag{2}$$

where Cn is the capacitance of the electrode in each cycle, and C1 is the capacitance of the first cycle. From figure (3-d), it can be seen that the PANI electrode loses its capacitance fast and gets 20% of the first cycle capacitance after 5000 cycles. This poor cycling stability is due to repetitive swelling and shrinkage of the polymer backbone structure during the charge-discharge process. While PANI/MWCNT of 1.5 wt. % MWCNTs nanocomposite electrode maintains its stability and saves more than 86% of its capacitance of the first cycle after 5000 cycles, which is a good and reasonable retention in the stability percentage compared with that of pure PANI electrode that suggests long term high stability for the charge/discharge applications.

The electrochemical impedance spectrum (EIS) is one of the main methods to inspect the fundamental behavior of the electrode materials for supercapacitors. Figure (5-a) shows a complex-plane plot of the impedance (Nyquist plot) of PANI and PANI/MWCNT-COOH of 1.5 wt.% MWCNTs loading electrodes in 1 M H_2SO_4 that were performed at an open circuit potential of 0.3V through a frequency range of 10 mHz -100 kHz. In the Nyquist complex-plane, the imaginary vertical component signifies the capacitive properties, while the real component signifies the ohmic features. Typically, the impedance spectrum of the supercapacitor consists of a semicircle arc at the high-frequency region and a straight line parallel to the imaginary axis at the low-frequency region. The semicircle represents the charge transfer limiting process resulting from the charge transfer resistance (R_{ct}) caused by the Faradic reactions at the interface between the electrode and the electrolyte solution in parallel with the double-layer capacitance (C_{dl}) [39, 40]. The R_{ct} can be measured from the semicircle diameter [39], where it is (152 Ω) for PANI, which is a high value that consistent with its insulating state. Whereas, the small semicircle diameter of the composite and its low R_{ct} (12.8 Ω) value, indicating that incorporation of MWCNTs is highly improving the charge transfer performance of PANI/MWCNT electrode, due to the significant enhancement of the electronic conductivity and porosity structure of the composite electrode.

The equivalent series resistance (ESR), which is the sum of the electrolyte resistance, the contact resistance of the active material/current collector and the resistance of the active electrode material, can be found from the intercept of the semicircle with the real axis of the Nyquist plot [39]. The ESR defines the Cs charged/discharged rate as an essential factor in the evaluation of the supercapacitor power density. From figure (5-a), the measured ESR value of the PANI electrode is (8.2 Ω), whereas it is (6.2 Ω) for the composite electrode of 1.5 wt.% MWCNTs loading. Therefore the use of the MWCNTs have relatively improved the power performance of the electrode due to the reduction in its internal electrical resistance, but the value of the ESR is still higher than the requirement to get a high power density, and it needs to be reduced. Furthermore, the impedance curve intersects the real axis at an angle of 45°, which is compatible with the porous structure of the nanocomposite electrode when saturated with the electrolyte [41].

At the low frequency region the composite electrode exhibits a nearly linear branch, which frequently departs from that predicted for ideal capacitor with a slope angle near to 90° due to the presence of "constant phase element" [42], representing a good capacitive behavior due to the reduced diffusion

IOP Conf. Series: Materials Science and Engineering **757** (2020) 012036 doi:10.1088/1757-899X/757/1/012036

resistance of the electrolyte ions in the nanostructure electrode. The transition point between the high and low-frequency regions being mentioned as the knee that represents the maximum frequency where the capacitive behavior is dominant and a sign of a supercapacitor power ability [43]. The knee frequency of the composite electrode is about (41 Hz), higher than that of pure PANI electrode, which is (8 Hz), indicating better power ability.

The energy density (E_s) and the power density (P_s) are given in equation (3) and (4), respectively:

$$E_S = \frac{1}{2} C_S . V^2 \tag{3}$$

$$P_S = \frac{E_S}{t} \tag{4}$$

They were calculated using the galvanostatic charge/discharge curves at a constant current density, where Cs is the specific capacitance, V is the charging voltage (potential operating window), and t is the time to charge or discharge [6, 41, 42, 44]. Figure (4) is a Ragone plot of PANI/MWCNT of 1.5 wt.% MWCNTs composite electrode. For the lower charging current density of 1.6 A/g, the energy density of the composite electrode of 1.5 wt. % MWCNTs is 34.8 Wh/Kg, and the power density is 1670 W/Kg. While, for the higher charging current density of 13.3 A/g, the electrode exhibited a good supercapacitance performance, as it can deliver an energy density of 12.8 Wh/Kg and a power density of 8381 W/kg.

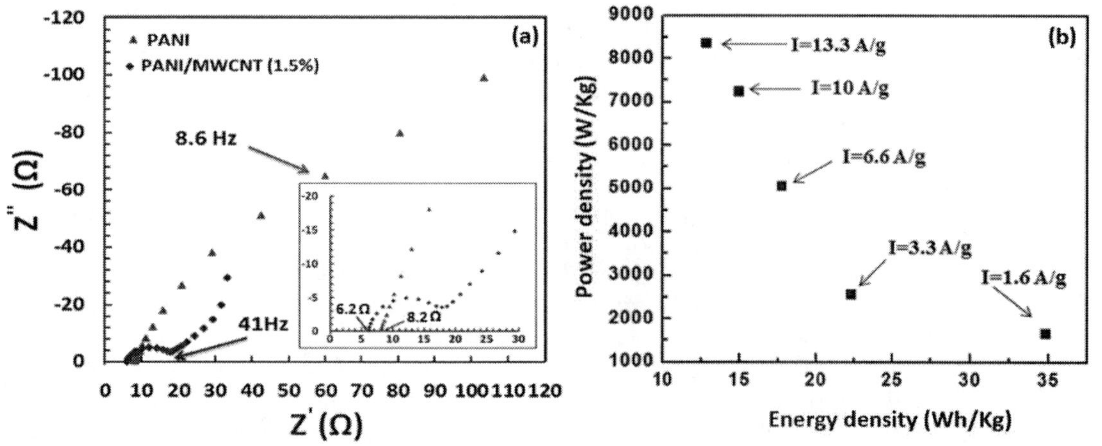

Figure 4. (a) AC impedance spectra of PANI and PANI/MWCNT of 1.5 wt. % MWCNTs composite electrodes that were performed at an open circuit potential of 0.3V through a frequency range of 10 mHz - 100 kHz. (b) Ragone plot of PANI/MWCNT of 1.5 wt. % MWCNTs composite electrode.

4. Conclusion

The possibility of using a microwave plasma in the polymerization of polymer composite materials as a powerful one-step method in the synthesis of the supercapacitor electrodes was demonstrated successfully in this work. Although the obtained results of the capacitance and the power density were nonoptimal, they can be optimized by better control of the plasma parameters and the deposition scheme that strongly affect the composite material thickness and structure. The SEM images didn't show a homogeneous coating of the CNTs with PANI, indicating that the polymer tends to grow both on the CNTs and on itself, creating micro-agglomerated polymerized islands. This growth may affect negatively on the three-dimensional nanopores structure of the CNTs that may be one of the reasons for the relatively low electrode capacitance values achieved by PANI/MWCNT composite. The other reason is may be due to the high composite thickness, which limits the accessibility of the electrolyte ions into the pores structure inside the electrode. So that, to increase the efficiency of the supercapacitor, the thickness of the electrode material should be reduced, and PANI coated layer on the nanotubes should be as thin as possible to earn the benefits of both the double-layer capacitance and pseudocapacitance performance of the composite constituents. In spite of the previous drawbacks, the achieved energy density crossed the threshold energy density limit of the battery [7,45], which is a favorable aim and meet the scientific ambition to use the supercapacitor as an independent source of energy by further enhancing of the energy density.

References

[1] Kim B K, Sy S, Yu A and Zhang J 2015 *Handbook of Clean Energy Systems* (John Wiley & Sons, Ltd.) **DOI:** 10.1002/9781118991978.hces112
[2] Wang G, Zhang L and Zhang J 2012 *Chem. Soc. Rev* .**41** 797-828. **DOI:** 10.1039/c1cs15060j
[3] Raza W, Ali F, Raza N, Luo Y, Kim K H, Yang J and Kumar S 2018 *Nano Energy* **52** 441–473. https://doi.org/10.1016/j.nanoen.2018.08.013
[4] Iro Z S, Subramani C and Dash S S 2016 *Int. J. Electrochem Sci.* **11** 10628 -10643
[5] Simon P and Gogotsi Y 2008 *Nature Materials* **7** 845-854. **DOI:** 10.1038/nmat2297
[6] Ramachandran R, Chen S and kumar G 2015 *Int. J. Electrochem. Sci.* **10** 10355 - 10388
[7] Saleem A M, Desmaris V and Enoksson P 2016 *Journal of Nanomaterials* **2016** 1-17 http://dx.doi.org/10.1155/2016/1537269
[8] Pandolfo A G and Hollenkamp A F 2006 *Journal of Power Sources* **157** 11-27 **DOI:**10.1016/j.jpowsour.2006.02.065
[9] Chen T and Dai L 2013 *Materials Today* **16** 272-280 **DOI:** http://dx.doi.org/10.1016/j.mattod.2013.07.002
[10] Pan H, Li J and Feng Y P 2010 *Nanoscale Res. Lett.* **5** 654-668 **DOI:** 10.1007/s11671-009-9508-2
[11] Ramya R, Sivasubramanian R and Sangaranarayanan M V 2013 *Electrochimica Acta* **101** 109-129 **DOI:** http://dx.doi.org/10.1016/j.electacta.2012.09.116
[12] Le T, Kim Y and Yoon H 2017 *Polymers* **9** 1-32. **DOI:** 10.3390/polym9040150
[13] Wang H, Lin J and Shen Z X 2016 *Journal of Science: Advanced Materials and Devices* **1** 225-255 **DOI:** http://dx.doi.org/10.1016/j.jsamd.2016.08.001
[14] Mi H, Zhang X, Xu Y and Xiao F 2010 *Applied Surface Science* **256** 2284-2288 **DOI:** http://dx.doi:10.1016/j.apsusc.2009.10.053
[15] Potphodea D D, Sinhaa L and Shirage P M 2019 *Applied Surface Science* **469** 162–172 **DOI:** https://doi.org/10.1016/j.apsusc.2018.10.277
[16] Sivakkumar S R, Kima W J, Choi J, MacFarlane D R, Forsyth M and Kim D W 2007 *Journal of Power Sources* **171** 1062–1068 **DOI:**10.1016/j.jpowsour.2007.05.103
[17] Shayeh J S, Ehsani A, Ganjali M R, Norouzia P and Jaleh B 2015 *Applied Surface Science* **353** 594-599. http://dx.doi.org/10.1016/j.apsusc.2015.06.066
[18] Marjanovic G C′ 2013 *Synthetic Metals* **177** 1-47 **DOI:** http://dx.doi.org/10.1016/j.synthmet.2013.06.004
[19] Choi H and Yoon H 2015 *Nanomaterials* **5** 906-936. **DOI:** 10.3390/nano5020906
[20] Liu D, Wang X, Deng J, Zhou C, Guo J and Liu P 2015 *Nanomaterials* **5** 1034-1047 **DOI:** 10.3390/nano5021034
[21] Ramana G V, Srikanth V V S S, Padya B and Jain P K 2014 *European Polymer Journal* **57** 137–142 **DOI:** http://dx.doi.org/10.1016/j.eurpolymj.2014.05.018

[22] Kumar A, Kumar V and Awasthi K 2018 *Polymer-Plastics Technology and Engineering* **57** 70-97 **DOI:**http://dx.doi.org/10.1080/03602559.2017.1300817
[23] Yasuda H K 1985 *Handbook plasma polymerization* (Academic Press Inc.: Orlando, FL.)
[24] Zhao X 2010 Iranian *Polymer Journal* **19** 823-841
[25] Lakshmi G B V S, Dhillon A, Siddiqui A M, Zulfequar M and Avasthi D K 2009 *European Polymer Journal* **45** 2873-2877 **DOI:** 10.1016/j.eurpolymj.2009.06.027
[26] Groenewoud LM H, Engbers G H M, White R and Feijen J 2002 *Synth. Met.* **125** 429-440 **DOI:** https://doi.org/10.1016/S0379-6779(01)00490-8
[27] Kumar A, Grant D, Alancherry S, Al-Jumaili A, Bazaka K and Jacob M V 2017 *Plasma Science and Technology for Emerging Economies* Springer 593-657 https://doi.org/10.1007/978-981-10-4217-1
[28] Oueiny C, Berlioz S and Perrin F -X 2014 *Progress in Polymer Science* **39** 707-748 **DOI:** https://doi.org/10.1016/j.progpolymsci.2013.08.009.
[29] Mittal G, Dhand V, Rhee K Y, Park S and Lee W R 2015 *Journal of Industrial and Engineering Chemistry* **21** 11-25 **DOI:** https://doi.org/10.1016/j.jiec.2014.03.022
[30] Zhou Y, Qin Z, Li L, Zhang Y, Wei Y, Wang L and Zhu M 2010 *Electrochimica Acta* **55** 3904-3908 **DOI:**10.1016/j.electacta.2010.02.022
[31] Ibrahim N I and Wasfi A S 2019 *Synthetic Metals* **250** 49-54 **DOI**:https://doi.org/10.1016/j.synthmet.2019.02.007
[32] Wasfi A S , Humud H R, Abed M A and Ismael M E 2017 *International Journal of Chem Tech Research* **10** 886-892.
[33] Wasfi A S, Humud H R and Ismail M E 2015 *Iraqi Journal of Physics* **13** 14-23
[34] Elnagga E M , Kabel K I, Farag A A and Al-Gamal A G 2017 *J Nanostruct Chem* **7** 75–83 **DOI:** 10.1007/s40097-017-0217-6
[35] Ramana G V, Padya B, Srikanth V V S S, Jain P K, Padmanabham G and Sundararajan G 2011 *Carbon* **49** 5239–5245. https://doi.org/10.1016/j.carbon.2011.07.041
[36] Reddy K R, Sin B C, Yoo C H, Sohn D and Lee Y 2009 *J. Colloid Interface Sci.***340** 160-165.
[37] Gull N, Khan S M, Islam A, Zia S, Shafiq M, Sabir A, Munawar M A, Butt M T Z and Jamil T 2016 *Materials Chemistry and Physics* **172** 39-46 **DOI:** http://dx.doi.org/10.1016/j.matchemphys.2015.12.048
[38] Wang Y, Wang Z and Xia Y 2005 *Electrochimica Acta* **50** 5641–5646 **DOI:** 10.1016/j.electacta.2005.03.042
[39] Guan D, Gao Z, Yang W, Wang J, Yuan Y, Wang B, Zhang M and Liu L 2013 *Materials Science and Engineering B* **178** 736– 743 **DOI:** http://dx.doi.org/10.1016/j.mseb.2013.03.010
[40] Li X and Wei B 2012 *Nano Energy* 1 479–487 **DOI:** 10.1016/j.nanoen.2012.02.011
[41] Xu G, Zheng C, Zhang Q, Huang J, Zhao M, Nie J, Wang X and Wei F 2011 *Nano Res.* **4** 870-881. **DOI:** 10.1007/s12274-011-0143-8
[42] Cheng Q, Tang J, Shinya N and Qin L 2013 *Journal of Power Sources* **241** 423-428 **DOI:** http://dx.doi.org/10.1016/j.jpowsour.2013.04.105
[43] Du C and Pan N 2006 *Nanotechnology* **17** 5314–5318. **DOI:** 10.1088/0957-4484/17/21/005
[44] Zhou G, Wang D, Li F, Zhang L, Weng Z and Cheng H 2011 *New Carbon Materials* **26** 180-186
[45] Miller J R and Simon P 2008 *Science* **321** 651-652 **DOI:** 10.1126/science.1158736
[46] M. Totolin, M. Grigoras, D. Conduruta, Plasma induced aniline polymerization, Journals of Optoelectronics and Advanced Materials, 9(2007) 881-883.

IOP Conf. Series: Materials Science and Engineering **757** (2020) 012037 doi:10.1088/1757-899X/757/1/012037

Effect of RTV Silicon Rubber on Some Mechanical Properties of Epoxy /Polyester Polymer Blends

Ahmed J Farhan[1] and Harith I Jaffer[2]

[1]Department of Physics, College of Science, University of Wasit, Kut, Wasit , Iraq.

kutpc19@yahoo.com

[2]Department of Physics, College of Science, University of Baghdad, Baghdad, Iraq.

harithibrahem@yahoo.com

Abstract: In this research the epoxy/ RTVsilicon rubber and unsaturated polyester / RTVsilicon rubber blends were prepared by adding varying weight (3,5, 7,10 and 20)% of (SR) to either of EP and UPE resins.The aim of research to analyze effect of RTV silicone rubber composition (3, 5, 7, 10 and 20) wt% of the mechanical properties of the epoxy and unsaturated polyester. Mechanical propertiessuch as impact strength, shore D hardness, young modulusand flexural strength were tested.The experimental results showed that the specimensof EP/SR and UPE/SR blends have a maximum values of impact strength (8.13 KJ/m^2 and 11.91 KJ/m^2) at the addition of 10% SR and 5% SR respectively.Hardness No. decreased slightly when the content of SR increased at a range of 3% to 20% in both of EP/SR and UPE/SR.Also flexural strength decreased rapidly when the content of SR increased at a range of 3% to 20% in both of EP/SR and UPE/SR.

Keywords: Polymer blends; Unsaturated polyester; Epoxy; RTV silicon rubber, Mechanical properties.

1.Introduction

Polymer-polymer blends have gained significant commercial growth, as it saves nearly 36% weight of the total polymer consumption without compromising weight. Epoxy is a versatile and widely accepted matrix material for fabrication of advanced composites, hardware components, electrical circuit board materials and missile equipments; because of its excellent bonding, physicochemical, thermal, mechanical, dielectric and aging characteristics. Epoxy resins belong to the most important thermosetting polymers with a high crosslinking degree and that contain many hydroxyl groups. Epoxy resins available in a wide variety of properties that are unattainable with other thermoset resins, they are easily cured without evolution of volatiles or by -products by a broad range of chemical species. These polymers have brilliant characteristics such as good chemical and moisture resistance and high dielectric properties, epoxy resins are widely working in modern industries, as adhesives, coatings and paints, and packaging materials for microelectronic devices[1,2].

Polyester resins (UPE)are comparatively hard and brittle materials and one of the important thermosetting materials used for the manufacture of glass-reinforced plastics and polymeric composites. These resins have wide spread use is due to their relatively low cost, comfortable of processing, high wetting properties with reinforcements, and good balance of properties . UPE are solutions of unsaturated polyester with unsaturated co reactant diluents like styrene [3]. Silicone sealant RTV (Room Temperature Vulcanizing) is a liquid form of adhesive. Typically, it looks, feels, and acts like a gel. It has a different chemical make-up from other organic polymer-based adhesives. Unlike other adhesives, silicone keeps its elasticity and stability in both high and low temperatures. Furthermore, silicone sealant is resistant to other chemicals, moisture, and weathering[4].

RTV silicone rubber is also one type of elastomer which have glass transition temperature, Tg, which is very low, have high elastic properties, hydrophobic properties, as well as good resistance to thermal and oxides [5].There are numerous numbers of previous studies related to polymer blends and their industrial applications, for examples:

Fahriadi Pakaya, Hosta Ardhyananta, and Sigit Tri Wicaksono [6]. Analyzed effect of RTV silicone rubber composition (0, 5, 10, 15, 20) wt% of the mechanical properties and thermal stability of thermoses epoxy. Testing and characterization conducted on thermoses epoxy by the addition of RTV silicone rubber. Results show that tensile strength, elongation at break and hardness has decreased, energy and impact strength increased maximum on the addition of 15% RTV silicone rubber respectively 0.294 J/ m2 and 6175 J / m2. The maximum degradation of temperature increase in the addition of 15% RTV silicone rubber is 328 and 349oC respectively at 5 and 10% degradation. Shatha H. Mahdi, Widad H. Jassim, Intisar A. Hamad and Kareem. A. Jasima[7]. Prepared epoxy/silicon rubber blends by adding the silicone rubber with varying weight (0.0, 0.25, 0.5, 0.75 and 1gm) to epoxy resin. Some thermal properties are determined for different samples of blends. The bulk resistivity and dielectric constant were evaluated and related to the neat epoxy. The results suggest that the inclusion of silicone epoxy effectively improved in the glass transition temperature (Tg) and the thermal insulation also improved the electrical properties like resistance and dielectric constant for using it as capacitor at high frequencies and in the high voltage strength applications.

2. Experimental Procedure

2.1. Materials

Epoxy resin (Sikadur-52) was used in this research; it is a low viscosity, free flowing and fast curing injection resin and primer/coating based on a two components solvent free epoxy resin; ideally suited to a wide range of building and civil engineering applications where highly penetrative material is required. Unsaturated polyester resin (UPE) is a liquid with moderate viscosity which can be cured to the solid state by adding (MethyleEthyleKeton Peroxide,MEKP) as a hardener, while cobalt octoate acts as a catalyst to accelerate the solidification process. SUPERMAX universal RTV silicon sealant made in U.A.E. was used in this research.

2.2. Material Preparation

2.2.1. Epoxy Resin preparation

A clean disposable aluminum container was put on sensitive electronic balance, and the required epoxy resin was then poured in. For100 gm of EP resin a sufficient amount of hardener (Sikadur-52-B) was added. The ratio of hardener to epoxy was approximately 1:2 by weight. The content was thoroughly mixed to be ready for further applications; blend specimens casting.

IOP Conf. Series: Materials Science and Engineering **757** (2020) 012037 doi:10.1088/1757-899X/757/1/012037

2.2.2. Unsaturated polyester Resin preparation

With the same above procedure the required unsaturated polyester resin was weighed. 100 gm of UPE resin were mixed with 0.5 gm of cobalt naphthenate as accelerator and 2 gm methyl ethyl keton Peroxide (MEKP)as hardener. The resin was mixed thoroughly with accelerator and hardener until a homogeneous state of the mixture evolved, and ready for blend specimens casting.

2.3. Polymer Blends Preparation

Epoxy/silicon rubber(SR) and unsaturated polyester/silicon rubber(SR blends are selected for investigation. Epoxy resin was prepared as in section (2.2.1), with weight fractions 3, 5, 7, 10 and 20% percentages of silicon rubber (SR). The epoxy and silicon rubber were added together and thoroughly prepared mixed to obtain the polymer blend required. Unsaturated polyester/silicon rubber(SR) polymer blend were prepared by the same manner.

3. Measurements

3.1. Impact test

Charpy impact test involve the use of hammer blow that will be delivered to the sample until reaches to breaking point. The sample is positioned in such a way that both of its ends are fixed in position and the blow isdielvered to the middle part. Samples of impact device has a dimensions of (10*55*10mm)according to ASTM (D4812). Figure1.shows the impact test specimens to carry the test.. The apparatus used in this test is manufactured by (Testing Machines,Inc , Amityville New York). The following equation can be used to calculate the impact strength (I.S.):[8]

I.S. =Uc/A

Uc: the fracture energy (Joule) from Charpy impact device.

A: the area of the samples (m^2) (cross-sectional)

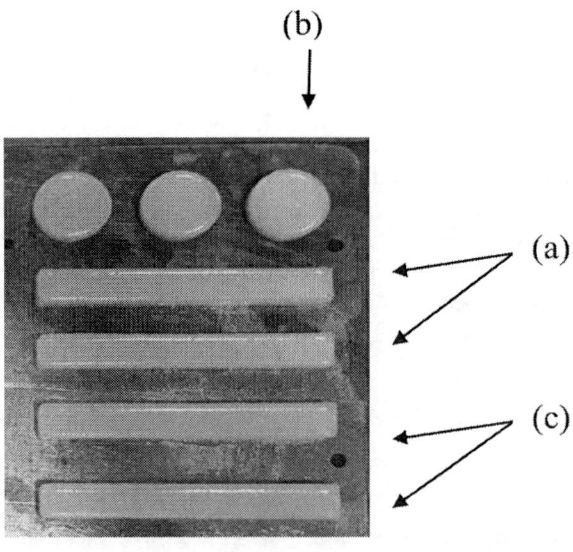

Figure 1.(a) Impact test,(b) Hardness test, and (c) Three – Point Bending testSpecimens.

3.2. Hardness test

Shore D was uitalized for calculating the hardness of the specimens, thickness of samples at least more than (3mm) which must have smooth surface and not be exposed to mechanical vibrations therefore the prepared sample has (5mm) thickness and (30mm) diameter with reference to ASTM (D 2240). Figure1.shows the hardness test specimens to carry the test. The shore D device involve a needle that is positioned vertically to the specimen for period about half a minute. To register an accurate reading, the average of three values measured from various sites of each specimen is estimated[9].

3.3. Three – Point Bending Test

Bending tests were carried out by using a Tinus Olsen machine model H50KT of (5kN) full scale load capacity, according to ASTM standard (D-790), this test method covers the determination of flexural properties of unreinforced and reinforced plastics in the form of rectangular bars molded directly or cut from sheets , plates or molded shapes. In this method the bar tests on two supports and is loaded by means of a loading nose midway between the supports. This method includes:

a) Preparation of test specimens. Number of specimens were cut from the prepared sheets, each specimen was in (125 mm) long, (96 mm) support span and (10 mm) width.Figure1.shows the bending test specimens to carry the test.

b) Procedure: The specimen is placed on two supports and a load is applied in center at constant rate of displacement (crosshead speed) equal to 0.5 mm / min. The loading at failure (psi) is the flexure strength [10].

4. Results and Discussion

4.1. Impact Behavior

The conventional charpy impact test is used to evaluate the impact strength of the blends that have (3,5,7,10 and 20)% of (SR) in either of EP and UPE. The results of this test are shown in Table1.and Figure 2.which show the effect of rubber content on the impact strength values of the prepared blends.

Table 1.clearly demonstrates the effect of blending (SR) with EP and UPE respectively on the impact strength of the resulting blends. In general, it is well known that the cured rubber-modified epoxy exhibits a two – phase microstructure which consisting of relatively small rubbery particles [11]. These particles are dispersed into the matrix of epoxy and the resulting phase separation nature will be related to the percentages of the rubber and the methodology of the preparation. Any increase in the rubber amount in the matrix would lead to an increase in the average size of the dispersed phase particles and eventually failing to produce a compatible and applicable blend [12].

The above-mentioned phenomenon is clearly noted in the present work, it can be seen that when the percentages of (SR) rubber in both of theEP and UPE blends increases to more than 20% wt., (Figure 2.), the rubber phase has notably separated from the epoxy or unsaturated polyester leading to decrease in the impact strength of the blends.

Figure 2.demonstrates that the Charpy impact strength of (thermoset /elastomer) blend is higher than that of brittle matrix itself, which implies the positive role of elastomer on the impact fracture toughness of brittle materials. Impact strength of (EP) can be maximized to almost (8.4) KJ/m2 on addition of (5) wt% of silicon rubber. The impact strength of (UPE) could be maximized to almost

(11.91) KJ/m2 on addition of about (5%) of silicon rubber. If the percentage of elastomer in the brittle matrix increased to more than the above-mentioned percentages, the impact strength would decrease to lower values. This indicates that the impact strength of binary blend is not simply an additive, and its dependence on blend composition reveals the influence of blend morphology, state of dispersion and any other structural parameter on impact toughening of this blend. Occurrence of a maximum in the impact strength at a particular blend composition may be attributed to the critical size and geometry of dispersed phase domains. Chiang and Song [13] have suggested that the rubber particles are often spherical with spherical inclusions of the brittle phase. At concentrations larger than the related maximum value of the rubbery phase, the dispersed particles tend to agglomerate or to form elongated rather than spherical particles, which reduce the impact strength.

Table 1. The effect of silicon rubber content (wt %) on the (I.S) of (EP/SR) and (UPE/SR) blends.

	Impact Strength (I.S), KJ/m^2					
	Silicon rubber (SR) content (wt) %					
Composition	0%	3%	5%	7%	10%	20%
EP/SR	7.3	6.16	8.4	6.07	8.13	6.4
UPE/SR	7.12	5.13	11.91	5.71	8.65	6.2

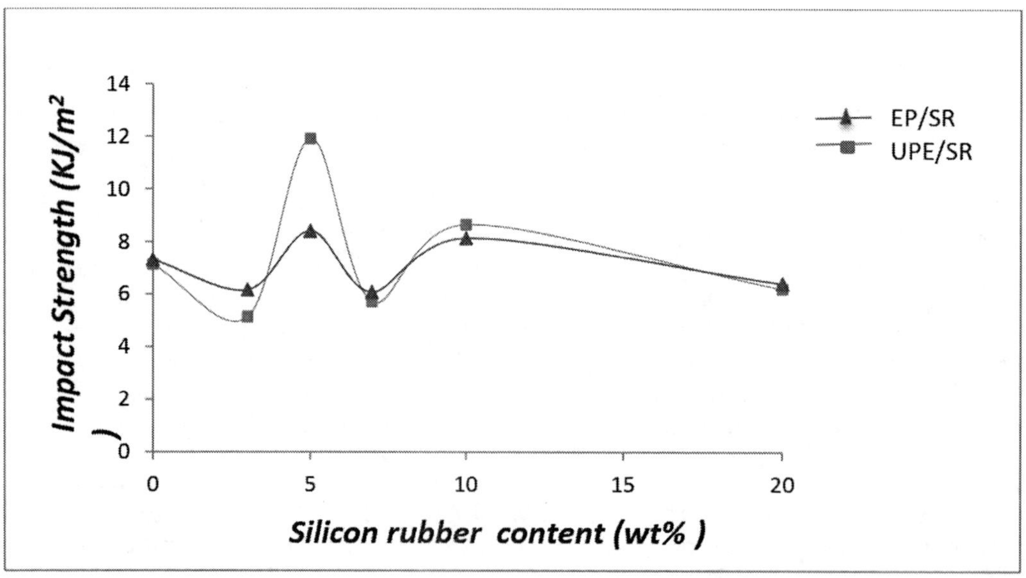

Figure 2. Charpy impact strength variation with SRrubber content in EP,UPE resins.

IOP Conf. Series: Materials Science and Engineering **757** (2020) 012037 doi:10.1088/1757-899X/757/1/012037

4.2. Hardness Test

The hardness is a measure of resistance to indentation and, hence, will not be greatly influenced by the matrix [14]. The shore D hardness is used to evaluate the hardness of the blends that have (3,5,7,10 and 20)% of (SR) in either of EP and UPE. The results of this test are shown in Table 2.and Figure 3.which show the effect of silicon rubber (SR) content on the Hardness No. values of the prepared blends.It can be noticed that the presence of silicon rubber would reduce the hardness of the blends and when increased thesilicon rubber (SR) ratio, the values of hardness No. are decreased for both epoxy and polyester blends. The reason for that is obvious, which is the elasticity of the silicon rubber.

Table 2. The effect of silicon rubber content (wt %) on the Hardness No. of (EP/SR) and (UPE/SR) blends.

	Hardness No.					
	Silicon rubber content (wt)					
Composition	0%	3%	5%	7%	10%	20%
EP/SR	79	63.9	60.2	67.2	60.4	66.6
UPE/SR	88.4	80.9	79.7	78.3	79.9	74.6

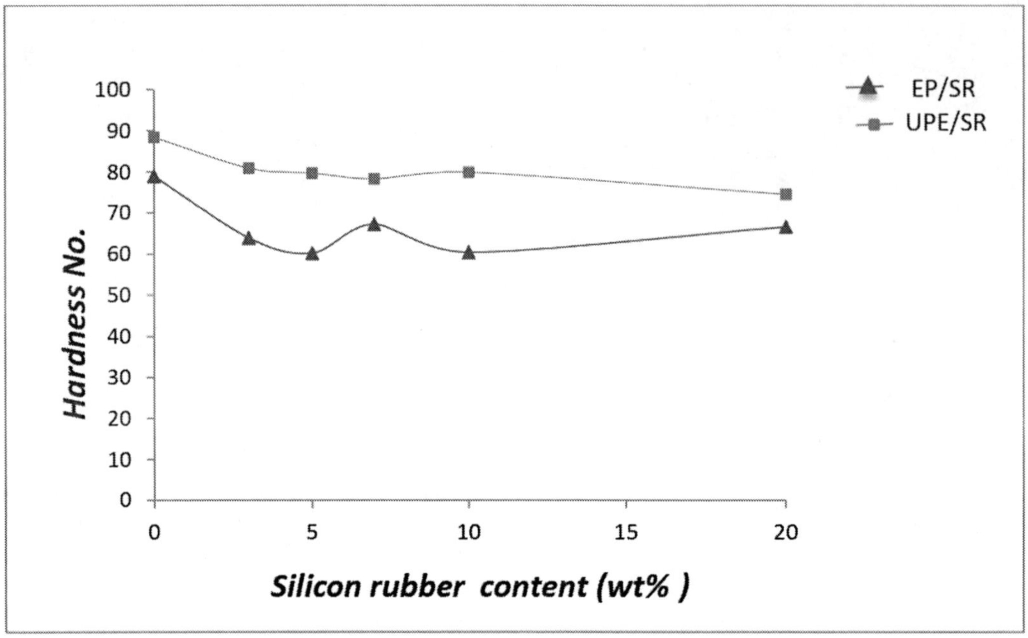

Figure 3. Hardness No.variation with SRrubber content in EP,UPE resins.

4.3. Bending Properties and Young's Modulus

The Young's modulus (E) is a measure of stiffness (rigidity) of the material. The greater the Young's modulus, the smaller the elastic strain resulting from the application of a given stress [19]. This reflects the importance of measuring the (E) of the blends that have (3,5,7,10 and 20)% of (SR) in either of EP and UPE, which were measured and are shown in table 3. This measurement was performed by using 3-point bending test method.

Table 3.states that the Young's modulus of (EP) and (UPE) has been reduced after blending them with the used silicon rubbers (SR). The reason for this reducing is the increasing in the flexibility of the polymer chains for the blends compared with that of pure (EP) or (UPE). This implies an increase in the strain rate and reduction in the stiffness (rigidity), which means, these blends will have new prospect of application. [15].

Table 3. The effect of silicon rubber content (wt %) on the Young's modulus (E) of (EP/SR) and(UPE/SR) blends.

	Young's modulus (E),MPa.					
	Silicon rubber content (wt) %					
Composition	0%	3%	5%	7%	10%	20%
EP/SR	1187	781	976	856	700	714
UPE/SR	3134	2573	1991	1605	1800	1294

4.4. Flexural Strength Behavior

Flexural strength is a measure of the resistance of material to be bent, when it goes under bending moment. During bending, the material undergoes gross flexural deformation before fracture, because the material is exposed to three types of stresses. The convex side of the material is extended (tensioned), while the concave side is compressed and, at the same time, the internal layers of the material shear each other [16].

The simple three-point bend method was used to calculate the flexural strength (F.S) for epoxy blends and unsaturated blends under study. According to the results shown in table4.and displayed in Figure 4.the flexural strength of (EP) and (UPE) are decreased when they are mixed with silicon rubber due to the dispersion phase between silicon rubber and polymer matrix. Any increase in the rubber amount in the matrix would lead to an increase in the average size of the dispersed phase particles and eventually failing to produce a compatible and applicable blend.

Table 4. The effect of silicon rubber content (wt %) on the Flexural Strength of (EP/SR) and (UPE/SR) blends.

	Flexural Strength, F.S (MPa)					
	Silicon rubber content (wt) %					
Composition	0%	3%	5%	7%	10%	20%
EP/SR	38.59	23.88	27.02	20.01	19.42	15.75
UPE/SR	105.4	48	48.31	40.22	46.39	28.14

IOP Conf. Series: Materials Science and Engineering **757** (2020) 012037 doi:10.1088/1757-899X/757/1/012037

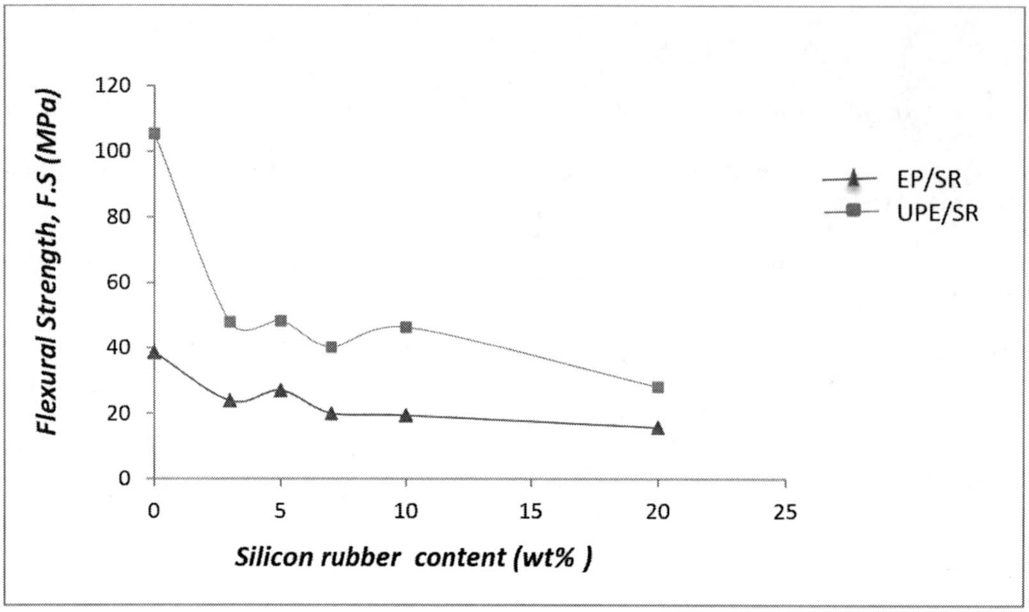

Figure 4. Flexural Strength variation with SRrubber content in EP,UPE resins.

5. Conclusions

The results showed that the specimens of UPE/SR blend have a highest values of impact strength at the addition of 5% SR .Hardness No. decreased slightly when the content of SR increased both of EP/SR and UPE/SR blends. Also flexural strength decreased rapidly when the content of SR increased in both of EP/SR and UPE/SR blends.

6. References

[1] Li J and Jian Guo Z 2011 *J. Polym. Plast. Tech. Eng.* **50** (10) 996–999.

[2] Swetha C and Kumar R J 2011 Materials and Design **32**4152–4163.

[3] Cherian A, Varghese L and Thachil A J 2007*J. European Polymer Journal*, **43** 1460–1469.

[4] Silicone Sealant Handbook Articles 2019A Handbook and Guide to Silicone Sealant Applications – Hotmelt.com. https://www.hotmelt.com/blogs/blog/silicone-sealant-handbook.

[5] MY S and Hs K 2009 *J. Corros. Sci.*, vol. **51**, no. 3, pp. 650–657.

[6] Fahriadi P, Hosta A, and Sigit T W 2017 *The Journal for Technology and Science* **28**(1).

[7] Shatha H M, Widad H J, Intisar A H and Kareem A J 2017 *International Conference on Technologies and Materials for Renewable Energy, Environment and Sustainability, TMREES* **17**, pp 21-24.

[8] Crawford R J 1987 *Plastics Engineering*, 2nd adition, (New York : Pergamon Press).

[9] Comyn J 1985 Polymer Permeability, *J. Springer*, chapter 1 pp 1–10.

[10]Annual Book of ASTM standards 1984 **08**.01.

[11] Kinloch A J, Shaw S J and Tod D A 1983 *Polymer* **24**(10) 1341-1354.

[12] Martuscelli E and Marchetta C 1987 *New Polymeric Materials*, (The Netherlands: VNU Science Press).

[13] Chiang W and Song L S 1988 *Journal of Applied Polymer Science* **36** 1685-1700.

[14] Agrawal B K 1994 *Introduction to Engineering Materials*, 4th edition, (New Delhi: Tota Mc Grew-Hill Publishing Company Limited).

[15] Hull D and Clyne T W 1996 *An Introduction to Composite Materials*, 2nd edition, (Great Britain: Cambridge University Press).

[16] James M G 2004 *Mechanics of Materials*, 6th edition, (Canada : Thomson).

Evolution and set up the maps for solar radiation of Iraq using Data observation and Angstrom model during monthly July2017

A f Abed1, E F Khanjer 1 and S A Abdullah1
1 Department of Astronomy, College of Science, University of Baghdad, IRAQ

Abstract.
The development that solar energy will have in the next years needs a reliable estimation of available solar energy resources. Several empirical models have been developed to calculate global solar radiation using various parameters such as extraterrestrial radiation, sunshine hours, albedo, maximum temperature, mean temperature, soil temperature, relative humidity, cloudiness, evaporation, total perceptible water, number of rainy days, and altitude and latitude. In present work i) First part has been calculated solar radiation from the daily values of the hours of sun duration using Angstrom model over the Iraq for at July 2017 . The second part has been mapping the distribution of solar radiation energy to Iraq using interpolation technique in Arc _GIs software Version 10.3.

Keyword: solar radiation, sun duration, Angstrom model, Arc _GIs

1. Introduction

The sun is a nuclear power, which is, generates power in form of radiation at the phenomenally high rate of 3.8x1023 kilowatts. An extremely small fraction of this is intercepted by the earth's surface, but this small fraction amounts to a huge 1.8x1014 kilowatt. On the average about 60% of this amount penetrates the atmosphere to reach the earth's surface (1.1x1014 Kw.). Of course this amount of power is distributed over the entire surface of the earth. To bring this number closer to home, consider that on bright sunny day each square meter of surface facing the sun receives about one kilowatt [1]. Energy form the sun is basis for life on earth, and the solar spectrum a mixture of radiative, visible, ultraviolet and infrared radiation has to be known to design any solar conversion system. Total daily solar radiation is considered as the most important parameter in the performance prediction of renewable energy system, particularly in sizing photovoltaic (pv) power and solar heating system. [1].In developing countries, such as Iraq, interest in solar energy applications has been growing in providing electricity and water supply in different areas. Understanding solar radiation data is essential for modeling solar energy systems. Solar radiation is used directly to produce electricity for photo voltaic (PV) systems and solar thermal systems.[2]

2. Solar Radiation

The radiation from the sun is the primary natural energy source of the planet Earth. Other natural energy sources are the cosmic radiation, the natural terrestrial radioactivity and the geothermal heat flux from the interior to the surface of the Earth, but these sources are energetically negligible as compared to solar radiation. When the spoken is of solar radiation, it means the electromagnetic radiation of the Sun. The energy distribution of electromagnetic radiation over different wavelength is called Spectrum. The electromagnetic spectrum is divided into different spectral ranges (Figure 1).

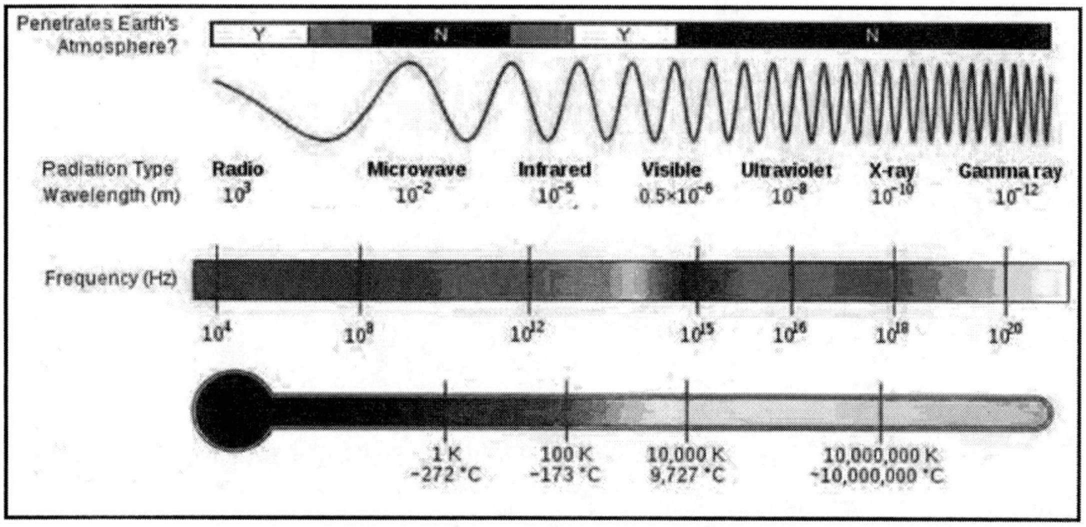

Figure 1. Spectral ranges of electromagnetic radiation, [3].

IOP Conf. Series: Materials Science and Engineering **757** (2020) 012038 doi:10.1088/1757-899X/757/1/012038

The quantity of solar radiation reaching the earth's surface varies dramatically as a function of changing atmospheric condition as well as the changing position of the sun through the day, [3].
Article I.

3.Sunshine Duration

Sunshine duration is the length of time that the ground surface is irradiated by direct solar radiation (i.e., sunlight reaching the earth's surface directly from the sun). Sunshine duration is also defined as the period during which direct solar irradiance exceeds a threshold value of 120 watts per square meter (W/m^2). This value is equivalent to the level of solar irradiance shortly after sunrise or shortly before sunset in cloud-free conditions. It was determined by comparing the sunshine duration recorded using a Campbell-Stokes sunshine recorder with the actual direct solar irradiance [4].

4.Angstrom Model

proposed first theoretical model for estimating global solar radiation based on sunshine duration, The fundamental Angstrom model can be considered as a base model, reconsidered this model in order to make it possible to calculate monthly average of the daily global radiation on a horizontal surface from monthly average daily total insolation on an extraterrestrial horizontal surface as per the following relation:[5]

$$\frac{H}{H_c} = a + b\frac{n}{N} \qquad \text{(1) [6]}$$

Where

H = Monthly average daily global radiation (*Wh/m²/day*).

H_c = Monthly average clear sky daily global radiation for the location solar radiation in a given day.

n = Actual sunshine duration in a day respectively (Hours).

N = Monthly average maximum possible bright sunshine duration in a day. Also known as monthly mean length of the day in hours.

a and b = Empirical coefficients. These coefficients are location specific coefficients referred to as fractions of extra terrestrial radiation on overcast days and on average days respect [7].

a= 0.25........................(2)

b= 0.50........................(3

N= (2/15) $\cos^{-1}$(- tanφ tanδ) ....(4)

The ratio *H/H0* called clearness index. Values of the monthly average daily extraterrestrial radiation (*Ho*) is calculated from the following equation, [8]:

$$H_o = \frac{24\times 3600\times I_{sc}}{\pi}\left[1+0.033\cos\left(\frac{360\times d}{365}\right)\right]\left[\cos\phi\cos\delta\sin\omega+\frac{\pi w}{180}\sin\phi\sin\delta\right] \text{....(5)}$$

Where:

I_{sc} = is the solar constant of a value 1367 Wm^{-2}

d = day of the year from January1 to December 31taking January 1st as 1.

Φ= latitude of location

ω= sunset hour angle

ω= $\cos^{-1}$(- tanφ tanδ) .... (6)

δ= declination angle

$$\delta = 23.45\sin\left[360\left(\frac{284+d}{365}\right)\right] \text{............(7)}$$

4.1 Interpolation techniques

Spatial interpolation is tool in GIS used to find the values of unknown points. It can be defined as a procedure of estimating the values of properties at unsampled locations based on the set of observed values at known locations. A large number of interpolation methods have been developed for use with point, line, and area data . No matter which interpolation technique is used, the derived values are only estimates of what the real values should be at a particular location. The quality of any analysis that relies on interpolation of observed data is, therefore, subject to a degree of uncertainty . Different interpolation methods can therefore generate different predictions at same locations. Many researchers have evaluated various methods for interpolation of different hydro climatic data. But accuracy of estimated values is varying from each methods based on the topography of the area, concentration and distribution of the measurement stations.So in this study some of these interpolation methods used in GIS were analysed and checked their accuracy based on the results produced.[9]

There are two main groupings of interpolation techniques: deterministic and geostatistical.

- **Deterministic interpolation techniques** create surfaces from measured points, based on either the extent of similarity (Inverse Distance Weighted) or the degree of smoothing (Spline). [9]
- **Geostatistical interpolation techniques** (kriging) utilize the statistical properties of the measured points[9]

4.2 Methodology

Using the data of the General Organization for Aerial Meteorology and Seismic Monitoring in Baghdad, the daily sun duration values were founvalues were found in the stations (Mosul, Kirkuk, Baghdad, Nasiriyah, Khanakin, Amara, Najaf, Hay, Diwaniya, Karbala) listed in below table(1) for the 2017.

Table 1...Represent values of the studied sun duration from Iraqi metrological Stations

Day	Badhdad	Mosoul	Nasriya	Alhay	kirkuk	Karbla	Khanqien	Dewania	Emara	Najaf
1	12.0	11.8	5.7	10.6	10.6	11.2	10.1	11.6	10.5	10.3
2	11.6	11.5	10.2	11.0	11.7	10.8	9.4	11.9	9.0	11.5
3	11.4	11.6	10.8	11.2	12.0	10.9	9.2	11.3	11.2	10.8
4	10.9	11.2	9.9	11.4	M	10.3	10.2	10.9	11.2	10.6
5	7.4	11.8	10.9	10.4	12.0	8.9	7.2	5.0	11.0	6.3
6	11.4	11.6	9.6	10.7	12.0	11.0	9.0	11.2	9.6	12.0
7	11.8	7.9	8.8	11.0	M	11.6	9.2	10.3	11.5	11.1
8	11.3	9.0	4.8	3.5	11.4	10.7	8.0	7.4	4.8	8.5
9	9.7	11.8	9.7	11.1	M	10.3	8.0	10.0	11.0	9.7
10	9.2	11.6	10.4	11.1	M	10.3	8.4	10.8	11.0	10.4
11	11.2	12.1	10.4	11.2	12.0	11.0	11.5	11.0	11.4	11.0
12	10.5	12.0	5.6	10.8	12.1	10.2	11.4	10.3	11.0	9.6
13	12.1	12.0	9.2	M	12.3	10.9	12.0	9.9	11.0	10.3
14	11.5	11.8	9.1	10.2	M	10.8	11.8	10.1	9.0	11.1

15	11.9	11.5	9.1	11.0	12.2	11.6	11.4	9.6	10.0	10.4
16	11.9	11.8	8.3	10.5	12.2	11.0	12.0	10.3	9.0	10.1
17	11.8	11.2	11.1	11.8	12.4	11.9	9.5	11.8	11.8	11.3
18	12.1	11.3	6.4	7.0	11.5	11.9	10.4	11.3	10.8	11.4
19	12.1	11.8	11.0	11.5	12.5	11.8	10.6	11.8	11.6	11.8
20	12.4	11.8	11.2	11.7	M	12.0	12.2	12.3	11.4	12.2
21	12.2	11.8	11.8	12.1	M	11.9	12.2	12.2	11.9	11.2
22	11.9	11.1	11.5	12.0	M	11.9	10.8	11.9	12.2	11.8
23	11.7	11.6	11.5	11.7	M	11.9	11.1	11.6	12.3	11.7
24	12.4	12.0	10.6	10.2	12.1	11.2	10.9	11.8	9.2	11.5
25	12.0	12.0	10.3	11.5	12.3	11.9	11.1	11.3	11.7	11.6
26	12.7	12.1	9.4	12.0	12.4	12.0	11.8	12.0	12.1	11.1
27	M	12	10.3	12.1	11.8	11.5	12.1	12.1	12.3	12.2
28	12.9	11.7	11.4	12.1	12.0	11.9	11.5	12.0	12.0	11.4
29	12.0	11.3	10.5	11.5	11.5	10.9	11.7	11.5	11.0	11.0
30	11.0	12.0	10.7	11.8	12.5	12.0	11.8	11.8	11.5	11.4
31	12.8	11.9	11.4	12.2	12.4	12.0	10.6	11.8	12.5	11.3

5.Experimental work

In this work we obtain daily solar radiation for July and August in year 2017 from daily values of sun duration using Angstrom Model and then show the results in map using interpolation techniques in ArcMap GIS 10.3 program.

Table 2. Represents estimation values of the solar radiation from sun duration for July

day	Badhdad	mosoul	Nasriya	Alhay	kirkuk	Karbla	Khanqien	Dewania	emara	najaf
1	27.94625	27.46217	18.73985	25.94418	25.79419	26.81684	25.12348	27.43508	25.80539	25.50474
2	28.05661	27.71435	26.02931	27.20851	28.08642	26.89663	24.71426	28.58799	24.19682	27.96558
3	28.56532	28.6722	27.72452	28.31322	29.36334	27.83646	25.12525	28.48912	28.32342	27.69415
4	27.98145	28.25083	26.50237	28.82012		27.08899	26.84585	28.0578	28.51769	27.5708
5	22.05658	28.55622	27.48405	26.68355	28.93899	24.3773	21.72413	18.41286	27.61308	20.40249
6	27.28318	27.37699	24.69708	26.30295	28.04043	26.73743	23.69814	27.06629	24.67746	28.2365
7	27.49547	21.68286	23.18618	26.38583		27.25304	23.66441	25.37848	27.12953	26.53492
8	27.21291	23.648	17.592	15.64873	27.23493	26.36893	22.28467	21.47265	17.58829	23.10286
9	25.61217	28.60097	25.69716	27.81733		26.57612	22.97552	26.1458	27.67495	25.6672

10	25.26479	28.77242	27.2305	28.2876		27.02463	23.9797	27.84148	28.14308	27.19548
11	28.03844	29.18284	26.90786	28.1062	29.12173	27.7839	28.43953	27.8211	28.42808	27.79994
12	26.13427	28.13688	18.86339	26.64609	28.37383	25.73474	27.41753	25.91845	26.95799	24.84943
13	27.86228	27.48264	23.72		28.00451	26.16532	27.65371	24.73402	26.33696	25.30183
14	27.13748	27.34194	23.70526	25.2932		26.16441	27.5141	25.16824	23.53386	26.62302
15	28.52364	27.67941	24.39022	27.23734	28.81687	28.1324	27.70482	25.13532	25.73175	26.32993
16	29.23081	28.81727	23.74712	27.13617	29.52795	27.89887	29.31432	26.85096	24.81337	26.51693
17	29.03533	27.85793	28.08314	29.12012	29.7894	29.25839	25.42308	29.15179	29.13592	28.34654
18	28.71679	27.26118	20.19892	21.08243	27.65383	28.48403	26.09473	27.62299	26.85005	27.74791
19	27.87075	27.17157	26.39145	27.07557	28.2827	27.49964	25.61424	27.54844	27.23814	27.5205
20	28.1293	26.99244	26.51992	27.1973		27.61808	27.76378	28.10849	26.77438	27.93226
21	28.46364	27.59207	28.01955	28.40823		28.09064	28.3853	28.59341	28.1264	27.06625
22	28.84309	27.34354	28.38575	29.0941		28.92045	27.08527	28.97683	29.421	28.79041
23	28.80106	28.35049	28.65322	28.89989		29.19067	27.7953	28.78154	29.85305	28.90443
24	29.31034	28.39614	26.72767	26.05718	28.66048	27.56472	26.95657	28.53846	24.54494	28.04657
25	27.78424	27.48017	25.42988	27.14661	28.02943	27.71769	26.38331	26.88738	27.46125	27.29904
26	28.31136	27.1359	23.67519	27.40238	27.67906	27.38147	26.92724	27.44182	27.56756	26.09806
27		27.33112	25.31151	27.90248	27.15853	26.99736	27.70597	27.94398	28.21818	28.05488
28	29.83927	27.70108	27.75509	28.74569	28.26965	28.41893	27.63633	28.63746	28.61546	27.68727
29	29.01372	27.60668	26.87992	28.35813	28.03276	27.40682	28.45734	28.40209	27.60465	27.58882
30	27.20861	28.38082	26.93291	28.54786	29.26494	28.83014	28.33198	28.59388	28.10965	27.93923
31	29.03889	27.35941	27.16028	28.27342	28.2187	27.95096	25.69624	27.72373	28.74424	26.93852

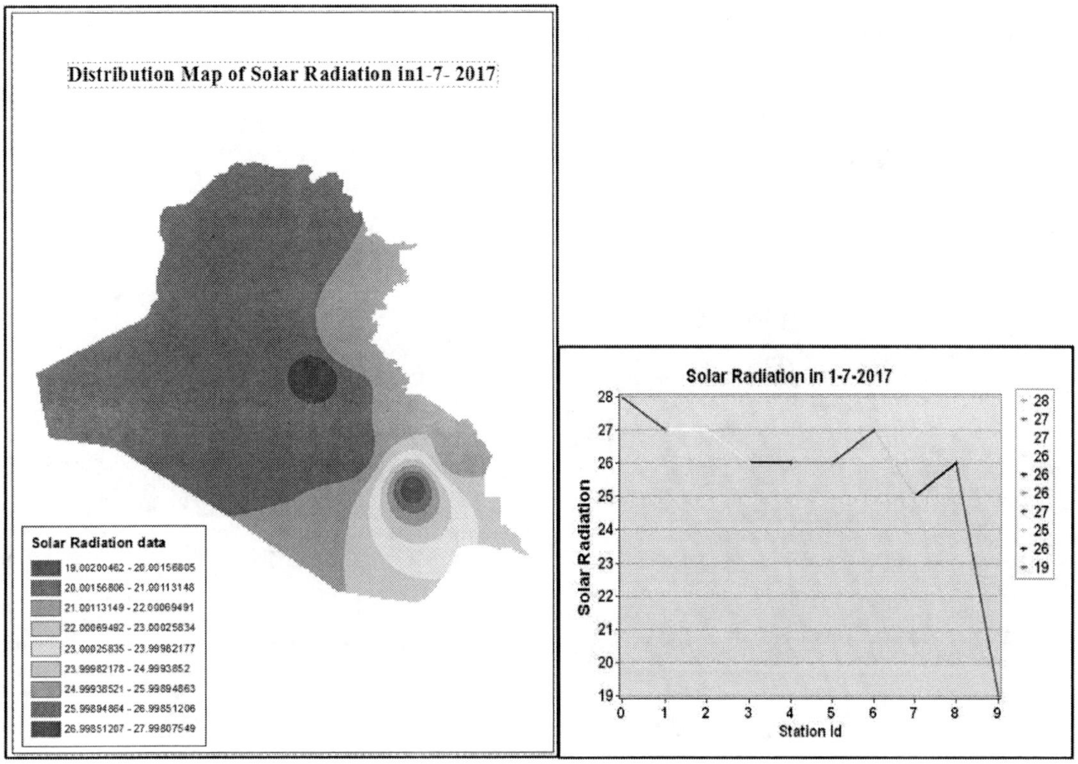

Figure 2. distribution map of solar radiation in 1-7-2017

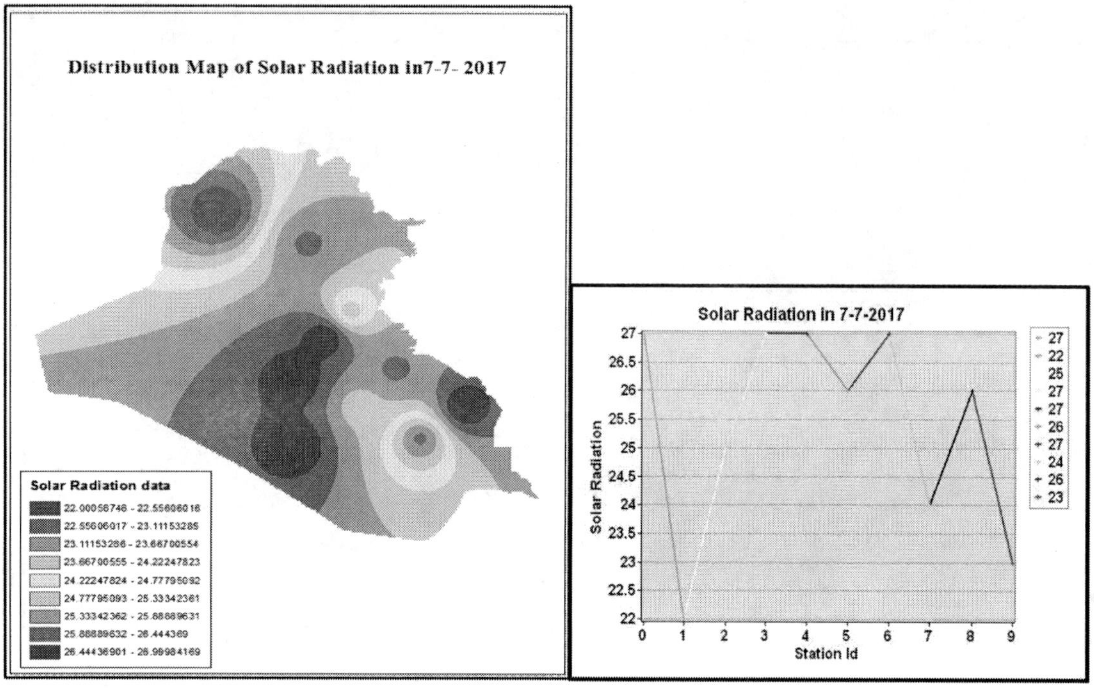

Figure 3.distribution map of solar radiation in 7-7-2017

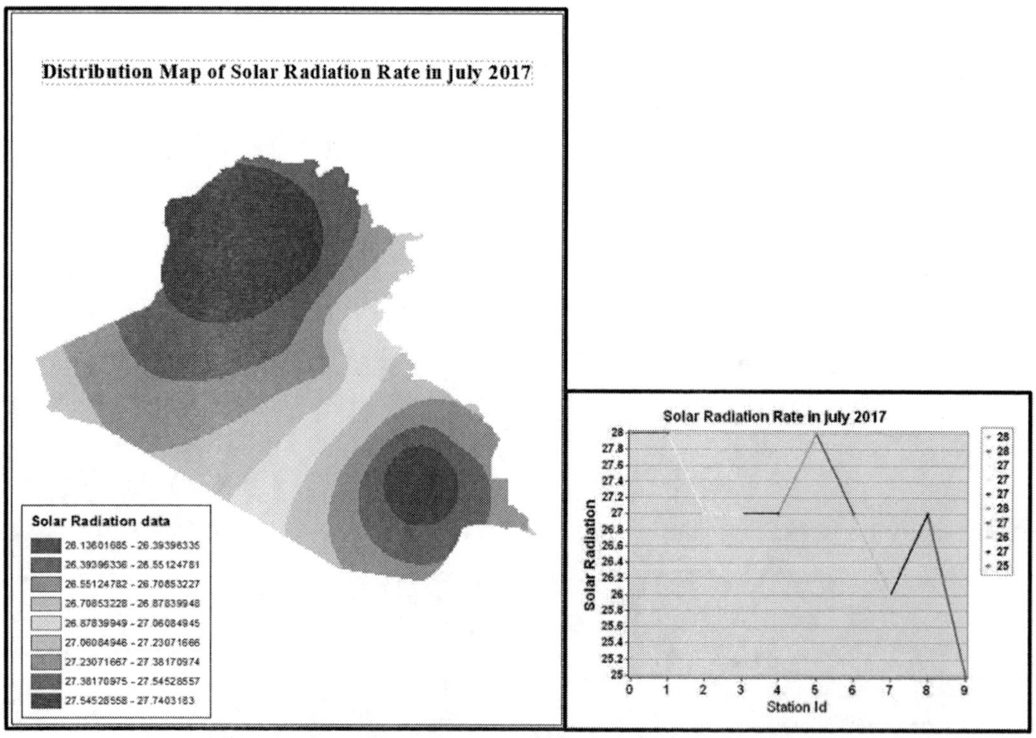

Figure 4. distribution map of solar radiation in July

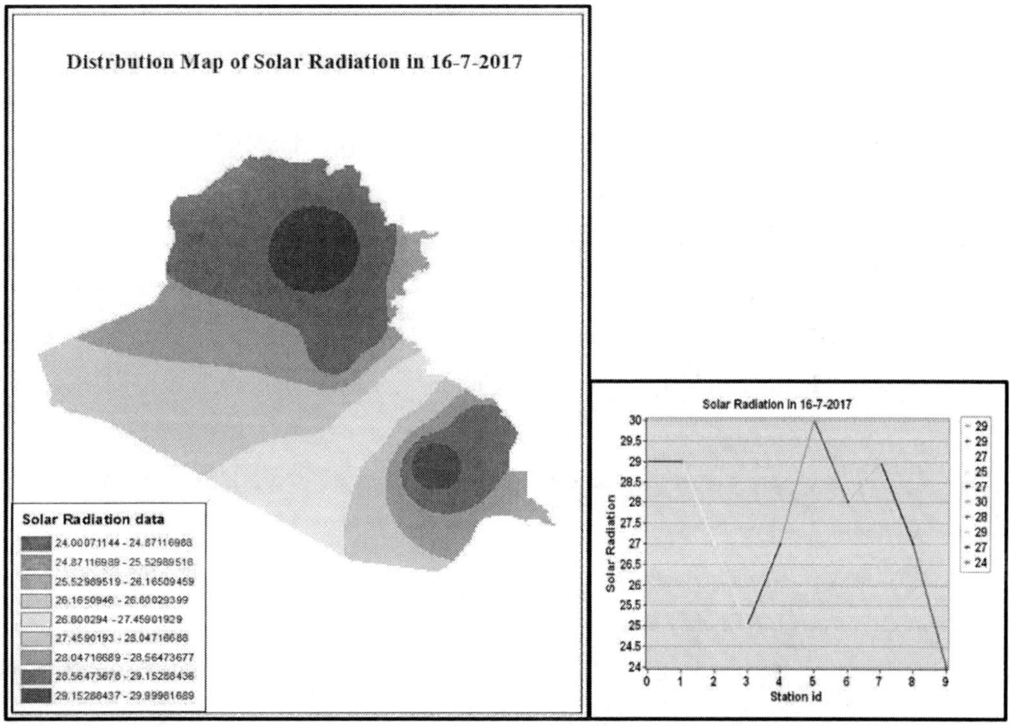

Figure 5. distribution map of solar radiation in 16-7-2017

6.Conclusion

Estimation of solar radiation from measured meteorological variables offers an important alternative in absence of measured solar radiation. The existing and developed models are comparatively studied and evaluated using monthly average daily solar radiation and measured meteorological data, monthly average daily sunshine duration.The most important outcomes of the study can be summarized as :the estimation of global solar radiation can be predicted with reasonable accuracy with all developed models considered in this study, due to its simplicity and better model performance evaluation.Therefore, it can be recommended that these models are used reasonably well for estimating the solar radiation in Iraq and possibly in its surroundings sites with similar climatic conditions if empirical coefficients are correctly calibrated

IOP Conf. Series: Materials Science and Engineering **757** (2020) 012038 doi:10.1088/1757-899X/757/1/012038

References

[1] Rabiu A B., **2012**. "Solar Radiation Models and Information for Renewable Energy Applications, pp:112 -125.

[2] Ahmed A, Hameed Neamah Mohsen Al-Fatlawy and Ali M. Al-Salihi, **2017** " Estimation Of Hourly Global Solar Radiation Incident On Inclined Surfaces In Iraq At Different Sky Condition ", International Journal of Research in Applied Vol. 5.

[3] Spokas and Forcella **2006** " Estimating Hourly Incoming Solar Radiation from Limited Meteorological Data" ,pp:182-189.

[4] Girma Dejene Nage **2018**. "Estimation of Monthly Average Daily Solar Radiation from Meteorological Parameters: Sunshine Hours and Measured Temperature in Tepi, Ethiopia",Vol. 3, No. 1, pp. 19-26, International Journal of Energy and Environmental Science,

[5] Didem VECAN **1940** " Measurement And Comparison Of Solar Radiation Estimation Models For Izmir/Turkey: Izmir Institute Of Technology Case",**2011**

[6] Prescott JA, "Evaporation from water surface in relation to solar radiation. Trans R Soc Austr", pp40-114.

[7] Rietveld& R **1978**. "A new method for estimating the regression coefficients in the formula relating solar radiation to sunshine. Agric. Meteorol. 19", pp 243–252.

[8] Fthenakis, Vasilis M. **2004** "Life cycle impact analysis of cadmium in CdTe PV production" ,Renewable and Sustainable Energy Reviews , vol **8** , pp 303–334

[9] Garnero G. and Godoneb D. **2013** "Comparisons Between Different Interpolation Techniques",The International Archives of the Photogrammetry, Remote Sensing and Spatial Information Science, Volume XL-5/W3.

Annealing temperature effects on the performance of the perovskite solar cells

A. K. Al-Mousoi, M. S. Mehde, A. M. Al-Gebori
Department of Applied Science, University of Technology, Baghdad, Iraq
*Corresponding author email: alialmousoi2006@yahoo.com

Abstract
In this work, perovskite solar cells were fabricated with one-step method by using spin coating technique. These solar cells were in the structure form of FTO/compactTiO_2/meso-porousTiO_2/$CH_3NH_3PbI_3$/Au. Different annealing temperatures were applied to investigate the effect of the annealing temperature on the properties of the fabricated solar cells since it lead to affect the crystallization degree of the $CH_3NH_3PbI_3$ layer. Three temperatures were used for annealing the perovskite layer typically: 60°C, 80°C and 100°C and the best result found at 100 °C where the efficiency was 5.68%, FF = 0.55, J_{sc}= 12.637 mA/cm^2 and V_{oc}=0.817 V.
Keywords: Perovskite solar cells, $CH_3NH_3PbI_3$, Lead Iodide perovskite.

1. Introduction

Nowadays halide organic-inorganic perovskite solar cell attracts a big deal of interest of solar cells researches society since incredibly development have been done in the solar cells efficiency from 3.8% to 22.1% since 2009 [1,2]. Perovskite materials earned great attention as an alternative to silicon solar cell, which took the most predominant position in the current solar cell market, of efficiency around 26% [3]. Perovskite solar cells attract the modern attention specifically from the researchers with experiences in dye sensitized-solar cell or organic-solar cell since various materials could be employed mutually in perovskite and organic-solar cell. The configuration form of perovskite solar cell is similar to that exist in the structure of dye sensitized solar cell [1]. The band gap of perovskite materials is tunable (e.g., lead halide perovskite has a band gap from 1.5eV to 2.3eV) [4] and moderately high absorption coefficient (about $10^4 cm^{-1}$) [5,6], which is closer to other thin film solar cell materials such as CdTe [7] and copper zinc tin sulfide (CZTS) [8]. The relatively low cost and simple fabrication technique play significant role to make a competitive device that made from silicon, which requires complicated and costly high-vacuum deposition methods. The efficiency of Perovskite solar cells that fabricated with spin coating method were found highly dependent on the deposition parameters such as the annealing temperature, spinning speed, spinning time and type of anti-solvent.

2. Experiment section

2.1. FTO glass preparation

The substrates used in this work were FTO glass in size of 1.5cm × 1cm with surface resistivity of 14Ω sq^{-1}, these FTO glass substrates were etched by a distance of 3mm along the substrate length. The etching process was done by using zinc powder and HCl (2M). The zinc powder was placed on the edge of the substrate while the rest of the substrate was covered by insulator tape then HCl was poured on it and waiting for 1 min. then the resulting paste was removed and the FTO glass soaked in deionized water and removing any remaining paste. The processes of substrates cleaning are extremely important since even small quantity of dust or any other material present on the FTO glass will affect the quality of the fabricated film. The cleaning was done by these steps: 1- the substrate were put in mixed of solution of soap and water in 100 watt ultrasonic device for 15min. 2- Immersing the substrate in deionized water and run the ultrasonic for 15min. 3- The substrates were washed in acetone ultrasonically for a period of 15min. 4- The substrates were putting in ethanol and ultrasonically were washed for 15 min. 5-Repeating the point number 2 for the same time. 6- Ultrasonically washing the substrates by immersing it in isopropanol for 15 min. 7-Drying the substrate by using compressed air. 8- Putting the substrate in oven at 100^0C for 15min. 9- Applying UV light to the substrate for 30 min.

2.2. Materials preparation

The chemicals used to prepare TiO_2 are listed in the table below

Table (1): Chemicals that used to synthesis compact TiO_2

Solvent	Stabilizer	Precursor
Absolute Ethanol	Hydrochloric acid (HCl)	Titanium isopropoxide (TIP)

Firstly we prepared a 2M of HCl from 12M HCl by using the relation M1V1=M2V2. Secondly absolute ethanol was put in two glass container each one have 5ml, then 0.635 ml of Titanium isopropoxide (TIP) was added to one container and 0.07 ml of HCl (2M) was added to the other container after that these two container were placed in an ice bath magnetic stirrer for 30 min. and then these two solution were mixed and run the ice bath magnetic stirrer for another 30min then the final solution was placed in the refrigerator. The final solution of compact TiO_2 has a colorless appearance. To prepare the meso-porous TiO_2, 5ml of absolute ethanol was used to solve 1 g of TiO_2 paste then the solution was placed on the magnetic stirrer at room temperature for 12 hours. The final solution was seemed to have a milky like appearance. The $CH_3NH_3PbI_3$ solution was synthesized by dissolving PbI_2 of mass 461mg in 0.635 ml of DMF and 0.071ml of DMSO solution and placed all these materials on magnetic stirrer at 70^0C for 30 min., after leaving this solution to cool to the ambient temperature, 159 mg of MAI was added to the solution and stirred at room temperature for 30 min. lastly we obtained a very clear yellow solution.

2.3. Device fabrication and characterization

The compact TiO_2 was deposited by using spin coater when the rotational speed is 4000 rpm and the time was sit to be 60 s. After the spin coating completed, the samples were put in the oven, at 100^0C for 30min and then annealed at 500^0C for 1 hour and leaved to cool to the ambient temperature for about 4 hours. The deposition of the meso-porous TiO_2 was done in the same condition of the compact TiO_2.

After the completion of the processes, the samples were left to cool to the ambient temperature gradually. The process of deposition of $CH_3NH_3PbI_3$ film was achieved by spin coating with pouring of anti-solvent (diethyl ether) in order to crystalize the perovskite material. After finishing this process the films were annealed in order to improve the crystallization. The deposition of Au was achieved by sputtering. The morphology of the films was examined using MIRA3 TESCAN scanning electron microscope (SEM). ARA AFM was used to study the topography of the films. The x-ray diffraction (XRD) patterns of substrate and perovskite films were obtained using Philips PW 1840 X – ray diffractometer of wavelength $\lambda = 1.5406°A$ from Cu - Kα. A Cecile CE 7200 Spectrophotometer supplied by Aquarius Company was used to record the optical transmission for the films (375-900 nm). Dark current – voltage measurements carried out by applying voltage supplied to the sample from a stabilized D.C. fine power supply, type L 30 – 2 Farnell of range (0.1 – 5)V. The current passing through the device was measured using a Keithley (2400) sourcemeter. The I-V measurements of the solar cells were done under 100mW/cm^{-2} light illumination power densities supplied by a Halogen lamp which is connected to a variable AC power supply and calibrated by Keithley 2400.

3. Results and discussion

All the characterization in this work was achieved to study the $CH_3NH_3PbI_3$ layer since it represents the important layer of the perovskite solar cells. The XRD patterns of the $CH_3NH_3PbI_3$ films formed by spin coating method on FTO glass substrates at different annealing temperatures, and make the other parameters fixed (spinning time = 60s , anti-solvent = Diethyl ether and rotation speed = 4000rpm), are shown in Figure 1. The patterns display seven diffraction peaks at 2θ values of approximately (14°.1′, 20°, 24 °.45′, 28°.4′,40°.45′ and 43°.05′) which correspond to the diffraction plans (110), (112), , (202), (220) , (400) and (411) crystalline planes of the $CH_3NH_3PbI_3$ tetragonal phase respectively. All the X-ray patterns show that the fabricated films are polycrystalline and it is clearly seen that diffraction intensities increased with annealing temperatures, which may due to the increasing of grain size and thickness.

The most preferred orientation found at 2θ values of approximately 14°.1′ and 28°.4′ that corresponding to (110) and (220) diffraction planes respectively, the lattice parameters which corresponding to these values are: a=8.87 A° and c=12.7 A°. Some of peaks appeared in the X-ray patterns that belong to PbI_2 (at 2θ about 12.6) and FTO, the peak of PbI_2 tend to be less with higher annealing temperature due to improvement in the reaction and crystallization degree. The intensities of the minor peaks are decreased with increasing of the annealing temperature as shown in Figure 1 [9]. The lattice constant (a and c) are calculated using the relation $\frac{1}{d^2} = \frac{h^2+k^2}{a^2} + \frac{l^2}{c^2}$ and $d sin\theta = n\lambda$ where hkl are miller indices, d is the inter planar spacing, n is a positive integer (usually equal to 1), λ is the x-ray wavelength and θ is the diffraction angle.

The close calculated value to the standard lattice constant (a_o≈8.8A° and c≈12.6 A°)[10] found at annealing temperature of 100°C.

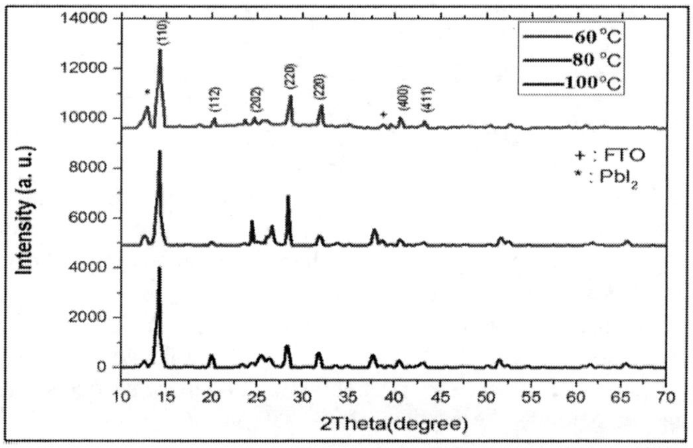

Figure 1: X-ray patterns of the $CH_3NH_3PbI_3$ films at different annealing temperatures: 60°C, 80°C and 100°C.

The surface morphology of films has been investigated using AFM images, which produces images of surfaces at very high magnification. Figure 2 shows the topography of the $CH_3NH_3PbI_3$ film deposited at different annealing temperature which be considered a significant parameter that influence the formation of the perovskite in one step method. The surface morphology has different shape for each perovskite film that prepared with a certain temperature. It is obvious that the particles, on the surface of the film, have different diameter values, however the $CH_3NH_3PbI_3$ film deposited at 60° C has average grain size of about 160nm, also the geometry of the surface has a high degree of regularity. As the annealing temperature increases the grain size will increase as obvious from images of samples prepared with annealing temperature of 80° C and 100° C which found to have average grain size of 187nm and 204nm respectively.

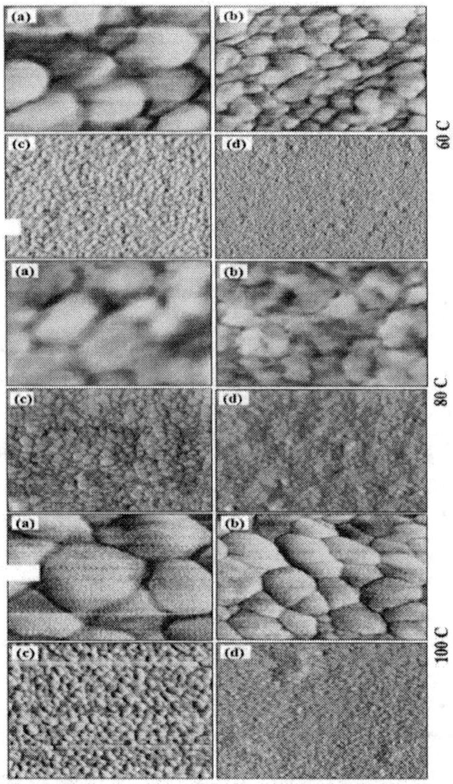

Figure 2: AFM images of $CH_3NH_3PbI_3$ film prepared with different annealing temperature, the letters here refer to the image scale a) 0.5 µm , b)1µm ,c) 5 µm and d) 10 µm

Figure 3 shows the SEM images of samples prepared with different annealing temperature to study the surface morphologies. It is clearly obvious that all samples are highly dense with full coverage. The surface of sample prepared with annealing temperature of 60 °C is not very smooth and its surface homogeneity is not uniform, this may due to less temperature established to fully formation of the perovskite layer, however, the average particle size is about 51nm and these particles found to be in group. By increasing the annealing temperature the surface of the samples begin to have a smooth and more homogeneity and the particles becomes to assemble in bigger crystals as shown in Figure 3. The average particle size for the samples prepared with annealing temperature 80 °C and 100 °C are 178 nm and 208 nm respectively.

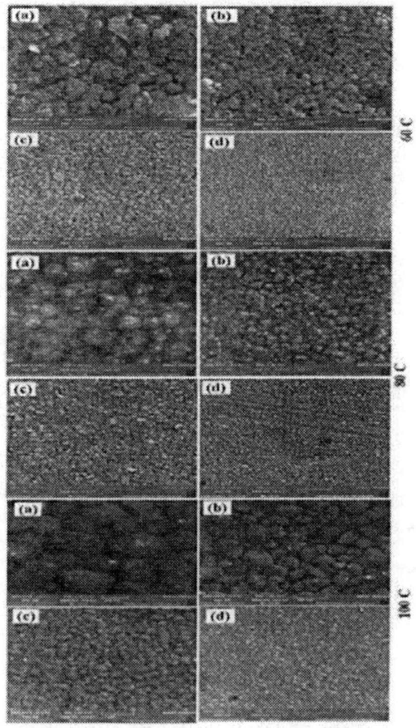

Figure 3: FE-SEM images of $CH_3NH_3PbI_3$ film prepared with different annealing temperature, the letters here refer to the image scale a) 0.2 μm , b) 0.5μm , c) 1 μm and d) 2 μm.

The transmittance spectra of the $CH_3NH_3PbI_3$ films deposited at different annealing temperatures are shown in Figure 4. We notice that the films have high transmission at long wavelengths approximately (60 – 70 %), and decreasing transmission to (20-35%) at short wave lengths (typically at about 750nm wavelength which corresponding to approximately 1.5 eV of energy gap).

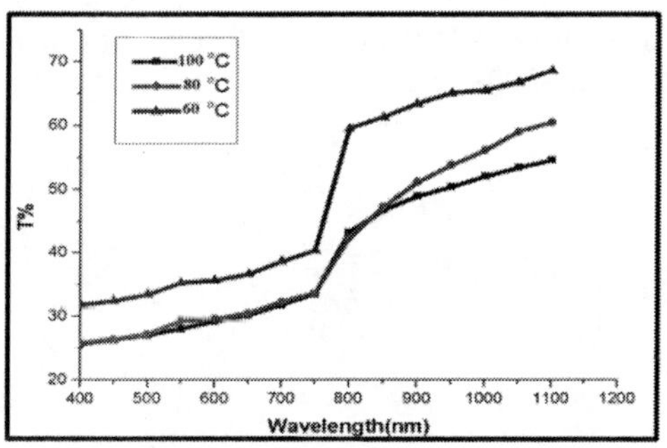

Figure 4: Transmittance as a function of wavelength for $CH_3NH_3PbI_3$ thin films deposited at different annealing temperatures.

It was observed that the transmission of the film decreased with increasing in annealing temperature, this may due improvement of the crystallization and orientation degree and.
Figure 5 shows the optical energy gap estimated by using Tauc plot, As expected the increasing the annealing temperature will lead to increase the value of energy gap due to crystallization improvement, the value of E_g of the sample annealed with 100° C is about 1.51 eV whereas there is a slightly difference in samples annealed with temperature of 80° C and 60° C which each has value of E_g about 1.5 eV and 1.25eV respectively, this differences of E_g of samples prepared with different annealing temperature may attributed to the different perovskite formation for each temperature.

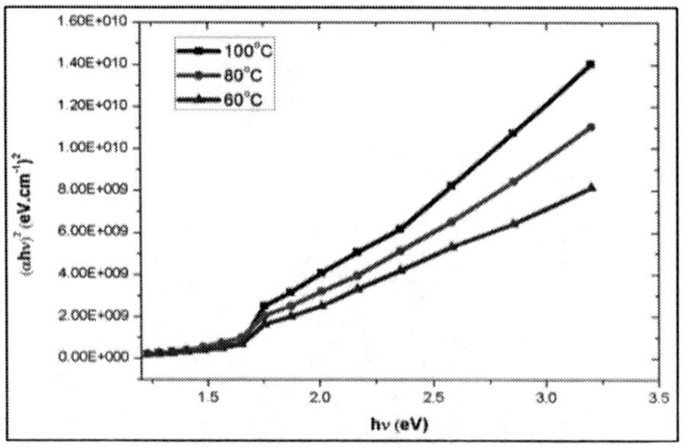

Figure 5: Shows the estimated energy gap of $CH_3NH_3PbI_3$ thin films deposited at different annealing temperature.

It can be noticed from the Figure 6 that the junction exhibits rectifying behavior, this rectifying behavior is attributed to heterojunction potential barrier at the $TiO_2/CH_3NH_3PbI_3$ interface. The formation of the heterojunction structure is referred to the difference in energy gap between the TiO_2 and $CH_3NH_3PbI_3$ [11]. Some of electrical characteristics of $TiO_2/CH_3NH_3PbI_3$ sandwich structure can be calculated from dark I-V measurements, such as ideality factor (n) and saturation current density (J_s). Both I_s and n provide important information on the predominant current transport mechanism in a device. There are a number of ways to obtaining these parameters, although the most consistent method involves their determination using the dark J-V curves. The slope of the J-V curve gives the n, whereas the y-intercept gives the J_s. They are usually considered together as there is great deal of correlation between these two parameters. It is desirable that J_s be as low as possible, so that higher voltages are required for the dark current to equal the light generated current. This increases the V_{oc} of the devices. A lower value of n will lead to reduction in V_{oc} value but that will help to increase the fill factors of the devices.

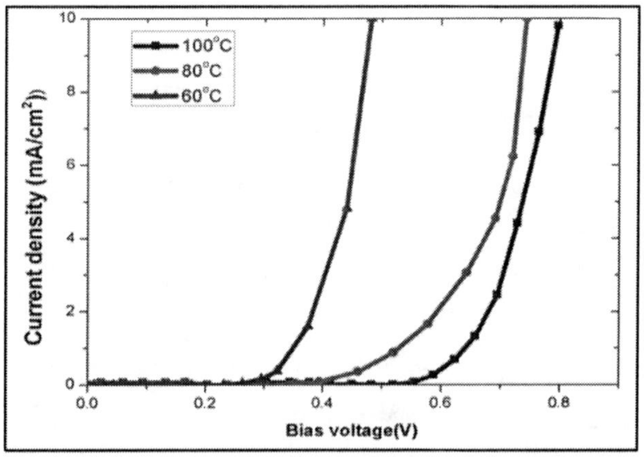

Figure 6: Dark I-V characteristics of FTO/c-TiO_2 /mp-TiO_2/$CH_3NH_3PbI_3$/Au solar cell at different annealing temperature of $CH_3NH_3PbI_3$ film.

Table (2): describes the ideality factor and the saturation current of TiO_2 / $CH_3NH_3PbI_3$ /Au solar cells prepared at different parameters of the absorber layer ($CH_3NH_3PbI_3$).

Parameters	Values	Ideality Factor	Saturation current (μA)
Annealing Temperature(°C)	60	3.47	2.71
	80	3.05	2.06
	100	2.18	0.731

The (J–V) characteristics of the solar cells under illumination of intensity power of 100 mW/cm^{-2} with FTO/c-TiO_2 /mp-TiO_2/$CH_3NH_3PbI_3$ /Au structure, at different annealing temperature of $CH_3NH_3PbI_3$ film, are shown in Figure (7). The effect of the annealing temperature on the solar cell parameters is very clear as Table (3) shows that, from the table one can notice that the increase in the J_{sc} and V_{oc} with increasing the annealing temperature, this may be due to improvement in the formation of the perovskite layer and this will cause high absorption of light which lead to high photocurrent generated. The efficiency of the solar cell is basically affected by series and shunt resistance which are established by the resistance of contact and leakage at the edge of device, the lower value of R_s and higher value of R_{sh} found to be in a solar cell prepared at annealing temperature of 100°C, the lower efficiency of the solar cell that fabricated at 60°C and 80°C may results from the incompleteness of the perovskite layer.

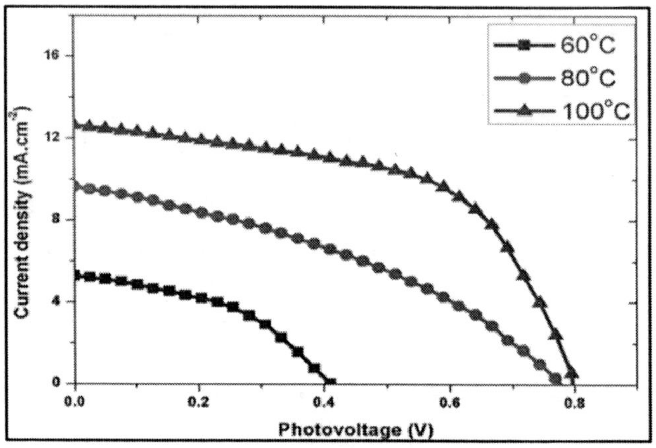

Figure 7: J-V characteristics of the solar cells under illumination of 100 mW/cm^{-2} with different annealing temperature.

Table 3: shows the parameters of the fabricated solar cells that prepared with different annealing temperature.

Temperature (°C)	V_{oc}(V)	J_{sc}(mA/cm^2)	FF	PCE (%)	R_s (Ω.cm^2)	R_{sh} (Ω.cm^2)
60	0.423	5.277	0.429	0.958	32.98	189.5
80	0.781	9.647	0.368	2.782	36.17	194
100	0.817	12.637	0.556	5.684	9.6	315.73

4. Conclusions

The different annealing temperatures play an important role on the quality of the perovskite films and lead to different influence of the fabricated solar cells. The best efficiency of the solar cells found at annealing temperature of 100°C. AFM and SEM analysis showed that there is big effect on the structure and morphology of the perovskite film that prepared with different annealing temperature; it is found that the grain size becomes larger when the annealing temperature increases. From the dark I-V measurements, the ideality factor and the saturation current will decrease with increase the temperature of annealing.

IOP Conf. Series: Materials Science and Engineering **757** (2020) 012039 doi:10.1088/1757-899X/757/1/012039

Finally the efficiency of the solar cell fabricated with annealing temperature of 100°C was found to have the maximum value among the other solar cells.

References

[1] Kojima A, Teshima K, Shirai Y, Miyasaka T 2009 Organometal halide perovskites as visible-light sensitizers for photovoltaic cells. *J. Am. Chem. Soc.* **131** 6050-6051
[2] Saliba M, Matsui T, Seo JY, Domanski K, Correa-Baena JP, Nazeeruddin MK Zakeeruddin SM, Tress W, Abate A, Hagfeldt A 2016 Cesium-containing triple cation PSCs: Improved stability, reproducibility and high efficiency. *Energy Environ. Sci.* **9** 1989-1997
[3] Yoshikawa K, Kawasaki H, Yoshida W, Irie T, Konishi K, Nakano K, Uto T, Adachi D, Kanematsu M, Uzu H 2017 Silicon heterojunction solar cell with interdigitated back contacts for a photoconversion efficiency over 26%. *Nat. Energy* **2** 17032
[4] Jeon NJ, Noh JH, Yang WS, Kim YC, Ryu S, Seo J, Seok SI 2015 Compositional engineering of perovskite materials for high-performance solar cells. *Nature* **517** 476-480
[5] Park NG 2015 PSCs: An emerging photovoltaic technology. *Mater. Today* **18** 65-72
[6] Green MA, Baillie AH, Snaith HJ 2014 The emergence of PSCs. *Nat. Photonics* **8** 506-514
[7] Rangel-Cárdenas J, Sobral H 2017 Optical absorption enhancement in CdTe thin films by microstructuration of the silicon substrate. *Materials* **10** 607
[8] Shi Z, Jayatissa AH 2017 One-pot hydrothermal synthesis and fabrication of kesterite Cu2ZnSn(S,Se)4 thin films. *Prog. Nat. Sci. Mater. Int.* **27** 550-555
[9] Leo K 2015 Perovskite photovoltaics: signs of stability. *Nat. Nanotechnol.* **10** 574-575
[10] Baikie T, Fang Y, Kadro JM, Schreyer M, Wei F, Mhaisalkar SG 2013 Synthesis and crystal chemistry of the hybrid perovskite (CH3NH3) PbI3 for solid state sensitised solar cell applications. *J. Mater. Chem. A.* **1** 5628
[11] Thjeel HA, Suhail AM, Naji AN, Al-Zaidi QG, Muhammed GS, Naum FA 2012 Fabrication and Characteristics of Fast Photo Response ZnO/Porous Silicon UV Photoconductive Detector. Adv. Mater. Phy. Chem. **101** 70-77

IOP Conf. Series: Materials Science and Engineering **757** (2020) 012040 doi:10.1088/1757-899X/757/1/012040

Hybrid SOLAR / DIESEL Low Power Generation Performance System for Local House in Baghdad City

[1]Ali S. Allw, [2]Falah I. Mustafa*
[1]Energy Engineering Dept, Engineering College Almusayb, Babylon University
[2]Solar Energy Research Center, Renewable Energy Directorate
[1,2]Higher Education and Scientific Research Ministry, Baghdad, Iraq
alihtallw@yahoo.com, falah_im@yahoo.com

Abstract. The national electrical energy situation report study today in Iraq represents that the value of the conventional electrical energy power is increasing day after day with increasing more demand on the electricity source in all sectors of life in Iraq. To avoid energy supply shortage, today should be using renewable energy sources with rise required management. A hybrid electrical energy system usually consists of two or more renewable energy sources used together with conventional electrical energy sources to extend increased system efficiency, but the design of hybrid power involves uneconomic and excessive capacities results in excess electric generation with increasing the initial costs. In this study, we find the optimization the best time required of sharing alternative between both of solar-diesel generator hybrid power system for 24 hours. The PV peak power was 1.3kW and diesel power 2.4kW with using both of theoretical calculation. The HOMER software was used to improve a simulation model for one day in Baghdad city in Iraq. The HOMER software is used to implement the optimization. The master objective is to optimize hybrid system component sizes, minimizing excess electricity generation, and set electric load and implement cost analysis based on life cycle price. Inputs data used for simulation consist of real solar radiation data (Averaged to monthly values), taken from the local weather station installed at Baghdad city. Based on maximum load, hybrid system parameters sizes considered and many simulation results are acquired.

Keyword: Hybrid system, Renewable energy, Homer software, solar energy, Efficiency.

1. Introduction

Hybrid renewable energy systems (HRES) is very common as stand-alone power system to provide national electric power in remote areas by reason of development in new renewable energy technology and high cost of all products produced from petroleum. The hybrid energy system typically involves two or more sources of renewable energy which were applied together to achieve high power system efficiency along with better balance in powered supply. Typically the hybrid energy systems involve two or more energy source; solar, wind, and hydro energy sources. Most of the renewable energy can be found including different forms of solar energy as photo voltage and thermal energy techniques [1-3].

A system that has all these types of energy sources have the advantage of the constancy and balance. The present study was performed by applying HOMER software which has been developed by the National Renewable Energy Laboratory (NREL). This software performs economic analysis of hybrid power systems [4]. Inputs to HOMER will provide an hourly simulation for each possible combination of components which were entered and the obtained results according to the system according to total Net Present Cost.[9-5]

In this paper, the simulation of a hybrid power generation system consists of DC electric power from PV panels and AC electric power from Diesel generator with Battery storage has a modeled and power management strategy. The system was designed and simulated by using two methods, PVGIS website and HOMER software simulation with comparison in the theoretical equation to check and prove the output data is the same for all in the same condition with our facts [10-11].

2. System Description

The hybrid power system was designed based on required peak load of present location in Baghdad city and to optimize the size and cost of components effectively. While the designing of power system solar radiation, load profile, size and cost considered components of the system were discussed in the next sections. The system design consists of the 2.4kW Diesel power generator (DG) and 1.3kW peak PV solar energy with charger and battery bank storage with an inverter from DC to AC power [12-15]. The block diagram of the Hybrid Solar Diesel Power System (HSDPS) as shown in figure 1.

2.1. Solar radiation Data

In this work, solar radiation was collected for two sources Davis instruments type Vantage Pro2 and PVGIS website. (Longitude 44.21° E, Latitude 33.21° N) the location from weather station installed at the ministry of science and technology, Jadryia, Baghdad of the year 2017, where data has been measured to monthly values to estimate the design of hybrid PV/Battery/DG power systems. First, we begin in Global Horizontal solar radiation for Iraq and especially in Baghdad city as shown in figure 2. Then Table 1. Shows monthly mean daily solar radiation ranges from 4.28 to 7.23 (kWh/m2/d), the total annual average radiation was 5.9kWh/m2/day and the monthly average solar radiation have maximum value in July 7.23 kWh/m2/day compared January 4.24 kWh/m2/day show in figure 3.

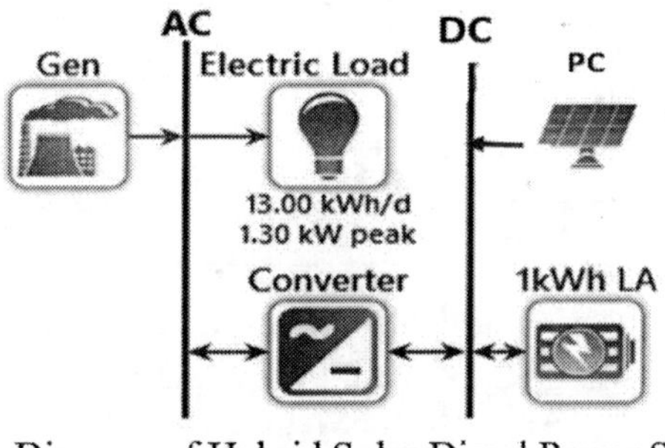

Figure 1. Block Diagram of Hybrid Solar Diesel Power System (HSDPS)

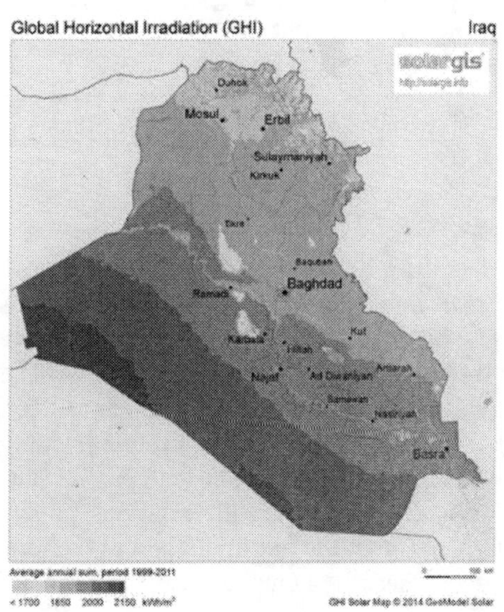

Figure 2. Iraq Map of Global Horizontal Irradiation (GHI)

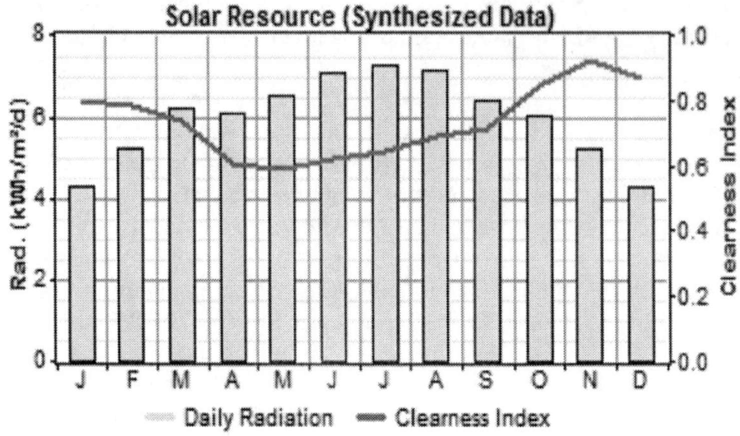

Figure 3. Monthly Average Solar Radiation in Baghdad city

Table 1. Annual average Radiation values in Baghdad city

Month Baghdad	Clearness Index	Average Radiation
		($kWh/m^2/day$)
Jan	0.798	4.24
Feb	0.787	5.24
Mar	0.738	5.7
Apr	0.607	6.1
May	0.586	6.51
Jun	0.617	7.10
Jul	0.641	7.23
Aug	0.687	7.15
Sep	0.715	6.40
Oct	0.843	6.04
Nov	0.920	5.19
Dec	0.868	4.30

2.2. Load Profile

Load is an important factor for any power generation system. HOMER software optimizes the system parameters based on load information of load profile. In this research load data have been provided for a simple Iraqi house by ministry of Electricity with the department of distribution company in Baghdad by consideration through meter reading instrument (MRI), the average of each 24 hour of the day in unit of kW. So that daily measured consumption is considered as 13 kWh/d in this study. The peak load was 1.3KW utilize to design the simulation model and it decides the size of system components as shown in figure 4. Shows Daily average load profile. The annual load is also a significant condition, for that HOMER measured seasonal load variations by simulating 12 months. The season load profile for the whole year shows that the minimum and maximum monthly average load is shown in figure 5 and Table 2. Represents the daily components of load demand.

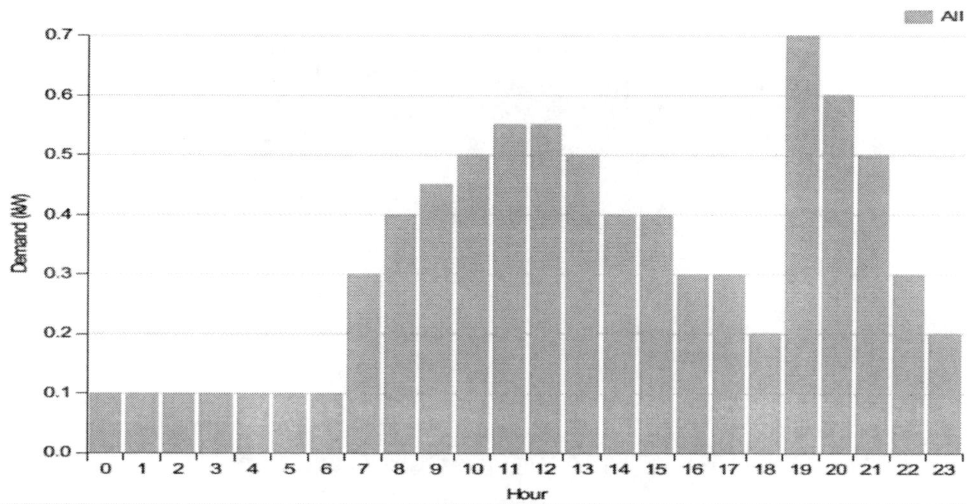

Figure 4. Daily Demand load profile

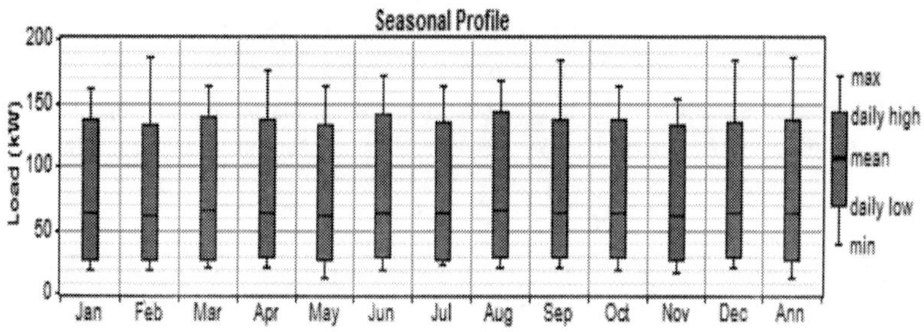

Figure 5. Annual load profile

Table 2. Daily components load demand.

no	name	Consumption Power W	Quantity	Load kW	Hour/ day	kWh/ day
1	Lighting	40	5	0.2	10	2
2	Fans	50	4	0.2	10	2
3	Refrigerator	250	1	0.25	24	6
4	TV	150	1	0.15	8	1.2
5	cooler	200	1	0.2	5	1
6	computer	100	1	0.1	6	0.6
7	printer	200	1	0.2	1	0.2
total				1.3		13

2.3. Diesel Generator

The diesel generator has mainly been used to provide power operates at night time, is of AC type. The generator have capacity 2.4kW is used in this study (China). The capital cost was considered is $200. The lifetime of the generator is rating at 5000 operating hours and the price of diesel is 0.35 $/liter this is the price in Baghdad Iraq. In the figure 6 shows that the working time table of hybrid Solar/Battery/diesel generator power systems that each of them working every day for one year appears that generator schedule depends on daylight and night hours that's mean working diesel 8 hours every day with PV solar 8 hours solar energy (from 8 clock morning time to 4 clock evening time) and 8 hours when used Battery bank storage.

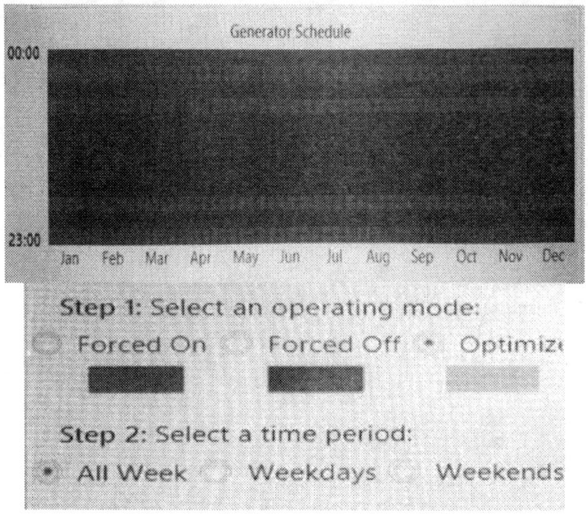

Figure 6. Generator Schedule

2.4. PV Solar Generator

The photovoltage modules type Silicon polycrystalline efficiency 17% the price here around 0.4$/W in Iraq with many companies specialize in Baghdad for solar energy components cost in the renewable energy market. That's mean the cost of 1.3kW peak power of PV price around 500$.

2.5. Converter

The converter function is rectifying like an inverter. Since 2kW output would be generated for designing simulation model with capital cost is assumed to be $200. The Lifetime of the converter is measured at around 10 years and an efficiency of 95% (inverter), 90% (rectifier).

2.6. Battery

The types of storage battery is a gel with capacity of 12V, 100Ah. Batteries used to provide about one hour of electricity, and arranged such that any string consist of two batteries, with total strings equal to six. Using 12 units of total batteries and use nominal voltage equal to 12V. This battery bank has the ability to provide about one hour of electricity.

3. Theoretical Calculations

To find the PV sizing system off-grid with peak load 1.3kW that mean the number of Module and number of Battery for 6A and three hours' time storage with using solar module 12V and battery 12V voltage with 200Ah capacity as show in equations (1-9). The peak sun hours in Baghdad Iraq from our calculations we found the P.S.H=5.9 and the efficiency of both of inverter, charger and battery, where:

$$\eta inv = \eta c\eta harg = \eta Batt = 0.9 \tag{1}$$

The safety factor=1.3.

The energy demand of load every day equal power load multiple hours of operation per day: that means,

$$El = (Energy\ of\ Load) = Power\ Load * Operation\ hours/day \tag{2}$$

$$El = PL * \frac{h}{Day} = 220V * 6A * \frac{3h}{day} = 3960Wh/day \tag{3}$$

$$Epv = Energy\ of\ Module = \frac{EL}{(\eta inv * \eta char)} \tag{4}$$

$$Epv = \frac{3960}{0.9*0.9} = 4888Wh \tag{5}$$

IOP Conf. Series: Materials Science and Engineering **757** (2020) 012040 doi:10.1088/1757-899X/757/1/012040

$$Ppv = \frac{(power\ of\ PV\ Module)}{(Peak\ Sun\ Hour)} = \frac{EL}{PSH} = \frac{4888}{5.9} = 828W \quad (6)$$

$$Actual\ Ppv = Ppv * 1.3(safety\ factor) = 828 * 1.3 = 1077 \cong 1100Wp \quad (7)$$

That means we required number of Modules equal to around ~ 1100 W that choosing 11 modules to have power 100watt, that means 11 Module connected in parallel or 120W, 9 panels connected in parallel depending on the designer.
The capacity battery required for storage power from the PV and diesel generator in a day when working that is:

$$Capacity\ Battery(C.B) = \frac{EL}{\eta Batt * DoD * Vbatt} = 3960/(0.9 * 0.7 * 12) \quad (8)$$

$$C.B = \frac{4400}{7.56} = 523Ah \sim 600Ah \quad (9)$$

That means we required about three numbers of Battery each capacity 200Ah with a total of three batteries connected in parallel each 12Volts.

4. Results and Discussion

The total price of complete PV solar / Diesel power generator system in this work consists of these items with specification as shown in table 3.

Table 3. Specification items with price list

Item, No.	power	price
PV panels, 10pics	1.3KW	500$
Gas Generator, 1	2.4KW	200$
Inverter , 1	2KW	200$
Charger , 1	50A	100$
Battery, 2	12V,200Ah	600$
Cables	6A, 220VAC	100$
structure	metal	100$
total		1800$

The hybrid power system HOMER software which is calculates each possible design of the present local location. It compose of PV array, diesel generator, converter, and batteries which were presented in the Performed simulation. In normal operation, PV supplies the load requirement. The exceeding energy of PV is stored in the battery until the capacity of the battery is full reached. The diesel generator is used as a backup and in night day, in the case when provide from PV is unavailable. The converter is used to convert DC to AC (as an inverter) and AC to DC (as rectifier or charger). After that we have two cases for simulation of power generation of PV / Diesel 1.3kW hybrid power generation performance system for a local house in Baghdad city؛

Case 1. HOMER simulation; first of all the HOMER software working of all the scale of low, medium and large power generation scale with here used around 1-2 kW, it's difficult to make simulation analyses for low scale. So we made more suggestions to analyses with adding another generator to give our study more choosing items. The results of the cost summary as shown in figures 7, 8, 9. The cost summary represents the t

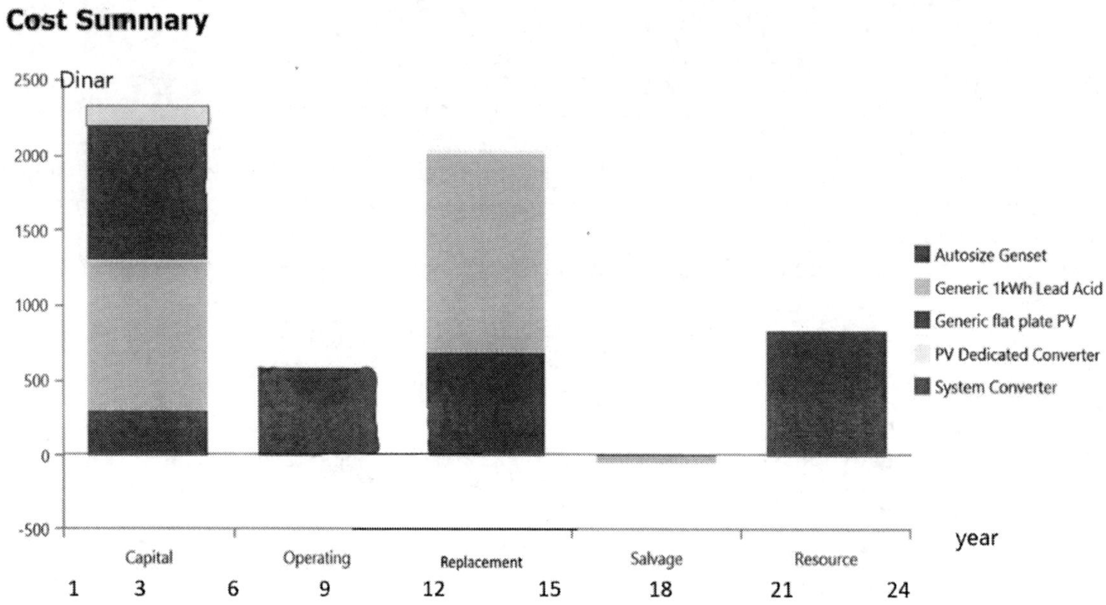

Figure 7. Cash flow analysis of NPC of the optimized hybrid energy system

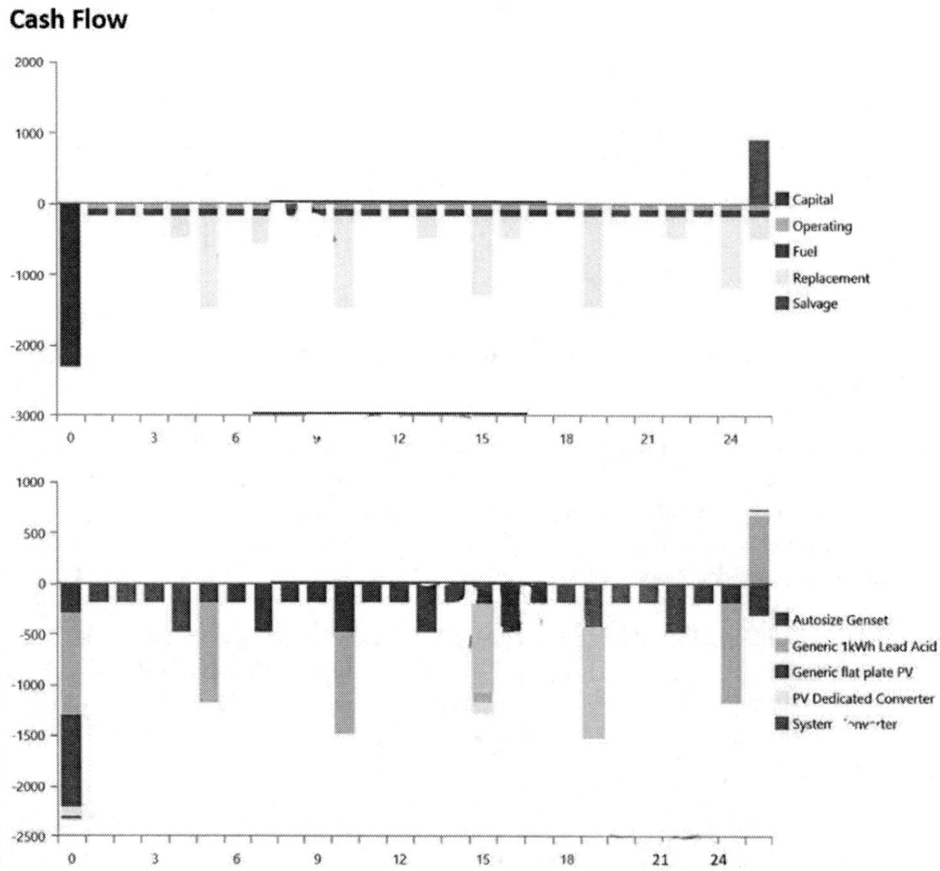

Figure 8. Cash flow

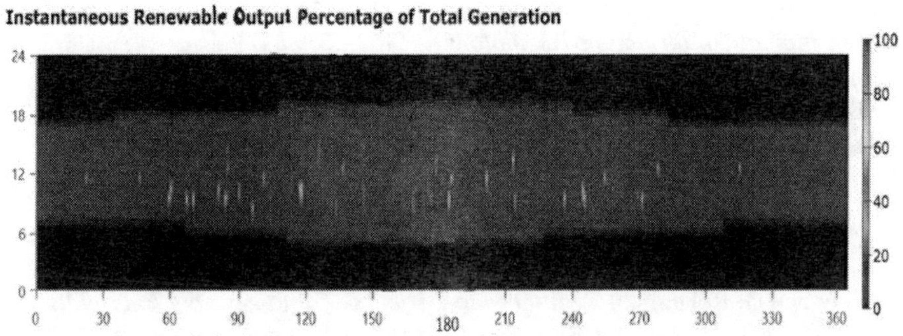

Figure 9. Instantaneous Renewable Output Percentage of total Generation

Case 2. Theoretical calculation;
In the theoretical calculation, it has put the total loss factor equal to 1.3 as a safety factor and calculated the number of PV power need for a local house in Baghdad we have found we need about 1100W and three batteries to build the hybrid power generation system.
From all simulation, we have found the difference not too much because the analyses between two systems different in basic, so any designer will go directly to Homer because more accurate in calculations.

5. Conclusions

In this study optimization of sharing alternative between both of solar-diesel generator hybrid power system, the PV peak power was 1.3kW and diesel power 2.4kW with using both of theoretical calculation and HOMER software were used to develop a simulation model for Baghdad city in Iraq. The number Module has 11 Modules connected in required about three numbers of Battery each capacity in parallel with three battery capacity has 200Ah connected in parallel each 12V. The total price of the complete PV/ Diesel power generator system was 1800$ USD. The saving energy has done about 30% every day. Hybrid Optimization Model for Electric Renewable (HOMER) software is used to implement the optimization. The main goal is to optimize hybrid system component sizes, minimizing excess electricity generation, and set electric load and implement cost analysis based on life cycle cost of sharing alternative between both of solar-diesel generator hybrid power system.

Acknowledgment
The authors would like to express the deep sense of appreciation to Solar Energy Research Center / Renewable Energy Directorate / Higher education and scientific research Ministry, for their invaluable support and constructive directions.

References

[1] Ahmed M. Kassem, Khairy Sayed, Emad El-Zohri, Amal M. Abd el-Hamed, 2015, *IOSR Journal of Electronics and Communication Engineering (IOSR-JECE)*.Volume 10, Issue 6, Ver. II, PP 63-70
[2] M. H. Albadi, R. S. Al Abri, M. I. Masoud, K. H. Al Saidi, A. S. Al Busaidi, A. Al Lawati, K. Al Ajmi,and I. Al Farsi, 2014, *International Journal of Smart Grid and Clean Energy,* vol. 3, No. 4, pp. 402_409
[3] B Kailash Krishna Prasad and K Rajesh, 2015, *European Journal of Advances in Engineering and Technology*, 2(6), pp 24-28.
[4] Kenneth E. Okedu*‡, Roland Uhunmwangho, 2014, *international journal of renewable energy research*, kenneth e. okedu et al. ,vol. 4, no. 2
[5] M. Alabdul Salam et al., 2013, *International Journal of Renewable Energy Research*, vol.3, no.2, pp. 301 – 307.
[6] HOMER, 2010, *National Renewable Energy Laboratory (NREL),* USA, V2.68.

IOP Conf. Series: Materials Science and Engineering **757** (2020) 012040 doi:10.1088/1757-899X/757/1/012040

[7] A. Al-Karaghouli, L.L. Kazmerski,, 2010, *Solar Energy*, Vol.84, pp. 710-714.
[8] Mohit Bansal, D.K.Khatod, R.P.Saini, 2014 *International Conference on Reliability, Optimization and Information Technology - ICROIT, India,* Feb, 2014, pp. 6-8.
[9] U. Suresh kumar, Dr. P.S. Manoharan, A.P.S. Ramalakshmi, 2012. *IEEE- International Conference on Advances in Engineering, Science and Management (ICAESM)* pp.30-31.
[10] A. V. Anayochukwu and E. A. Nnene, 2013, *Electronic Journal of Energy and Environment*, vol. 1, no. 1, pp. 57–70.
[11] S. M. Hussain and D. K. Sharma, 2014, *in Proceedings of the IOE Graduate Conference*, Oct. pp.420–428.
[12] R. Dufo-Lopez and J. L. Bernal- August, 2005, *Solar Energy,* vol. 79, no. 1, pp. 33–46.
[13] M. Nour and G. Rohani, 2014, *International Journal of Renewable Energy Research*, v4, no. 3
[14] ECOWAS, 2013, Center for Renewable Energy and Energy Efficiency (ECREEE), *HOMER Software for Renewable Energy Design*.
[15] K.R. Ajao, O. A. Oladosu and O.T. Popoola,2011, *IJRRAS*, vol. 7, no. 1, 2011, pp. 96-102.

IOP Conf. Series: Materials Science and Engineering **757** (2020) 012041 doi:10.1088/1757-899X/757/1/012041

A Novel Software Tool for Sizing Off-Grid PV System and PV/Diesel Hybrid System

Saja Majeed Hashim[1], Osamah Fadhil Abdulateef[2] and Falah I. Mustafa[3]

[1]College of Engineering, University of Baghdad, Baghdad, Iraq
[2]Al-Khwarizmi College of Engineering, University of Baghdad, Baghdad, Iraq
[3]Ministry of science and technology, solar energy research center, Baghdad, Iraq
E-mail:s.hashim1209@coeng.uobaghdad.edu.iq, drosamah@kecbu.uobaghdad.edu.iq, falah_im@yahoo.com

Abstract. The design of graphical user interface programming is for sizing the components of a stand-alone PV system and PV-Diesel hybrid power system based on Iraq conditions. This software program presents as a guideline tool for photovoltaic system integrated depends on the input data by the user to give correct results according to the weather condition of solar radiation of Iraq, with less time and effort. A stand-alone solar system design model applied firstly, on a software program is to provide the load with energy 300Wh/day. Secondly, as an experimental application, to identify the appropriate factors that must be entered in the program from average solar irradiation and safety factor, by test the performance of the photovoltaic solar system, where the generation energy rate by the solar module was 381Wh and the consumption energy rate was 250.33Wh during the working period. The system will suffer from a lack of power generation if solar irradiation is not high enough or in case of weather conditions change.

Keywords: Renewable energy, Stand-alone PV System, Hybrid PV-Diesel System, Design of Solar Power System, Solar System Design Program.

1. Introduction

Iraq is located in the Middle East between latitude (29° 5' and 37° 2' N), and longitude (38° 45' and 48° 45' E) [1]. Iraq receives more than 3000 hours of solar radiance per year in Baghdad [2]. The increase in the percentage of air pollutants and their negative effects led to the reliance on renewable sources of energy [3]. Solar energy is the most reliable and important energy resources to compensate for the restricted availability of the conventional energy source. Solar energy can be directly converted to electrical energy via photovoltaic (PV) technology and is considered as an environmentally friendly source of electricity. A standalone power system is independent on the main grid electricity .PV system is an interdisciplinary field, which involves a large variety of applications and can protect urban and rural areas from pollution by avoiding CO_2 emissions. Generally, a standalone photovoltaic solar system majorly included: photovoltaic solar modules, DC-DC solar charge controller, DC-AC inverter, storage batteries, loads, and other accessories like cables, frame for solar modules, connectors, etc., the previous components with adding diesel generator as a backup source will be a hybrid PV-Diesel system, see Figure 1 [4]. Using a diesel generator as a standby source will make utilization of hybrid systems more attractive [5]. The hybrid PV-Diesel system design provides power to the loads and to charging batteries when the solar irradiation is not available and the battery storage system is low and incapable to provide the loads [6]. There are many software and simulation programs is designed and most of these programs are complex and needs time to learn [7, 8]. In this paper, a new software program is designed to facilitate the sizing process of the system components (modules, storage, charge controller, DC-AC inverter, and diesel generator (DG), don't need more time to learn and more compatible with Iraq conditions.

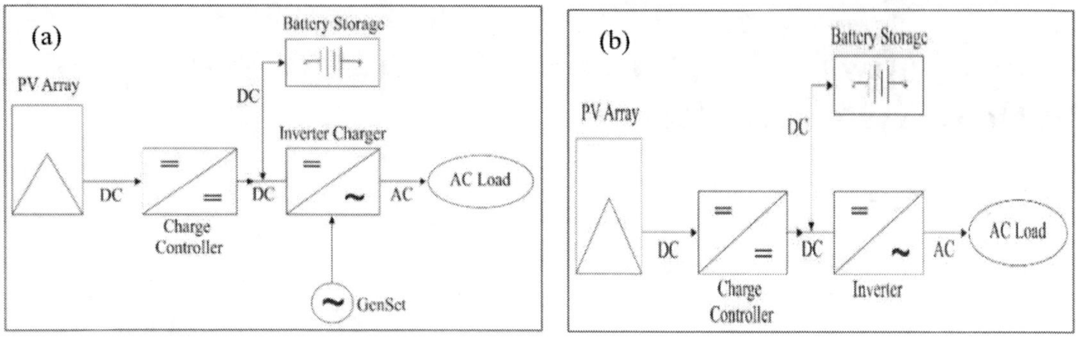

Figure 1. The block diagram of modes description for **(a)** Hybrid PV-Diesel system **(b)** Stand-alone PV solar system.

2. Software Design Methodology

The software program developed by employing GUI (graphical user interface) platform is based on visual studio and for creating a database by using Microsoft SQL Server. The overall process of input and output data for this software design tool summarized in the block diagram in Figure 2, explains the input and output information from this software program under the name of program is University of Baghdad_Solar System Design Tool (UOB_SSDT).

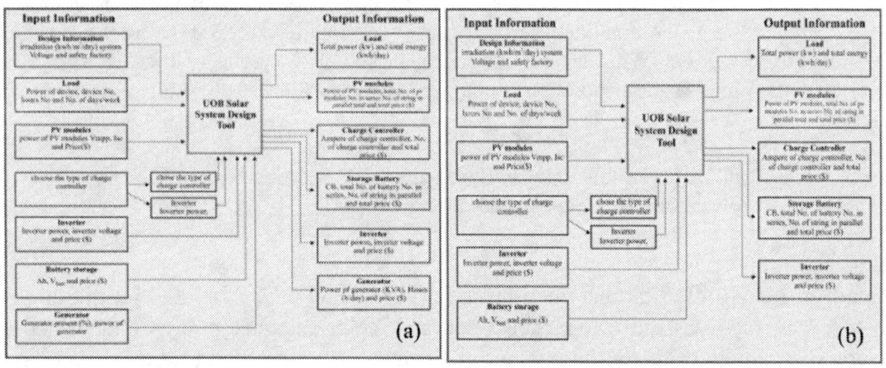

Figure 2. The block diagram of the followed methodology in UOB_SSDT for sizing implementation: **(a)** Hybrid PV-Diesel power system **(b)** Stand-Alone PV power system.

3. System Sizing

The properly photovoltaic system sizing is essential to achieve sustainable and reliable system design. The system sizing will be determined the size of each component existed in the photovoltaic system.

3.1 Load Assessment

Generally, load power refers to the electrical or thermal energy demand, Serving load is the primary purpose of any power system [9]. In the state of load analysis, a system designer should also discuss all the potential energy resources that can meet the capacity needs by the consumer and educate the customers about energy efficiency. The steps for load estimation are:

1. List all the alternative current (AC) electrical appliances that should be supplied by the designed power system.

2. Enter the items number of appliances.

3. Enter the operating wattage for one appliance.

4. Enter the number of hours per day for each item used.
5. Specify the number of days per week.
6. Multiply steps 2, 3 and 4 to calculate the daily energy requirement for 7 days a week.
7. Multiply steps 2, 3, 4 and 5 divided by 7 days a week to calculate the average daily energy requirement per day [10].

3.2 Photovoltaic Module Sizing
To estimate the actual power of the photovoltaic array by equation (1) [9].

$$P_{PV} = \frac{E_{L\_ave}}{PSH \times \eta_c \times \eta_{Inv} \times \eta_{Batt}} \times SF \quad (1)$$

Where the η_c, η_{Inv}, η_{Batt} is the efficiency of the charge controller, inverter and battery respectively, SF is the safety factor for compensating the losses due to temperature, dust, mismatch, and wire losses [17], E_{L_ave} is the average daily energy demand, and PSH is the peak sun hour.
In hybrid PV–Diesel system the energy demand will be multiplied by a fraction (n), this fraction can be defined by the PV to DG ratio [9]. The final equation (2).

$$P_{PV} = \frac{n \times E_{L\_ave}}{PSH \times \eta_c \times \eta_{Inv} \times \eta_{Batt}} \times SF \quad (2)$$

3.3 Charge Controller Sizing
The estimated solar charge controller rating (SCCR) depends on the output from PV array and battery voltage by equation (3) [11].

$$SCCR = Isc \times N_p \times SF \quad (3)$$

3.4 Battery Storage Sizing
The required storage batteries capacity is a function for many parameters like the maximum depth of discharge (DOD), autonomy days (AD) and the efficiency of an inverter and battery. The AD is the number of days that the batteries are able to supply the required load without recharging by photovoltaic modules or diesel generator. The battery capacity can be determined by using equation (4) [15].

$$C_{Bat\_nom} = \frac{E_{d\_daily} \times AD}{DOD \times SV \times \eta_{Batt} \times \eta_{Inv}} \quad (4)$$

Where Ed_daily is the daily energy demand and SV, is the DC-bus system voltage.

3.5 DC-AC Inverter Sizing
The calculated size of inverter in system design depends on the power of loads, the inverter size (IS) is calculated by equation (5) [16].

$$IS = \text{Total power} \times SF \quad (5)$$

Where the total power of loads will use to find the inverter size.

3.6 Diesel Generator Sizing
To obtain a good diesel utilization with long service life, the generator must be operated at optimum load. The estimating of diesel generator size (S_{dg}) can be found by the equation (6).

$$S_{dg} = S_{max\_dem} \times f_{over} \quad (6)$$

Where Smax_dem is the maximum required power during battery charging and f_{over} is oversize factor. The number of hours required in h/day for cover the load and charge the battery depends on PV to DG ratio by following equation (7) [9].

$$h = \frac{(1-n)\times E_{L\_ave}}{Total\ Power} \quad (7)$$

A model was applied to design a stand-alone PV solar system using UOB_SSDT, Figure 3 shows the main interface of the photovoltaic system design tool. It contains access information such as project name, designer or user name and the date. Also, it contains the type of system presented by the stand-alone PV system and hybrid PV-Diesel system. This window contains files that designed and saved for retrieval when needed to open it.

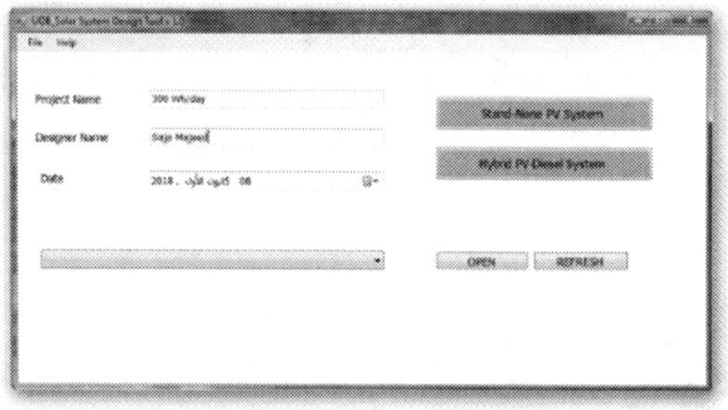

Figure 3. Main form in software program.

The second window has been designed as a home interface, it allows for easy navigation of the system components in this software design tool. It is containing the design information (Irradiation, System voltage and Safety factor). The peak sun hour (PSH) and SF in a software program used (5.7 kWh/m^2/day), and (1.15) respectively as default value. Also, it contains a diagram explain the type of current and includes components of the selected system such as the load analysis, PV Array, Inverter, Battery Bank, and DG for hybrid PV-Diesel system as shown in Figure 4,5. The solar irradiation data was obtained from Photovoltaic Geographical Information System (PVGIS) website. In Baghdad city, the suitable tilt angle for orientation PV array in the winter season is (45° – 48°), and for the summer season is 18°, the optimal tilt angle through the year is 33° for fixed photovoltaic modules and according to the latitude of Baghdad. In this program can input the average (daily, monthly and yearly) solar irradiation for system design.

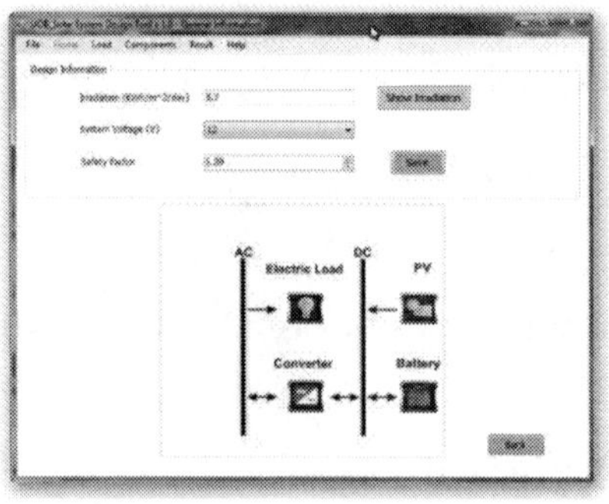

Figure 4. The home form in stand-alone PV system

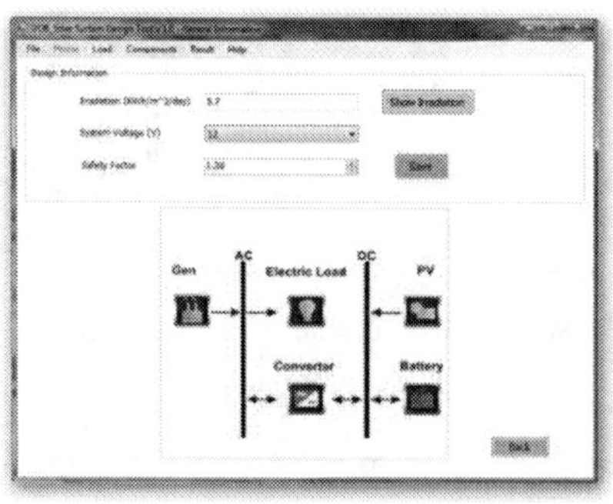

Figure 5. The home form in a hybrid PV-Diesel system.

Figure 6, shows the load profile window, the user can input the used loads power based on design system through input type of device, power of the device, quantity and the number of days in a week to view the actual load. From power and energy screen the user can know the values of the amount of energy demand and the power peak demand. These values are important in any sizing system.

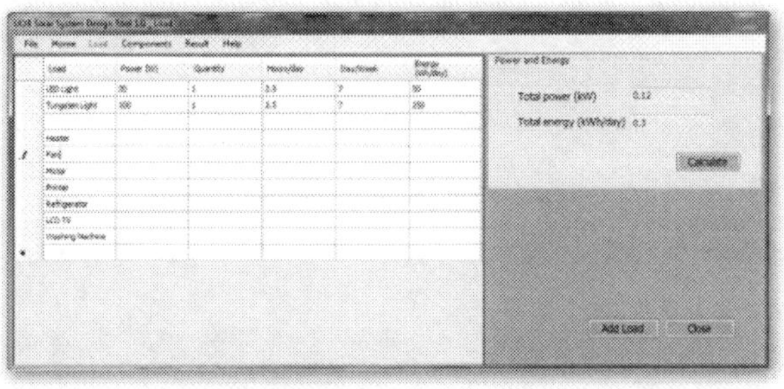

Figure 6. The load profile form.

In software program, there are types and models of photovoltaic modules available in Iraq stores and Iraq companies can be chosen from a combo box. For new data specifications and new types of module, the checkbox used as manual input for adding it manually as shown in Figure 7. The sizing solar charge controller, it is consist of two types of a charge controller are PWM and MPPT to calculate the current size. In addition, it contains the combo box to choose the charge controller model and the manual input property for adding a new model as shown in Figure 8.

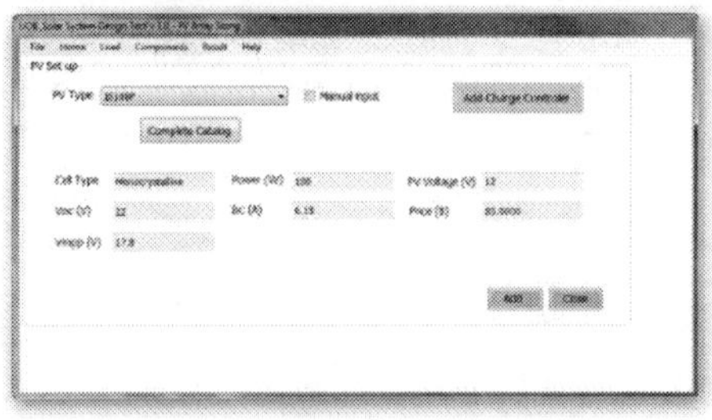

Figure 7. The PV modules form.

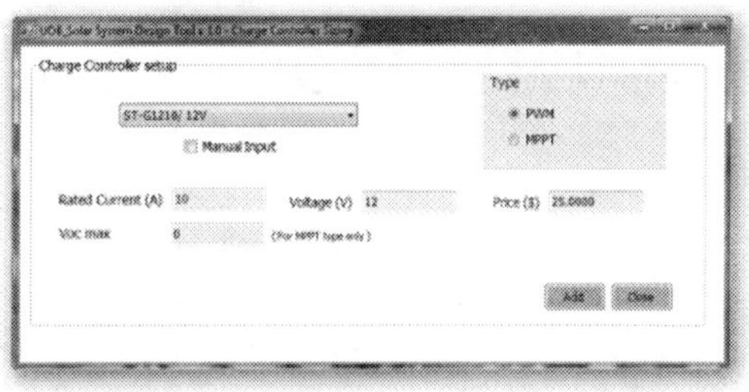

Figure 8. The charge controller form.

For sizing a battery storage capacity, it depends on the total daily energy consumption and battery capacity required. In addition, there is a list of batteries types available in Iraq stores and Iraq companies can be chosen it from the combo box or through enter the battery capacity at STC, battery voltage and it is price manually as shown in Figure 9. For inverter sizing, it includes three important parameters, firstly the short circuit current (I_{sc}) of the module which will be imported automatically when selecting the PV module, secondly is the safety factor for inverter and thirdly is the value of peak power required from the load power profile window. The three parameters determine the size and number of the inverter. In addition, the property of the combo box and manually input are existed in the interface as shown in Figure 10.

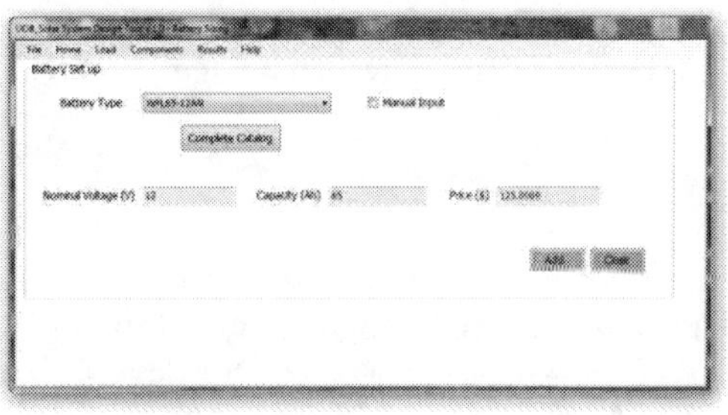

Figure 9. The battery storage form.

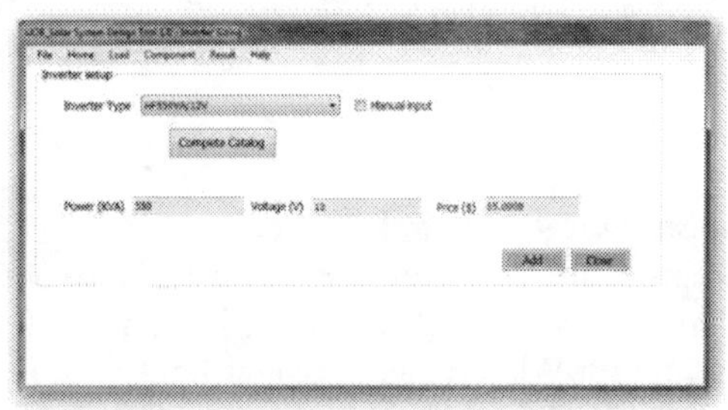

Figure 10. DC-AC inverter form.

When the selected system is hybrid PV-Diesel system the diesel generator interface will be. The fraction of energy contributed by photovoltaic (PV to DG ratio) and the power of diesel generator is presented in this interface as shown in Figure 11.

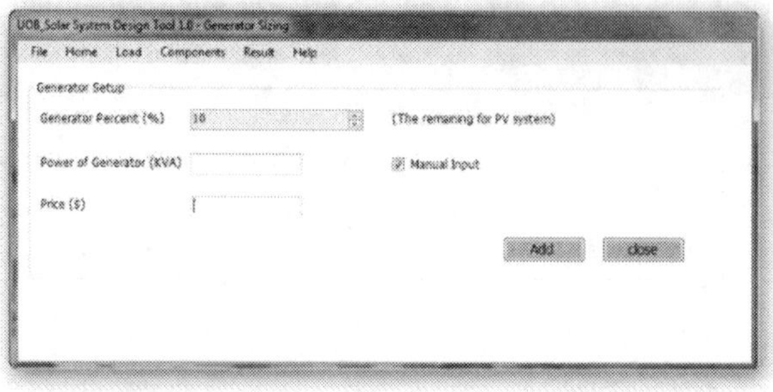

Figure 11. The diesel generator form.

Figure 12. show the result interface for stand-alone PV system in the software program, the result presented by the energy of loads, power of PV modules, total number of modules and how many series and how many parallels based on the system voltage. In addition, the program will calculate the charge controller current and the number of it, also, will calculate the power of inverter and the number of it. The program will calculate the total capacity of batteries in (Ah), the total number of batteries, how many in series and how many in parallel based on the system voltage. In addition, it contains the total cost of each component and initial capital cost as an approximate value. By press on button Calculate, all of the above results will calculate. The Figure 13, for hybrid PV-Diesel system results if it selected that will be influenced on the number of PV modules due to contribution ratio, also, it calculated the power of the generator, the number of hours to supply the loads and charging battery and diesel generator price to calculate the initial capital cost.

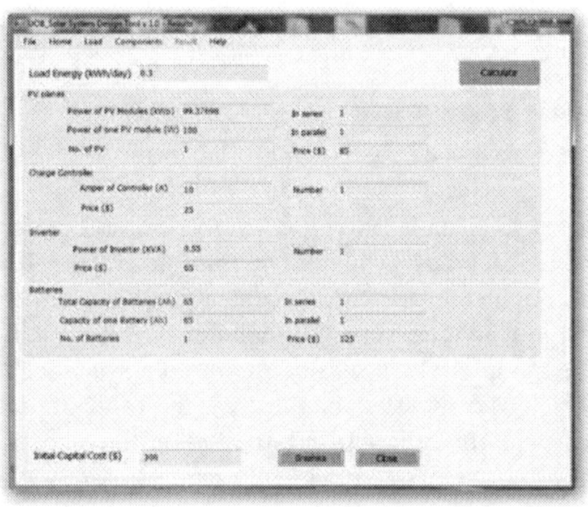

Figure 12. The final results form of the stand-alone PV system.

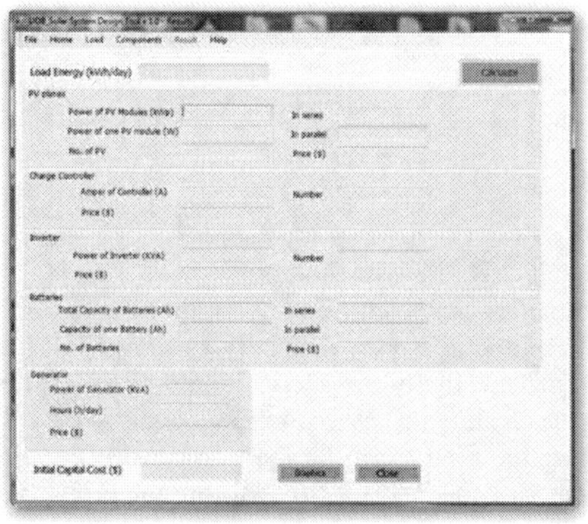

Figure 13. The final result form of the hybrid PV-Diesel system design.

4. Experimental System Design

An empiricist system for stand-alone photovoltaic system is designed to verify from the result of software program on between experimental work. The system is producing energy to turn on the lamps load with (300Wh/day) as the night light. The period of work was for a week from months (January, February, March, April). Table 1 shows the standard parameters adopted in the PV system design.

Table 1. The standard parameters adopted in the photovoltaic system design.

Peak sun hour (h)	5.7
DC voltage bus (system voltage) (V)	12
Safety factor	1.3
Efficiency of inverter	0.9
The efficiency of the charge controller	0.9
Efficiency of battery	0.85
Days of autonomy	1
Depth of discharge	50%

The structure is adjusted with the optimal tilt angle 33° to the south as an annual system, installing the PV module on the frame with the height of structure 1.06m, Figure 14 shows the installed PV module. Table 2 shows the specification of solar module. Then connect the other parts of the solar system charge controller, inverter and battery storage with loads board then connected it with the PV module (an improper sequence order can damage the controller). With apparatuses for measuring presented by: power meter, for measuring the kilo-watt-hour of loads, three apparatuses of digital multi meter for measuring the battery voltage, PV voltage, PV current. as shown in Figure 15, see Figure 1 (b) as a block diagram for system.

Table 2. The specification of mono-crystalline module at STC from ISTAR SOLAR manufacture.

Model	IS 100P
Peck Power	100W
Current At Maximum Power (Imp)	5.62
Short Circuit Current (Isc)	6.15
Voltage At Maximum Power (Vmp)	17.8
Open Circuit Voltage (Voc)	22
Normal Operation Cell Temperature	25℃
Dimensions	1200*540*40 (+/-2mm)
Power Tolerance	+/-3%

Figure 14. Installed PV module as a stand-alone PV system with solar power meter.

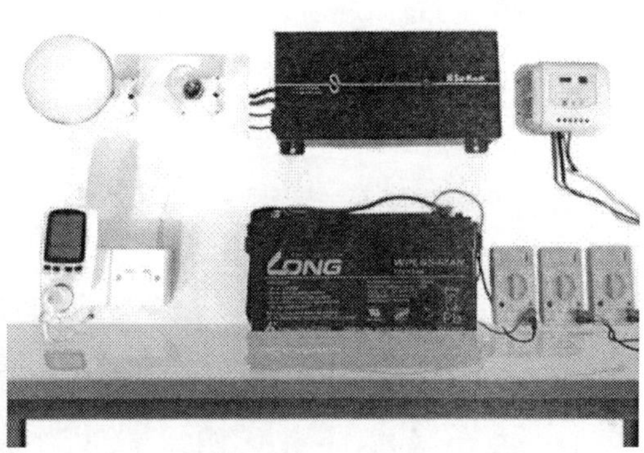

Figure 15. Other components of PV solar system with apparatuses for measuring.

To extraction the performance of the solar photovoltaic system more clearly, the load was fixed on (300Wh/day). In addition, the panel was cleaned daily to exclude losses due to dust. And to know the reason for the difference between actual production by PV solar module and design production with actual loads consumption, therefore the solar irradiation, produced a capacity of the solar module and energy consumption loads were measured.

5. Results and Discussion

The intensity of solar radiation for months (January, February, March and April) in work periods previously at a tilt angle (33°) degree was (2.96, 3.61, 4.62 and 5.25) respectively. The highest average daily solar irradiation during the work period was in April (5.25 kWh/m^2/day) due to the number of daylight hours during the week of this month was longer than the rest of the months, and the weather was sunny with few clouds in most days. The average of solar radiation through four months was (4.11 kWh/m^2/day), as shown in Figure 16.

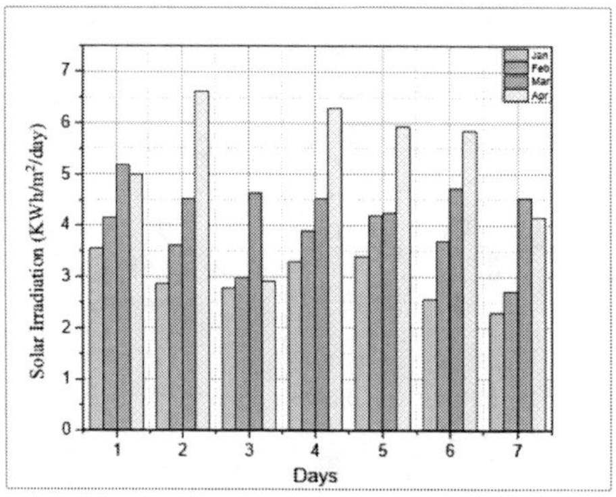

Figure 16. The intensity of solar radiation for months with work periods.

The produced power from the PV module changes directly proportional with the value of solar irradiation. Most of the energy generated from the solar panel in the three months of February, March and April was within design limits and covered the demand energy to loads, the Figure 17 and Figure 18 summarized the data of experimental work for generation and consumption energy during the week respectively. The load energy of the solar system has been proven to be (300Wh/day) throughout the working period as a constant daily load. The highest rate of generated energy during the week was (473.8Wh) on March due to the increase the weekly rate of solar irradiation, while the highest rate of consumption energy was (292Wh) on April, because of the compensation the shortage of generation due to clouds and rain to meet the required load for energy by energy stored in batteries in days at high solar radiation rate.

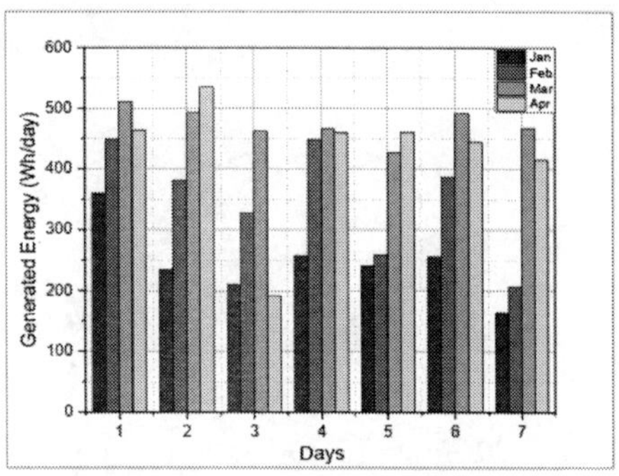

Figure 17. Generation energy for a week per month of four months.

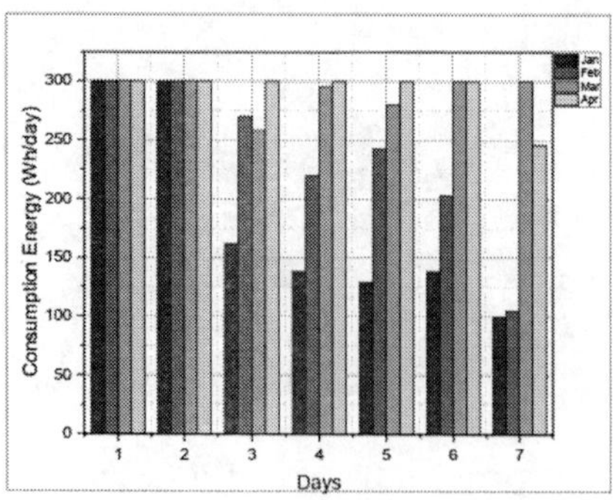

Figure 18. Consumption energy for a week per month of four months.

From Table 3, the difference between practical and theoretical in generating power due to losses caused by heat losses and DC cable losses. In addition, to decrease the solar irradiation due to weather conditions. Moreover this effects on the difference between the daily required load in design (300Wh) and actually average consumption load during the months (250.33Wh) as well as equipment losses represented by charge controller, storage battery and DC-AC inverter or due to defects in components. The conversion efficiency is the output consumption energy to input generation energy and the rate of DC-AC energy conversion efficiency was (0.665) very close to the theoretical efficiency which comprises the efficiency of charge controller, inverter and storage battery (0.9, 0.9 and 0.85) respectively with error percentage equal to -3.5%.

To evaluate the photovoltaic installation performance and it is involved losses like (components losses, temperature losses, DC and AC cable losses, dust and snow losses, losses at weak radiation) by using the performance ratio (PR). The performance ratio for the designed system was approximately 61% that's mean 39% is wasted energy, the incident solar radiation is not converted to useful energy due to heat losses, conduction losses, and other components losses. With taking the solar module efficiency (0.153) the performance ratio is calculated from formula in equation [18].

$$\mathrm{PR} = [(UsefulEnergy\ (kWh))/(Irradiance\ (kWh/m2) \times PVarea\ (m2) \times PVmoduleEffeciency))] \times 100 \qquad (8)$$

Table 3. The weekly average of ambient temperature, experimental and theoretical generated energy in watt-hour, error percent, consumption energy in watt-hour, and with conversion efficiency respectively arrangement in the table.

Month	Ta (C°)	Eexp. (Wh)	Etheo. (Wh)	Error (%)	Econs. (Wh)	$\eta_{conv.}$
Jan.	11.6	246.3	327.7	24.8	181	0.735
Feb.	9.11	379.7	400.8	5.2	234	0.616
Mar.	14	473.8	440	-7.7	294.3	0.621
Apr.	20.29	424.2	458.3	7.4	292	0.688

6. Conclusion

The software program is developed to acts as a guideline tool the final result depends on data entered by the user. Also, it facilitates the design process for systems with the possibility of saving and printing the results. Due to changes in weather conditions throughout the year, especially solar radiation and temperature, are heading towards increasing solar panels in stand-alone PV system to increase the generation capacity to cover the loads during the interruption of power processing during winter months and to avoid increasing the depth of battery discharge more than 50% to avoid costly losses. Or the user can use an alternative source of renewable energy as a backup generator as the hybrid system to cover this gap in the working of the solar system during the winter months and to charge the battery storage.

References

[1] Mahmood F H Al-Hassany G S 2014 *Iraqi J. of Sci.* **55** 1663-74
[2] Kazem H A Chaichan M T 2012 *Rene. and Sustainable Energy Reviews* **16** 6007-6012
[3] Perera F 2017 *Int. J. of Environ. Res. and Public Health* **15** 16
[4] El Shenawy E T Hegazy A H Abdellatef M 2017 *Int. J. of Appl. Eng. Res.* **12** 10433-10446
[5] Ismail M S Moghavvemi M Mahlia T M I 2012 *J. of Asian Scientific Res.* **2** 599

[6] Olatomiwa L Mekhilef S Huda A S N Sanusi K 2015 *Energy Sci. and Eng.* 271-285
[7] Hussein B N Essa I A Hussein D J 2013 *Energy Procedia* **36** 852-861
[8] Lalwani M Kothari D P Singh M 2010 *Int. J. of Appl. Eng. Res.* **1** 585-601
[9] Mohamed N S S Adek N A B M Hamid N F A 2018 *IOP Conf. Ser.: Materials Sci. and Eng.* **318** 012042
[10] Ibrahim U H Aremu D A Unwaha J I 2013 *Int. J. of Scientific & Technol. Res.* **2** 187-195
[11] Hussein B N Essa I A Hussein D J 2013 *Energy Procedia* **36** 852-861
[12] Ibrahim M 2005 *Germany. 20 th-EU-PVSEC, Barcelona. Spain.* 6-10
[13] Omar M I A A 2007 Computer a aided design and performance evaluation of PV-Diesel hybrid system
[14] Mohanty P Sharma K R Gujar M Kolhe M Azmi A N 2016 *Solar Photovoltaic System Appl. Springer. Cham.* 49-83
[15] Hansen A D Sørensen P E Hansen L H Bindner H W 2001 Models for a stand-alone PV system.
[16] Kassem A M Sayed K El-Zohri E el-Hamed A M A 2015 *IOSR J. of Electronics and Commun. Eng. Ver. II* **10** 2278-2834
[17] Mathema P 2011 Optimization of integrated renewable energy system-micro grid (IRES-MG) Doctoral dissertation Oklahoma State University
[18] Al Dulaimi N H Alkhalidi A 2017 M.Sc. Thesis German Jordanian University School of Natural Resources Engineering and Management, Department of Energy Engineering

IOP Conf. Series: Materials Science and Engineering **757** (2020) 012042 doi:10.1088/1757-899X/757/1/012042

Fabrication and Development Low Cost Dual Axis Solar Tracking System

Ahmed Jadah Farhan

Department of Physics, College of Science, University of Wasit, Kut, Wasit /Iraq

aalomairi@uowasit.edu.iq

Abstract. Solar energy is gaining importance day by day as an energy resource. By sun tracking we can increase the electric energy of solar panel. When the solar panel is perpendicular to the sun's rays it receives more sunlight. This paper about follow the sunlight direction across the sky by using a DC motor(satellite motor), the LM324 IC control the movement of DC motor by this the solar panel will change their position, solar panel detects the sunlight using four light dependent resistors (LDRs). The objective is to design and implement an automatic solar-tracking mechanism using an embedded system design with minimum cost and reliable structure. The results presented in this review confirm that the azimuth and altitude dual axis tracking system is more efficient compared to other tracking systems.

Keywords: LM324 IC, solar-tracking, solar panel, solar energy, Dc motor.

1. Introduction

The increasing demand of energy, the depletion of fossil fuel reserves, the unexpected events taking place on the international scene (local armed conflicts, natural disasters like earthquakes, tsunamis, floods, hurricanes, etc. That have the potential to partially cripple the energetic systems, proves that the energy security and diversity is a serious aspect that mankind should seriously consider when deciding the short and middle term energy policy. The general opinion shared by the most part of the specialists supports the idea that the exclusive dependence on the energy produced from fossil fuels (coal, oil, nuclear, etc.) is hazardous, unsustainable and harmful for the environment. In this context, many developed countries (e.g. USA, Germany, Spain, Denmark, France, Italy, etc.) launched in the last Decade's ambitious programs for supporting the rapid development of alternative energetic technologies based on solar energy, wind energy, tidal and wave energy, biomass etc. One of the most promising renewable energy sources characterized by a huge potential of conversion into electrical power is solar energy. The conversion of solar radiation into electrical energy by Photo-Voltaic (PV) effect is a very promising technology, being clean, silent and reliable, with very small maintenance costs and small ecological impact. The interest in the PV conversion systems is visibly reflected by the exponential increase of sales in this market segment with a strong growth projection for the next decades. According to recent market research reports carried out by European Photo Voltaic Industry Association (EPIA), the total installed power of PV conversion equipment increased from about 1 GW in 2001 up to nearly 23 GW in 2009 [1].

Different researches estimate that covering 0.16% of the land on earth with 10% eicient solar conversion systems would provide 20 TW of power, nearly twice the world's consumption rate of fossil energy [2]. his proves the potential of solar energy which in turn points out the necessity of tracking mechanism in solar systems. The tracking mechanism is an electromechanical system that ensures solar radiation is always perpendicular to the surface of the photovoltaic cells(solar cells) which maximizes energy harnessing [3]. Over the years, researchers have developed smart solar trackers for maximizing the amount of energy generation. Before the introduction of solar tracking methods, static solar panels were positioned with areas on cable tilted angle based on the latitude of the location. In this competitive world of advanced scientific discoveries, the introductions of automated systems improve existing power generation by 50% [4]. There are mainly two types of solar trackers on the basis of their movement degrees of freedoms. these are single axis solar tracker and dual axis solar tracker. Again these two systems are further classified on the basis of their tracking technologies. Active, passive, and chronological trackers are three of them [5, 6]. Previous researchers used single axis tracking system which follow sonly the sun's daily motion [7]. But the earth follows a complex motion that consists of the daily motion and the annual motion. he daily motion causes the sun to appear in the east to west direction over the earth whereas the annual motion causes the sun totiltata particular angle while moving along east to west direction [8].

2. Solar Tracker

One of the factor that can improve the efficiency of the photovoltaic system is the availability of solar tracking mechanism in the system [9]. There are typically two types of solar tracker designs available i.e. one and dual axis solar tracker. The one axis solar tracker consists of a horizontal-axis tracker which always oriented along East-West or North-South direction; tilted-axis tracker which is tilted from the horizon by an angle oriented along North-South direction; and vertical-axis tracker also known as an azimuth sun tracker as shown in figure 1. The dual axis solar tracker consist of azimuth-elevation and tilt-roll sun tracking systems, follow the sun in the horizontal and vertical plane as shown in figure 2. A solar tracker is a device that orients a solar panel toward the sun. The solar tracker techniques can be used as a sun tracking algorithm or light tracing sensors [10].

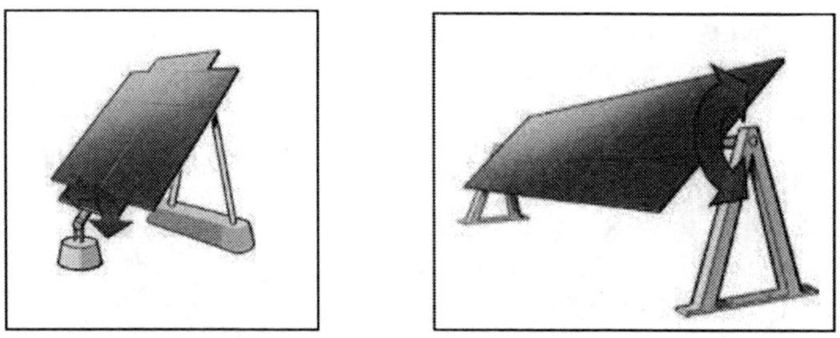

(a) (b)

Figure 1. The single solar tracker: (a) Vertical –single axis tracker

(b) horizontal-single axis tracker.

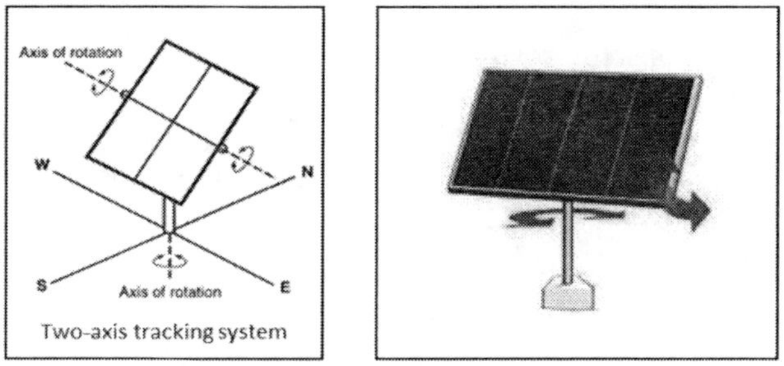

Figure 2. The dual axis solar tracker.

3. Experimental part .

Mechanical Tracking Design

The first part fabricated is the frame of the structure as following

1. Metal pipes are cut using a special metal saw to the required lengths:

   (60 cm* 2, 100 cm * 3, 90 cm * 1, 15 cm * 1).

2. Metal pipes shall be welded to form the solar cell-bearing base.
3. The metal base above is welded to the fixed metal stand or rack.
4. Connect the motors between the base and the metal stand, as well as in Figure 3:

Figure 3. Mechanical tracking design.

DC Motors

The motors used to rotate the axes are satellite motors, figure 4. They are pretty known for their long service life and their wind and rain with standing potential. The tight seal will prevent water from penetrating to the inside. The rugged aluminum casting maintains the greatest strength in high stress areas. Their large bearing surfaces and strongly reinforced mast support lateral loads. Ball bearing is provided to hold the thrust (vertical) loads. The more important features of these motors are their high torque capabilities and precision in positioning. Almost all houses in cities, where satellite broad casting is available, have one setup using this type of motors. So usually it is cheap and abundant in the market and people can easily maintain it.

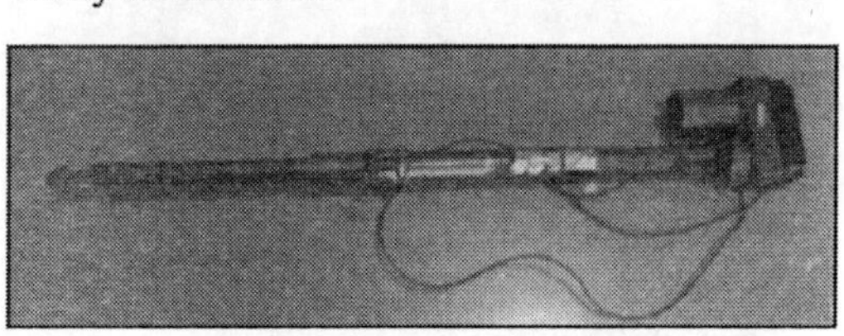

Figure 4. Satellite motor.

3. 3. Electronic circuit

There are many electronic circuits for solar pursuits, simple and complex ones, some of which are made up of known electronic components such as the timer 555 [11]. The Arduino or Microcontroller, which can be programmed and erased several times, has developed and manufactured a simple electronic circuit and simple and inexpensive components where anyone can manufacture it. An electronic circuit for the single axis tracker and another electronic circuit that is similar to the Horizontal-single axis tracker as shown in Figure 5.

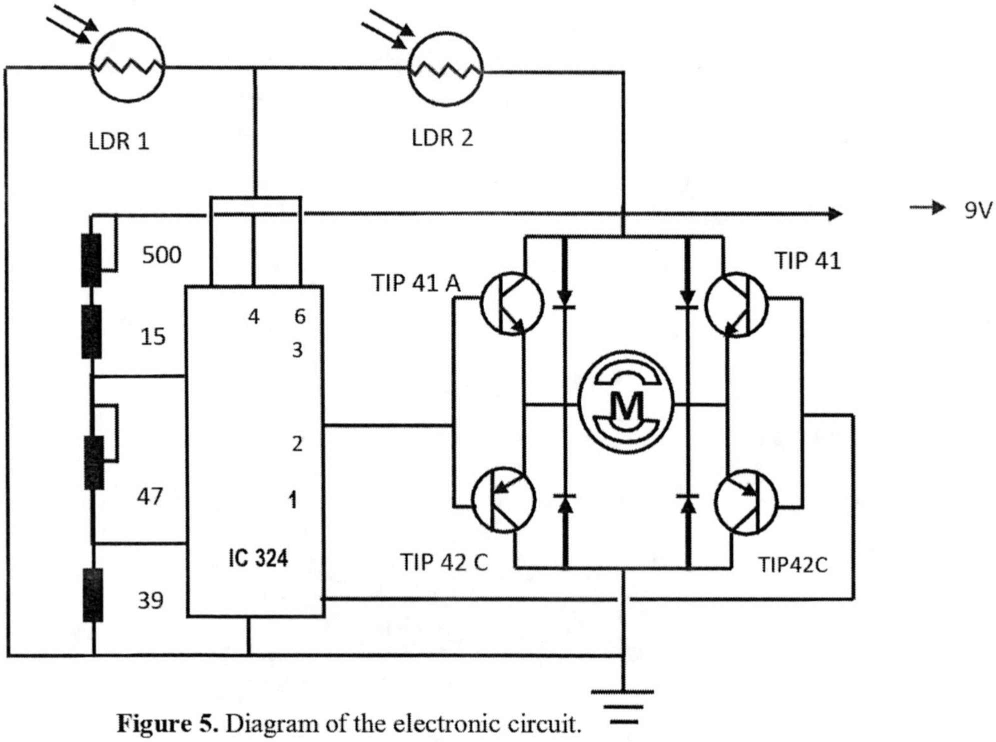

Figure 5. Diagram of the electronic circuit.

Light sensors (R_{LDR}).

The tracker tracks the sunlight by four light-based resistors (LDR) at the top of the solar cell panel looking for sunlight and identifying it on either side (east, west, north or south) and then moving to find the best point of sunlight) as shown in figure 6. A light dependent resistor (LDR) is a resistor whose resistance decreases with increasing incident light intensity. Figure 7. shows the Basic diagram of LDR circuit [12].

Figure 6. Four R_{LDR} kits.

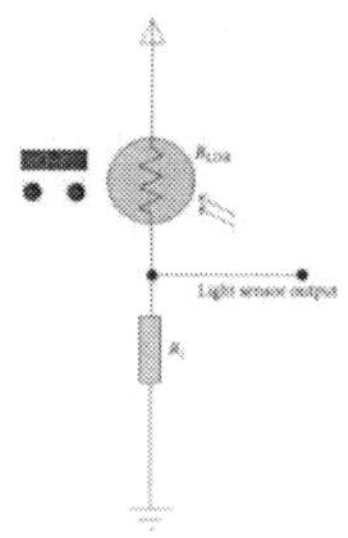

Figure 7. Basic diagram of LDR circuit.

IC. LM324

LM324 as denoted in figure 8. is a 14 pin IC consisting of four independent operational amplifiers (op-amps) compensated in a single package. Op-amps are high gain electronic voltage amplifier with differential input and, usually, a single-ended output. The output voltage is many times higher than the voltage difference between the input terminals of an op-amp. These op-amps were operated by a single power supply LM324 and need for a dual supply is eliminated. They can be used as amplifiers, comparators, oscillators, rectifiers, etc. The conventional op-amp applications can be more easily implemented with LM324 [13].

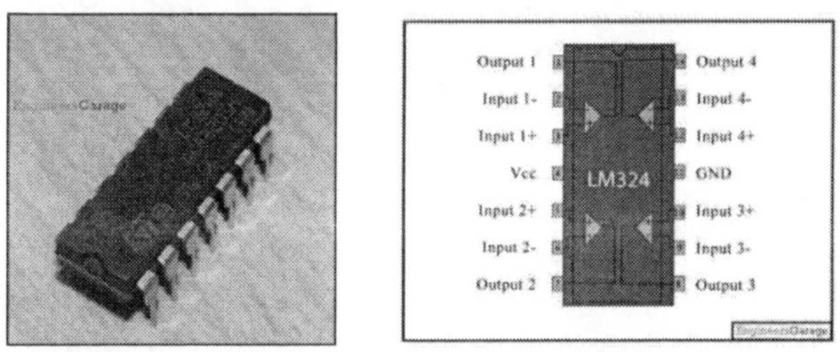

Figure 8. The basic LM324 circuit.

4. Test angles axles.

The angle of rotation of the horizontal axis (330 °) of solar panel, it is moving from the extreme right to the far left to trace the motion of the sun from east to west. The angle of the vertical axis is (from 0 to 90 degrees) with any horizon moving from a level parallel to the horizon to a vertical level with the horizon to track the movement of the sun from north to south as shown in figure 9.

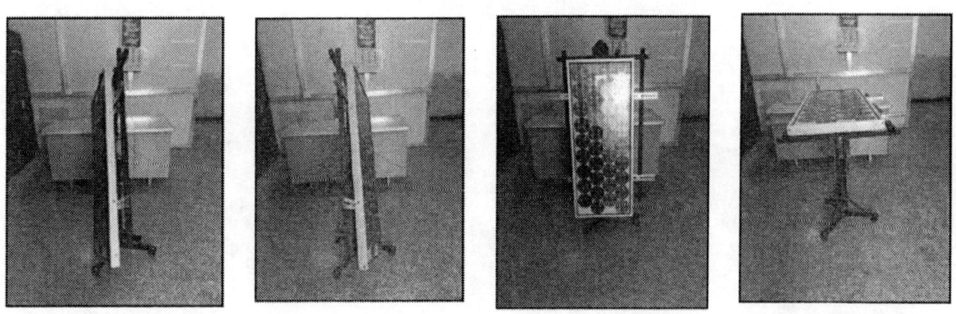

Figure 9. Solar cell movement to the extreme right, the extreme north and solar cell movement with the horizon line.

IOP Conf. Series: Materials Science and Engineering **757** (2020) 012042 doi:10.1088/1757-899X/757/1/012042

5. Experimental and method.

Experimental Setup.

1. Solar Panel: A solar cell or photovoltaic cell is a device that converts solar energy into electricity by the photovoltaic effect. The collection of solar cells is called a solar panel or solar array, which has 18 volts. Solar panel is placed on top of the tracker board which absorbs the maximum sunlight which is used as supply to the motors or stored in the battery for future use.
2. Checking the battery (voltage display).
3. Connect the wires to the DC motor and power source.
4. Reference to optical resistance (LDR)
5. Rotating Variables Resistors (Trimming Tool)
6. Connect the power cord and turn DC motor.
7. Control the variable resistance (trim pot, 500 K) clockwise. Note the position.
8. Continue to rotate (trim pot) clockwise until the DC motor works in the opposite direction. Note the position.
9. Split the angle between positions in step 6 and step 7 and set the trim pot at that point. (This trim pot determines the equilibrium point)
10. Rotate (trim pot 47K) clockwise until the DC motor begins to search.
11. Pull back in a clockwise direction until the engine takes a break. (This trim pot determines sensitivity).
12. Check the correct operation by shading each LDR individually and in turn checking the rotation DC motor and vice versa for the other direction.

Results of solar panel at month of September.

Solar panel is placed with an angle of 45 degrees with the horizon on top of the fixed system. solar panel is directed towards the base using the electronic compass. Solar panel with the stationary base is moved to an open place, for example, on a building surface, and begins by measuring the voltages of the solar panel by using an avow meter every hour from sunrise to sunset. The voltages are repeated twice per hour at the first time and the cell is stationary without the second time when the circle of the solar tracker is directed to direct the solar cell directly towards the sun so that the sun is vertical on the solar panel. Table 1. shows the fixed and the tracking systems voltage values per each hour from 6am to 6pm for one solar panel (for September).

Table 1. Electrical properties of fixed and tracking systems data (for the month of September).

Time (hr)	Fixed-axis Voltage (V)	Dual-axis Voltage (V)
6	17	18
7	18	18.6
8	18.32	18.6
9	18.33	18.6
10	18.6	18.6
11	18	18.6
12	17.94	18.6
13	17.75	18.6
14	17.6	18.6
15	17.55	18.6
16	16.6	18.6
17	15	17
18	12.55	14.88
Average Voltage per Hour (V/hr.)	**14.54**	**18.14**

Figure10. show that the relationship between time and voltage of the fixed and dual solar systems of the solar panel (for the month of September). It is quite obvious that the percentage increase in power is focused at the early and afternoon hours. After all, the fixed panel at these periods of the day will not be erected in the optimum irradiation receiving the position. Similar results are obtained for different days of the year and for different weather conditions (hot, and cold) where maximum voltage gain reached, with this type of tracking mechanism, is 5% more than a fixed one.

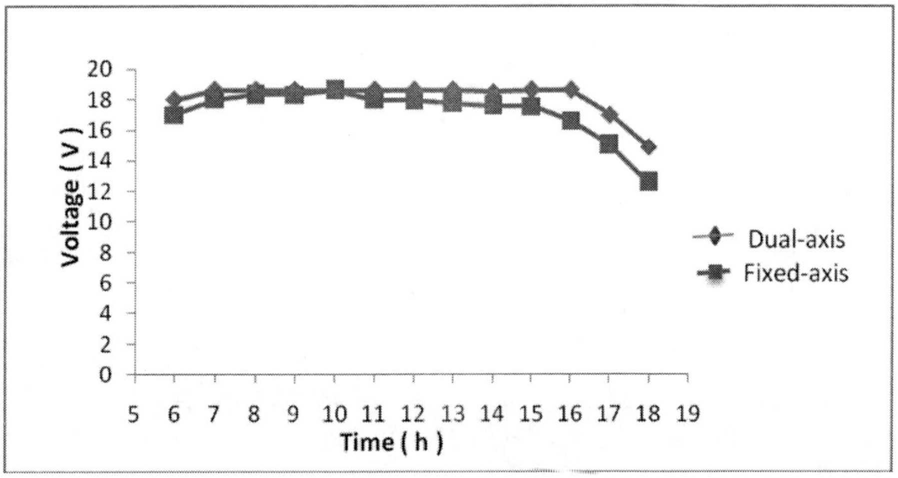

Figure 10. The relationship between time and voltage of the fixed and mobile solar system of the solar panel (for the month of September).

The results of solar panel at month of January.

Figure 11. and table 2. show that the measured voltage values of the solar tracker are higher than the voltage values of the fixed system for the same solar daylight hours. That means the solar tracker improved an efficiency of the solar panel compared with a fixed axis system. Where the maximum voltage gain reached, with this type of tracking mechanism, is 16% more than a fixed one.

Table 2. Electrical properties of fixed and tracking systems data (for the month of January).

Time (hr)	Fixed-axis Voltage (V)	Dual-axis Voltage (V)
6	16	20
7	16.3	20.2
8	16.7	21.2
9	19.6	20.7
10	20	20.3
11	20.2	20.3
12	20.5	20.5
13	20	20
14	19.9	20
15	19.9	19.9
16	19.6	19.6
17	18.6	19.2
18	15	17
Average Voltage per Hour (V/hr.)	**17.11**	**19.92**

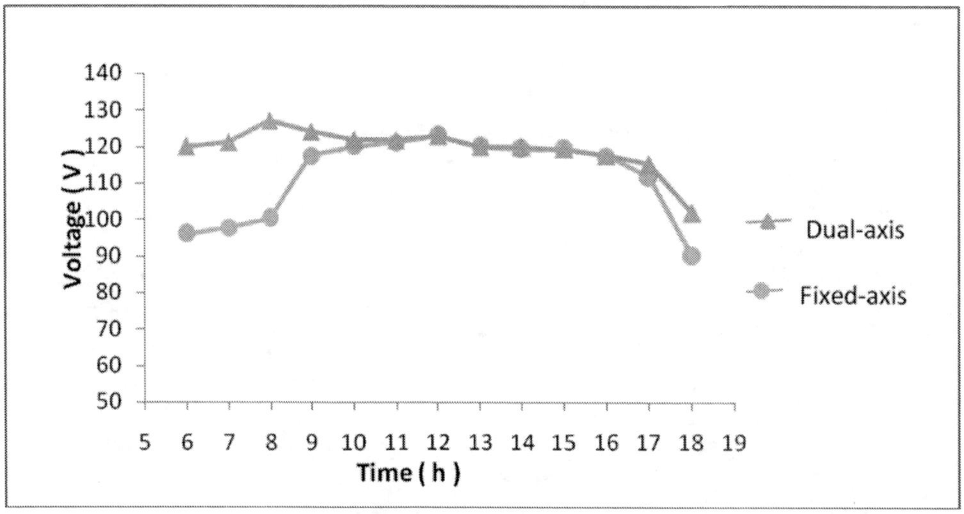

Figure 11. The relationship between time and voltage of the fixed and mobile solar system of the solar panel (for the month of January).

6. Conclusions

A solar tracking system can easily increase the solar energy absorbed compared to a fixed system. However, to carry on a fair comparison with a fixed system, the inclination angle of the fixed panel must be considered and the orientation of the fixed panel should be changed once or twice a season according to the summer and winter sun arch. The power loss due to the control electronic circuit and the antenna motors is very little compared to the power gained from the tracking system. When comparing the price of PV solar panels, the cost of tracking solar energy will be relatively low. Solar Tracker reduces the space requirement of the solar power plant, and it keeps the same output. Electronic Circuit Tracker is a simple to use and inexpensive circuit.

7. References

[1] Tiberiu T, Constantin D and Liviu K 2012 *U.P.B. Sci. Bull., Series C,* **74**(1), PV PANEL.

[2] H Mousazadeh, A Keyhani, A Javadi, H Mobli, K Abrinia, and A Sharii *Energy Reviews*, **13**(8)1800–1818.

[3] M Benghanem *Applied Energy*, **88**(4) 1427–1433.

[4] C Praveen *Proceedings of the International Conference on Computing and Control Engineering (ICCCE '12).*

[5] D F Fam, S P Koh, S K Tiong, and K H Chong, 2012 *Rese J of Rec Sci* **1**(9) 74–78.

[6] A M Sharan and M Prateek, 2006, *Journal of the Indian Institute of Science*,**86**(5) 415–437.

[7] C Alexandru and M Comsit, Virtual Prototyping of the Solar Tracking Systems, Department of Product Design and Robotics, *University Transilvania of Bras* , ov, Brasov, Romania.

[8] A Hsing, 2010 Solar Panel Tracker, *Senior Project, Electrical Engineering Department, California Polytechnic State University*, San Luis Obispo, Calif, USA.

[9] K Chong, C Wong, U Tunku and A Rahman 2014 General Formula for On-Axis Sun-Tracking System, *Universiti Tunku Abdul Rahman Malaysia*, Chapter 3, pp. 263–291.

[10] M A Ghias, Kh S Karimov, S I A Termizi, M J Mughal, M A Saqib and I H Kazmi 2007 A Photo-Voltaic System with Load Control, *Proceedings of the International Conference on Electrical Engineering (ICEE),* Lahore (Pakistan).

[11] Shrishti R 2013, I. J. T. R. A. e-ISSN: 2320-8163, **1**(4) 77-85.

[12] CdS Photoconductive Photocells, Advanced Photonix, http://www.cooking-hacks.com/skin/frontend/default/cooking/pdf/ LDR-Datasheet.pdf.

[13] LM324 IC Pin Diagram and Description LM324N Datasheet, https:// www. Engineers garage. com/electronic.../lm324n-datasheet.

IOP Conf. Series: Materials Science and Engineering **757** (2020) 012043 doi:10.1088/1757-899X/757/1/012043

Measurement of the distance to the central stars of Nebulae by using Expansion methods with Alladin Sky Atlas

Sundus A. Abdullah Albakri[1] , Mohamed Naji Abdul Hussien[1] , Huda Herdan[1]
[1]Department of Astronomy and Space,College of Science, University of Baghdad

[1]Corresponding author: Sundusalbakri70@yahoo.com

Abstract. The usual methods of distance determination in Astronomy parallax and Spectroscopic with Expansion Methods are seldom applicable to Nebulae. In this work determination of the distances to individual Nebulae are calculated and discussed. The distances of Nebulae to the Earth are calculated. The accuracy of the distance is tested by using Aladin sky Atlas, and comparing Nebulae properties were derived from these distance made with statistical distance determination .The results showed that angular Expansions may occur in a part of the nebulae that is moving at a velocity different than the observed velocity . Also the results of the comparison of our spectroscopic distances with the trigonometric distances is that the spectroscopic distances are 55% larger. Since using trigonometric parallaxes with large relative measurement errors can introduce systematic errors, we carried out a Monte Carlo simulation of the biases introduced by selection effects and measurement errors. It turns out that a difference between both distance scales of the observed size is expected for the present day data if the underlying distance scales are identical.

1.Introduction

Assumption of Zero-Age Main Sequence Stars Spectroscopic parallax distance estimates involve comparison of (de-reddened) apparent magnitudes with absolute magnitudes, which are assumed based on stellar evolutionary status, to derive a distance modulus. Several early works reported distances ranging from 185 to 2000 pc (falling off the range of distance versus publication year. In the first modern set of distance estimates to the Orion Nebula region, derived a distance modulus (DM) of 8.5 magnitudes by comparing absolute magnitudes [1] along the main sequence with the de-reddened apparent magnitudes of B stars in the cluster. His method would be repeated by numerous authors in later studies[1] re-calculated the total absorption towards the three brightest stars in the Trapezium, used a different absolute magnitude relation[2] and derived a much smaller distance modulus, 7.38 magnitudes [1,2].

Eight methods are considered as independent ways of determining the distance. Some of the methods are reliable than others, but most have a limited application. Each of the methods ,their expected accuracy and some results for the more reliable methods will be discussed in turn [3].

(1) Trigonometric parallax. Unfortunately, none of the nebulae are close enough to allow distances to be reliably determined. A possible exception is NGC 7293 for which parallax of 0."020+_.006 has been given, corresponding to a distance of 50 pc. This value is very uncertain.
(2) Spectroscopic parallax. Some planetary nebulae are excited by stars which have binary companions. If the spectral type and luminosity class of the companion(which is presumably a 'normal' star since it has a 'normal' spectrum) can be measured, a spectroscopic distance for the system can be determined this method may have a wide application, since it is estimated that at

IOP Conf. Series: Materials Science and Engineering **757** (2020) 012043 doi:10.1088/1757-899X/757/1/012043

least 10% of the exciting stars are binaries .the method may be applied both to visual binaries(where both stars are seen separately)and to spectroscopic binaries where the 'normal' star dominates the spectrum. Only a very limited number of such cases have been well studied. These nebulae are all nearby. This is consistent with the fact that the nebulae all have large angular diameters.

(3) Expansion distances. Radial velocity meassurments of the nebulae show a splitting of emission lines. This is interpreted as an expansion of the nebulae , with velocities typically of 20 km s^{-1} . this expansion velocity may be compared with angular expansions derived from studying the location of knots, filaments, edges, and other features seen on both old and new photographs [3].

2.Basic Consideration and Equations

2.1 Distance modulus:

The distance modulus μ=m-M **is the difference between the** apparent magnitude) **and the** absolute magnitude **M of an** astronomical object. **It is related to the distance d or D in parsecs by [3,4]:**

.

$$\log_{10}(d) = 1 + \frac{\mu}{5}$$

$$\mu = 5\log_{10}(d) - 5 \qquad \text{............ (1)}$$

m-M=5log10(D(pc))-5 ....(2)

Where : m= Apparent magnitude ; M= Absolute magnitude ; D= the distance in parsecs.

This definition is convenient because the observed brightness of a light source is related to its distance by the inverse square law (a source twice as far away appears one quarter as bright) and because brightness's are usually expressed not directly, but in magnitudes.

Absolute magnitude M is defined as the apparent magnitude of an object when seen at a distance of 10 parsecs. Suppose a light source has luminosity L(d) when observed from a distance of d parsecs [4].

2.2 Spectroscopic distances

Spectroscopic distances can be derived from the model atmosphere analyses described in [5] and offer an approach to the PN distance scale independent from the properties of the nebulae. After the stellar mass is estimated from a comparison with evolutionary models [5] the distance can be calculated in a straightforward way from effective temperature, gravity, and the reddened apparent magnitude of the stars[5].

1. The stellar radius is computed from surface gravity and mass.
2. The surface flux in the V band is calculated from model atmospheres, with the appropriate parameters (thus bypassing the uncertain bolometric corrections),
3. and finally the absolute magnitude MV is derived and compared with the red end V0 magnitude of the star.

4. This procedure can be condensed into one formula [6]:

$$D=0.111\frac{\sqrt{M*Fv}10^{0.4vo}}{g} \qquad ....(3)$$

Where : $M\star$=is the solar mass in $M_\odot$; g =is the surface gravity in cm/s^2 ; F_V= is the model atmosphere flux in $10^8 erg\ cm^{-2}s^{-1o}A^{-1}$.

2.3 Expansion distances

Distances for PNe can also be determined by the parallax expansion method as is usually done for resolved novae. As a nebula expands its rate of motion appears as a proper motion measurable in arc seconds per year. Spectra obtained at the same period of time give the expansion velocity directly in kilometers per second. If the expansion is uniform then the proper motion corresponds to the same velocity. The geometry of the expansion allows to find the distance D to the nebula as[7]:

$$Dpc=211\frac{Vkm/sec}{\ddot{\theta}mas/yr} \qquad ....(4)$$

Where V= is the radial velocity ; $\ddot{\Theta}$=is the proper motion.

This method has been applied both at optical and radio wavelengths to determine distances to PNe [7]

3 . Determine the physical size of the PN using the small-angle formula

Explains the derivation of the small-angle formula [8]

$$l=\frac{r\theta}{206265} \qquad ....(5)$$

Where l: is the physical size, r: is the distance (note that l and r must be expressed in like units Mpc for galaxies) as shown in figure 1. θ: Angular size of nebulae in arcs units

206265= is the number of arc seconds in a radian [7,8]

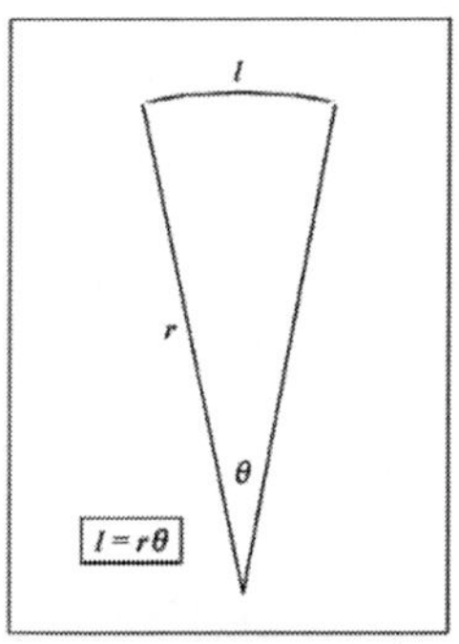

Figure1. the physical size and angular size

4. Aladin(atlas sky)

Aladin is an interactive sky atlas developed and maintained by the Centre de Donnie's astronomies' DE Strasbourg (CDS) for the identification of astronomical sources through visual analysis of reference sky images. Aladin allows the user to visualize digitized images of any part of the sky, to superimpose entries from the CDS astronomical catalogues and tables, and to interactively access related data and information from various data servers [7,8].

In this case we use Aladin in the undergraduate mode (developed in the frame work of the EuroVO-AIDA European project) [8,9]

edit -> user preferences -> profile ->

IOP Conf. Series: Materials Science and Engineering **757** (2020) 012043 doi:10.1088/1757-899X/757/1/012043

undergraduate.
Restart Aladin in order to validate the new Configuration [9,10] Open the server selector window as shown in figure 2:
File -> Load astronomical image

-> Aladin image server [8,11]

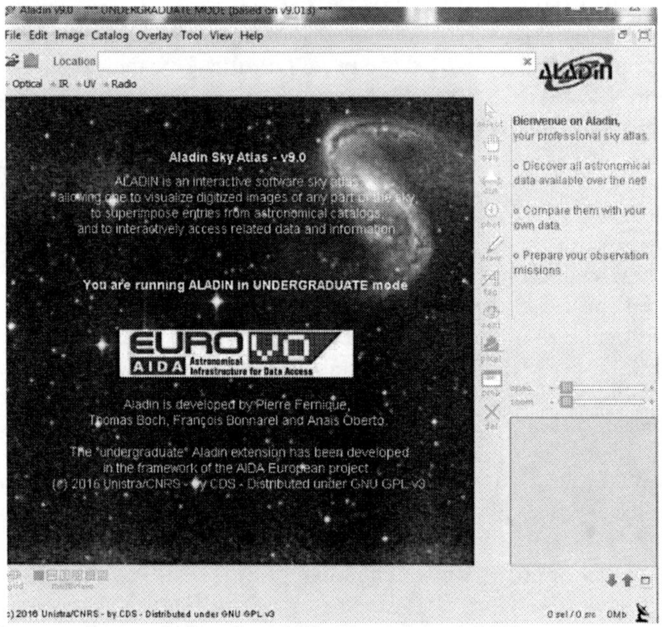

Figure 2. Server selector window for Aladin Sky Atlas-v9.0 [8]

5. The Results

In this work the spectroscopic distances were calculated by using the Equation (1) and compared with observed distance as shown in table 1:

Table 1. Represents the values of Spectroscopic distances of nebulae with extinction E(B-V).

NP	SPECTRAL TYPE [13]	mv	Mv	E(B-V)	Observed D(pc) [13]	The calculated D(pc)
NGC 245	KO V	14.3	5.9	0.01	470	**478**
NGC 1514	A0 III	9.42	-0.2	0.045	400	**470**
NGC 2346	A2 V	11.12	1.4	0.22	640	**690**
NGC 3132	AO V	10.06	0.7	0.07	670	**744**
A CO 35	G8 III-IV	9.63	1.9	0	360	**346**

According to the relation(1) had been showed between (D) and (E(B-V)) , as along as the distance increased the Extinction E(B-V) increased too, leads to conclude that when the nebulae has largest obseved distance (D) the value of E(B-V) stron and that will increase as shown in figure 3,A. The values of the calculated D is less than the values in observed D . the relation between Calculated D and extinction E(B-V) shows a big range as(400-700) in the first interval . The distance increases the extinction increase too, as shown in figure 3,B

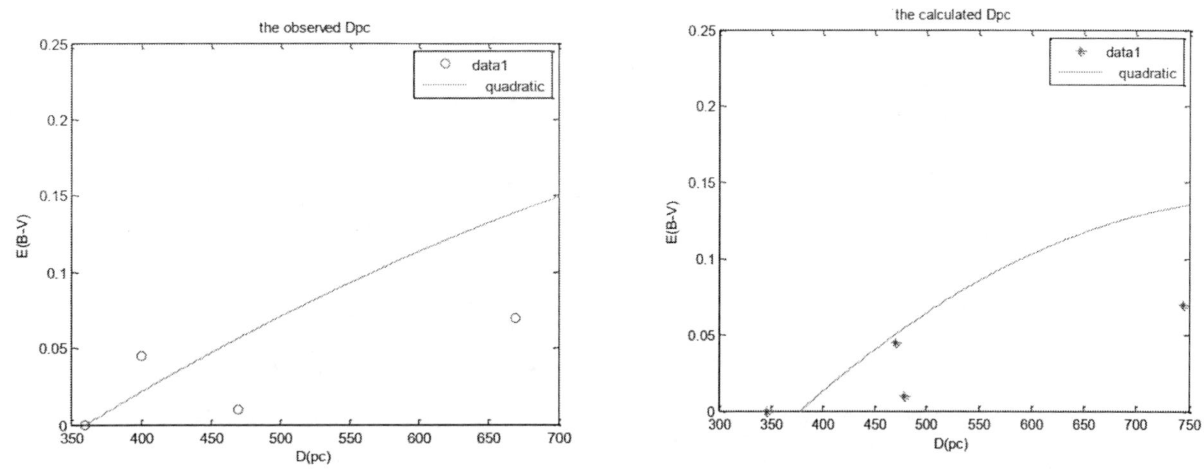

A **B**

Figure3. **A**- Represents the relationship between the Observed distance and the extinction E(B-V) , **B**- Represents the Calculated distance in this work.

The angular Expansion distances were calculated by using the Equation 3 and compared with observed distances as shown in table 2, also the angular expansion were be calculated by using equation 4 for the adopted samples as estimated in table 3. The results compared with the observed values.

Table 2. Represents the values of the angular expansion rate and resultant distance.

NP	**$\dot{\theta}$ ("/100years)**	**θ"**	**V(KMS^{-1})**	**Observed D(pc)**	**The calculated D(pc)**
NGC246	1.4+0.5	100	38	620	**682.9**
NGC3242	0.83+0.25	15	30	760	**767.5**
NGC3587	2.0+1.0	99	41	430	**453.4**
NGC6572	0.81+0.10	49	16	420	**418.7**
NGC6720	0.9+0.1	40	30	700	**705.109**
NGC7009	0.75+0.3	14	21	600	**596.8**
NGC7662	1.0+0.6	10	26	550	**560.9**
	0.6+0.17	10	26	900	**917.4**
NGC2392	0.72+0.06	6	18	530	**528.5**

Where V: the radial velocity of the nebula in km/s unit.

Table 3 . Represents the values of angular sizes and physical size with the observed and calculated distances for the adopted sample.

V	$\boldsymbol{\theta}$(arcsecond) [19]	L(Jy)	D(pc)Observed [19]	D(pc)Calculated
A30	63.5	0.023	2020	1120
A31	486.0	0.155	240	214
A33	134.0	0.014	1060	794
IC4637	9.3	0.132	1500	1525
KI-14	23.0	0.001	3500	3299
KI-22	90.5	0.011	1130	1033
Mz3	12.7	0.649	1440	916
NGC1535	9.2	0.160	2310	1646
NGC3132	22.5	0.230	770	1250
NGC7008	14.0	0.217	870	1233
SP3	17.8	0.061	2380	1965
A21	307.5	0.327	540	278
A35	368.0	0.255	230	255
NGC1514	80.O	0.288	480	540
NGC6853	165.0	1.324	380	310
NGC7296	300.0	1.291	210	220
PHL932	135.0	0.0100	310	540

As the angular sizes and physical size increase the distance (D) decrease. Stretch a large $\boldsymbol{\theta}$ be a close distance as shown in figure 4 in a good agreements with the observed values for the angular size and distances as shown in the figure 5. The results showed the same behaviors and agree with observed values.

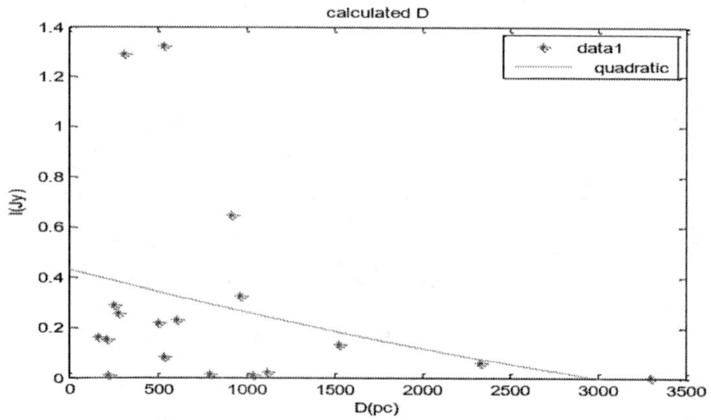

Figure 4. The angular size with calculated distances for the sample that adopted in table (3)

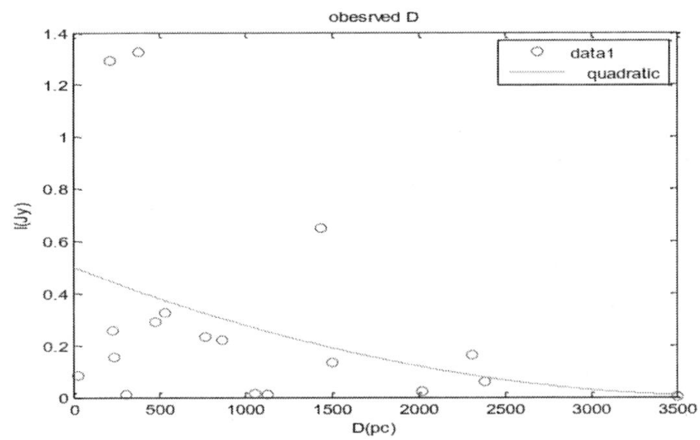

Figure 5. The angular size with observed distances for the sample that adopted in table 3.

In this work the Aladin sky Atlas were be used to calculate the distance of the nebulae through Atlas Aladan program. Where some samples will be adopted for the nebulae and we introduced these samples in the program as the results show:

To find the distance by using spectra method we will do the following steps in Aladin Atlas:

1- Choose the objects that we need to find the D(pc).

2-Determind the location related to our sample.

3-The simplest method to obtain the linear size of the nebula is to multiply the expansion velocity by the time passed since the explosion.

4- Need only to know the year the image has been taken. And calculate the time passed since the Explosion .

5- Determine the Angular size of the nebula:

In order to compute the angular size of the nebula we use the tool "dist" and track a vector from the centres of the nebula to the outermost visible part. The outer parts of the nebula are very faint, so we adjust the intensity distribution by increasing the contrast of the image. The centre of the nebula matches with the pulsar (marked on the image).

6-now have both the linear and the angular size of the nebula: by comparing them we can compute its distance from any object near it .In order to obtain the distance in parsec we have to express the sizes in the proper units: the linear size in parsec and the angular size in radians. 1 pc = 3.085 · 1016 m, 1 rad = 206264"

7-finally compute the distance to the nebulae by using the following relation [4] :

d = R / r (pc).

Where R is represents the radius of Nebula in parsec unit.

5.1 Linear size of the nebula .

The simplest method to obtain the linear size of the nebula is to multiply the expansion velocity by the time passed since the explosion. Observations performed several years apart reveal the slow expansion of the nebula, telling us the expansion is accelerating. In this tutorial we analyses an image of the Nebula as recent as possible and consider the expansion constant. This simplification gives us an approximated result, but enough accurate for our aims.

Start Aladin and switch to the
undergraduate mode from the menu:
edit -> user preferences -> profile ->

IOP Conf. Series: Materials Science and Engineering **757** (2020) 012043 doi:10.1088/1757-899X/757/1/012043

undergraduate.
Restart Aladin in order to validate the new configuration.
Open the server selector window:
File -> Load astronomical image
-> Aladin image server
In the field "target" enter "NGC246" and press "submit". The list of available images appears (figure 6): look at the dates of the images and load in Aladin the most recent one (taken in 1994). We use this the Skull (NGC246) Nebula image in the next step[11], when we will compute its projected angular size.
Now we need only to know the year the image has been taken. We calculate the time passed since the explosion (1785):

$1yr=3.154*10^{7}s.$

$(1994 - 1785)\ yr = 209\ yr = 6.591*10^{10}\ s.$

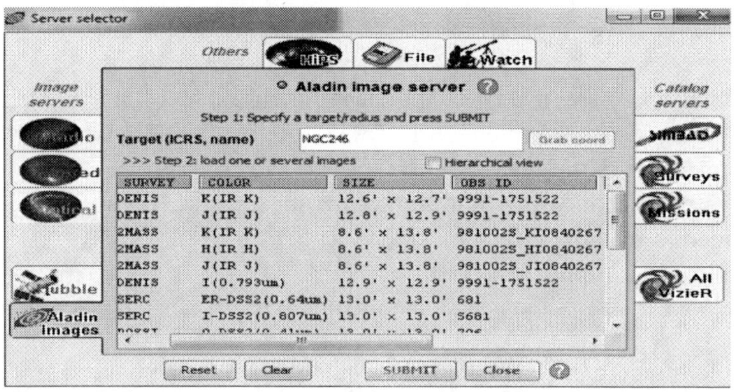

Figure 6. The server selector window with the list of available image .

We multiply this time by the velocity and obtain the linear size (the radius) of the nebula:

$R = 6.591*10^{10}\ s \cdot 38\ km/s =$
$= 2.54 \cdot 10^{13}\ km.$

5.2 Angular size of the nebula

The modulus of the projected angular distance vector 7.6'. In the case of figure 7. it is $r= 4.103'$. the angular size in radians:

$1pc=3.085 \cdot 10^{16}\ m$

$R=2.54 \cdot 10^{13}\ km.= 2.54 \cdot 10^{16} m=0.8233\ pc.$

Then

$1rad=206265''$

$r= 4.103'=246.18''=1193 \cdot 10^{-3}\ rad.$

finally compute the distance to the Crystal Ball Nebula.

$d = R / r =0.8233pc /1193 \cdot 10^{-3}= 690\ pc$

.

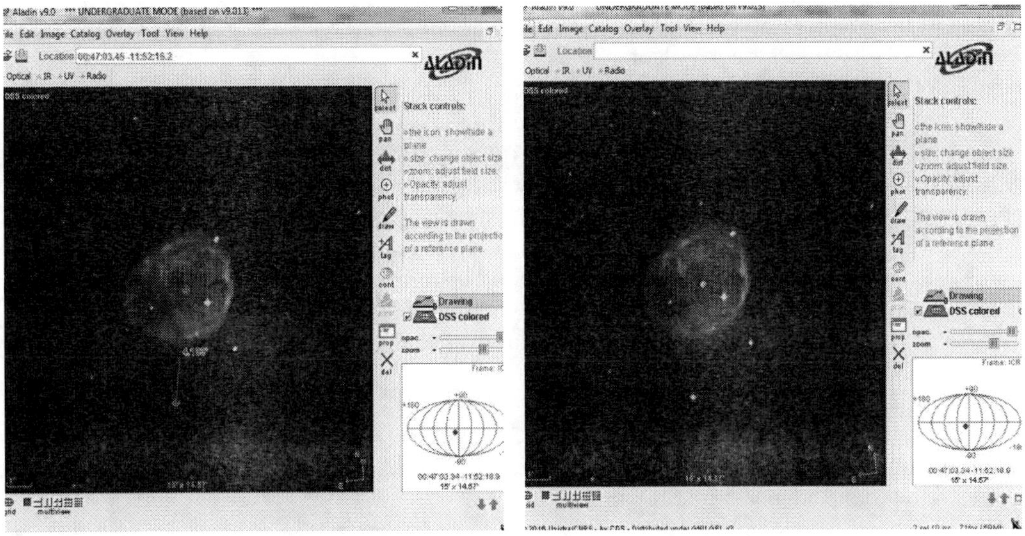

Figure 7. The distance vector for computing the angular size of the nebulae

In this sample, we calculated the distance from the nearest star through it is (UCAC2 27592058(12.949,Star).
The correct value of the distance to the Skull Nebula is about 620 pc .very close to the value obtained with our simple analysis. In particular our value is a particular, we have High value slightly because of the approximations of our procedure.

6. *Conclusions*

The distances of planetary nebulae are discussed as derived from their angular expansions and radial expansion velocities. An assessment is given of distances derived by this method both at optical and radio wave lengths. The recent VLA radio data show promise in establishing a more accurate distance scale for planetary nebulae.
Both the linear and the angular size of the nebula are comparing and computed their distance from us in parsec unit.
we have to express the sizes in the proper units: the linear size in parsec and the angular size in radians.

The distance calculated in with spectroscopic less distance in with expansion where the highest value in the spectroscopic is 879 pc. While the expansion is 917.4 pc , so note that the increase occurring in expansion.

7.References

[1] Trumpler R.J 1930 , Preliminary results on the distances, dimensions and space distribution of open star clusters. Lick Obs. Bull. Vol XIV, No. 420.
[2] Sundus A. A 2012, Distances Scale Determination for a Planetary Nebula in Galactic , Bulge International Journal of Astronomy and Astrophysics 3(02):81-86

[3] Wilson R.E. 1940.conntr.mount Wilson obs.631,110.
[4] Stuart R. Pottasch1.983," Planetary nebulae,Astudy of late stages of stellar Evolution" .p100,101.
[5] Zelik Gregory and Simith , Introductory Astronomy and Astrophysics(1992,Thomson learning).
[6] Heber .U.,Hunger k., Jonas G., Kudritz kiR.p1984 A&A130 ,119.
[7] Laty pov.A.A., 1957,publ.Astro.Obs.Tashkent 5,31.
[8] http://cosmo.phy.tufts.edu/uzirbel//laboratories/Astrometry prelab.pdf.
[9] lafrate G.,M.Ramella INAF-Astronomical observatory of Trieste."Distace to the crab nebula".
[10] Lizia S. Richared and V.Eva,"The magellanic cloud calibration of the calactic planetary Nebulae Distance scale".Astro physical ,Journal vol 80.2088,pp.1129.1150.
[11] Sun Kwok, "The Origin and Evolution of Planetary Nebulae", (Cambridge Astrophysics) ,(2007).

Acknowledgments
Authors wishing to acknowledge assistance or encouragement from colleagues, special work by astronomy department or financial support from organizations should do so we would like to thanks dr.sun kwok for assistance to perform our paper.

IOP Conf. Series: Materials Science and Engineering **757** (2020) 012044 doi:10.1088/1757-899X/757/1/012044

Evaluation of some properties of polyetheretherketone composites used for biomedical applications

A A Mohammed[1,2], E S Al-Hassani[1], and J K Oleiwi[1]

[1]Department of Materials Engineering, University of Technology - Iraq.

Abstract. Polyetheretherketone is considered to be a prime candidate with the potential to replace biomedical metallic materials as an orthopedic and dental implant on account of its elastic modulus similar to that of human cortical bone. This study presents the effect of polyetheretherketone specimens incorporated with different nano titanium dioxide and nano hydroxyapatite loading up to (1.5 wt%) have been prepared through a series of processing stages involving melt compounding, granulating and hot pressing. Following analysis of mechanical and chemical interactions by flexural, surface roughness and fourier transform infrared spectroscopy respectively. The results showed that the nano titanium dioxide and nano hydroxyapatite particles were uniformly disperse into polyetheretherketone, so the flexural modulus, flexural strength and surface roughness of polyetheretherketone biocomposites obviously increased with the addition of nano powder. On the other hand, break strain decreased with the rise of the total nano titanium dioxide and nano hydroxyapatite content. Furthermore, the best properties obtained when reinforced with nano hydroxyapatite.

Keywords: Polyetheretherketone, Hydroxyaptite, Titanium Dioxide, Polymer composites, dental implants.

1. Introduction

A dental subgingival implant is a fixture, surgically placed into the alveolar bone, which functions as an artificial root that can stabilize and support a fixed or removable prosthesis [1-2]. The implant material, usually titanium and its alloys [3], zirconia [4], or, as a potential future material, fiber reinforced composite (FRC) [5]. Over the last several decades, commercially pure grade 2 or 4 titanium and its alloys have been the material of choice for endosseous implants [6]. However, titanium has been shown to exhibit a variety of problems. Because of the high modulus of elasticity of the titanium alloys, dental implants made from the material can cause stress-shielding [7] which may lead to periodontal bone loss [8-9]. Polyether-ether-ketone commonly referred to as PEEK was first introduced in the late 1990s as a high performance thermoplastic polymer for replacing metal implant components. The material was initially used in orthopedics for the development of "isoelastic" hip stems and fracture fixation plates, with stiffness comparable to bone [10]. By April 1998, PEEK was commercially offered as a substrate for an implant material [11]. Poly-ether-ether-ketone (PEEK), a unique high performance semi-crystalline polymer, is one of the most potential thermoplastics for medicine, automotive and aerospace applications, with the advantage of good impact performance, chemical inertness, heat resistance and biocompatibility [12-16]. Hydroxyapatite (HAp) is an interesting biomaterial with potential orthopedic, dental, and maxillofacial applications due to its excellent biocompatibility, bioactivity, and osteoconductivity [17-18]. Some structural features of HA are presented in Figure 1 and Table I. This structure consists of Ca^{2+} ions surrounded by PO_4^{3-} ions and OH- with a Ca/P ratio of 1.6 [19-20].

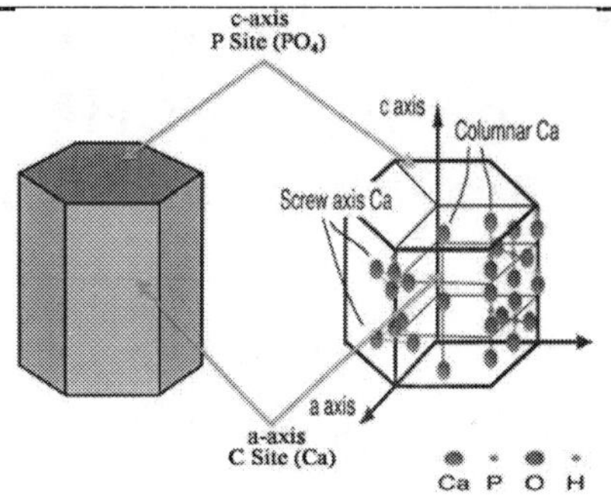

Figure 1. Hydroxyaptite hexagonal crystal structure [20].

Table 1. Hydroxyaptite crystallographic data[19].

Mineral	apatite
General group formula	$A_5(BO_4)_3X$ -
	Where A=Ca, Sr, Ba, Fe, Pb, Cd and rare earth;BO4=PO_4 $^{3-}$, VO_4^{3-}, SiO_4^{4-}, AsO_4^{3-}, CO_3^{2-}; X=OH^-, Cl^-, F^-, Co_3^2
Formula	$Ca_{10}(PO_4)_6(OH)_2$
Crystale structure	Hexagonal
Symmetry group	P_{63}/m
Class	6/m
Lattice parameters	a=b=9.432A°, c=6.881 A°;$\alpha=\beta=90°$, $\gamma=120°$
Ca/P ratio	1.2-1.6667

Titanium dioxide (TiO_2) material has been demonstrated with good antimicrobial activity [21-24], biocompatibility, bioactivity [25-28] hydrophilicity [29], and corrosion resistance [30]. Titanium dioxide occurs as two important polymorphs, the stable rutile and metastable anatase. Figure (2) shows the geometrical structures of the unit cells of anatase and rutile TiO_2. The unit cells contain four Ti atoms and eight O atoms for anatase TiO_2 and two Ti atoms and four O atoms for rutile TiO_2, respectively [31].

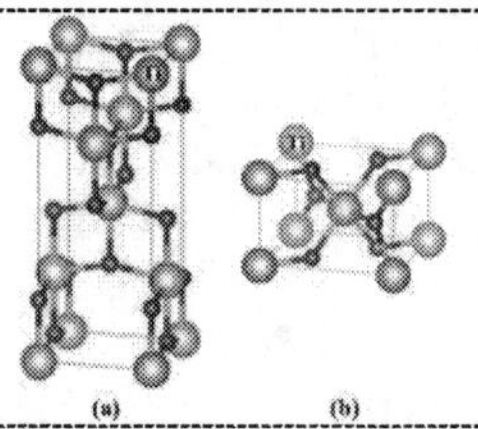

Figure 2. Unit cells of anatase TiO_2 (a) and rutile TiO_2 (b). Large light-blue and small red spheres are Ti^{4+} and O^{2-} ions, respectively [31].

The basic properties of antase and rutile are show in table (2):

Table 2. Table properties of anatase and rutile [32].

Property	Anatase	Rutile
Crystal structure	Tetragonal	Tetragonal
Lattice parameters	a = 0.3785	a = 0.4594
	c = 0.9514	c = 0.29589
Density (kgm^{-3})	3894	4250
Hardness (Mohs)	5.5-6	6-6.5
Bulk modulus (GPa)	183	206

Tang et al. investigated cyclic load on the PEEK/HA composite with different content of HAp showed the HAp/PEEK composite is a promising fatigue-resistant material for biomedical applications [33].
Wong et al. illustrated the bending modulus of strontium-containing hydroxyapatite/polyether ketone (Sr-HA/PEEK) increased with increasing the volume fraction of Sr-HA. The elastic modulus of 25 vol% and 30 vol% Sr-HA reinforcement showed 113% and 136% increase, respectively, in comparison with pure PEEK. The bending strengths of 25 vol% and 30 vol% Sr-HA reinforcement showed 25% and 29% decrease, respectively, in comparison with pure PEEK [34].
Kim et al. incorporated $30CaO \cdot 70SiO_2$ (CS) microspheres into PEEK to obtain a bioactive bone-repairing material with mechanical properties similar to those of the natural bone. It was observed the formation of hydroxyapatite on the surfaces of all CS-PEEK composites after immersion into simulated body fluid (SBF), indicating a potentially bond to the natural bone. Additionally, the bending stiffness and strength increased in 38% and in 6%, respectively, with the addition of 20 vol% CS particles. It was not observed a significantly decrease of the mechanical properties of the composites after immersion in SBF [35].
Wang et al. incorporated nano-FHA into PEEK in order to enhance its antibacterial property and osseointegration of implant materials. Mechanical tests revealed that nano-FHA acts as reinforcement filler in PEEK considering that both E and tensile strength increased in 210% and 67%, respectively, with the incorporation of 40 wt% nano-FHA into PEEK. Furthermore, microbiological assays indicated that the composite possesses an antibacterial activity. The in vivo assays showed high bioactivity, osseointegration and bone-implant contact [36].
The main aim of this study was to fabrication of poly(ether ether ketone) (PEEK) composite containing nanohydroxyaptite and nano titanium dioxide at different weight fraction (0, 0.5, 1, 1.5)% for two types of reinforcement and assess the some mechanical properties and chemical analysis was used

2. Materials and Methods

2.1. Sample Preparation

PEEK(551G) granule, Nano hydroxyapatite and Nano titanium dioxide were purchased from (Jilin Joinature Polymer Co., Ltd, china), (N & R Industries, Inc, China) and (Hangzhou Union Biotechnology Co., Ltd, china) respectively with purity 99.5%, 99% and > 99.8% respectively. The two nano particles have 20nm particle size. The biopolymer composites containing (0, 0.5, 1, 1.5) wt% for two types of nano powder were fabricated via a series of processes melt compounding, granulating and hot pressing process. PEEK granule were dried in an oven 80 °C for 1 h prior to compounding and hot pressing. In brief, nano powder were dispersed in ethanol alcohol using ultrasonic mixer to obtain a homogeneous mixture and then mixing with PEEK granule for 20 min. After well dispersed, the mixture was dried in a oven at 90 °C for 24 h to remove the excess ethanol alcohol.
In the compounding process, Compounding was carried out in an internal mixer (AHAAKE BUCHLER PRODUCT, HBISYSTEM 90, USA) at temperature of 360 °C and mixing speed at 90 rpm. The press was done by using hot press type (TOYOSEIKI, Japan). Temperature in hot press was adjusted at 360°C when reaching this temperature the pressure was applied for (15) min and its value equal (15MPa). The mould was then removed from hydraulic press and put in cooling system cooled with water jet (5 L/min) at temperature room temperature. The mould was opened, and then the composite sheet was removed and cutting the specimens by CNC machine for each test.

2.2. Characterization

The flexural property of synthesized composite [width (b)~13 mm, thickness (d)~4 mm and length~160 mm] was measured by three-point bending using Universal testing machine (SANTAM, Type STM-50, Iran) followed by ASTM D790-03[37].The support span was 128 mm with the cross head velocity at 5 mmmin-1. The flexural strength, flexural modulus and flexural stain at break of composites were calculated by using the given equations [38]:

$$\text{Flexular Strength} = \frac{3FL}{2bd^2} \qquad (1)$$

$$\text{Flexural Modulus} = \frac{L^3M}{4bd^3} \qquad (2)$$

$$\text{Flexural Strain \%} = \frac{6Dd}{L^2} \qquad (3)$$

Where, F is the maximum flexural load, L is the support span mm, b= width of the sample, d= thickness of the sample, D maximum deflection of samples and m= slope of the linear portion of the flexural load- deflection curve, respectively.

The measurement of roughness surface was done by fixing the sample on the machine (TR200, type TA620 stand and column, Time group Inc.), and applying the needle of the device perpendicular to the sample and taking the reading on different places on the surface of sample. The prepared samples were characterized by attenuated total reflection Fourier-transform infrared (ATR-FTIR) spectroscopy followed by ASTM E1252 [39] using (BRUKER-type I11316, Germany) FTIR spectrophotometer for 32 scans over the range of 700–4000 cm−1 at a resolution of 4.0 cm−1. For these analyses the samples were mixed crystals with KBr (potassium bromide).

3. Results and Discussions

3.1. Flexural Test

Figure (3) shows that the variation in flexural strength of the composite specimen increases with an increasing percentage of two type of reinforcement (from 0 to 1.5%) as they resist the deformation of the composite material. The flexural strengths of the sample were found to be about 7% higher than that of pure polyetheretherketone. The plastic region of sample depends on ductility of the PEEK matrix and interparticle distance of reinforcements. At higher % reinforcing, the interparticle distance increases make matrix hard to undergo yielding, increasing the flexure strength [40]. The highest flexural strength reached a value of 163.431 MPa for (PEEK+1.5%HAp) and showed good bend properties. While about 162.321 MPa for sample reinforced with 1.5%TiO2. Also the same trend was observed for flexural modulus in figure (4) were improved for all composites when adding nano particles. The highest flexural modulus of was obtained for the 1.5% composite reinforced because of excellent particle dispersion and distribution. On the other hand, while total reinforcement raises from 0 wt% to 1.5 wt% the flexural break strain % results of HAp/PEEK and TiO2/PEEK separately decreased as seen in figure (5). Such a results reveals that the existence of ceramic powder in biocomposites could importantly delay the plasticity to brittleness transmission of the polymer composites which rise with increase particle reinforcement [41].

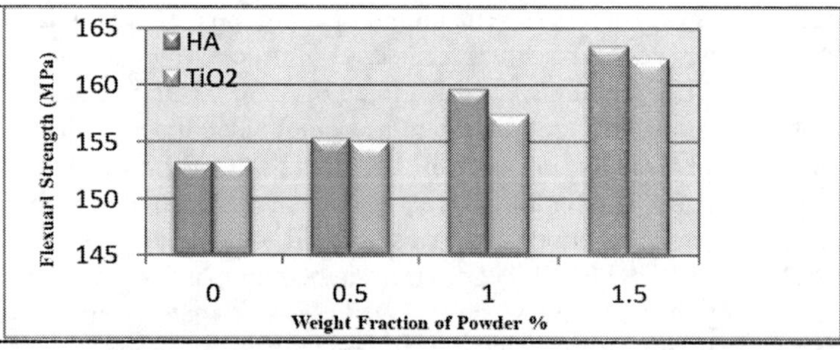

Figure 3. Results of flexural strength for polymer composites with two types of reinforcement.

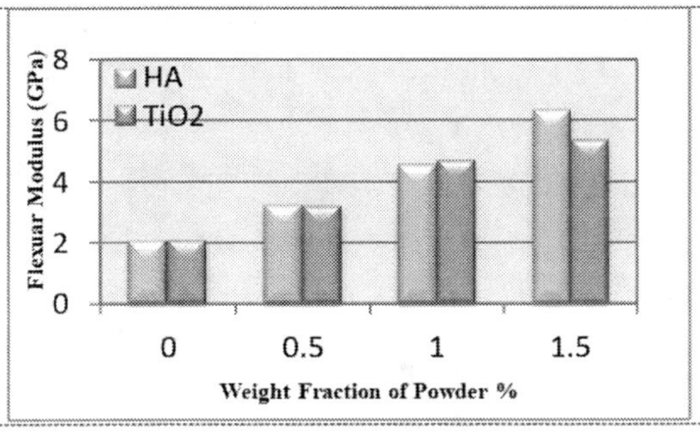

Figure 4. Results of flexural modulus for polymer composites with two types of reinforcement.

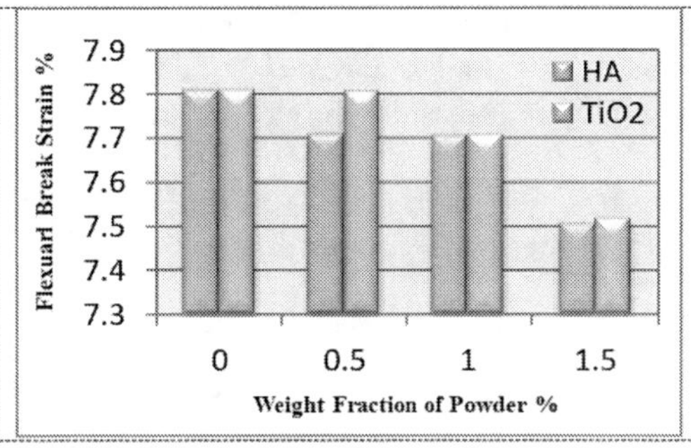

Figure 5. Results of flexural break strain% for polymer composites with two types of reinforcement.

3.2. Surface Roughness

Figure (6) shows the effect of nano particles on surface roughness of nanocomposites, this figure indicate the reinforcing leads to a clear increase in surface roughness. From this figure it can be noticed that the maximum surface roughness was obtained when using HAp as reinforcement material. The increased in surface roughness was reached to (1.253μm) for (1.5%HAp+PEEK) while reached to (1.0163 μm) for (1.5%TiO2) compared with value of (0.478 μm) for neat PEEK.

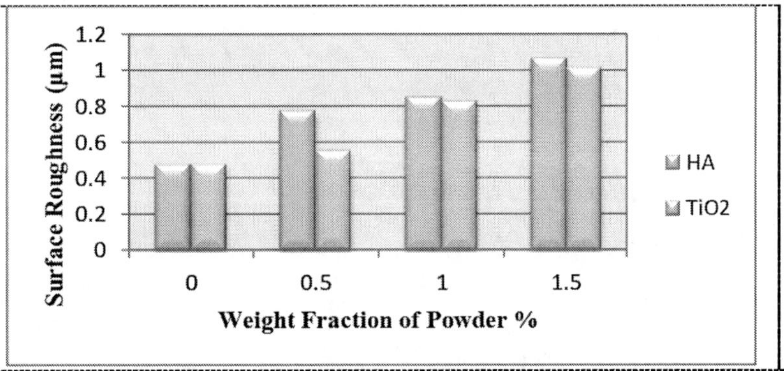

Figure 6. Results of surface roughness for polymer composites with two types of reinforcement.

3.3. FTIR Analysis

Figure (7) is the IR spectrum of HAp, In thise figure, IR characterization curve for HAp the absorption peak at 3443.41cm^{-1} is corresponding to the contracting vibration peak of O-H in H_2O; the absorption peaks at 1036.90cm^{-1} are corresponding to the key band of PO_4^{3}-; the absorption peak at 790.89cm^{-1} is corresponding to the key band of PO_4^{2-}; the absorption peaks at 568.10 cm^{-1} and 603.51cm^{-1} belong to the bending vibration absorption peak of PO_4^{3-}.

The spectrum of TiO_2 show in figure (8) exhibits an intense and wide band centered at ~3426.99 cm^{-1} attributed to the O–H stretching, as well as peaks at 1654.66 and 1043.50 cm^{-1} arising from the bending vibration of coordinated H_2O and Ti–OH [42]. The peak at ~566.42cm^{-1} is related to the Ti–O–Ti stretching, and that at 1465.75 cm^{-1} has been assigned to TiO_2 lattice vibrations [43].

FTIR spectra Figure (9) revealed bands attributed to the PEEK main chain in it, the stretching vibration peak for C=O is at 1648.10 cm^{-1}; the aromatic ring framework vibration peak is at 1593.49 cm^{-1} and 1485.73 cm^{-1}; asymmetric stretching vibration peak for R-O-R is at 1217.06cm^{-1}; symmetric stretching vibration peak for R-CO-R is at 925.29cm^{-1}; bending vibration absorption peaks for C-H out of the benzene ring plane is at 835.27cm^{-1} and 765.07cm^{-1}, meanwhile, at 836.92cm^{-1} is the characteristic peak of aromatic ring para-position substitution [44]. It can be again deduced that the chemical structure of PEEK substrate is not observably altered by the addition of nanoparticles bio composite samples as can be seen in figures (10 &11). Also it was clear that no new peaks and manifestly peaks shift were observed.

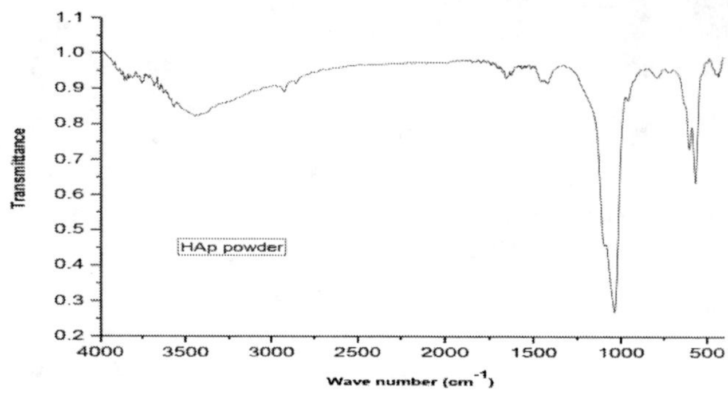

Figure7. FTIR analysis for HAp powder.

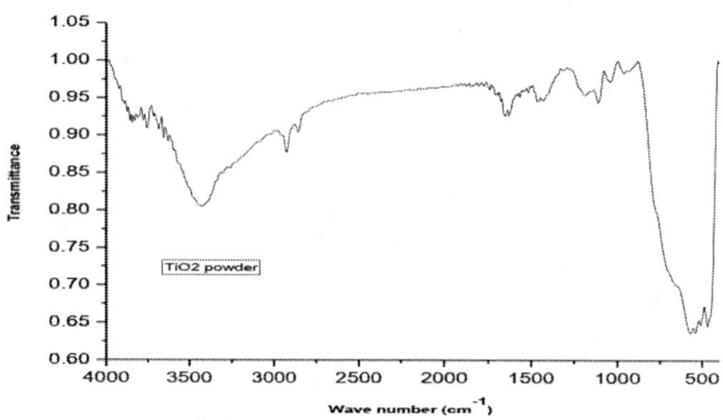

Figure 8. FTIR analysis for TiO_2 powder.

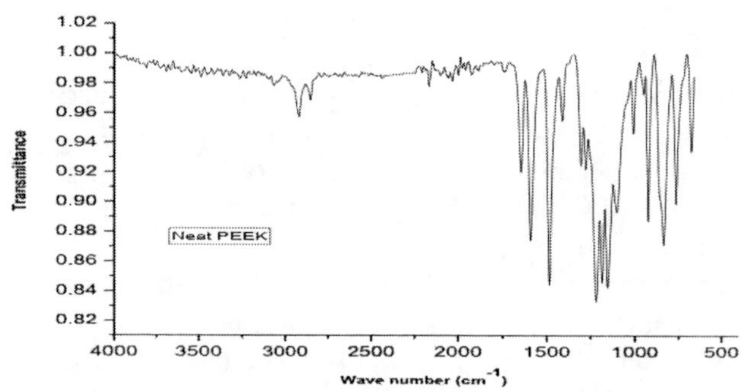

Figure9. FTIR analysis for pure PEEK.

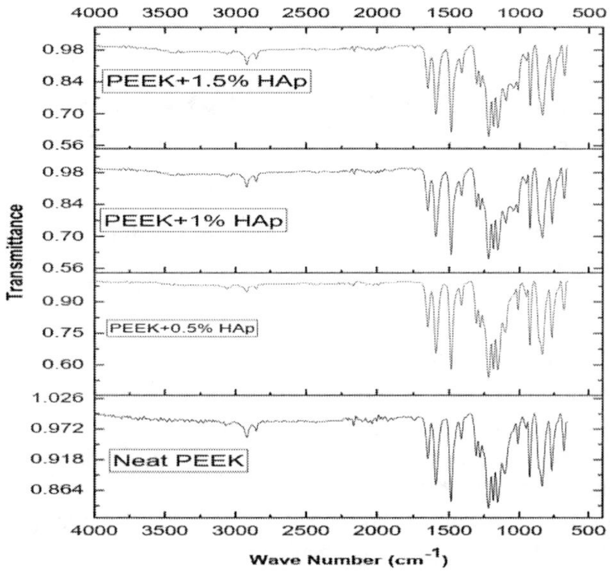

Figure10. FTIR analysis for polymer reinforced with HAp.

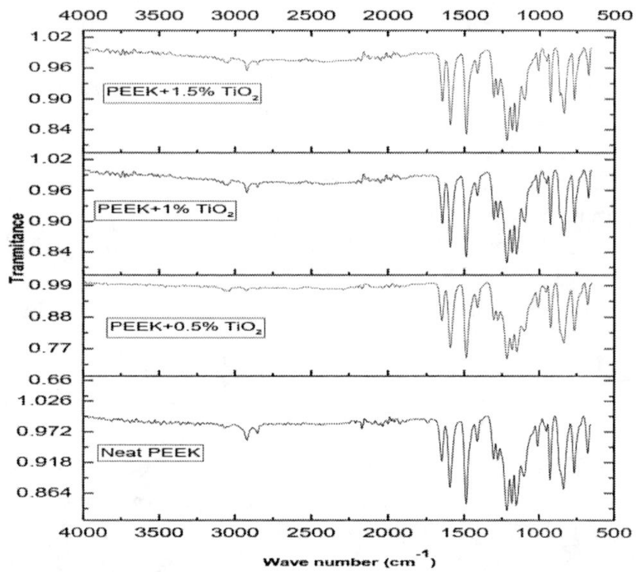

Figure11. FTIR analysis for polymers reinforced with TiO_2.

4. Conclusions

In this paper, the flexural behavior, surface roughness and FTIR analysis of nHAp/PEEK and $nTiO_2$/PEEK composites were investigated through, three-point flexural tests, surface roughness test and FTIR analysis it was found that:

1. The flexural behavior of polymer composite improved by the increase weight fraction of powder but the higher values obtained at (1.5wt%) for two types of powders. While the flexure break strain% showed opposite behavior.

IOP Conf. Series: Materials Science and Engineering **757** (2020) 012044 doi:10.1088/1757-899X/757/1/012044

2. The surface roughness of polymer samples increased with increase weight fraction of powder but the higher values obtained at (1.5wt%) for two types of powders.
3. The best properties obtained when reinforced with (nHAp) than reinforced with ($nTiO_2$).
4. The chemical structure of PEEK substrate is not observably altered by the addition of nanoparticles bio composite samples.
5. The above results suggest that 1.5% HAp +PEEK has clinical potential as an effective material for dental applications.

5. Acknowledgements

The Authors gratefully acknowledge the University of Technology-Iraq, Iran polymer and petrochemical institute for their support and performing this work.

6. References

[1] Awad M A, Rashid F and Feine J S 2014 Clinc. Ora. Imp. Rese. 25 46–51.
[2] Turkyilmaz I, Company A M and McGlumphy E A 2010 Gerodont. 27 3–10.
[3] Guo C Y, Tang A T H and Matinlinna J P 2012 J. Adhes. Sci. Tech. 26 189–05.
[4] Liu D, Matinlinna J P and Pow E H N 2012 J. Adhes. Sci. and Tech. 26 1249–65.
[5] Zhang M and Matinlinna J P 2012 Silic. 4 73–78.
[6] Br°anemark P-I, Breine U, Adell R, Hansson B O, Lindstr¨om J and Ohlsson A 1969 Scand. J. Plast. Reconstr. Surg. 3 81–00.
[7] Sarot J R, Contar C M M, Cruz A C C D and De Souza Magini R 2010 J. Mater. Sci Mater. Med. 21 2079–85.
[8] Huiskes R, Weinans H and Rietbergen B V 1992 Clin. Orthop. Relat. Res. 274 124–34.
[9] Najeeb S, Khurshid Z, Matinlinna J P, Siddiqui F, Nassani M Z and Baroudi K 2015 Int. J. Dent. 1-7.
[10] Kurtz S M 2012 PEEK Biomaterials Handbook (USA: William Andrew Elseveir Inc.).
[11] El Awadly T A, Radi I, EL Khadem A and Osman R 2017 J. Oral Implantol.43 456-61.
[12] Garcia-Gonzalez D, Rodriguez-Millan M, Rusinek A and Arias A. 2015 Com. Struc. 133 1116–26.
[13] Rae P J, Brown E N and Orler E B 2007 Poly.48 598–15.
[14] Zheng B, Wang H, Huang Z, Zhang Y, Zhou H and Li D 2017 Poly. Test. 63 349–59.
[15] Kurtz S M and Devine J N 2007 Biomaterials 28 4845–69.
[16] Zheng B, Deng T, Li M, Huang Z, Zhou H and Li D 2019 Pol.11 1-12.
[17] Hench L L 1998 J. Ameri. Cer. Soc. 81 1705–28.
[18] Pramanik N, Mishra D, Banerjee I, Maiti T K, Bhargava P and Pramanik P 2009 Int. J. Biomater. 1-8.
[19] Kay M I, Young R A and Posner A S 1964 Nature 204 1050.
[20] Ortega Clemente A I, Domínguez-Crespo M A, Torres-Huerta A M and Onofre-Bustamante E 2011 ECS Trans. 36 187-96.
[21] Chung C J, Lin H I, Tsou H K, Shi Z Y and He J L 2008 J. Biomed. Mate. Res. B 85B 220-24.
[22] Chung C J, Lin H I and He J L 2007 J. Mater. Res. 22 3137-43.
[23] Chung C J, Lin H I and He J L 2007 Surf. Coat. Technol. 202 1302-07.
[24] Chung C J, Lin H I, Chou C M, Hsieh P Y, Hsiao C H, Shi Z Y and He J L 2009 Sur. Coat. Tech. 203 1081-85.
[25] Xie D, Weng Y J and Huang N 2007 J. Func. Mater. 38 518.
[26] Lin C M and Yen S K 2006 Mater. Sci. Eng. C-Biomimetic Supramol. Syst. 26 54.
[27] Wu J M, Xiao F, Hayakawa S, Tsuru K, Takemoto S and Osaka A 2003 J. Mater. Sci.-Mater.Med. 14 1027.
[28] Harle J, Kim H W, Mordan N, Knowles J C and Salih V 2006 Acta Biomater. 2 547.
[29] Hata A, Kai Y, Yamanaka I, Oosaki H, Hirota K and Yamazaki S 2000 JSAE Rev. 21 97.
[30] Shan C X, Hou X and Choy K L 2008 Surf. Coat. Technol. 202 2399.
[31] Jia J, Yamamoto H, Okajima T and Shigesato Y 2016 Nano. Res. Lett. 11 1-9.
[32] Hanaor D A H and Sorrell C C 2011 J. Mater. Sci. 46 855–74.

[33] Tang S M, Cheang P, AbuBakar M S, Khor K A and Liao K 2004 Int. J. Fat. 26 49–57.
[34] Wong K L, Wong C T, Liu W C, Pan H B, Fong M K, Lam W M, Cheung W L, Tang W M, Chiu K Y, Luk KD and Lu W W 2009 Biomater. 30 3810–17.
[35] Kim I Y, Sugino A, Kikuta K, Ohtsuki C and Cho S B 2009 J. Biomater. Appl. 24 105–18.
[36] Wang L, He S, Wu X, Liang S, Mu Z and Wei J 2014 Biomater 35 6758-75.
[37] ASTM International, D 790–2003a.
[38] Askeland D R and fulay P P 2009 Essential of Materials Science and Engineering (USA: Cengage Learning).
[39] ASTM International, E 1252 – 98, 2002.
[40] Yunus M and Alsoufi M S 2018 Int. J. Biomater. 1-8.
[41] Pan Y, Chen Y and Shen Q 2016 J. Mater. Sci. Tech. 32 34-40.
[42] Colthup N B, Day L H and Wiberley S E 1990 Introduction to Infrared and Raman Spectroscopy (San Diego: Academic Press, Inc.).
[43] Nyquist R A and Kagel R O 1971 Infrared Spectra of Inorganic Compounds (New York: Academic Press, Inc.) p 214.
[44] Jin Y S, Bian C C, Zhang Z Q, Zhao Y and Yang L 2017 IOP Conf. Series: Mater. Sci. Eng. 16 71-6.

Gliding arc discharge for water treatment

T A Hameed1, S J Kadhem2
1, 2, Physics Department, College of Science, University of Baghdad, IRAQ
Corresponding author: E-mail: tamaraabood.d@gmail.com

Abstract

The significant shortage of usable water resources necessitated the creation of safe and non-polluting ways to sterilize water and rehabilitate it for use. The aim of the present study was to examine the ability of using a gliding arc discharge to inactivate bacteria in water. Three types of Bacteria satisfactory were used to pollute water which are Escherichia coli (Gram-negative), Staphylococcus aurous (Gram-positive) and salmonella (Gram-negative). A DC power supply 12V at 100 Hz frequency was employed to produce plasma. pH of water is measured gradually during the plasma treatment process. Contaminated water treated by gliding arc discharge at steadying the gas flow rate (1.5 l/min) and changing the exposure time of the polluted water to the plasma during periods of 10, 20 and 30 min.The bacteria which used show different responses when expose to produced plasma, most of them inactivated when treated with plasma for 30minutes.That's means Survival rate decreased with treatment time. Results show that gliding arc plasma is a powerful and green tool to treatment water without generating any byproducts.

Keywords: Gliding arc discharge, Escherichia coli.Salmonella,Staphylococcus aureus,Water treatment

1. Introduction

Plasma treatment is a committed ,non –chemical technique for inactivation of bacteria in water [1].The plasma which used in this work is ionized gas own concentrated energy with highly localized temperature increases close the arc of the gliding arc discharge. It is known, that plasma has a good ability to generate highly reactive species. The gliding arc discharge is able to generate a considerable amount of active plasma species because of its exceptional plasma properties. The collision of the energetic electrons produced in the discharge with gas atoms or molecules is the main reason for the formation of highly reactive species. The reactive species which are formed in electrical discharges in liquid or gas-liquid medium, which causes the degeneration of the biological pollutants, are ozone O_3, hydrogen peroxide H_2O_2, and hydroxyl radicals. Also reactive oxygen species (ROS) and reactive nitrogen species (RNS) are generated in the plasma like superoxide, atomic oxygen, hydroperoxyl radicals, nitrogen oxides, nitrites, nitrates, proxy nitrites as well as UV and electric fields. All of these plasma species may play a role in the inactivation of bacteria. Most of the active plasma varieties have very short life live on the order of μs or fewer [2-5].The exceptions to these are ozone and H_2O_2, which have comparatively long half-lives, and as

a result, may be particularly useful in the treatment of contaminated water [6].On the other hand, can be formed directly in water through the dismantling of water molecules by plasma discharges [7].
Gliding arc discharge plasma is an exemplary source of thermal plasma, however in some environments it may also produce non-thermal plasma (NTP). This discharge usually burns between two separating electrodes blown by injected gas. The discharge started in the narrowest inter-electrode area is wafted by flowing gas external the diverging area. The gliding arc collects the advantages of (thermal and non-thermal) plasma, i.e., non-thermal plasma conditions at a higher power [1].The exploratory, we find that the gliding arc discharge plasma is in contact with water by feeding argon gas, for inactivation of three sorts of bacteria (*E-Coli, salmonella, and Staphylococcus aureus).* The cell dike of these bacteria is so thick and develops on a rich medium. *Staphylococcus aureus* is answerable for a large range of infections like mild skin contagion (folliculitis, impetigo). *Salmonella* which can be transported by drinking or eating dirtied water or nourishment, in explicit. Salmonella passes through the body and simply make its way out of gorge acid, also accountable for many diseases including puke and fever [8]. These microorganisms in water may transport to the body [9].

2. Experimental part

The experimental setup used in the present work is illustrated in Figure (1). The gliding arc plasma system consists of four main parts: Power supply, DC/AC converter, coil, and two Electrodes. The gliding arc was generated using a dc power supply 12V connected to the DC/AC converter circuit. DC to AC converter circuit is designed and implemented. The DC/AC converter circuit is consisting of two major parts; the first part is the pulse generator circuit part while the second part is the high voltage circuit part. The purpose of the DC to AC converter circuit is to convert the 12 volts DC power supply to AC voltage equal to about 220 volts. DC/AC converter circuit is joined to the coil which is used to raise the output voltage 3000 to 13000 volt. The output of this circuit is attached to two electrodes. The gliding arc discharge (GAD) reactor consists of two stainless steel diverging electrodes, 1 mm thick, located under a feeding gas nozzle. These electrodes attached to ceramic support placed between two thin rectangular glass plates. The discharge is produced by two knifes-shaped electrodes 2.5 cm radius and thickness 1mm.The maximum gap between two electrodes is 2cm and the minimum is 4mm. The distance between the connecting points of the electrodes is 1 cm. The AC high voltage circuit provided 13kV at a frequency of 50 Hz. The gas Ar, (99.998% purity) is at a total flow of o.5, 1and 1.5 L/min.The arc is pushed away by the supply gas flow, carding, along the increasing electrode cavity and breaks in a large plasma plume. This system has been used to treat bacteria-contaminated water.
Three types of bacteria were used in this work. Two of them are Gram-negative (*salmonella* and *Escherichia coli*) and the other Gram-positive (*Staphylococcus aureus*). The streaking method was used for isolation pure colony of each bacteria. Bacterial was activated by culturing it on nutrient broth for 24h at 37°C and a second day takes a touch of activated bacteria by a loop and streaks it on nutrient agar .The concentration of peroxide (H_2O_2) was measured using test strips as follows 40 mL samples of treated water by gilding arc system were collected and a test strip was dipped into each water sample for about 20 seconds. After extraction strip of water, observed a change in color of a test strip and compared with the colors from the calibration standard data provided by the manufacture company. Also the acidification of samples was measured.

2.1 Treatment method

To test for inactivation of bacteria, 3 samples of distilled water (40 ml for each sample) were taken using a sterile pipette and placed in a petri dish and contaminated with three types of pathogenic bacteria (*salmonella, Escherichia coli and Staphylococcus aureus).* 20 ml of the contaminated water is put in a petri dish and placed on a magneticstirrer. The contaminated water is exposed to the plasma generated from the arc discharge and at a distance of 15mm between the surface of the water and the plasma. The contaminated water is treated by exposing it to plasma within a period of 10 minutes and changing the flow rate of argon gas (0.5,1 and 1.5) l/min. It was also treated by stabilizing the gas flow rate (1.5 l/min) and changing the exposure time of the contaminated water to the plasma during periods of (10,20 and 30) min these times were chosen to get a gradual killing rate .After treatment completed, 5 microliters of contaminated water were pulled and placed over Macconkey agar, then the bacteria are distributed by using sterile swab over the agar. Then, the plates were incubated at 37°C for 24h and bacterial colonies were counted.

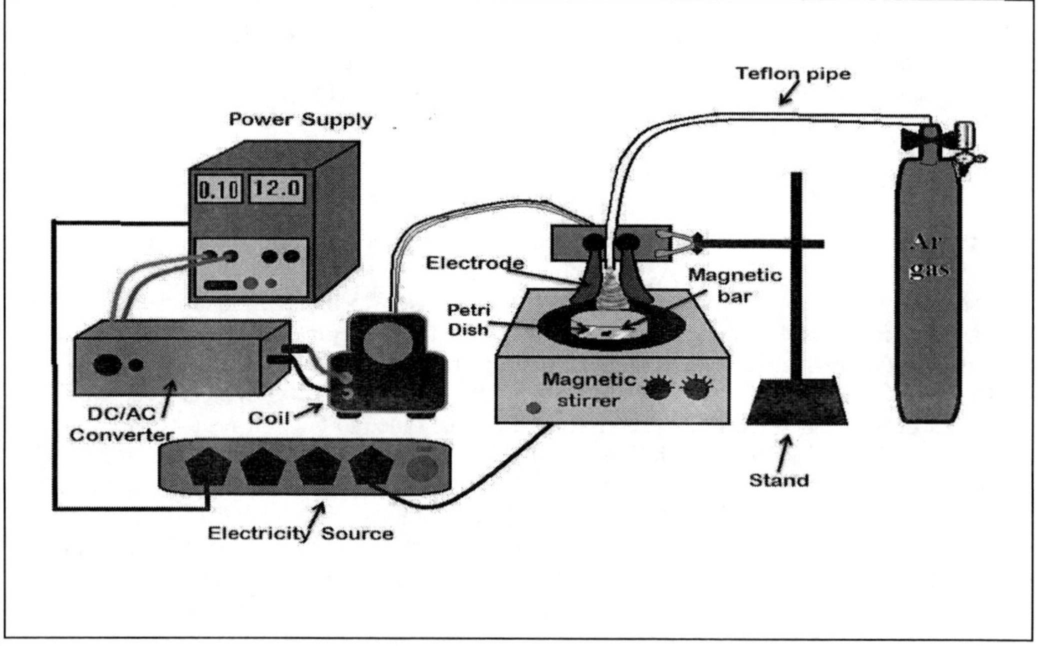

Figure 1. Schematic diagram of gliding arc discharge plasma system

3. Results and discussion

The acidity of the water was tested before and after the water treatment for all samples. It was observed that the pH of the sample gradually decreases when the exposure time increases as a result of nitric and nitrate products such as HNO_3, NO_2, and HNO_2 dissolving in treated water. Figures 1, 3 and 3 represent the change of pH of plasma treated water during different treatment periods which have been previously contaminated by the three types of bacteria (*salmonella and Escherichia coli and Staphylococcus aurous*). This study discusses the effect of changing the concentration of H_2O_2 on the sterilization of contaminated water by the gliding arc discharge system. Experimental results indicate that the gliding arc discharge reacts with water. It causes decreasing the pH of water but it generates H_2O_2. When water is exposed immediately to the Gilding arc discharge, the reactions begin with the dissociation of water molecules as follows [10].

$e + H_2O \rightarrow H + OH + e^{\cdot}$.......... (1)

e + $H_2O \rightarrow$ H- + OH .... (2)

M+ + $H_2O \rightarrow$ H2O+ + M .... (3)

H_2O+ + $H_2O \rightarrow H^+(H_2O)$ + OH .... (4)

After that H_2O_2 is produced from the recombination of hydroxyl radicals.

OH + OH + M $\rightarrow H_2O_2$ + M .... (5)

Figure 5 shows the relationship between the peroxide concentration and the exposure of the contaminated water to the gilding arc discharge.

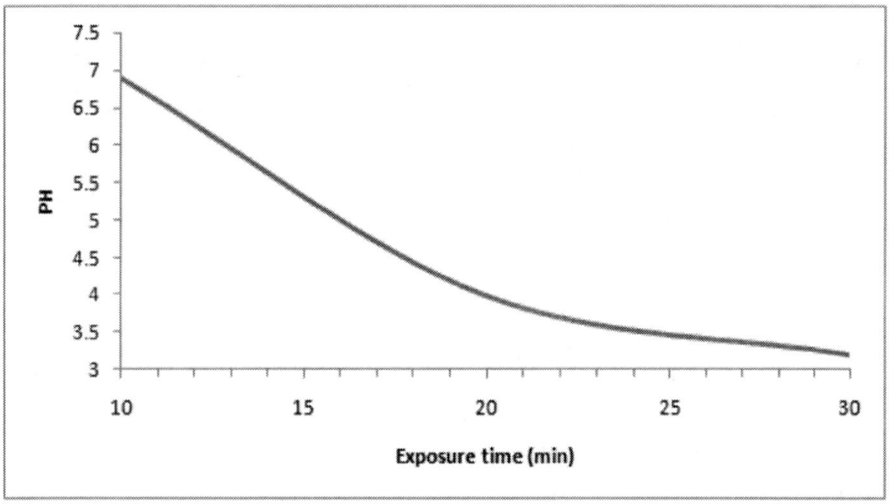

Figure 2. PH changing of water polluted with staphylococcus bacteria before and after treatment by Gilding arc plasma.

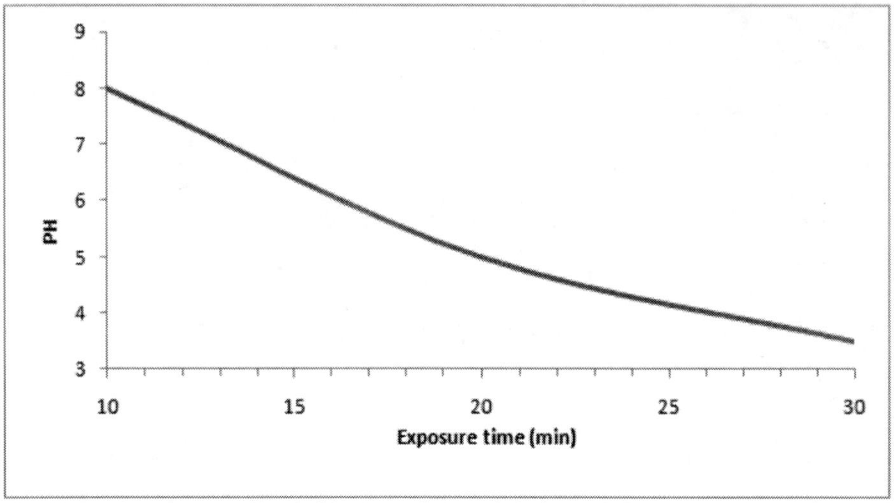

Figure3. PH changing of water polluted with salmonella bacteria before and after treatment by Gilding arc plasma.

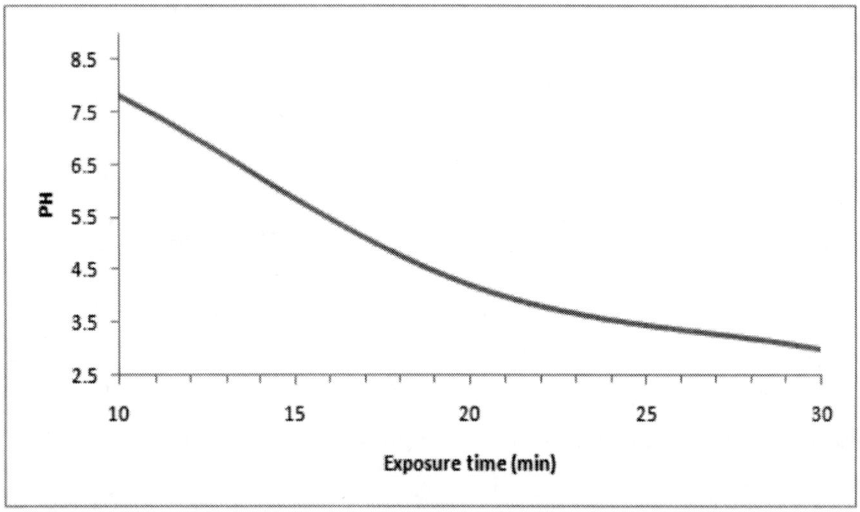

Figure 4.PH changing of water polluted with E-coli bacteria before and after treatment by Gilding arc plasma.

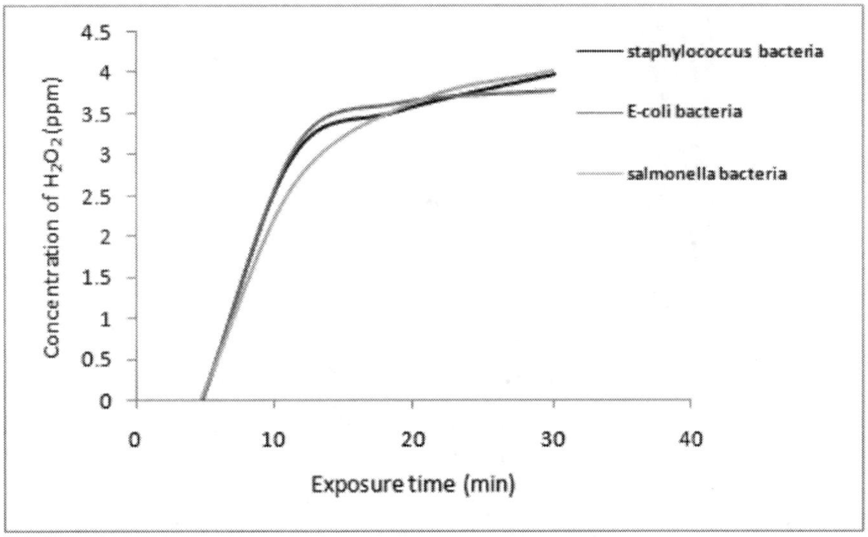

Figure 5. Concentration of peroxide of treated water.

3.1Inactivation of Bacteria

'Figure (6)', 'Figure (7)', 'Figure (8)', 'Figure (9)', 'Figure (10)', 'Figure (11) 'respectively show the effect of treatment by Gliding arc discharge on Escherichia coli, Staphylococcus aureus and Salmonella bacteria colonies, respectively. One can be observed from the figures that the rates of killing increase with an increase in the exposure time of the polluted water to the plasma formed by the Gliding arc discharge. These Figures refer to the high efficiency of the gliding arc discharge plasma water treatment method to inhibit bacteria. When H_2O_2 form in low pH water, it becomes a powerful oxidizer and an active tool for the inhibition of bacteria. It has an immediate effect on the outer membrane of bacteria because of the peroxidation of cell membranes.
When increasing the exposure time, the number of bacteria colonies was decreases. E-coli bacteria are more resistant than the other types of bacteria, where at 20 minutes a few colonies were killed. While salmonella bacteria were the most affected compared with other types of bacteria. At 20 min, there is no trace of salmonella colonies appears.

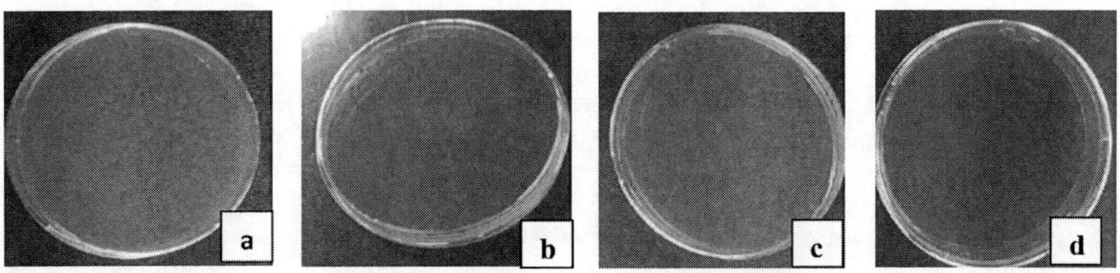

Figure 6: E-coli bacteria inactivation at different time: a) control, b)10, c) 20min, d) 30min

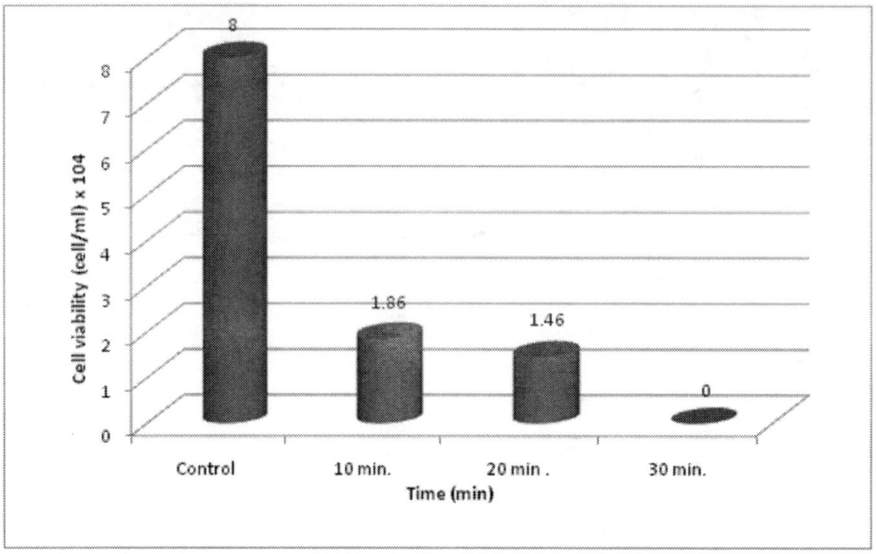

Figure 7. Histogram effect gliding arc plasma of different time in cell No. viability of Escherichia coli. All data are expressed as a mean± standard deviation LSD test (ANOVA).Statistical significance was considered as P<0.01.

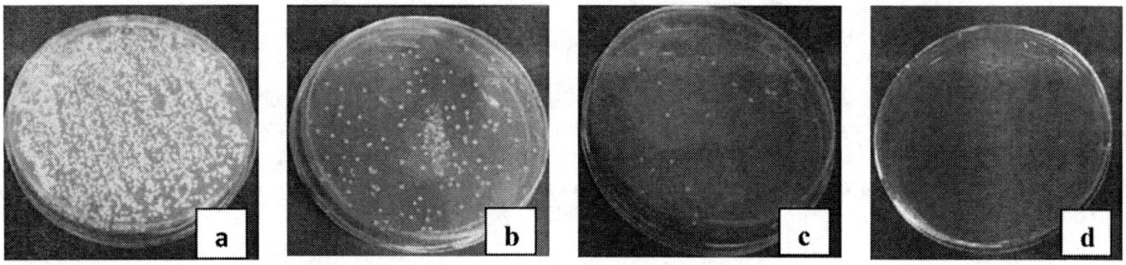

Figure 8. Photograph of the Staphylococcus aureus bacteria inactivation at different time :a) control, b) 10min, c) 20min, d) 30min

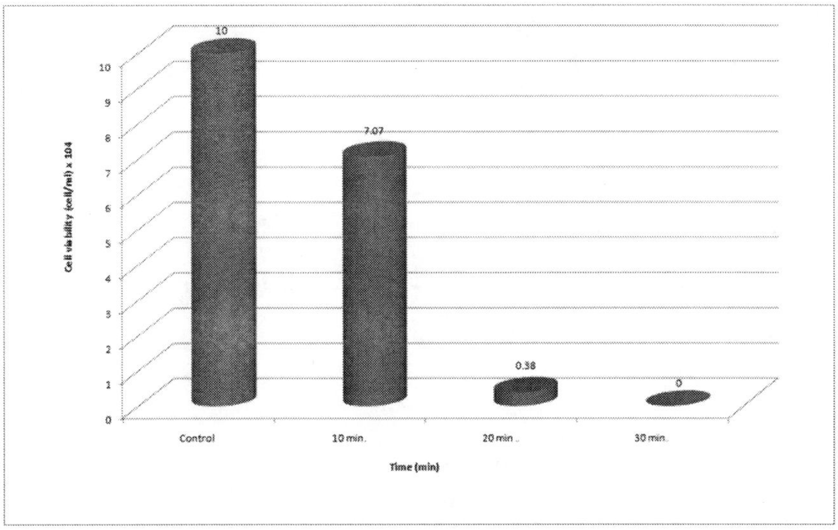

Figure 9. Histogram effect gliding arc plasma of different time in cell No. viability of Staphylococcus aureus. All data are expressed as a mean ± standard deviation LSD test (ANOVA) .Statistical significance was considered as P<0.01.

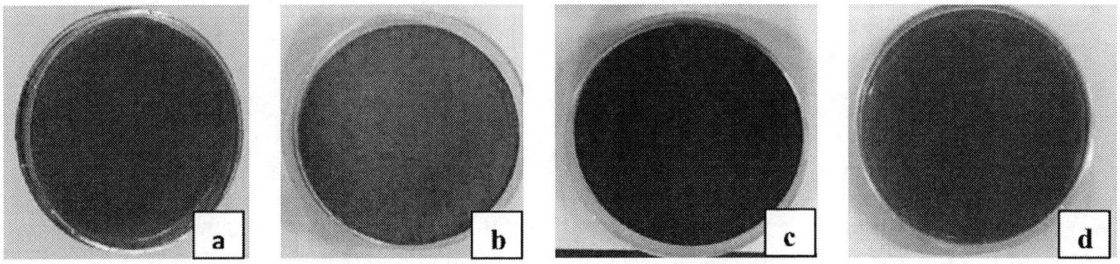

Figure 10. Photograph of the Salmonella bacteria inactivation at different time :a) control, b) 10min, c) 20min, d) 30min

IOP Conf. Series: Materials Science and Engineering **757** (2020) 012045 doi:10.1088/1757-899X/757/1/012045

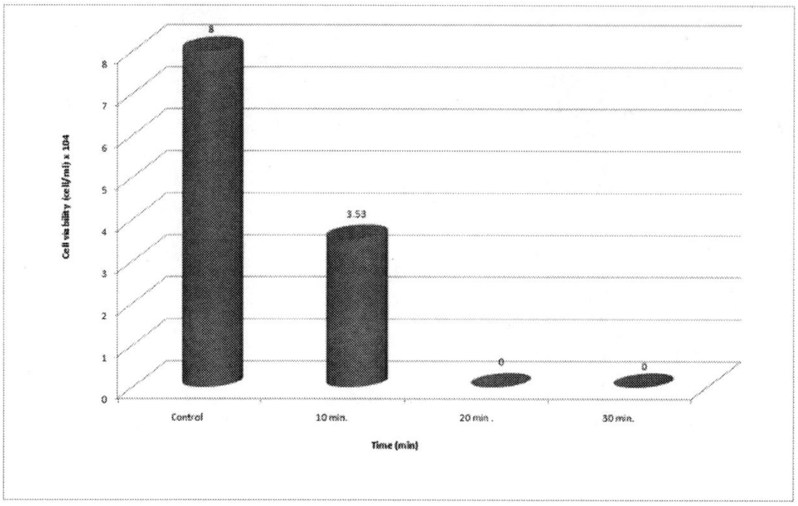

Figure 11. Histogram effect gliding arc plasma of different time in cell No. viability of Salmonella. All data are expressed as a mean ± standard deviation LSD test (ANOVA).Statistical significance was considered as $P<0.01$

4. Conclusion

The gliding arc discharge method is very efficient in the treatment of biologically contaminated water. A large volume of water can be treated by this method. The pH value of treated water a bit decreases with increasing treatment time, and the product of H_2O_2 in water increases safely with increasing treatment time. The colonies density decreases gradually with increasing exposure time and there is no trace of bacteria (*salmonella*) at 20 min but the (Escherichia coli and Staphylococcus aureus) killed at 30 min. When peroxide is formed in a low pH water, it becomes a strong oxidizer and has high efficiency in inhibiting bacteria.

Acknowledgment
We would like to present our thanks to the staff of biological and thermodynamics laboratory in the University of Baghdad, college of science, physics department for their kind support.

Reference

[1]Yang Y, Hyung- soon K, Andry S, Yong Cho, Alexi F. 2011 Rev. Sci .Instrum. 82 1-10.

[2] Alexi F.2008, Cambridge Univ. Press, Cambridge, UK. 71 1619-1625.

[3] Helmut S.1993 Eur. J.Biochem. 215 213–219.

[4] William A. Pryor, 1986, Annu. Rev Physiol. 48 657–667.

[5] Sies H, Stahl W and Sundquist A.R.1992 Annu.NY Acad Sci. 669 67–20.

[6] Oehmigen k, Hhnel M, Brandenburg R, Wilke C, Weltmann K.D and Woedtke T.V.2010

IOP Conf. Series: Materials Science and Engineering **757** (2020) 012045 doi:10.1088/1757-899X/757/1/012045

. Plasma Process Polym. 7 250–257.

[7] Forni L, Bahnemann D, Hart E.J.1982 , J. Phys chem. 86 255–259.

[8] Mehranz G, Gharagozalian, Davoud D and Mahmood Ghoranneviss.2017, J Theor Appl Phys 11 171-180

[9] Andersen B, Frisvad J, Søndergaard C, Rasmusseni, and Larsen I.S. 2011Associations

. Appl. Environ. Micro. 12 4180–4188.

[10]Hyoung- Sup K., Kamau C. W, In-Hwan H., Dong- Hwan L., Alexander R., Alexander F,

and ChoY. I.2011, Plasma Medicine. 1 295–307.

IOP Conf. Series: Materials Science and Engineering **757** (2020) 012046 doi:10.1088/1757-899X/757/1/012046

Preparation of Hydroxyapatite/ Zirconia Porous Composites via Polymeric Sponge Method and Study the Physical and Bioactivity Properties

M Mohammad[1], A Taha[1], S Shahatha[1*]

University of Technology/ Baghdad/ Iraq[1]

*sarahamid_91@yahoo.com

Abstract. In this study, Hydroxyapatite/ Zirconia porous composites at different ratio were fabricated by polymeric sponge method. HAp were produced via solid state reaction in molar ratio (3:2) between TCP and $Ca(OH)_2$. The crystal structure of the samples were characterized using X-ray diffractometer (XRD), scanning electron microscopy (SEM). Furthermore, the porosity and rheological properties of the samples were also tested. It's found that the compression strength with the highest value (13.2 MPa) obtained when the composition ratio is (30:70 wt %) and porosity (65.7%). Incubated for 30 days in Simulation Body Fluid, a bone like apatite layer was completely formed on the surface of the porous composite.

Keywords. Hydroxyapatite, Zirconium oxide, Polymeric sponge method, mechanical properties, simulation body fluids, degradation.

1. Introduction

Hydroxyapatite (HAp) ($Ca_{10}(PO_4)_6(OH)_2$ is a promising material for bone reconstruction based on its chemical structure. It is very much the same to the mineral parts of bone and tooth and shows excellent biocompatibility and osteoconductive properties [1, 2]. Additionally, HAp materials do not exhibit any cytotoxic effects [3]. Yet, the application of HAp is restricted for use as implant material because of its poor mechanical properties [4, 5]. Many studies focused at improving the mechanical properties of HAp, for example making them as composite materials [6]. Addition of second-phase ceramic materials (e.g. ZrO_2, TiO_2 or Al_2O_3) into the HAp matrix for improving strength and toughness [7].

Zirconia (ZrO_2) is a well-known reinforcing factor as a result of it's comparatively "high mechanical strength, toughness, wear resistance, corrosion resistance and biocompatibility" [8, 9]. At ambient pressure, ZrO_2 polymorphs occur in three forms; monoclinic (M), tetragonal (T), and cubic (C) depending on the temperature [9, 10]. As a result of phase transformation, pure and single-phase ZrO_2 tends to fracture even at room temperature, leading to deterioration in mechanical properties and consequently making it unsuitable for clinical applications. Therefore,

ZrO_2- based composites possess enhanced toughness following controlled phase transformation of stabilized ZrO_2 doped with yttria (Y_2O_3) (Y-TZP) [10]. Theoretically, ZrO_2 can be utilized as a composite material because of its stable-biocompatibility and proceed superb mechanical properties [11]. Furthermore, ZrO_2 is also very appropriate to be utilized as a bone and dental implant material because of its aesthetic quality properties [12].

Many techniques have been sophisticated to manufacturing porous HAp /ZrO_2 composite scaffolds, generally including of organic particles in the HAp/ZrO_2 composite powder, direct foaming, and polymeric sponge method [13, 15]. Porous ceramic which prepared by the replication method have a "controllable pore size, interconnected pores, and desired geometry" but a weak mechanical properties for using in load-bearing [16].

Previous studies focusing on the use of other materials to form HAp composites which have better compressive strength. Wang et al. studied 3%yttria-stabilized 20wt% zirconia (ZrO2) to get porosity between (87%~35%) and a compressive strength of (12 MPa) [17].

Here, we report on zirconium oxide (ZrO_2) particles that used for reinforcing hydroxyapatite particles for orthopedic implant considerations. Specifically, the effects of ZrO_2 particle on bioactivity property of hydroxyapatite for bone regeneration. The porosity and viscosity were determined by Archimedes method and viscometer at spindle speed of (100 rpm). The Crystallographic & morphological studies were performed using optical micrograph, X- Ray diffraction and "Scanning electron microscope", compression strength of the templates were tested by compression strength test "ASTM C77-88 (1999)" using Instron machine "model 1195" at crosshead speed of (1 mm/min).

2. Experimental Procedure

2.1 Preparation of the Samples

The fabrication of high porous HAp/ ZrO_2 biocompatibility ceramic foam was done. Initially HAp powder was fabricated by solid state synthesis the starting material that have been used were commercial tricalcium phosphate TCP (Himedia, India) and calcium hydroxide, $Ca(OH)_2$ (Himedia, India). The powders mixed in 3:2 molar ratio and were mechanically dispersed in de-ionized water at solid loading 55%. Thereafter adding 3% yttria-stabilized zirconia with ratios of (5, 10, 15, 20 and 25 wt%) from the total solid loading , loaded and 4% (poly vinyl alcohol); PVA were added as a binder and magnetic stirred for 2 hours until the PVA dissolve and the slurry with (pH= 12) is ready for the dipping process .

Polymeric sponges that used each cut into a dimension of 1 cm^3. The viscosity of the ceramic slurry was analyzed utilizing viscometer. The templates were squeezed by hand many times then put in a "desiccator" and evacuated to (10 $^{-1}$ bar0 to take off the restricted air and make certain that the templets is fully filled with the ceramic slurry. The excess ceramic slurry was extracted by squeezing the sponge through a roller. The filled samples were dried in "microwave oven" for (10 to 25 minute) according to viscosity of solid loading. The Samples were sintered at 1250 °C with heating rate of 5 °C/min and three hours of soaking time.

2.2 Immersion in SBF Solutions

Samples (HAp+ ZrO_2 porous composites) (1 cm^3) were put in plastic containers with (500 ml) of (SBF x5 gm/l). The containers were sealed with an airtight and the temperature stay constant at 37 °C. After (30 days) in touch with SBF to perform many characterizations rinsing them with distilled water and then drying them with microwave to prepared it for XRD and SEM analysis. Table (1) shows the composition of simulation body fluids.

Table 1. Composition of SBF and concentration of (SBF×5) [17].

ITEM	SBF×5 (gm/l)
NaCl	40.18
$NaHCO_3$	1.76
KCl	1.125
K_2HPO_4	1.15
$MgCl_2.6H_2O$	1.555
$CaCl_2$	1.465
$NaSO_4$	0.36

3. Results and Discussion

3.1 Morphological Characterization

Representative SEM and optical pictures of HAp/ZrO_2 samples sintered at 1250℃ are respectively showed in Fig (1). As the HAp/ZrO_2 concentration was "increased", the pores became "interconnected with more dense and thicker pore walls", which are very important factors that enhancing the mechanical properties of the porous materials.

At low concentration of HAp/ZrO_2 which is (35%), the "polyurethane foam structure" was completely replicated with fully interconnected open pores. Yet, when the HAp/ZrO_2

concentration reached to (55%), the pores were slightly blocked as an effect of thicker walls. A elevated densification regime is indicated after sintering at (1250 °C) which showed that the sintering process almost complete and subsequently grain growth occurred at temperature (1250°C) with minimum decomposing of HAp to Tri calcium phosphate.

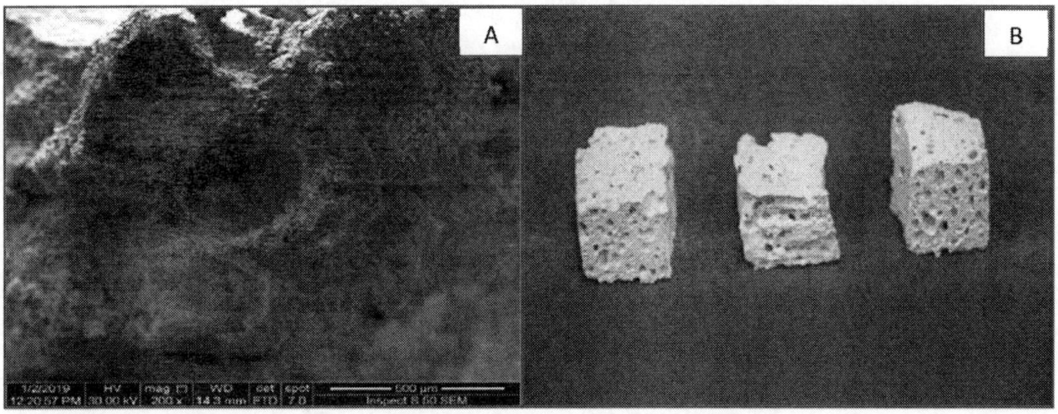

Figure (1). A. SEM image of hydroxyapatite/ZrO_2 porous ceramic sintered at 1250 °C, B. Optical micrograph of ceramic composites foam.

3.2 Crystallographic Investigation

The results of the crystallographic analysis of the produced samples, carried out with "X-ray diffraction", are presented in Fig. 2. X-ray diffraction patterns of HAp/ZrO_2 composite samples reveal peaks associated with HAp (Card no. 01-074-977). From the XRD it can be seen that at temperatures of (1250 °C), a continued shrinkage in the yield of HAp phase and an increased yield of (b-TCP) is can be seen. HAp is decomposed by heat into (b-TCP), ($Ca_3(PO_4)_2$), ($CaZrO_3$) and (water vapor). The ($CaZrO_3$) released stabilizes the (m-ZrO_2) by solid solution which create the formation of (t-ZrO_2). This is lead to the fact that "porous HA/ZrO_2 scaffolds" have a better mechanical properties compared with that of pure HAp porous scaffolds. The following chemical reaction is suggested:-

$Ca_{10}(PO_4)_6(OH)_2 + ZrO_2 \rightarrow 3Ca_3(PO_4)_2 + CaZrO_3 + H_2O \uparrow$...(1)

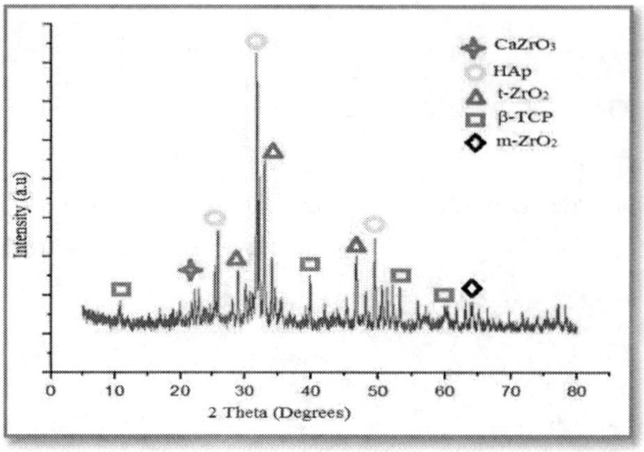

Figure (2). XRD patterns of HAp/ ZrO2 powder.

3.3 Physical Properties Measurement

Fig.3 (A) showed the porosity over the change of the ratios of (HA/ZrO_2) samples as an action of (HA/ZrO_2) concentration. The porosity "decreased with increasing ($HA\text{-}ZrO_2$) concentration". The obtained porosity ranging from (65.7 to 84.4%), are predictable to be sufficient to allow the tissue ingrowth and anchor the prosthesis to the surrounding bone in addition to provide blood and nutrients to the "bone-like vascular canals". This behavior could be understood in term of the rheological properties where viscosity of the slurry is one of the most important factors that control the porosity. Since solid loading is responsible for open and close porosity, clearly increasing viscosity will greatly decreases the physical properties as shown in Fig.3 (B), because it will block the open porous, therefore, decreasing the solid loading usually enhances the physical properties but decreases the mechanical one [19].

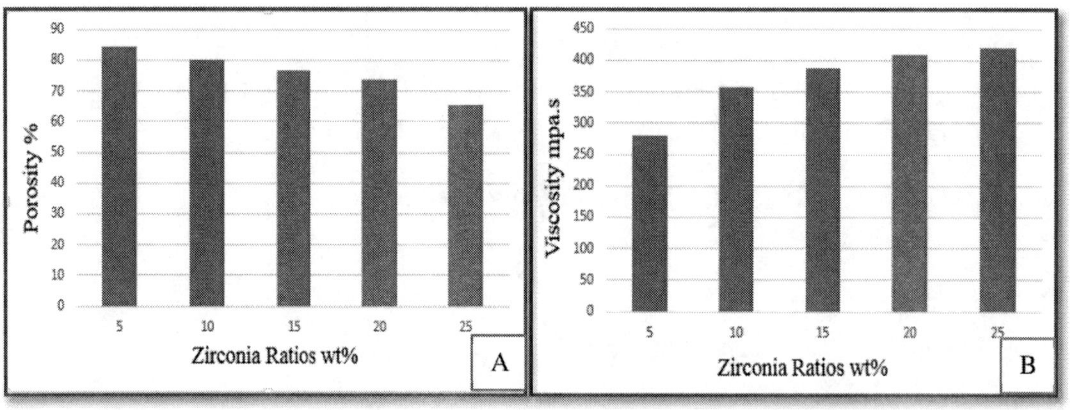

Figure (3).A. Porosity as % of hydroxyapatite /ZrO_2 porous ceramic at (5, 10, 15, 20, 25%) from total solid loadings, B. Viscosity of hydroxyapatite /ZrO_2 porous ceramic at (5, 10, 15, 20, 25%) from total solid loadings.

3.4 Compression Strength Measurement

Fig. 4 shows the compressive strength HAp composites (5, 10, 15, 20 and 30wt% ZrO_2) prepared under vacuum and dried by microwave oven. The highest value for compressive strength was obtained for samples prepared at (25 wt. %) (13.2 MPa), (20wt %) (10.1 MPa), (15wt %) (8.3 MPa), (10wt %) (6.25 MPa) and finally the lowest at (5 wt. % ZrO_2) (4.5 MPa). This was attributed to highest density, the higher result of porosity percentage contributed to lower strength of samples. This result due to cell wall thickness that increased with increasing zirconia ratio. This was mainly because of the porosities decreased as the weight fraction of the zirconia increased. The strength is a property which is strongly dependent on the porosity and microstructural defects of the samples.

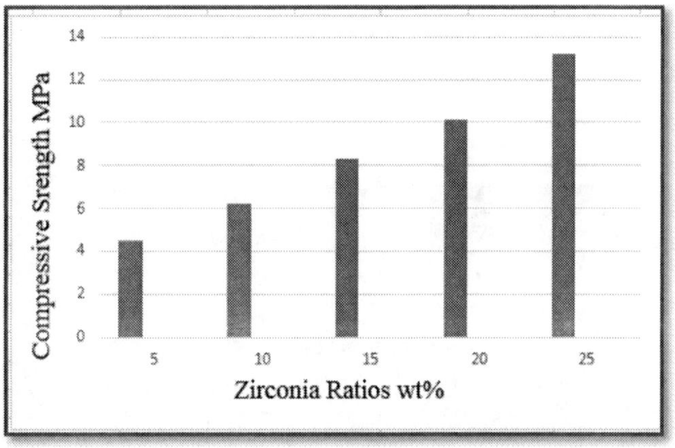

Figure (4). Compressive strength of hydroxyapatite /ZrO_2 at (5, 10, 15, 20, 25 wt. %) from total solid loadings.

3.5 Bioactivity Tests

In order to investigate the bioactivity behavior of the specimens and, in vitro test was carry out. In (vitro test) was carried out in “simulated body fluid (SBF)”, prepared according to the “Kokubo prescription” [20, 21] under a stable temperature (37 °C) and pH (7.2). Templets were extracted from the solution after (4 weeks). Bioactivity tests were done on the templates which showed highest physical and mechanical properties. Starting from the first week, intensive “apatite layer” formation for all samples were determined as shown in Fig. 5 which show the SEM image. When the immersion period in SBF was “increased”, the “size and quantity of the apatite layer” increased, too, as predictable. Unexpectedly, the size and quantity of the apatite layer also increased with increasing firing temperature. This is perhaps based on a quiet “smooth morphology, well-arranged grains, and presence of the glassy phase” which improving the apatite formation at higher firing temperatures. At the end of (4 weeks), the samples' surfaces were completely covered by the "apatite layer". For the (HZC composites0 sintered at (1250 °C), the present apatite nucleation was seen only around ZrO_2 grains in the "first week". At the end of (3 weeks), the apatite accumulation also happen on the ZrO_2 grains.

The results in Fig. 6 of XRD analysis of sample (75% HAp + 25% ZrO_2) composites, soaked in SBF solution for 30 days and if compared between A and B its can been seen that all the peaks referred as $Ca_3(PO_4)_2$, $CaZrO_2$ are completely disappears and HAp phase had also changed and become less in height in some positions.

Also the Ca/P ratio change before the immersion was (1.67) and after (1.25) for the selected sample. This decrease may be attributing to precipitation of a non-stoichiometric HAp from SBF. This result is in a good agreement with the earlier report that had written by Kaygili et al. [22] and also agree with Bulut et al. as they stated in their paper the apatite nucleation was also observed on (ZrO_2 grains). At the end of the 'four weeks", the matrix was fully coated with an appetite layer. [23]

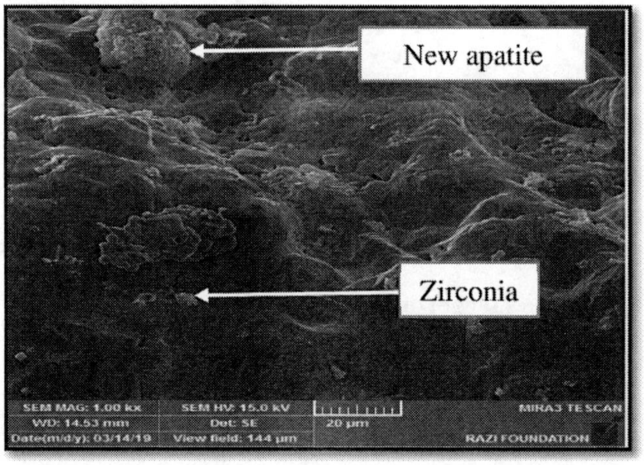

Figure (5). Scanning electron micrographs of HAp/ZrO_2 porous composites after 30 days of immersion in 5xSBF at 37 ℃.

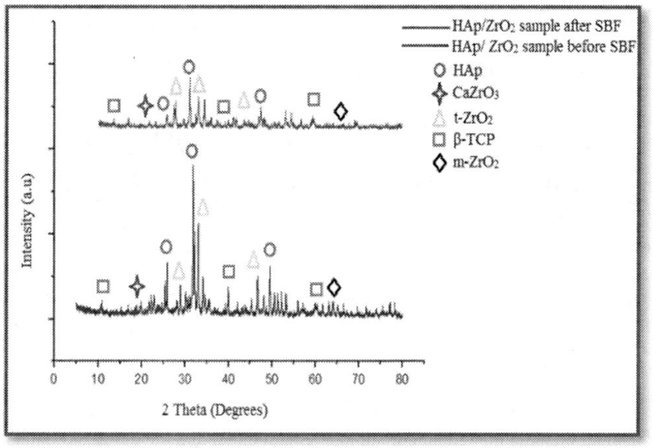

Figure (6). Show the XRD analysis before and after SBF soaking for 30 days at 37 ℃.

4. Conclusion

HAp/ZrO_2 porous composites were successfully fabricated by mixing Tri Calcium Phosphate and Calcium Hydroxide in "molar ratio (3:2)" via polymeric sponge methods. The porosity in the porous ceramic can be controlled by changing (HAp/ZrO_2) concentration and the porosity is ranging (65.7- 84.4%). A compressive strength of (10.1 MPa) for) (HAp/ZrO_2) porous scaffolds with (HAp/ZrO_2) concentration of (25 wt% zirconia+ 75wt% HAp) is the best because it's similar to cancellous bone, the composites having the best physical and mechanical properties also enhanced bioactive properties which is very comparable to that of cancellous bone. Overall, this study introduces a new method to create (HAp/ZrO_2) porous ceramic composites and its can be used as promised candidate for bone replacement and regeneration.

Acknowledgement

Authors would like to thank Ceramic lab. /applied sciences department/ University of Technology for their help to accomplish this work.

5 References

[1] Hannora, A.E. 2014 Preparation and characterization of Hydroxyapatite/ Alumina nanocomposites by high-energy vibratory ball milling J. Ceram. Sci. Tech. 5 293 298.

[2] Descampsa, M, Boilet, L, Moreaua, G., Tricoteaux, A., Lud, J., Leriche, A., Lardot, V., Cambier, F. 2013 Processing and properties of biphasic calcium phosphates bioceramics obtained by pressureless sintering and hot isostatic pressing J. Eur. Ceram. Soc., 33, 1263 1270,

[3] Brown, O., McAfee, M., Clarke, S., Buchanan, F. 2010 Sintering of biphasic calcium phosphates J. Mater. Sci. – Mater. M. 21 2271 – 2279.

[4] Kim, S.J., Bang, H.G., Song, J.H., Park, S.Y. 2009 Effect of fluoride additive on the mechanical properties of Hydroxyapatite/ Alumina composites Ceram. Int. 35,1647 – 1650.

[5] Goller, G., Demirkiran, H, Oktar, F.N., Demirkesen, E. 2003 Processing and characterization of bioglass reinforced hydroxyapatite composites Ceram. Int.,29 721 – 724.

[6] Demirkol, N., Oktar, F.N., Kayali, E.S. 2013 Influence of niobium oxide on the mechanical properties of hydroxyapatite Key Eng. Mat. 529 – 530 29 – 33.

[7] Rao, R.R., Kannan, T.S. 2002 Synthesis and sintering of hydroxyapatite- zirconia composites Mater. Sci. Eng. 20 187 – 193.

[8] Hisbergues, M., Vendeville, S., Vendeville, P. 2009 Zirconia: established facts and perspective for a biomaterial in dental J. Biomed. Mater. Res. B., 88, 519 – 529.

[9] Daou, E.E. 2014 the zirconia Ceramic Strengths and weaknesses Open Dent. J., 8, 33 – 42.

[10] Evis, Z. 2007 Reactions in hydroxyapatite-zirconia composites Ceram. Int. 33 987 – 991.

[11] Boslama N, Chevalir Y, Bouaziz J, and Ben Ayed F, 2013 Influence of the sintering temperature on Young's modulus and the shear modulus of tricalcium–phosphate–fluorapatite composites evaluated by ultrasound techniques Mater. Chem. Phys. 141 289–297.

[12] Sharmin N, Hasan M S, and Parson A J 2015 Cytocompatibility mechanical a dissolution properties of high strength boron and iron oxide phosphate glass fiber reinforced bioresorbable compactJ. Mech. Biomed. Mater. 59 41–46.

[13] H.R. Ramay, M. Zhang 2003 Biomaterials. Vol.24, p.3293.

[14] A.M. Li, K.N. Sun, Y.S. Yin, et al. 2004 Acta Mater. Composite Sinica. Vol. 5, p. 98.

[15] K.R. Roop, M. Wang 2002 Mater. Lett. Vol.7 p.133.

[16] D.M. Liu, T. Troczynski, W.J. Tseng 2001 Biomaterials Vol. 22, p.1721.

[17] Li-li WANG, Xiu-feng WANG, Hong-tao JIANG and Cheng-long YU 2010 Preparation of Porous Hydroxyapatite-Zirconia Composite Scaffolds by Combination of Gel-Casting and Polymer Sponge Methods Advanced Materials Research Vols. 105-106 pp 616-619.

[18] Hasa, A.F., A.H.K. Ltaief, and Hamodi, J.F. 2015 The electrodeposition of strontium-hydroxyapatite (Sr-HA) on TiO_2 nanotubes layer fabricated by anodizing of Ti metal and Ti-6Al-4V alloy to increase the bonding with bones Engineering and Technology Journal 33(6 Part (B) Scientific) p. 1003-1018.

[19] W. Barsoum. Michel,"fundamentals of ceramics 1997 ISBN 0-07-114184-7, 24, P9.

[20] Kokubo, T.: Bioactive glass Ceramics: properties and applications, Biomaterials, 12, 155 – 163.

[21] Maniatopoulos, C., Sodek, J., Melcher, and A.H.1990 Bone formation In Vitro by stromal cells obtained from bone marrow of young adult rat Cell Tissue Res. 254 317 – 330, (1988).

[22] O. Kaygili, et al. 2016 Progress in Biomaterials 5(3-4), 173.

[23] B. Bulut, Z.E. Erkmen, E.S. Kayali, 2016 Biocompatibility of Hydroxyapatite-Alumina and Hydroxyapatite-Zirconia Composites including Commercial Inert Glass (CIG) as a Ternary Component J. Ceram. Sci. Tech. 07 [03], pp 263-276.

IOP Conf. Series: Materials Science and Engineering **757** (2020) 012047 doi:10.1088/1757-899X/757/1/012047

Corrosion resistance of Ti6Al4V alloy by Radio Frequency Technique used for Coating Deposition of multilayer (HA/TiN/Ti6Al4V-substrate) for Optimization power

M. A. ulkareem [1]*, F. T.M.Noori[2], M. K. Khalaf[2]

Department.of Physics. Baghdad Factuality Science, Iraq, Baghdad

Applied Physics Center, Materials Research Directorate

*Corresponding author email: mustafaphphy@gmail.com

Abstract: The objective of this study is to perform physical / chemical characterization and to determine the corrosion resistance of calcium-phosphate (HA), TiN-based bio-ceramics and thin metal plating processed by a Ti-6Al-4V process. TiN and HA coating maximum power 150 watts. The presence of TiN200 Watt TiN s coating resulted in electrochemical corrosion. High current cathodic and passive anodic polarization shows an active surface. EIS shows the active behavior assumed by the active cathodic currents imposed. And HA on the sample surface, some noise was observed during both polarization and EIS tests.
Key word: sputtering, HA, Corrosion, polarization, TiN, Electrochemical and power.

Keywords:sputtering, HA, Corrosion, polarization, TiN and Electrochemical.

1. Introduction

The choice of biocompatible materials depends on their biological compatibility and vital functions. The first characteristic refers to the effect of implantation on surrounding tissues and the second to its ability to tolerate the function of the biological system being replaced. In practice, both properties are closely interrelated and correlated with corrosion resistance. If implant rejection due to mechanical defects is relatively rare, most failures are due to lack of biological compatibility which is directly related to the corrosion resistance of the implant [1], [2]. From biocompatibility and from the perspective of biomechanics, thin coatings may represent advantages over thick PSHA coatings [3]. However, morphology, reduced dimensions and process-dependent configuration may lead to significantly different electro-chemical behavior during bone interaction in the physiological environment. The objective of this study was the physical / chemical characterization and determination of the corrosion resistance of a thin layer of biochemical ceramics based on calcium and P-varying on titanium alloys (Ti-6Al-4V) using electrochemical methods [4]. Excellent tissue compatibility, better corrosion resistance, and use of pure commercial titanium (CPTi) in transplantation processes, but materials were inhibited due to lack of strength and extremely weak corrosion resistance. Interest in the use of Ti-6Al-4V for prostheses [5]. It is evident from the literature that the α cpTi phase, vanadium is added to 4% by weight, which is enough to stabilize at

10% β phase volume at room temperature and also adds 6wt% of aluminum in cpTi. Although starting α+ β alloy Ti-6Al-4V [6].

2. Experimental work

Ti-6Al-4V alloy samples were used in plasma sputtering (3 x 10 x 10) mm using a diamond cutter (Struers, Denmark). Ti-6Al-4V alloys were manufactured by grinding and polishing mirror respectively, using the Struers-RotoPol-21 system, Denmark. The samples were grinded by SiC sheet in the flowing steps: (120, 180, 240, 320, 500, 600, 800, 1000 and 1200μm of grain size). Samples are polished with a DP- suspension for (15, 9, 6, 3, 1μm), and the use of the DP-cloth. The fine polishing of samples was made by the same machine with (0.51μm) alumina powder and DP-Lubricant red. The polished samples are cleaned with a triachloroethyl acetate solution. The diluted samples were washed with deionized water, dried and stored in a dissociator above the silica gel pad and were used in plasma treatments, microscopic development and electrochemical investigation. The system was pumped down to the base pressure (about 10-4 mbar), Argon gas was presented into the chamber. DC discharge of argon fluorescence (argon plasma) was neglected and applied to each sample for one hour. The sample has been cleaned with a break until all the small arcs have disappeared and a uniform glow can be seen throughout the sample. Plasma treatment of argon was used only as a standard measure to clean the surface of Ti alloy samples, reducing and removing the original oxide and the contaminated layer. Argon gas was introduced into the evacuated chamber and the flow rate was adjusted until the pressure was stabilized at the desired pressure, and the droplets were cooled in the same gas. Cooling is performed in the same environment mostly to prevent oxidation of samples. The plasma samples of Ti-6Al-4V were treated under pressure N2 (1-3mbar). Figures were obtained from the circular Ti alloy of diameter 1.5 cm and thickness of 2 mm for polarization and EIS tests in the simulated body fluids (SBF). The samples were prepared for polarization investigations by leaving an area of 1 cm2 exposed and the other sides covered with a special varnish.The test solution was to (SBF) after its composition is shown in Table 1.

Table 1. Composition of Simulated Body Fluid

Reagents	g/l
NaCl	8.035
$NaHCO_3$	0.355
KCl	0.225
$K4HPO_4.3H_2O$	0.231
$MgCi_2.6H_2O$	0.311
$CaCl_2$	0.392
Na_2SO_4	0.072
$((HOCH_2)_3CNH_2)$	6.118
1.0 M HCl	Appropriate amount for adjusting the pH 7.4

Using the EG & G electroelectric system, the polarization and ELS tests were achieved: Model 273 Potentiostat (EG & G) + Model 1260 Solartron FRA, in three schematic electrodes where the samples were connected to a working electrode, the Pt wire was used as a resistor for Ag / AgCl as a reference electrode. Tafel polarization tests were performed at a 0.5 mv / sec scan rate and ELS tests were performed by scanning a sine wave (m 5mV) from 100 KHz to 10 MHz. All Nyquist and Bode schemes are recorded in one chart to easily compare. All tests were conducted at room temperature. It should be noted that because of the surface coatings presence, some samples demonstrated severe noise in current and potentials that interfere with the measurements. Below, the electrochemical test results are reported. Figures include "Tafel" polarization followed by EIS schemes from Nyquist and Bode where the results of simulated circuits are collected.

3. Results and discussion

3.1 Corrosion test

IOP Conf. Series: Materials Science and Engineering **757** (2020) 012047 doi:10.1088/1757-899X/757/1/012047

The results of electrochemical tests are reported below. The figures include Tafel polarization followed by EIS plots of Nyquist and Bode on which the results of simulated equivalent circuits is superimposed Figure 1(a) result of sample M1 pure Ti6Al4V alloy E $_{O.C.P}$ = -105 mv in Solution: Saturated Body Fluid (SBF) .

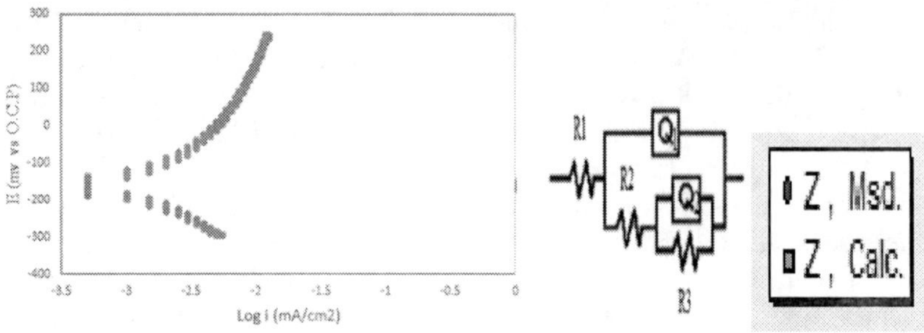

Figure.1 Electrochemical tests results on sample No.1 in SBF: **a.** Tafel polarization graph of sample No.1 in SBF.

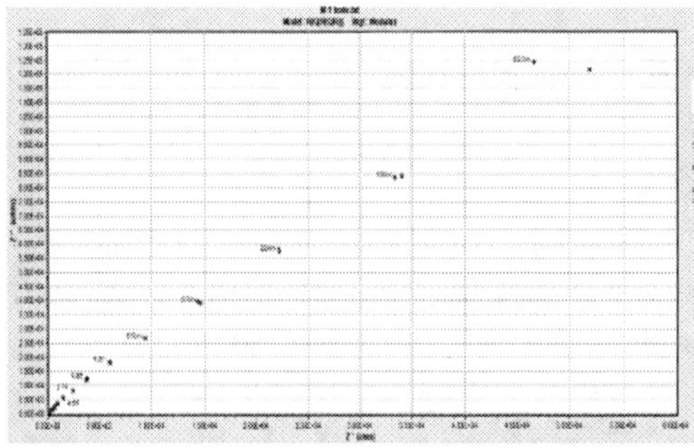

Figure.1 b. EIS Nyquist plot of sample No.1 and superimposed simulated graph obtained from the proposed equivalent circuit.

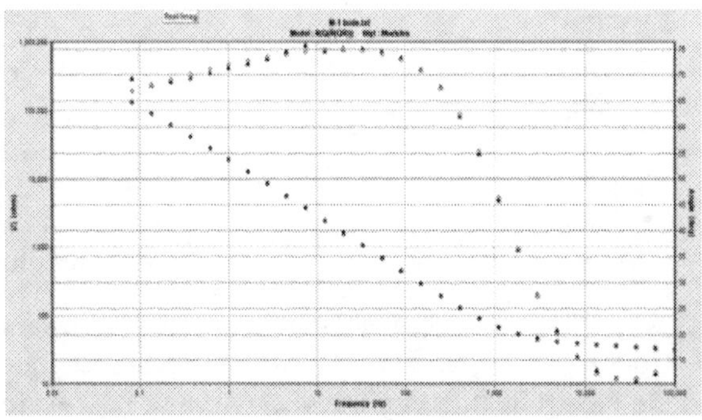

Figure.1 c. Bode plot and superimposed simulated graph of sample No.1

Figurl (b and c) shows Tafel polarization test and EIS diagrams of sample No.1.The polarization shows high cathodic and anodic currents which is typical for Ti alloys. The EIS shows two semi circles one at very high frequency that is due to the adsorbed ions on the surface and the other at medium frequency shows the charge transfer resistance, Rct.

Figure 2(a) result of sample M2 pure Ti6Al4V alloy E $_{O.C.P}$ = -397 mv in Solution: Saturated Body Fluid (SBF).

IOP Conf. Series: Materials Science and Engineering **757** (2020) 012047 doi:10.1088/1757-899X/757/1/012047

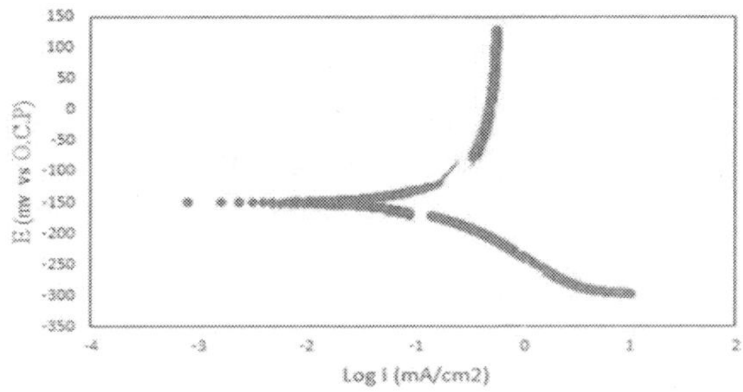

Figure.2 Electrochemical test results on sample No.2 in SBF: a. Tafel polarization graph of sample No.2 in SBF.

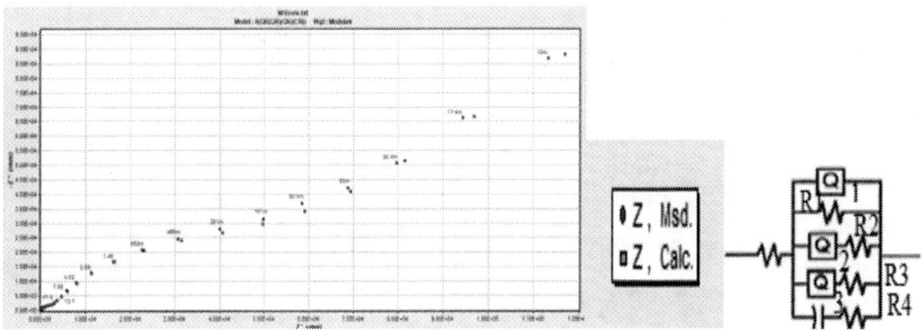

Figure.2 b. EIS Nyquist plot of sample No.2 in SBF and superimposed simulated graph obtained from the proposed equivalent circuit.

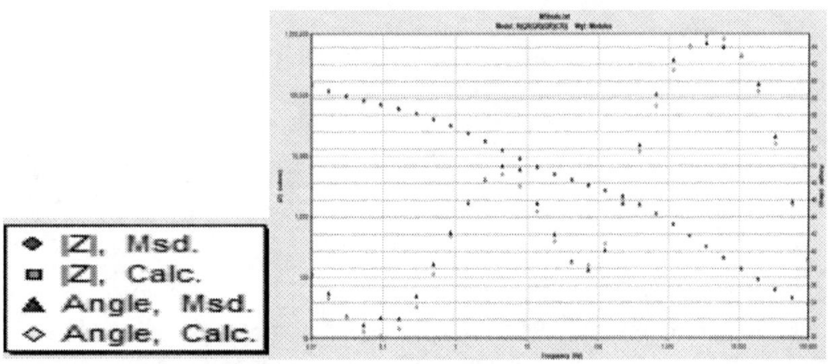

Figure.2 c. Bode plot and superimposed simulated graph

Figure 2(a and b) indicates corrosion tests of specimen No.2 with 200 Watt TiN coating. High cathodic current and passive anodic polarization shows active surface. The EIS shows active behavior presumably due to the active cathodic currents. Two semicircles one for R_{ct} for cathodic current and the next is for TiN coating reactivity or transfer of ions through it with Warburge diffusion.

Figure 3(a) result of sample M3 pure Ti6Al4V alloy $E_{O.C.P}$ = +83 mv in Solution: Saturated Body Fluid (SBF).

IOP Conf. Series: Materials Science and Engineering **757** (2020) 012047 doi:10.1088/1757-899X/757/1/012047

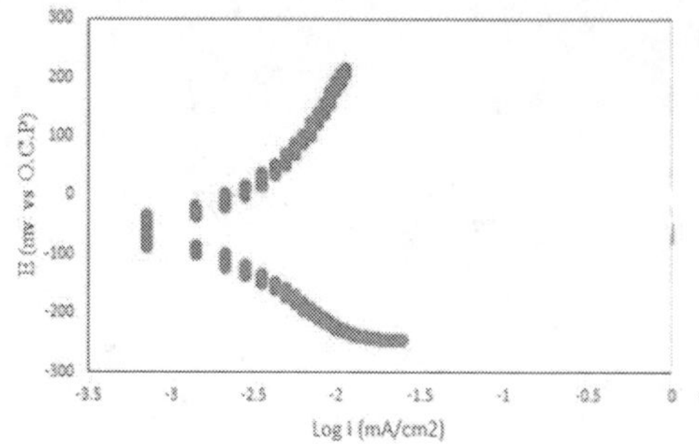

Figure.3 Electrochemical test results on sample No.3 in SBF: a. Tafel polarization graph of sample No.3 in SBF

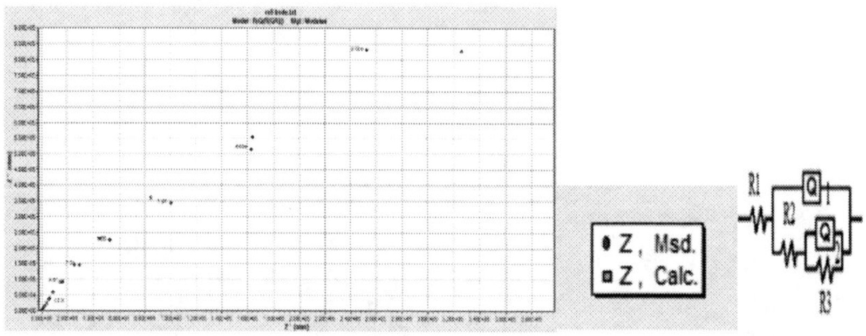

Figure.3 b. EIS Nyquist plot of sample No.3 and superimposed simulated graph obtained from the proposed equivalent circuit.

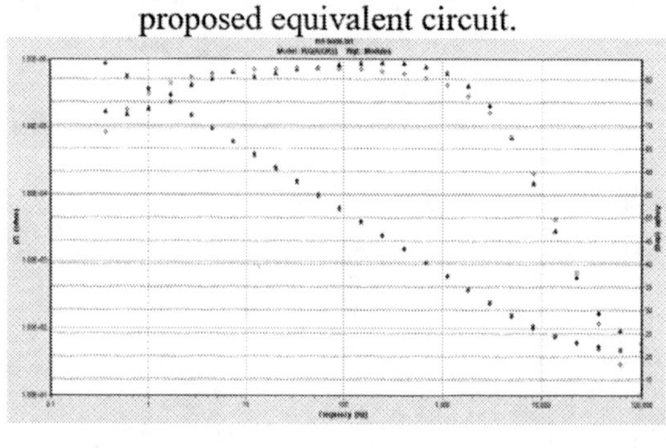

Figure.3 c. Bode plot of sample No.3 in SBF and superimposed simulated graph

Figures 3 (b and c reflects the behavior of sample No.3 and shows unreactive surface because the E_{OCP}=+88mv and polarization indicates similar results as before. The C_{dl}=48 μF but $C_{coating}$=300 μF.

Figure 4(a) result of sample M4 5 hours nitriding coating $E_{O.C.P}$ = +170 mv in Solution: Saturated Body Fluid (SBF).

IOP Conf. Series: Materials Science and Engineering **757** (2020) 012047 doi:10.1088/1757-899X/757/1/012047

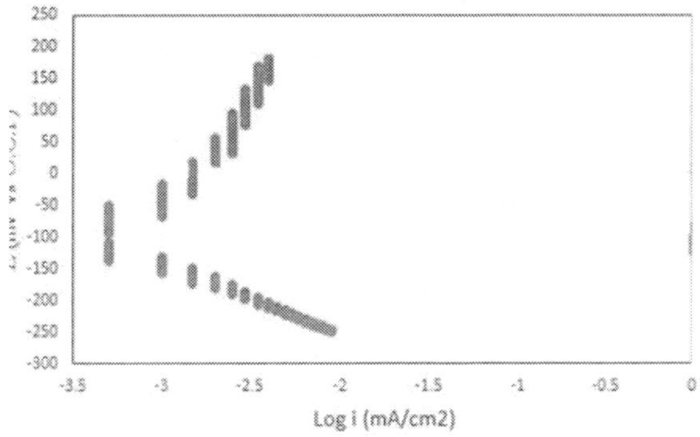

Figure.4 Electrochemical tests results on sample No.4 in SBF: a. Tafel polarization graph of sample No.4 in SBF

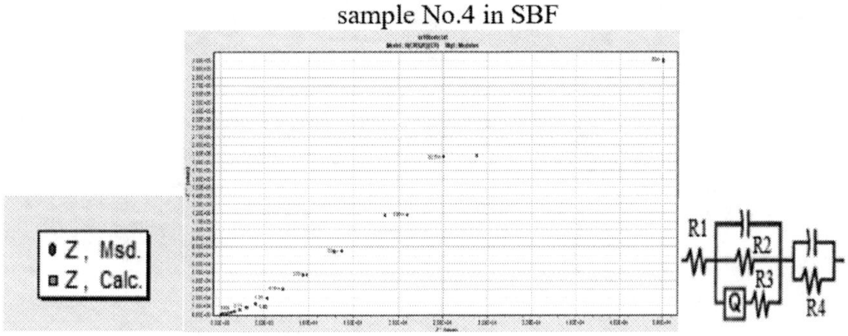

Figure.4 b. EIS Nyquist plot of sample No.4 in SBF and superimposed simulated graph obtained from the proposed equivalent circuit.

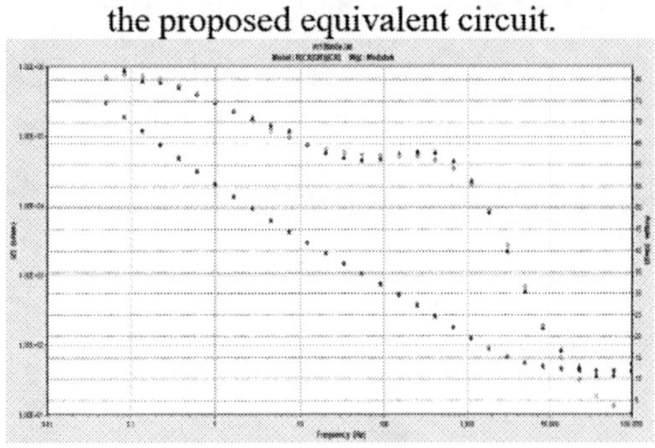

Figure.4 c. Bode plot of sample No.4 in SBF and superimposed simulated graph

Figures 4 (b and c) shows Tafel polarization and EIS for specimen No.4 with 5 hours nitriding process. The E_{OCP}=+170mv showing more noble surface compared to others. But the Rct in EIS spectra is low presumably due to some weak points on the surface which is not converted to nitride. Figure 5(a) result of sample M5 100watt 2h HA on Ti_6Al_4V $E_{O.C.P}$ = +58 mv in Solution: Saturated Body Fluid (SBF).

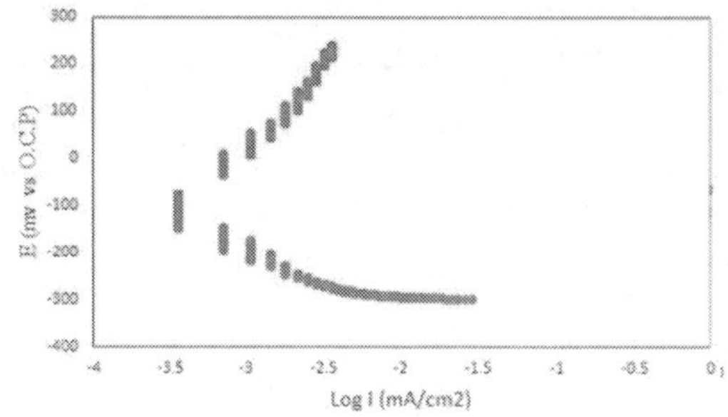

Figure.5 Electrochemical tests results on sample No.5 in SBF: **a.** Tafel polarization graph of sample No.5 in SBF.

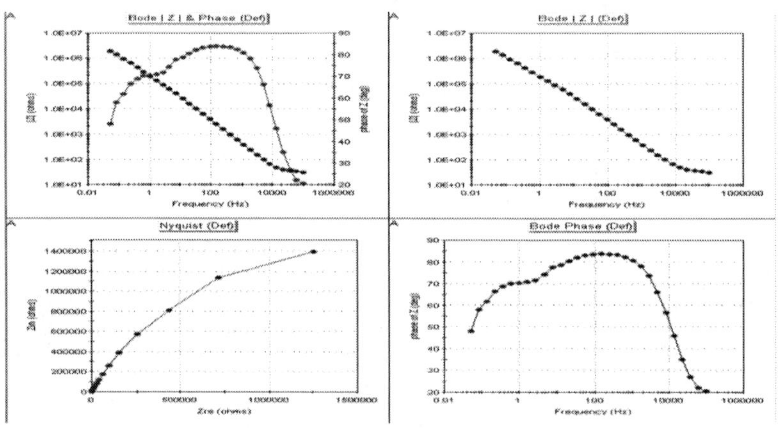

Figure.5 b. EIS graphs of sample No.5 in SBF

The polarization curve shows high anodic + cathodic polarization with E_{OCP}=+58mv. The presence of HA has protected the surface but H_2 formation is a weakness for such coatings. EIS spectra show two semicircles one for Rct with high resistance and the next for HA coating.

Figure 6(a) result of sample M6 150 Watt 3h/Thickness/HA on TiN T=100 °C E O.C.P = +93 mv in Solution: Saturated Body Fluid (SBF).

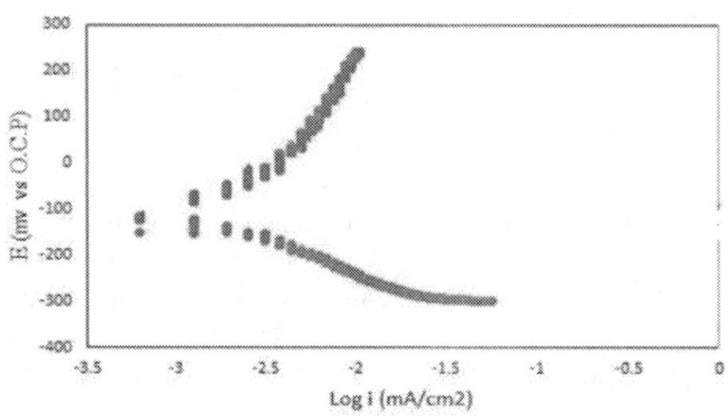

Figure.6 Electrochemical tests results on sample No.6 in SBF: **a.** Tafel polarization graph of sample No.6 in SBF.

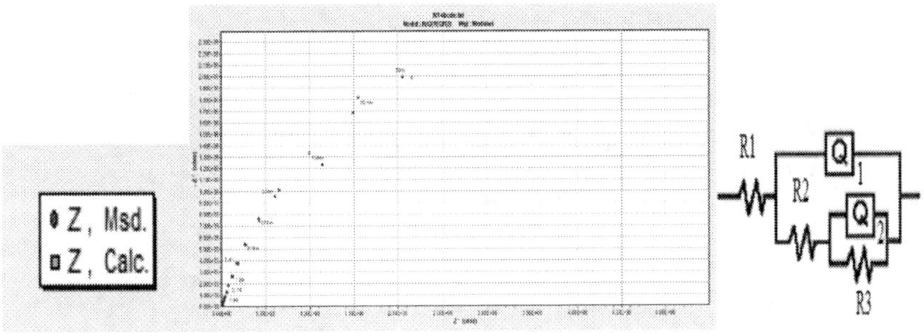

Figure.6 b. EIS Nyquist plot of sample No.6 in SBF and superimposed simulated graph obtained from the proposed equivalent circuit.

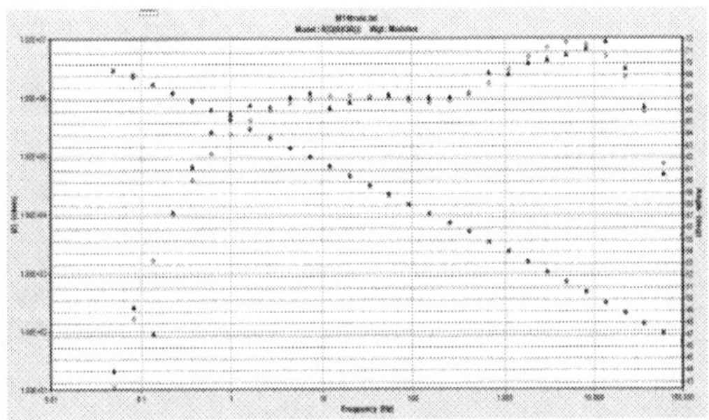

Figure.6 c. Bode plot of sample No.6 and superimposed simulated graph

Figures 6(b and c) show the Tafel and EIS of 3 hours at 150-watt TiN processed coating + HA at T=100C shows again reactive surface in polarization but two semicircles; one at high frequencies for the coatings and one at medium frequencies for the Rct. The resistance of the Rct indicates the presence of a resistive coating.

The overlapped EIS and polarization graphs of Ti alloy with different coating and treatments are reported in below. Figure 7 shows the EIS data from Ti_6Al_4Vcoated with TiN at different powers. It can be seen that at 50 and 100 Watts better coatings are obtained.

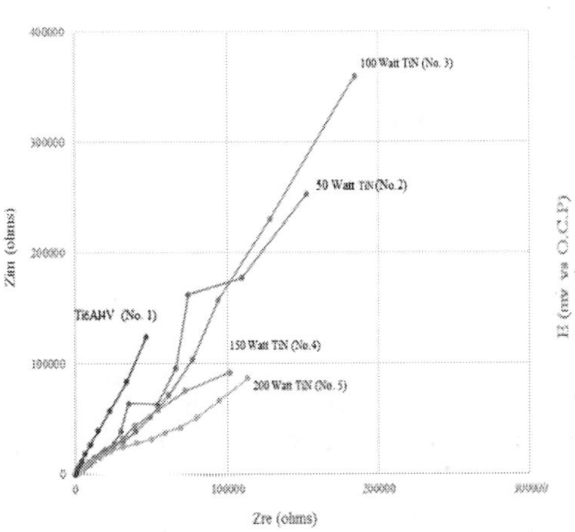

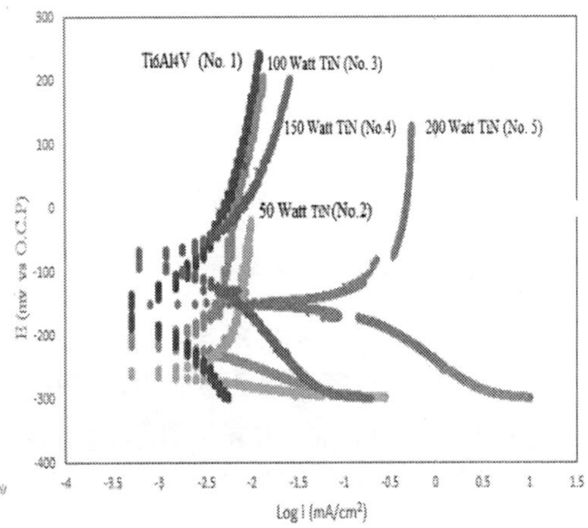

Figure. 7 Nyquist Plot in SBF for Ti6Al4V with TiN coatings at different conditions. **Figure. 8** Polarization curves in SBF for Ti6Al4V and TiN coated Ti6Al4V

The polarization graphs in Fig.8 confirms the better quality of the coating at lower powers.

In Figure 9, HA-coated samples are compared in which HA with medium power provided better quality coating because it has lost its hydrated water.

The polarization graphs in Figure10 compares the anodic behavior of the coating in which sample No.5 gives better behavior.

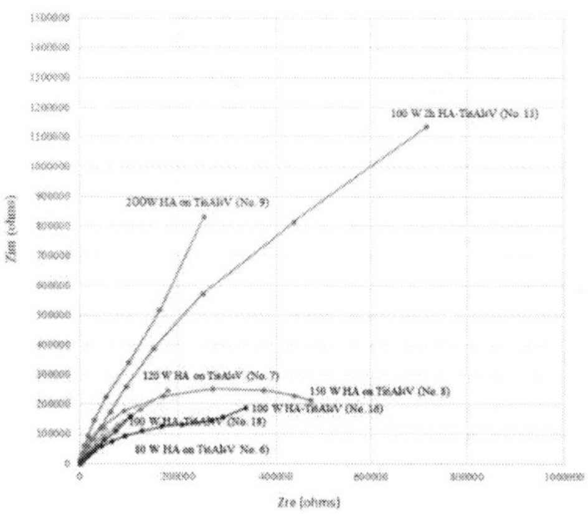

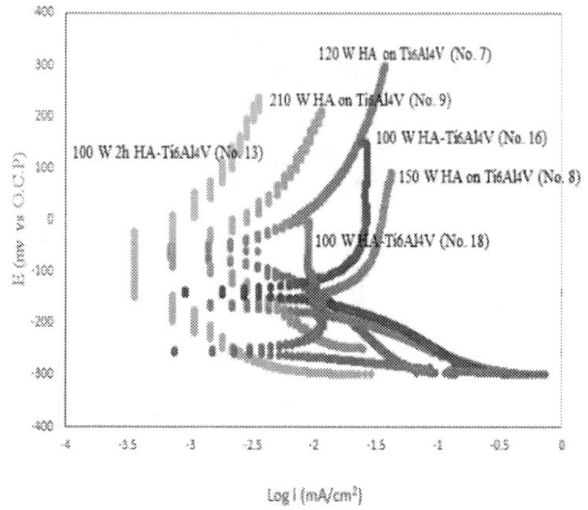

Figure. 9 Nyquist Plots in SBF for Ti alloy samples coated with HA at different conditions **Figure. 10** Polarization curves in SBF for HA-coated samples.

Figure11 shows the effect of nitriding process on the corrosion behavior of Ti alloy. Here again nitriding at lower time (5Hours) showed better corrosion resistance in SBF both in EIS and Polarization tests.

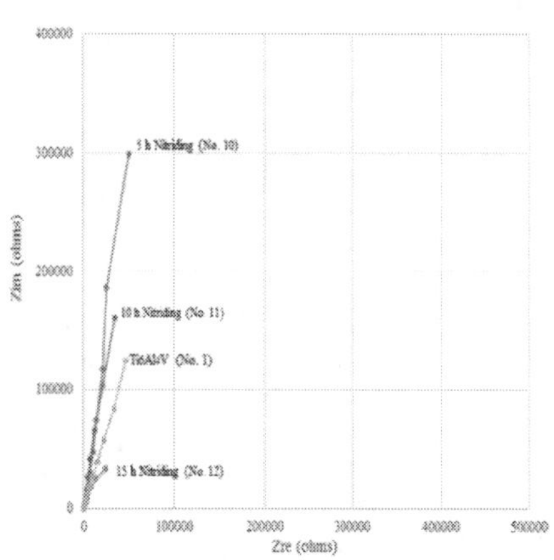

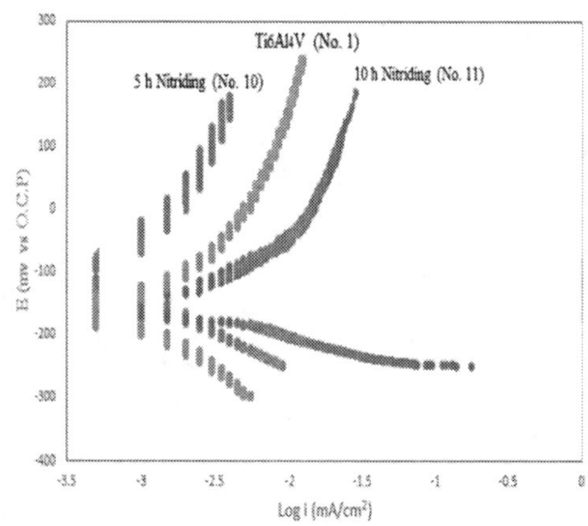

Figure. 11 Nyquist Plots in SBF of Ti alloy with nitride coating at different duration times. **Figure.12** Polarization curves in SBF for Ti alloy with nitride coating at different duration.

Figure 13 and 14shows the effect of HA-TiN-coatings on Ti alloy at different conditions. High power gives better HA coating and better corrosion resistance.

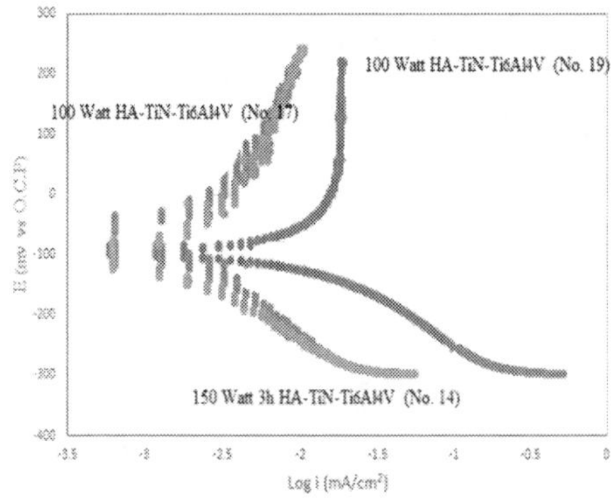

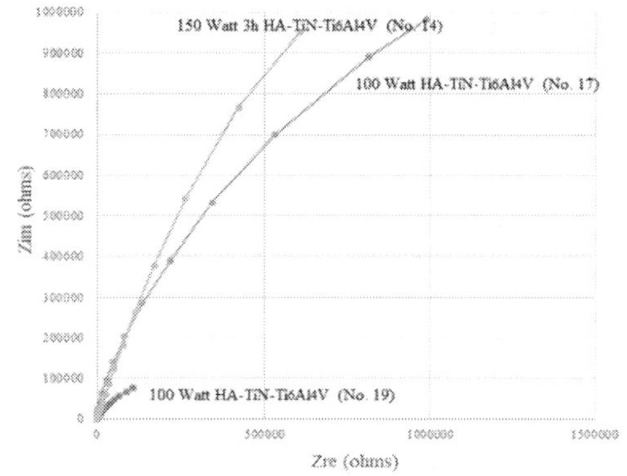

Figure. 13 Nyquist plots of HA-TiN-coated Ti samples with different power and different time.
Figure. 14 Polarization curves in SBF of HA-TiN-coated samples with different power and different time.

4. Conclusion

Due to the presence of Ti/ TiN and HA on the surface of the samples, some noises appeared during both polarization and EIS tests. The corrosion rate could not be measured because the anodic and cathodic lines of Tafel polarization were not homogenous. The corrosion rate can be measured by immersion tests. Morphological investigation needed to report the way corrosion occurred on the surface. The sample size was not large enough for wire connection and we could not weld or solder any wire for such connections because it may damage the surface. The coatings looked homogenous but need SEM analysis for detecting weak points. Hydrogen formation is a main problem with implants which the cathodic Tafel lines were showing. It is important also to measure the coatings thickness which may reflect its durability. What we report is only our observations and do not judge about the performance of the samples.

5. References

[1] E. Kelesoglu, C. Mitterer, M. Ürgen, 2002 Surf. Coat. Technol. **160**, 82.
[2] Belov M. Balaceanu, V. Braic, D. Macovei, M. Genet, A. Manea, D. Pantelica, M. Braic, F. Negoita, J. Optoelectron. 2002, Adv. Mater. **4**(1), 107.
[3] D. Muster, Biocompatibilite ET, Paris 2000 evolution de l`utilization des biomateriaux Meta liques, in Actualites en Biomateriaux, vol. V, Ed. Romillat.
[4] S.B. Al-Baghdadi, F.T.M. Noori, W.K. Ahmed, A.A. Al-Amiery: 2016, Journal of Advanced Electrochemistry 2(1) 67–69.
[5] Farah T. M. Noori: Feb 2016, Experimental, Kinetics, Isotherms and Quantum Chemical Calculations; Corrosion Inhibitor for Mild Steel in Acidic Medium:Wulfenia journal Klagenfurt austia Vol 23, No. 2.
[6] Buddy D.Ratner, Ph.D. 2004 Biomaterial science, Elsevier Inc, 2nd Edition.

IOP Conf. Series: Materials Science and Engineering **757** (2020) 012048 doi:10.1088/1757-899X/757/1/012048

Friction and Wear Testing of Lubricated Joint Implant Material using AMTI Ortho-POD

A. I. Shamki *, S. J. Abass

Biomedical Engineering Department, Al-Nahrain University, Baghdad, Iraq

*Corresponding author email: alialamiry85@gmail.com

Abstract

Friction and wear are caused by complex interactions between surfaces that are in mechanical contact and sliding against each other. The goal of this study was to determine the effect of different lubricants on the wear and friction of most commonly used implant material ultrahigh molecular weight polyethylene (UHMWPE), and to compare the results of UHMWPE with another materials polytetrafluoroethylene (PTFE). To accomplish this goal, pin-on-disk cyclic testing by AMTI Ortho-Pod pin on disk was used with lubricant. The lubricant used to mimic synovial fluid was bovine albumin, hyaluronic acid, and the artificial lubricant (silicone oil) were used. Gravimetric analysis was used to characterize wear after 700 thousand cycles at normal load 300 N, sliding velocity 87.92 mm/sec with different wear paths. Using these data, pin-on-disk wear tests were conducted to explore that circular path had a higher mean coefficient of frictions and specific wear rate than a line path. UHMWPE had a lower mean coefficient of frictions and specific wear rate than PTFE.

Keywords: Pin on disk, Wear rate, Friction coefficient, Lubricant, UHMWPE, and PTFE.

1. Introduction.

During the service life of the biomaterials face relative motion on micro- or macro-levels, this can bring a tribological damage. Besides, the surrounding biological environments can also contribute to that damage by having a corrosive effect. All of the combination and synergy of tribology as well as the corrosion and the biological environment are extremely complex and have a great importance [1]. Therefore, tribology has an important role in a number of medical devices and implants, like the artificial joints, artificial teeth, dental implants, orthodontic appliances, cardiovascular devices, contact lenses, artificial limbs, and not to forget the surgical instruments [2]. It can be defined as the study of friction, wear and lubrication in the design of bearings, and it can be said that it is the science of interacting surfaces in relative motion [3]. However, friction is a common phenomenon in mechanical systems which may be increased or decreased basing on the sliding pairs and operating parameters [4]. Well, wear can be defined as 'the progressive loss of substance from the surface of a body brought about by mechanical action', besides, it is one of the most common causes of joint replacement failure which in turn causes aseptic loosening and then can result in chronic inflammation at the implant–bone interface, it is not surprising that all implants for total joint replacement is subjected to wear [5].

Mechanical wear assessment also includes the use of a pin and disk method in order to create wear in the material and see the result. One of the common screening tools is Pin-on-disk wear testing which is used to quantify the wear rate of polymer material that is used in orthopedics. Both friction coefficient and wear rate can differ significantly because of introducing an external lubricant to the polymer tribo- system. The external lubricant can define as a substance which can alter the nature of the surface interaction of the contacting solids. The main way a lubricant helps influences friction and wear of thermoplastics comes by reducing adhesion of the polymer to the counter face.

Today, polymers occupy a major position in the bearings field. Thus, the polymeric biomaterials main advantages compared to the metal or ceramic materials are seen ease of manufacturability to produce various shapes (latex, film, sheet, fibers, etc.), and ease of secondary process ability, however, a reasonable cost, besides the availability with desired mechanical and physical properties [6].

2. Materials and Methods:

2.1 *Testing Machine*

The Ortho-POD is a six-station pin-on-disk machine that can provide suitable test motions, environments, and loads to produce wear rates similar to the encountered in joints simulators. However, three independent servo-motors the engine has which control motions corresponding to the pin rotary motion, the plate with the disks rotary motion, and a constant normal load on the six pins. Each one of the six pin drive axes are centered on a planet gear which is driven by a single sun gear controlling all pins to move with the same rotary motion. Each of the pin load actuators provides an identical load, but any one of those pins can be turned off individually. Disk from AMTI lab, Watertown, USA. AMTI multi-component force sensors can measure the three force components (Fx, Fy, and Fz) at the pin-sample disk interface. The machine is attached with a PC so, the machine also contains a waveform board which has fixed microprocessor. The PC is for programming a desired waveform that is then be moved to the waveform board. However, circular wave that was used in the tests is programmed by 36 points and (7 mm) radius after being converted to the template file "wav". The controller reads are embedded by waveform generators, the template data and performs an interpolation algorithm in order to provide smooth real time wave generation. The interpolated digital signals are fed to a digital to analog converter that provides the desired time which varying analog

IOP Conf. Series: Materials Science and Engineering **757** (2020) 012048 doi:10.1088/1757-899X/757/1/012048

output voltages or the waveforms. These analog waveforms then being fed to the servo control loop circuitry that is, along with the proper feedback signals, are conditioned to provide the driving signals to both the force and the rotation actuators

2.2 *Sample Preparation*

The size of all pins is Ø9.525mm x 30mm (.375 in x .787 in) to fit existing Ortho-POD pin holders, for each of the tests a group of six samples with same materials being used. Pins will be reused during the different tests. Every pin and pin holder are identified with the same distinct number in all tests. The cleaning and weighting of the samples should be done after every 100000.

2.3 *Fluid Test Medium*

The counter face surfaces of the components will be kept immersed in a lubricating solution at 33°C ± 3 degrees during the test. Validation of the lubrication solution temperature can be controlled by the temperature of the external base plate and external circulating temperature control unit. Each pin on disk specimen should have its own environmental chamber for every separate station of that machine, which will contain both the specimen and lubricating solution. Any fluid lost by evaporation during the test is to be compensated by adding extra amount of fluid. The fluid test medium is to be replaced fully at each stopping point for gravimetric weighing.

2.4 Test Fluids

Synovial fluid is considered as the best lubricant for a natural hip or knee joint. The Synovial fluid has unique lubricant properties that protect cartilage and bone surfaces from extremely high contact pressures during the patient's life span. There are four major biological components can be found, which make synovial fluid an efficient lubricant, hyaluronic acid (HA), albumin, mucinous glycoproteins (mainly lubricin), and globulin [7].

Different test lubricants were used for wear tests in which synovial fluid components and an artificial oil was used.

2.4.1 Bovine Albumin.

Diluted filter-sterilized bovine serum 20 ±2 g/L protein concentration similar to the protein content found in a healthy human body with 1% Sodium Azide to retard bacterial growth. The bovine serum is stored in the fridge until ready for use.

2.4.2 *Hyaluronic Acid.*

Diluted hyaluronic acid with deionized water 3 g/900ml which within the normal range of hyaluronic acid concentration in the synovial fluid [8], the fluid is stored in the fridge until ready for use.

2.4.3 *Artificial Lubricant Silicone Oil.*

Silicones are polymers consists mainly macromolecular compounds of organic silicone polysiloxanes. That silicone oils kinematic viscosity being proportional to the degree of polymerization (i.e. proportional to polymers molar mass) [9]. Silicone fluids are mainly clear, colorless, bland, and tasteless, and they are produced in a wide range of viscosities. They are so much inert and are remarkably stable, bringing a little, change in physical properties over a wide temperature range [10].

2.5 *Experimental polymer materials*

2.5.1 *Ultra-High Molecular Weight Polyethylene (UHMWPE).*

Ultra-high molecular weight polyethylene (UHMWPE) is considered to be a unique polymer with both stellar physical and mechanical properties. Most amazing is its chemical inertness, lubricity, impact resistance, and also abrasion resistance. However, polyethylene consists only of carbon and hydrogen. In the first place, the UHMWPE should be polymerized from ethylene gas. Then, the polymerized UHMWPE, as a powder, has to be consolidated and machined into the required shape. UHMWPE was utilized in orthopedics as a bearing material in artificial joints for the past 45 years. The UHMWPE that was used in orthopedic implant is a type of polymer generally classified as a linear homopolymer [11].

2.5.2 *Polytetrafluoroethylene PTFE.*

Polytetrafluoroethylene (PTFE), known as Teflon®, is a fluorocarbon solid, and is a high-molecular-weight compound consisting wholly of carbon and fluorine.

2.6 *Load Profile, Wear Path Procedure, and Wear Measurement*

A waveform was programmed to apply constant loading for all pins with a peak load of 300 N (maximum contact pressure of 4.205 MPa) which is well within the physiological range [12], and a 14mm diameter of circular wear path was conducted at a frequency of 2 Hz, sliding periodically on its own 316 stainless steel disk.

The setup of these wear tests was unique to this study, where each test was further divided into three subtests. Every subtest utilized a certain wear testing lubricant composition for the set number of the intervals. Using this setup, the effect of different lubricant compositions on UHMWPE and PTFE wear could be successively shown. Each polymer submitted to 0.7 million cycles with each lubricant of the three lubricants, when the lubricant type changed at the end of 0.7 million cycles other parameters (circular wear path, load & chambers temperature) are kept constant.

Pins are changed when the previous material is tested with three lubricants. At regular intervals (every 0.1 million cycles) during the wear tests, the pins were detached from the experimental apparatus, wiped clean and weighed using a high-precision balance which reads 0.0000g. The weight loss of each pin due to wear was determined by calculating the difference in the weight of the sample between two measurement points. The volumetric wear (Δv) in $(mm)^3$was estimated by taking the change in mass of the specimen and dividing it by the density of the specimen material as shown in equation (1)

$$\Delta v = \frac{w_1 - w_2}{\rho} \tag{1}$$

w_1 The weight loss in gram of the specimen before test.
w_2 The weight in gram of the specimen after test.
ρ The experimental density of the specimen, which is for UHMWPE and PTFE 0.00093, 0.00214 ($g\ /\ mm^3$) respectively.
The Specific wear rate (Ws) of the specimen is calculated as in equation (2).

$$W_s = \frac{\Delta v}{F_n S_s} \tag{2}$$

Where S_s is Sliding distance (m), F_n is Normal load (N) [13].
The data acquisition system of the AMTI Ortho-POD collects multi-component force data and friction coefficients. Force sensors in those three vertical legs of that machine measures three of the force components. The nine outputs are summed to deliver a single value for each of the forces Fx, Fy, and Fz where x and y are the transverse direction and the perpendicular direction is z. Each test runs for 0.7 million cycles, friction was collected every 50,000 cycles. The sample rate was 200 data sets/second and its duration were ten seconds. Data were measured and collected for the last ten cycles. The friction coefficient was determined by the ratio of the transverse force to the vertical load, equation (3).

$$cf = \frac{\sqrt{(fx)^2 + (fy)^2}}{(f_z)} \tag{3}$$

fx, fy and fz computed by AMTI Ortho-Pod machine.

3. Results and Discussion

Figures 1 shows the average weight loss results obtained from the test for the six pins PTF components with bovine albumin over 0.7 million cycles. The average volumetric wear rate of all pins at the end of each 100000 cycles is illustrated in figure 2. This configuration of figures is repeated for all tests.

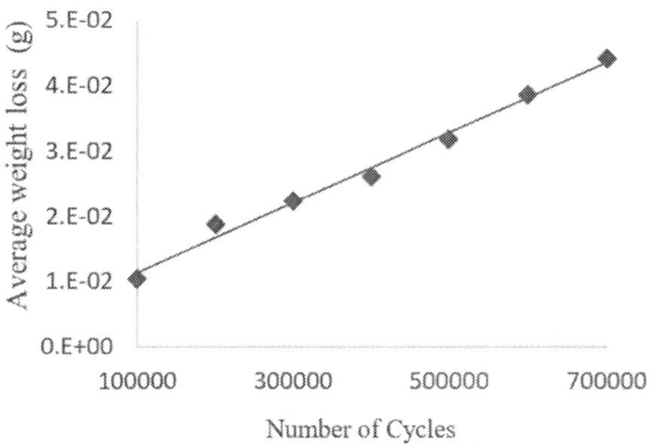

Figure 1. The average weight loss with number of cycles of PTFE with bovine albumin.

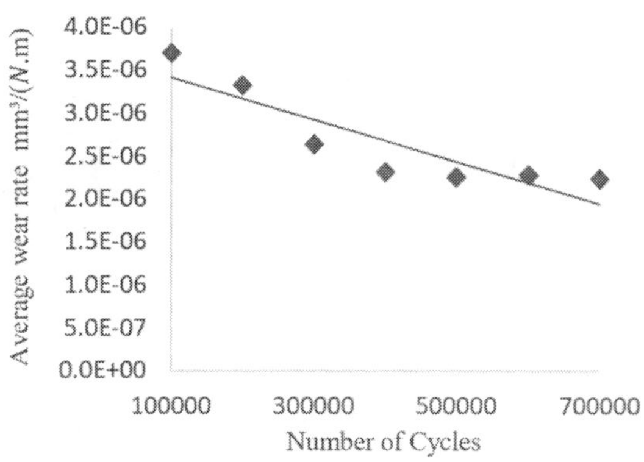

Figure 2. The average wear rate with number of cycles of the PTFE with bovine albumin.

Because, all pins of PTFE, and UHMWPE were not presoaked for all tests, that caused increasing in weight due to lubricant absorption specially when HA was the lubricant medium, because the HA more aqueous than others lubricants (see figures 3, 4, 5, 6, 7, 8, 9 & 10) show that UHMWPE more Absorbent than PTFE. When the test was conducted, they absorbed the lubricant causing their weight to be increased. Silicone oil is more viscous than HA, however, it showed greater wear rates than those for HA, that is due to the amount used of SO in the test which was less than that used for HA (see table 1) which results in less absorption of lubricant, another cause to that is changing the HA and BA each 50000 cycles, whereas SO was changed after 300000 cycles for both PTFE and UHMWPE that resulted in more wear debris accumulation within SO solution which acted as third particles causing abrasive wear. Wear results of PTFE with HA was more than those of UHMWPE, this is due to high coefficient of friction of PTFE which caused increasing liquid temperature leading to weakening of the interactions among the polymer chains which form HA leading to a decrease in its viscosity.

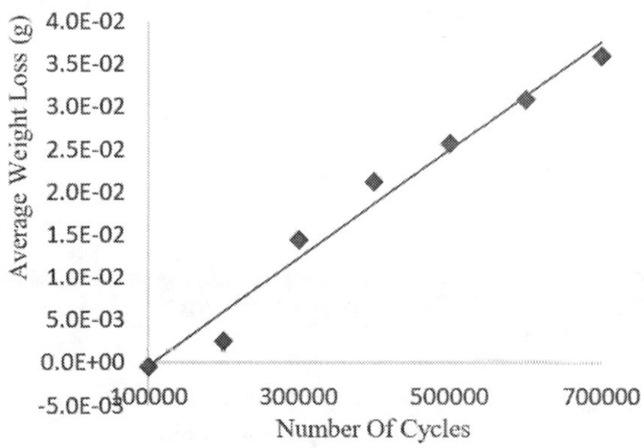

Figure 3. The average weight loss with number of cycles for PTFE with hyaluronic.

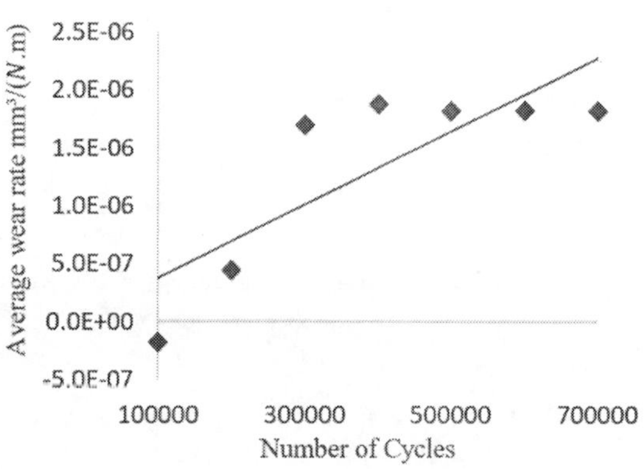

Figure 4. The average wear rate with number of cycles for PTFE with hyaluronic acid.

IOP Conf. Series: Materials Science and Engineering **757** (2020) 012048 doi:10.1088/1757-899X/757/1/012048

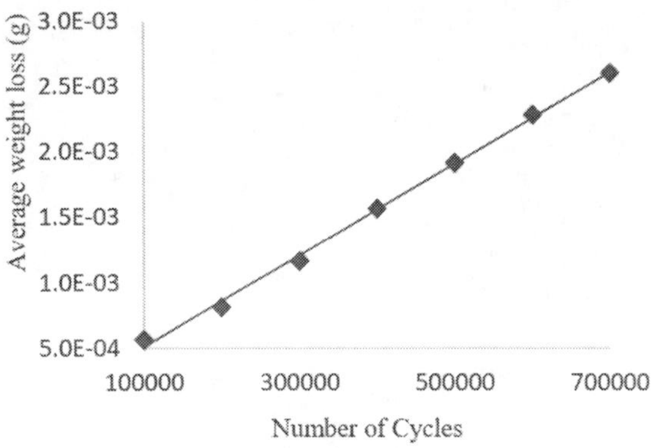

Figure 5. The average weight loss with number of cycles of PTFE with silicone oil.

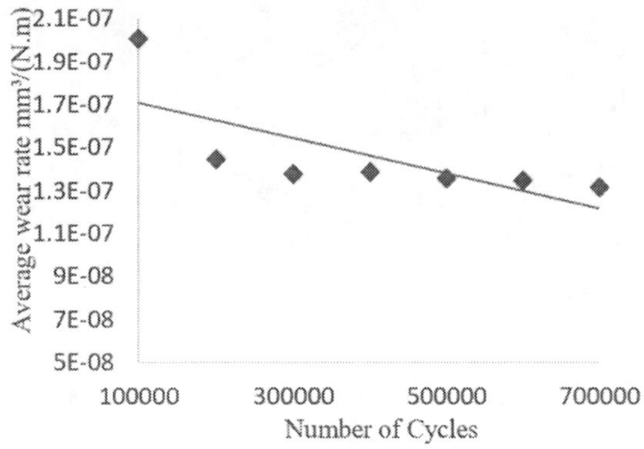

Figure 6. The average wear rate with number of cycles of PTFE with silicone oil.

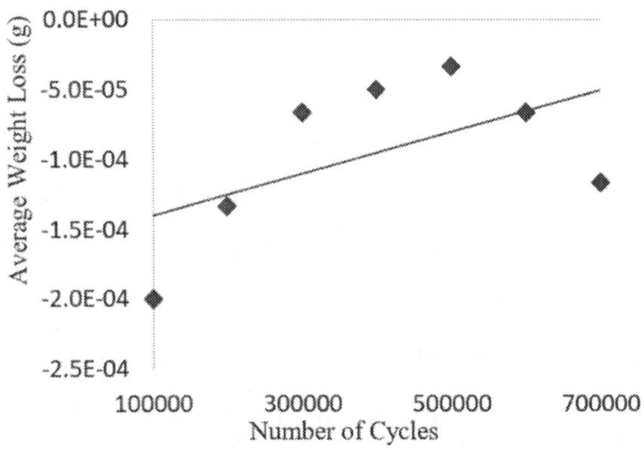

Figure 7. The average weight loss with number of cycles for UHMWPE with hyaluronic acid.

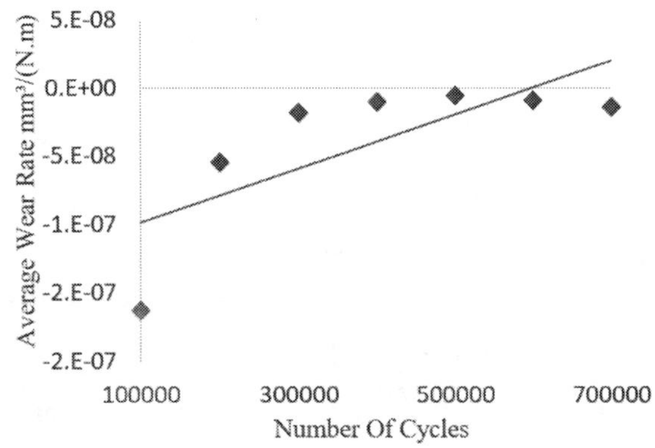

Figure 8. The average wear rate with number of cycles for UHMWPE with hyaluronic acid.

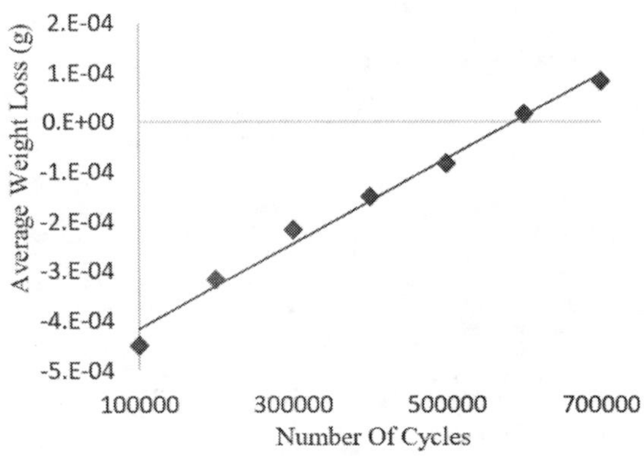

Figure 9. The average weight loss with number of cycles for UHMWPE with silicone oil.

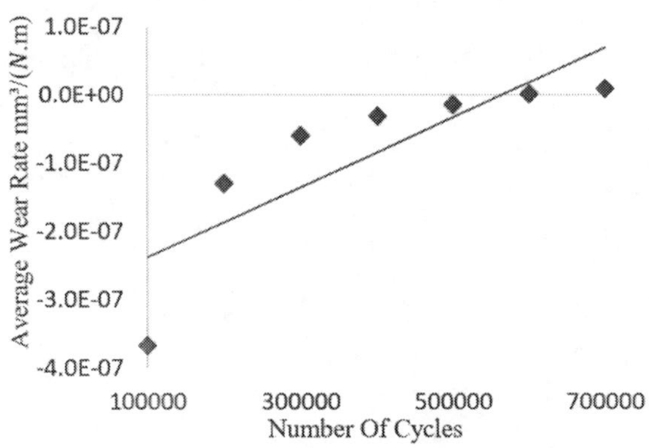

Figure 10. The average wear rate with number of cycles for UHMWPE with silicone oil.

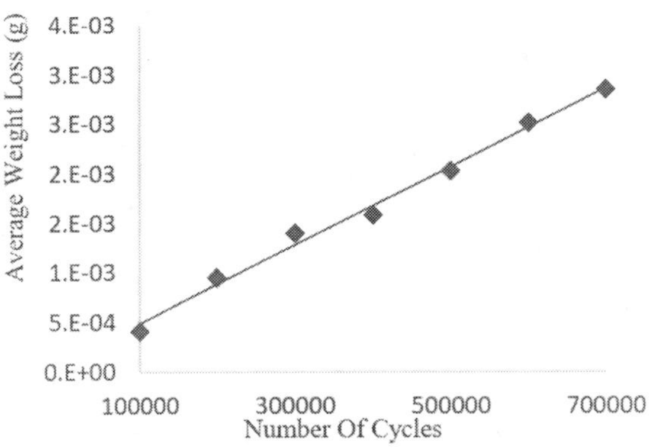

Figure 11. The average weight loss with number of cycles of UHMWPE with bovine albumin.

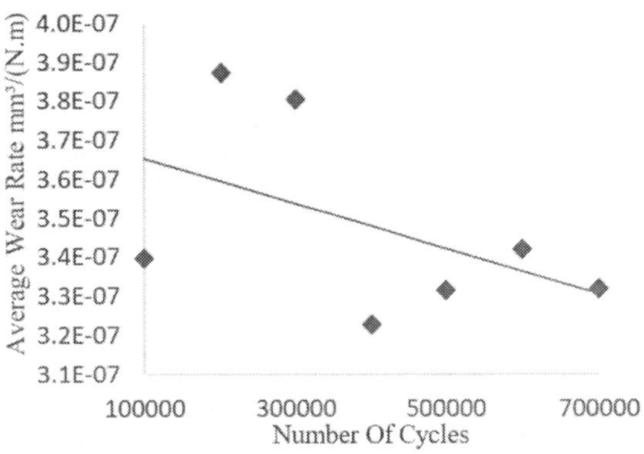

Figure 12. The average wear rate with number of cycles of UHMWPE with bovine albumin.

Figures 11 & 12 show the average weight loss and the average wear rate of UHMWPE with BA and it is noticed that the average wear rate started with high magnitude then it decreased with number of cycles, that can be explained by the fact that the specimens were not soaked before the test.

Pin Material	Load (*N*)	Wear Path	Lubricants	Lubricant Temperature (°*C*) ±2	Lubricant Volume (*ml*)±1	Average *cf*
PTFE	300	Circle	Bovine albumin	36	4	0.2237
PTFE	300	Circle	Silicone Oil	33	2	0.2515
PTFE	300	Circle	Hyaluronic Acid	33	5	0.3681
UHMWPE	300	Circle	Bovine albumin	30	4	0.0969
UHMWPE	300	Circle	Silicone Oil	31	2	0.1845
UHMWPE	300	Circle	Hyaluronic Acid	32	5	0.0956

Table 1. The coefficients of friction (cf) during different tests with sliding distance 30772 m for each test steady velocity 87.92 mm/sec.

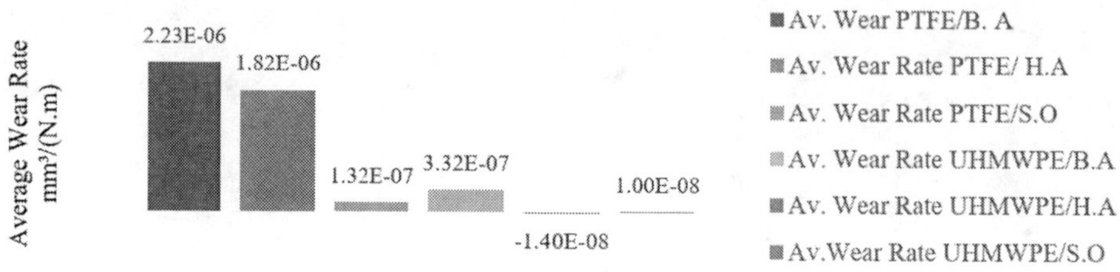

Figure 13. The comparison between wear rate of PTFE and UHMWPE with different lubricants

Regarding the comparison shown in figure 13, the volumetric wear rate of Polytetrafluoroethylene (PTFE) with Bovine Albumin (BA) is higher than PTFE with Hyaluronic Acid (HA) or PTFE with Silicone Oil (SO) that is because the PTFE wear rate was increased due to the increase in temperature on the polymer counter surface, where the (BA) temperature reached 36°C (see table 1), increasing temperatures in polymer contact surface results in the softening of polymer surface which leads to severe wear. Both polymers, the disks and the hollow acrylic plastic cylinder surround the samples were not sterilized that led to bacterial contamination of the serum lubricant (BA), which causes protein cleavage due to inflammatory response of the protein as a result of increase the friction and wear. Cleavage of proteins entails scission of the peptide chain, leading to smaller peptide fragments with exposed hydrophobic and hydrophilic moieties. Albumin affinity for the hydrophilic metal surface exposes the hydrophobic groups that can bond with the hydrophobic polymer surface. This results in the attachment of the albumin molecule to both metal and polymer surfaces. This interaction results in reduced friction by separating the surfaces, but also to increased wear due to effective removal of polymer where polymer–protein–metal adhesion has occurred. This adhesive wear effect was more pronounced against the smoother metal disk. Although the HA had higher protein concentration than BA, HA solutions display progressively increased viscosity and viscoelasticity.

4. Conclusions

The choice of lubricant can significantly affect the results and validity of in-vitro testing. The results from different laboratories are hardly ever comparable as, for example, the protein concentration and the protein composition of the liquids vary, and the differences in the roughness of counter surfaces, and the chambers temperature. The viscosity, the test environment and the chemical reactions of these test does not match the more viscous synovial fluid, biological environment and chemical reactions, which surrounds the polymer, so the in-vitro results still differ from in-vivo results.

The wear rate was higher for BA at all loads. When the hyaluronic acid and silicone oil were lubricants, PTFE and UHMWPE were at their lowest wear-magnitudes, due to the high viscosity of HA and silicone oil. The effect of lubricant makes the difference in friction coefficient and wear rate values

IOP Conf. Series: Materials Science and Engineering **757** (2020) 012048 doi:10.1088/1757-899X/757/1/012048

based on their lubricating ability. The results show that BA and silicone oil exhibited stable wear rate with the passage of time.

Reference

[1] Y. Yan, Bio-tribocorrosion in biomaterials and medical implants. 2013.

[2] Z. M. Jin, J. Zheng, W. Li, and Z. R. Zhou, "Tribology of medical devices," Biosurface and Biotribology, vol. 2, no. 4, pp. 173–192, Dec. 2016.

[3] Z. Jin and J. Fisher, "Tribology in joint replacement*Note: This chapter is an updated version of Chapter 2, from the first edition of Joint replacement technology, edited by P. A. Revell and published by Woodhead Publishing, 2008*," in Joint Replacement Technology, Elsevier, 2014, pp. 31–61.

[4] U. S. Tewari, S. K. Sharma, and P. Vasudevan, Polymer Tribology, vol. 29, no. 1. 1989.

[5] S. Sivananthan, S. Goodman, and M. Burke, Failure mechanisms in joint replacement. 2014.

[6] J. B. Park and J. D. Bronzino, Biomaterials - Principles and Applications - J. Park, J. Bronzino. 2003.

[7] A.-M. Trunfio-Sfarghiu, Y. Berthier, M.-H. Meurisse, and J.-P. Rieu, "Multiscale analysis of the tribological role of the molecular assemblies of synovial fluid. Case of a healthy joint and implants," Tribol. Int., vol. 40, no. 10–12, pp. 1500–1515, Oct. 2007.

[8] F. Duff, "Hyaluronic Acid in Synovial Fluid. I. Molecular," vol. 10, no. 4, 1967.

[9] T. Kałdoński and P. Wojdyna, "Liquid lubricants for space engineering and methods for their testing," J. KONES, vol. 18, no. 1, pp. 163–184, 2011.

[10] B. HELAL and B. S. KARADI, "Artificial Lubrication of Joints," Rheumatology, vol. 9, no. 8, pp. 334–340, 2007.

[11] S. M. Kurtz, UHMWPE Biomaterials Handbook. Elsevier, 2016.

[12] G. Wang, W. Huang, Q. Song, and J. Liang, "Three-dimensional finite analysis of acetabular contact pressure and contact area during normal walking," Asian J. Surg., vol. 40, no. 6, pp. 463–469, 2017.

[13] G. Agarwal, A. Patnaik, and R. K. Sharma, "Parametric Optimization and Three-Body Abrasive Wear Behavior of Sic Filled Chopped Glass Fiber Reinforced Epoxy Composites," Int. J. Compos. Mater., vol. 3, no. 2, pp. 32–38, 2013.

Au–Ag–Cu nanoparticles alloys showed antifangal activity against the antibiotics- resistant Candida albicans

H. R. Humud 1 , T. K. Al-Kafiji 2 , A. M. Al-araji 3, L. F. Ali 4

1, 2 Department of Physics, College of Science, University of Baghdad, Baghdad, Iraq
3, 4 Department of Biology, College of Science, University of Baghdad, Baghdad, Iraq
Corresponding author: E-mail: dr.hammad6000@yahoo.com

Abstract
Formation of Au–Ag–Cu ternary alloy nanoparticles (NPs) is of particular interest because its trimetallic process that includes miscible (Au–Ag and Au–Cu) and immiscible (Ag–Cu) system. So there is a possibility of phase segregation in this ternary system. At this challenge, it was an attempt to use the synthetic technique to generate such trimetallic alloy nanoparticles by exploding wire technique. This work aims to develop an effective antifungal treatment against the pathogenic fungus Candida albicans using Au–Ag–Cu nanoparticles alloys. In this work, three metals alloys (A, B, and C) containing Au-Ag-Cu nanoparticles were prepared as a physical methods by exploding wire with different ratio of each elements (A: 50% Au 25% Ag 25% Cu; B: 25% Au 50% Ag 25% Cu; C: 15% Au 60% Ag 25% Cu). The antifungal activity of the alloy nanoparticles was tested against pathogen fungi, Candida albicans grown on a solid medium using disk diffusion method. The highest antifungal activity against Candida albicans was observed on alloy B that was prepared by 160 A and concentrations 12.5 μg.ml-1. The treatment with this alloy showed more than 85% inactivation of the fungus growth.

Keywords: Exploding Wire, Au–Ag–Cu alloy nanoparticles, Candida albicans, Antibiotic- resistant, Growth rate.

1. Introduction

Nanotechnology is a multidisciplinary and fast growing scientific field in physics, chemistry, and biotechnology sciences [1]. Hybrid and multifunctional nanoparticles that constructed from more than one metallic phase have attracted much interest because of their unique catalytical, optical, electrical, and magnetically properties compared to the individual monometallic metals and bimetallic [2]. This notion leads toward the concept and possibility of personalized medicine for the potential of early detection of diseases and most importantly, molecular targeted therapy. The application of nanotechnology to biomedicine or biological designated as nanomedicine has dramatically accelerated the diagnosis, imaging, and treatment of many diseases.

Living organisms are built of cells that are typically 10 μm across. However, the cell parts are much smaller and are in the sub-micron size domain and proteins typical size is just 5 nm which is comparable with the dimensions of the smallest humanmade nanoparticles [3]. Functionalities can be added to nanomaterials by interfacing them with biological molecules or structures. Because the size of nanomaterials is similar to that of most biological molecules and structures, nanomaterials can be useful for both in vivo and in vitro biomedical research and applications. Thus far, the integration of nanomaterials with biological processes has led to the development of physical therapy applications, and drug delivery vehicles [4]. The nanoparticles are produced by using various techniques like chemical methods, physical methods, and mechanical methods. One of the physical processes is the electrical explosion wire (EEW) technique for producing nanoparticles, which has recently gained immense importance [5]. EEW method has become one of the promising ways for producing metal nanoparticles because it is ecologically safe, simple, effective, and cost-effective. [6,7]. Therefore, these nanoparticles can be used in almost any environment with no risk of rapid chemical reactions. Furthermore, it has a great potential in a wide range of applications with great interest from both scientific and technological points of view [8]. It is important to note that the metallic nanoparticles are most promising application as they show good antimicrobial properties and therefore it has a great interest from researchers in the field of microbial and fungi resistance against metal ions and antibiotics [9].

Candida albicans is one of the many fungi which live in and on all of us [10]. It is most common fungal pathogen presents in the oral cavity and is best known for causing thrush in the mouths of babies. It also causes the sore white and moist plaques in the mouth, on the tongue and the upper respiratory tract, as well as in some parts of gastrointestinal, blood, and genital tracts of the human body. In certain conditions, this fungus causes many types of diseases, for example, dermatocandidiasis, and may also lead to morbidity and mortality of immuno-compromised (e.g., via HIV infection) patients [11]. Recently, the development of antibiotic-resistant pathogens has become a world concern. For instance, Candida albicans has become resistant to antifungal drugs such as fluconazole and azole [12]. Thus, the current treatment of Candida albicans with the commonly used antifungal agents such as fluconazole or other azoles has become unsuccessful [13]. For this reason, new antifungal therapies are of utmost importance for effective treatment of Candida albicans infection. Importantly, using the trimetallic Alloy nanoparticles such as (Ag, Au, Cu) as antifungal agent has attracted much interest [14]. Therefore, this study was designed to investigate the antifungal treatment of Au–Ag–Cu nanoparticles alloys against the pathogenic fungus Candida albicans.

2. Experimental Setup and Procedures

2. 1 produce the nanoparticles

In this work the process was achieved by exploding wire (EEW) technique of producing nanoparticles as we described previously [15].

2. 2 Fungi Preparation

Candida albicans. The fungi isolate were obtained from Central Teaching laboratories in the City of Medicine, Teaching Hospital, Baghdad. The culture of *Candida albicans* was prepared under the following conditions: First, the fungi were cultured in the Sabouraud Dextrose Agar Medium that was prepared according to manufacturer protocol. In brief, the amount of 65g agar (40 g dextrose, 10 g peptone,15g agar) was dissolved in 1L distilled water. The medium PH was adjusted to 6.8 and sterilized by autoclaving. It was cooled down to 45-50°C then poured in sterile Petri dishes and left for 30 min at 37 °C to solidify. The most viable cellular colonies selected for the treatment (inoculum) by a sterile loop.

IOP Conf. Series: Materials Science and Engineering **757** (2020) 012049 doi:10.1088/1757-899X/757/1/012049

The sterile loop was streaked on the whole surface of the medium plates containing Sabouraud Dextrose Agar, and then kept stored in the refrigerator at 4°C and used as needed.

2 .3 Antifungal activity test of alloys nanoparticles by Disk-Diffusion Method

The antifungal activity of synthesized in a distilled water solution of different alloys (A,B,C) NPs where don by exploding wire a different current (100A,160A) and with four concentration (6.25, 12.5, 25 and 50 μg ml^{-1}) are tested against pathogen fungi of *Candida albicans*. Were examined using the disk diffusion method as (NCCLS document M44-A) [16] per the modified method of against of *Candida albicans* were taken from fresh stock cultures (24 hours) by the loop, then suspended in normal saline, and adjusted by comparison using standard 0.5 McFarland turbidity (5x106 cell ml^{-1}) tubes. These pathogens were absorbed on the surface of sterile cotton swab by dipping into the suspension. The sterile cotton swab was streaked on the whole surface of the medium plates containing Muller Hinton Agar. Then each test sample (50 μ ℓ) was soaked separately on a paper disc (6 mm diameter) made up from Whatman filter paper, and should wait until it completely dry .These impregnated paper discs were ascetically transferred to the surface of the petri dish media. The inoculated plates were then safely incubated at a temperature of 35 °C for 24 hours in the case of fungi. . The clear zone around the disc was considered as the zone of inhibition of microbial growth. The diameter of inhibition zones was estimated using a meter ruler and the average value for each organism was recorded. All tests were carried out in triplicate. Plates containing Muller Hinton Agar with fungi cultures(untreated) were used as a control. The diameter of the inhibition zone was precisely determined to calculate the antifungal activity of the test samples.

3. Results and discussion

3. 1 Inhibition of Fungus Growth by applying different alloys NPs

The antifungal activity of different alloys NPs (A, B, C) was tested against the pathogen fungi *Candida albicans*. Figures 1,2 and 3 shows the inhibition zone (IZ) of Candida albicans grown in solid medium (Muller Hinton agar plates) using disk diffusion methods. Different alloys nanoparticles (A, B and C) were applied at increasing concentrations (6.25, 12.5, 25 and 50 μg ml^{-1}). Figures 1,2 and 3 illustrated the inhibition zone at a varied concentration of nanoparticles ranged from 6 mm to 40 mm. The best inhibition of growth rate was observed after applying (160 A with 12.5 μg.ml^{-1}) to alloy B, that produced 40 mm zone diameter as showed in Figure 2. It is most likely that the small size of NPs compared to fungi facilitates the penetration through the cell wall and causes uncontrolled mass transfer through membranes. The cell wall is the primary target of the nanoparticles whereby the small size of these particles facilitates the penetration of these particles through the cell wall and interferes with the intracellular processes. Another effect of the nanoparticles as antifungal is the ability to induce cellular toxicity through the release of reactive oxygen species (ROS) [17]. Additionally, to achieve higher antifungal efficacy of nanoparticles, moderation, attenuation or erosion of the cell wall probably will enhance the effect of nanoparticles and facilitate their interaction with cellular components [9].The antifungal effect of alloy NPs could be due to NPs toxicity. The high antifungal activity of NPs could be through a combination of membrane damage and oxidative stress. The antimicrobial mechanisms of NPs can be explained in three stages: Primary NPs fungi contact, membrane perturbation, and membrane oxidation in an electronic structure[18]. These results indicated that direct physical membrane disruption of the fungi cells by the NPs is an important cytotoxic mechanism for effective fungi cell inactivation. Besides, metals alloy have also been demonstrated to be active agents in killing fungi cells through the generation of ROS that can induce oxidative stress within the pathogenic cells [9, 19]. The nanoparticles are likely to improve the directionality of the propagation of nanoparticles towards diseased sites within a tissue. Furthermore, the Au-Ag-Cu alloy consists of three elements, it is unlikely the fungus produced a resistant mutant to the nanoparticles compared with the single element nanoparticle.

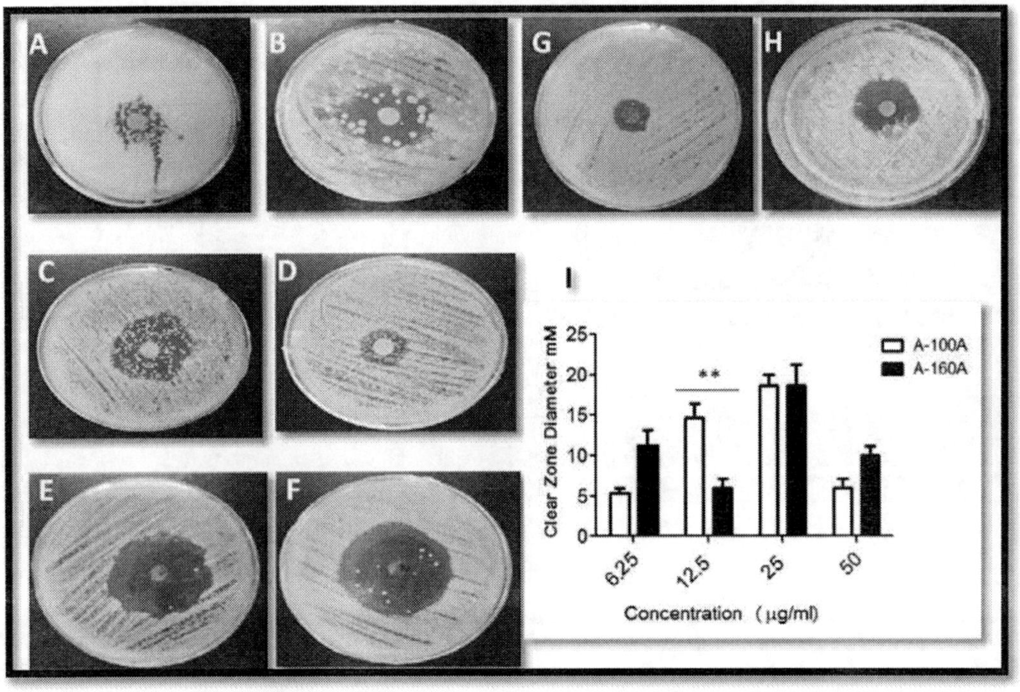

Figure 1. Growth inhibition of fungal *Candida albicans* grown in solid medium, samples in Petri dishes. (A) A-100A and (B) A-160-A at concentrations of 6.25 $\mu g.ml^{-1}$. (C) A-100A and (D) A-160A at concentration of 12.5 $\mu g.ml^{-1}$. (E) A-100A and (F) A-160A at concentration of 25 $\mu g.ml^{-1}$. (G) A-100A and (H) A-160A at concentration of 50 $\mu g.ml^{-1}$. The alloy (A) nanoparticles where done by exploding wire currents (100A,160A) respectively. (I) The statistical analysis of the differences between A-100 and A-160 showed significant difference at the concentration 12.5 µg/ml. Two-way ANOVA with Bonferroni post-test. $^{**}p<0.01$.

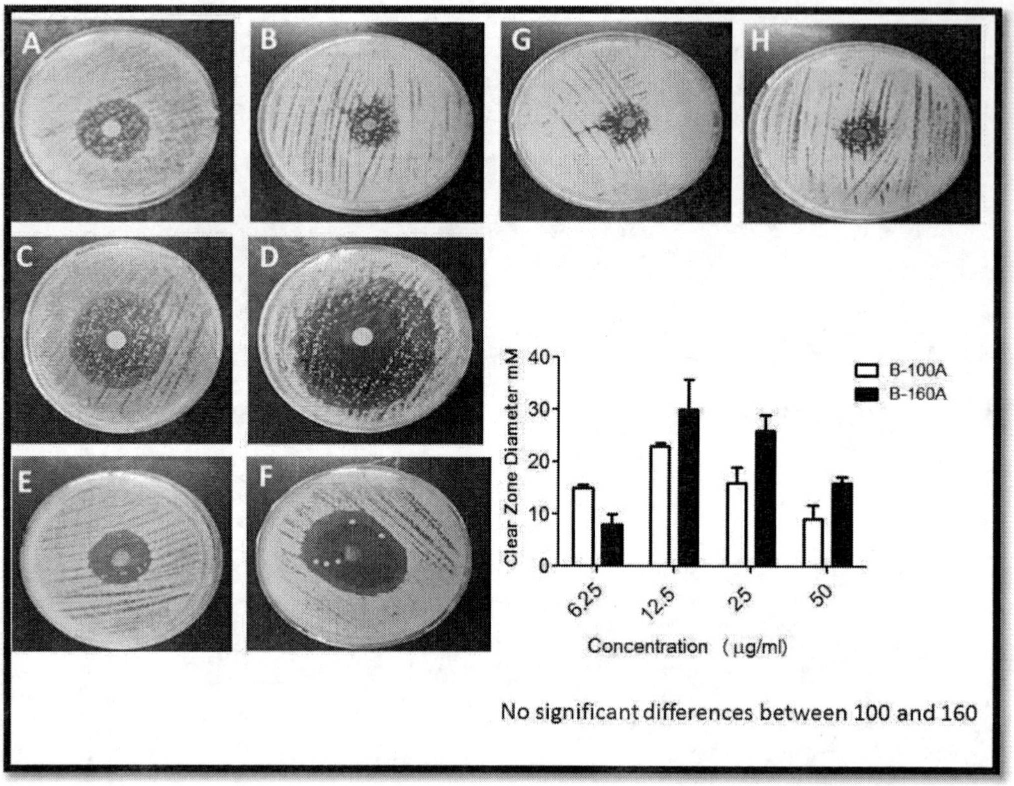

Figure 2. Growth inhibition of fungal *Candida albicans* grown in solid medium, samples in Petri dishes. (A) A-100A and (B) A-160-A at concentrations of 6.25 µg.ml^{-1} . (C) A-100A and (D) A-160A at concentration of 12.5 µg.ml^{-1}. (E) A-100A and (F) A-160A at concentration of 25 µg.ml^{-1}. (G) A-100A and (H) A-160A at concentration of 50 µg.ml^{-1}. The alloy (B) nanoparticles where done by exploding wire currents (100A,160A) respectively. (I) The statistical analysis of the differences between A-100 and A-160 showed no significant differences between A-100 and A-160. Two-way ANOVA with Bonferroni post-test. **p<0.01.

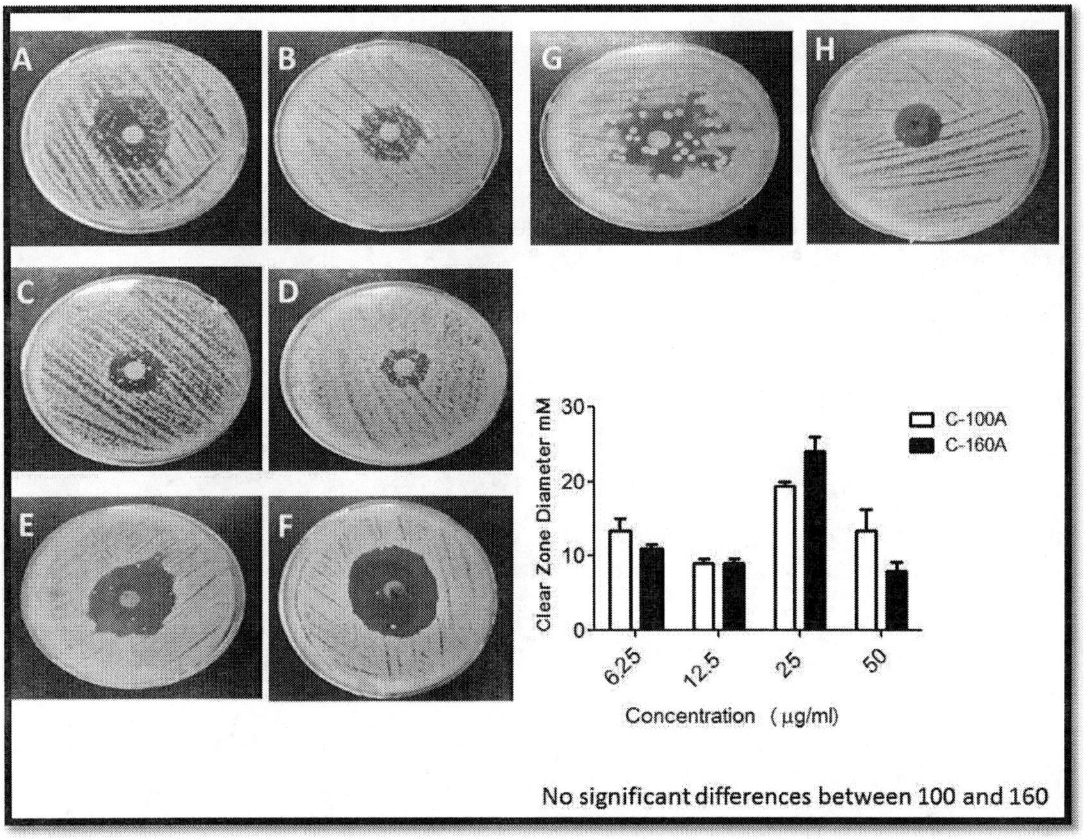

Figure 3. Growth inhibition of fungal *Candida albicans* grown in solid medium, samples in Petri dishes. (A) A-100A and (B) A-160-A at concentrations of 6.25 μg.ml^{-1} . (C) A-100A and (D) A-160A at concentration of 12.5 μg.ml^{-1}. (E) A-100A and (F) A-160A at concentration of 25 μg.ml^{-1}. (G) A-100A and (H) A-160A at concentration of 50 μg.ml^{-1}. The alloy (C) nanoparticles where done by exploding wire currents (100A,160A) respectively. (I) The statistical analysis of the differences between A-100 and A-160 showed no significant differences between A-100 and A-160. Two-way ANOVA with Bonferroni post-test. **p<0.01.

4. Conclusions

The outcome of this study showed that the alloys nanoparticles inhibited the growth rate of the pathogenic fungus *Candida albicans*. The best antifungal activity against pathogen fungi of *Candida albicans* was for a concentrations 12.5 μg.ml^{-1} of alloy B prepared by 160 A that showed more than 85% growth inhibition after the treatment. The enhanced penetration by nanoparticles offers a spatial selectivity that could imrove their safe use as a therapeutic strategy.

Acknowledgment

We would like to present our thanks to the staff of biological laboratory in the University of Baghdad, college of science, physics department for their kind support.

REFRENCES

[1] Mohd H, Arvind K., Aftab A, Mohammed F A, Mohammed B, Shah Al K and Mohd M, 2017 Tropical Journal of Pharmaceutical Research, **16** 12:2967-2976.

[2] Masaharu T, Mika M, Hisayo K, Masatoshi O, Sachie H, Yuki Y and Toshitaka I , 2013 Journal The Royal Society of Chemistry, Cryst Eng Comm, **15**:1345–1351.

[3] Dunn W, 2004, Applications of nanoparticles in biology and medicine, Journal of Nanobiotechnology, 2, 3-10.

[4] Boisseau Pand Loubaton B, 2011 Nanomedicine, nanotechnology in medicine, Comptes Rendus Physique, **12** 7:620–636.

[5] Wankhede P, Sharma P K , and Jha A K, 2013 J. Eng. Res. Appl., 3:1664–1669.

[6] Masaharu T, Mika M, Hisayo K, Masatoshi O, Sachie H, Yuki Y and Toshitaka I , 2013, journal The Royal Society of Chemistry, CrystEngComm, **15**:1345–1351.

[7] Rahman – L-U., Afzal S, Sher B K, Abdullah M A, Hidayat H, Changseok H, Rumana Q, Muhammad N, A, Muhammad A Z, Muhammad I,and Heinz-Bernhard K, 2015 J Appl Electrochem, **45** 5: 463-472

[8] Chen C, Gabriella R, Sravan T, and Jing Z, 2017, Au–Cu–Ag Nanorods Synthesized by Seed-Mediated Coreduction and Their Optical Properties, Part. Part. Syst. Charact, 34, 1600384.

[9] Salama A O, Amira A. El-Adly, and Abdel-Aleam H M, 2015 Journal of Medical Microbiology, **64**:1151–1161.

[10] Sayyada G N, Amna S, Shazia T Hakim, Yasmeen A, and Shahana U, 2013 Open Journal of Medical Microbiology, **3** 185-192.

[11] Xiong Z, Lu X P, Feng A, , Pan Y , and Ostrikov K, 2010 Physics of plasmas, **17** (123502), 1–6,.

[12] Nelson D , Priscyla D M , Priscyla D M , Oswaldo L A and Elisa E, 2007 Journal of Biomedical Nanotechnology, **3** 2: 203–208.

[13] Peng S , Yi S, Haiyan W, Weidong Z, Jose L L , Wei L J Z , Ruoyu L , and Jing F, 2011 Applied Physics Letters, **98** 021501:1-3.

[14] Isam M Ib , Iftikhar M A, Batol I Dh , Qays A A, Asmeit R, Eisa M H, and Aljameel A.I, 2017 Materials Science and Engineering, **73**: 665–669.

[15] Thikra K Al-K and Hammad R H, 2018, Iraqi Journal of Physics, **16** 39 :81-92

[16] NCCLS., 2004 Method for Antifungal Disk Diffusion Susceptibility Testing of Yeasts; 24, NO. 15, ISSN 0273-3099, Approved Guideline. NCCLS document M44-A, NCCLS, Wayne, Pennsylvania 19087-1898 USA,

[17] Aswathy R, Preethy C, S Sudheer K, 2013 Colloids and Surfaces B: Biointerfaces, 105:342– 352 .

[18] Mohammad J, Katharina M , Ali Akbar A , Dorleta J , Idoia R, Teofilo R, Vahid S , Wolfgang J and Morteza M, 2012 Trend in Biotechnology , **30** 10:499-511.

[19] Shu F S and Anna Cristina S S, 2014 Appl. Materials and Interfaces, **6** :20154 – 20163.

Building a spin coater device for thin-film preparation

F H Ali

Department of Physics, College of Science, University of Baghdad, Baghdad, Iraq

Corresponding author email: **falahhasan@scbaghdad.edu.iq**

Abstract

A low cost, simple design of the rotary spindle (spin-coater) is described in this research. This simple coating system is used as thin films deposition device for microscope slide (75*25) mm . The easily system can be built by electronics and machine. Designed with a brushless engine with Arduino Uno, designed for speed control. The rotational speed ranges from 350 to 10,000 rpm to give us many thickness according to the materials used. In our design, the rotational speed is controlled manually by the variable resistor.

Key words: low cost, spin coater, thin-film

1. Introduction

The technology of thin film plays an important role in high-tech industries [1]. The thin film technology has been enhanced mainly for the need of integrated circuits industry. Demand for the development of smaller, smaller, faster-speed devices, especially in the next generation of integrated circuits, requires sophisticated materials and new processing technologies suitable for integration technology [2]. In this regard, the technology and physics of thin films can play a chief part to reach this goal. Thin film production has been developed for the past 40 years. Thin film as a two-dimensional system is of unlimited importance to many problems of real world [3]. The costs of their materials are very small compared to the matching bulk material and achieve the same function when it arises to surface operations. Thus, knowledge and design can be used for the nature of films, their functions and their new features to develop new technologies for future applications [4].

Thin films are thin layers of materials ranging from parts of nanometers to several micrometers in thickness [5]. A thin film can be applied uniformly through various techniques: chemical vapor deposition (CVD) and PVD [Hussein, 2011]. Another way is rotational films (spin-coater). Spin-coater is a quick and easy way to create thin and homogeneous films [6].

Spin coating is the procedure used to apply thin films uniform on a flat substrate. The machine used to build thin film is called spin-coater, or simply rotor. In short, a limited amount of solution is deposit on the substrate, which is then rotated at high speed for propagation, liquid by centrifuge force. Manufacturing techniques are comparatively simple because equipment is available. One of the mainly important factors in rotational coatings is repetition. Several ways to develop composite inorganic semiconductors play vital roles in advancing technologies. For performance of thin films based on film, it is necessary to control film properties. Therefore, many advanced techniques are used, such as molecular beam epitaxy, laser assisted evaporation, sol-gel, ion beam radiation, thermal evaporation,

chemical vapor deposition, vacuum deposition, chemical bath deposition, etc. These processes are intensive of energy and include high temperature, pressure and vacuum. But spin coating is one of the most cost-effective, fast and simple ways to deposit thin films and the way to prepare single and heterogeneous films [7].
In general, the thickness of the spindle film is relative to the square root of rotation speed as in the equation below where t is the thickness and (ω) is the angular velocity:

$$t \propto \frac{1}{\sqrt{\omega}}$$

This means that the film, which is woven at a rate of four times the speed, will be half its thickness. The spin curve of this equation can also be calculated as shown below.

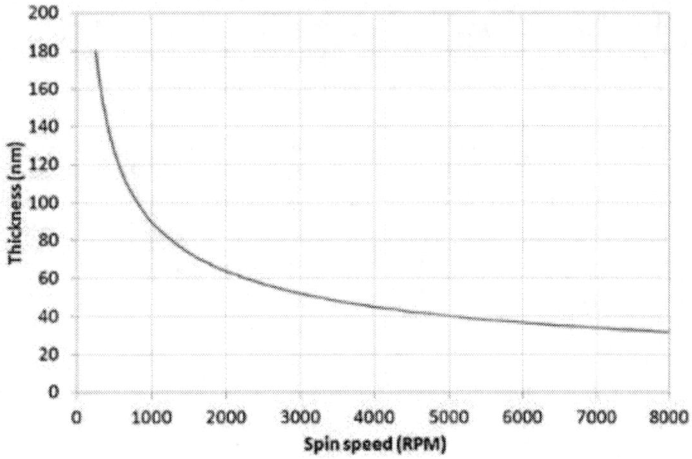

Figure 1. Example spin curve for a solution.

In this paper, a cost-effective and very easy spin coater has been planned with their modify base for spin holder. Our future design is approximately 75% cost-effective than the marketplace available spin coating system. There is no time extreme stay and the device is easy to clean after each process.

Furthermore, we use the basics of the physics in the design and construction of a low-cost spin-coater that can be programmed by computer using Arduino Uno card and brushless-motor.

2. Experimental

This part displays the hardware and software, adopted in this work, such as Arduino Uno brushless-motor and substrate holder.

2.1. Arduino Uno

Arduino UNO (Arduino, Italy)
Arduino UNO is a microcontroller based on ATmega328.For analog inputs, there are 6 very useful analog pins and for digital: It consists of 14 digital I / O pins where the number of PWM pins is 6. [10].

Figure 2. The Arduino UNO board

The stand for our device, a microcontroller needed to read data from our sensors and we work with Arduino because it is easy to utilize and has an dynamic community with lots of free codes and help to find when wanted [10]. In this project, we worked with the Arduino Uno board based on the 8-bitmica328p microcontroller. The panel has an internal clock frequency of 16 MHz and memory of 32 KB ashes and a number of analog and digital screws. It operates at a voltage of 5 volts and is easier to operate via USB, which also provides a serial line for communication and programming from a computer [11].
In the final product, the ATmega328p processor was used, which was flashed using an Arduino boot loader and an internal 8MHz oscillator rather than the entire motherboard, make the device smaller and easier to handle. For most programming and testing, the UNO board was used.

2.2. Brushless-Motor

In conventional DC motors, mechanical change is applied using brushes that lead to mechanical friction, noise, electric spark and wireless interference. These defects can be overcome by Brushless DC Motors (BLDC), as shown in fig.3. Due to the presence of permanent magnets in the rotor, they are also known as DC motors (PMBLDC) free of permanent magnet brushes. The (BLDC) engine was developed based on brushed DC motors.

Brushless motor controlled by 30A ESC controller in this work.

IOP Conf. Series: Materials Science and Engineering **757** (2020) 012050 doi:10.1088/1757-899X/757/1/012050

Figure 3. The brushless motor& 30A ESC Controller

2.3. Substrate holder:

The sample holder is made locally as shown in figure 4.

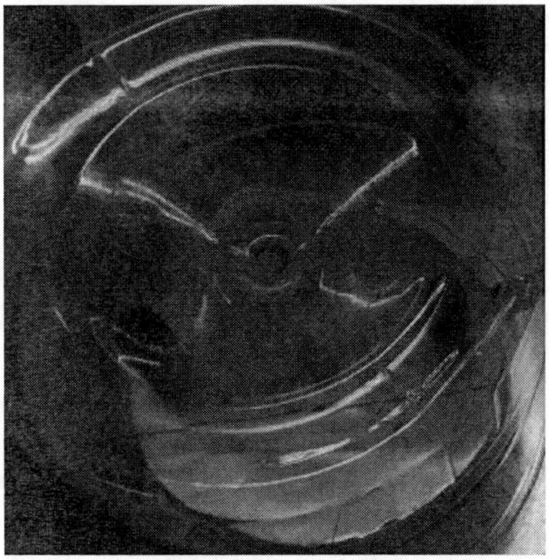

Figure 4. Substrate holder

2.4. Device assembly

Figure 5. illustrate the low-cost spin coating device after connecting the hardware and uploading the code.

IOP Conf. Series: Materials Science and Engineering **757** (2020) 012050 doi:10.1088/1757-899X/757/1/012050

Figure 5. Spin-coating device

2.5. *The software*

The following steps represent the used code with illustration each step after double slash .

```
#include <Servo.h>// servo library was used to control ESC

Servo esc; //Create of a servo class with name as esc

void setup()

{

esc.attach(8); //state the esc signal pin,Here as D8

esc.writeMicroseconds(1000); //initializing of the signal to 1000

Serial.begin(9600);

}

void loop()

{

intval; //Creating a variable val

val= analogRead(A0); // input Reading from analog pin a0 and store in val

val= map(val, 0, 1023,1000,2000); //mapping of val (adjust if desired)

esc.writeMicroseconds(val); //using of val as the signal to esc
```

IOP Conf. Series: Materials Science and Engineering **757** (2020) 012050 doi:10.1088/1757-899X/757/1/012050

}

3. Preparing of samples

Preparing samples that coated by manufactured device are shown in figure 6.

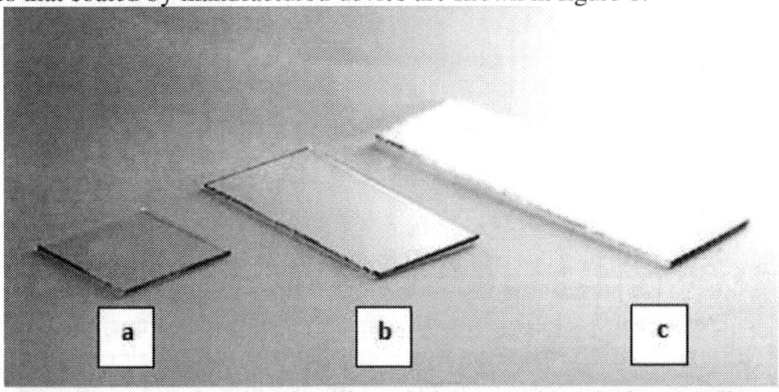

Figure 6. Prepared Samples with different thicknesses: a) 85 nm b)105 nm c)150nm.

4. Conclusion

Brushless motor is a good motor that we can use for build low cost spin coater.
High homogenized thin film is get by this manufactured device. Wide range spin-speed is gated by using brushless motor (350-10000) RPM giving us many thickness according to the materials used ,and in this work was about (1000,1500 and 2000)RBM.

5. References

[1] Ohring M 1992 *Academic Press limited*, 24-28 Oval Road, London, NW1, 7DX, UK, XIII.
[2] Kashyap S C and Wuttig M 2003 *World Scientific Publishing Co. Pvt Ltd.* **5** Toh Tuck Link, Singapore, 596224 *Tehran, Iran, IX.*
[3] Seshan K 2002 2nd ed Norwich, New York, U.S.A.: Noyes publications, William Andrew publishing.
[4] Aguilar R G and López J O 2011 *Latin American Journal of Physics Education*, 5 (**2**), 368-373.
[5] Smirnov M A Rambu P Baban, C and Rusu G I 2010 *Journal of Advanced Research in Physics*, 1 (**2**), 021011.
[6] P Claude Sevvanthi A Jayanthi C and Poiyamozhi A 2012 *Advances in Applied Science Research*, 3 (**6**), 3573-3580.
[7] Hussein H F Shabeeb G M and Hashim S S 2011 *J. Mater. Environ. Sci.* 2 (4), 423-426)."

[8] Dunbar P and Birnie III 1987 "The Key Stages in Spin Coating", *J. Imaging Technology*, 13, 122.

[9] Schubert D W and Dunkel T 2003 "Spin-Coating"*J. Appl. Phys.*, 49 (**7**), 3993.

[10] M. Margolis 2011 "*Arduino cookbook*: O'Reilly Media, Inc.

[11] C Yi and F Bai *Journal of Changchun University of Science and Technology (Natural Science Edition)* **4** 016, 2011.

IOP Conf. Series: Materials Science and Engineering **757** (2020) 012051 doi:10.1088/1757-899X/757/1/012051

Synthesis of copper oxide thin films by electrolysis method based on Porous Silicon for Solar Cell Applications

M. H. Oudah[(1)] ,M. H. Hasan[(1)] ,A. N. Abd[(2)]

Physics Department, College of Science, University of Anbar, Anbar, Iraq [(1)]

Physics Department, Faculty of Science, University of Al- Mustansiriyah, Baghdad, Iraq [(2)]

Abstract. Copper oxide thin film was successfully prepared by electrolysis method and deposited on glass and porous silicon substrates. The main goal of this paper is to study the properties of copper oxide thin films and prepare a solar cell of low cost and good efficiency. These properties were optimized at annealing temperature (200) °C. The transmission spectrum were studied for the prepared thin film showed good transparency in the visible region. The structure of films was characterized by X-ray diffraction. The average grain size of the thin film from AFM analysis is about 46.3 nm. The energy bandgap is found to be (3.8) eV this wide bandgap is due to the quantum size effect. The solar cell efficiency is found to be (12.9%).

Keywords: copper oxide, thin film, solar cell, electrolysis

1. Introduction

Solar cells are future energy resources. Silicon is a conventional semiconductor material used for solar cells. The reduction in the cost of solar cells is the most important issues. [1]. Copper has two major oxide phases, Cu_2O and CuO. Both are p-type semiconductors. The energy bandgap for bulk Cu_2O is 1.8–2.5 eV and for bulk, CuO is 1.2–1.9 eV. They are promising semiconductors for solar cell applications due to their electrical and optical properties, as well as nontoxicity, natural abundance and inexpensive [2]. Thin films Copper Oxide have been prepared using various techniques such as sputtering process, chemical vapour deposition, electro-deposition, vacuum evaporation, thermal oxidation, chemical solution deposition, electron-beam Evaporation [3–6].

In this work, we discuss the effect of annealing on the properties of copper oxide thin film, photovoltaic properties and its applications in the solar cell this thin film were prepared by electrolysis method. The film was characterized using UV-Vis spectrophotometer, (AFM), (XRD) and I-V characteristics at dark and illumination conditions.

2. Experimental:

The steps for preparing copper oxide thin film and porous silicon were summarized in this section

2.1. Preparation of substrates

In this work, 2 x 2 cm^2 and glass an n-type silicon (1- 10) Ω .cm, (2×1.5) cm dimensions substrates were used with a thickness of 0.1 cm. the substrates were cleaned using methanol and deionized water in the ultrasonic bath in order to get a clean surface contact for the next process.

2.2. Preparation of copper oxide solution using the electrolysis method

Copper nanoparticles have been prepared by electrolysis cell. The electrodes are copper as a positive electrode and gold as a negative electrode. Water was used with Hydrochloride (HCl) in ratio about 8:1 as an electrolyte liquid. The applied voltage is (6) Volt.

2.3. Preparation of porous silicon substrate

The porous Silicon (PS) layers were fabricated by anodic etching with an electrochemical cell where 0.785 cm^2 etched area substrate was placed in the Teflon etching cell using an admixture of 1:1 ethanol (purity 99.99%) and aqueous hydrogen fluoride (50% purity). The sample was anodized at a current density of (15) mA/cm^2 for (15) min etching time. No further thermal or chemical treatment was carried after etching.

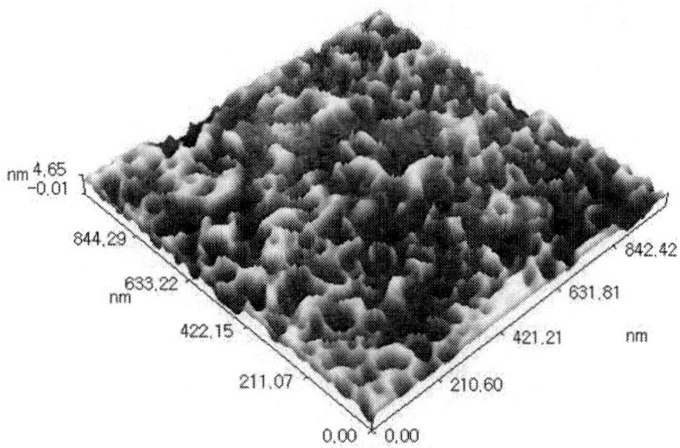

Figure1. AFM images of the porous silicon (PS)

2.4. Preparation of copper oxide thin films deposition by Drop-casting method

The drop-casting method has been used to deposit a Cu nanoparticles solution on glass and PS samples. The annealing process at (200) °C is then applied for (90) minutes to the samples to get the copper oxide. Thin film thickness is found to be (200±10) nm the thickness of the film is measured using (Fizeau Interferometer Method). The bottom of the (PS) sample is coated with thick aluminium of very high purity (99.99%) in a vacuum. At a pressure of (5x10^{-6}) mbar. In addition, silver paste was added on the top layer of the copper oxide to measure the electrical properties.

2.5. Samples Characterization

The crystal structure was studied at the following operation conditions-: Source Cu-Kα radiation of wavelength (λ=1.5405 Å), Voltage =4 kV, Current =20 mA, Scanning speed = 5 cm/min by XRD 6100 Shimadzu. Optical transmittance measurements were complete with (UV/Visible 1800 spectrophotometer). The energy bandgap of the thin film was specified from the optical transmittance spectrum (300-1100nm). The shape and size of copper oxide were investigated by using XRD and AFM (AA 3000 Scanning Probe Microscope).

3. Results and Discussion

The obtained results from the experiment are discussed in this section.

3.1. Structural properties of copper oxide thin films

Fig.2 shows the diffraction pattern of copper oxide film deposited on glass substrates annealed at (200) °C. Analysis of diffraction spectra indicates two peaks positions for Cu_4O_3 at crystal planes, 47.47°, 57.74°. In addition, there are two peaks of CuO_2 at crystal planes, 28.97°, and 39.93°. CuO_2 is not stable and it will decay to one of the stable forms of copper oxide. Also, the figure shows two peaks of Cu_2O at crystal plane, 27.91° and 42.16° and one peak of (Au_2O_3) at a crystal plane, 27.91°. These peaks were associated with the reflection of the peaks of (COD) cards no. (96-210-5391, 96-152-1321, 96-900-7498, 96-900-5770 and 96-900-0604) [7]. The (Au_2O_3) peak is present as a result of using a gold electrode in the electrolysis cell. When saturation state accrues a reverse current may flow and this will lead to getting Au nanoparticles. Au_2O_3 is a narrow bandgap semiconductor (0.83) eV [8]. According to the pattern, results have a small peak and the film shows a good crystallinity.

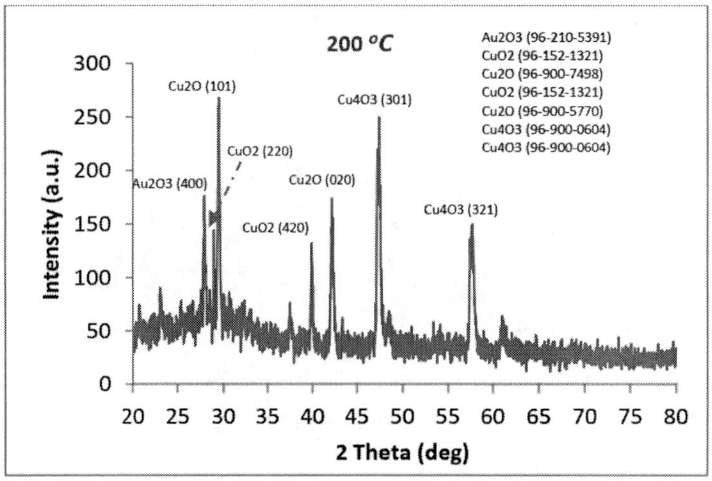

Figure2. XRD pattern for copper oxide thin film annealed at 200 °C.

The grain size was calculated using Debye-Scherrer's equation [9]:

$$(G_S) = \frac{0.94 \times \lambda}{\beta \cos\theta} \qquad (1)$$

Where (λ) represents the wavelength of X-ray radiation, (G_S) represents the grain size, (θ) is the angle of diffraction and (β) is the full width at half maximum.

the value of dislocations density (δ) and strain value (ε) can be calculated by using the next relations [10]:

$$\delta = \frac{1}{(G_S)^2} \qquad (2)$$

$$\varepsilon = \frac{\beta \cos\theta}{4} \qquad (3)$$

Fig. 3 shows a (3-D) AFM image of the thin film. AFM image explains that the grains are distributed uniformly within the scanning area (1600×1600nm) with individual straight grains growing upwards. The average grain size is calculated from AFM analysis using the software is found to be (46.3) nm.

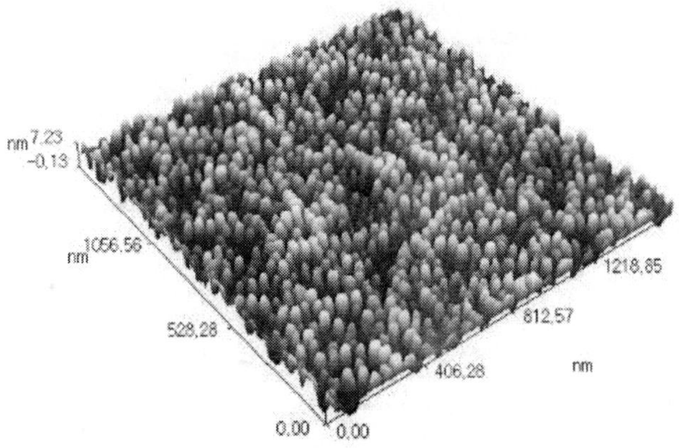

Figure3. AFM images of copper oxide thin film annealed at 200 °C

3.2. Optical properties of copper oxide thin films

The transmittance as a function of wavelength is shown in Fig. 4. It is found that the transmittance of the film increase with the increasing of the wavelength. At UV-region, the transmittance is sharply decreased as a result of the wide of absorbed particle size.

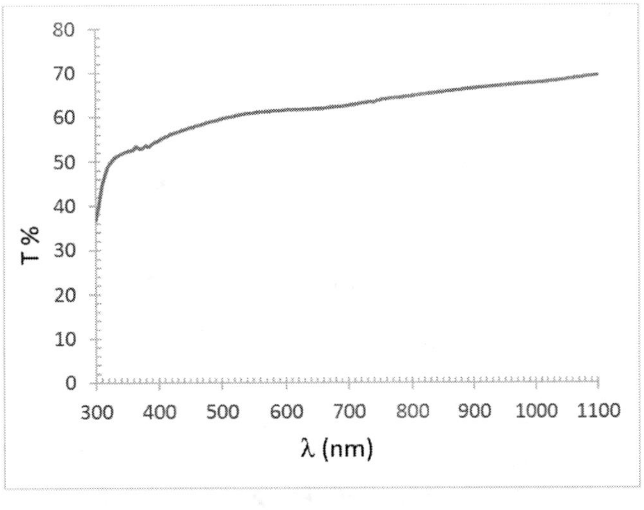

Figure 4 .Transmission spectra for copper oxide thin film annealed at 200 °C

Fig. 5 shows the energy bandgap of copper oxide measured from the plotting $(\alpha h\upsilon)^2$ versus the photon energy ($h\upsilon$) where (α is the coefficient of absorption). By taking, a straight line from the linear section of the curve toward the x-axis, which represents the photon energy. (Eg) is found to be 3.8 eV, this results in good agreement with [11]–[14]. The wide bandgap is due to the quantum size effect.

IOP Conf. Series: Materials Science and Engineering **757** (2020) 012051 doi:10.1088/1757-899X/757/1/012051

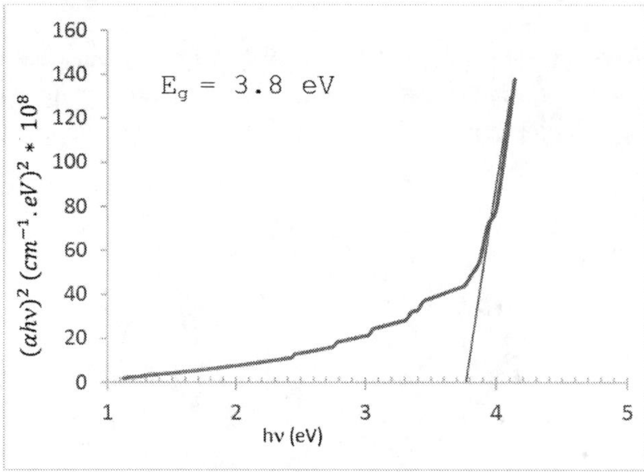

Figure 5. $(\alpha hv)^2$ versus E_g of copper oxide thin film annealed at 200 °C

3.3. Electrical properties of the heterojunction

Fig. 6a and Fig. 6b Shows the I-V characteristics at the dark condition and in the presence of light respectively which confirm the heterojunction behaviour of (p-copper oxide /n-PS) junction and Fig. 7 Shows The measurement of the open-circuit voltage, short-circuit current, fill factor and the efficiency all data are explained in Fig 7 under (40 μwatt) light power. This figure represents the I-V curve at the light condition with the existence of variable resistance. This experiment is a simple method to measure the efficiency of a solar cell. There are two factors we need to find, short-circuit current (I_{SC}) occurs when $R = 0$ so that $V = 0$ and The second factor is the open-circuit voltage (V_{OC}) occurs when $R \rightarrow \infty$, in this case, the net current is zero. From these two points, we can find the maximum power rectangle and calculate the efficiency of the solar cell using equation (4). All results of the heterojunction p-copper oxide/n-PS confirm that this heterojunction could be used in solar cell applications [15].

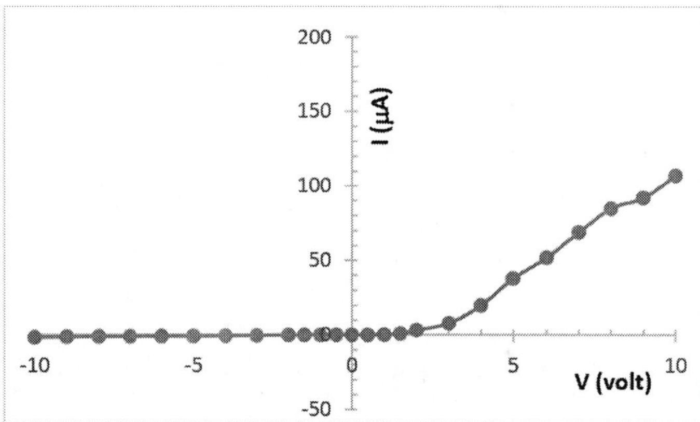

Figure 6. a) the I-V characteristics in the dark condition of the (p-copper oxide/n-PS) heterojunction annealed at 200 °C

The solar cells conversion efficiency depends on the morphology of interfaces.

$$\eta = \frac{V_m I_m A_2}{P_{in} A_1}\ 100\ \% \qquad (4)$$

(η) is the efficiency of the solar cell, (V_m) is the maximum output voltage, (I_m) is the maximum output current, (P_{in}) is the light source power which is equal to (40) μwatt in our experiment. (A_1) is the area of the solar cell sample of radius (0.5) cm. (A_2) is the area of the electric contacts which have a value of (0.04) cm^2.

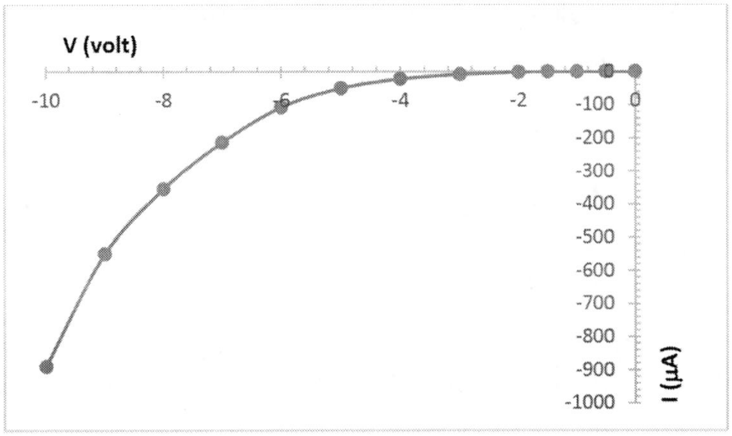

Figure 6. b) the I-V characteristics in the light condition of the (p-copper oxide/n-PS) heterojunction annealed at 200 °C

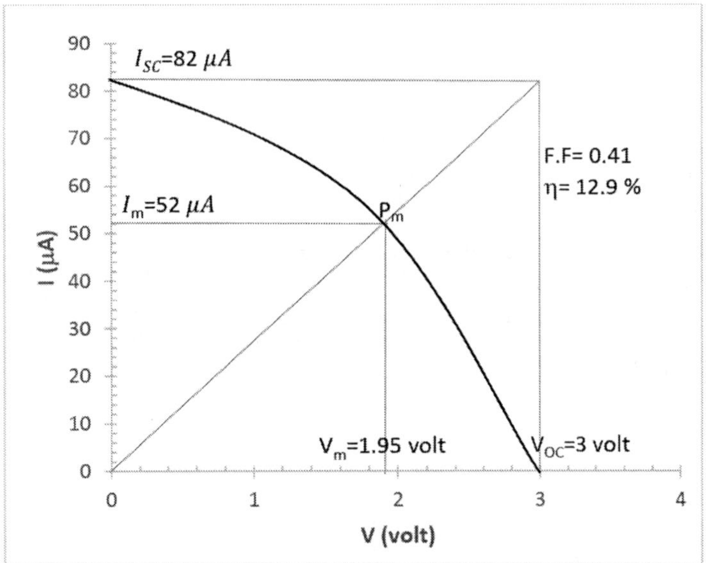

Figure7. open-circuit voltage of the (p-copper oxide/n-PS) heterojunction annealed at 200 °C

4. Conclusion

In the present work, copper oxide thin films prepared by electrolysis method for solar cell applications has been studied. The prepared thin film shows good transparency in the spectral range (350-1100) nm the energy bandgap is found to be (3.8 eV). This wide bandgap due to the quantum size effect. The performance of the solar cell heterojunction has been improved by the porosity of Silicon and the existence of the gold oxide that form sublevels inside the bandgap of copper oxide that helps to increase the charge carriers that share in the conduction process. The solar cell efficiency is (12.9%).

IOP Conf. Series: Materials Science and Engineering **757** (2020) 012051 doi:10.1088/1757-899X/757/1/012051

References

[1] L. C. Olsen, R. C. Bohara, and M. W. Urie 1979 *Appl. Phys. Lett.* **34** 1:47–49.

[2] F. Gao, X. J. Liu, J. S. Zhang, M. Z. Song, and N. Li 2012 *J. Appl. Phys.* **111** 8: 2–6.

[3] Y. F. Lim, C. S. Chua, C. J. J. Lee, and D. Chi 2014 *Phys. Chem. Chem. Phys.* **16** 47: 25928–25934.

[4] C.-Y. Lin, Y.-H. Lai, D. Mersch, and E. Reisner 2012 *Chem. Sci.* **3** 12 :3482–3487.

[5] L. Liao *et al 2009 Nanotechnology* **20** 8 :85203, 2009.

[6] A. Paracchino, N. Mathews, T. Hisatomi, M. Stefik, S. D. Tilley, and M. Grätzel 2012 *Energy Environ. Sci.* **5** 9 :8673–8681.

[7] S. Gražulis *et al.* 2012 *Nucleic Acids Res.* **40** D1 :420–427.

[8] H. Shi, R. Asahi, and C. Stampfl 2007 *Phys. Rev. B* **75** 20.

[9] J. M. Rzaij 2016 *Iraqi J. Phys.* **14** 31 :1–12, 2016.

[10] Y. Shen 2014 "Development of Thin Film Photovoltaic Cells based on low cost metal Oxides." University of Bolton.

[11] M. Dhaouadi *et al.* 2018 *Am. J. Phys. Appl.* **6** 2 43.

[12] K. S. Wanjala, W. K. Njoroge, N. E. Makori, and J. M. Ngaruiya 2016 *Am. J. Condens. Matter Phys.* **6** 1 :1–6 .

[13] H. Hashim, S. S. Shariffudin, P. S. M. Saad, and H. A. M. Ridah, 2015 *IOP Conf. Ser. Mater. Sci. Eng.*, **99** 1.

[14] H. Kidowaki, T. Oku, T. Akiyama, A. Suzuki, B. Jeyadevan, and J. Cuya 2012, *J. Mater. Sci. Res.* **1** 1 :138.

[15] T. Oku, R. Motoyoshi, K. Fujimoto, T. Akiyama, B. Jeyadevan, and J. Cuya 2011 *J. Phys. Chem. Solids*, **72** 11 :1206–1211.

Structural and Morphological Characteristics of Organic and Inorganic Heterojunction Blend NiPcTs/ CdS

A. S Hassan[1] , I. H Shallal[2] and M. A Aboode[3]*

[1,2] Department of Physics, College of Education for Pure Science /Ibn Al-Haitham, Baghdad University, Baghdad, Iraq
[3] Solar Energy Research Center, Renewable Energy Directorate
[1, 2,3] Higher Education and Scientific Research Ministry, Baghdad, Iraq
*Corresponding author:email: morooj_a@yahoo.com

Abstract.
In the present study NiPcTs, CdS thin films, and Blends of NiPcTs:CdS were prepared with 1:2 content mixing ratio of NiPcTs to CdS solutions. Cadmium chloride and thiourea were used as the essential materials for deposition CdS thin films while using organic powder of NiPcTs to deposit NiPcTs nanostructure films. The spin-coating technique was employed to fabricate the NiPcTs , CdS films and NiPcTs-CdS blend. Structural properties of films have been investigated via X-Ray diffraction(XRD),and show that thin films of NiPcTs, and CdS have monoclinic and polycrystalline hexagonal structure respectively while the blend has two polycrystalline structure with cubic and hexagonal phases.
Atomic force microscope (AFM) confirmed that the surface of all samples are quite smooth and they are comprised of spherical numerous nanoparticles with diameter less than 70 nm. Scanning Electron Microscopy (SEM), and enery dispersive spectroscopy (EDS)analyzer has been achieved to determine the chemical composition of the molecular materials, which exhibit the existence of all essential elements of thin films and blend hetrojunction (BHJ) of NiPcTs –CdS. The preparation of chemical solutions,deposition of NiPcTs, CdS thin films and the blend hetrojunction BHJ of NiPcTs –CdS have been characterized.

Keywords: Thin Film, Organic Solar Cell, Structural; FTO Conductive Polymers, blend heterojunction

1. Introduction

Over the past few decades, considerable attention has been paid towards the organic materials being used in fabrication of organic solar cells (OSC). This was obviously seen in metal phthalocyanines (MPc) which are molecular organic semiconductors. Organic photovoltaic devices including organic solar cells, gas sensors, field effect transistors, optical data storage have wide range of industrial applications; due to their unique optical and electronic properties as well as their chemical and thermal stability. In addition the low cost and simple device preparation via solution processing makes them more interested in some technology applications [1-2]. The power conversion efficiencies of solar cells were clearly improved from 0.001% to5.5% within the years of (1975 -2006).Development and new inventions in polymer science in parallel with participate significantly to the amount of investigations about OSC[3-4]. Possibly the development in efficiency will make them a competitive alternative to inorganic solar cells in the near future [5-8].Organic thin films deposition techniques like roll-to-roll processing, ink-jet printing or spin casting are widely used, due to their cost effectiveness, and less material consumption can be obtained from the wet processing method to deposition of thin films[9-12].CdS is one of compound belong to II-VI semiconductor with an energy gap ranging from (2.82- to 2.5)eV. The chalcogenide families are used as window and buffer layer for CdS/CdTe solar cells, and continue to be a good choice to enormous researches due to their high-efficiency in solar devices[13-14]. Examples on the photoelectronic devices, like transducers laser materials, photo-sensors, optical wave guides and non-linear integrated optical devices. [15-16]

In this work synthesis and physical characteristics of organic NiPcTs inorganic CdS, and Blends of NiPcTs -CdS semiconductors thin films, have been studie**d.**

2. Device structure and experimental method

The studied thin films of Nickel Phthalocyanine Tetrasulfonic Acid Tetrasodium Salt NiPcTs films were synthesized on pre-patterned (Fluorine-doped Tin Oxide) FTO coated glass substrates of thickness (3.2mm) and (10×10 cm) dimensions

The substrates were cleaned in acetone, ethanol, and deionized water for 10 min respectively. After drying in 70°C, the substrates were cut into four equal pieces with dimensions (2.5x2. 5) cm.

Thin films of NiPcTs prepared by sol-gel spin coating methods, with thickness of (130 ± 7) nm, this is achieved by dissolving (0.65gm) organic powder of (NiPcTs), which was provided by (Sigma- Aldrich) in (10ml) deionized water. the solution has been deposited on FTO glass substrate ,and, leave samples for 60 minutes in spin coater type [Laurell-WS-650Mz-23NPP]with speed of (1500) rpm ,then drying process in (70°C) temperatuer.

Two (Al) electrodes of 200 nm thickness were deposited through a shadow mask onto glass substrate using thermal evaporstiion method with 40 ± 5 μm gaps between them. The evaporation rate of (0.3 nm/s) was choosing to deposit electrode films, and the Vacuum chamber was maintained at a pressure of (5×10^{-5}) mbar throughout the evaporation process. CdS films were deposited on glass substrates by spin-coating method. CdS solutions were prepared by dissolving (1.83gm) cadmium chloride ($CdCl_2$), with (25ml) of ethanol, purchased from (Edwards), (50ml) of ethanol, placed in a glass baker with a piece of magnet bar was used to mix the solutions over 60 minutes. A specialsyringe was used for depositing the materials on FTO glass substrate that has been left for 10 minutes.. Then, (0.76gm) of Thiourea was dissolved in 25ml of ethanol, then dropped wisely to the mixture under stirring for 5 hours. The sol solution was dropped onto the glass substrate about 10 cycles at speeds of 1400 rpm for 30 Sec. After deposition, the films were dried in air at 150 °C for one hour.Blends of NiPcTs -CdS were fabricated with 1:2 content concentrations ratio of CdS solutions. The preparation procedure as follows: The organic NiPcTs and CdS materials were dissolved separately with concentrations of 1:2 g/L for NiPcTs -CdS respectively then starrier over one hour on the hot plate with elevated temperature of 70°. A solution of CdS was filtered using polytetrafluoroethylene (PTFE) filter with pore size of 0.2 μm. 0.5 ml of the final solutions for NiPcTs and CdS were dropped onto the FTO glass substrate with the thickness of 130 nm The film was left about 10 cycles at speeds of 1500 rpm for 30 Sec.

After deposition, the films were dried in air at 70 °C for one hour.Film thicknesses were examined using Dektak profilometer instrument. Structural characterization for both of thin films and blends were identified using XRD, atomic force microscopy, EDS and SEM measurements

3. Results and discussion

3-1 X-Ray Diffraction Technique (XRD):

The crystalline of NiPcTs, CdS thin films and heterojunction blends of NiPcTs-CdS were investigated using x-ray diffraction type (Shimadzu XRD- 6000) ,with source of radiation Cu-Kα of wavelength, 1.5418Å, measurements recorded in the range of 2θ between (0 - 80°). As shown in figure (1-a),NiPcTs thin film has polycrystalline structure, with monoclinic phase. The XRD patterns of NiPcTs film has shown two predominant sharp peaks belonging to the reflection surfaces (100) and ($\bar{1}$02) at diffraction angles (6.90^0)and (8.96^0)respectively ,this result matched with the standard JCCDD card No.(11-0744),with simple difference. CdS thin film has polycrystalline hexagonal structure, with unsharp peaks belonging to the reflection surfaces (100),(002),(101),(110),(103),(112) at diffraction angles 24.75°, 26.465° , 28.1781°, 43.5010°, 47.7362°, 51.7811° respectively; as compared with the standard JCCDD card No. (96-101-1055), and the best growth at orientation (101) as shown in figure. (1- b). The NiPcTs/CdS blend heterojunction(BHJ)had two polycrystalline structure, the first one was cubic structure which was predominant and belong to the organic semiconductor (NiPcTs) , and the other one was the hexagonal structure which was belong to inorganic (CdS) semiconductor , as shown in figure (1-c),also one can notice that predominant crystalline planes (101) and (102).
The value of the crystallite size was estimated using the relation[4].

$$L = \frac{K\lambda}{\beta cos\theta} \ldots\ldots\ldots\ldots (1)$$

Where, L is the mean crystallite size, K is the Scherer constant with value 0.94, λ= 1.542 Å of the X-ray source wavelength and β is the full width at half maximum (FWHM) , value. The FWHM,interplane distance (d) and crystallites size (G.S) of NiPcTs, CdS thin films and The NiPcTs/CdS blend heterojunction(BHJ)are listed in Table 1.
One can observe from table (1) that the average crystallite size is small This may be due to the effect of temperature throughout the deposition and preparation method. The crystallite size of organic semiconductor (NiPcTs) was decreased while the grain size of CdS increased in the blend structure as compared with thin films structure, this might be attributed to the interaction between organic and inorganic semiconductors. The experimental values for interplanar spacing are in good matching with the joint committtee on powder diffraction standards(JCPDS) with numer (96- 101-1055)

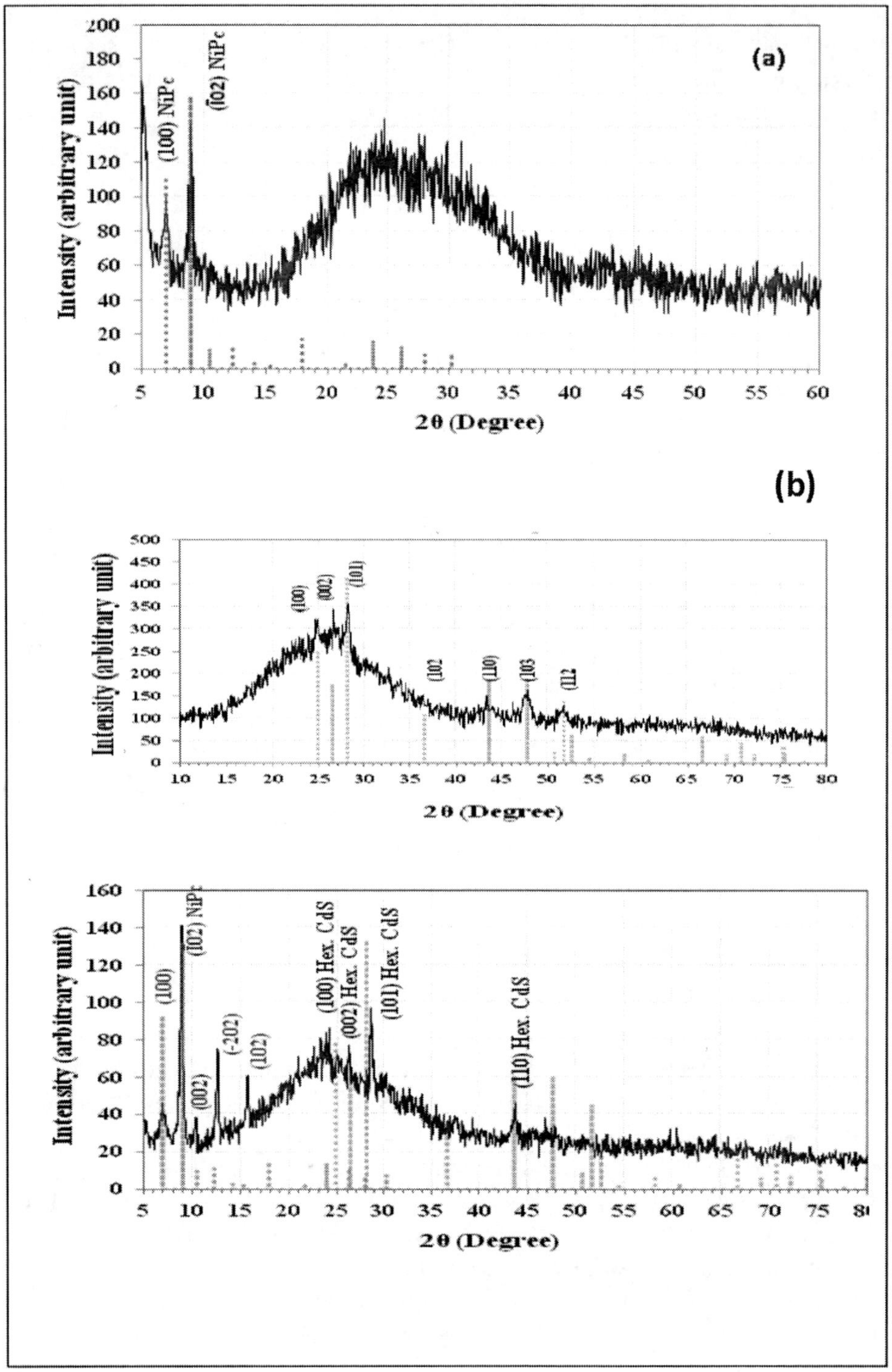

Figure (1): XRD pattern of (a): NiPcTs thin film. (b): CdS Thin Film
(c): Blend hetrojuncction NiPcTs -CdS.

Table (1): XRD results for NiPcTs , CdS thin films and blend hetrojuncction (BHJ)NiPcTs –CdS.

Sample	2θ(Deg)	FWHM	$d_{hk\ell}$ (Exp) (A)	C.S (nm)	$d_{hk\ell}$ (stdand.)	hkℓ	phase
NiPcTs	6.9000	0.2400	12.8005	33.2	12.6519	)100(	Monoclinic.CuPc
	8.9600	0.1866	9.8616	42.7	9.8591	)Ī02(	Monoclinic.CuPc
CdS	24.7519	0.6186	3.5941	13.1	3.5940	(100)	Hex.CdS
	26.4650	0.6187	3.3652	13.2	3.3685	(002)	Hex.CdS
	28.1781	0.6186	3.1644	13.2	3.1710	(101)	Hex.CdS
	43.5010	0.7138	2.0787	12.0	2.4577	(110)	Hex.CdS
	47.7362	0.9041	1.9037	9.6	2.0750	(103)	Hex.CdS
	51.7811	1.0469	1.7641	8.4	1.9045	(112)	Hex.CdS
Blend hetrojuncction(BHJ) NiPcTs -CdS	7.0019	0.6983	12.6519	11.4	12.6145	(100)	Monoclinic.CuPc
	8.9106	0.4190	9.8591	19.0	9.9161	(Ī02)	Monoclinic.CuPc
	10.3538	0.3724	8.4792	21.4	8.5370	(002)	Monoclinic.CuPc
	12.6350	0.3259	7.2111	24.5	7.0003	(-202)	Monoclinic.CuPc
	15.7076	0.3258	5.7751	24.6	5.6372	(102)	Monoclinic.CuPc
	24.1806	0.4190	3.5940	19.4	3.6777	(100)	Hex.CdS
	26.2291	0.3724	3.3685	21.9	3.3949	(002)	Hex.CdS
	28.6499	0.4656	3.1710	17.6	3.1133	(101)	Hex.CdS
	43.7803	0.4656	2.0750	18.4	2.0661	(110)	Hex.CdS

3-2 Atomic Force Microscope *(AFM)*

The AFM surface images of spin coated nanostructures NiPcTs ,CdS thin films and Blend Hetrojunction NiPcTs –CdS respectively are analysed and shown in Figure 2 It well defined that Topography with sample morphology obtained by AFM give good information about surface features like the root mean square surface roughness, ten point heigh, and average diameter of grain size.These images confirm that the films are homogeneous without any defects like cracks and continuous with very well assembled grains to uniformly covering all surface of the substrate with a good adherence. The surface of all samples are quite smooth and they are comprised of spherical numerous nanoparticles with diameter less rhan 70 nm, and one can observe that the height of each surface was in the range of (6–12) nm, which is advantage of the spin coating technique for preparation smooth surfaces .The results of AFM measurement were listed in **Table (2)..** The root mean square of the NiPcTs, CdS films was 1.74 nm and 2.22 nm respectively and 3.6nm for blend hetrojunction (BHJ) NiPcTs –CdS. These results confirm the XRD results about the nanostructure of our samples.

IOP Conf. Series: Materials Science and Engineering **757** (2020) 012052 doi:10.1088/1757-899X/757/1/012052

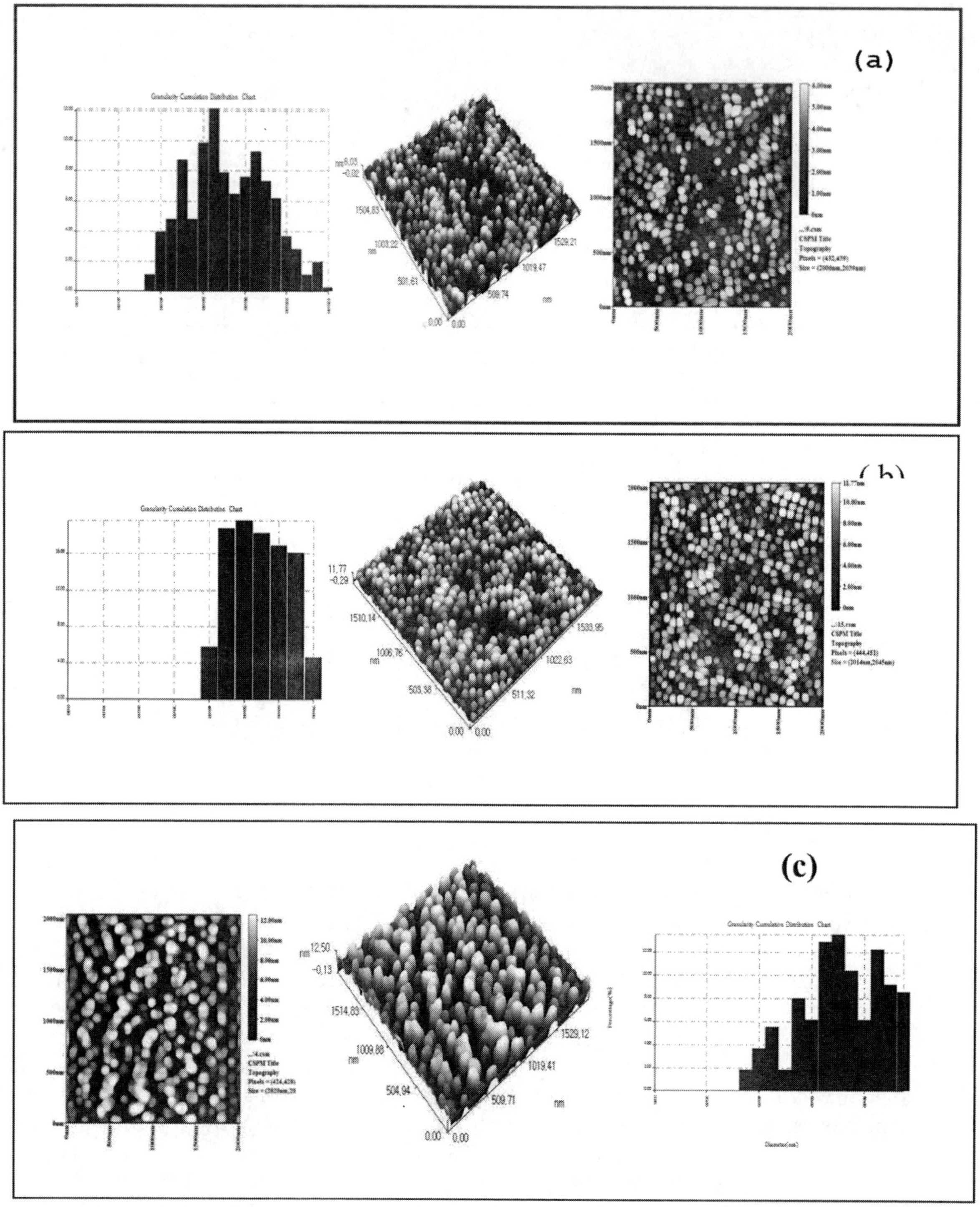

Figure (2):AFM images for(a): NiPcTs thin films. (b): CdS thin films.
(c): Blend hetrojunction of NiPcTs -CdS.

Table(2): AFM parameters of NiPcTs,CdS thin films and blend hetrojuncction(BHJ) NiPcTs -CdS.

Samples	Root Mean Square (RMS) (nm)	Roughness (nm)	Ten point Height (nm)	Average diameter (nm)
NiPcTs	1.74	1.51	6.03	69.79
CdS	2.22	1.93	7.7	51.99
Blend hetrojuncction(BHJ) NiPcTs -CdS	3.6	3.1	12.6	68.4

3-3 EDS Spectrum Measurement:

A qualitative elemental analysis was carried out to investigate the chemical composition of the novel molecular materials of the nanostructure NiPcTs , CdS films and for hetrojunction blend NiPcTs – CdS. For this purpose an EDS analyzer which installed with scanning electron microscope was made for 150 second and measured peaks were compared with the database to distinguish the chemical elements. Figure (3 a-c) shows EDS spectra for NiPcTs , CdS and for hetrojunction blend NiPcTs – CdS respectively.As seen from this figure spectra,(a),(b)and (c) , the existence of essential elements of thin films and HJB of NiPcTs –CdS.

3-4 Scanning Electron Microscopy (SEM)

Figure (4-a-c) represent the Scanning electron micrographs images of as-deposited NiPcTs, CdS, and for hetrojunction blend NiPcTs –CdS respectively. It is seen that no particular crystal orientations in the all samples. Scanning electron micrographs of NiPcTs, CdS thin films are It can be noticed a polycrystalline structure, consist of many smaller crystallite of the size about few tens of nanometers.SEM images also indicated some uncovered parts of the substrate are evidently granular structure with low homogeneity and the distribution of the grains is not uniform.The morphology of BHJ NiPcTs –CdS is present in figure 4 (c) indicated a network of nanotube.

IOP Conf. Series: Materials Science and Engineering **757** (2020) 012052 doi:10.1088/1757-899X/757/1/012052

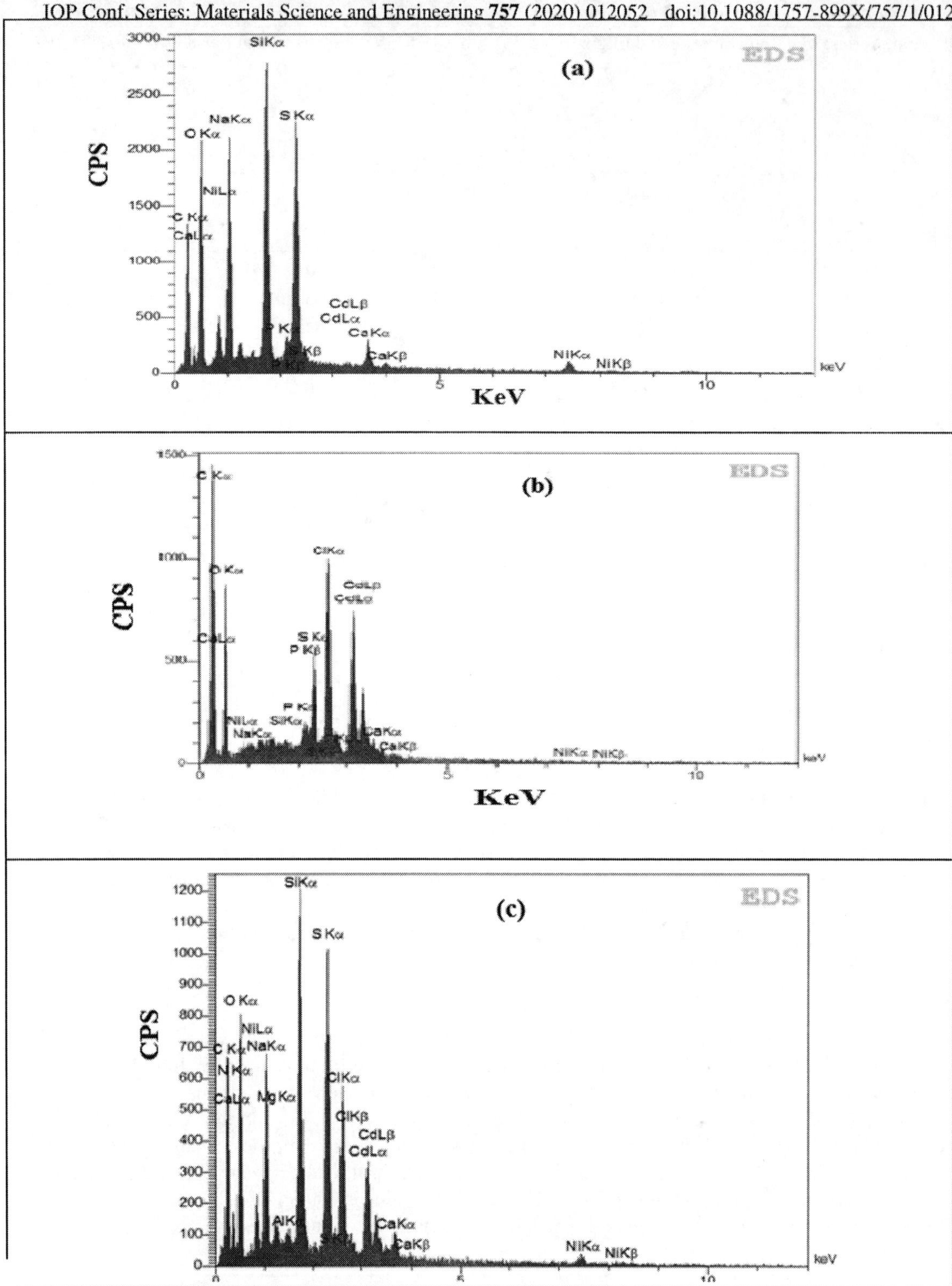

Figure (3): EDS Spectra of (a): NiPcTs Thin Films.(b): CdS Thin Films (c): Blend hetrojuncction NiPcTs -CdS.

Figure (4): SEM surface images for (a): NiPcTs thin films. (b): CdS thin films.
(c): BHJ of NiPcTs -CdS.

4. Conclusions:

Thin film of NiPcTs ,CdS were deposited successfully on the FTO substrates and hetrojunction blend NiPcTs- CdS was fabricated using spin coating technique with 130 nm thickness.All structural measurements show that the films and blend hetrojunction have polycrystalline structure and EDS analysis examine the chemical elements oncentration of the initial materials.The surface morphology measurement displays that the NiPcTs ,CdS and blend hetrojunction NiPcTs- CdS exhibits excellent performance and the grains uniformly covering all surface of the substrate with a good adherence.SEM images show many smaller crystallite of the size about few tens of nanometers .

References:

[1] Thomas Kietzke 2007 "Recent Advances in Organic Solar Cells ",Hindawi Publishing Corporation Advances in Optoelectronics ,Volume 2007, Article ID 40285, p-15.

[2] Shota Mochizuki, Nobuo Satoh, and Shigetaka Katori 2018"Nanoscale investigation of bulk heterojunction organic solar cell by scanning capacitance force microscopy", Japanese Journal of Applied Physics 57.

[3] Ameer F. Abdulameer, Izzat M. Al-Essa, Mahdi Hasan Suhail 2015"Temperature effect on optical properties of nickel (ii) phthalocyanine tetrasulfonic acid tetrasodium salt (NiPcTs) organic thin films", Iraqi Journal of Physics, Vol.13, No.28, PP. 44-51.

[4] Nada K. Abbas1, Ameer F. Abdulameer , Atyaf H. Kadhum1 2017" The structure and optical properties of organic semiconductor bulk hetrojunction blend (NiPcTs/Alq3) thin films", Iraqi Journal of Physics, Vol.15, No.33, PP. 40-48.

[5] Adel H.Omran ALKhayatt, Ghofran Mohammed Hussein AL-Haddad 2016"Synthesis and Structural Characterization of Sol-Gel Deposited Fluorine Doped Nanocrystalline A CdS Thin Films",International Journal of Research in Applied, Natural and Social Sciences 2347-4580 Vol. 4, Issue 5, May ,125-136.

[6] Jurchescu, Oana Diana 2006" Molecular organic semiconductors for electronic devices", Solid State Chemistry of the Materials Science Centre at the University of Groningen, the Netherlands.Ph.D. thesis .

[7] G.Khrypunov, S. Bereznev, A. Meriuts, G. Kopach, N. Kovtun, N. Deyneko, 2010 "Development Organic Back Contact for Thin-Film CdS/CdTe Solar Cell" , Physics and Chemistry of Soild STtate, № 1 C. 248-251 V. 11, P. 248-251.

[8] Mohan V. Jacob 2014" Organic Semiconductors: Past, Present and Future", Electronics, 3, 594-597; doi:10.3390/electronics3040594.

[9] Pang-Leen Ong and Igor A. Levitsky 2010" Organic / IV, III-V Semiconductor Hybrid Solar Cells",Energies, 3, 313-334.

[10] L. Leontie , I. Druta, R. Danac, M. Prelipceanu, G.I. Rusua 2005" Electrical properties of some new high resistivity organic semiconductors in thin films",Progress in Organic Coatings 54 ,175–181.

[11] Husham N. Noori and Lamiaa K. Abbas 2018" Annealing effect on the optical properties of organic semiconductor Alq3: C60 blend thin films",Iraqi Journal of Physics, Vol.16, No.39, PP. 1-10.

[12] Shota Mochizuki,, Nobuo Satoh, and Shigetaka Katori 2018"Nanoscale investigation of bulk heterojunction organic solar cell by scanning capacitance force Microscopy",Japanese Journal of Applied Physics 57, 08NB05.

[13] Ulrich Hörmann 2014" Voc from a morphology point of view: On the open circuit voltage of polycrystalline organic heterojunction solar cells",Ph.D. thesis , Faculty of Mathematics, Natural Sciences and Technology the University of Augsburg.

[14] Serbenta, Almis 2016" Organic donor-acceptor systems Charge generation and morphology ", Ph.D. thesis, University of Groningen.

[15] Bjoern Nlesen 2012" Performance improvement of organic solar cells by metallic nanoparticles ", Ph.D. thesis , department of t of Electrical Engineering, Arenberg Doctoral School of Science, Engineering & Technology.

[16] Yasin Arslan 2015 Nano Material Doped Organic Solar Cell Produced and Characterization ", thesis in Master of Science, Fatih University.

Acknowledgement

The authors thank the College of Education for Pure Science /Ibn Al-Haitham, Baghdad University, Iraq, and Solar Energy Research Center, Renewable Energy Directorate, especially Dr. Falah I. Mustafa Alattar, for providing financial support.

The Influence of x ratio and Annealing Temperatures on Structural and Optical Properties for $(CuO)_x(ZnO)_{1-x}$ Composite Thin Films Prepared by PLD

E M Nasir[1], **M F A Alias**[1*] **and A N Mohammed Ali** [1]

[1] Department of Physics, College of Science, University of Baghdad, Baghdad, Iraq.

*Corresponding author: may20131313@yahoo.com

Abstract. Thin films of $(CuO)_x(ZnO)_{1-x}$ composite were prepared by pulsed laser deposition technique and x ratio of $0 \leq x \leq 0.8$ on clean corning glass substrate at room temperatures (RT) and annealed at 373 and 473K. The X-ray diffraction (XRD) analysis indicated that all prepared films have polycrystalline nature and the phase change from ZnO hexagonal wurtzite to CuO monoclinic structure with increasing x ratio . The deposited films were optically characterized by UV-VIS spectroscopy. The optical measurements showed that $(CuO)_x(ZnO)_{1-x}$ films have direct energy gap. The energy band gaps of prepared thin films decreased as x ratio increased, while they increased with increasing annealing temperatures. Also the optical constants such as refractive index, extinction coefficient and dielectric constants have been calculated.
Keywords: $(CuO)_x(ZnO)_{1-x}$ films, Pulsed laser deposition, Annealing temperatures, Structural properties, Optical properties.

1. Introduction

Zinc Oxide (ZnO) is n-type semiconductor with a very useful and important material for multiple applications in both microelectronic and optoelectronic devices. Zinc oxide is a group II-VI wide direct band-gap semiconductor (3.4 ev at room temperature) that has been extremely investigated for many years [1]. Copper oxide (CuO) is p-type semiconductor with band gap energy of about 1.51-1.74 ev [2, 3]. CuO is especially used in solar cell technology and photovoltaic applications [2]. The composite of n-type ZnO and p-type CuO were expected to show the reduced sensitivity, due to their opposite response to gases. Doped CuO was shown to reduce the electrical conductivity of CuO as gases sensors. This was proposed to be responsible to increase sensitivity of CO gas for ZnO-CuO composite by increasing the current through p-n function [3]. The conductivity is controlled by ZnO-CuO boundaries [4], due to their superior properties such as high crystalline quality, large aspect ratio, and quantum confinement effects. ZnO nanostructures have attracted great research interest. Different techniques such as molecular beam epitaxy, sputtering, sol-gel processing, vapor deposition, and electrochemical deposition have been employed to fabricate ZnO nanowires and nano rods. These have been widely used in laser devices, gas sensors, ultraviolet–visible light emission devices, and many other applications [5]. CuZnO (CZO) with different forms (e.g., thin films, nano rods and powder which is combined by ZnO and CuO has been extensively studied, owing to its potential applications in many fields, including gas sensors and solar cells [6]. In this study the effects of X ratio and annealing temperatures on the structural and optical properties of thin $(CuO)_x(ZnO)_{1-x}$ films prepared by pulsed laser deposition technique have been investigated.

2.Experimental

Zinc oxide and copper oxide powder with 99.999% purity were used as a source materials to prepare $(CuO)_x(ZnO)_{1-x}$ thin films by pulsed laser deposition technique with thickness 100nm on glass substrate at room temperature (RT), different x ratios (x= 0, 0.2, 0.4, 0.6, and 0.8) and annealing at 373 and 473K. These materials were mixed in gate mortar for one hour. After that, the mixture was pressed at 5 Ton to form a target with 2.5 cm diameter and 4 cm thickness. Finally, the targets were sintering at 873K temperature to ensure the homogenous of the materials. The target must be as dense and homogenous as possible to ensure perfect quality of the deposit films. The pulsed laser deposition experiment is carried out inside a vacuum chamber generally at (10^{-3}Torr) vacuum conditions, at low pressure of a background gas for specific cases of oxides and nitrides. Nd:YAG laser focusing on the target surface has been done using a convex spherical quartz lens with focal length 10cm and the lens has been placed inside the vacuum chamber. This laser focusing is necessary in order to exceed the ablation threshold energy density when the laser hit the surface of the target material. The specifications of laser are Nd:YAG laser with pulse energy of 800 mJ and at 6 Hz of frequency and 1064 nm wavelength, and number of shots equals to 600 pulse. Structure properties of $(CuO)_x(ZnO)_{1-x}$ thin films were determined by X-ray diffraction using Shimadzu 6000, with Cu-K_α radiations operated at 40kV and 30mA, with scanning angle 20°-60°. The optical properties of the deposited films were measured in wavelength range of 190 to1100 nm using UV-VIS spectrophotometer (Optima-3000).

3.Results and Discussion

The structure of the prepared $(CuO)_x(ZnO)_{1-x}$ films has been investigated by using XRD to ensure the stoichiometry of material. Figure 1 and 2 show the XRD patterns of the deposited $(CuO)_x(ZnO)_{1-x}$ films on glass substrate with different x ratio (0, 0.2, 0.4, 0.6 and 0.8) and annealed at (373 and 473) K. These figures show polycrystalline structure of all the prepared samples, also from these figures, the phase change from ZnO hexagonal to CuO monoclinic with increasing x ratio from 0 to 0.8. Tables 1 and 2 give the inter planer distance (d), intensity (I/I_o) and crystallite size (D) for the prepared samples in comparison with the standard value as in JCPDS card. The values of d and 2θ are nearly similar to that in the JCPDS cards (96-900-4180 for ZnO and 96-410-5686 for CuO) as listed in these Tables. Similar result has been emphasized by Buchholz et al [7]. Figure 1 shows the X-ray diffraction patterns for $(CuO)_x(ZnO)_{1-x}$ films at different x ratio where annealed at 373K. This figure illustrates that all peaks in the first and second patterns which correspond to x= 0 and 0.2 are identically with the $(CuO)_x(ZnO)_{1-x}$ standard peaks. There are four peaks for polycrystalline hexagonal ZnO phase for x= 0 and 0.2 located at 2θ equal to 31.8 °, 34.5°, 36. 25° and 56.6° with hkl (100),(002), (101) and (110) respectively, Also there are four peaks for polycrystalline hexagonal ZnO and mono clinc CuO for x= 0.4 located at 2θ equal to 35.6 °, 36.45°, 38. 9° and 56.85 ° with hkl (11-1), (101),(111) and (110) respectively, while at x=0.6 the peak for (110) plane disappear compare for ZnO and another peak appear at 2θ equal to 48.70 for (20-2) for mono clinc CuO. Finally, at x=0.8, it is observed that the peak for (101) plane was disappeared compare for ZnO at x=0.6 as shown in Figure 1 and Table 1.The figure shows that all peaks in the first and second patterns which correspond to x= 0 and 0.2 are identically with the $(CuO)_x(ZnO)_{1-x}$ standard peaks. It is observing four peaks for polycrystalline

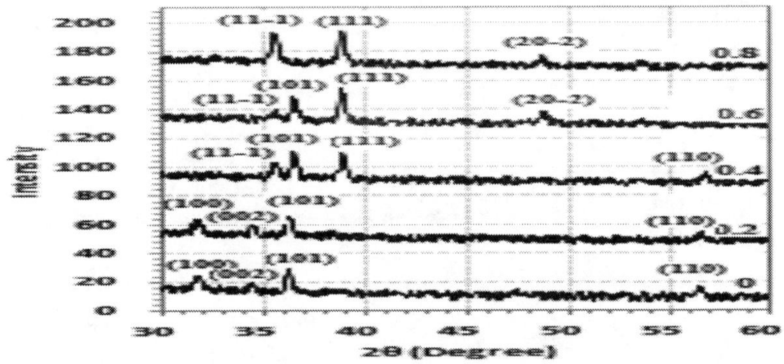

Figure 1: X-ray diffraction pattern for $(CuO)_x(ZnO)_{1-x}$ with different x ratio and annealed at 373K.

Table 1: Structural parameters: 2θ, d_{hkl}, (hkl), FWHM and D of $(CuO)_x(ZnO)_{1-x}$ films with different x ratio and annealed at 373K.

Cu content	2θ (Deg.)	d_{hkl} Exp. (Å)	d_{hkl} Std. (Å)	Int (Arb. Unit)	FWHM	D (nm)	hkl	phase
pure	31.8	2.8117	2.817	25	0.4103	20.132	(100)	ZnO
	34.5	2.5976	2.603	19	0.5275	15.769	(002)	ZnO
	36.25	2.4761	2.478	25	0.4689	17.826	(101)	ZnO
	56.6	1.6248	1.626	16	0.7034	12.827	(110)	ZnO
0.2	31.8	2.8117	2.817	10	0.35087	23.541	(100)	ZnO
	34.35	2.6086	2.603	9	0.4093	20.312	(002)	ZnO
	36.2	2.4794	2.478	14	0.4678	17.864	(101)	ZnO
	56.6	1.6248	1.626	9	0.2923	30.857	(110)	ZnO
0.4	35.6	2.5198	2.521	16	0.3508	23.779	(11-1)	CuO
	36.45	2.4629	2.478	20	0.4093	20.431	(101)	ZnO
	38.9	2.3133	2.315	20	0.3508	24.011	(111)	CuO
	56.85	1.6182	2.626	11	0.4093	22.067	(110)	ZnO
0.6	35.6	2.5198	2.521	15	0.4093	20.382	(11-1)	CuO
	36.45	2.4629	2.478	10	0.2923	28.604	(101)	ZnO
	38.9	2.3133	2.315	22	0.3508	24.011	(111)	CuO
	48.7	1.8682	1.708	9	0.2923	29.822	(202)	CuO
0.8	35.65	2.5161	2.521	18	0.4093	20.385	(11.1)	CuO
	38.95	2.3104	2.315	21	0.3508	24.015	(111)	CuO
	48.75	1.8664	1.867	10	0.4678	18.642	(20-2)	CuO

hexagonal ZnO phase for x= 0 and 0.2 located at 2θ equal to 31.65°, 34.5°, 36. 25° and 56.65° with (100), (002),(101) and (110) planes respectively. Also five peaks for polycrystalline hexagonal ZnO and monoclinic CuO appeared for x= 0.4 was located at 2θ equal to 35.6°, 36.45°, 38. 85°, 48.65 ° and 56.9°

IOP Conf. Series: Materials Science and Engineering **757** (2020) 012053 doi:10.1088/1757-899X/757/1/012053

with hkl (11-1), (101),(111), (20-2) and (110) respectively, whereas at x=0.6 the peak for (110) plane disappeared compare for ZnO at x=0.4. Finally at x=0.8 the peak for (101) plane disappear compare for ZnO at x=0.6 as shown in figure 2 and Table 2.

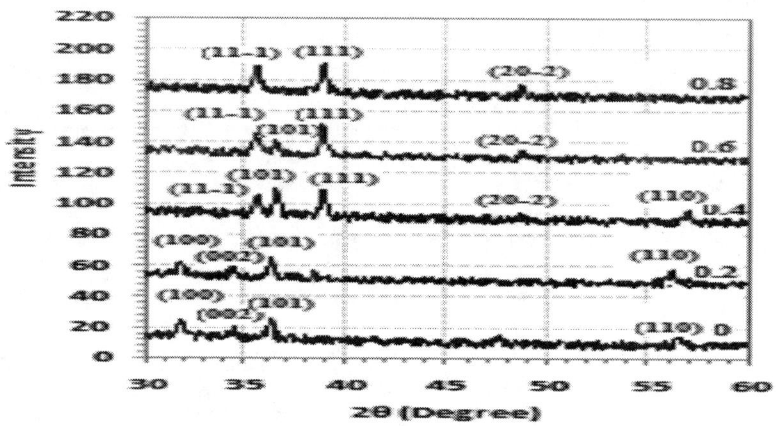

Figure.2: X-ray diffraction pattern for $(CuO)_x(ZnO)_{1-x}$ with different x ratio and annealed at 473K.

Table2: Structural parameters: 2θ, d_{hkl}, (hkl), FWHM and D of $(CuO)_x(ZnO)_{1-x}$ films at different X ratio and annealed at 473

x ratio	2θ (Deg.)	FWHM (Deg.)	Int. (Arb. Unit)	d_{hkl} Exp. (Å)	d_{hkl} Std. (Å)	D (nm)	hkl	Phase
pure	31.65	0.4778	14	2.8196	2.817	17.2807	(100)	ZnO
	34.5	0.2986	9	2.6095	2.603	27.8379	(002)	ZnO
	36.25	0.2986	19	2.4801	2.478	27.9812	(101)	ZnO
	56.65	0.4181	10	1.6274	1.626	21.5669	(110)	ZnO
0.2	31.75	0.3584	11	2.8248	2.817	23.0376	(100)	ZnO
	34.5	0.2389	9	2.6095	2.6037	34.7973	(002)	ZnO
	36.3	0.2986	15	2.4841	2.478	27.9764	(101)	ZnO
	56.45	0.5376	9	1.6243	1.626	16.7837	(110)	ZnO
0.4	35.6	0.3584	16	2.5204	2.521	23.2783	(11-1)	CuO
	36.45	0.2986	20	2.4605	2.478	28.0052	(101)	ZnO
	38.85	0.2986	20	2.3146	2.315	28.2051	(111)	CuO
	48.65	0.7168	7	1.8713	1.867	12.1602	(20-2)	CuO
	56.9	0.4181	10	1.6164	1.626	21.6094	(110)	ZnO
0.6	35.55	0.2389	10	2.5286	2.521	34.9059	(11-1)	CuO
	36.55	0.2986	19	2.4644	2.478	28.0003	(101)	ZnO
	38.85	0.2986	25	2.3180	2.315	28.2000	(111)	CuO
	48.75	0.5376	11	1.8670	1.867	16.2213	(20-2)	CuO
0.8	35.5	0.2986	23	2.5286	2.521	27.9247	(11-1)	CuO
	38.9	0.41816	26	2.3180	2.315	20.1428	(111)	CuO
	48.75	0.4678	10	1.8664	1.867	18.642	(20-2)	CuO

The crystallite size (grain diameter D) of the $(CuO)_x(ZnO)_{1-x}$ films has been calculated by using the Scherrer's formula [8]:

$$D = \frac{K\lambda}{\beta \cos\theta} \tag{1}$$

where D is the crystallite size, λ the X-ray wavelength used, β the angular line width of half maximum intensity, θ the Bragg's diffraction angle and *K* some constant with a value of 0.9, which can be deduced from the lower full width at half maximum (FWHM) of diffraction peaks associated these processes as shown in Tables1 and 2. It is observed that the crystallite size in general increases with increasing of annealing temperatures and decreases with increasing X ratio .

In addition, it is observed from these Tables a decreasing in d_{hkl} with increasing X ratio. This is nearly in agreement with Chowa et al [6].The intense and sharp peak in X-ray diffraction pattern reveal the good crystallinity of the films.

The optical absorption of thin $(CuO)_x(ZnO)_{1-x}$ films was studied in the wavelength (λ) range from 200 nm to 1100 nm. Figure 3 shows the transmittance for $(CuO)_x(ZnO)_{1-x}$ films deposited on glass at different X ratio s at RT(293K) and both annealing temperatures (373 and 473)K .In general, It can be observed from this figure that the transmittance decreases with increasing of X ratio which means increase in the reflection and absorption. The onset of absorption edge for films became less sharp which is due to the fact that bigger crystalline sizes are deposited, and because in the case of more atoms are present in the film so more states will be available for the photons to be absorbed. The shifts of transmittance spectrum to longer wavelengths (lower energies) indicating decreasing of E_g with increasing x ratio [9,10]. Also it can be seen from figure 3 that the transmittance increases with increasing of annealing temperatures. The shifts of transmittance spectrum to shorter wavelengths (higher energies) with the increase of T_a may be attributed to the crystallization of film structure by increasing the crystallite size. This result in agreement with Chih Hung Hsu et al[11].

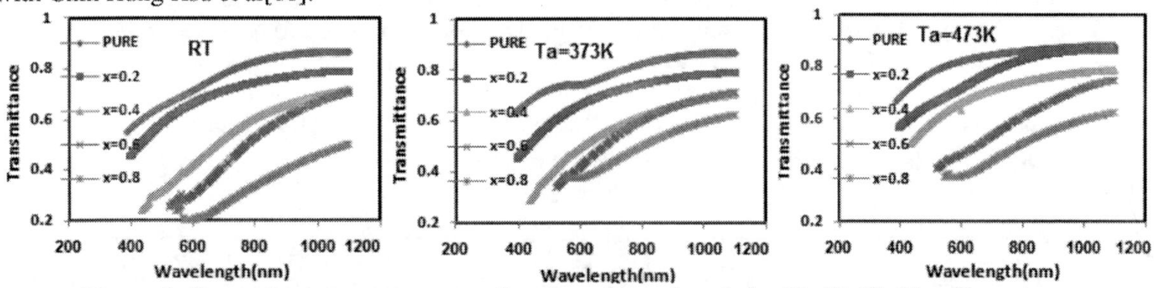

Figure 3: Transmittance spectrum as a function of wavelength for $(CuO)_x(ZnO)_{1-x}$ films at different x ratio and annealing temperatures.

The absorbance spectrum for $(CuO)_x(ZnO)_{1-x}$ films with different x ratio and annealing temperatures was recorded in the wavelength range 190-1100nm. Figure 4 shows that the absorption spectrum within the visible wavelength region in range (400-500) nm for $(CuO)_x(ZnO)_{1-x}$ thin film spectrum. It is obvious that the absorbance spectrum decreases with increasing of annealing temperature. This may be due to improve the crystallite size and increasing the transmittance which is led to increase in the energy gap. Also it can be observed from Fig.4 that the absorbance increased with increasing of x ratio . This is due to create the localized states in the energy gap lead to decrease in the transmission and then decrease in the energy gap.

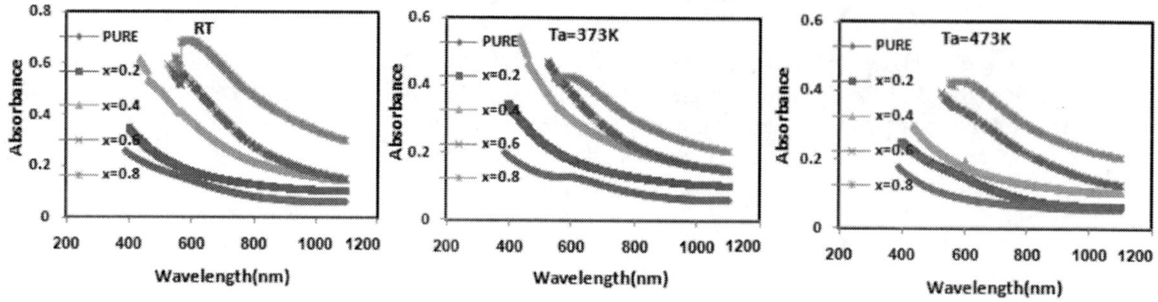

Figure 4: Absorbance spectrum as a function of wavelength for $(CuO)_x(ZnO)_{1-x}$ films at different x and T_a.

The absorption coefficient (α) was measured from the region of high absorption at the fundamental absorption edge of the film from equation [10]:

$$\alpha = 2.303\,\frac{A}{t} \qquad (2)$$

where A is the absorptance and t is the thickness. Figure 5 shows the variation of the absorption coefficient versus the wavelength at both annealing temperatures (T_a) and x ratio of $(CuO)_x(ZnO)_{1-x}$ films. It can be noticed that α decreases with increasing of T_a for all x ratio as listed in Table 1 and figure 5. This is due to increase in energy gap with T_a due to improvement in crystal structure with increasing annealing temperatures and decreasing the absorption [11]. Also, the value of α increases with increasing x ratio for all samples due to create of localized states in the gap (as mentioned previously). In the shorter wavelength the absorption coefficient exhibits high values of α ($\alpha > 10^4$) which means that there is a large probability of the allowed direct transition. The values of the absorption coefficient are nearly in agreement with values reported by other researchers [10-12].

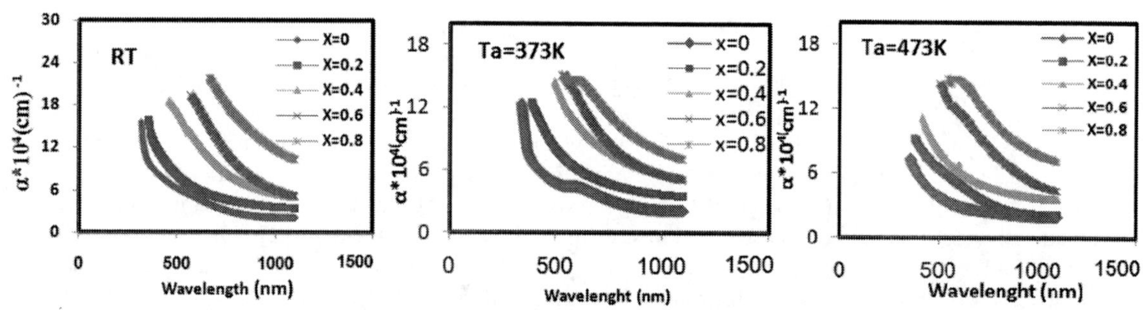

Figure 5: Absorption coefficient as a function of wavelength for $(CuO)_x(ZnO)_{1-x}$ films at different x and T_a.

The optical energy gap values (E_g) for $(CuO)_x(ZnO)_{1-x}$ films with different x ratio (x= 0, 0.2, 0.4, 0.6 and 0.8) and both annealing temperatures (373 and 473)K have been determined by using Tauc equation[13]:

$$\alpha(h\upsilon) = B\,(h\upsilon - E_g)^r \qquad (3)$$

where B is inversely proportional to amorphosity, r is a constant depending on the nature of type optical transition, by plotting the relations of $(\alpha h\nu)^2$ versus photon energy (hv) and selecting the optimum linear part, as shown in Figure 6. It is shown that the values of E_g have direct transitions, and it is increased with increasing T_a for all prepared samples. This is due to the growth of grain size and the decrease in defect states near the bands and these in turn increase the value of E_g [14]. Also the optical energy gap decreases with increasing x ratio as given in Table 3, E_g decreases about half value from 3.0 to 1.3 eV and from 3.2 to 1.4 eV and from 3.3 to 1.5eV for samples prepared at RT, 373 and 473K respectively, when x ratio increases from 0 to 0.8. This is may be attributed to fact the addition of CuO to the ZnO made the material more opaque (more absorbance). The decrease of E_g when x ratio is added to the ZnO is in agreement with Chowa et al [6] and Thakur et al[14].

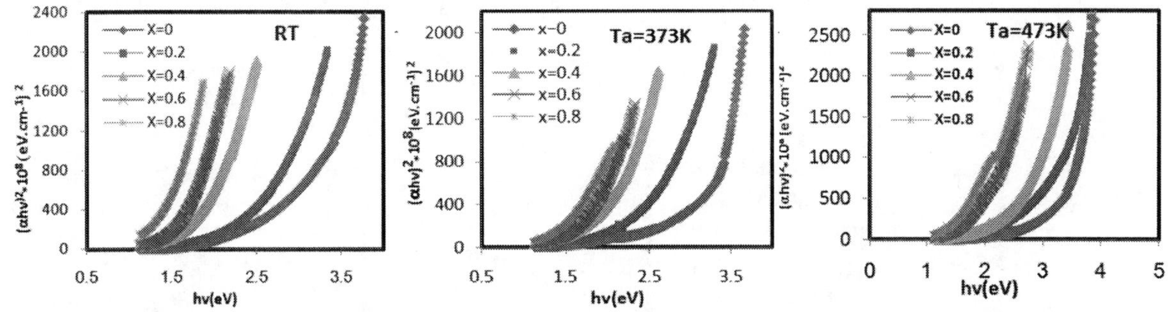

Figure 6: $(\alpha h\nu)^2$ as a function of hv for $(CuO)_x(ZnO)_{1-x}$ films at different x and T_a.

The variation of the refractive index as a function of the wavelength in the range 190-1100 nm for different x ratio and annealing temperatures is shown in Figure 7. The refractive index for $(CuO)_x(ZnO)_{1-x}$ films has been calculated from the equation [15]:

$$n = \left(\frac{4R}{(R-1)^2} - k^2\right)^{\frac{1}{2}} - \frac{(R+1)}{(R-1)} \quad (4)$$

where R is the reflectance and n is refractive index. It can be observed from these figures and Table 3 that the refractive index decreases with increasing of T_a which means decreasing of the reflection because the increase in energy gap is due to the expansion of the lattice and growth of the grain size. Whereas the value of refractive index increases with increasing of x ratio , from 2.336 to 2.613, 2.25 to 2.612 and 2.02 to 2.61 for samples annealed at 293, 373 and 473 K respectively, when x ratio increases from 0 to 0.8. This attitude may be due to the change in the bond length due to the decrease in the defect density which means decreasing of the reflection where the refractive index depends on it. This result is close to this obtained by Cheng et al [16].

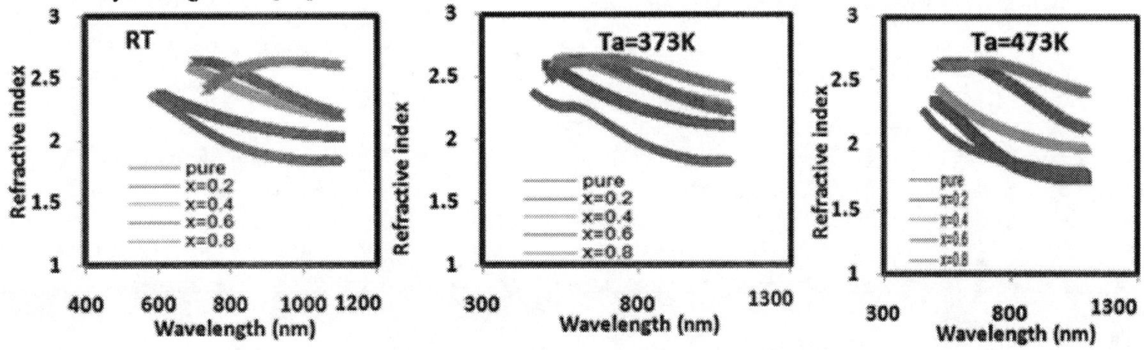

Figure 7: The refractive index as a function of wavelength for $(CuO)_x(ZnO)_{1-x}$ films at different x and T_a.

Figure 8 shows the variation of extinction coefficient (k) as a function of wavelength at different x ratio and annealing temperatures. The values of extinction coefficient are determined by using equation [15]:

$$k = \frac{\alpha\lambda}{4\pi} \qquad (5)$$

The extinction coefficient decreases from 0.2 to 0.14 with increasing annealing temperature from RT to 473K due to decrease the absorption coefficient as shown in Table 3, whereas it increases (0.2-1.07, 0.19-0.67, 0.14-0.67) with increasing of x ratio at RT, 373 and 473K respectively due to more grain boundaries with addition of CuO to ZnO and decreased the energy gap as a result to increasing the absorption with increasing x ratio , that was shown by Austin and Mott [17].

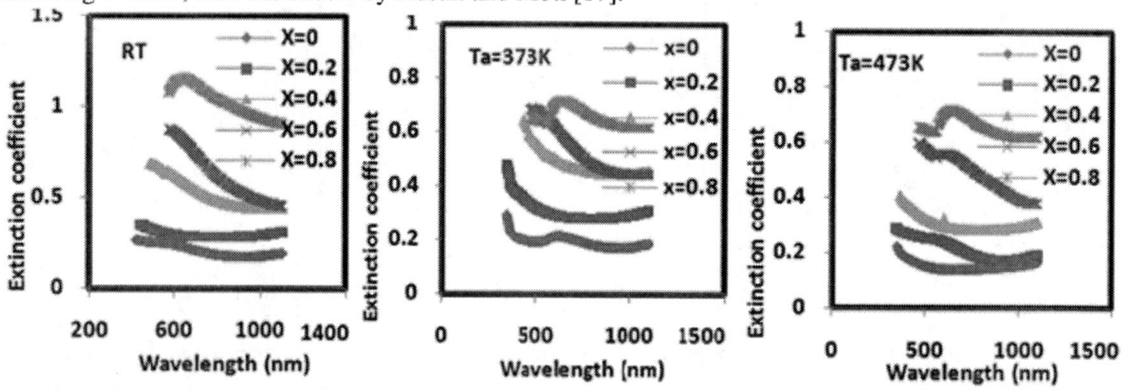

Figure 8: Extinction coefficient as a function of wavelength of $(CuO)_x(ZnO)_{1-x}$ films at different x and T_a.

The plots of real and imaginary (ε_r and ε_i) parts of the dielectric constant with wavelength in the range of 190–1100nm for $(CuO)_x(ZnO)_{1-x}$ thin films deposited at different x ratio and annealing temperatures are shown in Figures 9 and 10. Real and imaginary dielectric constants were calculated using equations [12,13]:

$$\varepsilon_r = n^2 - k^2 \qquad (6)$$

$$\varepsilon_i = 2nk \qquad (7)$$

The behavior of ε_r is similar to that of refractive index because smaller value of k^2 compared with n^2 and ε_i mainly depends on the k values. It is observed that ε_r and ε_i increased with increasing of x ratio , while ε_r and ε_i decrease with increasing of annealing temperatures as shown in Table 3. This behavior is in agreement with Bahsi and Oral results [18].

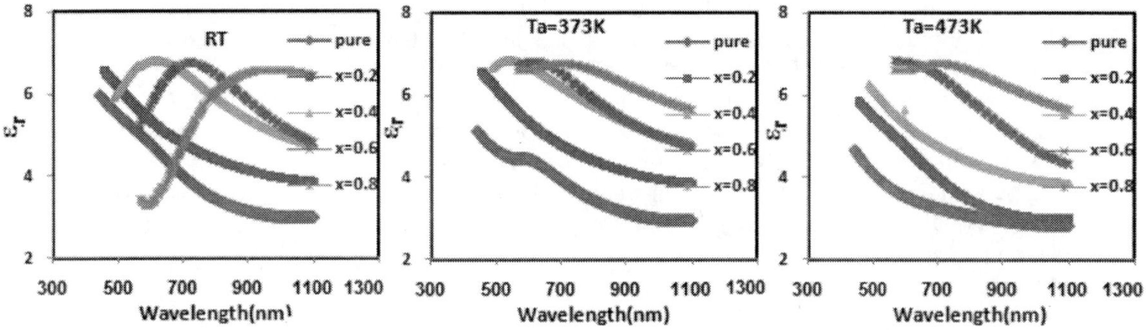

Figure 9: Real part of dielectric constant as a function of wavelength for $(CuO)_x(ZnO)_{1-x}$ films at different x and T_a.

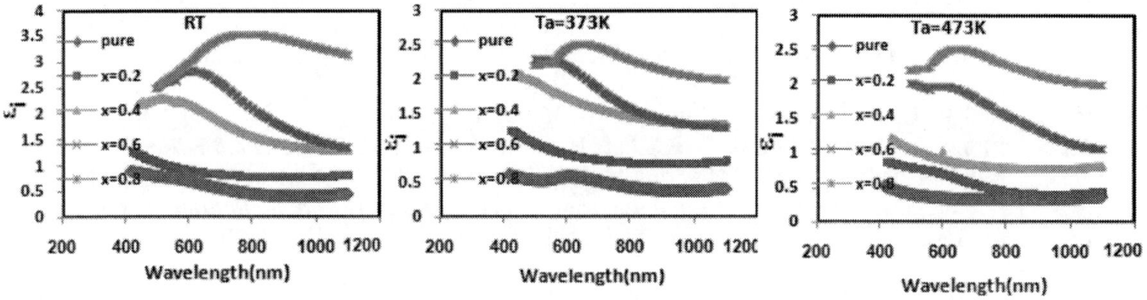

Figure 10: Imaginary part of dielectric constant as a function of wavelength for $(CuO)_x(ZnO)_{1-x}$ films at different x and T_a.

Table 3: Optical parameters of $(CuO)_x(ZnO)_{1-x}$ films at different x ratio and T_a.

T_a (K)	x ratio	E_g (eV)	$10^4\alpha$* $(cm)^{-1}$	n	k	ε_r	ε_i
	pure	3.0	3.15	2.15	0.150	4.600	0.646
	0.2	2.3	3.85	2.28	0.183	5.164	0.837
RT	0.4	1.7	8.16	2.64	0.389	6.834	2.059
	0.6	1.5	11.26	2.52	0.537	6.067	2.711
	0.8	1.3	14.88	2.11	0.710	3.966	3.004
	pure	3.2	2.78	2.07	0.133	4.281	0.551
	0.2	2.4	3.85	2.28	0.184	5.164	0.838
373	0.4	1.9	6.54	2.59	0.312	6.604	1.616
	0.6	1.6	8.16	2.64	0.389	6.834	2.059
	0.8	1.4	9.18	2.63	0.438	6.739	2.308
	pure	3.3	1.86	1.84	0.089	3.409	0.327
473	0.2	2.7	3.15	2.15	0.150	4.600	0.646
	0.4	2.4	4.26	2.35	0.203	5.463	0.954
	0.6	1.7	7.30	2.62	0.349	6.773	1.831
	0.8	1.5	9.20	2.63	0.438	6.739	2.308

4.Conclusions

A polycrystalline structure for the $(CuO)_x(ZnO)_{1-x}$ films prepared successfully by pulsed laser deposition at different x ratio (x= 0, 0.2, 0.4, 0.6 and 0.8) and annealing temperatures (373 and 473) K. The phase change from hexagonal ZnO to monoclinic CuO with increasing x ratio . Also the prepared films have better crystalline with annealing temperatures. The optical transitions in $(CuO)_x(ZnO)_{1-x}$ film is direct transition and the value of optical energy gap increases with increasing of annealing temperatures, and decreases with increasing x ratio . The optical constants increase with increasing of x ratio , while they decrease with increasing annealing temperatures. Thus the characteristic of $(CuO)_x(ZnO)_{1-x}$ thin films prepared by pulsed laser deposition depend strongly on the prepared condition like annealing temperatures and x ratio .

5.References

[1] Jonda C, Mayer A, Stolz U, Elschner A and A Karbach, A 2000 *Journal of Materials Science* **35** 5645.
[2] Özyurt Kus F and Serin T N 2009 *Journal of Optoelectronics and Advanced Materials* **11**(11)1855.
[3] Basu S and Dutta A 1994 *Sensor and Actuators* B **22** 83.
[4] Malyshev V V 1992 *Sensor and Actuator* B **10** 11.
[5] Lung-C C, Cheng-A H and Xiuyu Z 2014 *Materials* **7** 7304.
[6] Chowa L, Lupana O, Chaia G, Khallafa H and Onoa L K 2013 *Sens. Actuators A Phys.* **189** 399.
[7] Buchholz D B, Chang R P H , Song J H ,and Ketterson J B 2005 *Applied Physics* **87**(8)1.
[8] Zhou H, Dan-qing Y, Zhi-ming Y, Lai-rong X, and Jian L 2007 *Thin Solid Films* ***515*** 6909.
[9] El Sayed A M, Said G, Taha S, Ibrahim A and Yakuphanoglu F 2013 *Super lattices and Microstructures* **62** 47.
[10] Ahmed M N, Nadia A A, Nariman S and Farid El-Tantawy 2014 *Journal of Applied Physics* **6**(4)17.
[11] Chih-H H, Lung-Chien C and Xiuyu Z 2014 *Materials* **7**(2)1261.
[12] Pankove J I 1971 *Optical Process in Semiconductors* Dover Publications New York
[13] Burns G 1985 *Solid State Physics* Academic Press Inc., Harcourt Brace Jovanovich, New York.
[14] Thakur S and Sharma N 2014 *CODEN* (USA) **5**(4)18.
[15] Taus J 1974 *Amorphous and Liquid Semiconductor* Plenums Press New York and London.
[16] Cheng W X, Ding A L, Zheng X S, Qiu P S and He X Y 2009 *Journal of Physics*: Conference Series, **152** 12.
[17] Austin I G and Mott N F 1969 *Polarons in crystalline and non-crystalline materials* **18** 71.
[18] Bahsi Z and Oral A 2007 *Opt. Mater* **29**(6) 672.

Study Optical Properties of Copper and Silver for Astronomical Solar Filters Prepared by DC Sputtering Plasma

S. M. Abd Al Kareem and W. I. Yaseen*
Dept. of Astronomy and Space, College of Science, University of Baghdad, Iraq
*Corresponding author email: waleedib1972@yahoo.com

Abstract. This study presented a technique to sputter high strength, highly conductive copper and silver astronomical solar filter on glass substrates at room temperature. In this work, the thicknesses of copper are (110 and 230) nm and for silver are (230 and 400) nm using power 25 and 50 W respectively have been sputtered on glass substrate by DC sputtering plasma technique at room temperature. The base pressure of the plasma sputtering system found that it is very good at 0.1 mbar. The thickness and DC power sputter effect on the optical properties were studied by using UV-Visible spectrophotometer. The transmittance for all coated layers is increases rapidly as the wavelength increases in the range of (350-750) nm. The absorbance decreases rapidly at short wavelengths corresponding to the energy gap of the layer and the range of optical band gap values between (2 and 2.6) eV,

Keywords. Astronomical solar filter, optical properties UV-Visible spectrophotometer, DC sputtering plasma.

1. Introduction

Solar filters are very important for observing the Sun safely with your telescope. A filter can greatly enhance the human eye's perception of small details on solar system and deep sky objects. Correct use of photo-visual filters can greatly enhance an observing. Filters work by blocking a specific part of the color spectrum, thus significantly enhancing the remaining wavelengths. Colored filters work by absorption/transmission, and immediately telling which part of the spectrum they are reflecting, and therefore transmitting. The so-called light-pollution reduction and nebulae filters are very selective in the wavelengths they transmit. [1]. Pure metals like copper, silver and aluminum are very conductive and very soft. The coating layer for these metals become stronger by various approaches like grain size reduction, solid solution strengthening, cold working leads to high strengths but low electrical conductivity. For instance, Copper, which has been reinforced two times or three times, has been lost conductivity by 50 to 60 percent in comparison with pure copper. There are three different types coating processes:

1- vapor phase, which includes physical vapor deposition (PVD) (a) Thermal evaporation (b) Electron beam-PVD (c) plasma Sputtering (PS).
2- Chemical vapor deposition (CVD) (a) Low pressure CVD (b) Plasma enhanced CVD (c) Photochemical and laser-CVD.
3- Liquid phase, which includes painting, dipping, and electroplating.
4- Solid phase, which includes (a) plasma spry processes (b) laser cladding [2].
After deposition the different deposition methods (PVD, CVD and PS) different parameters and annealing are lead to different coating layer properties like texture, grain size and grain boundary misorientation [3].

IOP Conf. Series: Materials Science and Engineering **757** (2020) 012054 doi:10.1088/1757-899X/757/1/012054

Physical sputtering is a process in which atoms are ejected from a source material due to impinging energetic particles. Sputter deposition takes place in a chamber at a reduced pressure. Low background pressure is realized by a tightly sealed system consisting of a vacuum chamber, gas supply lines and vacuum pumps [4]. DC or Direct Current Sputtering is a coating layer Physical Vapor Deposition (PVD) Coating technique where a target material to be used as the coating is bombarded with ionized gas molecules causing atoms to be "Sputtered" off into the plasma. These vaporized atoms are then deposited when they condense as a coating layer on the substrate to be coated. DC Sputtering is the most basic and inexpensive type of sputtering for PVD metal deposition and electrically conductive target coating materials. Two major advantages of DC as a power source for this process is that it is easy to control and is a low cost option if you are doing metal deposition for coating [5]. The growth of coating layer copper and silver coating layer on glass is a subject of great interest, as many optical filters, low emissivity coatings, and sun protection coatings are made of coating layer metallic and dielectric layer [6]. In general, one speaks about solar coatings when the optical layer properties are adapted to the solar spectrum. The considered solar range may vary with application. A weak or near-zero absorptance is a requisite if only interference effects should occur inside the layer. Interference coating layer consist of alternating high- and low-refractive index materials [7]. Optical experiments provide a good way of examining the properties of coating layers. Especially measuring the absorption coefficient for various energies gives information about the band gaps of the material. Knowledge of these band gaps is very important for understanding the electrical properties of a semiconductor, and is therefore of great practical interest [8]. The UV-Visible spectrophotometer uses two light sources, a deuterium (D2) lamp for ultraviolet light and a tungsten (W) lamp for visible light. After bouncing off a mirror, the light beam passes through a slit and hits a diffraction grating which can be rotated allowing a specific wavelength to be selected. At any specific orientation of the grating, only monochromatic light successfully passes through a slit. A filter is used to remove unwanted higher orders of diffraction. The light beam hits a second mirror before it gets split by a half mirror. One of the beams is allowed to pass through a reference cuvette (which contains the solvent), the other passes through the sample cuvette. The intensities of the light beams are then measured. As a result, there is a sharp increase in absorption at energies close to the band gap energy that manifests itself as an absorption edge or reflection threshold in the UV-V region is absorbance spectrum. Having the values of Absorbance (A), wavelength (λ), and Thickness (t). Other optical parameters can be calculated using the following relationships [9]:

$$T = \frac{1}{\exp(2.303 * A)} \quad (1)$$

where T is Transmittance A is absorption

$$R = 1 - (A + T) \quad (2)$$

where R is Reflectance

$$\alpha = 2.303 \frac{A}{t} \quad (3)$$

where α is Absorption coefficient

$$k = \frac{\alpha \lambda}{4\pi} \quad (4)$$

Where k is Extinction coefficient

$$n=\left(\left(\frac{4R}{(R-1)^2}\right)-K^2\right)^{1/2}-\frac{(R+1)}{(R-1)} \quad (5)$$

where n is Refractive index

The band structure of coating layers can be probed directly by measuring their absorption spectrum. For direct band gap [9]:

$$\alpha h\nu \propto \sqrt{h\nu - E_g} \quad (6)$$

where α is the absorption coefficient, $h\nu$ is the energy of incident photons and E_g is the band gap energy of the coating layers, E_g is the intercept of the straight line obtained by plotting $(\alpha h\nu)^2$ vs. $h\nu$. For indirect band gap semiconductors [9]:

$$\alpha h\nu = B(h\nu - E_g)^2 \quad (7)$$

where B is inversely proportional to amorphousity.

2. Experimental Setup

DC plasma sputtering technique has been used to sputter the Copper and silver as layers on glass substrate. The glass substrate were used for copper deposition and washed in ultrasonic device and dry in air. Plasma sputtering system apparatus was used at a rate of 10kV DC and current 500 mA for copper and silver sputter. Plasma chamber was made from Pyrex glass. The length, thickness, and diameter of the cylinder chamber are 30cm, 5mm, and 20cm respectively. Two plane-parallel stainless steel electrodes were used as electrode. Copper and silver are used as a target for sputtering system. The photography picture for plasma chamber is illustrated in Figure (1). The discharge was fed with argon at variable flow rate (100-600) sccm.

Figure (1): DC plasma sputtering system.

The base pressure of the plasma sputtering system was better at 0.1 mbar. The thickness for coated layer was measured by using Spectroscopic Reflectmeter SR300 systems. The optical properties were measured using UV-Visible spectrophotometer. The images of copper and silver astronomical solar filter illustrated in Figure (2)

IOP Conf. Series: Materials Science and Engineering **757** (2020) 012054 doi:10.1088/1757-899X/757/1/012054

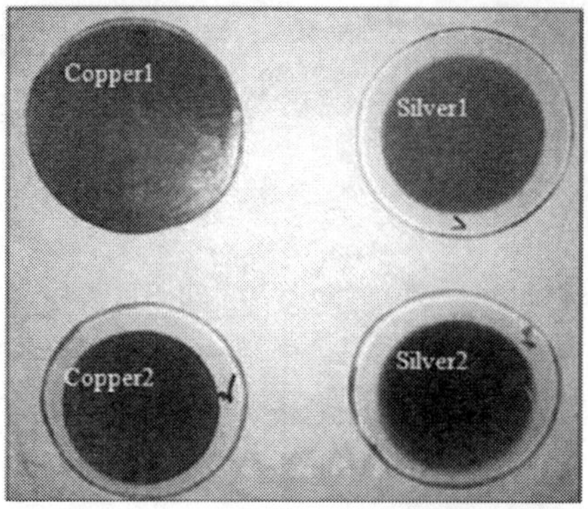

Figure (2): Copper and silver astronomical solar filter.

3. Results and Discussion

The treatment was performed for different time intervals. After this treatment of the class samples copper and silver was deposited stepwise on the sample by DC sputtering from a pure copper and silver target with argon as processing gas. The total pressure during the copper and silver deposition was 0.1 mbar. The deposition rate of the silver was controlled by time. Determine of th thickness of coated layer using Spectroscopic Reflectometer. The thickness for copper and silver was measured for each sputtering power as shown in Table (1). The optical properties of copper and silver astronomical solar filter is grown on glass substrate which involve the transmission, refractive index, absorption coefficient and the optical energy gap (E_g).

Table (1): Thickness of copper and silver coated layer.

Sputtering power W	Coated layer	Thickness (nm)	Sputtering time (mint)
25	Copper1	110	30
50	Copper2	230	30
25	Silver1	230	30
50	Silver2	400	30

3.1 Transmittance

The spectral dependence of the transmittance (T) for copper and silver astronomical solar filter deposit at conditions (d =5cm, power = (25 and 50) W, sputtering time =30 minutes and 0.1 mbar pressure). Figure (3) illustrate the transmittance versus wavelength for those prepared astronomical solar filter deposited on glass slides.

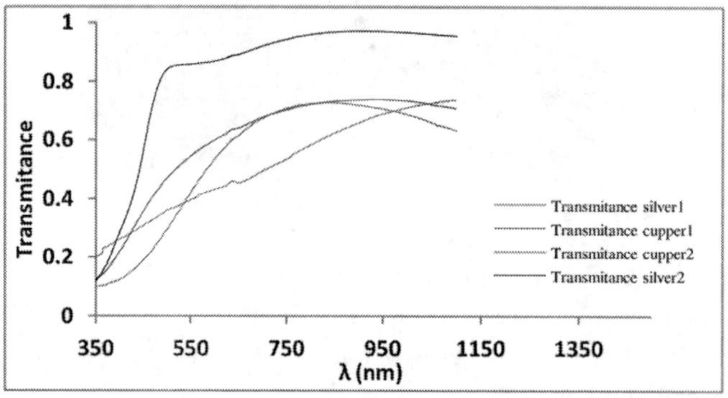

Figure (3): Transmittance as a function of wavelength for sputtering process copper and silver astronomical solar filter on glass.

The transmission of astronomical solar filter deposited in the UV-Visible region (300-600 nm) on glass slides has been examined and found that all astronomical solar filter prepared by DC sputtering method has high transmission at longer wavelengths. The transmittance (T) can be calculated using equation (1) shows the relation between transmittance and wavelength for copper and silver layers.

It can be noticed that the transmittance for all coated layers increases rapidly as the wavelength increases in the range of (350-750) nm, and then increases slowly at higher wavelengths. Maximum value in transmission at 0.7232 for 1048 nm of silver 1 solar filter at power 25 W and silver 2 at 0.97 for wavelength 858 nm at power 50 W. Copper 1 at power 25 W a maximum value in transmission at 0.739 for wavelength 930 nm and copper 2 at power 50 W 0.724 for wavelength 802 nm. Optical transmission measurements depend strongly on the power DC sputtering; this result is in agreement with [10].

3.2 Absorbance

Figure (4) demonstrate the optical absorbance as a function of wavelength of copper and silver coated layers. The absorbance decreases rapidly at short wavelengths corresponding to the energy gap of the coated layers. This clear increase of energy is product from the interaction of the material electrons with the incident photons which have enough energy for the occurrence of electron transitions. The increase in energy deposition of silver and copper leads to increased absorption, this result is in agreement with [10].

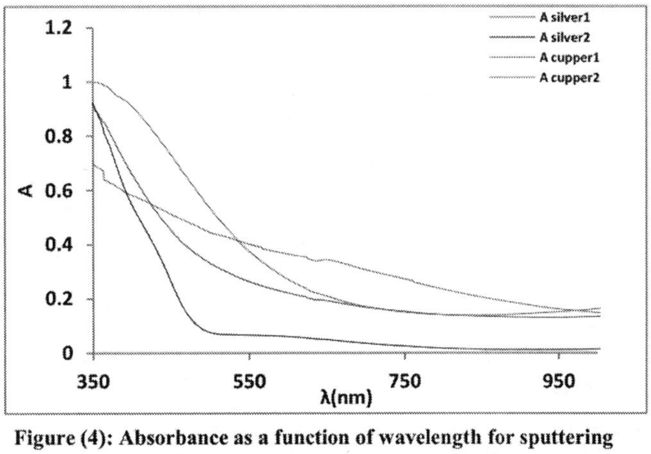

Figure (4): Absorbance as a function of wavelength for sputtering process copper and silver astronomical solar filter on glass.

3.3 Refractive Index (n)

The refractive index (n) calculated from Equation (5). The refractive index of the copper and silver coated layers samples prepared under different conditions (d =5 cm, applied power (25 and 50) W, sputtering time =30 mints and 0.1 mbar pressure) as represent in Figure (5). From this figure, the refractive index increases in the visible region in the shorter wavelength at maximum value n= 2.5553 λ= 649 nm for silver1, silver2 maximum value n= 2.537 λ= 433 nm, copper1 n= 2.611 λ=478 nm and copper2 n= 2.554 λ= 556. This increase goes back to the film densification and improvement in the crystalline of the coated layer, this result is in agreement with [11].

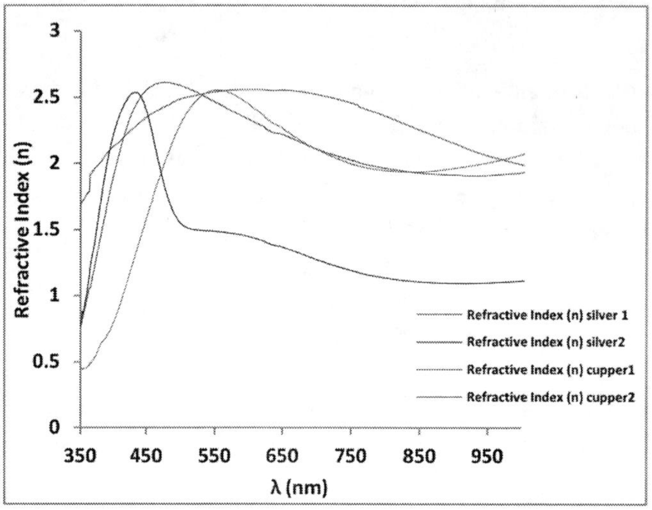

Figure (5): Refractive index as a function of wavelength for sputtering process copper and silver astronomical solar filter on glass

3.4 Absorption coefficient (α)

The absorption coefficient is variation with wavelength of power (25 and 50) W for copper and silver astronomical solar filters are shown in Figure (6).

This figure shows that in the region of short visible wavelength, absorption coefficient takes high values and then decreases significantly with increasing wavelength and in the long visible wavelength the absorption coefficient takes low values, this is because the absorption coefficient is directly dependent on the absorption and inversely with the thickness according to the equation (3). Copper1 it have maximum value greater than other films where reach α= 208153.85 cm-1 λ=329 nm, copper2 α= 100130.43 cm-1 λ=354 nm, silver1 α= 99214.60 cm-1 λ=312 nm and silver2 α= 56562.42 cm-1 λ=335 nm these results are consistent with [12].

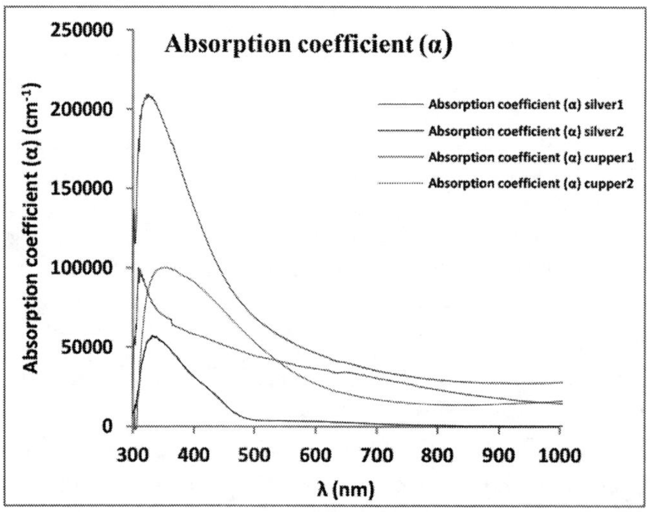

Figure (6): Absorption coefficient as a function of wavelength for sputtering copper and silver solar filter on glass with DC power 25W for sputter silver1 and copper1 and 50 W for silver2 and copper2.

3.5 Optical Energy Gap (E_g)

Study and estimate the average value of the energy gap is very necessary to understand the optical properties of the astronomical solar filter manufactured by DC sputtering technique. The coated layers structure and arrangement and the distribution of atoms in the crystal lattice are very important for value of energy gap. To calculated the value of band gap (Eg,) the graph of $(\alpha h\nu)^{1/2}$ vs. hν has been plotted for different preparation conditions (d =5cm, applied power (25 and 50) W, , sputtering time =30 mints and 0.1 mbar pressure) as demonstrated in the Figure (7). To determine the band gap (Eg) of the copper and silver solar filter can be used data point near the absorption edge.

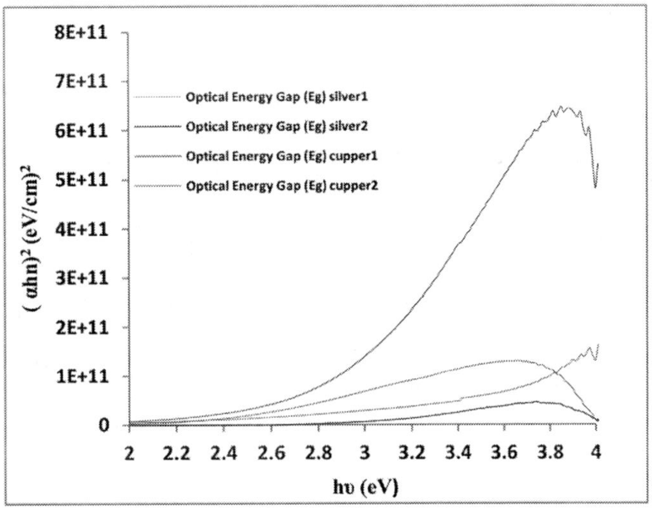

Figure (7) :($\alpha h\nu)^2$ as a function of hv for sputtering process copper and silver astronomical solar filter on glass.

The extrapolating the straight line portion of the graph of the energy axis is use to determined band gap. From energy gap values will notice that the increase in energy deposition leads to increase the thickness and density of formed solar filter, where for copper filter it will notice when sputtering energy 25 W produces energy gap 2.45 eV and energy 50 W the energy gap equal to 2.2 eV. For silver filter at sputtering energy 25 W energy gap equal to 2.6 eV and at energy 50 W energy gap become 2 eV.

Energy gap in the copper and silver solar filter which, is deposited in this work is low value due to from impurities these results are consistent with [12].

4. Conclusions

Copper and silver astronomical solar filters were deposited on glass substrates by DC sputtering technique successfully. The optical properties variation is depending on the amount of sputtering power. Were found to have high absorbance in the ultra violet region and scale down as the wavelength increased. They have high transmittance in the visible / near infra-red region. It has high refractive index. The energy gap for the fabricated copper and silver solar filter was found to be 2, 2.45, 2.2 and 2.6 eV.

IOP Conf. Series: Materials Science and Engineering **757** (2020) 012054 doi:10.1088/1757-899X/757/1/012054

References

[1] Knisely D, 1998, Pulcherrima Productions, 1-10.
[2] Singh J, Wolfe D, 2005, Journal of Materials Science 40, 1– 26.
[3] Koetter T, Wendrock H, Schuehrer H, Wenzel C, Wetzig K, 2000, Journal Microelectronics Reliability, 1295-1299.
[4] Liljeholm L, 2012, Uppsala: Acta University Upsaliensis, 52-54.
[5] Hughes M, 2016, Semicore Equipment, 1-5.
[6] Mack I, 2008, Thesis University Basel, 14-15.
[7] Mertin S, 2015, Thesis Federal School of Lausanne, 32-33.
[8] Benno G, Joachim K, 2003, Perkin-Elmer-Sputter-Sputtering-Deposition-System, 1-2.
[9] Khorsand A, Manaf A, and Darroudi M, 2014, Journal Springer, Vol. 9, 399.
[10] Nabeel A. Bakr*, Ziad T. Khodair, Hussein I. Mahdi, 2016, International Journal of Materials Science and Applications, Vol. 5, No. 6, 261-270
[11] Ezenwa I, Okereke A., Egwunyenga N, 2012, International Journal of Science and Technology, Vol. 2, No. 3, 101-106.
[12] Amer Hassan M, 2012, Eng. & Tech. Journal, Vol. 30, No. 14, 2421-2430.

Study The Structural Properties of Nanostructure (TiO_2: Cu) Thin Film Prepared By RF Magnetron Sputtering

Khalowd D Khemkheem[1], Najwa J Jubier[2] and Abdul_Hussain K Elttayef[3]

[1,2]Department of Physics, College of Science, Wasit University, Kut, Iraq

kutpc19@yahoo.com

[3]Ministry of Science and Technology, Baghdad, Iraq

Abstract. In this research, TiO_2: Cu thin films deposited on a glass substrate with different thicknesses (100) and (200) nm by the radio frequency (R.F.) magnetron sputtering process using target TiO_2: Cu. The sputtering deposition was performed by using the power of 100 Watt. Crystalline structure, surface morphology, and electronic structure were studied using X-ray diffraction(XRD).XRD patterns demonstrate that TiO_2 films deposited on a glass substrate at 500 C° are observed to be brookite (orthorhombic) polycrystalline phase with preferred orientation along (111). The average surface roughness (Ra) measured with atomic force microscopy (AFM) was 0.532 nm, and the minimum grain size was 40.5 nm for TiO_2 doped 5%Cu.

Keywords: TiO_2: Cu , thin films, magnetron sputtering, grain size.

1. Introduction

Thin films are more advantageous part for gas sensing applications because of their higher surface to volume ratio and controlled surface morphology over their bulk counter [1-3]. Titanium dioxide (TiO_2), especially in thin film form, is widely preferred as a semiconductor material because of its unique material. In view of its versatile properties which comprise the high refractive index, wide bandgap, and resistance to chemical and physical impacts [4, 5].Titanium dioxide TiO_2 (titania) is a cheap, non-toxic and one of the most efficient semiconductor photocatalysts for extensive environmental applications [6].TiO_2 has also been preferred as a material for gas sensors due to its low resistivity to the reduced driving power of the sensor [7]. TiO_2 exists in three polymorphic phases: rutile, anatase, and brookite, in nature rutile is the most common crystal phase conversely, anatase thermodynamically not stable as rutile [8,9], both anatase, and rutile have tetragonal crystal structures but belong to different space groups [10]. In addition to that TiO_2 doped Cu are widely used in many fields including in the field of medicine and industrial applications as well as in decontamination as these particles are available in a very low cost, TiO_2 thin films are very good preservatives in the ultraviolet ray area due to a tape. It has been observed that copper-coated titanium membranes improve the properties as a photocatalyst in the visible region [11]. In this research, the nanostructure TiO_2: Cu films were synthesized using R.F. sputtering and annealed at 500 °C. The structural properties of these films were investigated.

IOP Conf. Series: Materials Science and Engineering **757** (2020) 012055 doi:10.1088/1757-899X/757/1/012055

.2. Experimental Details

Thin films of TiO_2: Cu was deposited on glass substrates by RF plasma sputtering using TiO_2: Cu target (diameter 5 cm and thickness 0. 3 cm) prepared from TiO_2 and Cu powders with a purity of 99.99% from (Merck, Germany) (Figure 1. illustrate the sputtering system). The film thickness was measured using the thickness monitor in the sputtering system and by applying, a 10-ton compression then dried at 80 OC. Before starting deposition, the glass substrates were cleaned thoroughly. The process of cleaning substrates begins by immersing in a cleaning solution to remove any oil or dust can be found on the surface and then placed under tap water for 15 minutes. The substrates are then immersed in the pure acetone solution, which reacts with the contamination such as grease and some oxides. Finally, the substrates are re-washed in an ultrasonic bath for 15 minutes, and then dried by hot air. The clean substrates and the TiO_2: Cu target are fixed in the chamber of sputtering system. High-purity argon was used as the sputtering and reactive gas. Then deposition of TiO2: Cu films were carried out at 100 °C under a constant dc power of 100 W. The deposition time was set at 180 min. The base pressure of the deposition chamber was kept at 2-3 × 10^{-5} Torr, and sputtering pressure $2.5*10^{-3}$Torr and annealing thin films at 500°C .X- ray diffraction studies were carried out using type (SHIMADZU Japan) XRD600, wavelength 1.541 A° using CuKα radiation. AFM were studied. The morphological characteristics of the surfaces were studied using AFM (model A3000 Angstrom Advance Lns.)

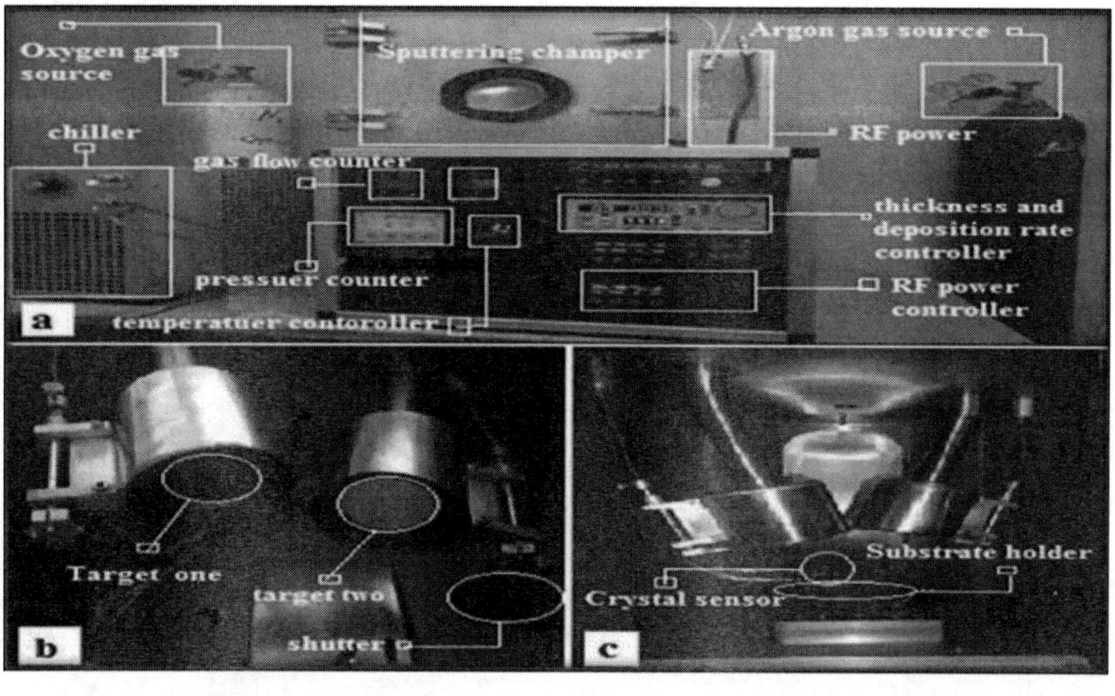

Figure 1.a,b&c Illustrate the sputtering system and the components used in this work.

IOP Conf. Series: Materials Science and Engineering **757** (2020) 012055 doi:10.1088/1757-899X/757/1/012055

3. Results and discussion

3.1. Thin films Characterization:

The structural properties of copper-doped Titanium oxide thin films with weight percentages (3%, 5%, and 7%) were studied using x-ray diffraction and atomic force microscopy.

3.2. X-ray diffraction results

The crystalline structure of the TiO_2: Cu thin film can be identified by studying the phase of that substance. When a beam of X-ray incident on the surface of the thin film, the peaks appear on specific angles due to the reflection of Bragg on the surface of the parallel crystal. The diffraction of the X-ray of the pure TiO_2 thin film at room temperature with different thicknesses (100 and 200) nm is shown in Figure 2, and after the annealing process at 500 °C, is shown in Figure 3. It is clear from these two figures that the crystalline structures of the prepared thin films were amorphous and become crystallized of brookite (orthorhombic) polycrystalline phase after annealing especially for TiO_2 doped with 5% Cu and the preferred peak (111) at the angle of 31.7°as in Figure 4.Obtained results have a good agreement to the standard results according to the data file [13463-67-7]. The full –width at half maximum (FWHM) of (111) of brookite peak were also evaluated to analyze the variation of grain size in the films doped. The average crystalline size of the thin films was obtained using Scherer's formula [12]:

$D= [(k\lambda/\beta Cos\theta)]$ .... (1)

Where D is the size of the crystal, k is the constant number of 0.9, λ is the x-ray wavelength ,θ is the angle of Bragg in degree ,and β is the full width at half the maximum (FWHM) of the selected peak . The average crystallization size calculated after annealing at thickness (100 and 200) nm for pure TiO_2 thin films were 59.82 nm and 59.996 nm and for TiO_2: 5%Cu were found to be 58.8 and 98.17 nm as illustrate in Table1.

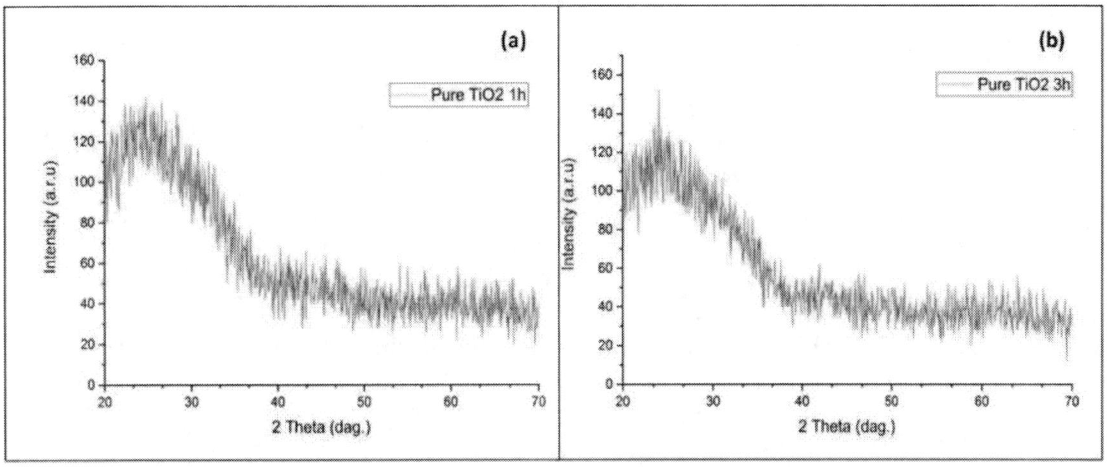

Figure 2. X-ray Diffraction of pure thin film TiO_2 at room temperature with thickness a: (100nm) and b: (200 nm).

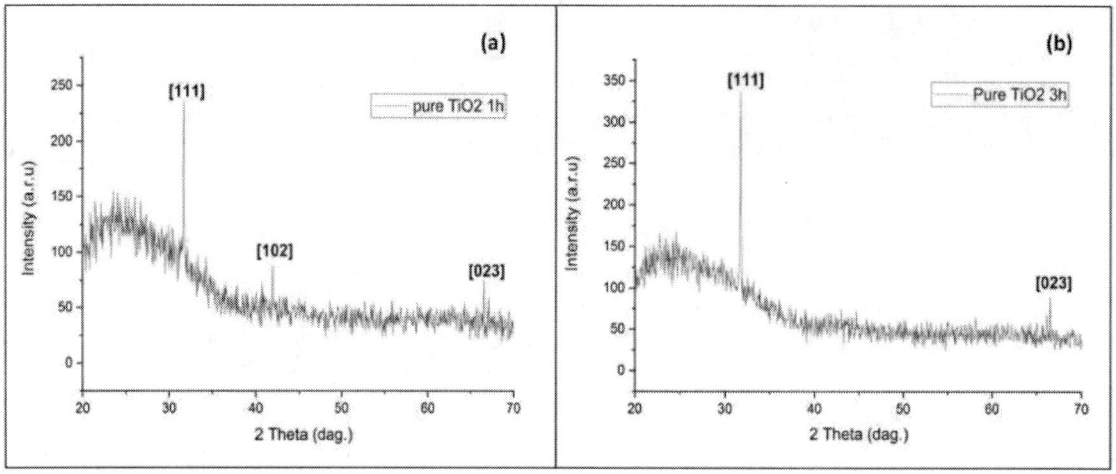

Figure 3. X-ray Diffraction of pure thin film TiO_2 at temperature500°C with thickness a: (100nm) and b: (200 nm).

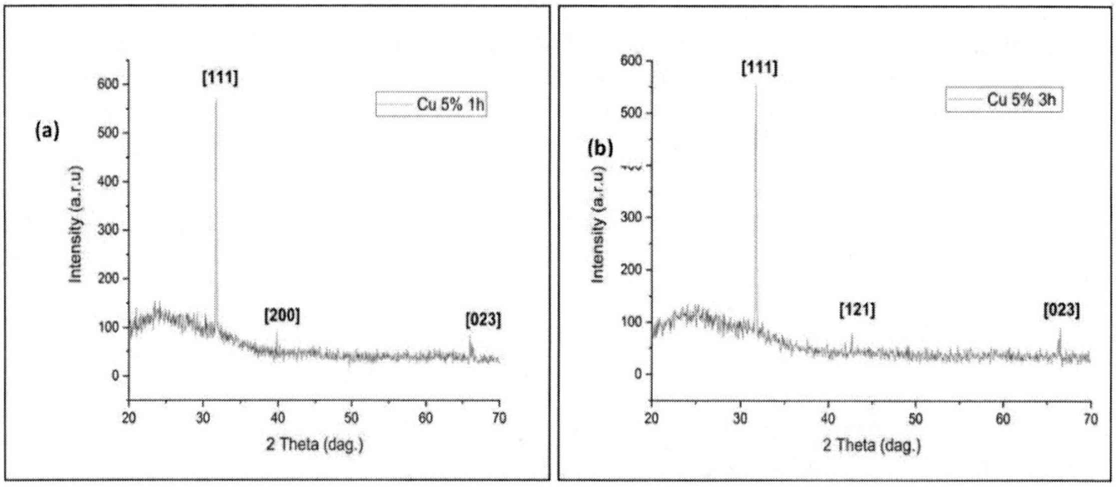

Figure 4. X-ray Diffraction of TiO_2:5%Cu thin film at temperature500°C with thickness a: (100nm) and b: (200 nm).

Table 1. Some Synthetic Properties of TiO_2 Thin Films.

Sample(at 500°C)	2θ(deg.)	hkl	d(Å)	FWHM (deg.)	D(nm)	a	b	c
(Stander)	31.46	[111]	2.766	-	-	4.53	5.4	4.9
TiO_2 Pure (100nm)	31.728	[111]	2.817	0.138	59.82	4.52	4.78	5
TiO_2 Pure (200nm)	31.724	[111]	2.818	0.1376	59.996	3.9	8.05	4.39
TiO_2 :5% Cu (100nm)	31.7278	[111]	2.815	0.14	58.8	4.36	5.08	4.89
TiO_2 :5%Cu (200nm)	31.744	[111]	2.816	0.084	98.17	4.23	5.62	4.72

IOP Conf. Series: Materials Science and Engineering **757** (2020) 012055 doi:10.1088/1757-899X/757/1/012055

3.3. Results of the atomic force microscopy (AFM) Morphology

Surface morphology of all TiO_2: Cu thin films with different concentration (0, 3, 5, and 7) % of Cu are measured by Atomic force microscopy (AFM) , the best results were noticed at 200nm thickness as in Figure (5- a, b, c, d).It is clear that all TiO_2: Cu films have a smooth surface with regular crystalline grains and a nanoscale structure. Through surface morphological study, it was noticed that the grain size of TiO_2: Cu is a nanostructure. It was observed that the surface of the thin film is rougher when increasing the copper doping in TiO_2 films, this result agrees with [13]. Also it was noticed the particle size increases as increasing the concentration of dopant Cu in the thin film and after the annealing process. The increase in grain size reduced the grain boundaries and thus reduced the barrier between them. The results are shown in Table 2.

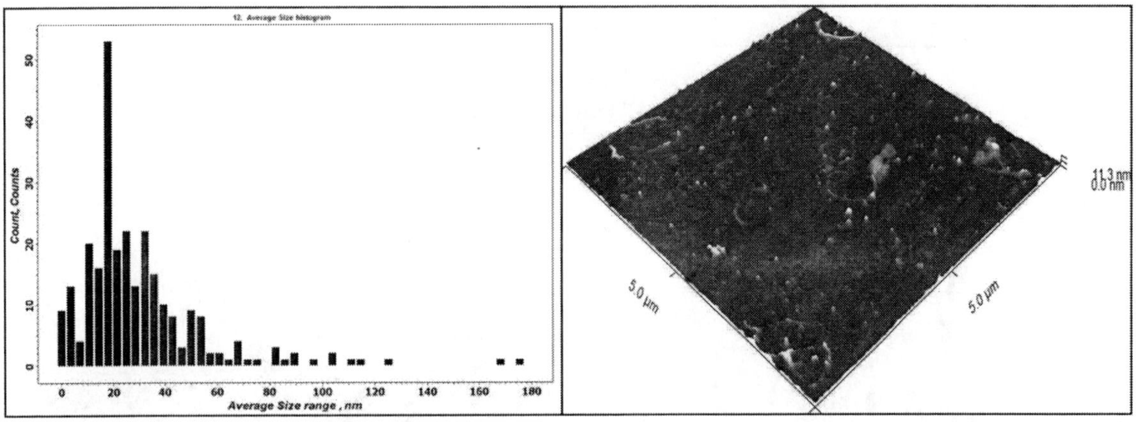

Figure (5-a). AFM image for Pure TiO_2 thin film.

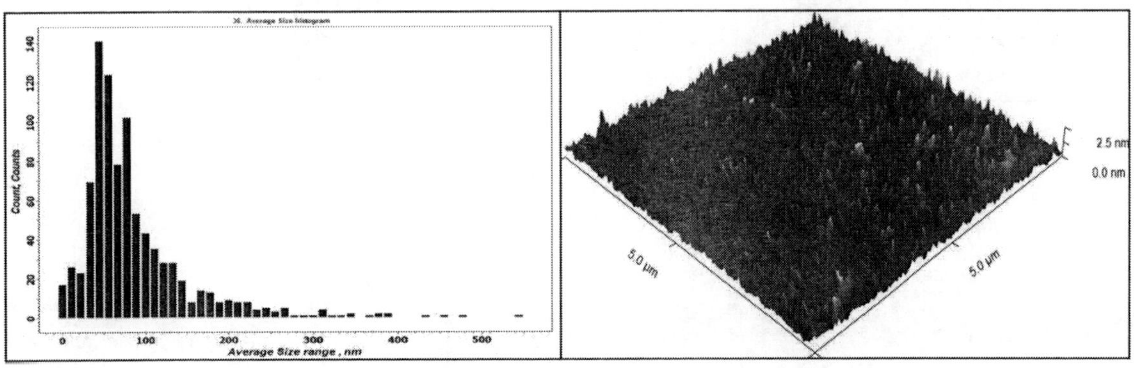

Figure (5-b). AFM image for TiO_2:3%Cu thin film.

IOP Conf. Series: Materials Science and Engineering **757** (2020) 012055 doi:10.1088/1757-899X/757/1/012055

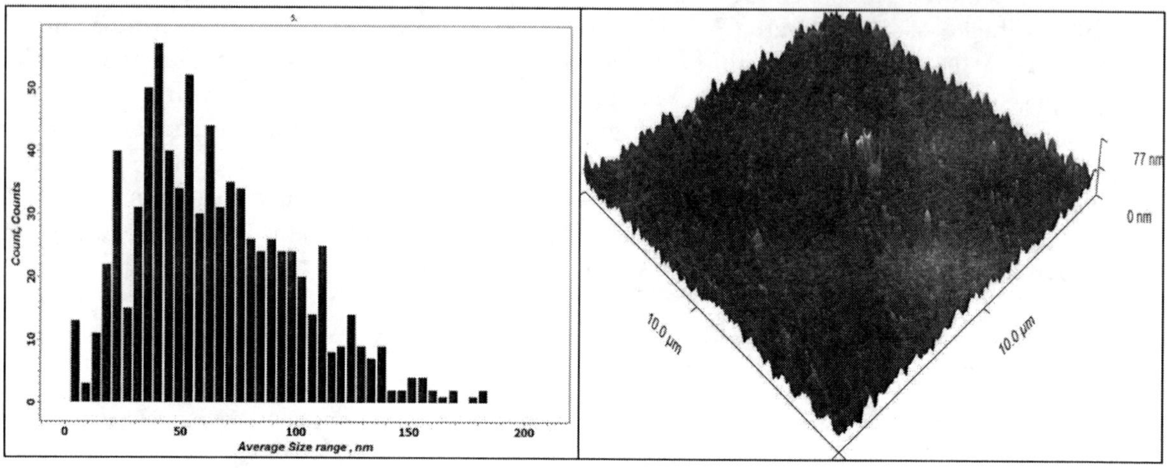

Figure (5-c). AFM image for TiO2:5%Cu thin film.

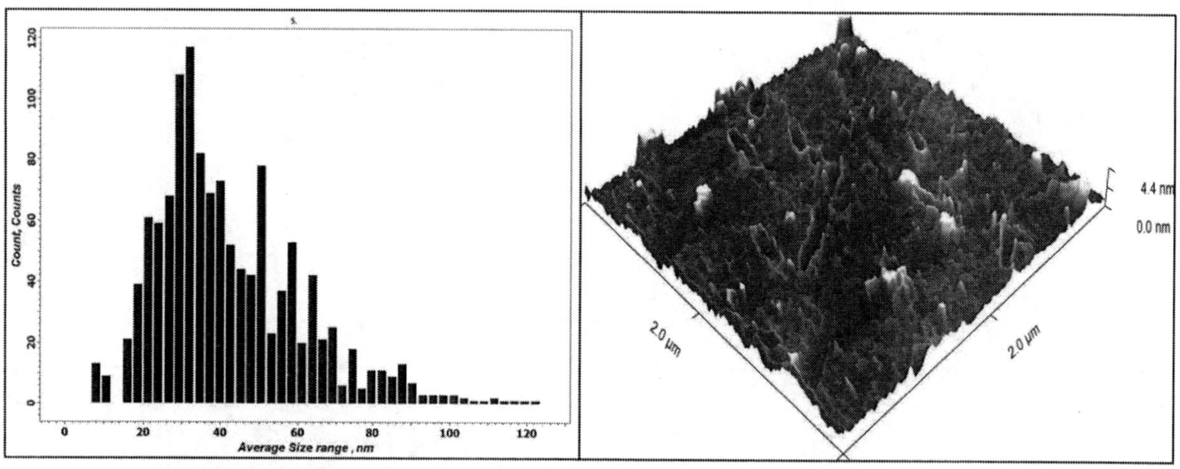

Figure (5-d). AFM image for TiO2:7%Cu thin film.

Table 2. The values of surface roughness root mean square and grain sizes for TiO_2: Cu thin films.

Thin Film	Surface Roughness (nm)	RMS (nm)	Average diameter (nm)
Pure TiO2	0.416	0.073	102.3
TiO_2+3%Cu	0.532	0.226	40.5
TiO_2+5%Cu	0.793	0.353	62.2
TiO_2+7%Cu	0.895	0.421	86.8

4. Conclusion

1-The radio frequency (R.F.) magnetron sputtering technique was used in this work to prepare thin films of TiO_2: Cu with various Cu contents (0, 3, 5and 7) %.
2-XRD analysis showed a crystalline structure to be a brookite polycrystalline phase of pure TiO_2 when annealed at 500 °C with preferred orientation along (111). The same structure also observed for TiO_2 doped with 5%Cu.
3-AFM analysis showed a smooth surface with regular crystalline grains with nanoscale structure films .It was noticed that the average surface roughness (Ra) measured was 0.532 nm, and the minimum grain size was 40.5 nm for TiO_2 doped 5%Cu.

5. Reference

[1] Dainius P and Ludwig J,2005 *Journal of Electroceramics*, **14**.111–103.

[2] Chowdhuri A, Gupta V, Sreenivas K, 2003 *Sensor Actuators B*, **93** 579–572.

[3] Haridas D, Sreenivas K,and Gupta V, 2008 *Sensor Actuators B* **133**, 270-275 .

[4] Chopra K,1969 *Thin Films Phenomena* (New York: McGraw – Hill Book Company).
[5] Lever k D, 1971*Thin Films* (London: Wykeham Pub p.32).
[6] Peter C S, 2004 CdO thin films deposited by spray pyrolysis, *Electronics*, **Vol.15**, pp. 22-24.
[7] Gurumuru K, Mangalaraj D, Narayandass S K, NakanishiY and Hatanaka Y, 1997 *Applied surface science,* **Vol.113**,pp.422-425.
[8] Adawiya J H, Raad M S AL-Haddad, and khaled Z Yahya, 2011, *IJAPVOL.*7,NO.2 pp .27-31.
[9] Kingon A I, Maris J and Stiffer S, 2000 Materials Fundamentals of Gate Dielectrics *Nature* **406**, p 1032.
[10] Fries M and Simko R, 2012 Nano-Titanium dioxide (part1) Basics, Production, Applications *Nano trust dossiers,* **33**.
[11] Tribble A, 2002 *Electrical Engineering Material and Devices*, University of lowa.
[12] Cullit B D, 1978 *Elements of X-Ray Diffraction* (Adison: Wesley).

[13] Sreedhar M, Reddy I N, Bera P, Amachandran K D, Saravanan G, Ms Rabel A, Anandan C, Kuppusami P and Brijitta J, 2015 *Applied Physics A, Materials Science & Processing*, Springer-Verlag Berlin Heidelberg.

The role of aluminum doping on structural and optical properties of ZnO thin films prepared by PLD

R A Rasheed and M F A Alias*
Physics Department, College of Science, University of Baghdad, Baghdad, IRAQ
* corresponding author email: may20131313@yahoo.com

Abstract.
In this work, pure and doped ZnO thin films with different Al wt% concentration were prepared using pulsed laser deposition technique on clean glass substrates at room temperature. Then investigate the effect of Al concentration on the structural and optical properties for the as deposited ZnO thin films. The structure of the prepared ZnO powder and thin films were examined by X-ray diffraction, whereas the optical properties of prepared thin films were characterized and investigated using UV-VIS-NIR spectrophotometer. The structure of powder and prepared pure and doped ZnO thin films were polycrystalline. Powder, pure and Al doped ZnO thin films were synthesized in hexagonal phase and the dominate orientation is (101). The crystallite size for (101) plane is increased with increasing Al wt%. The optical energy gap of doped ZnO thin film increased with increasing doping weight concentration. These make the prepared films as good candidate for gas sensor.
Keywords:

1. Introduction

Zinc Oxide (ZnO) films have become technologically important because of their range of electrical and optical properties, together with their high chemical and mechanical stabilities, which make them suitable for a variety of applications such as flat panel display electrodes ,gas sensors and antibacterial agent [1,2]. ZnO is an n-type semiconductor from II-VI group having wurtzite crystal structure, with wide and direct band gap of 3.37 eV. It is generally used in optoelectronic devices especially for blue and UV emissions, like laser diodes [3, 4]. There are many elements doped with ZnO were used different techniques for various applications [5-8]. The doping of ZnO films with the group III components can increase the conductivity of the films, by comparison with other elements and aluminum is the best dopant [7, 9-12]. There are several techniques used for fabricating ZnO films, for example chemical vapour depositions [8], pulsed laser deposition, dc reactive sputtering [6], spray pyrolysis and sol–gel process [7, 13]. Among these techniques, PLD has unique advantages such as: stoichiometric deposition, easy control of experimental parameters and a naturally clean process. Through the preparation of ZnO thin films, the growth condition such as substrate type, deposition temperature, background gas pressure and laser energy density, have great influence on the crystal quality of ZnO film and cause a change in its structural and optical properties [14].

In this paper, pure and doped ZnO thin films with different Al concentration were grown using pulsed laser deposition technique from prepared powder, then study the role of Al wt% concentration on structural and optical properties of all prepared thin films.

2. Experimental Work

2.1 Sample preparation

The targets were composited of pure (ZnO) and doped (ZnO: Al) powder with high purity (99.999%).Different Al concentrations (3,5,7,10) wt.% were mixed together and prepared by pressing them under 5 Ton for time of 10 minutes to form a target of 2.5 cm diameter. ZnO:Al thin films were prepared on clean substrate using pulsed laser deposition (PLD) method (Nd: YAG (Huafei Tongda Technology—DIAMOND-288 pattern EPLS) (SHG Q-switching beam) as shown in figure (1) which housing inside a vacuum chamber of 10-3Torr.

2.2 Sample Characterization

The structural properties of prepared films are examined using X-ray diffraction. The optical measurements (transmittance and absorbance spectra) were recorded, in the wavelength range (190-1100) nm using UV-VIS-NIR spectrophotometer at room temperature. The XRD parameters of ZnO powder and films, transmittance and energy gap were calculated for ZnO thin films as a function of different Al concentrations.

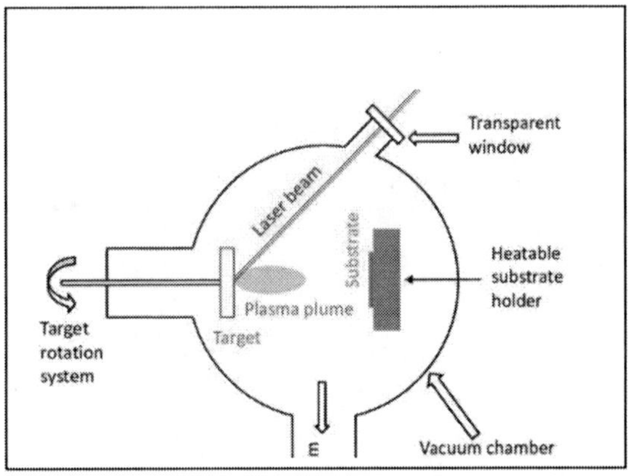

Figure 1. Schematic diagram for experimental set out for pulsed laser deposition technique.

3. Results and Discussion

The crystal structure and various structural parameters of the prepared powder, pure and Al doped ZnO thin films with different concentration (3, 5,7,10) wt% were analyzed from X-ray diffraction (XRD) patterns. Figure 2 display the X-ray diffraction patterns of ZnO powder. X-ray diffraction peaks showed that all samples had a polycrystalline structure .Eleven peaks were observed at 2θ=31.77°,2θ =34.43°,2θ =36.26°,2θ =47.58°,2θ=56.62°,2θ =62.86°,2θ =66.40°,2θ =67.97°, 2θ =69.10°, 2θ =72.57° and 2θ = 76.98° corresponding to the wurtzite hexagonal phase of ZnO matching to the lattice planes of (100) , (002), (101) ,(102), (110),(103) , (200) ,(112), (201),(004) and (202) plans respectively. These values are matched with values in the standard card number (96-901-1663), (96-901-1664) and (96-901-1665). The dominate peak is for the plan (101) orientation. The average crystallite size (C.S Av) is 46.28 nm. Table (1) illustrates the X-ray parameters for the ZnO powder.

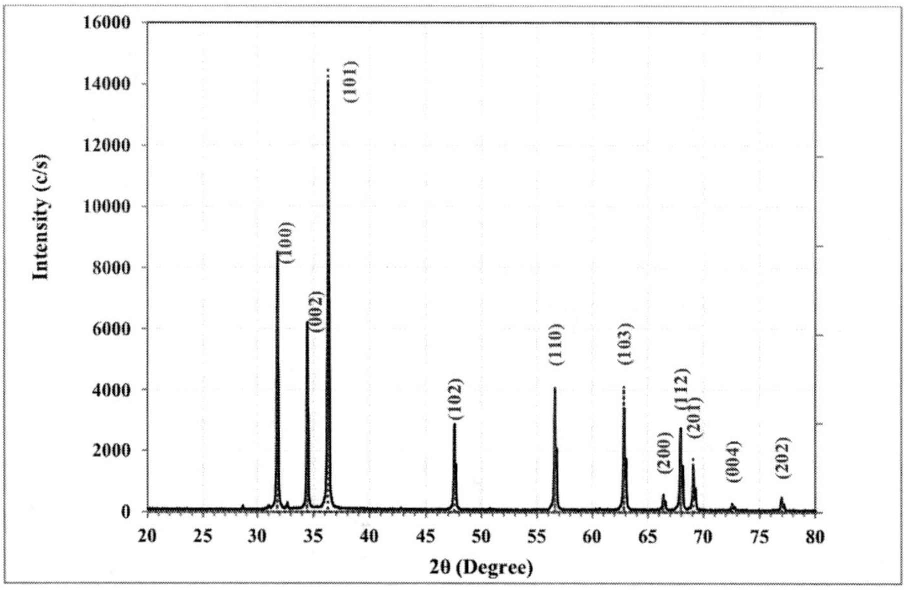

Figure 2. X-ray patterns of ZnO powder.

Table 1. XRD parameters for ZnO powder.

$2\theta_{Exp}$ (Deg.)	$d_{Exp.}$ (Å)	$d_{Std.}$ (Å)	FWHM (Deg.)	C.S (nm)	$C.S_{Av}$ (nm)	hkl	Phase	Card No. ASTM
31.7757	2.8138	2.8137	0.2103	39.3	46.28	(100)	Hex.	96-901-1663
34.4393	2.6021	2.6035	0.1752	47.5		(002)	ZnO	
36.2617	2.4754	2.4754	0.2103	39.8		(101)		
47.5818	1.9095	1.9110	0.2454	35.4		(102)		
56.6238	1.6242	1.6245	0.2103	42.9		(110)		
62.8621	1.4772	1.4772	0.2102	44.3		(103)		
66.4019	1.4067	1.4069	0.1753	54.2		(200)		
67.9790	1.3779	1.3782	0.2103	45.6		(112)		
69.1005	1.3582	1.3582	0.2103	45.9		(201)		
72.5701	1.3016	1.3017	0.1752	56.3		(004)		96-901-1664
76.9860	1.2376	1.2377	0.1753	57.9		(202)		96-901-1665

Figure 3 display the X-ray diffraction patterns of pure and doped ZnO thin film with different concentration of Al metal. Nine peaks were observed at $2\theta=31.77°$, $2\theta=34.41°$, $2\theta=36.23°$, $2\theta=47.51°$, $2\theta=56.54°$, $2\theta=62.80°$, $2\theta=66.27°$, $2\theta=67.87°$ and $2\theta= 68.98°$ corresponding to (100), (002), (101), (102), (110), (103), (200) ,(112) and (201) plans for ZnO thin films respectively. These values are matched with values in the standard card number (96-901-1663). Two peaks were disappeared for (004) and (202) orientation as compared with that was shown in figure (2) for ZnO powder. The samples which

IOP Conf. Series: Materials Science and Engineering **757** (2020) 012056 doi:10.1088/1757-899X/757/1/012056

investigated were prepared at room temperature without any treatment, so, the samples have polycrystalline structure.

The prepared films exhibit a dominant peak corresponding to the (101) plan of ZnO. Presence of various peaks corresponding to plans (100), (002), (101), (102), (110), (103), (200), (112) and (201)suggests polycrystalline nature of the prepared thin film. One can observe that phases for all peaks were hexagonal.

Another noticeable remark is that the angle value is slightly shifted towards lower angle when ZnO thin film doped with Al for (101) orientation which is due to stresses within the lattice caused by doping and crystalline defects [15]. This indicates that the lattice parameters of the Al doped ZnO films are larger than that of pure ZnO thin films (Table 3). Such an increment in lattice parameter can be attributed to the different ionic radius of Al3+ (0.053 nm) comparing to that of Zn2+ (0.074 nm)[16]. The variation in lattice parameter signifies the incorporation of Al3+ due to doping also introduces stress in the film. Substitution replacement of smaller size Al3+ was resulted in a tensile stress [17, 18].

The average value of crystallites size (C.S) was measured by using Scherrer equation [19]:

$$C.S = k\lambda/\beta \cos\theta \qquad (1)$$

In this relation, k is the shape factor having value 0.94, λ is wavelength of the Cu Kα (1.5414 Å), β is full width of half maximum (FWHM) in radian and θ is the Bragg's angle in degree.

Table 2 illustrates the values of Bragg's diffraction angles (2θ), full width of half maximum (FWHM), inter-planer distance (dhkl), the crystallite size (C.S), and the corresponding Miller indices of the standard cards (hkl) for pure and doped ZnO with different Al concentrations. .

The full width of the peaks changed due to the change in strain applied on lattice due to the particle size change (eq.1). The crystallite size at (101) orientation is increased with increasing Al, whereas the average crystallite size value varied unsystematic with increasing Al wt% concentration from 0 to 10.

IOP Conf. Series: Materials Science and Engineering **757** (2020) 012056 doi:10.1088/1757-899X/757/1/012056

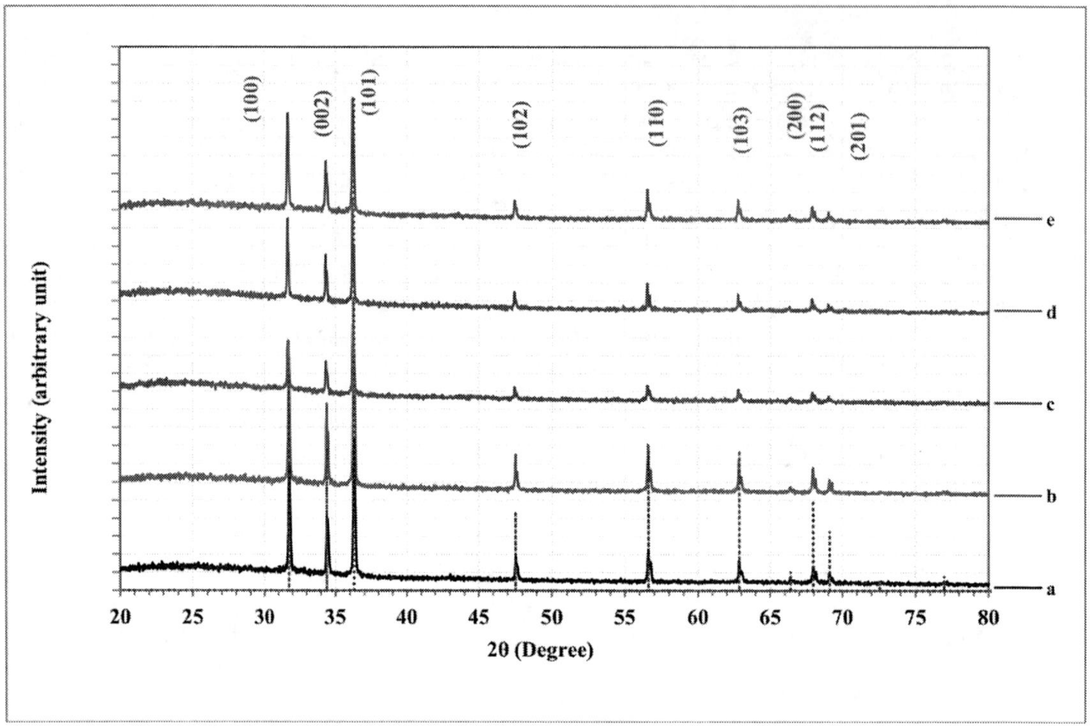

Figure 3. X-ray patterns of pure ZnO films and doped ZnO thin film with different Al wt% concentrations (a) 0,(b) 3,(c)5,(d)7,(e)10.

Table 2 XRD parameter for pure and doped ZnO thin film with different Al wt% concentrations (3, 5,7 and 10).

Al wt%	2θ (Deg.)	d_{hkl} Exp.(Å)	d_{hkl} Std.(Å)	FWHM (Deg.)	C.S (nm)	$C.S_{Av}$ (nm)	(hkl)	Phase ZnO	Card No. ASTM
	31.7750	2.8139	2.8137	0.1629	50.7	50.08	(100)	Hex.	96-901-1663
	34.4190	2.6035	2.6035	0.1698	49.0		(002)	Hex.	96-901-1663
	36.2390	2.4768	2.4754	0.1686	49.6		(101)	Hex.	96-901-1663
0	47.5110	1.9122	1.9110	0.2127	40.8		(102)	Hex.	96-901-1665
	56.5400	1.6264	1.6245	0.1415	63.8		(110)	Hex.	96-901-1663
	62.8020	1.4784	1.4772	0.1825	51.0		(103)	Hex.	96-901-1663
	66.2760	1.4091	1.4069	0.2004	47.4		(200)	Hex.	96-901-1663
	67.8720	1.3798	1.3782	0.2216	43.2		(112)	Hex.	96-901-1663
	68.9830	1.3603	1.3582	0.1745	55.3		(201)	Hex.	96-901-1663
	31.7350	2.8173	2.8137	0.1377	60.0	73.48	(100)	Hex.	96-901-1663
	34.3810	2.6063	2.6035	0.1471	56.6		(002)	Hex.	96-901-1663
	36.2190	2.4782	2.4754	0.1562	53.5		(101)	Hex.	96-901-1663
3	47.4910	1.9130	1.9110	0.1270	68.4		(102)	Hex.	96-901-1665
	56.5200	1.6269	1.6245	0.0879	102.6		(110)	Hex.	96-901-1663
	62.7820	1.4789	1.4772	0.0881	105.7		(103)	Hex.	96-901-1663
	66.2970	1.4087	1.4069	0.1854	51.2		(200)	Hex.	96-901-1663
	67.8520	1.3802	1.3782	0.0998	96.0		(112)	Hex.	96-901-1663
	69.0030	1.3599	1.3582	0.1365	70.6		(201)	Hex.	96-901-1663
	31.6940	2.8209	2.8137	0.1615	51.2	48.95	(100)	Hex.	96-901-1663
	34.3410	2.6093	2.6035	0.1700	48.9		(002)	Hex.	96-901-1663
	36.1590	2.4821	2.4754	0.1548	54.0		(101)	Hex.	96-901-1663
5	47.4300	1.9153	1.9110	0.1387	62.6		(102)	Hex.	96-901-1665
	56.4800	1.6280	1.6245	0.2154	41.9		(110)	Hex.	96-901-1663
	62.7210	1.4801	1.4772	0.2022	46.0		(103)	Hex.	96-901-1663
	66.2560	1.4095	1.4069	0.3234	29.3		(200)	Hex.	96-901-1663
	67.8120	1.3809	1.3782	0.2286	41.9		(112)	Hex.	96-901-1663
	68.9430	1.3610	1.3582	0.1487	64.8		(201)	Hex.	96-901-1663
	31.6740	2.8226	2.8137	0.1364	60.6	60.63	(100)	Hex.	96-901-1663
	34.3200	2.6108	2.6035	0.1474	56.4		(002)	Hex.	96-901-1663
	36.1380	2.4835	2.4754	0.1471	56.8		(101)	Hex.	96-901-1663
7	47.4100	1.9160	1.9110	0.1341	64.7		(102)	Hex.	96-901-1665
	56.4590	1.6285	1.6245	0.1704	52.9		(110)	Hex.	96-901-1663
	62.7210	1.4801	1.4772	0.1158	80.4		(103)	Hex.	96-901-1663
	66.2360	1.4099	1.4069	0.1702	55.7		(200)	Hex.	96-901-1663
	67.8120	1.3809	1.3782	0.1411	67.9		(112)	Hex.	96-901-1663
	68.9230	1.3613	1.3582	0.1915	50.3		(201)	Hex.	96-901-1663
	31.6740	2.8226	2.8137	0.1451	56.9	54.34	(100)	Hex.	96-901-1663
	34.3410	2.6093	2.6035	0.1685	49.4		(002)	Hex.	96-901-1663
	36.1590	2.4821	2.4754	0.1711	48.9		(101)	Hex.	96-901-1663
10	47.4300	1.9153	1.9110	0.1845	47.0		(102)	Hex.	96-901-1665
	56.4590	1.6285	1.6245	0.1034	87.2		(110)	Hex.	96-901-1663
	62.7210	1.4801	1.4772	0.1954	47.6		(103)	Hex.	96-901-1663
	66.2360	1.4099	1.4069	0.1819	52.2		(200)	Hex.	96-901-1663
	67.8120	1.3809	1.3782	0.1589	60.3		(112)	Hex.	96-901-1663
	68.9430	1.3610	1.3582	0.2437	39.6		(201)	Hex.	96-901-1663

Table 3 and figures 4 (a and b) show the variation of lattice constants for hexagonal pure and doped ZnO thin films. The lattice constant (a) reduced to a minimum value of 3.2533Å with increase of Al to 3% for the prepared thin films and then increase by increasing Al up to7 % after that it remain constant. On the other hand, it's observed that the lattice constant (c) has apposite behavior with this parameter, i.e., it increases with increasing Al concentration and the maximum value of lattice constant (c) is 5.2216 Å at 7% Al then it decreases to 5.2185nm at 10 %.

Table 3. Lattice constants for pure and Al doped ZnO thin films at different concentrations.

Al wt %	a (Å)	c (Å)	Average C.S (nm)
0	3.2542	5.2071	50.1
3	3.2533	5.2127	73.4
5	3.2551	5.2185	48.9
7	3.2559	5.2216	60.6
10	3.2559	5.2185	54.3

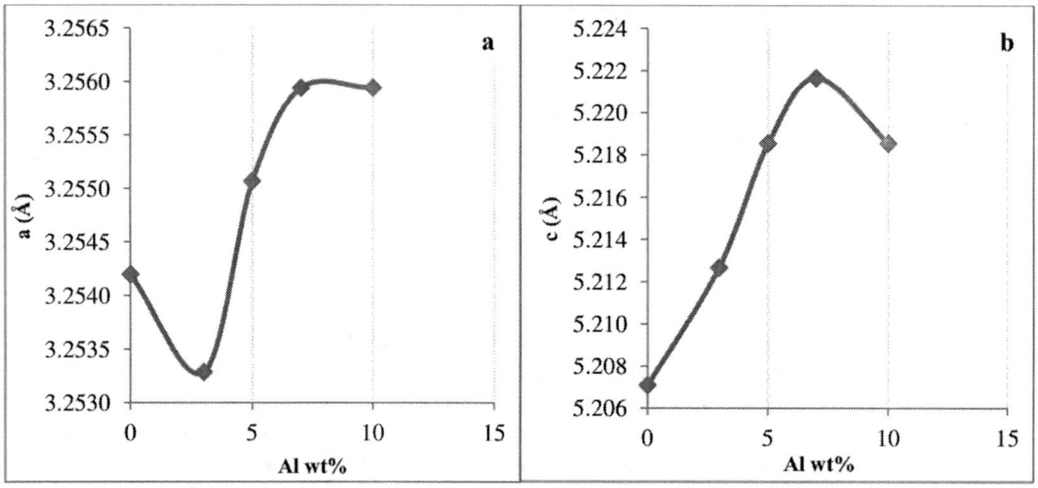

Figure 4. Variation of lattice constants vs Al wt% concentration for pure and doped ZnO thin films (a) **a** and (b) **c**.

Figure (5) shows the transmittance (T) spectrum for pure and doped ZnO thin films through which study the effect of the Al on their optical properties. It can be observed that the optical transmittance is increased suddenly at wavelength more than 130 nm then become nearly constant at wavelength ≥180 nm, while having low values at short wavelength because only photons with energy equal to or greater than the energy difference between the valence and conduction bands interact with valence electrons creating a pair of electron-hole. It is also observed that the transmittance has maximum value for ZnO doped with 7% Al concentration, whereas the minimum value is at ZnO doped with 3%. Another noticeable remark is that the value of T decreases with other Al doping (5 and 10) %. This may be due to the change in the crystallization rate or because of the change in the value of the energy gap.

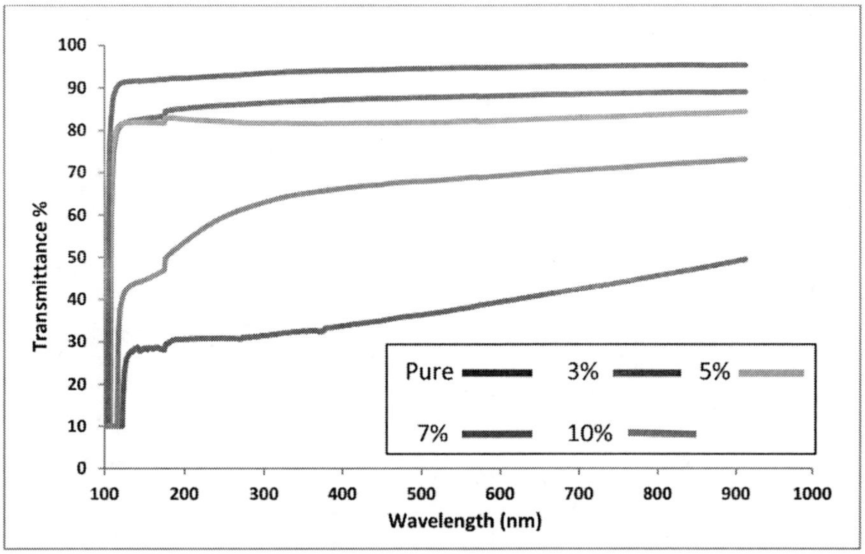

Figure 5. Transmittance vs. wavelength for pure and doped ZnO thin films with different Al wt% concentrations.

The amount of energy gap is one of the most important parameters that determine thin films applications. The direct optical gap of the samples was calculated using the Tauc formula [20], where the optical energy gap was calculated from the graphing between (αhυ)2 as a function of the photon energy (hυ). Determining the energy gap (Eg) from the tangent intersects point of the linear region with the x-axis according to equation [20]:

$$(\alpha h\upsilon)^2 = B\,(h\upsilon - E_g) \qquad (2)$$

where α is absorption coefficient and B is factor inversely proportional to amorphousity.
Figures 6 (a-e) show the change in the value of the direct optical energy gap with the various Al wt% concentrations in zinc oxide. On can observe there are two energies gap which is confirming the polycrystalline structure of prepared films.
These figures show that increasing the Al concentration (0,3,5,7 and 10 wt%) has increased its transparency and energy gap values (Eg2) (3.35, 3.70, 3.87 and 3.98) eV respectively .Whereas the lower energy gap (Eg1) is varied (2.0, 2.3, 2.0, 3.1 and 2.78) eV with increasing Al concentration from 0 to 10% respectively . The low values of the band gap (Eg1) may be due to the existence of tail states within the band gap and the absorption edge is not sharp .There are several factors that change the value of the optical band gap, such as the difference in distance between the atoms due to lattice distortion. The difference in the values of the energy gap from the standard values is due to preparation conditions.

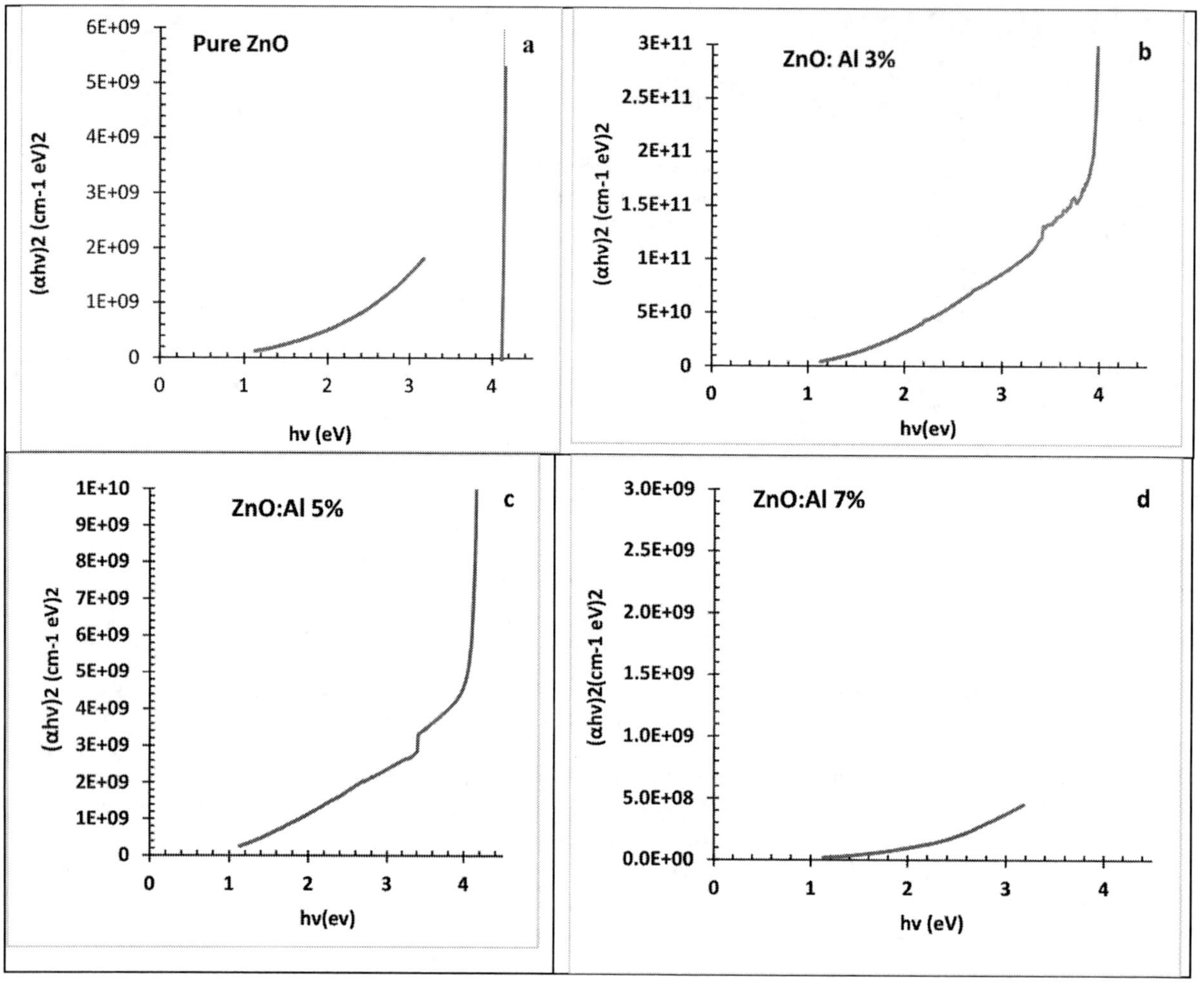
Pure ZnO
a
6E+09
5E+09
4E+09
3E+09
2E+09
1E+09
0
(αhv)2 (cm-1 eV)2
0
1
2
3
4
hv (eV)
ZnO: Al 3%
b
3E+11
2.5E+11
2E+11
1.5E+11
1E+11
5E+10
0
(αhv)2 (cm-1 eV)2
hv(ev)
ZnO:Al 5%
c
1E+10
9E+09
8E+09
7E+09
6E+09
5E+09
4E+09
3E+09
2E+09
1E+09
(αhv)2 (cm-1 eV)2
hv(ev)
ZnO:Al 7%
d
3.0E+09
2.5E+09
2.0E+09
1.5E+09
1.0E+09
5.0E+08
0.0E+00
(αhv)2(cm-1 eV)2
hv (eV)

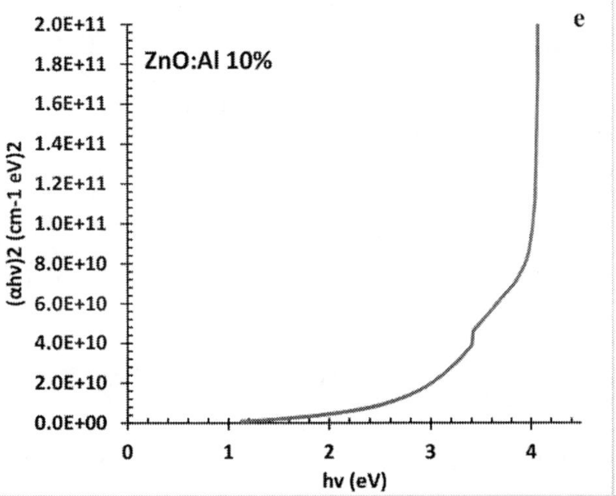

Figure 6. $(\alpha h\upsilon)^2$ versus (hυ) for doped ZnO thin films at different Al wt% concentration (a) pure (b) 3 (c) 5 (d) 7 (e) 10.

4. Conclusions

The results of examination of as deposited ZnO thin films prepared on glass substrates at different Al wt% concentration using PLD technique are summarized as following:

Powder, pure and Al doped ZnO thin films were synthesized a polycrystalline structures with hexagonal phase. The best sample crystallinety appeared for (101) orientation for pure and doped ZnO thin films.

The crystallite size for all prepared thin films at (101) orientation is increased with increasing Al concentration, whereas the average crystallite size value varied unsystematic with increasing Al concentration from 0 to 10%. The increasing Al concentration in ZnO thin films leads to increase transparency and in turn increase energy gap with values more than standard value for bulk samples.

References

[1] Selmi M, Chaabouni F, Abaab M and Rezig B 2008 Superlattices and Microstructures 44 268.

[2] Ayeshamariam, A, Kashif M, Vidhya V Sankaracharyulu S, Swaminathan MGVand Bououdina V, 2014 Journal of Optoelectronics and Biomedical Materials 6 85.

[2] X. M. Fan, J. S. Lian, Z. X. Guo and H. J. Lu 2005 Applied Surface Science 239 176.

[3] W. Zhaoyang, S, Liyuan and H. Lizhong 2010 Vacuum 85 397.

[4] Silva RF, Maria ED, Zaniquelli 2004 Thin Solid Films 449 86.

[5]Mai M.A.Ahmed , Wael Z.Tawfik , M.A.K.Elfayoumi, M.Abdel-Hafiez and S.I.El-Dek 2019 Journal of Alloys and Compounds 791 586.

[6] Saurabh Kunj and K.Sreenivas 2016 Current Applied Physics 16 748.

[7] Tanusevskia A and Georgieva V 2010 Appl Surface Sci. 19964.

[8] Muhammad R.Islam MukhlasurRahman S.F.U.Farhad and Jiban Podder 2019 Surfaces and Interfaces 16 120.

[9] J.N. Zeng, J.K. Low, Z.M. Ren, T. Liew, and Y.F. Lu 2002 Appl. Surf. Sci. 197 362.

[10]M.F.A.Alias, R.M.Aljarrah, H.Kh.Al-Lamy and K.A.W.Adam 2013International Journal of Application or Innovation in Engineering and Management 2 198.

[11]M.F.A.Alias and Kh. M. Rashid, 2014 International Journal of Advanced Scientific and Technical Research 4 627.

[12]M. F. A. Alias, Kh. M. Rashid and K. A. Adem 2014 International Journal of Innovative Research in Science, Engineering and Technology 3 15538.

[13]Salah. Q. Hazaa Manar Th. Mansor and Najiba A. Al-Hamadani 2016 International Journal of Latest Research in Engineering and Technology 02 01.

[14] K. Siraj, M. Khaleeq-ur-Rahman, S. I. Hussain, M. S. Rafique and S. Anjum 2011 Journal of Alloys and Compounds 509 6756.

[15] Aleksander D. Pogrebnjak, Abid A. Muhammed, Emad T. Karash, Nawfal Y. Jamil and Janusz Partyka 2013 Przeglad Elektrotechniczny 89 315.

[16] K.M. Sandeep, S. Bhat and S.M. Dharmaprakash 2018 Opt. Laser Technol. 102 147

[17] M.L. Addonizio, C. Diletto 2008 Sol. Energy Mater. Sol. Cells 92 1488.

[18] G.G. Rusu, A.P. Rambu,V.E. Buta, M.Dobromir, D. Luca, M. Rusu 2010 Mater. Chem. Phys.123 314.

[19] B. D. Cullity 1978 Element of X-ray diffraction 2nd Edition, Addison-Wesley, Philippines.

[20]V. P. Kunets, N. R. Kulish, V. P. Kunets and M. P. Lisitsa 2002 Semicond. Physics Quantum *Electron. Opto*. 59.

IOP Conf. Series: Materials Science and Engineering **757** (2020) 012057 doi:10.1088/1757-899X/757/1/012057

The Impact of Silver: Gold Nanoparticles Ratio on its Structural, Morphological and Optical Properties

M. F. A. Alias* and KH. M. Rashed
Department of Physics, College of Science, University of Baghdad,
*Corresponding author: may20131313@yahoo.com

Abstract. The rapid development of nanomaterial applications has prompted the researchers to explore the possibility of controlling their properties. In this work, synthesize of silver: gold alloys (Ag: Au) nanoparticles by pulsed laser ablation in distilled water. Changing the ratio of the used target alloys (Ag: Au) lead to control the band wavelength of the plasmon frequency resonance. Atomic analysis shows decrease in the percentage of gold than its initial values using energy dispersive x-ray spectroscopy. The absorption spectrum illustrates the plasmon peaks in the visible range changes from 404 nm to 524 nm by changing the ratio of Ag: Au from 100:0 to 0:100 using UV-VIS spectrophotometer. The scanning electron microscopy indicates a change in the size of the NPs and achieves regular volume, especially for 50:50 ratios.

Keywords: Metal nanoparticles Ag: Au, Plasmon effect, Elemental analysis, Scanning electron microscopy.

1. Introduction

Recently, nanoparticles (NPs) have been used in many bio-researches. Some researches included the use of NPs in diagnosis and treatment due to their distinctive and varied characteristics. These are due to their large surface area relative to the size of the sample, high stability and accessibility within the cell and interaction with its components. Among the applications that have been invested on their optical properties, which are characterized by plasmon resonance phenomenon, the most common nanoparticles in this field are gold and silver [1].

The optical properties of metal NPs strongly, dependent on particle size and its composition. To understand these dependences of the optical properties of metal NPs and nanostructures, one has to consider the impact of spatial plasmon resonances confinement. Metal particles are particularly interesting because they are easily formed and modified [2].

Nanoparticles can be formed by various techniques such as chemical [3], physical [4], biological [5], etc. Pulse laser ablation technique is one of the simplest techniques for the production of nanoparticles and easy control of their shape and size with different parameters, which has been of great importance in recent years [6].

In this work Ag: Au alloys and its nanoparticles were prepared by laser ablation technique in liquid. We did study the effect of Au: Ag ratio on metal nanoparticles properties using various standard characterization techniques such as energy dispersive X-ray spectroscopy (EDX), X-ray diffraction (XRD), scanning electron microscopy (SEM) and Uv-Vis spectrophotometer.

2. Experimental work

Five samples of metal targets were prepared from gold, silver and their alloys of both, with different ratios of Ag: Au such as 100:0, 75:25 50:50, 25:75 and 0:100. The alloys were melted in graphite crucibles at sufficient melting point for both metals, then poured molten into small molds and left to cool in air.

The metal nanoparticles were prepared by bombarding the metal target by Nd: YAG Pulsed laser with 1064 nm wavelength and 900 mJ pulses at 9 ns pulse duration and 10 Hz repetition frequency in distilled water. The nano-particles were collected by centrifuge and dropped on glass slides to dry at 353 K under 0.1 mbar vacuum for thirty minute.

The prepared nanoparticles were characterized using energy dispersive X-ray spectroscopy (EDX), X-ray diffraction (XRD), UV-visible spectrometer and scanning electron microscopy (SEM).

3. Results and Discussion

From the elemental analysis for prepared alloys samples, shown in Fig. 1. It is observed that the silver (Ag) peaks located around 3 and 3.2 keV corresponding to L_α and L_β transition respectively. The gold (Au) peaks at 2.2 and 9.7 keV belong to M_α and L_α transitions with intensities depend on their ratio in sample. The ratio of weight and atomic percentage are shown in Table 1, were calculated by the software attached to the examination device, using peak intensities for K_α transition for elements.

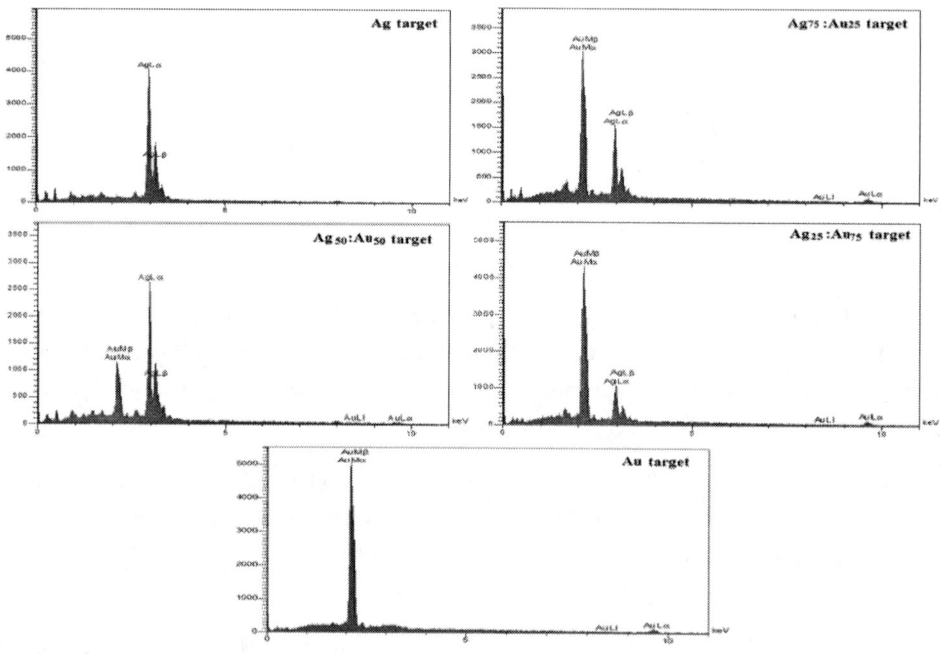

Figure 1: EDX for Ag, Au and their alloys target at different Au ratio.

IOP Conf. Series: Materials Science and Engineering **757** (2020) 012057 doi:10.1088/1757-899X/757/1/012057

Fig.2 shows the elemental analysis of nano Ag: Au samples deposited on glass substrates. The same peaks observed of Ag and Au as in alloy samples, with intensities depend on Ag: Au ratios in sample, in addition to very small peaks come from glass substrate especially for silicon peak at 1.74 keV.

Table 1 displays the Ag, Au alloys values and their nanoparticles samples and composite targets for prepared at different initial Au ratio. It seems that the gold ratio in films is nearly same as in alloy samples.

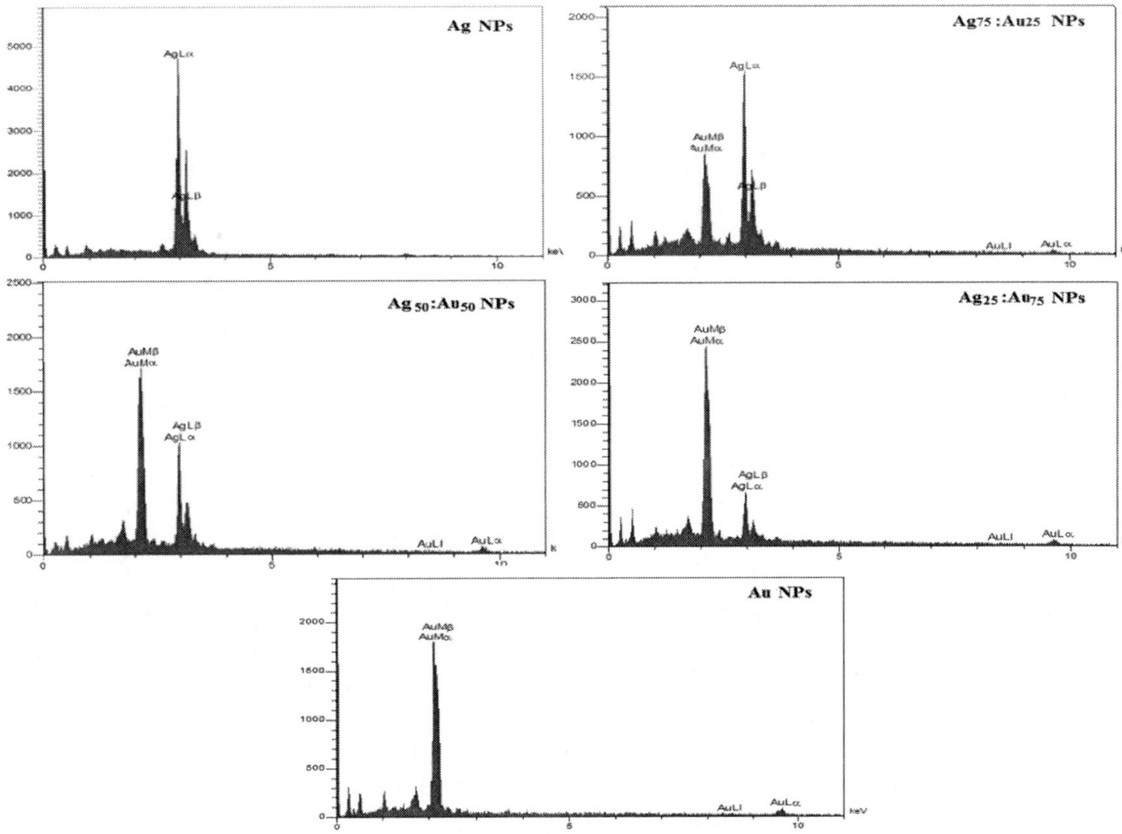

Figure2 : EDX for AgNPs, AuNPs and their alloys nanoparticles at different ratio.

Table 3: Elemental analysis for Ag, Au and their composite targets and nanoparticles at different Au ratio.

Ag: Au	Element	Alloy sample		Nano sample	
		Wt%	atm%	Wt%	atm%
Pure Ag	Ag	100.00	100.00	100.00	100.00
	Au	00.00	00.00	00.00	00.00
75:25	Ag	79.22	87.44	77.29	86.14
	Au	20.78	12.56	22.71	13.86
50:50	Ag	53.55	67.79	55.31	69.33
	Au	46.45	32.21	44.69	30.67
25:75	Ag	35.88	50.54	34.86	49.43
	Au	64.12	49.46	65.14	50.57
Pure Au	Ag	00.00	00.00	00.00	00.00
	Au	100.00	100.00	100.00	100.00

Fig.3 shows the X-ray diffraction patterns for Ag NPs, Au NPs and their composite at different ratio prepared by pulsed laser. This figure shows polycrystalline structures for all prepared samples with peaks appear at $2\theta = 37.2785^0$, 43.3584^0 and 63.0386^0 corresponding to (111), (200) and (220) planes for Ag or Au structures, whereas for Au, the peaks appears $2\theta = 37.3200^0$, 43.3200^0 and 62.9600^0 corresponding to the planes (111), (200) and (220) respectively. Ag and Au have cubic structure with small variation in their dimensions (a=4.1720 Å for Ag and a=4.1450 Å for Au) [7], which identical according standard card No. (96- 901- 3054 and 96-901-3045). These results are in a good agreement with reported results of Chen *et al*[8].

The XPowder software was used to calculate the peaks width at half maximum is shown in Fig.4. This figure shows that the peak width along (111) and (200) directions increased with increasing Au ratio, while irregular variation along (220). In general, the average crystallite size decreases with Ag decreasing in Ag: Au ratio samples. Table 1 shows the structural parameters for prepared metal nanoparticle including Bragg angle (2θ) , full width at half maximum (FWHM), experimental Inter-planar spacing [9], crystallite size [10] and the corresponding planes.

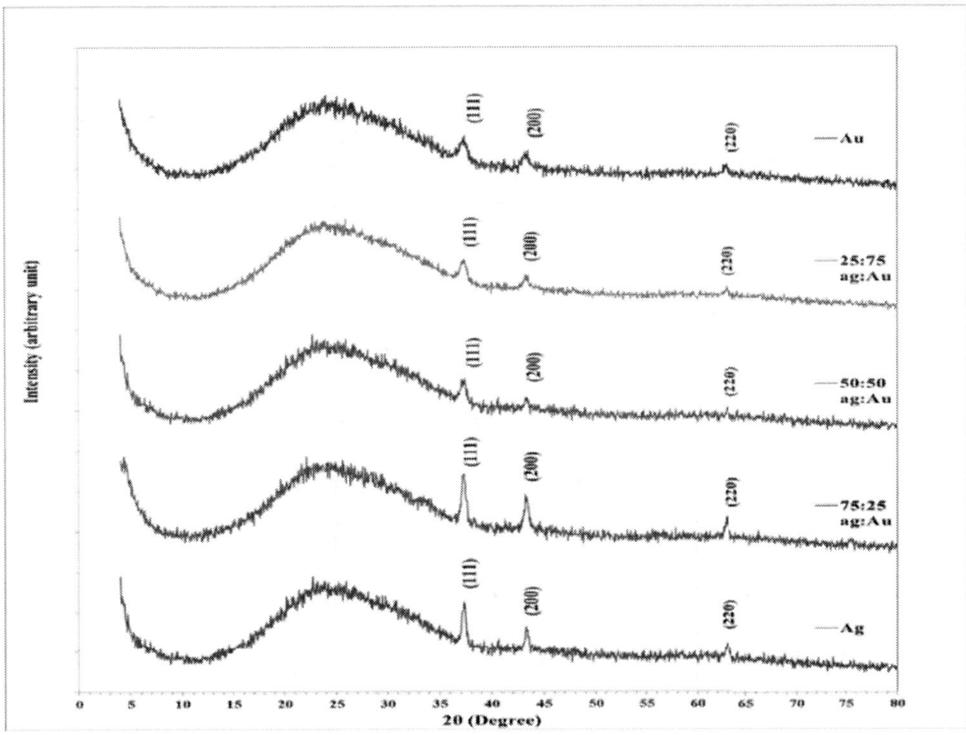

Figure 3: X-ray diffraction patterns for AgNPs, AuNPs and their composite NPs.

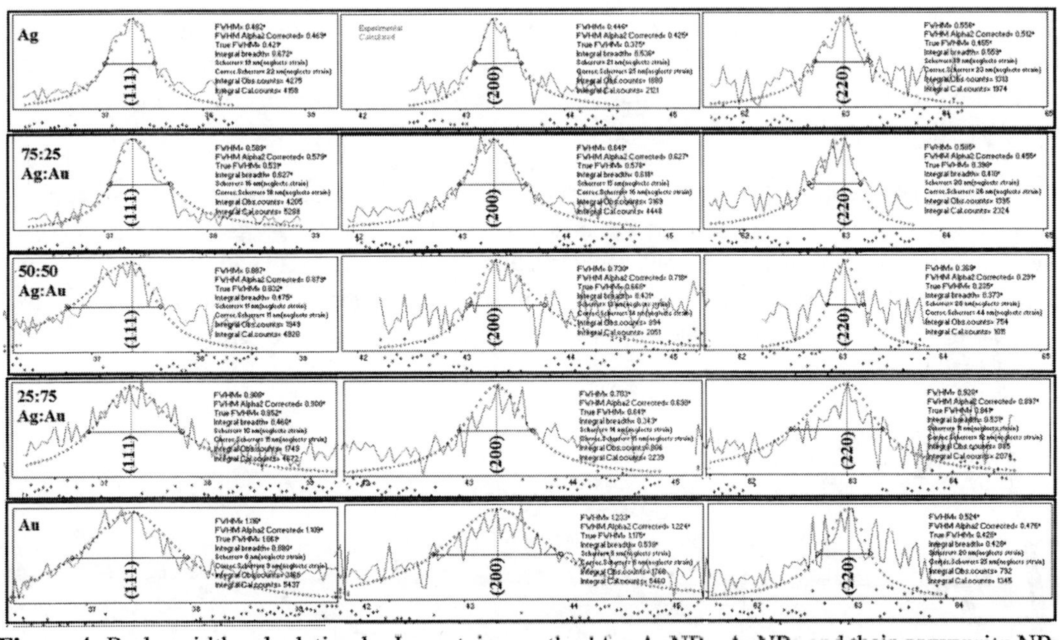

Figure 4: Peaks width calculating by Lorentzian method for AgNPs, AuNPs and their composite NPs.

Table 2: Structural parameters for AgNPs, AuNPs and their composite NPs alloys

Ag: Au ratio %	2θ (Deg.)	d_{hkl} Exp.(Å)	hkl	FWHM (Deg.)	C.S (nm)	Average C.S (nm)
	37.2785	2.4101	(111)	0.4820	17.4	
100:0	43.3584	2.0852	(200)	0.4460	19.2	17.8
	63.0386	1.4735	(220)	0.5560	16.8	
	37.2800	2.4100	(111)	0.5895	14.2	
75:25	43.3600	2.0851	(200)	0.6408	13.3	15.3
	63.0400	1.4734	(220)	0.5055	18.4	
	37.4000	2.4026	(111)	0.8871	9.5	
50:50	43.3200	2.0870	(200)	0.7300	11.7	15.5
	63.0800	1.4726	(220)	0.3693	25.2	
	37.4400	2.4001	(111)	0.9075	9.2	
25:75	43.3300	2.0865	(200)	0.7027	12.2	10.5
	63.0000	1.4743	(220)	0.9204	10.1	
	37.3200	2.4076	(111)	1.1159	7.5	
0:100	43.3200	2.0870	(200)	1.2330	6.9	10.7
	62.9600	1.4751	(220)	0.5240	17.8	

Fig.5 illustrates the UV-visible absorption spectra of the aqueous solution for producing AgNPs, AuNPs and their composite at different percentage ratio prepared by pulsed laser with infrared wavelength of 1064 nm (900 mJ) in distilled water.

The peak appeared at 405 nm is the specific peak of AgNPs plasmon resonance, whereas the peak at 526 nm is the specific peak of AuNPs ,which is due to the excitation of the quantum size confinement effect, which have resonance frequency owing to interaction between electromagnetic wave frequency with electron resonance inside a sub of wavelength dimension. These peaks confirm the formation of NPs in the solution [11].

Increasing the gold ratio leads to shift of absorbance peaks toward higher wavelength in spectra (red shift) corresponds to variation in plasmon resonance. This sift is due to the variation of the nanoparticle material which cause changing its resonance location. This result agree with previous researches reported in Ag: Au nanoparticle composites [12]. Table 3 shows the wavelength values of plasmon resonance peak for Ag: Au nanoparticles. These absorbance peaks can use as a selective absorbance peak for irradiation treatment.

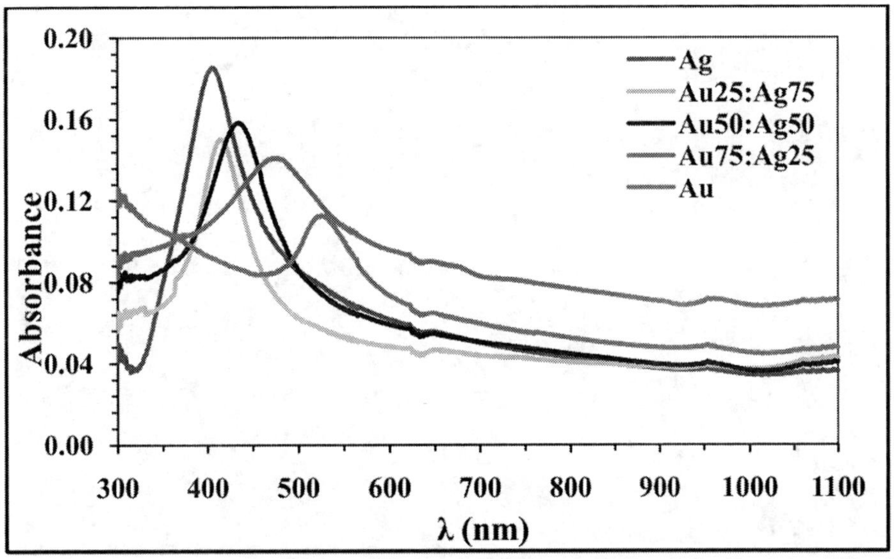

Figure5: UV-VIS spectra for AgNPs, AuNPs and their composite at different Au ratio in distilled water prepared by PLD.

Table 3: Wavelength at maximum absorbance values corresponding to plasmon resonance for AgNPs, AuNPs and their composite at different Au ratio.

Ag: Au ratio	Plasmon peak (nm)
100:0	404
75:25	414
50:50	434
25:75	477
0:100	524

The typical scanning electron microscope images, for Au, Ag, and Au: Ag nanoparticles at three different ratios are shown in Fig.6. The pure Ag sample appeared with spherical shape various in diameter from 60 to 710 nm, for this reason the plasmon peak appears with a broad feature as shown in figure 5. The bimetallic nanoparticles samples appeared as spherical shape with less average diameter. The different in diameters is smaller especially for Ag25:Au75 sample with diameters in the range 224 to 277 nm. The Au NPs sample appeared with the smallest regular shape NPs, with diameter varied from 26 to 91 nm.

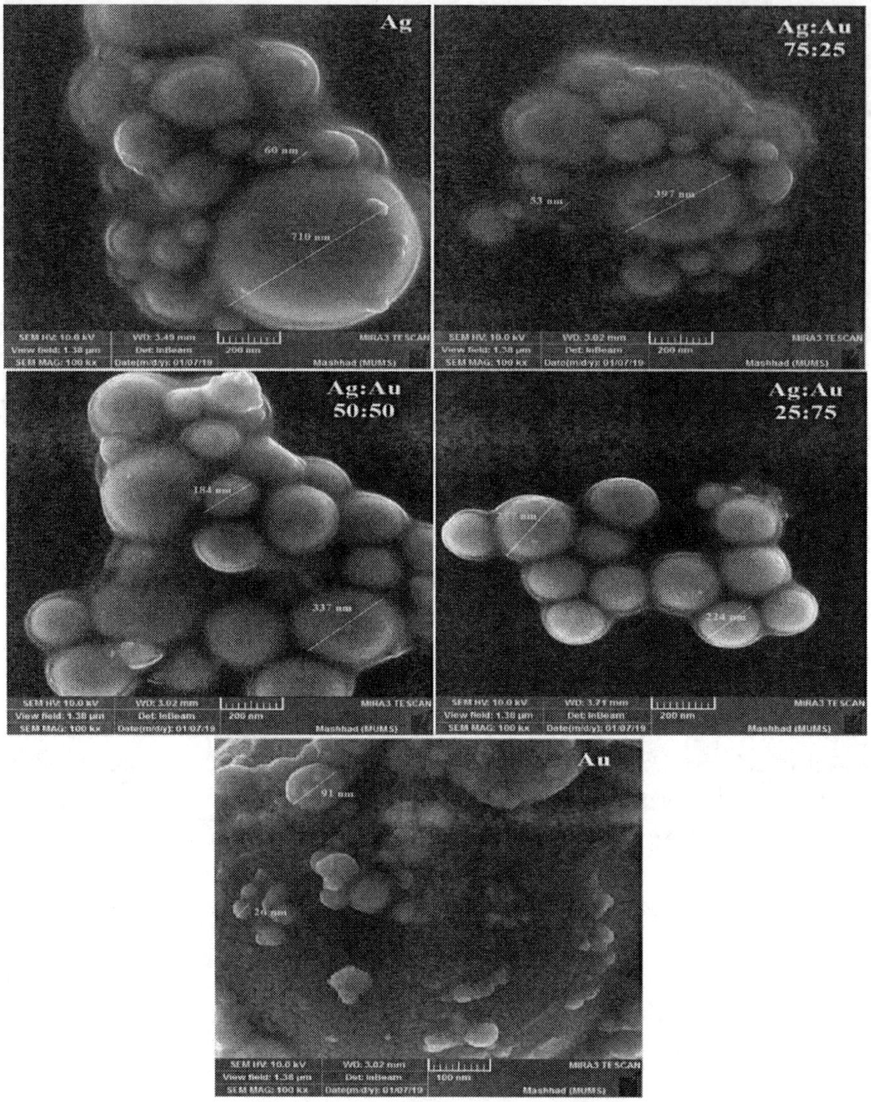

Figure 6 : SEM images for Ag, Au and their alloys nanoparticles at different Au ratio.

4. Conclusions

Synthesize and characterize Ag: Au alloys and their nanoparticles at various ratios using pulse laser ablation in distilled water were investigated. The outcome of this study can be summarized as follows: Elemental analysis of the alloys and nanoparticals show that the result ratios of gold less than theoretical one. Polycrystalline structure with cubic phase is shown for all prepared samples. The crystallite size increased with increasing Au ratio. SEM images indicate the decrement of nanoparticle with decreasing Ag:Au ratio. UV - visible measurements showed tunable resonance peak from 404 to 524 nm by varying the Ag: Au ratio between 100:0 to 0:100.

References

[1] P. Tiwari, K. Vig, V. Dennis, and S. Singh 2011 *Nanomaterials* **1** 1.
[2] R. G. Finke, 2002 *Metal Nanoclusters: Synthesis, Characterization and Applications*. USA: Marcel Dekker, Inc.
[3] Y. Wang, Dong Yan Zhang,Qi Hao,Jie Liu,Yu,Sheng,Shi,Jun Zhu,Li Su and Yan Wang 2018 *Sep. Purif. Technol.* **192**.
[4] N. M. Al-hada, E. Saion, Z. A. Talib, and A. H. Shaari, 2016 *Polymers (Basel).* **8** 113 1.
[5] P. Beukes and M. Maaza 2015 *J. Alloys Compd.* **646** 1043.
[6] P. Wankhede, P. K. Sharma, and A. K. Jha2013 *J. Eng. Res. Appl.*, **3** 6, 1664.
[7] Suh I.-K., Ohta H., Waseda Y. 1988 *Journal of Materials Science* **23** 757.
[8] D.-H. Chen and C.-J. Chen 2002 *J. Mater. Chem.* **12** 5 1557.
[9] W. H. Bragg and W. L. Bragg 1918 *X Rays and Crystal Structure*. London: G. Bell and Sons,LTD.
[10] P. Scherrer 1918 *Göttinger Nachrichten Gesell* **2** 98.
[11] Kh. M. Rashed, M.F. A. Alias and M.M.F. Al-Halbosiy,Inpress in Indian Journal of Public Health Research and Development 2019.
[12] G. Compagnini, E. Messina, O. Puglisi, R.S.Cataliotti, and V. Nicolosi, 2008 *Chem. Phys. Lett.* **457** 386.

IOP Conf. Series: Materials Science and Engineering **757** (2020) 012058 doi:10.1088/1757-899X/757/1/012058

The Role of Tl Substitution on Superconducting Properties for Hg-1223 System

M.F.A.Alias* , B. A. Aljurani, and G.Y.Hermiz
Department of Physics, College of Science, University of Baghdad,
*Corresponding author: may20131313@yahoo.com

Abstract. High temperature superconductors with a nominal composition $Hg_{1-x}Tl_xBa_2Ca_2Cu_3O_{8+\delta}$ for ($0\leq x\leq 0.4$) were prepared by two- step solid state reaction in air. The effect of the substitution of Tl for Hg, oxygen content (δ) and sintering time (20-200h) on the Hg-12223 superconductivity system has been investigated to obtain the optimum conditions for the formation and stabilization of the high critical temperature (T_c).The results showed that the optimum sintering temperature for the prepared samples was equal to 880 °C, whereas the sintering time is equal to 100 h.The quantity of the element content for $Hg_{1-x}Tl_xBa_2Ca_2Cu_3O_{8+\delta}$ system were carried out using x-ray fluorescence (XRF). X-ray diffraction (XRD) analysis showed a tetragonal structure with a high T_c phases (1223) as a dominant phase and low T_c phase (1212) addition to exist of some impurity phases for all samples. It was found that the change of the Tl concentrations of prepared samples produced a change in the lattice parameters, density ρ_m and c/a ratio. The highest T_c for $HgBa_2Ca_2Cu_3O_{8+\delta}$ composition was 118 K. The results for the doped samples, represented highest T_c equal to 124 K for $Hg_{1-x}Tl_x Ba_2Ca_2Cu_3O_{8+\delta}$ sample with x=0.4.The oxygen content was not systematic changed but it played a great role in the evaluation of T_c.

Keywords: Hg-1223, Tl substitution effect, Structural properties, HTSC

1. Introduction

After the discovery of high temperature superconductivity in Bi-Sr-Ca-Cu-O (BSCCO) system [1] Putilin et al.[2]in 1993 was first observed superconducting in mercury-based cuprates $HgBa_2Ca_{n-1}Cu_nO_{2n+2+\delta}$ [n: the number of consecutive Cu-O layers] with the compound $HgBa_2CuO_{4+\delta}$ (Hg-1201, T_c= 94 K). The compound extended to a homologous series of Hg-1212 (n=2), Hg-1234 (n=4) and Hg-1245 (n=5) having T_c= 120, 126 and 101 K, respectively [2, 3]. In the same year the highest critical temperature [n=3, T_c =133 K] among oxide superconductors has been reported by Schilline et al.[3].Application of pressure at 29 GPa [4] can increase the critical temperature of the second member to 154K and that of the third to 164K at 31 GPa [4,5]. While partial substitution of Hg^{+2} by ions of various radii and valance such as Tl^{+3} in $HgBa_2Ca_2Cu_3O_{8+\delta}$ compound may affect the high temperature superconductore HTSC phase formations, chemical stability and superconducting properties[6].
Die et al. [7] showed that the transition temperature T_c increased to 138 K for a nominal composition of $Hg_{0.8} Tl_{0.2}Ba_2Ca_2Cu_3O_{8.33}$, while Wagner et al. [8] obtained T_c around 135 K and δ = 0.18 for Hg-1223 .They studied the tetragonal phase transformation by neutron powder diffraction technique founding that the lattice parameter was around a = 3.8478Å and c = 15.7782Å. Whereas Pandy et al. [9] synthesized the Tl - doped Hg - based HTSC bulk material corresponding to $Hg_{1-x}Tl_xBa_2Ca_2Cu_3O_{8+\delta}$ with x = 0.2, 0.4, 0.5, 0.6 and 0.8. They found the variation of T_c from ~ 106 to ~ 133 K for under doped samples and from ~ 106 to ~ 129 K for over doped samples. Microstructure investigations revealed the presence of ordered and /or disordered stacking faults and the accompanying poly type like regions.
Giri et al. [10] studied structural / microstructures variations of the as synthesized $(HgTl_{0.2}Bi_x)Ba_2Ca_2Cu_3O_{8+\delta}$ HTSC phases with x = 0.05, 0.1, 0.15) and different nominal compositions employing transmission electron microscopic (TEM) technique. The TEM revealed the presence of defect

substructure and staking faults. The as synthesized samples exhibited stability and have T (onset) of 133 K. They found that the phase $(HgTl_{0.1}Bi_{0.1})$-1223, had the highest density of ordered defect substructure and exhibited the highest integrain critical current density (J_c= $6.2\times10^6 A/cm^2$). Powder diffraction data for high temperature superconducting compound $Hg_{0.8}Tl_{0.2}Ba_2Ca_2Cu_3O_{8+\delta}$ at pressure range 0–20 GPa and temperature interval 100–300 K were studied by S. Titova et al. [11]. They found that Ba-atoms do not shift towards HgO_x-planes and splitting of BaO-planes do not decrease with external pressure.

The effect of neutron irradiation on the microstructure properties of $Hg_{1-x-y}Tl_xPb_yBa_2Ca_2Cu_3O_{8+\delta}$ for $0\leq x\leq 0.4$ and $0\leq y\leq 0.4$ were investigated by Ghazala et al.[12]. They found that the microstructure for most samples before irradiation was to be dense and the morphology of these samples shows needle – like and plate –like layered for different compositions. While irradiation with fast and thermal neutrons induced more voids and defect in most of the samples.

In this paper, we studied the effect of Tl substitution on the superconducting properties such as electrical resistivity, transition temperature and structure of HBCCO system.

2. Experimental work

Appropriate weights of pure materials Hg_2O, $BaCO_3$, CaO, CuO, and Tl_2O_3 in proportion of their molecular weights were used in order to prepare the samples by two step solid state reactions. During the first step, mixing the oxides, and carbonates of Ca, Cu, and Ba developed $Ba_2Ca_2Cu_3O_8$ precursor. The mixture homogenization takes place by adding a sufficient quantity of 2-propanol to form a paste, during the process of grinding for about (50-60) min. Measuring the weight of the dried mixture (w1), and put it in an alumina crucible, calcinied in a tube furnace in air that has programmable controller type (Eurptherm818) for 24 hours at 1073K with a rate of 275K / min. In the second step, the $Ba_2Ca_2Cu_3O_8$ precursor was regrind again and mixed with Hg_2O and $Tl_2\,O_3$ to obtain the nominal compositions $Hg_{1-x}Tl_xBa_2Ca_2Cu_3O_{8+\delta}$, where x varied from 0 to 0.4. This mixture was then pressed into pellets 1.3 cm in diameter and (0.2 – 0.3) cm thick, using hydraulic type (SPECAC), under pressure of 0.7GPa.

The pellets were pre sintered at 1153K for 100 h with a rate of 393K /h and then cooled to room temperature by the same rate of heating. The x-ray fluorescence (XRF) was used to determine the quantity of the element content for $Hg_{1-x}Tl_xBa_2Ca_2Cu_3O_{8+\delta}$ system. The structure of the prepared samples was obtained by using X-ray diffractometer (XRD) type (Philips) with the $Cu_{K\alpha}$ source. A computer program was established to calculate the lattice parameters a, b, and c. The program is based on Cohen's least square method [13].

Four probe dc method at temperature range (77-300) K was used to measure the resistivity (ρ) and to determine the critical temperature (T_c). The amount of oxygen content (δ) was determined from a simple chemical experiment method called iodometric titration[14] for the samples $HgBa_2Ca_2Cu_3O_{8+\delta}$, and $Hg_{1-x}Tl_xBa_2Ca_2Cu_3O_{8+\delta}$ with different nominal composition. The extra oxygen content (δ) is directly associated with the average oxidation state of copper ($v_{av\ Cu}$). Assuming $v_{Hg} = v_{Ba} = v_{Ca} = +2$, $v_O = -2$, one can obtain for Hg – based superconducting cuprates the following correlation between δ and $\mathbf{v_{av\ Cu}}$[15,16]:

$$\mathbf{v_{av\ Cu} = (2 / n) \times (n + \delta)} \text{.........(1)}$$

where n is the number of CuO_2 planes. Density (ρ_m) of unit cell of the samples were calculated using the following equation [17]:

$$\rho_m = \frac{W_m}{N_a V} \quad \text{......................(2)}$$

where W_m is the molecular weight, N_a is Avogadro's number and V is the volume of the unit cell.

3. Results and Discussion

The quantity test of the element content for $Hg_{1-x}Tl_xBa_2Ca_2Cu_3O_{8+\delta}$ system were carried out by XRF at different nominal composition (x= 0 and 0.2) which sintered in air .These data illustrated in Figs. (1a and b) which showed that the peaks belonging to the superconducting phase consist of Hg-Ba-Ca-Cu-O,and Hg-Tl-Ba-Ca-Cu-O, respectively. This result agree with the results reported by Xu et al.[18].

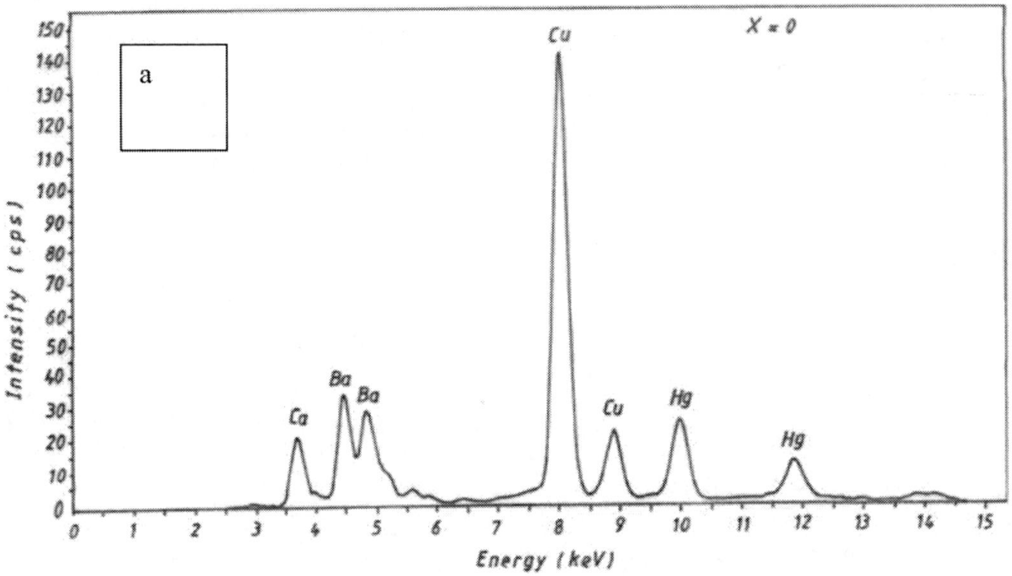

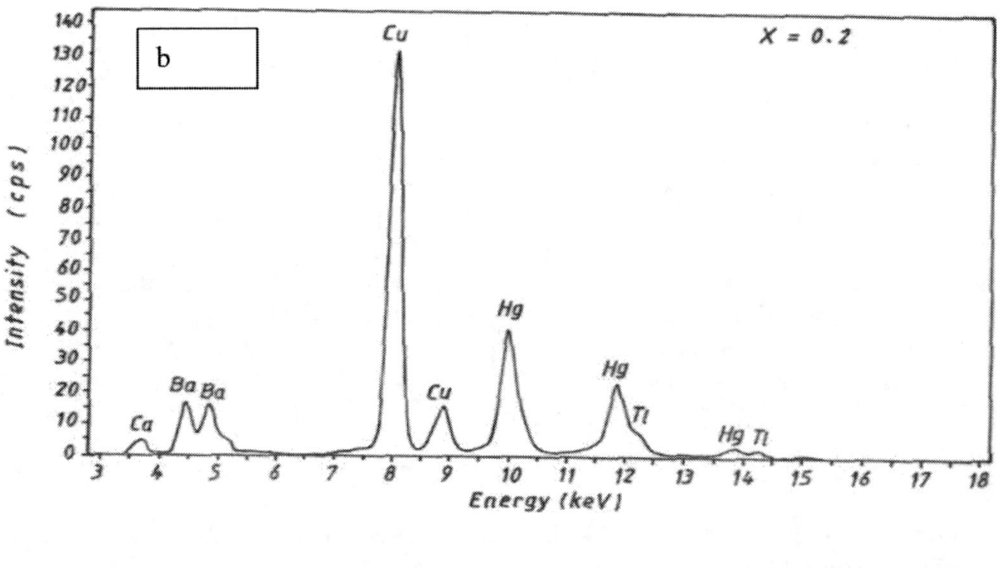

Figure 1. X-ray fluorescent patterns for $Hg_{1-x}Tl_xBa_2Ca_2Cu_3O_{8+\delta}$ samples with (a) x = 0 and (b) x = 0.2 .

A typical XRD pattern for the Tl free samples $HgBa_2Ca_2Cu_3O_{8+\delta}$ prepared by solid state reaction in air method is shown in Fig.(2), indicating the dominant phase of Hg-1223 together with amount of low-phase Hg-1212 and Hg -1234 with some impurity phase like CuO at 38.8^{0}and $CaHgO_2$ at 15.6^0 and 31.2^0

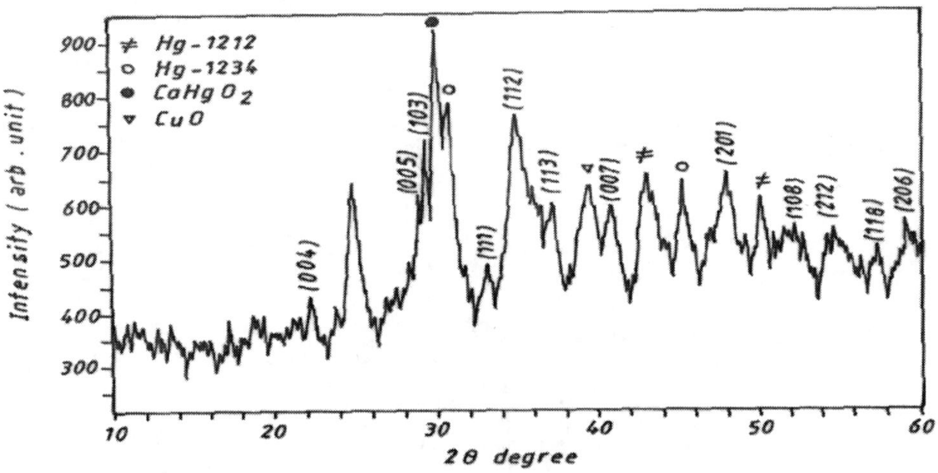

Figure 2.X-ray diffraction patterns for $Hg_{1-x}Tl_xBa_2Ca_2Cu_3O_{8+\delta}$ sample for x = 0 .

X- ray diffraction patterns for $Hg_{1-x}Tl_xBa_2Ca_2Cu_3O_{8+\delta}$, samples in Figs. (3a-c) revealed the presence of tetragonal Hg,Tl-1223 structural phase as the dominant phase .However, some peaks attributable to impurities e.g. $CaHgO_2$ and CuO and mercury phase e.g.Hg-1212,1234 (low phases) are also observed in Fig.(3c). Our results presented are in agreement with Dai et al.[7] and Pandey et al.[9] results. The lattice parameters of the $(Hg_{0.8}Tl_{0.2})$-1223 were slightly less than that of the pure Hg-1223 sample. Also they revealed that samples have a multiphase with the presence of Hg,Tl-1223 as the dominant phase. The addition of thallium introduces more oxygen atoms and stronger bonding into HgO planes, raising the vibration frequency of oxygen atoms and hence stabilizing the crystal structure of the Hg-1223 phase [17,19]. The difference in lattice parameter, c/a ratio, and density of unit cell, were illustrated in Table (1) and shown in figures. (4a and b), that are obtained from XRD patterns. Table (1) showed that the Hg-1223 has tetragonal structure with the lattice parameter a=3.851 Å ,c=15.813 Å of Hg-1223 , which affects the volume of the unit cell and then causes an increase of the density from 6.163 to 6.210 g/cm^3.

IOP Conf. Series: Materials Science and Engineering **757** (2020) 012058 doi:10.1088/1757-899X/757/1/012058

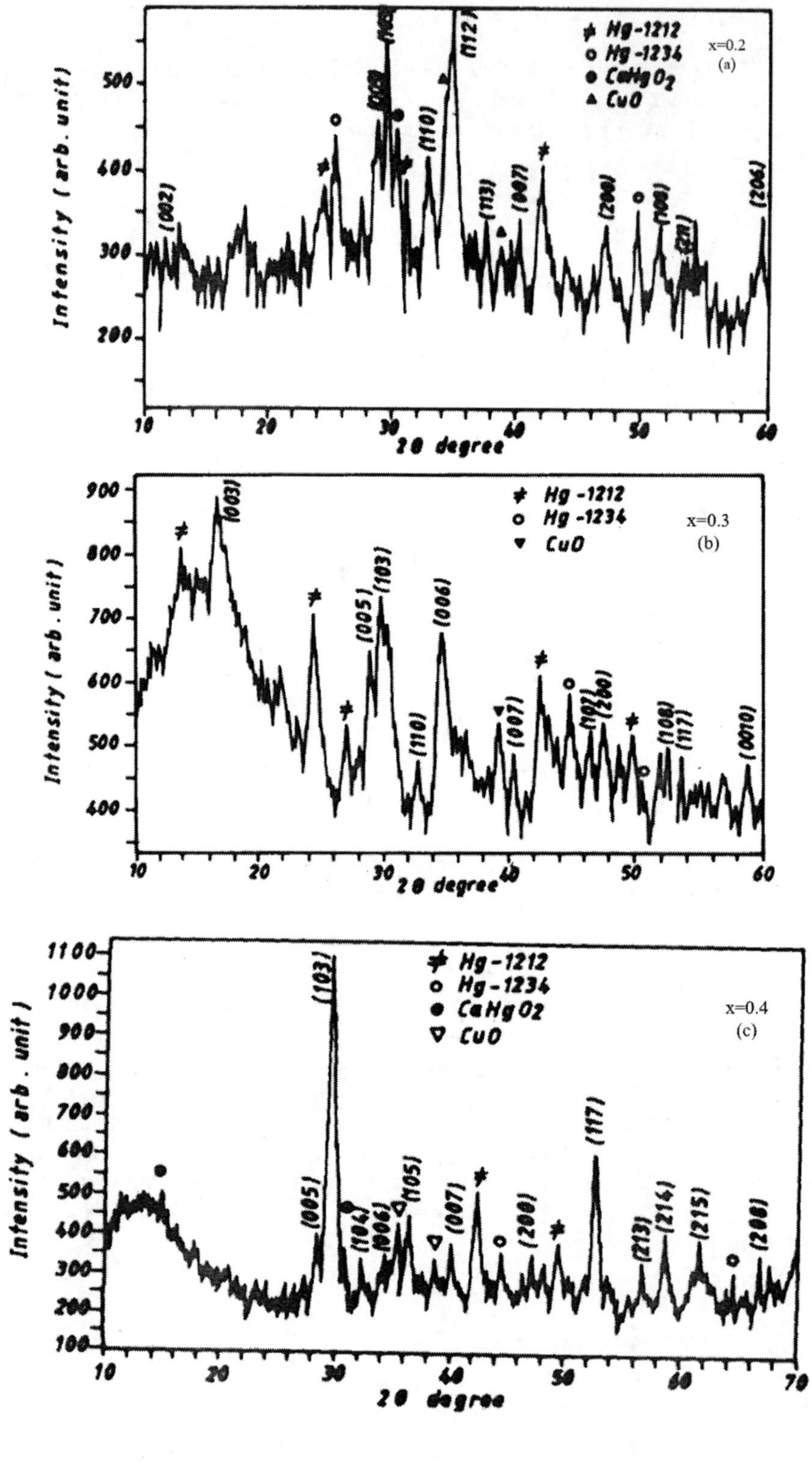

Figure 3. X-ray diffraction patterns for $Hg_{1-x}Tl_xBa_2Ca_2Cu_3O_{8+\delta}$ samples at (a) x = 0.2 (b) x =0.3 (c) x = 0.4 .

Table 1 .Lattice parameter a, c, c/a ratio, density (ρ_m) and volume of unit cell at different composition of $Hg_{1-x}Tl_xBa_2Ca_2Cu_3O_{8+\delta}$.

x	a (Å)	c (Å)	c/a	ρ_m (g/cm^3)	V (Å)^3
0.0	3.851	15.813	4.106	6.163	234.509
0.1	-	-	-	-	
0.2	3.847	15.765	4.097	6.226	233.312
0.3	3.842	15.786	4.108	6.236	233.016
0.4	3.849	15.801	4.105	6.210	234.088

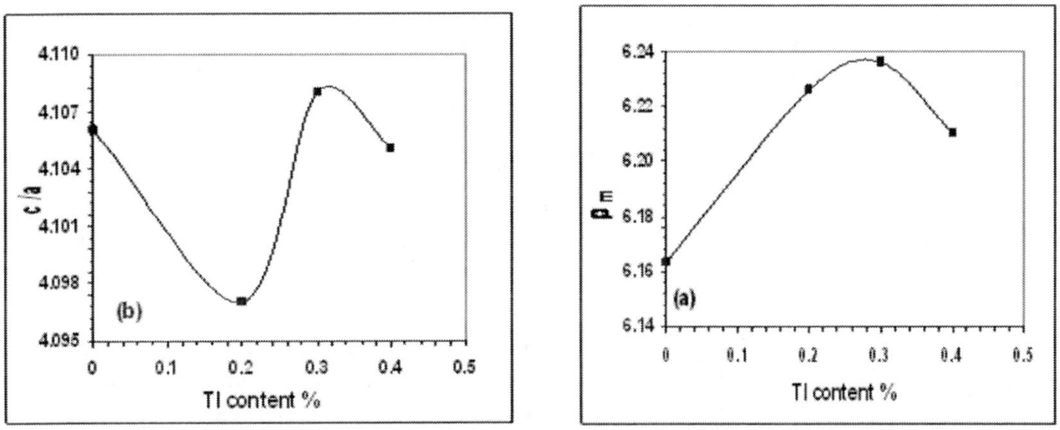

Figure 4. Variation of (a) density of unit cell,(b)c/a ratio as a function of Tl concentration for $Hg_{1-x}Tl_xBa_2Ca_2Cu_3O_{8+\delta}$ samples sintered in air.

In order to clarify the effect of sintering time on the electrical d.c. resistivity and the critical temperature, the samples of nominal composition $HgBa_2Ca_2Cu_3O_{8+\delta}$, and $Hg_{0.6}Tl_{0.4}Ba_2Ca_2Cu_3O_{8+\delta}$, were sintered in air at 880^{0}C for periods of (20-200)h, as shown in Figs.(5 a and b). It is found from these figures that prolonged sintering time up to 100 h could improve the superconductor behavior and the T_c of stoichiometric composition. Also it raised T_c from 100 K to 118 K for the composition x= 0 as the time increases from 80 h to 100 h, at the same sintering temperature , while T_c increased to 124K for the composition of x=0.4. The increased values of T_c could be attributed to the increase of contact areas between grains during the sintering process and in other words, decrease of the porosity. Another feature was found that when the sintering time increase above 100h (140h and 200h) the critical temperature value decrease for all the prepared samples .

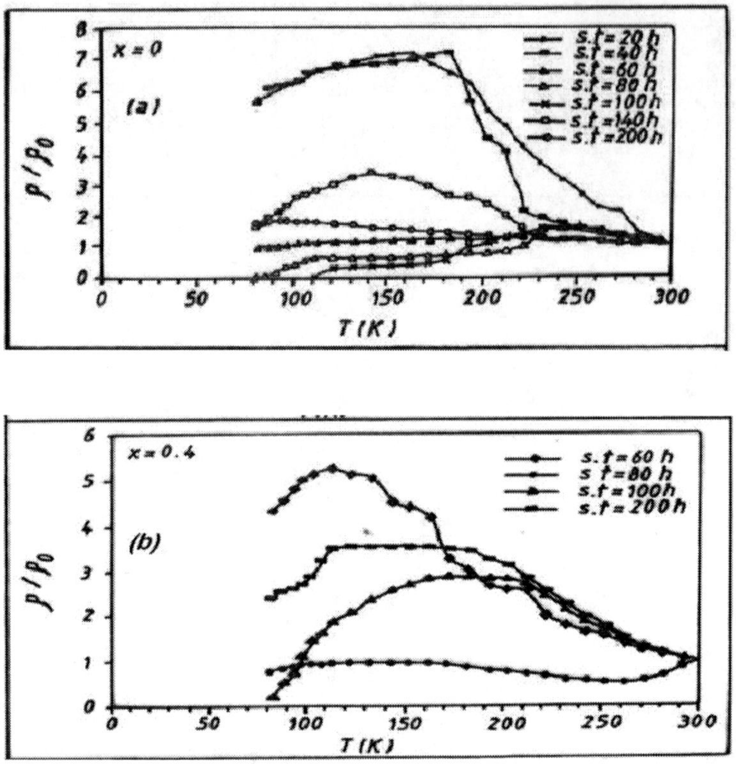

Figure 5. Temperature dependence of normalized resistivity for $Hg_{1-x}Tl_xBa_2Ca_2Cu_3O_{8+\delta}$ with (a)x=0 (b) x=0.4 sintered in air at different periods of time.

It is noticed from Fig. (6) and Table (2) that the $Hg_{0.6}Tl_{0.4}Ba_2Ca_2Cu_3O_{8+\delta}$ sample has the highest T_c of 124 K. Pandey et al.[9] found that T_c increased when Tl concentration increased for the as-synthesized samples .This result is in agreement with our result of the samples as shown in Fig.(7) which indicates that the variation of transition temperature with Tl content for $Hg_{1-x}Tl_xBa_2Ca_2Cu_3O_{8+\delta}$ samples (x=0.2,0.3 and0.4).One can observe that Tl content plays a major role on the formation of high T_c-phase.

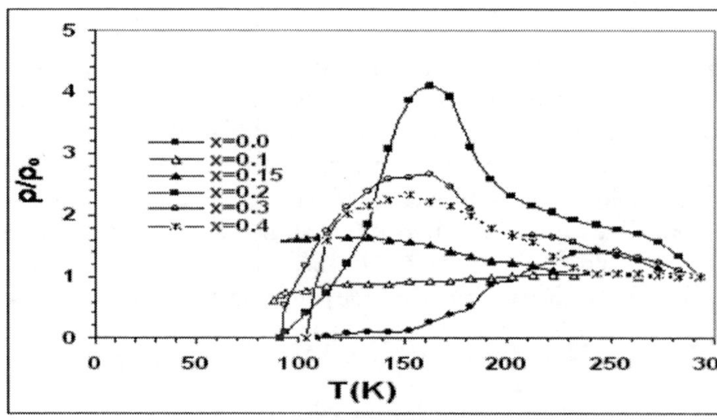

Figure 6. Temperature dependence of normalized resistivity of $Hg_{1-x}Tl_xBa_2Ca_2Cu_3O_{8+\delta}$ at different x value.

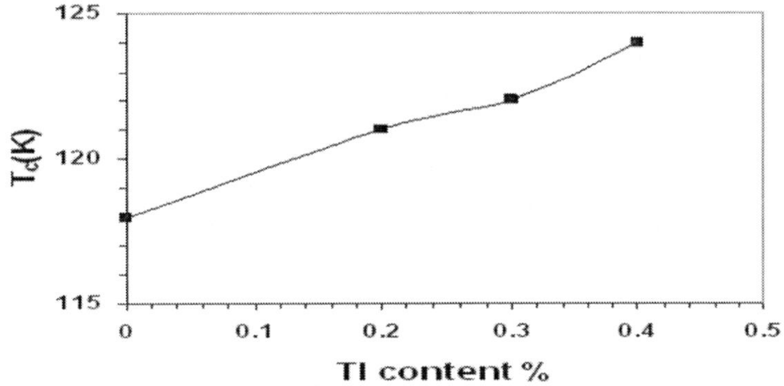

Figure 7. Transition temperature as a function of Tl content for $Hg_{1-x}Tl_xBa_2Ca_2Cu_3O_{8+\delta}$ samples.

Table 2. Oxygen content, the average Cu valence and transition temperature for different composition of $Hg_{1-x}Tl_xBa_2Ca_2Cu_3O_{8+\delta}$.

X	δ	v_{Cu}	T_c(K)
0.0	0.30	2.20	118
0.1	0.38	2.25	<77
0.2	0.42	2.28	121
0.3	0.44	2.29	122
0.4	0.48	2.32	124

The relationship between transition temperature T_c and oxygen content δ for $Hg_{1-x}Tl_xBa_2Ca_2Cu_3O_{8+\delta}$ with $0 \leq x \leq 0.4$ can be seen in Table (2). It is to be noted from this Table that δ=0.48 for the sample with x=0.4 to have the highest T_c of 124 K is with optimum carrier concentration.

In pure Hg-1223, the interstitial oxygen occupancy δ=0.30 led to a mean value of copper valency +2.20 (see Table 2) .Because of the +2 valency of Hg, all the oxygen sites was not filled as compared to Tl-1223 compound. The O (4) site is partially filled with interstitial oxygen. Due to this fact, there is a great flexibility in the HgO with regard to the introduction of extra oxygen atoms through oxygenation or through introduction of other suitable elements with higher number of associated oxygen atoms in place of Hg. Thus, substituting Hg^{+2} by Tl^{+3} will bring in greater number of associated oxygen atoms in the metal oxide HgO layer, hence disturbing the charge balance. In an analogous manner to Tl-1223 compounds, this will lead to a change in doping level of CuO_2 layer by creating extra holes to offset the charge misbalance in HgO layer [7].

Fig. (8) demonstrated the variation of the average Cu valance obtained by equation (1) and oxygen content as a function of Tl concentration for $Hg_{1-x}Tl_xBa_2Ca_2Cu_3O_{8+\delta}$ samples sintered at 100h. One can observe that the average Cu valance and oxygen content increase with increasing Tl content.

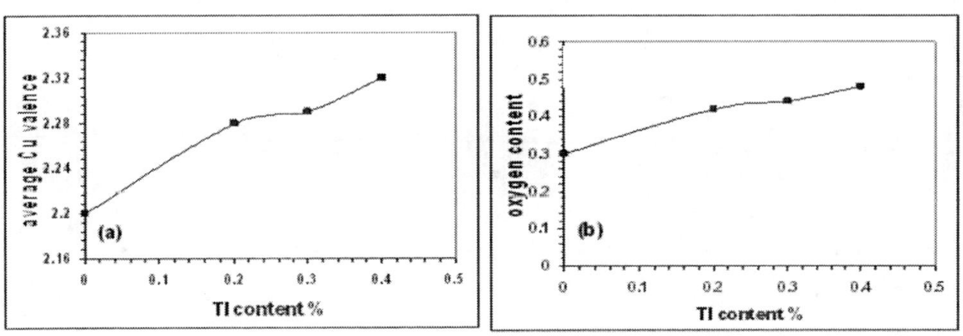

Figure 8. (a)Variation of average Cu valence v_{Cu} and (b) oxygen content δ as a function of Tl content for $Hg_{1-x}Tl_xBa_2Ca_2Cu_3O_{8+\delta}$ samples sintered at 100h.

IOP Conf. Series: Materials Science and Engineering **757** (2020) 012058 doi:10.1088/1757-899X/757/1/012058

4. Conclusions

The conclusion from the results for preparing $Hg_{1-x}Tl_x\ Ba_2Ca_2Cu_3O_{8+\delta}$ for $0\leq x \leq 0.4$ system can be summarized as follows:$HgBa_2Ca_2Cu_3O_{8+\delta}$, and $Hg_{1-x}Tl_x\ Ba_2Ca_2Cu_3O_{8+\delta}$ for $0\leq x \leq 0.4$ synthesizing with high-T_c phase and the optimum sintering time was 100 h .The substitution of Tl in Hg for the compound $Hg_{1-x}Tl_x\ Ba_2Ca_2Cu_3O_{8+\delta}$ with x=0.4 has maximum value of $T_c \approx 124$ K .X-ray diffraction analyses for all samples have shown tetragonal structure and there is a change of lattice constants with the increase in Tl concentration.

References

[1] H. Maeda, Y. Tanaka, M. Fukutomi and T. Asano 1988 *Jpn.J.Appl.Phys.***27** L209.
[2] S.N.Putilin, E.U.Antipov, O.Chmaissem and M.Marezio, 1993 *Nature* **362** 226.
[3] A.Sehilling, M.Cantoni, J.D.Gou and H.R. Ott 1993Nature **363** 56.
[4] L.Jansen and R. Block 1994 *Physica A* **212** 143.
[5] I.Kirschner,R.Laiho,P.Lukács,A.C.Bodi,Dimitrova- Lukacs,E.Lähderanta and G.Zsolt 1996 *Phys.B* **99**,501.
[6] J.Z,Liu,C.Chang,M.D.Lan,P.Klavins and P.N.Shelton 1994 *High-T_c* **8** 3.
[7] P.Dai,B.C.Chakoumakes,G.F.Sun,K.W.Wong,Y.Xin and D.F.Lu 1995 *Physica C* **243** 201.
[8] J.L.Wagner,B.A.Hunter,D.G.Hinks and J.D.Jorgensen 1995 *Phys.Rev.B* **15** 15407.
[9] A.K.Pandey,G.D.Verma and O.N.Srivastara 1998 *Physica C* **306** 47.
[10] R.Giri,G.D.Verma,R.S.Tiwari and O.N.Srivastava 2003 *Cryst.Res.Technol* **38** 760.
[11] S.Titova,S.Pryanichnikov,V.Balakirev,Y.Ohishi and T.Adchi 2009 *Phys:Conf.Ser.***150**.
[12] G.Y.Hermiz, M.F.A.Alias and B.A.Aljurani 2017 *Indian Journal of Natural Sciences* **8** 12642.
[13] I.F.Ferguson and A.H.Rogrson: 1984 *Computer Phys.Communication* **32**.95.
[14] D.A.Paviov 2004 *Synthesis and Properties of Substituted Hg-Based Superconductors* PhD Stockholm University, Department of Inorganic Chemistry.
[15] H.M.Sao, C.C.Lam, P.C.W.Fung, X.S.Wu, J.H.Du, G.J.Shen,J.C.L.Chow,S.L.Ho, K.C.Hung and X.X.Yao 1995 *Physica C* **246** 207.
[16] V.Z.Kresin and S.A.Wolf 1990 *Fundemantals of superconductivity* Plenum press, New York.
[17] Q.L.Xu, F.Foong,S.H.Liou,L.Z.Cao and Y.H.Zhang 1997 *Supercon.Sci.Technol.***10** 218.
[18]A.Manthiram , J.S.Swinnea, Z.T.Sui,H.Steinfink and J.B.Goodenough 1987 *J.Am.Chem.Soc.* **109** 6667.
[19]A.K. Jassim, M.N.Makadsi and M.F. Alias 2010 *Iraqi J. Phys*. **8** 11 53.

IOP Conf. Series: Materials Science and Engineering **757** (2020) 012059 doi:10.1088/1757-899X/757/1/012059

Preparation and characterization of equiatomic NiTi shape memory alloy

M A Mohammed[1], A A Aljubouri[1], S H Mohammed [1*]
[1]Department of applied sciences, University of Technology, Baghdad, Iraq
[*] Corresponding author e-mail: safa_ha_90@yahoo.com

Abstract .The present study investigated the vacuum induction melting of equiatomic NiTi (Ni 50 at. %, Ti 50at. %) shape memory alloy using graphite crucible. The cast alloy heat-treated at 865 °C for 15 mints, followed by quenching the sample in icy water. The microstructure and micro hardness of alloy were investigated before and after heat treatment by using an optical microscope and Vickers micro hardness tester. Transformation temperatures upon heating/ cooling were investigated by using differential scanning calorimeter (DSC). Scanning electron microscope (SEM) supported with energy dispersive X-ray (EDX) and X-ray diffraction analysis (XRD) were carried out to investigate the chemical composition and the phases present in the alloy after heat treatment. The results show that the vacuum induction melting process in a graphite crucible provides binary NiTi shape memory alloys with good chemical homogeneity. The microstructure becomes more homogenous and the Vickers's micro hardness increases from 186.06 to 238.74 after heat treatment with highly controlling on the composition of the alloy. The DSC curve shows that the R- phase exists upon the heating and cooling process. In addition the precipitate phase Ti_2Ni was also exist.

Keywords: equiatomic NiTi alloy, NiTi shape memory alloys, DSC, XRD.

1. Introduction

Shape memory alloy is a type of smart material that has two distinct properties such as superelasticity (SE) and superior thermal shape memory effect (SME). SMAs are the materials that can achieve high deformation and regain the original shape upon unloading or after heating above a certain temperature. Shape memory effect refers to the phenomena where the SMA will automatically come to its original position after heating and super elasticity effect is where SMAs may undergo non-linear deformation and comes back to its original shape upon unloading. Due to these amazing properties, SMA's have found various applications in the biomedical field as well as in industrial sectors such as for making fasteners, coupling, actuators, etc., [1-3]. Shape memory behavior detected in 1965 by Buehler and Wiley of the U.S. Naval Ordnance Laboratory (NOL), received a United States patent for a group of NiTi alloys, which is called 55-Nitinol. These compounds have chemical compositions in the range of (53-57) wt. %, corresponding to a (48-52) at. %, which implies they are near equiatomic alloys. NiTi shape memory alloys known as (Nitinol) alloys according to their chemical element symbols and the Naval Ordnance Laboratory acronym [4]. NiTi Shape memory alloys contain two phases, first is the martensite phase which is stable at low temperatures formed by rapid cooling is defined according to the German physicist Adolf Martens, and it has an eccentric and highly crystalline phase, the second phase is the

IOP Conf. Series: Materials Science and Engineering **757** (2020) 012059 doi:10.1088/1757-899X/757/1/012059

Austenite phase which is stable at high temperatures, which is named according to the English physicist Charles Austen; it has a face-centered cubic structure geometry [4]. In near equiatomic Nitinol alloys, the thermo elastic martensitic transformation between B2 austenite phase and B19′ martensite phase gives an increase to the shape memory effect (SME) and super elasticity (SE). Furthermore, high mechanical strength and ductility are noticed in Nitinol alloys [5, 6]. Vacuum induction melting (VIM) is a common method used for producing NiTi alloys, because of the electrodynamics' forces that give excellent stirring of the melt and produce a greater chemical and microstructural homogeneity in the alloy [7-9]. This experimental work deals with, preparation of equiatomic NiTi ($Ni_{50}Ti_{50}$) shape memory alloy using vacuum induction melting method, and investigates the phase transformation temperature, micro hardness, and microstructure of binary NiTi shape memory alloy.

2. Experimental work

2.1 Materials

Nickel plate with high purity 99.9 wt. % and titanium wire 98.66 wt. % were used to produce NiTi alloys. The constituent elements of the alloy, titanium wire and nickel plate were dipped in acetone and alcohol in an ultrasonic bath, then they were rinsed with distilled water and dried by pressurized air before melting.

2.2 Melting and Homogenization

The alloy composition was equiatomic NiTi alloy 50 % at. Ni and 50 % at. Ti (which is equivalent to 55 % wt. Ni and 45 % wt. Ti). The ingredient elements nickel and titanium are charged in a graphite crucible and melted in high-frequency induction furnace type (EQ-SP-25TC-15AB) made in the U.S.A. the furnace evacuated to 10^{-3} mbar and after that purged with high purity argon. The melting process was carried out under argon atmosphere. After that, the melting process repeated three times to obtain a good homogenization of the chemical composition. The sample was heat-treated at 865 °C for 15 min, in a programmable furnace. Followed by quenching the samples in icy water. The sample were poured with dimensions (25mm×20mm), and then it is sectioned into same dimensions for different examinations. Differential scanning calorimetry (DSC) produced by SETARAM, type 131 EVO, made in French. A different temperature range from (-100°C – 300 °C) with heating/cooling rate 10°C/min in nitrogen atmosphere was used to investigate the thermal behavior of prepared alloy. Scanning electron microscope SEM supported with the energy dispersive x-ray analysis unit EDX model (VEGA3LM), is used to investigate the chemical compositions of samples. The analysis of the different phases of the heat-treated sample, x-ray diffraction test was carried out using Shimadzu device, made in japan. Microstructure study was carried out at various stages of processing. The sample was grinded and polished followed by etching using (10 ml HF, 20 ml HNO_3, and 30 ml H_2O) solution for 10 sec. Optical microscope with magnification of (4X) was used to investigate the microstructure of samples. For hardness measurements, the test was carried out at room temperature using micro hardness Vickers tester type (Laryee model HVS-1000) made in china, under 300 gf. load for holding time 10 sec.

3. Results and discussion

Figure 1 shows the DSC profile of binary $Ni_{50}Ti_{50}$. The transformation temperatures, austenite start temperature (As), austenite finish temperature (Af), martensite start temperature (Ms), martensite finish temperature (Mf), R- phase start temperature (Rs) and R- phase finish temperature (Rf) are given in (Table 1). The alloy was found to undergo a multi-stage martensitic transformation. There are four peaks in the DSC curve, two exothermal peaks appear during the cooling process which refers to the transformation from the austenite phase (B2) to the R- phase and from R- phase to martensite phase (B19'), respectively [10]. Also, two endothermic peaks are revealed during heating which indicates the transformation from the martensite phase (B19') to the

R- phase and from the R- phase to the austenite phase (B2), respectively [10]. This phenomenon is different from most of the results that characterized in previous works of literature, where R-phase transformation occurs only during cooling [1]. This behavior may be predicated to the formation of coherent Ti_2Ni precipitates, which produces local stress fields [12, 13]. This transformation behavior is most alike to that presented by Niraj N. et al [10].

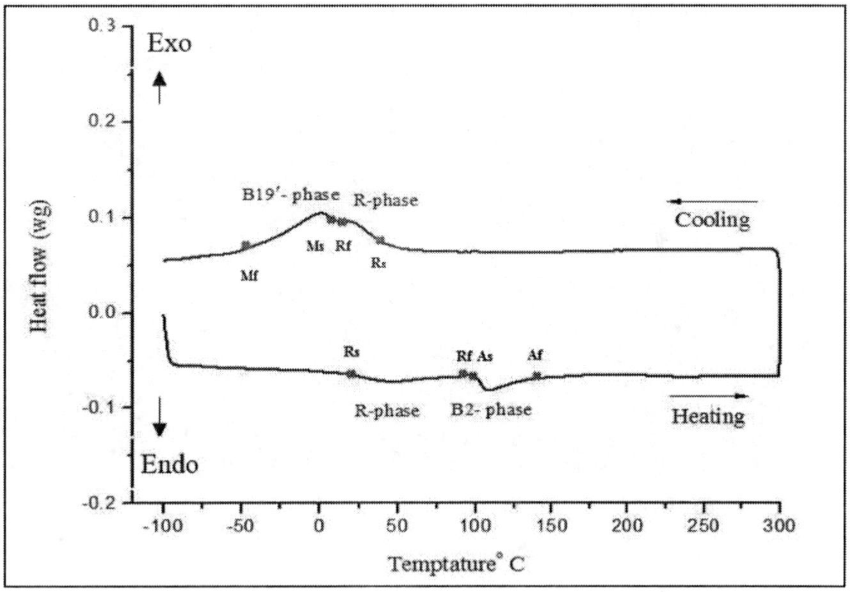

Figure 1. DSC profile of heat treated $Ni_{50}Ti_{50}$ shape memory alloy.

Table 1. The transformation temperatures of heat treated $Ni_{50}Ti_{50}$ shape memory alloy.

	Transformation temperatures °C					
	Martensite phase		R- phase		Austenite phase	
	start	finish	Start	finish	start	Finish
Heating	-	-	14.4	89.5	100.1	137.2
Cooling	7.6	-42.6	39.7	12.7	-	-

Figure2 show the microstructure of the $Ni_{50}Ti_{50}$ alloy by SEM analysis, the microstructure of the alloy consist of NiTi phase (matrix) presented by dashed pointer and precipitates of intermetallic compound of Ti_2Ni phase presented by solid pointer. Table 2 show the microanalysis, by energy dispersive X-ray spectrometry, of different phases found in the microstructure of the investigated alloy that presented in figure 2. It can be noticed that, the matrix has approximately the same content of Ni and Ti. The matrix phase has the composition of Ni 49.8% at. And Ti 50.2% at. , (which is equivalent Ni 54.9 % wt. and Ti 45.1 % wt.), which is very close to the percentage of Nickel and Titanium elements that are mixed before melting, these results indicate to the successful mixing process during melting. Ti_2Ni phase has higher Ti percentage, almost double,

IOP Conf. Series: Materials Science and Engineering **757** (2020) 012059 doi:10.1088/1757-899X/757/1/012059

than Ni in its precipitates. Ti content in the latter phase reaches around, 68.28 % at while Ni content is about 31.72 % at (equivalent to Ti 63.71%wt and Ni 36.29%wt).

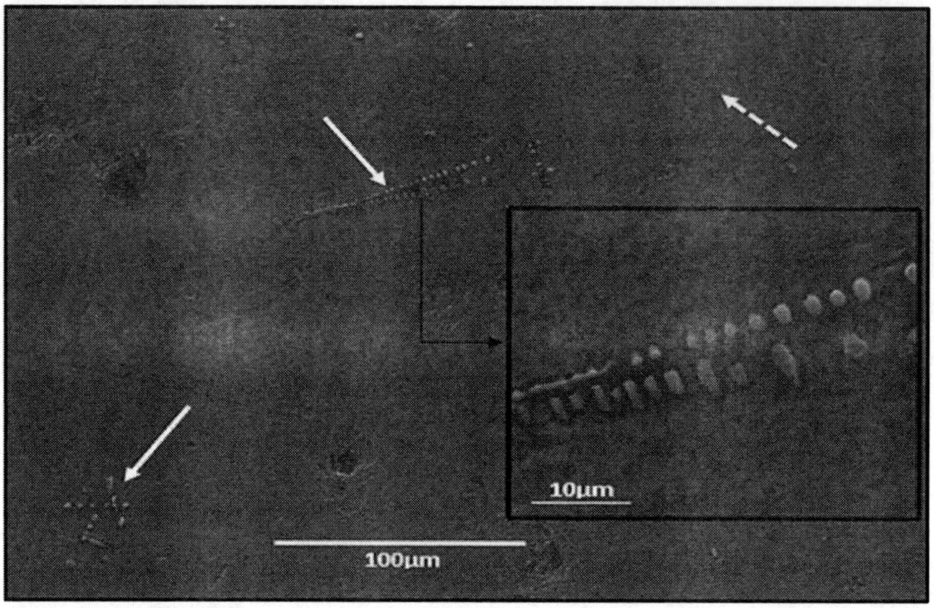

Figure 2. SEM micrograph showing the microstructure of heat treated $Ni_{50}Ti_{50}$ Shape memory alloy.

Table 2. Chemical composition of different phases in heat treated binary $Ni_{50}Ti_{50}$ Shape memory alloy

Phases	Ti % at.	Ni % at.
Matrix (NiTi)	50.2	49.8
Ti_2Ni	68.28	31.72

The crystalline phases of the heat-treated $Ni_{50}Ti_{50}$ alloy were investigated by X-ray diffraction (XRD) at room temperature, Figure 3 shows that the $Ni_{50}Ti_{50}$ alloy contains both R- phase (rhombohedral) and martensite (B19′) phase (monoclinic). In addition to diffraction peaks of Ti_2Ni phase are observed in the alloy. Coexist of R- phase and martensite phase because the martensite finish temperature (Mf) of the $Ni_{50}Ti_{50}$ SMA is lower than the room temperature, which leads to incomplete transformation of $Ni_{50}Ti_{50}$ SMA from martensite (monoclinic) to R- phase (rhombohedral) at room temperature.

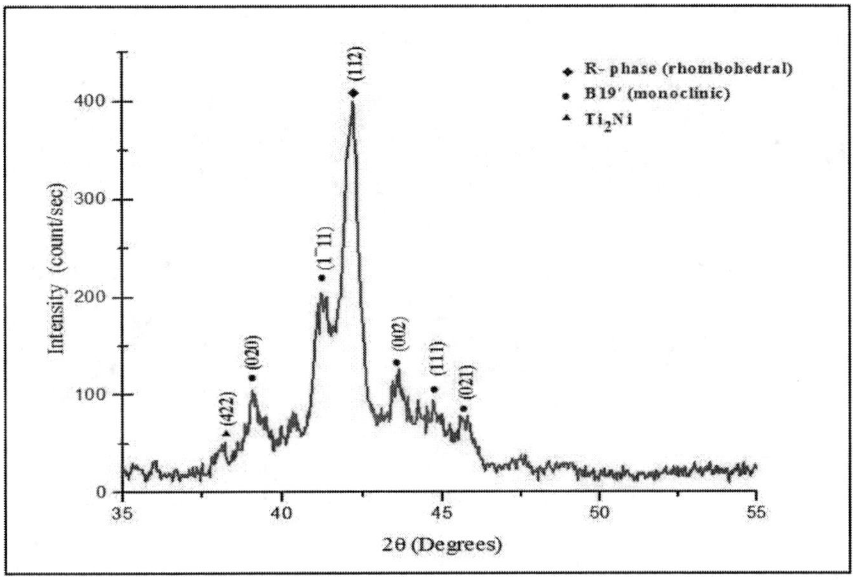

Figure 3. X-ray diffraction (XRD) of heat treated $Ni_{50}Ti_{50}$ shape memory alloy.

The microstructures of $Ni_{50}Ti_{50}$ SMA before and after heat treatment determined by the Optical Microscope are presented in Figure 4. From the images found that the microstructures of the equiatomic NiTi SMA after heat treatment at 865 °C is more homogenous because of solution treatment dominated by equiaxed grains rather than dendrites. From figure 4 the grain size after heat treatment seems to be smaller than the grain size before the heat treatment (grain refinement) Also, there are precipitates Ti_2Ni in the grain interior as well as at the grain boundary, represented by a solid pointer in figure 4b. The matrix NiTi phase represented by dashed pointer in figure 4b.

The results of micro hardness test for $Ni_{50}Ti_{50}$ are shown in Table 3. The optioned values was determined by taking the average of the seven measurements on the specimen. By comparing the micro hardness values of the specimen, before and after the heat treatment at 865 °C in normal atmosphere, it is obviously, an appreciable increase in micro hardness after heat treatment because the structure becomes more homogenous, and also due to the refinement of the grain size, with grain refinement, that's mean there are more and more grains that are nearby are turned to have a different orientation. This makes it difficult for a dislocation to cross between grains, also, introduces more and more grains boundaries which effect the movement of dislocations and that's make the material harder and stronger.

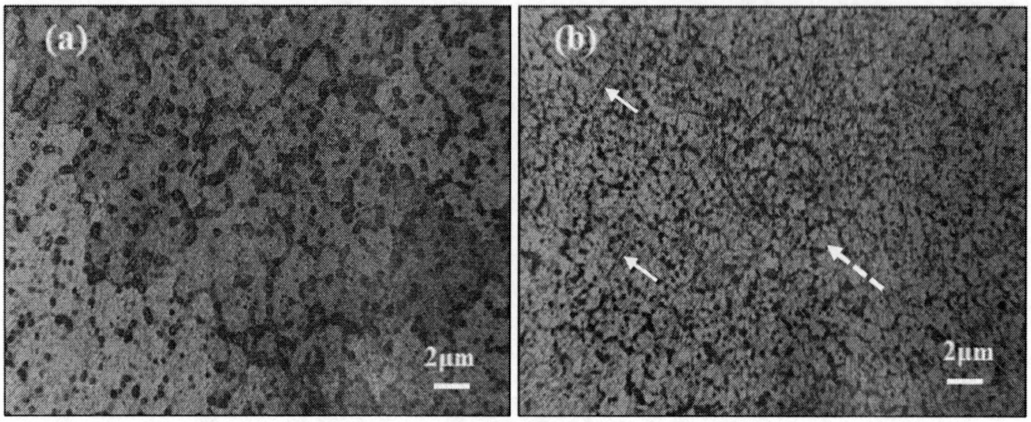

Figure 4. Optical Micrograph showing the microstructure of Ni50Ti50 shape memory alloy: a) before heat treatment, b) after heat treatment at 865 °C for 15 min.

Table 3. Micro hardness of $Ni_{50}Ti_{50}$ sample before and after heat treatment.

Sample	Before heat treatment	After heat treatment
$Ni_{50}Ti_{50}$	186.06	238.74

4. Conclusions

It was proved by EDX that the composition of the alloy could be controlled by using a vacuum induction melting process for preparation NiTi shape memory alloy. Binary NiTi shape memory alloy with two-steps martensitic transformation during cooling and heating can be produced. XRD patterns revealed that the strongest peak belongs to the R- phase (rhombohedral) which is a form of martensitic phase. The optical micrographs show homogenous structure of NiTi alloy after heat treatment. There is a noticeably increasing in hardness value after heat treatment at 865°C.

Acknowledgment

The authors would like to thank the University of Technology/ applied sciences department, and specially the metallurgy and material properties laboratory.

References

[1] Birnbaum, Andrew J., and Y. Lawrence Yao 2010 *J. Manuf. Sci. Eng.* **132** 4 041002.
[2] Wu, Ming H., and L. M. Schetky 2000 *SMST conf.* **vol. 171** (California) P.171-182.
[3] Lagoudas, Dimitris C 2008 *Springer Science & Business Media.*
[4] Lecce, Leonardo 2014 Elsevier (USA).
[5] Moberly, W. J., J. L. Proft, T. W. Duerig, and R. Sinclair 1990 *Acta metallurgica* **38** no. 12 2601-2612.
[6] Adharapurapu, R. R., and K. S. Vecchio 2007 *Experimental mechanics* **47** no. 3 365-371.
[7] Badakhshan Raz, S., and S. K. Sadrnezhaad 2004 *J. Mater. Sci. Technol.* **20** no. 5 593-598.
[8] Suzuki, Y., K. Otsukaand, and C. M. Wayman 1998 133-148.
[9] Frenzel, J., Z. Zhang, K. Neuking, and G. Eggeler 2004 *J of Alloys Compd.* **385** no. 1-2 214-223.

[10] Nayan, Niraj, C. N. Saikrishna, K. Venkata Ramaiah, S. K. Bhaumik, K. Suseelan Nair, and M. C. Mittal 2007 *Mater. Sci. Eng. A* **465** no. 1-2 44-48.
[11] Nagase, Takeshi, Atsushi S, Hiroyuki Y. Yasuda, Hirotaro M, Tomoyuki T, and Tomoyuki Ka 2011 *Intermetallics* **19** no. 9 1313-1318.
[12] Nam, Tae-Hyun, Jae-Hoon K, Mi-Seon C, Hee-Woo L, Yeon-Wook K, Hee-Joong I, Jeung-Sun A and Tadaoki M 2002 *J. mater. Sci. Lett.* **21**, no. 10 799-801.
[13] Liu, Yinong, Melina B, Geraldine T, J. I. Kim, and Miyazaki S 2006 *Mater. Sci. Eng. A* **438** 617-621.

The effects of additives in to heat treatment temperature and time on the crystallinity of lithium silicate glass-ceramic

S. A Zaidan[1], M. F Al-Hilli[2] and D. F Mahdi[3*]
[1]Applied Science Department-University of Technology, Baghdad, Iraq
[2]Department of Physics-College of Science-University of Baghdad, Baghdad, Iraq
[3]Applied Science Department-University of Technology, Baghdad, Iraq
[*]Corresponding author E-mail: drshihab1969@gmail.com,

Abstract .The glass powders were prepared from binary system of 45 wt.% Li_2O and 55 wt.% SiO_2 and melted in platinum crucible (90% Pt- 10% Rh) at 1195 ^oC for 1:15 hr. The glass molten was rapid cooled in cold water at 3^oC and the produced frit was milled in agate mortar and four sample groups were prepared. In the first group the temperature changes of heat treatment 750, 800, 850, 900 and 950^oC was studied. Second group the time changes of heat treatment 2, 4, 6, 8 and 10 hr. was examined. Third group the addition effect of P_2O_5 4, 7, 11, 14 and 18 wt.% as nucleating agent was studied. Fourth group the addition changes of nucleating agent mixing P_2O_5 and TiO_2 (quarter of P_2O_5 percentages as same above) was inspected. XRD, AFM and optical microscope had been done for all four group specimens. Crystalline phase percentages were calculated from XRD histogram by GSAS (General Structure Analysis System) and EXPGUI. The grain size distribution and average diameter of the crystalline phases for all four group specimens were calculated by AFM measurement. The density and mass fraction of crystalline phases were calculated by optical microscope image analyses by image J 1.52i. The results indicated good crystalline phase percentages, phases density, mass fraction and average diameter on nano-scale for 4 wt.% of (3 wt.% P_2O_5- 1 wt.% TiO_2), heat treated at temperature 950^oC for 6 hrs.

Keywords: Glass-Ceramic, Powder Technology, Lithium Silicate, GSAS, Image J 1.52i.

1. Introduction

Glass-ceramics are inorganic materials consisting of crystalline phases and amorphous phase (glass phase). It is interesting materials due to its unique mechanical, thermal, chemical and electrical properties. It can be designed for specific application by employing its special properties. It may be produced from a close control of molten glass with seeds (Nucleating agents) like P_2O_5, TiO_2, ZrO_2, MgF....etc., or cooling rate; the crystalline phases is growing in glass matrix [1].

Lithium silicate is first glass-ceramic developed by Stookey (1953, 1959), it produced from binary Li_2O-SiO_2 glass base. The main crystalline phase formed in this glass base could be $Li_2Si_2O_5$, Li_2SiO_3, Li_4SiO_4, ...etc; which depends on particular ratio of Li_2O and SiO_2 and the type of components that introduced the glass base composition and nucleating agents. One of the important properties of lithium silicate is good chemical durability which makes it suitable for biomaterial in human medicine and restorative material in dentistry particularly. The additional properties of lithium silicate such as high flexural strength, fracture toughness, young modulus values are 120.1, 110.7 and 126.7 GPa., along x-, y- and z-axis, electrical resistivity, band gap 5.36 ev. , and low loss factor have increased the demand for this material in electrical applications. Relatively, high thermal expansion coefficient render it proper material for sealing substrates in the electrical industries [2].

The glass powders sintering are one of the glass-ceramic formations in which the glass molten is rapidly cooled in cold water or cold metal rollers. The produced frit may be milled to 5 μm particle size and compacted to bulk material, then put in furnace for firing in which the sintering and crystallization process occur in one firing cycle. It is quite simple forming method and mixing different glass powders to make glass-ceramic composites. The closed pores may still exist in the material prepared by this method which reduces the mechanical strength, dielectric properties such as dielectric breakdown strength. However, we obtain high quality surface finish by this method [3].

2. Materials and Methods

2.1. Base Glass preparation

Base glass is binary system of Li_2O (45%wt) and SiO_2 (55%wt) in which the starting materials were Lithium carbonate Li_2CO_3 (BDH Chemicals Ltd, England, quality 99.5%) and Silicon dioxide SiO_2 (Thomas Baker Chemicals Ltd, India, quality 99%). Starting materials were mixed by using magnetic heating mixer and diluted with water for 1:30 hr. After drying, the mixture was calcined in crucible at 700°C for 1 hr. It was milled in agate mortar for half hour and compressing to cylinder shape with 1.8 diameter × 2.4 heights mm by applying 5 ton pressure and put in platinum crucible (90% Pt-10% Rh). The base glass mixture was melted in furnace at 1195°C for 1:15 hr., then base glass molten rapidly cooled in cold water at temperature of 3°C producing a frit.

2.2. Specimen preparation

The produced frit was milled in agate mortar for 1 hour to get a fine glass powder. A 5 g specimen in the shape of a disc (10 mm diameter by 2.5 mm thickness) was prepared by using biaxial pressing. Four groups were studied in this work; in the first group (G1) the change of heat treatment temperatures as 750 °C, 800 °C, 850 °C, 900 °C and 950 °C was studied. In the second group (G2) the change of heat treatment times as 2, 4, 6, 8 and 10 hrs was inspected. In the third group (G3) the change of addition percentage of P_2O_5 (Fluka AG, Buchs SG, quality 99%) 4, 7, 11, 14 and 18% as nucleating agent was examined. In the fourth group (G4) the change of addition percentage of mixing P_2O_5 and TiO_2 (BDH Chemicals Ltd, England, quality 98%) as nucleating agents was studied. The corundum (α-Al_2O_3) (BDH Chemicals Ltd, England, quality 99%) as reference material for XRD specimens, table (1) shows all four groups and their changes that studied in this work.

2.3. Testing Methods and Measurements

XRD was done by X-ray diffractometer (XRD-6000 Shimadzu, German) with measurement conditions as shown in table (2). In this work, general structure analysis system (GSAS & EXPGUI) which is comprehensive system for the refinement of structural models for both single crystals and powder diffraction data was employed [6]. The grain size measurements for all four groups in this work were used atomic force microscope analysis (AFM) (type SPM-AA3000, Angstrom, USA). Optical microscope and image J 1.52i software were used to measure the crystalline phases density and mass fraction for four group specimens. The optical microscope was polarizing microscope model MT9430 transmitted illuminator, MEIJI Techno Co. LTD, Japan with digit camera of type canon DS126431, Canon Inc., Japan connecting to computer.

IOP Conf. Series: Materials Science and Engineering **757** (2020) 012060 doi:10.1088/1757-899X/757/1/012060

Table 1. Four groups and their changes.

First Group G1	G11		G12		G13		G14		G15	
Temperatures ^{O}C	750		800		850		900		950	
Time hr	4		4		4		4		4	
Nucleating Agent %	0		0		0		0		0	
Second Group G2	G21		G12		G23		G24		G25	
Temperatures ^{O}C	800		800		800		800		800	
Time hr	2		4		6		8		10	
Nucleating Agent %	0		0		0		0		0	
Third Group G3	G31		G32		G33		G34		G35	
Temperatures ^{O}C	800		800		800		800		800	
Time hr	4		4		4		4		4	
Nucleating Agent P_2O_5 %	4		7		11		14		18	
Forth Group G4	G41		G42		G43		G44		G45	
Temperatures ^{O}C	800		800		800		800		800	
Time hr	4		4		4		4		4	
Nucleating Agent	P_2O_5	TiO_2	P_2O_5	TiO_2	P_2O_5	TiO_2	P_2O_5	TiO_2	P_2O_5	TiO_2
%	3	1	5.25	1.75	8.25	2.75	10.5	3.5	13.5	4.5

Table 2. XRD Measurement Conditions.

X-ray Tube	
Target	Cu
Voltage KV	40
Current mA	30
Wavelength ^{O}A	1.5418
Scanning	
Scan Range 2θ (deg)	10-80
Scan Mode	Continuous Scan
Scan Speed (deg/min)	8
Sampling Pitch (deg)	0.02

3. Results and discussion

Figure (1) is showing XRD histogram of base glass, G12, G32 and G42, there is 7.5 wt.% crystalline phase in base glass XRD histogram due to the ratio of Li_2O to SiO_2 is non-stoichiometric. A phenomenon occurred in this state called "Phase Separation" render the base glass certain opacity and phase separation is playing a part in the nucleation of lithium silicate glass-ceramics [4]. Primary phase 5.3 wt.% Li_4SiO_4 and secondary phase 2.2 wt.% SiO_2 in base glass, they are considered as nuclei for growing the crystalline phases in heat treatment later.

It is obvious from figure (1); the base glass shows lower intensity values which mean low reflections from low content of crystalline phases, while G32 and G42 showed higher intensity values from high reflections of high content of crystalline phases [5].

IOP Conf. Series: Materials Science and Engineering **757** (2020) 012060 doi:10.1088/1757-899X/757/1/012060

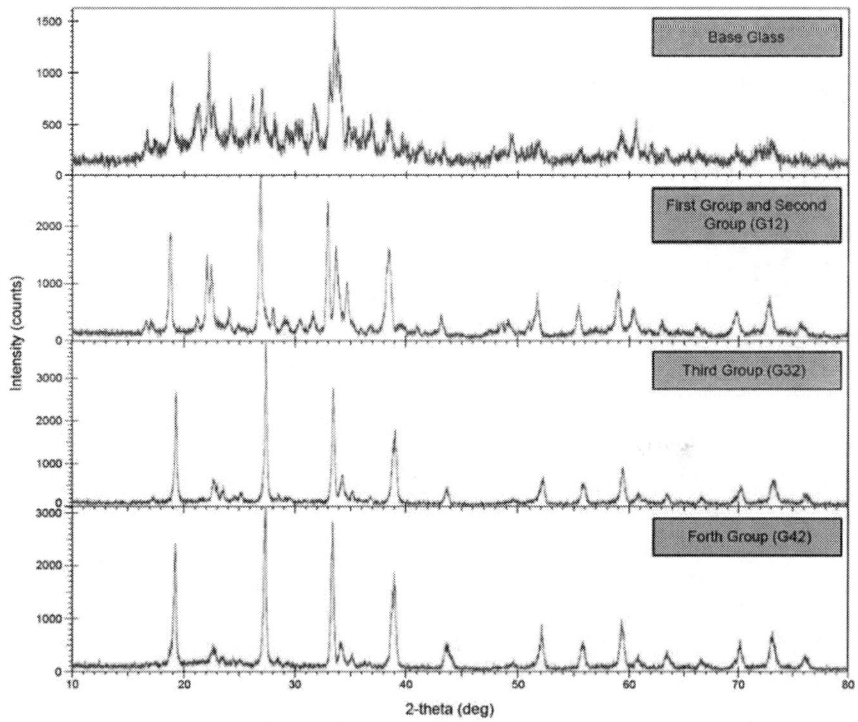

Figure 1. XRD histogram of base glass, G12, G32 and G42.

Corundum (α-Al_2O_3) was used as reference material for crystalline phase percentage calculations from EXPGUI and which add 1g for each specimen and done by [7]:

$$P_f = \frac{C_A}{C_m} \times P_m \tag{1}$$

Where P_f is crystalline phase fraction, C_A is Addition fraction of Corundum in specimen, C_m is Mass fraction of Corundum that calculated from EXPGUI and P_m is Mass fraction of crystalline phase that calculated from EXPGUI.

$$G_f = 1 - \sum_{i=0}^{i} P_f \tag{2}$$

Where G_f is Glass fraction.

Figure (2) shows the changes of heat treatment temperatures, time and nucleating agent percentages with crystalline phase percentages. The increasing of heat treatment temperature and time lead to increasing in crystalline phase percentages as seen in figure (2) a and b. Figure (2) c shows the crystalline phases increasing with adding percentage of nucleating agent which has been done for a certain limit. This behavior is attributed to increasing interfacial energy (γ) or energy barrier value which is restricting the crystalline phase growing and depending on viscosity of the glass-ceramic composition [8]. The crystalline phase percentages in forth group (G4) higher than third group (G3) because of producing viscosity depending on the type and amount of nucleating agents added to the base glass.

IOP Conf. Series: Materials Science and Engineering **757** (2020) 012060 doi:10.1088/1757-899X/757/1/012060

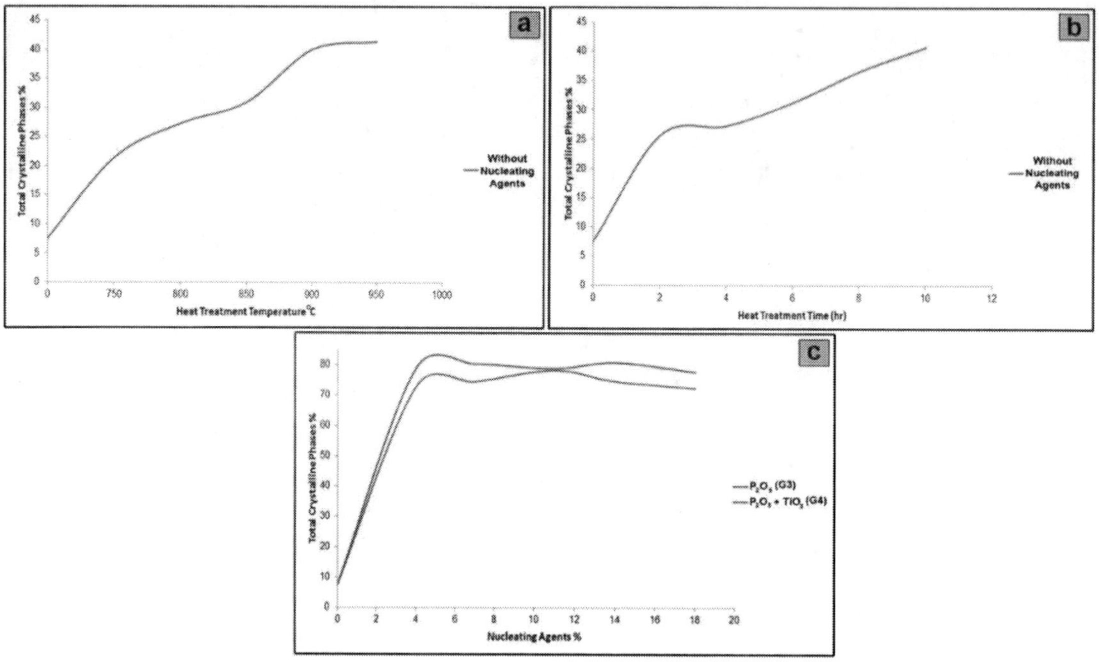

Figure 2. Heat treatment temperature, time and nucleating agent percentage.
(a) First group (G1), **(b)** Second group (G2), **(c)** Third (G3) and fourth (G4) group.

Figure (3) shows the growing crystalline phases in base glass have primary phase and secondary phases. The primary phase may be low or high percentage but still dominant phase and secondary phase percentages less than primary phase. The primary phase which limits the glass-ceramic properties which are depending on chemical composition and crystalline structure of the primary phase and secondary phases are making differences in the glass-ceramic properties [8].

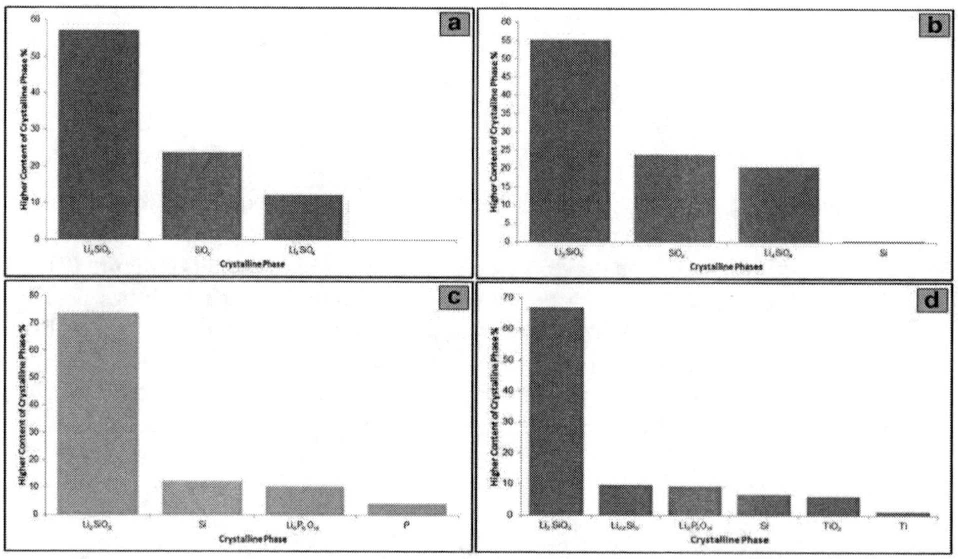

Figure 3. Primary phase and secondary phase percentages. **(a)** First group (G1), **(b)** Second group (G2), **(c)** Third group (G3), **(d)** Fourth group (G4).

The primary phase in this work is lithium silicate Li_2SiO_3 has an orthorhombic structure with lattice constants (a, b, c) = (9.396, 5.396, 4.661 ^{O}A)[space group $Cmc2_1$ (No.36)]. It has four molecular units (24 atoms) per cell, consisting of one type of Li atoms with site of 8b, one type of Si atoms with site of 4a, and two types of O atoms with site of 8b and 4a, respectively. Li_2SiO_3 have number of properties such as low electric conductivity, high mechanical resistance, high thermal stability, mechanically stable structure and good young modulus values [9]. Figure (3) indicates existence of crystalline phases in a and b and vanishing in c and d like Li_4SiO_4 and SiO_2, appear new crystalline phases in c and d while absence in a and b like $Li_6P_6O_{18}$, $Li_{22}Si_5$, TiO_2...etc. This means the secondary phases is metastable phases which their existence or vanishing is depending on the chemical composition, mixing and addition percentage of base glass and nucleating agents, temperature and time of the heat treatment. Whereas primary phase Li_2SiO_3 exists in a, b, c and d of figure (3) which meaning stable phase.

The grain size measurement range on all specimens is 25 × 25μm and choice has been done for one specimen from each group as shown in Figure (4).

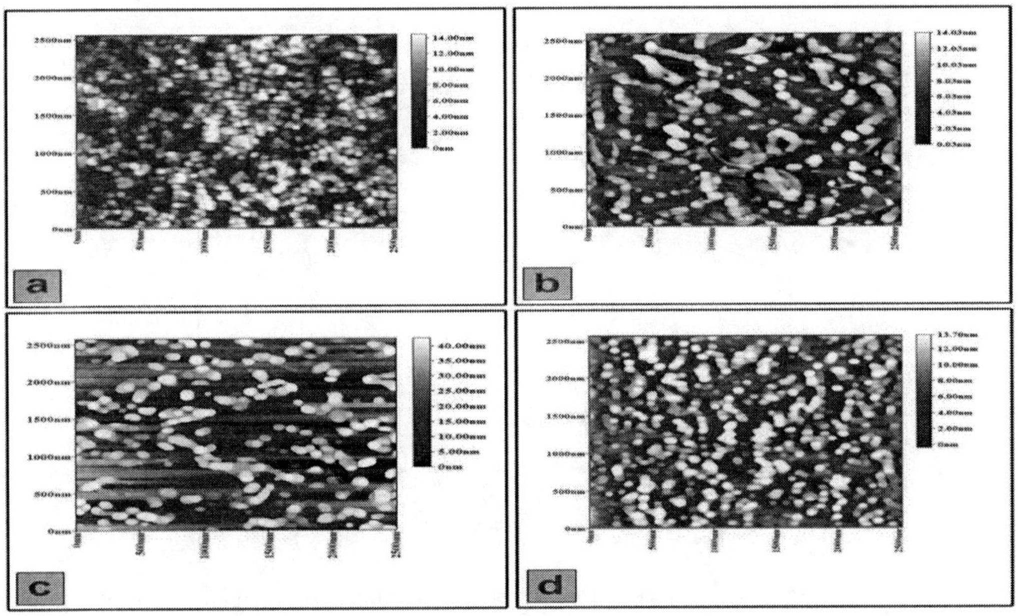

Figure 4. Grain size distribution for selecting specimen. **(a)** G12, **(b)** G23, **(c)** G32 and **(d)** G42.

The grain size results of crystalline phases in lithium silicate glass-ceramic for all four group specimens are nano-scale, average diameter from 64 nm to 108 nm. As seen in figure (4), c is high grain size distribution (0<ps<40 nm) because of P_2O_5 helps in growing of the crystalline phases and has high energy of reaction with others oxides to creating new compounds of crystalline phases. The rest of figure (4), a, b are (0<ps< 14 nm), (0.03<ps< 14.03 nm) and d show low grain size distribution (0<ps< 13.7 nm) in above figure, TiO_2 was reduced from P_2O_5 activity to larger crystalline phases in diameter but increased the density of crystalline phase with 13.7 nm diameter as comparison with c.

Figure (5) shows the temperature, time of heat treatment and addition percentage of nucleating agent changes with average diameter was studied. It may be seen an increasing to a certain limit and begun to decrease except for G3 (P_2O_5) curve in c, the average diameter increases with addition percentage of nucleating agents.

IOP Conf. Series: Materials Science and Engineering **757** (2020) 012060 doi:10.1088/1757-899X/757/1/012060

A wide range at high temperature, the glass is semi-liquid solid material and crystalline phases are lookalike islands in the middle of glass. The large crystalline phase (low energy) is encountering glass barrier (high energy) leading to broken parts of crystalline phase (island) and small pieces of crystalline phase are begin to grow up [10]. As mention before, P_2O_5 has high energy to reaction and get attracting the ions in glass and building order arrangement by increasing the diffusion rate of ions in glass.

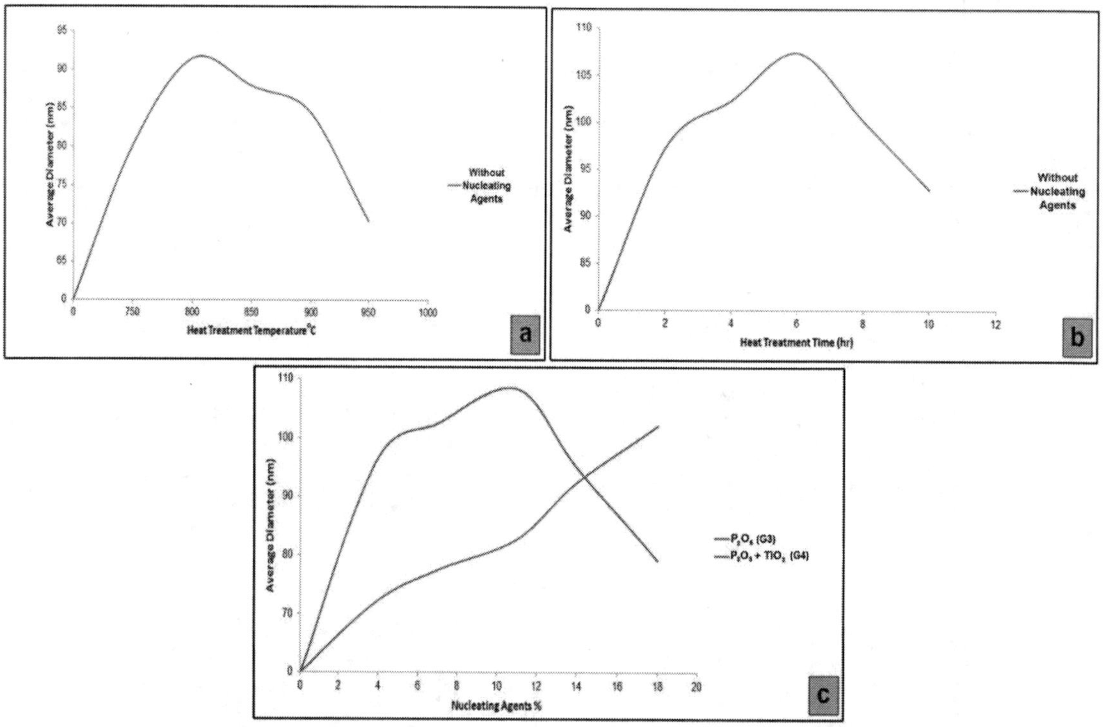

Figure 5. Relationship between average diameter and **(a)** Heat treatment Temperature (G1), **(b)** Heat treatment time (G2) and **(c)** Addition percentage of nucleating agents (G3 and G4).

The specimens were grinded with silicon carbide paper 1200 mesh and 1500 mesh and polishing with cloth and alumina paste for half hour. Etching has been done by dilute solution with 3% HF concentration for 2 min; the optical microscope images were analyzed by image J 1.52i. The optical microscope images were magnified force (100X) and image size 1×1mm.The analysis of images had been done by image J1.52i indicating that black dots are representing the crystalline phases in white area which is glass phase. The calculation of crystalline phase's number in 1 mm^2 area which represents density of crystalline phases in image J 1.52i, the mass fraction calculation of crystalline phases had been done by limiting all crystalline phases (black dots). The image J 1.52i is calculating the total area (pixel) of the crystalline phases and by dividing on image area (pixel) can got on mass fraction of crystalline phases in 1 mm^2 area [11].

The density and mass fraction of crystalline phases were studied with temperature, time of heat treatment and addition percentage of nucleating agent changes, as shown in figure (6) and (7). Figure (6) a and b are shows increasing of crystalline phases density with temperature and time of heat treatment. Figure (6) c increasing of crystalline phase's density reaches to a certain value and begins to decreasing for G3 and G4 curves. This due to increase in the viscosity with nucleating agent addition, but G4 curve shows higher values than a, b and G3 curve in c. The obvious of figure (7), a was little increased of mass fraction with heat treatment temperature up to 900°C, sharp increase was begun due to the reducing viscosity of base glass composition and diffusion rate increased.

Figure 6. Crystalline phases density (No. /mm^2) with **(a)** Heat treatment temperature (G1), **(b)** Heat treatment time (G2) and **(c)** Addition percentage of nucleatin agents (G3) and (G4).

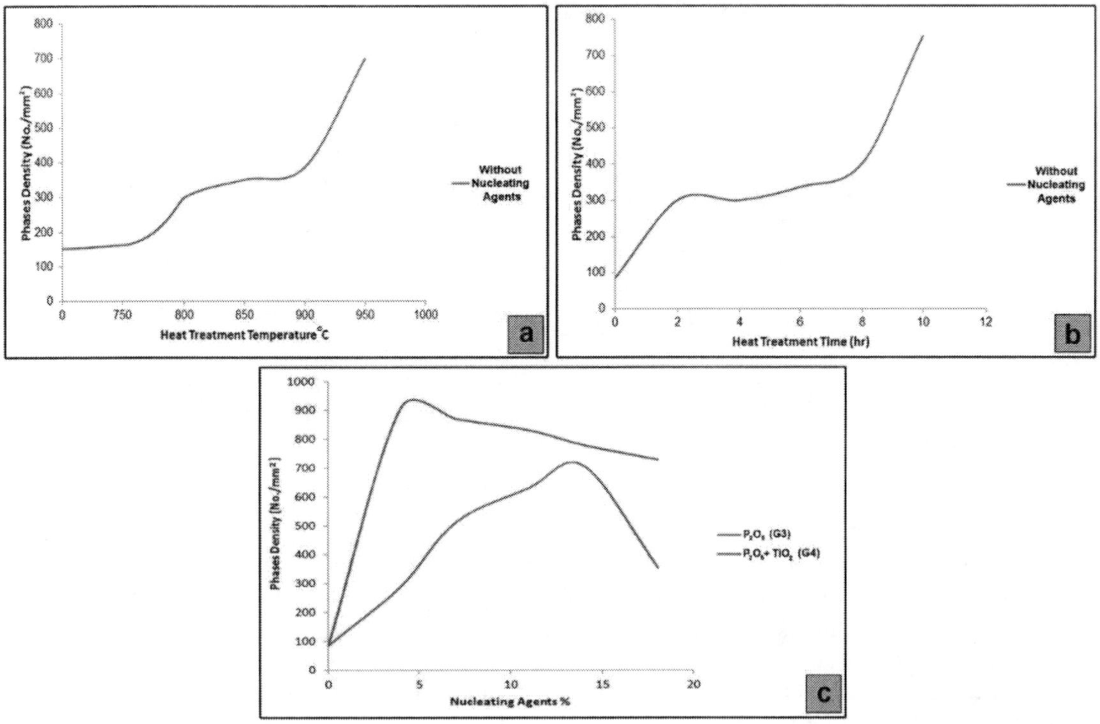

Figure (7) b and c were increased of mass fraction with heat treatment time and addition percentage of nucleating agents until 8 hrs and 4% in G4 curve and 11% in G3 curve of c are begin to decrease. This related to breaking the crystalline phases to tiny pieces and reaction with softening base glass. Generally, crystalline phases were increased with temperature and time of heat treatment and nucleating agent percentages comparison to base glass. The mass fraction values are so little with comparison to high values of crystalline phases density, this meaning the particle size for all crystalline phases on small scale. Small scale is nano-scale and down and these results provide good agreement and confirmation with AFM results.

IOP Conf. Series: Materials Science and Engineering **757** (2020) 012060 doi:10.1088/1757-899X/757/1/012060

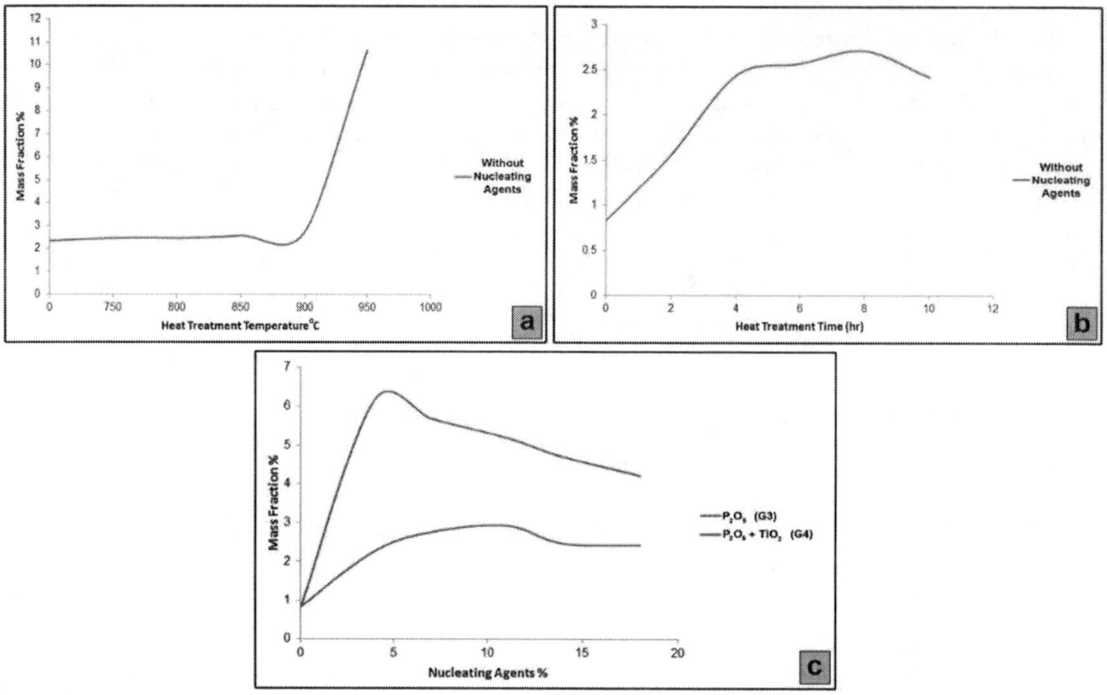

Figure 7. Mass fraction of crystalline phases (Percentage/mm^2) with **(a)** Heat treatment temperature (G1), **(b)** Heat treatment time (G2) and **(c)** Addition percentage of nucleating agents (G3) and (G4).

4. Conclusion

1- The non- stoichiometric ratio (Li_2O:SiO_2) can used to making glass-ceramics without adding nucleating agents due to phase separation phenomena that creating crystalline phases which consider nuclei for crystalline phases growing at heat treatment. This state is need for high temperature and more time of heat treatment.
2- Chemical composition of crystalline phases are coming from chemical composition of base glass, so it can governing of the crystalline phase types of the glass-ceramic by type of oxides that forming base glass and their percentage.
3- P_2O_5 is very active nucleating agent for crystalline phases growing of lithium silicate free-alumina glass-ceramic or with little alumina content.
4- A little addition of TiO_2 as nucleating agent with P_2O_5 get higher values for crystalline phase percentages, average diameter and mass fraction when using P_2O_5 alone, i.e. good work for a certain value, but then decreasing is occur.
5- The best results from point view of the work is 4% nucleating agent P_2O_5 and TiO_2, heat treatment temperature is 950 ^oC and heat treatment time is 6 hr. The glass-ceramic is become good percentage with crystalline phases, high phases density and good mass fraction with interesting in nano-scale.

References

[1] Schmitt, T.J., Tomiyama, M., Folgueras, M.V. and Bahr, F.L., 2012. *Materials Science Forum* **727**:1308-1312. Trans Tech Publications.

[2] Holand, W. and Beall, G.H., 2012. *Glass ceramic technology*. John Wiley & Sons, pp. 75-80.
[3] Ponsot, I., Detsch, R., Boccaccini, A.R. and Bernardo, E., 2015 *Advances in Applied Ceramics*, **114**:S17-S25.
[4] Hijiya, H., Kishi, T. and Yasumori, A., 2009 *Journal of the Ceramic Society of Japan*, **117**:1361:120-126.
[5] Kniess, C.T., de Lima, J.C. and Prates, P.B., 2012. IntechOpen, pp. 295-297.
[6] RBV, L.A.D., 2004. *General structure analysis system (GSAS)*. Los Alamos National Laboratory Report LAUR: 86-748, pp. 51-60.
[7] Gualtieri, A.F., 2003. A guided training exercise of quantitative phase analysis using EXPGUI. *GSAS Tutorials and Examples*.
[8] Prado, M.O., Nascimento, M.L.F. and Zanotto, E.D., 2008 *Journal of Non-Crystalline Solids*, **354** 40-41:4589-4597.
[9] Tang, T. and Luo, D.L., 2010. *J. At. Mol. Sci*, **1**, pp.185.
[10] Karmakar, B., Rademann, K. and Stepanov, A. eds., 2016. *Glass nanocomposites: synthesis, properties and applications*. William Andrew, pp. 13-19.
[11] Ferreira, T. and Rasb, W., 2012. ImageJ user guide: IJ **1.46 r**, pp. 29-34.

Determination of The Weibull Modulus by Electrical Breakdown and Mechanical Strength for Soda Lime Glass

S A Zaidan[1*] and A Z Al Ani[1]
[1]Applied Science Department-University of Technology, Baghdad, Iraq
[*]Corresponding author E-mail: drshihab1969@gmail.com,

Abstract. The interrelationships between mechanical and electrical properties are important to characterize one signifies the other, especially for brittle materials such as glass. The relationship between Weibull modulus from electrical breakdown and Weibull modulus from three-point bending test and other mechanical properties has been studied for soda lime glass slides. The results show that the electrical breakdown and Weibull modulus increased with the rate of rising voltage increases from 20.1-24.7 kV/mm and 10.9-13.78 ,respectively. The value of Weibull modulus was calculated by using mechanical strength is (12), this result identified the conditions necessary to achieve the same results in electrical breakdown strength test. Therefore, the best rate of rising voltage in the electric breakdown test is 2 kV/s which gives the same results as the mechanical tests. This means that it is possible in the future to evaluate the dielectric products electrically in terms of matching mechanical strength specifications using electrical tests.

Key words: soda lime glass, dielectric, bending, breakdown, Weibull modulus.

1. Introduction

Glass is a poor conductor that conducts electricity at normal temperatures. In this regard, it considers insulator materials. The resistance of glass to electricity varies with its composition. Glass withstand current will lower the temperature, but humidity will increase the conductivity of the current. The rough surface of the glass reduces the resistance to electrical current. [1]

Soda lime glass used in almost all electrical and electronic industries, and insulators are used for light sources (light envelopes) and tubes, cathode ray tubes, capsules and for holding microcircuit components. Glass has high resistance and high electrical breakdown as well as low dielectric constant and low dielectric losses. [2]

When the strength of a series of equivalent glass samples measures, usually find that the results are quite scattered. The reason is the scale-distributing behavior that leads to failure. This behavior is very different from the behavior of metals. Therefore, when we use ceramics, we must adopt different design methods.[3]

Weibull probability of survival analysis has been developed as, among other things, an engineering design method for components made from such materials as ceramics. Failure loads in ceramic components are determined by defect size, and are thus characteristic of the specimens, not the material. The defects have a size distribution and thus the failure loads are variable. An alternative way of describing this situation is that at any given load, a fraction of the specimens tested will survive. Weibull described this fraction, the survival probability (Ps), at any tensile stress as (Eq. (1)):[4]

$$P_s = exp\left\{-\left(\frac{\sigma_{max}}{\sigma_o}\right)^m\right\} \ldots..(1)$$

where σ_{max} is the maximum strength (either $\sigma_{max}=\sigma_{Emax}$ for electrical strength or $\sigma_{max}=\sigma_{Mmax}$ for mechanical strength) , σ_o = the characteristic strength for which the survival probability is 0.37 (1/e) (so, either $\sigma_o=\sigma_{oE}$ for electrical strength or $\sigma_o=\sigma_{oM}$ for mechanical strength). To determine Weibull modulus it must take a natural logarithm of both sides of equation (1)

$$\ln\left(\ln\left|\frac{1}{P_s}\right|\right) = m \ln \sigma_{max} - m.\ln\sigma_o \ldots..(2)$$

Now if we plot ln(ln|1/P$_S$|) versus ln σ we will get a straight line of slope (m). The higher the Weibull modulus the lower is the variability of strength. [5]

The Connectivity between mechanical characteristics and electrical characteristics is an urgent necessity for material evaluation.

Electrical breakdown can be classified to two types: 1. Volume breakdown 2. Surface breakdown [6,7]. Dielectric breakdown is the failure of a dielectric to withstand the applied electrical field[8]. Dielectric strength can be measured from the electrical field σ_{Emax} represents the field in which the dielectric material fails:

$$\sigma_{Emax} = \frac{V_{br}}{h} \ldots..(3)$$

Where $\mathbf{V_{br}}$ is the maximum voltage applied to the dielectric and **h** the thickness of the dielectric materials.

There are many types of electrical breakdown: (a) Intrinsic breakdown, (b) Electromechanical breakdown, (c) Electrothermal breakdown, (d) Erosion breakdown, (e) Streamer breakdown[9].

The three-point bending test was performed by the bending device by this test calculate the strength (σ_{Mmax}) for each slide according to equation:

$$\sigma_{Mmax} = \frac{3PL}{2bh^2} \ldots..(4)$$

Where P is the maximum applied load, L is the distance between the supports, b is the width of the slide and h is the thickness of the slide.[10]

2. Materials and Methods

2.1 Materials

Soda lime glass used as samples, where supplied by SUPERIOR\MARIENFELD company-Germany and its classified as brittle materials. The same glass slides were used in mechanical and electrical strength tests. The typical composition of soda lime glass slides (by %weight) are: 72.3% SiO_2, 13.3% Na_2O, 8.8% CaO, 4.3% MgO and 1.3% of (Al_2O_3+K_2O+Fe_2O_3). The dimension of the samples is fixed (length =76 mm, width =26 and thickness tolerances 0.95–1.05 mm).

2.2 Dielectric strength measurement

The dielectric strength tested by using high voltage supplier (BAUR - PGO - S -3) Germany with a range (0-60 kV) and frequency (50 HZ).The breakdown has been measured for five rate of rising voltage

(0.5, 1, 2, 3, 5 kV/s). For each rate of rising voltage take 10 point of failure and the calculate the average. The electrical field was found from Equation 3.

2.3 Mechanical strength measurement

The strength has been measured by three-point bending test for 10 samples with the same dimension and take the average values. The samples dimension (76×26×1mm), and the distance between the two supports of devise is 50 mm. The device gives value of maximum force, and the strength measured by equation 4.

3. Results and discussion

The rate of rising voltages (ramping speed) is one of the most parameters in the electrical breakdown strength test of dielectrics. So, that it can determine the value of electric breakdown and types, as well as explain the causes of change in values.

Fig. 1 shows increasing the electrical strength of the dielectric with rate of rising voltages. The low rate of rising voltage means increasing the heat between the electrodes (inside the contact area of the sample) due to increased leakage currents. In addition to the cumulative effects of electrochemical transformations and erosion, which breaks the structure of the material and accelerated the breakdown of the dielectric material as a result of heating processes. The values of the electric breakdown taken at low rate of rising voltage was represented the operational values of the sample, in other words when used as high-voltage insulators. As for the values at which the electric breakdown occurs at high rate of rising voltage, it represents the moment when the circuit is opened or closed.[11]

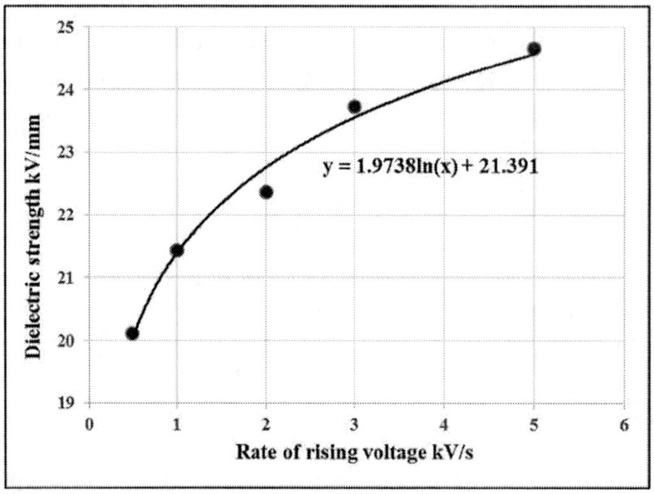

Figure 1. Variations of dielectric strength with rate of rising voltage.

The electrical breakdown at a low rate of rising voltage is called "Electrothermal" breakdown, while the electric breakdown at high rate of rising voltages is called "Pure Electrical Breakdown".[12] In both cases, damage on the contact point of the sample and between electrodes occurs, and it is clearly visible by the eye or using a simple microscope. The damage is accompanied by the appearance of microcracks as shown in Fig. 2, showing that the extent of damage at the point of breakdown increases in two ways: the first involves electrochemical changes and this is apparent at 0.5, 1 and 2 kV/s, accompanied by limited microcracks and cannot cause fractures in the sample.

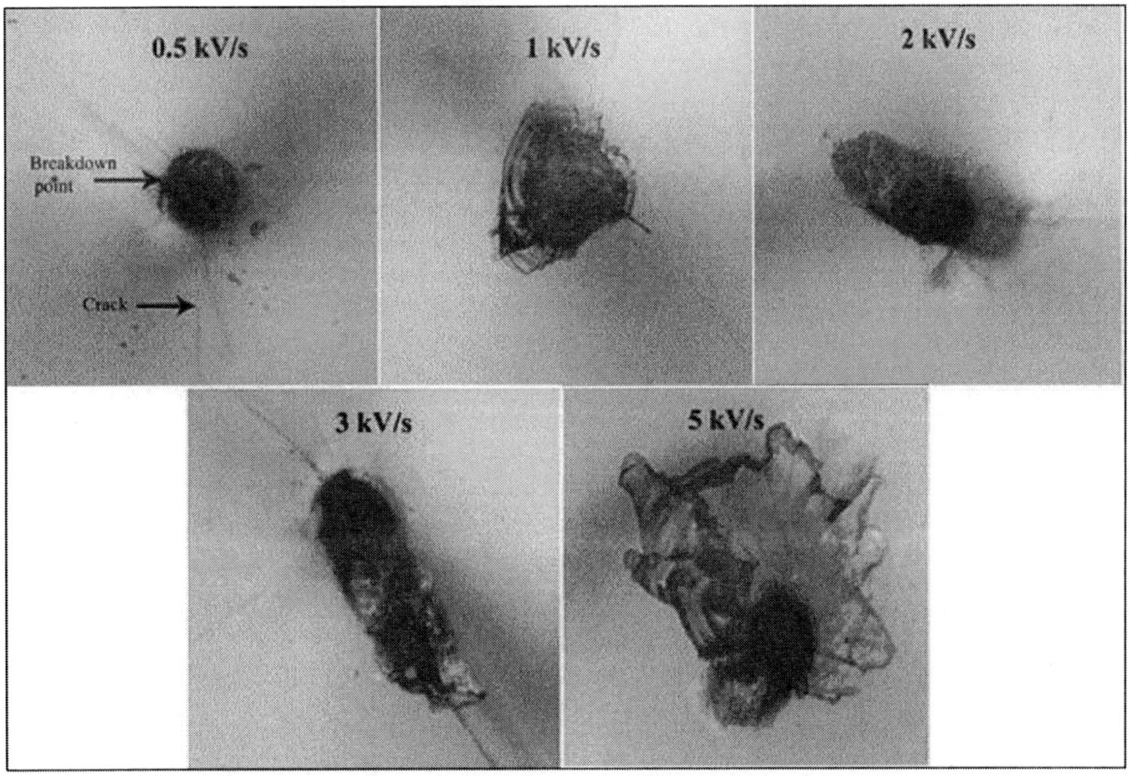

Figure 2. Optical microscopy images for electrical breakdown zone and for different rates of rising voltage.

The second represents an increase in the rate of rising voltage up to 3 and 5 kV/s, where the extent of damage increases significantly, turning into the other type of effects, called "Electromechanical Effect"[13]. This type may cause a sample fraction or the emergence of cracks extending long distances from the edges of the breaking point, and if it reaches the tip of the sample can get a break of the sample. The electrical breakdown strength is exponentially changing with the rate of rising voltage, as in Fig. 1, which means that in high rate of rising voltage the effect on the electrical breakdown is less. The amount of electrothermal breakdown is 20.12 kV/mm at 0.5 kV/s and increased up to 24.66 kV/mm at 5 kV/s which represents pure electrical breakdown.

Since glass is exposed to many factors, leading to the emergence of structural defects such as the residual stresses due to manufacturing processes and annealing, dislocations, nonhomogeneous distribution of glass compounds and the amorphous structure of glass, All of these factors lead to different results of the electrical breakdown and mechanical strength tests. It is therefore necessary to establish a statistical mechanism by which to assess the degree of discrepancy in the results and the evaluation of the glass products.

The Weibull modulus is one of the best statistical methods to describe this randomization, and it is possible to find variations in measurements. Fig. 3 shows the change in the values of the electrical breakdown strength (x-axis), which represents the logarithm of the greatest electrical stress ($\ln|\sigma_{Emax}|$) where the electrical breakdown occurs. The y-axis represents the logarithm-logarithm inverted probability of survival of the sample without breakdown ($\ln\ln|1/P_s|$), note that the probability of failure can be found from the relationship ($P_f = 1 - P_s$).

Fig. 3(a) shows a change in the probability of survival with electrical breakdown strength at the low rate of rising voltage (0.5 kV/s), that mains at the electrical breakdown associated with the electrothermal changes associated with the breakdown due to the high temperature associated with applied voltage. The Weibull modulus was calculated from the slope of the straight line, and its value (m=10.875) and the characteristic strength value of electrical filed is σ_{Eo}= 21MPa.

When increasing the rate of rising voltage up to 2 kV/s, the Weibull modulus increases to m=11.934, as shown in Fig. 3(b). This increase in the Weibull modulus means that the duplicate values of results is large, in other words, the measured values of the maximum electrical field intensity are less variable in the glass samples.

The Weibull modulus is continues in increase with the rate of rising voltages up to m=13,683 at the 5 kV/s, as shown in Fig. 4, and σ_{Eo}= 25.5 MPa, which can be considered as the optimum value for the characterization of electrical strength of soda lime glass samples, as shown in Fig. 3(c,d and e).

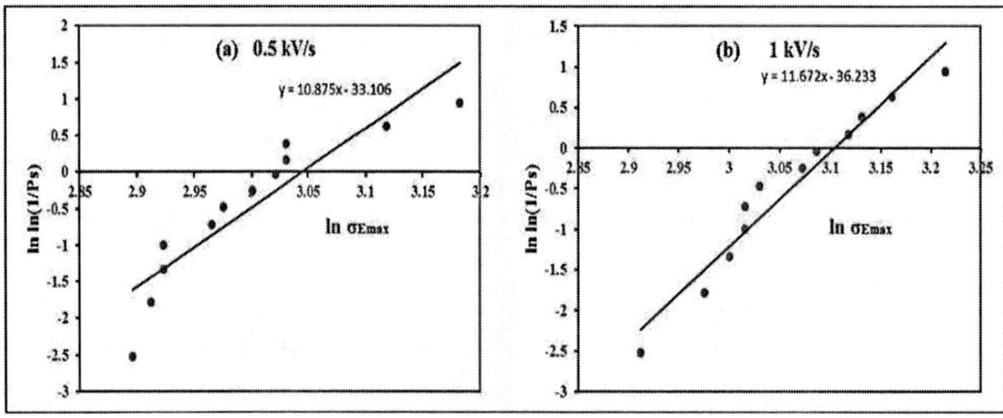

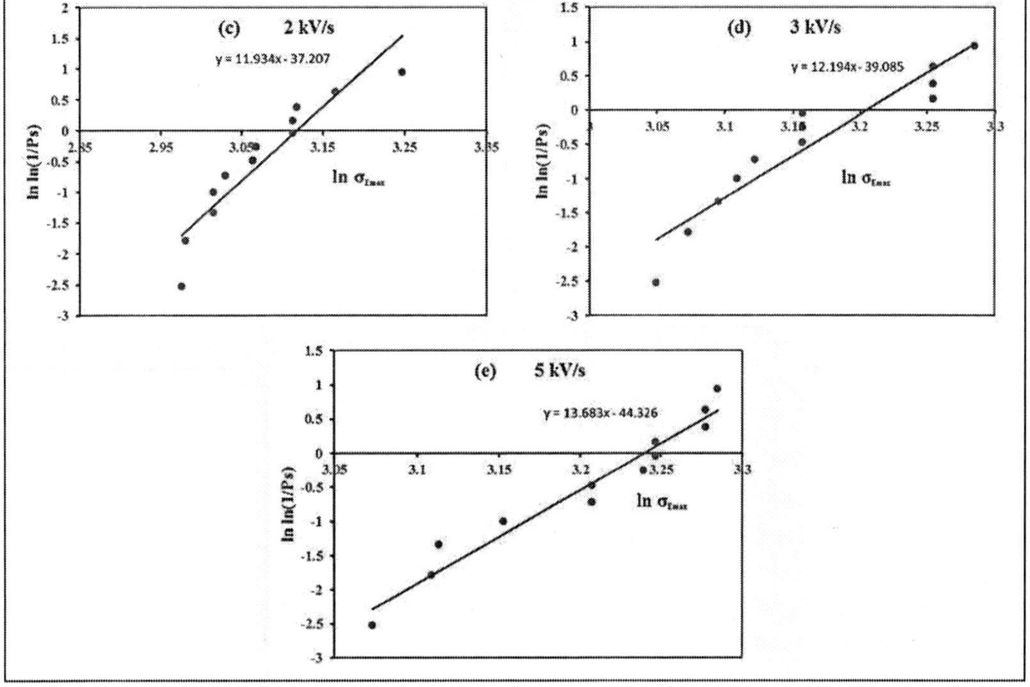

Figure 3. Weibull distribution of the electrical breakdown strength with different rate of rising voltage: (a) 0.5 kV/s, (b) 1 kV/s, (c) 2 kV/s, (d) 3 kV/s and (e) 5 kV/s.

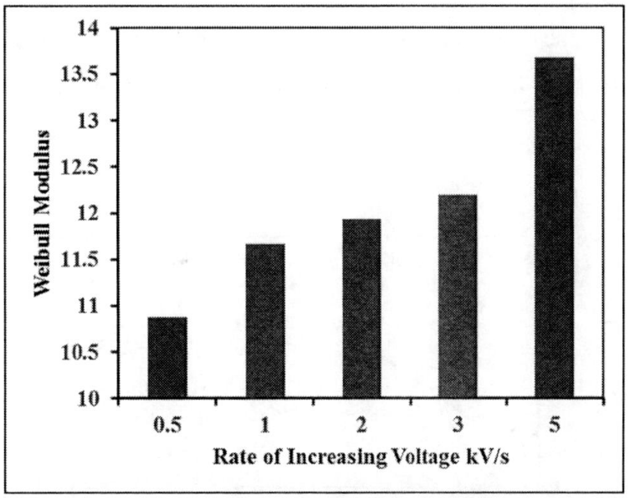

Figure 4. The Weibull modulus dependence of rete of rising voltage.

Fig. 5 shows the relation between the characteristic strength value of electrical filed and the rate of rising voltages, representing the high values of the rising which are the most accurate in electromechanical properties.

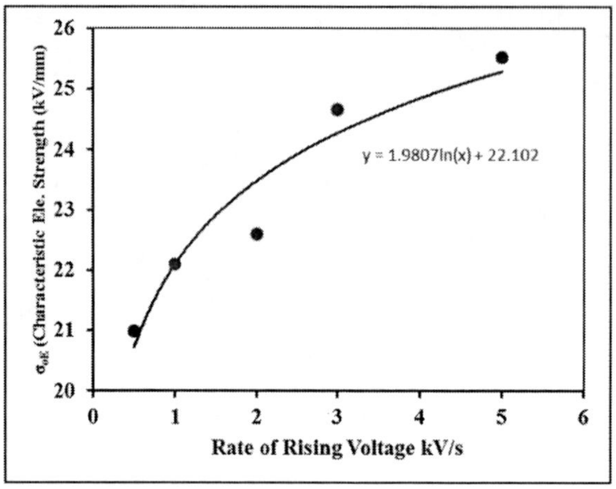

Figure 5. Variations the characteristic electrical strength with rate of rising voltage.

Glassware is subject to mechanical failure theories for brittle materials. Microcracks and structural defects are the most important factors determining mechanical strength. Bending strength is one of the mechanical strength types, and characteristics by which resistance is evaluated because it simulates the weakness of these materials, especially tensile strength[14]. Fig. 6 shows the test results of a set of soda lime glass slides, and according to relationship 2. The Weibull modulus can also be found in the same way as in electrical breakdown tests, but replacement the electrical stress with mechanical stress. The slope of the straight line represents the Weibull modulus m=12, and the characterization of electrical strength value is σ_{Mo}=97.52 MPa. This is the optimum value for bending strength by three point test.

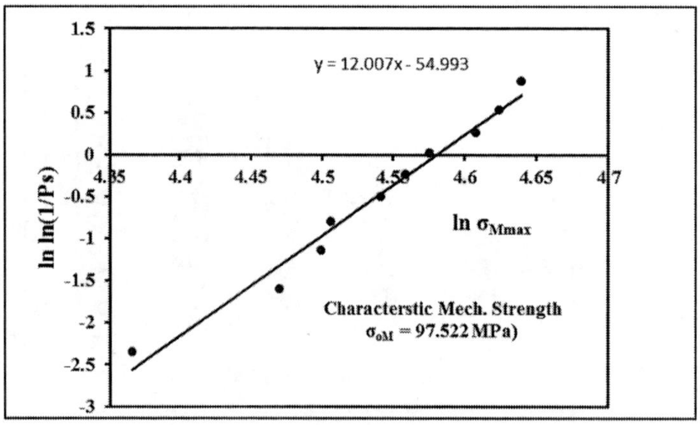

Figure 6. Weibull distribution of the bending strength.

4. Conclusions

The rate of rising voltages is determining the mechanism of electrical breakdown strength. At low rates, thermal emission increases and electrothermal breakdown occurs, and at high rates, pure electrical breakdown occurs, and often causing it to fracture of sample. In other words, the size of the damage in the area of breakdown increases with increasing the rate of rising voltages as well as extending the microcracks. Increasing the Weibull modulus with the rate of rising voltages increases means obtaining less difference in the results of electrical strength measurements. The Weibull modulus calculated from the mechanical bending test corresponds to the calculated electrical breakdown test at the value of m=12. This corresponds to the rate of rising voltage is 2 kV/s. These conditions determine the adoption of the electrical breakdown tests as an alternative to the mechanical in finding the Weibull modulus to evaluate brittle dielectric products such as glass.

References

[1] Shelby, J.E., 2007. *Introduction to glass science and technology*. Royal Society of Chemistry.

[2] Pye, L.D., Stevens, H.J. and LaCourse, W.C., 1972. Introduction to Glass Science. *The Vitreous State Plenum Press, New York.*

[3] Carter, C.B. and Norton, M.G., 2007. *Ceramic materials: science and engineering*, New York: springer **716**, 712.

[4] Burrow, M.F., Thomas, D., Swain, M.V. and Tyas, M.J. 2004 *Biomaterials*, **25** (20) pp.5031-5035.

[5] Sorensen, C.D., 1992. Measuring the Weibull modulus of microscope slides.

[6] Carabajar, S., Olagnon, C., Fantozzi, G. and Le Gressus, C., 1995, October. Relations between electrical breakdown field and mechanical properties of ceramics. In *Proceedings of 1995 Conference on Electrical Insulation and Dielectric Phenomena* (pp. 278-281). IEEE.

[7] Rizk, F.A. and Trinh, G.N., 2014. *High voltage engineering*. CRC Press.

[8] Miller, H.C., 1989 *IEEE transactions on electrical insulation* **24** (5) pp.765-786.

[9] Zaidan, S.A. and Mohammed, A.Z., 2017. *Ibn AL-Haitham Journal For Pure and Applied Science* **30** (1) pp.339-354.

[10] Li, D., Li, W., Wang, R. and Shen, X., 2015 *Journal of Non-Crystalline Solids*, **409**,126-130.

[11] Mieller, B., 2019 *Journal of Advanced Ceramics*, pp.1-9.

[12] Rajput, R.K., 2004. *A textbook of electrical engineering materials*. Firewall Media.

[13] Chen, J., Yao, M. and Yao, X., 2016. *Journal of Electronic Materials* **45** (2) pp.892-898.

[14] Biolzi, L., Cattaneo, S. and Rosati, G., 2001 *Rock mechanics and rock engineering* **34**(3) pp.217-233.

IOP Conf. Series: Materials Science and Engineering **757** (2020) 012062 doi:10.1088/1757-899X/757/1/012062

Using Remote Sensing and GIS in Measuring Vegetation Cover Change from Satellite Imagery in Mosul City, North of Iraq

M. F. Allawai[1] , B. A. Ahmed[2*]

1. Baghdad University, College of Science, Physics Department
2. Baghdad University, College of Science, Remote Sensing and GIS Department

*Corresponding author email: b_ali_ahmed@yahoo.com

Abstract

The aim of the study is the assessment of changes in the land cover within Mosul City in the north of Iraq using Geographic Information Systems (GIS) and remote sensing techniques during the period (2014-2018). Satellite images of the Landsat 8 on this period have been selected to classify images in order to measure normalized difference vegetation index (NDVI) to assess land cover changes within Mosul City. The results indicated that the vegetative distribution ratio in 2014 is 4.98% of the total area under study, decreased to 4.77% in 2015 and then decreased to 4.54 % in 2016, after then decreased to 3,59% in 2017,then increased to 4.39% in 2018. Land cover change of the area was identified using Normalized Difference Vegetation Index (NDVI) technique. Highest NDVI value was found in 2015 (6.26%) which denotes presence of moderate- high vegetation cover at that time period. After 2015, highest NDVI value was found following a decreasing trend (6.05% in 2016 and 5.96% in 2018) which clearly represents the vegetation cover change in the study area, also Green Normalized . Difference Vegetation Index (GNDVI) is studied in this paper. Statistical properties for NDVI and GNDVI were illustrated has been computed. From the results of this study one can clearly notice that there are the spatial variation in the vegetal cover from 2014 to 2018 in Mosul city, north of Iraq.

Keywords: Land Cover Change, Remote Sensing, NDVI, GNDVI, Mosul City, Landsat Satellite Imagery.

1. Introduction

Satellite imagery is an excellent way to monitor the earth's surface cover and the ongoing changes. Quick monitoring requires the application of certain techniques to distinguish the image style Technologies are the use of change detection techniques. Change detection is the process of distinguishing differences in the case of objects or phenomena by observing them in different periods of time, including the ability to determine temporal effects using a multi-temporal data set [1]. Typically, change detection procedures must include data generated by the same mobile sensor on the satellite (like that) with the same spatial accuracy, spectral accuracy, and spectral radiation and to some extent at the same time of day [2]. The Normalized Difference Vegetation Index (NDVI) is the most accurate method of digital processing of spatial visualization in the presentation of vegetation. Where it is based on the fact that plants Show a high reflectivity in the wavelength range near the red wavelength range [3]. The values of NDVI have a range of (-1 to +1). In general, the result is positive, indicating that the cell has a plant cover, and the higher the resulting positive value, the greater the plant's greenery and density and vice versa with regard to the negative values that indicate the non-green surface parameters. Therefore, the evidence of vegetative differences is used to distinguish between diseased and healthy plants [4]. The major objectives of the study are to assessment vegetal cover changes of the studied area using Normalized Difference Vegetation Index (NDVI), Green Normalized Difference Vegetation Index (GNDVI) for the period (2014-2018).

2. Description of the studied area

The area of study in this research included the city of Mosul as shown in figure 1, which lies within the boundaries of the province of Nineveh, Mosul is located 400 km (250 miles) north of Baghdad, located on the west bank of the Tigris River, the number of the population of Mosul, according to estimates in 2003 about 1.4 million people, while believed that the current number of approximately 1.376 million people. In this research, Landsat 8 satellite data were used for the Operational Land Imager (OLI) and eleven spectral channels and during the time period of 2014/6, 2015, 2015, 2016, 2017 and 2018 within coordinates 424949'19. E, 363737'48 " N.

IOP Conf. Series: Materials Science and Engineering **757** (2020) 012062 doi:10.1088/1757-899X/757/1/012062

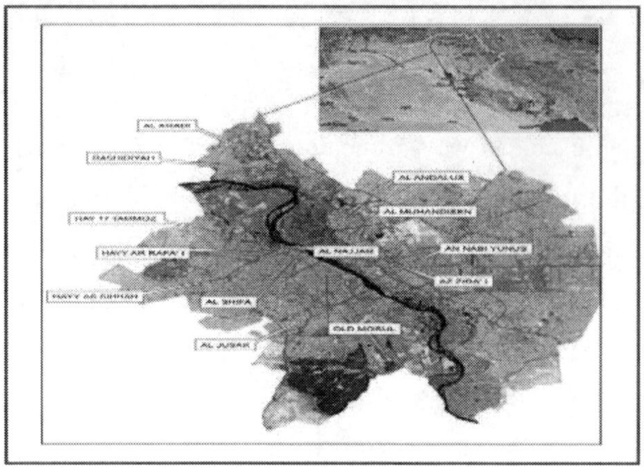

Figure 1. The study area (Mosul city)

3. Data and Methodology

The basic concepts, available imagery sources and classification techniques of remote sensing imagery related to vegetation mapping were introduced, analyzed and compared. Vegetation cover mapping were provided to iterate the importance of thorough understanding of the related concepts and careful design of the technical procedures, which can be utilized to study vegetation cover from remote sensing imagery [5]. Table1 illustrated Landsat 8 provides metadata of the bands such as type of sensors, channel spectral etc., which have been used to obtain the results of this study.

Table 1. Some of the information and Landsat -8 imagery used in this study.

Type of sensor	Channel Spectral	Georeferencing	Date
Operational Land Imager (OLI)	4	PATH=170, ROW =037	2014/06/09
Operational Land Imager (OLI)	5	PATH=170, ROW =037	2014/06/09
Operational Land Imager (OLI)	4	PATH=170, ROW =037	2015/06/12
Operational Land Imager (OLI)	5	PATH=170, ROW =037	2015/06/12
Operational Land Imager (OLI)	4	PATH=170, ROW =037	2016/06/14
Operational Land Imager (OLI)	5	PATH=170, ROW =037	2016/06/14
Operational Land Imager (OLI)	4	PATH=170, ROW =037	2017/07/19
Operational Land Imager (OLI)	5	PATH=170, ROW =037	2017/07/19
Operational Land Imager (OLI)	4	PATH=170, ROW =037	2018/07/17
Operational Land Imager (OLI)	5	PATH=170, ROW =037	2018/07/17

The ratio between the spectral reflections at the near-infrared wavelength (0.85-0.88) μm and the red wavelength (0.64-0.67) μm on their total is calculating using the following equation [6]

NDVI= (NIR- visible Red)/ (NIR+ visible Red)..................... (1)

IOP Conf. Series: Materials Science and Engineering **757** (2020) 012062 doi:10.1088/1757-899X/757/1/012062

Figure 2 shows that NDVI is calculated from the visible (red) and near-infrared light reflected by vegetation. Healthy vegetation (left) absorbs most of the visible light that hits it, and reflects a large portion of the near-infrared light. Unhealthy or sparse vegetation

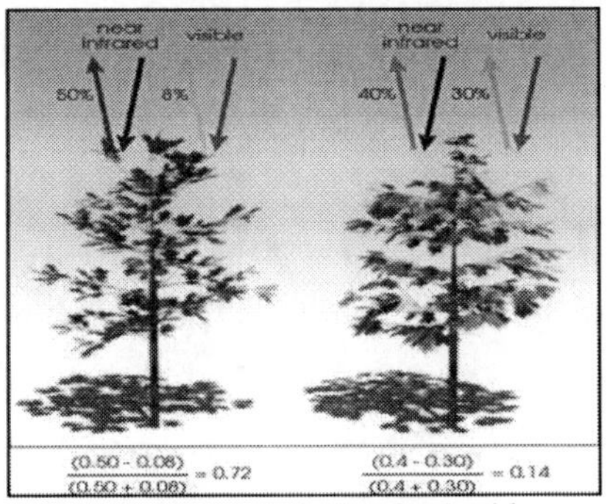

Figure 2. Show that NDVI healthy and unhealthy vegetation [7]

(right) reflects more visible light and less near-infrared light. The numbers on the figure above are representative of actual values, but real vegetation is much more varied. Figure 2. Shows the scientific meaning of the vegetation cover guide [7].

In this study, the ratio of vegetative cover was determined in the city of Mosul based on satellite images of Landsat 8, and figure 3 illustrates the work steps followed in this research.

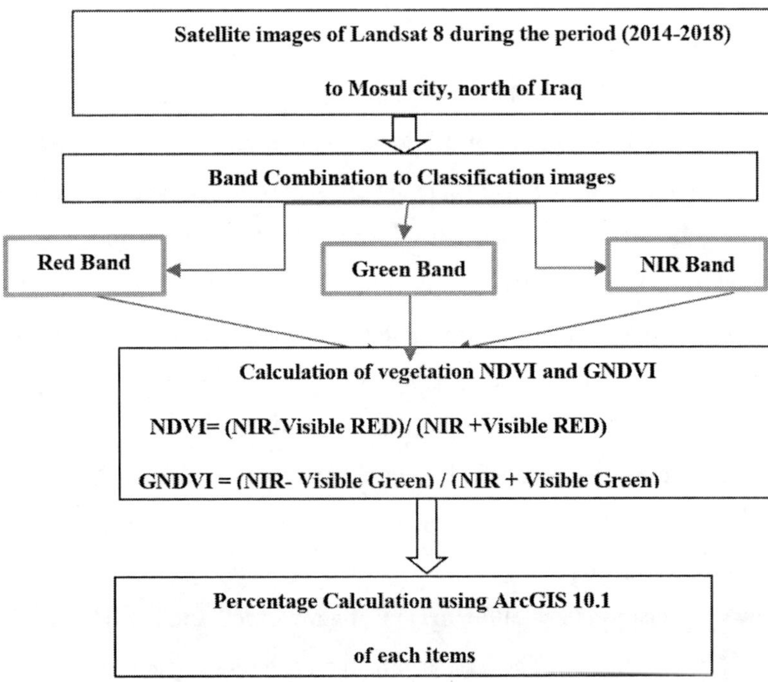

Figure 3. Show the flowchart of the methodology

3.1 Image Classification

Reflections measured by satellite sensors depend on the local characteristics of the earth's surface. Multi-spectral image classification is an important step in the extraction of thematic information from satellite images [8]. The aim is to automate this as much as possible by using suitable image processing and image analysis software. A variety of classifications methods exists. The general purpose of digital classification is to automatically make all pixels in land use categories or categories. Multi-spectral data are usually used for this classification. The digital classification process is an important basis in the study of land use and land cover. The classification process is based on the numerical values of the multiple domains and is based on two types of classification: supervised classification and unsupervised classification, in each method, there is a special method of training, the process by which the computer operates with classification controls such as the number of categories and the statistical criteria for each category [9]. The use of the method used to carry out the classification process requires the Training samples for each species expected to be in the area under study. In this method, each unit of the image is classified on the basis of the degree of convergence and matches with the training samples in terms of spectral response and some statistical calculations. There are several methods of the categorization of the vector, has been used in this research one of these methods called maximum likelihood classification that method of classification The amount of correlation and variation of the spectral response of the training samples is calculated together, and on the basis of which the unknown pixels are distributed to those known categories. In order to do this properly, the method is based on the assumption that modules of training are distributed systematically. That the regular imposition in the distribution of these units must be for all spectral channels used, thus, the spectral value and matrix of difference and correlation can be calculated for each set of training samples used. The many calculations that this method relies on to classify each unit of the image is the only thing in this way that makes it slow to implement and expensive compared to other methods, yet it achieves high accuracy results [10]. The classification process is a representation of the data of the studied phenomenon (vegetation) according to its real location and depends greatly on the extent of similarity or difference between the values of the elements of the phenomenon, i.e. the extent of self-correlation between the elements of the phenomenon. The function of spatial analysis of the important functions in the information system, this function is used to create maps showing the vegetation cover in the study area [11].

Figures (4, 5) show classification of satellite image using Maximum likelihood classifier for a period (2014-2018) in Mosul city. The Landsat images were classified into four classes represent water, vegetation, soil and urban areas.

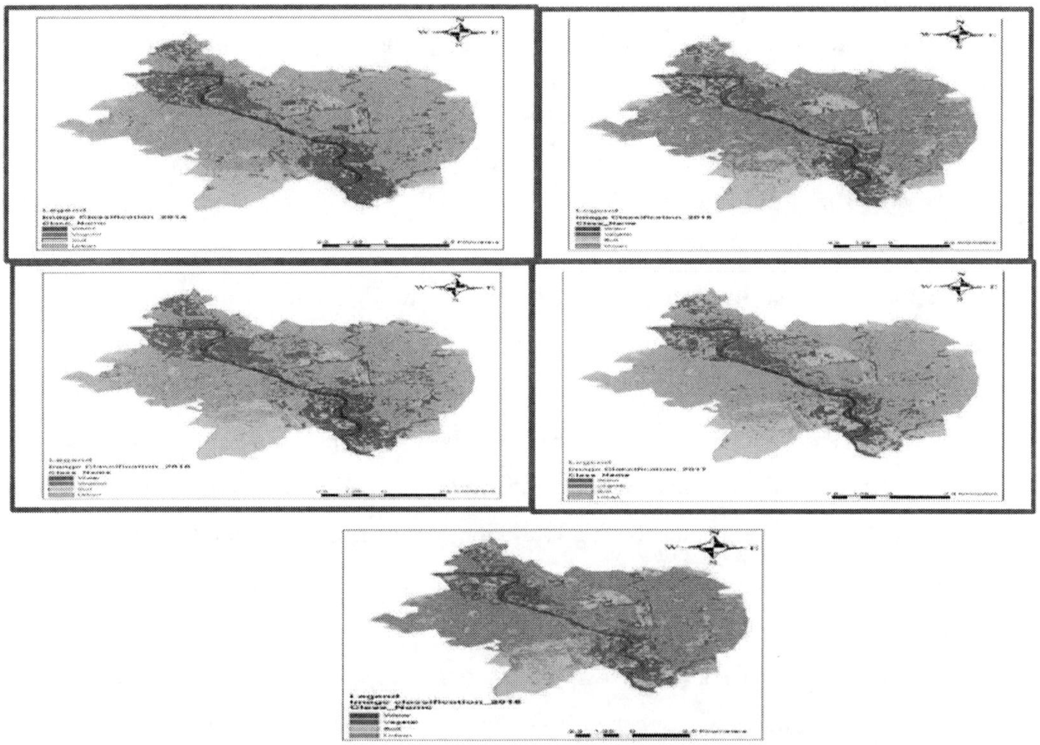

Figure 4. Shows classification of satellite image using Maximum likelihood classifier for a period (2014-2018)

From the previous figures, the blue color represents the water (the water bodies and the rivers), the green color represents the vegetation, the yellow color represents the soil and the gray color represents the urban.

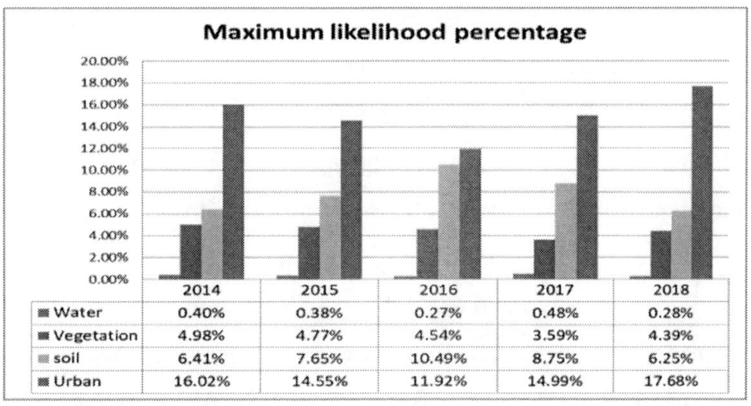

	2014	2015	2016	2017	2018
Water	0.40%	0.38%	0.27%	0.48%	0.28%
Vegetation	4.98%	4.77%	4.54%	3.59%	4.39%
soil	6.41%	7.65%	10.49%	8.75%	6.25%
Urban	16.02%	14.55%	11.92%	14.99%	17.68%

Figure 5. Show the land-cover classification for Mosul city for period 2014 to 2018

Figure (4, 5) shows that the results of the vegetation index calculation for the period (2014-2018). The areas falling to the north-west and the south –west of the study area was a good plant density (large vegetation cover) with values ranging from 4.98% to 4.39%. From this figure, it is found that high density was only present in 2014. After that due to anthropogenic disturbances or other related reasons, this area started to lose the vegetation cover, from 2015 to 2017 caused losses in soil productivity (decline of soil quality) is cumulative. Thus, the highest value found was 4.39% for 2018. This clearly shows the conversion of moderate-deep vegetation cover to low or low – moderate vegetation cover from 2014-2018 time period.

3.2 Normalized Difference Vegetation Index

The normalized difference vegetation index (NDVI) is the most extensively used satellite index of vegetation health and density [12, 13].The NDVI is an index used to identify vegetation and its health through the levels of chlorophyll detected in the leaves. Healthy vegetation absorbs most of the incoming visible, and reflected a large portion (about 25%) of the near infrared (NIR) light, but a low portion in the red band (RED).Unhealthy or sparse vegetation reflects more visible light and less NIR light [14]. In this section, the NDVI technique is used for extracting the various feature presented in the three band satellite image (GREEN, RED and NIR), the NDVI is a simple numerical indicator that can be used to analyze the remote sensing measurements, from a remote platform and assess whether the target or object being observed contains live green vegetation or not, from the equation (1). Vegetation index is calculated by using the reflectance in the red and near infrared bands [15]. The range is theoretically between -1 and +1, where increasing positive values indicate increasing green vegetation and negative values indicate non-vegetated surface features such as water, barren land, ice, snow, or clouds [16].It should be noted, that the information obtained from NDVI data is mostly used qualitatively, in relation to other NDVI data, e.g.to monitor vegetation over time, or to compare different regions with similar characteristics. The magnitude of NDVI is indicative of the level of photosynthetic activity in the vegetation being monitored. The basic NDVI image gives an indication of where vegetation is present and active, but has limited value beyond that [8].

The results of NDVI have a range from (- 1 to 1). In general, the result is positive, indicating that the cell has a plant cover. The higher the resulting positive value, the greater the vegetation and density of the plant. The opposite is true for the negative values that indicate the non-green surface parameters. Therefore, the evidence of vegetative differences is used to distinguish diseased and healthy plants. Figure (6) shows NDVI images for study area. These images display blue area (low value NDVI) which represents no vegetation such as water and other region; while green area (high value NDVI) represent high vegetation such as agriculture land.

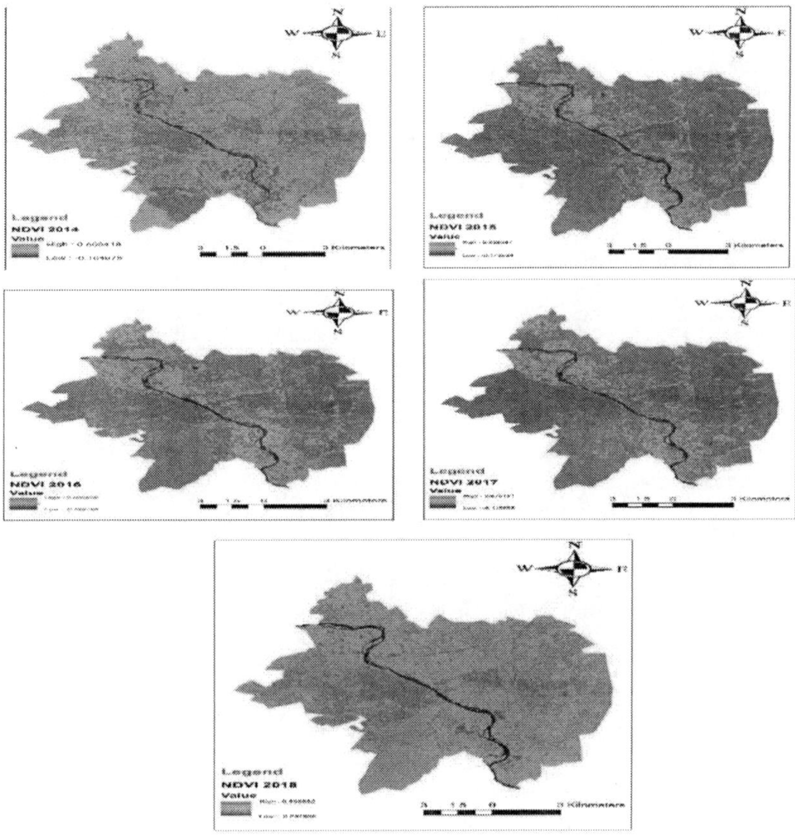

Figure 6. The image resultant by applying NDVI method from 2014 to 2018

3.3 Green Normalized Difference Vegetation Index

Green Normalized Difference Vegetation Index (GNDVI) a variation of NDVI has been developed which uses the green band on of the electromagnetic spectrum rather than the red band. This is referred to as Green Normalized Difference Vegetation Index (GNDVI), defined by [17]:

GNDVI = (NIR-Green) / (NIR + Green) …………….. (2)

The benefit of GNDVI over NDVI is that the green band can cover a broader range of chlorophyll in plants than the red band. This applies to mature plants and can be useful for monitoring yield crop in the late growing season.Figure 7 shows the green vegetation index to highlight the change in vegetation. It was noted that there is a difference between the values of the green vegetation index and values of classification of the satellite images taken with the dates mentioned above.

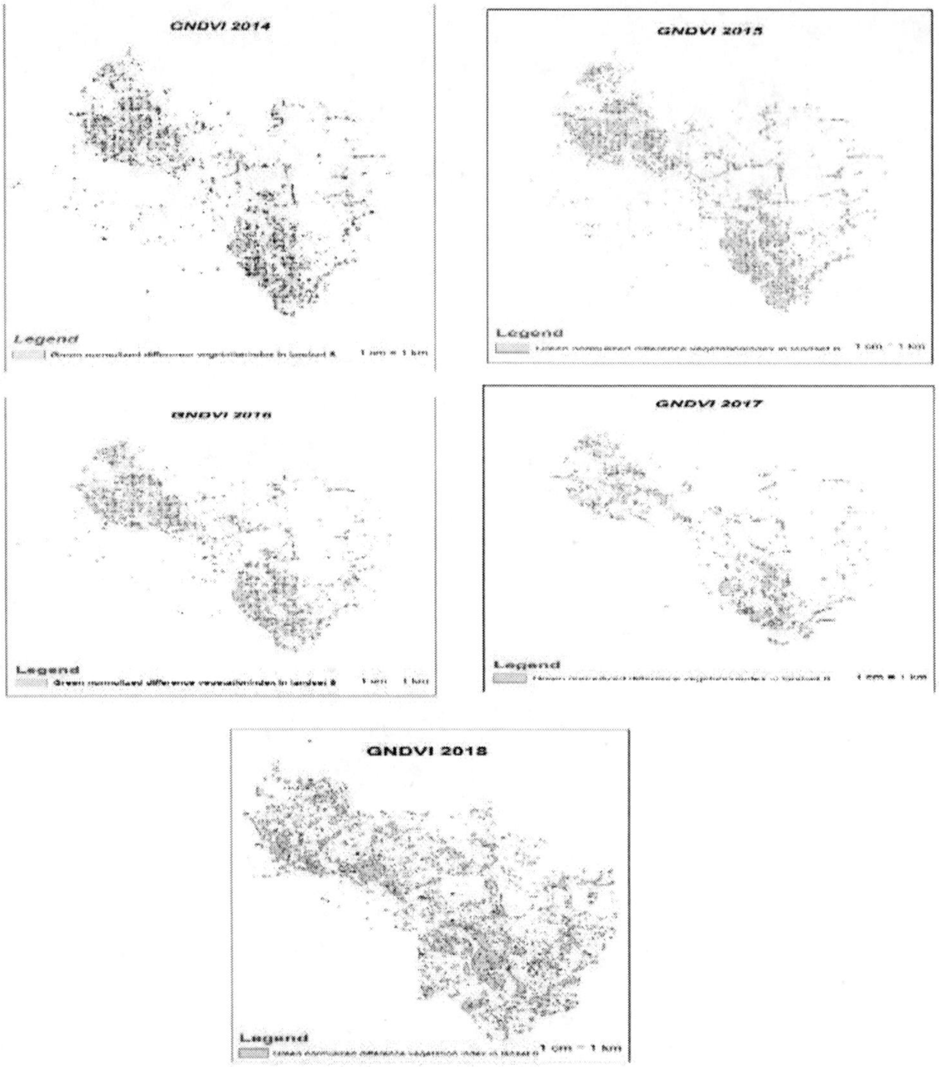

Figure 7. The image resultant by applying GNDVI method from 2014 to 2018

Figure 8 shows the statistical comparison between NDVI and GNDVI for the period (2014-2018).

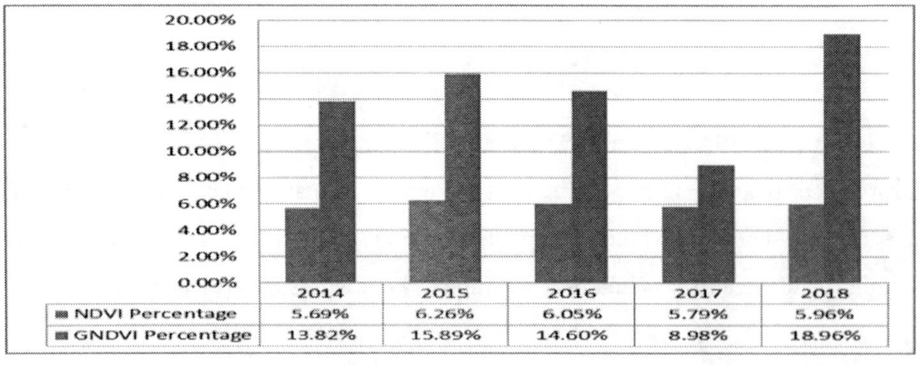

Figure 8. The statistical of NDVI and GNDVI percentage results of the vegetal regions.

IOP Conf. Series: Materials Science and Engineering **757** (2020) 012062 doi:10.1088/1757-899X/757/1/012062

The distributions of NDVI and GNDVI in Mosul city for the period 2014-2018 showed the variations of density vegetation due to change in surface area of the water body of the study area represented by Mosul Lake.

4. Conclusions

From the results of this work, several concluding remarks can be drawn and are summarized as follows: The results of satellite image using Maximum likelihood classifier show that vegetation cover decreased from 4.98% in year2014, 4.77% in year 2015, 4.54% in year 2016, 3.59% in year 2017and then increased to 4.39% in 2018;. While water decrease about 0.4%, 0.38%, 0.27%, 0.48%, and 0.28% for years 2014, 2015, 2016, 2017 and 2018 respectively. Urban area decreased about (16.02 -11.92) % during the period (2014-2016), after that increase when re-reconstruction of building in Mosul city. The (NDVI) and (GNDVI) has been found to be a good indicator for vegetation and land use/ land cover changes, and gives the same behavior about the increasing of vegetal cover but the value of the NDVI is greater than the value of GNDVI. This increase can be attributed to interest of people in the cultivation of arid lands and increases the water of the Mosul city for these years.The statistical relationship between NDVI and water area reflected by significance the effective of NDVI as an indicator of vegetation-moisture conditions for Mosul city. Normalized difference vegetation index (NDVI) depends of spectral reflectance of land use cover so the low value of NDVI which represents no vegetation such as water, soil and urban area ; while green color area represent high vegetation such as agriculture land and vegetation.

References

[1] Hassan E. 2014 . *Journal of Kufa- Physics*.**6** 2.

[2] Anupam K., Manpreet K. 2013. *International Journal of Engineering Trends and Technology*. **4** 1.

[3] Gilabert M.A. , Gonza´lez J. P., Melia J. 2002.*Remote Sensing of Environment* **82** 304.

[4] Arsalan A. O., Ahmad K. K. 2013.*Saudi Society for Geosciences*. **10** 1007.

[5] XIE, Y. 2008. *Journal of Plant Ecology*. **1** 9-23

[6] M. Gandhi.G, S.Parthiban, T. Nagaraj, and A. Christy. 2015. *Procedia Comput. Sci.* **57** 1199-1210.

[7] Murat O,Bilgili,Cemil&Salici,Aylin.2015.*Journal of environmental protection and ecology* **16** 264-273.

[8] Schultz .G. A. and Engman. E. T. 2000.*Ruhr-Universitat Bochum NAS A/Goddard Space Flight Center*.

[9] Saleh. M. A. 2013. *International Journal of Emerging Tec hnology and Advanced Engineering*.**3**. 389.

[10] Bogus. O.Y.1988.*Mosul University for obtaining the MS degree in remote sensing*.

[11] Khalid A. H. 2015. *AL-Mostansiriyah journal for arab and international studies* **50** 161-177.

[12] Bellone. T. P, Boccardo, and Perez, F. 2009. *Am. J. Environ. Sci.***5** 460-466.

[13] Pierce. F. J . and Clay D. 2006 .*GIS Applicatoins in Agriculture*. 41-42.

[14] Ahmed M. A. and Ahmad W. A. 2013.*Iraqi J. Sci.***54** 895-901.

[15] Chen x., Vierling L. and Deering D. 2005. *Remote Sens. Environ.***98** 63-79.

[16] Yacouba D., Guangdao H and Xingping W. 2010. *Rep. Opin.***2** 7- 16.

[17] Salman S. S. 2016. *Ph. D., University of Baghdad.*

IOP Conf. Series: Materials Science and Engineering **757** (2020) 012063 doi:10.1088/1757-899X/757/1/012063

Investigation In to Some Electrical Properties of Ytteria – Silicon Carbide ,Ceramic Composites

I A. Mohammed * A A. Ibrahim

Department of physics ,Ibn AL Haytham , College of Science , ,University of Baghdad, Baghdad, IRAQ

* corresponding author: Email: intehaa2019@yahoo.com

Abstract: Ytteria ,silicon carbide powders of (99.99) purity having (40) μm grain size were used to form different combinations of the system Y2O3- SiC . Silicon carbide particles was added in weight percent ranging (5,10,15and 20), dry milling were done for all the combinations at 8 hours .Samples of disc shape were fabricated and sintered at different temperature (800,1000,1200, 1400 and 1500) °C ,under static air and for three hours as soaking time .Porosity for all sintered samples were measured . SEM after sintering were detected too .Electrical properties were studied to find the dielectric constant , breakdown voltage and the capacitance were don. Improvement in its electrical properties were achieved at all the combinations after adding the dispersed Sic particles . And especially at the combination (Y2O3 80-SiC20)Wt.% , that sintered at1500 °C.

Keywords: ceramic composites , Sintering , Dielectric constant , Brake down voltage ,Capacitance

1.Introduction

The composite materials mainly consist of two or more than material can be intimately bonded to have a new material with unique properties compared with their consistent .As all composites ceramic materials ,ytteria –matrix composites can be tailored . By addition of high strength ,high temperature resistance ,high hardness to get attractive properties with the ability to operate under high temperature .Silicon carbide is a filler material can be used successfully due to its excellent properties ,such as high thermal conductivity ,low coefficient of thermal expansion ,high modules and low density [1] . Chemical properties reflect its high stability during synthesis under severe conditions .SiC particles can be used with other materials ,like metals Al-SiC metal matrix composites .Ytteria in recent twenty years received many scientific attention in many fields , because of its excellent properties such us high dielectric constant ,low absorption in broad range (near UV to IR),superior electrical brake down voltage and low leakage current [2] . Yttrium oxide is one of the most important compounds of yttrium and accounts forth largest use. It is widely used in making YVO4 europium, and Y2O3 europium phosphors to give the red color in color television tubes. Many hundreds of thousands of pounds are now used in this application. Yttrium oxide also is used to produce yttrium-iron-garnets, which are very effective microwave filters. It also has potential use in ceramic and glass formulas, as the oxide has a high melting point and imparts shock resistance and low expansion characteristics to glass. Hard, dense layers are deposited by electron-beam evaporation .In the current investigation our goal is to study the effect of adding silicon carbide on the microstructure and electrical properties of ytteria after sintering proses .

IOP Conf. Series: Materials Science and Engineering **757** (2020) 012063 doi:10.1088/1757-899X/757/1/012063

2. Experimental details

2.1 preparation of *Y_2O_3- SiC composites ceramic material .*

High purity ytteria and silicon carbide of 40 μm , were used to fabricate the composites ceramics .Different weight percentage ranging (5,10,15 ,20) SiC were dry mixed with the matrix ytteria at soaking time 8 hrs. .Discs of 2 cm in diameter were compacted to be ready for sintering process . Different sintering temperature used under static air ,starting step by step to get the sintering behavior , gradually from (800,100,1200, 1400and 1500) ˚C. The sintered body of three samples from each combination are ready to be test.

2.2 Characterization of samples

The fabricated samples(Disc shape) , from each combination after sintering was measured its porosity ,by using Archimedes' method. .Scanning electron microscopy were applied to study its microstructure after sintering .And by applied the required electrical conditions to calculate the di electrical constant ,brake down voltage and capacitance . The tests were done according to the ASTM D374/374M [3] , by using high accurate instruments having a well calibrated test (± 0.01) tolerance in measurements .

3.Results and discussions

3.1Characterization

The composites ceramic after sintering proses its sintering activity reflect the behavior of its grain size and porosity distribution .This can be seen from the figure 1 , which shows the relation between the sintering temperature and porosity for all the combinations .The silicon carbide act as sintering aid [4] , by reducing the porosity with increasing its weight percentage added to the matrix ,this behavior was noticed at all sintered temperature having maximum value of (68) and reached minimum value (50), for the silicon carbide 5 weight percent sintered at temperature 800 ˚C. While at the weight percentage 20 from silicon carbide the maximum value is (52) and reduced to be (12) as a lowest value reached after sintering temperature at 1500 ˚C .

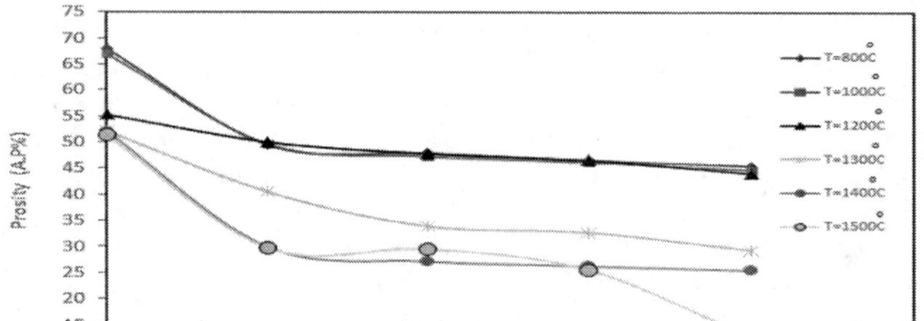

Figure 1. The Porosity behavior with respect of Wt.%SiC , at different sintering temperatures .

The well homogenization between the silicon carbide particles and yttrium oxide particles lead to get this improvement in porosity distribution compared with the matrix [5]. More over the formation SiO_2 ,liquid phase [6] lead to increase the sintering density [7], and reduce the porosity .The results found in porosity distribution correlated to the changes in grain size after sintering [8] , as shown in figure 2(A,B) the scanning electron micrographs of ceramic composites representative by the combination (Y_2O_3 95 Wt.% - SiC 5 Wt.%) ,sintered at 1500 °C ,showed the gain size distribution with open porosity figure 3 (A) .And in figure2 (B) ,the SEM micrographs of the combination (Y_2O_3 80 Wt.% - SiC 20 Wt.%) sintered at 1500 °C ,the grain growth exist with mass transportation forming the closed porosity [9].

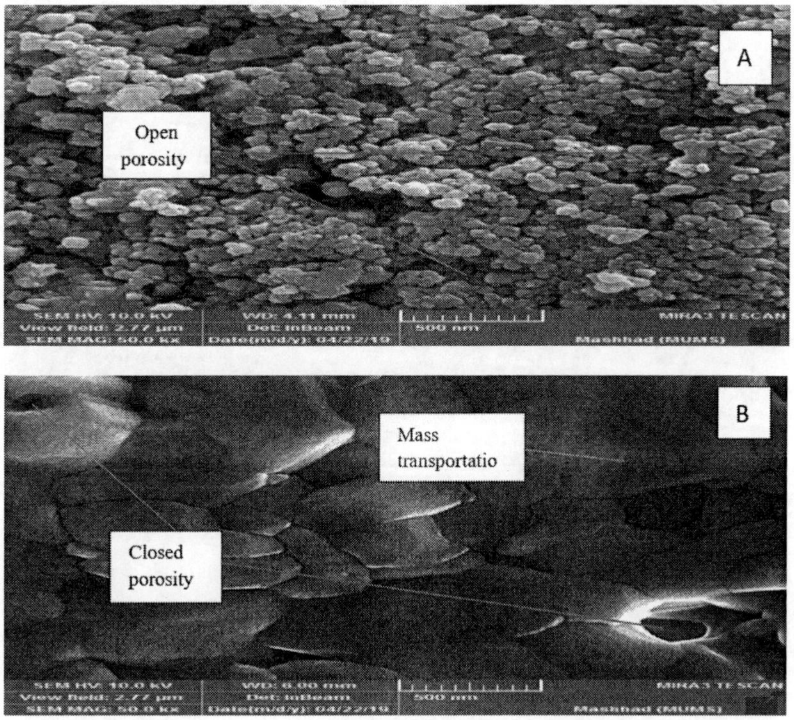

Figure 2. SEM micrographs ,a) Y2O3 95 Wt.% - SiC 5 Wt.% ,b) Y2O380 Wt.% - SiC 20 Wt.% , Sintered at 1500 °C .

The di electric constant as a function of silicon carbide added to the ytteria was tested for all combinations sintered at different sintering temperatures as shown in figure 3. The general behavior shows the decreasing in K value with respect to the silicon carbide added and increasing in sintering temperature .The highest value signed at sintering temperature 800 °C , for the 5 weight percent SiC added .And lowest one at sintering temperature 1500 °C ,for the 20 weight percent SiC added . This behavior focus on the microstructure changes [10] and porosity distribution that drive the space charge in addition to the surface roughness [11] .The capacitance calculations reflect and matched the same results with K and for all combinations as shows in figure 4 .

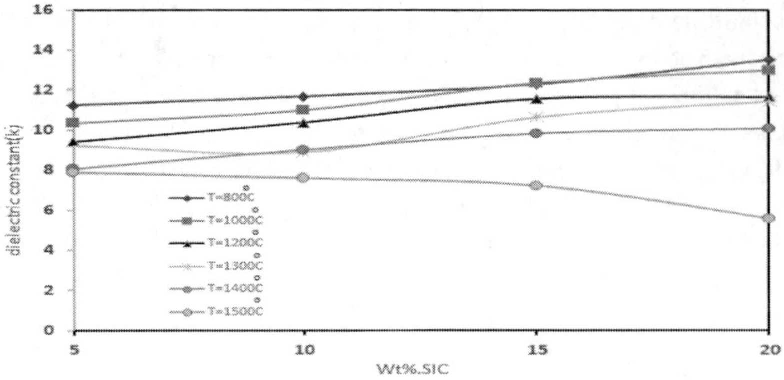

Figure 3. The di electric constant with respect to the SiC weight present , sintered at different temperatures .

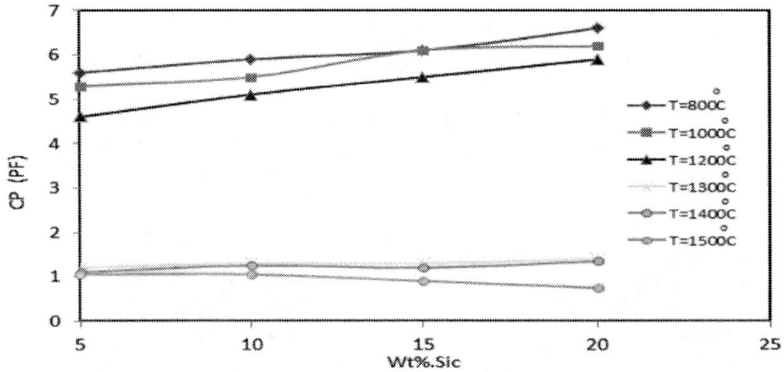

Figure 4. The Capacitance of the composite ceramics with respect to the SiC weight present , sintered at different temperatures .

The break down strength analysis for all combinations after sintering shown in figure 5, the increasing in ability to stand the voltage applied was noted with the increasing in silicon carbide added and with increasing in sintering temperatures having minimum value of 1.5KV/mm for 5 weight percent SiC added ,sintered at 800 °C, and maximum value of 6.5KV/mm for 20 weight percent SiC added ,sintered at 1500 °C . Based on this data the thickness dependence explain the increasing in break down strength by impeded electrons or hole through the test samples that deeply effected by the surface roughness and porosity distribution [12].

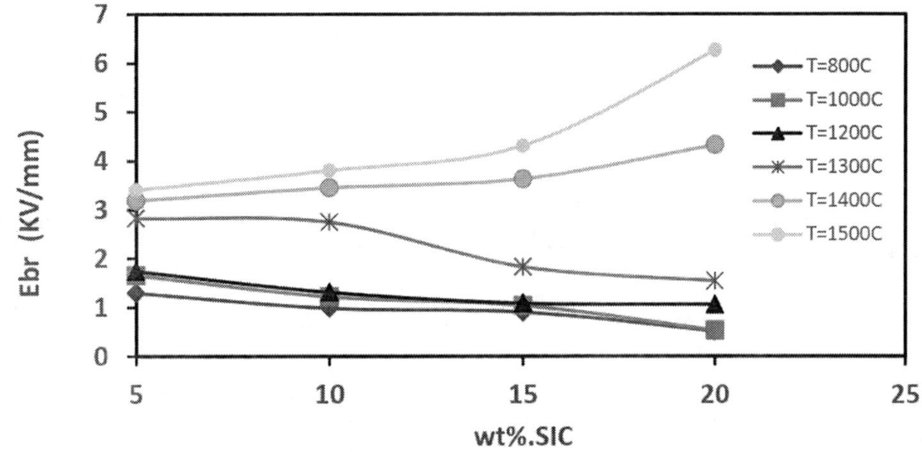

Figure 5. The Brake down strength of the composites ceramics with respect to the SiC added, sintered at different temperatures .

4. Conclusions

The effect of added dispersed silicon carbide particles of (40) µm was shown at all the fabricated compacted composites ceramics sintered at different sintering temperatures ranging from(800 ,1000,1200,1400,1500) °C , under static air for 3 hours .Clear improvement in sintering density was noticed by the microstructure and porosity distribution .Di electric constant ,capacitance and brake down voltage were improved comparing with the matrix . The improvement signed at the combination (Y_2O_3 80 Wt.% - SiC 20Wt.%) , sintered at1500 °C, under static air for 3 hours .

Acknowledgments

Authors wishing to acknowledge assistance from the state commission of research and development chemical research centre, material lab, ministry of industry .And the university of technology, materials science department .

5. References

[1] Wen, L., Sun, X., Lu, Q., Xu, G. and Hu, X 2006 *Opt. Mater.* **29**:239–245.

[2] Malec, D., Bley, V., Lebey. T., Talbi, F., Lalam, F. 2005 IEEE Annual Report Conference on Electric Insulation and Dielectric Phenomena, , 63-66.

[3] Owate, I.O., Freer, R. 1992 Sixth International Conference on Dielectric Materials, Measurements and Applications 443-446.

[4] H. Liang, X. Yao, J. Zhang, X. Liu and Z. Huang 2104 Journal of the European Ceramic Society **34**: 831-835.

[5] J. Ihle, Hermann M. and J. Adler 2005 Journal of the European Ceramic Society 25 **7**:1005-1013.

[6] R. M. German, P. Suri and S. J. Park,. Journal of Materials Science **44** **1** : 1-9.2008.

[7] D. Galusek, K. Ghillanyova, J. Sedlacek, J.Kozankova and P. Sajgalik. 2012 Journal of the European Ceramic Society **32**: 1965-1970..

[8] J. Ihle, Hermann M. and J. Adler 2005 Journal of the European Ceramic Society **25** **7**:1005-1013..

[9] Wang, L., Jiang, W. and Chen, L., J. Mater. Sci., 2004 **39**, 4515–4519.

[10] Haddour, L., Mesrati, N., Goeuriot, D., Tréheux, D., J.Eur 2009 Ceram. Soc **29**:2747 – 2756.

[11] Gerson, R., Marshall, T. C. 1959 J. Appl. Phys. **30:**1650-1653.

[12] Chen G., Zhao J., Li S., Zhong L. 2012 Appl. Phys. Lett. **100**: 222904-1 - 222904-4.

Silicon Carbide particles dispersion Effect on the sintering behavior and damage characterization of Yttrium Oxide Composites Ceramics

Intehaa A. Mohammed * Ayser A. Ibrahim

[1] College of education for pure science, Ibn AL-Haytham , University of Baghdad, Baghdad, IRAQ

* corresponding author: Email: tarik@scbaghdad.edu.iq

Abstract: Yttria, silicon carbide powders of (99.99) purity having (40-60) µm grain size were used to form different milling were done for all the combinations at 12 hours .Samples of disc shape were fabricated and sintered at different sintering temperature (800,1000,1200 and 1400) °C, respectively under static air and for three hours as soaking time . XRD test reflect the formation of new phase which is Y_5Si_3C (Unique). SEM test shows the sintering behavior after the sintering proses. Sintered samples were mechanically tested by hardness and fracture strength. Improvement was noticed for all the combinations after adding the dispersed SiC particles detected especially at the combination (Y_2O_3 80-SiC20) Wt.%, sintered at 1400 °C.

Keywords: ceramic composites, Sintering , Hardness , Fracture strength .

1.Introduction

The composite materials mainly consist of two or more than material can be intimately bonded to have a new material with unique properties compared with their consistent. As all composites ceramic materials ,yttria –matrix composites can be tailored . By addition of high strength ,high temperature resistance ,high hardness to get attractive properties with the ability to operate under high temperature .Silicon carbide is a filler material can be use successfully du to its excellent properties ,such as high thermal conductivity ,low coefficient of thermal expansion ,high modules and low density [1] . Chemical properties reflect its high stability during synthesis under severe conditions. SiC particles can be used with other materials ,like metals Al-SiC metal matrix composites [2].Yttria in recent twenty years received many scientific attention in many fields , because of its excellent properties such us high dielectric constant ,low absorption in broad range (near UV to IR),superior electrical brake down voltage and low leakage current [3] .

Yttrium occurs in nearly all of the rare-earth minerals. It is recovered commercially from monazite sand, which contains about 3%, and from bastnasite, which contains about 0.2%. The metal is now produced mainly by reduction of the fluoride with calcium metal. Yttrium oxide is one of the most important compounds of yttrium and accounts for the largest use. It is widely used in making YVO_4 europium, and Y_2O_3 europium phosphors to give the red color in color television tubes. Many hundreds of thousands of pounds are now used in this application. Yttrium oxide also is used to produce yttrium-iron-garnets, which are very effective microwave filters.

It also has potential use in ceramic and glass formulas, as the oxide has a high melting point and imparts shock resistance and low expansion characteristics to glass. Hard, dense layers are deposited by electron-beam evaporation sputtering. Typical applications are protection of aluminum and silver mirror coatings, intermediate layer in wide band visible AR coatings and for XeCl (308 nm) laser AR and dielectric mirror designs.

Yttrium oxide is the thermodynamically most stable oxide and is suitable for extreme operation conditions. It is used as a coating in metal production. Yttria is often used with zirconia to form yttria Stabilized Zirconia (YSZ).

Yttria has the common characteristics of ceramics that are not found in metallic or organic materials, including high hardness (Next to Diamond), high mechanical strength, high temperature stability, chemical resistance, erosion resistance, low electrical conductivity.
Many researchers study the sintering behavior of yttria ,E.KOSTIC ,M.M.RISTIC , mentioned the powder activity from 1000 °C to 1400 °C ,while Marlowe and Wildwer study the densification from 670 to 1600 °C [4].
However, as a ceramic material, both yttria and SiC is brittle and its mechanical strength is considered to be less than that of other metal carbides. Aside from that, SiC is susceptible to chipping and fracture when introduced to large mechanical stress or shock . Another disadvantage of SiC is that the material is difficult to be manufactured in any required shapes especially for component design for certain application. Moreover, it is difficult to density the SiC without the use of suitable additives and external pressure [5]. This is mainly due to the covalent nature of silicon carbon (Si-C) bonding together with its low self-diffusion coefficient. Among common sintering aids to obtain a dense silicon carbide are aluminum oxide (Al_2O_3), yttrium oxide (Y_2O_3) and boron oxide (B_2O_3).
As experimentally investigated by Omori and Takei [6], β-SiC could be densified by sintering at the temperature of 2100 °C using yttrium oxide and aluminum hydroxide. In the present work our goal is to study and improve the sintering behavior and the mechanical properties for the yttria after adding silicon carbide as dispersed particles under elevated temperature.

2. Experimental details

2.1 preparation of *Y_2O_3- SiC composites ceramic material .*

High purity yttria and silicon carbide of 40 µm , were used to fabricate the composites ceramics .Different weight percentage ranging (5,10,15 ,20) SiC were dry mixed with the matrix yttria at soaking time 8 hrs. .Discs of 2 cm in diameter were compacted to be ready for sintering process . Different sintering temperatures used under static air , gradually from (800,100,1200,and 1400)°C .The sintered bodies of three samples from each combination are ready to further tests .

2.2 Characterization of samples

The fabricated samples were measured its sintered density(Bulk density) , by the geometrical method using high accurate electronic balance and high accurate digital micrometer and according to the ASTM C29/C29M-17a .X-ray diffraction have been done by using (6000-Shimadzu) with wavelength λ=0.15418 nm. Scanned from the two theta degree (20-70) ,to identify the resulting crystal structure .

The damage characterization for each combination after sintering was observed by applied the Vickers hardness test and fracture strength respectively .By using high accurate instruments having a well calibrated test (± 0.1) tolerance in measurements.

3.Results and discussions

3.1Characterization

The investigated XRD pattern for each combination after sintering process was discussed .The most attractive and unique one shown in figure 1 , which indicate the domain phase No.1 ,belong to the yttria [7],while the most second domain phase is No.3 ,belongs to the new phase formation Y_5Si_3 (File 00-038-0794).And the silicon carbide phase [8],was also detected and signed by No.2,in all sintering temperature

.More over silicon oxide phase [9], also signed by No.4 Wt.%, at sintering temperature 1200 °C and1400 °C. The new phase No.3 ,was generated at sintering temperature 1400 °C, and it was very clear can be signed in all the combinations ,and its intensity increased as increasing in the SiC added weight percentage ,reaching its maximum at the 20 Wt.% .

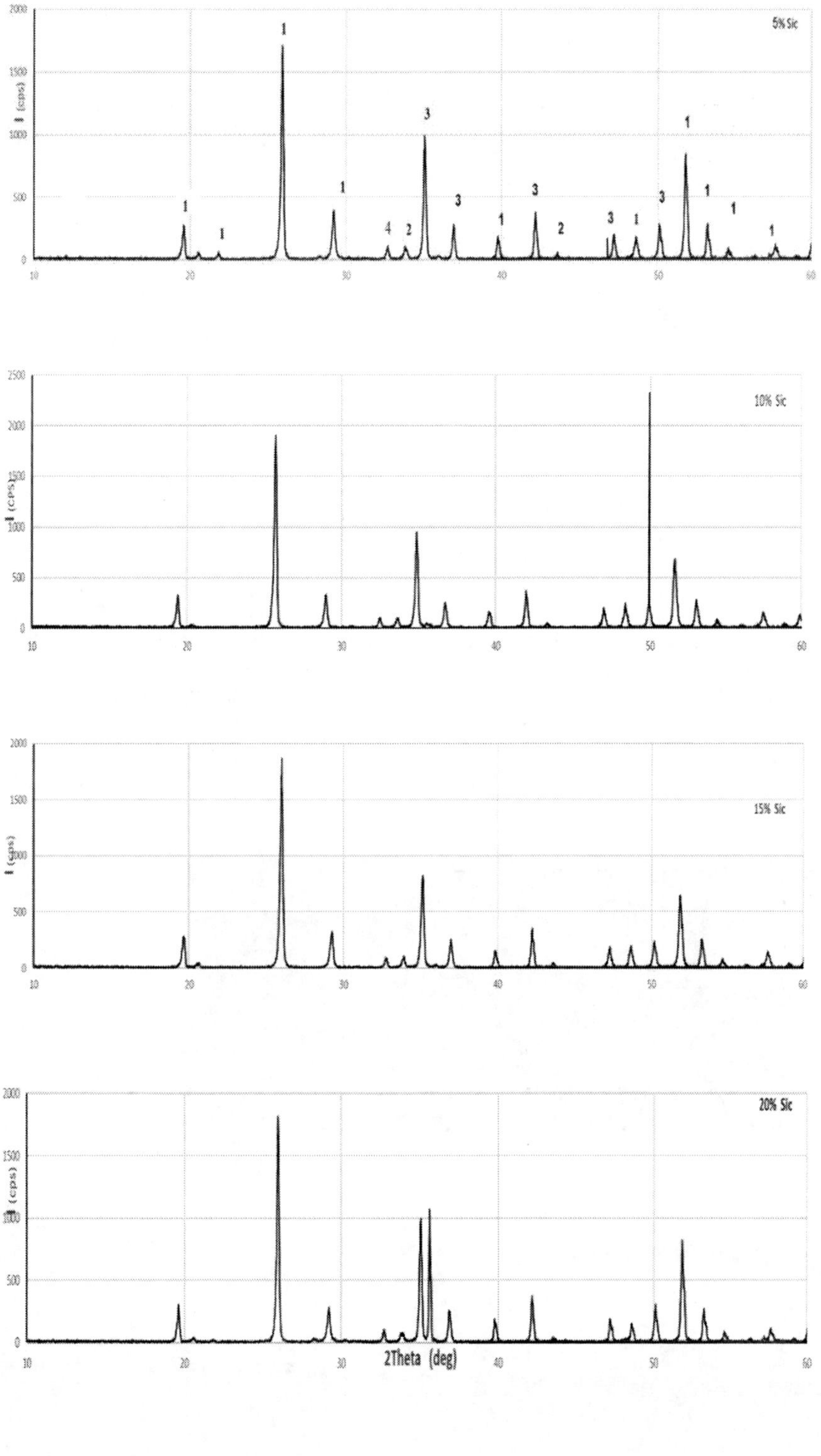

Figure1: XRD pattern for the ceramic composite materials ,sintered at 1400 ˚C, under static air for 3 hours , no.1 (Y_2O_3) , no.2 (SiC) ,no.3(Y_5Si_3) , no.4 (SiO_2) .

IOP Conf. Series: Materials Science and Engineering **757** (2020) 012064 doi:10.1088/1757-899X/757/1/012064

The sintered density relation with the added silicon carbide shown in figure 2.The increasing in sintering density was noticed for all combinations comparing with its value regarding the matrix (Yttria) .Starting from (SiC 5 Wt.%) with highly increased value reflect the first sintering stage and overlapping with the second sintering stage at the (SiC 10,15 Wt.%) .While the best results can be shown at the sintering temperature 1400 °C , and especially for the combination (Y_2O_3 80-SiC20)Wt.% , reached the maximum value 2.570(gm./cm3) which represent the final sintering stage .This increasing in sintered density is due to the best homogenization done between the yttira and silicon carbide powder [10] ,which lead to increase the powder activity of the matrix [11],and the formation of liquid phase [12] .

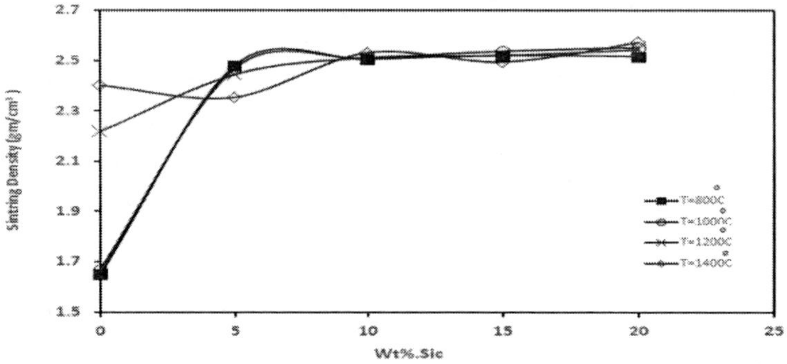

Figure 2: Sintered density for the composites ceramics (Y_2O_3-SiC) sintered at different temperatures under static air for 3 hours .

at sintering temperature 1200 °C as shown in figure 3,the scanning electron micrograph which indicate the SiO_2 as the liquid phase [13].And at 1400 °C , the three sintering stages[14] as shown in figure 4 (,A ,B, C) the scanning electron micrographs of the composites ceramic material. Where (A) is the first stage of grain contacting , (B) the second stage of grain growth with open porosity and (C) ,the third stage(final stage) of mass transportation with closed porosity .

Figure 3: SEM micrograph for the composites ceramics (Y2O3 80 SiC 20)Wt.% , sintered at 1200 °C under static air for 3 hours.

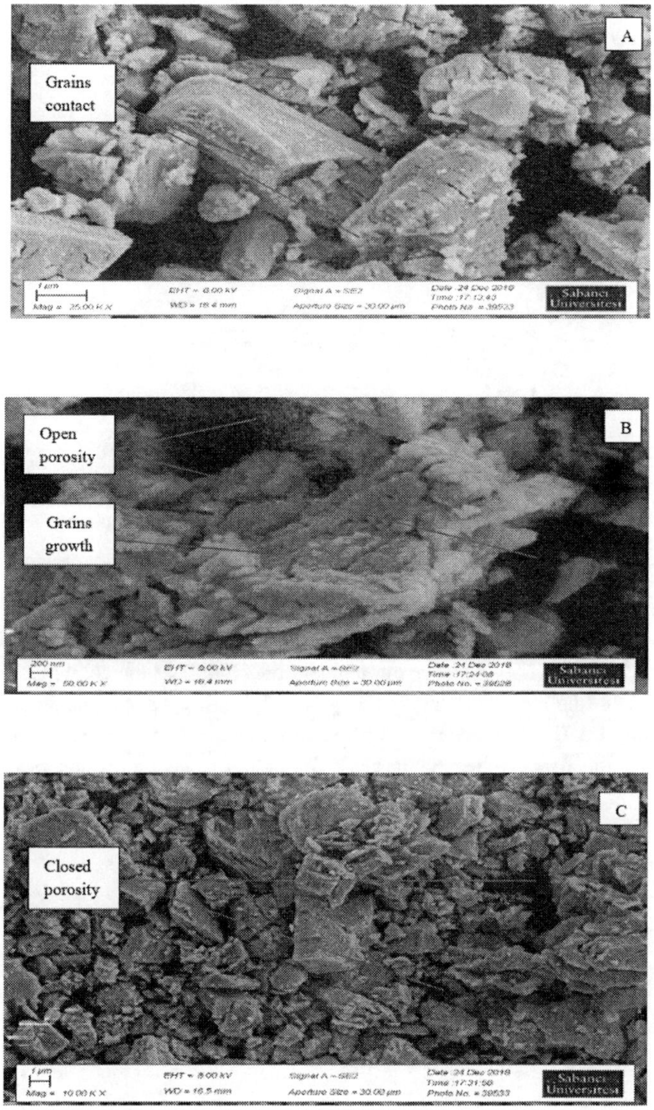

C

Figure4: SEM micrograph for the composites ceramics (Y2O3 80 %-SiC 20) Wt.% ,sintered at 1400 ˚C under static air for 3 hours ,A) First stage ,B) Second stage ,C) Third stage of sintering .

hardness test can clearly be shown for all the fabricated composites ceramic Y_2O_3- SiC ,the increasing in hardness after adding the dispersed silicon carbide particles sintered at different sintering temperature under static air was observed at all the combination and at all sintering temperature reaching its maximum value of 380Mpa at the combination (Y_2O_3 80 - SiC 20) Wt.% ,sintered at 1400 ˚C as shown in figure 5 .The increasing in hardness can be translated to the improvement in the sintered density coupled with increasing in silicon carbide dispersed particles , that reflect the matching in two materials to have the final product with hard and uniform shape after the mass diffusion settled during the final stage of sintering [15]

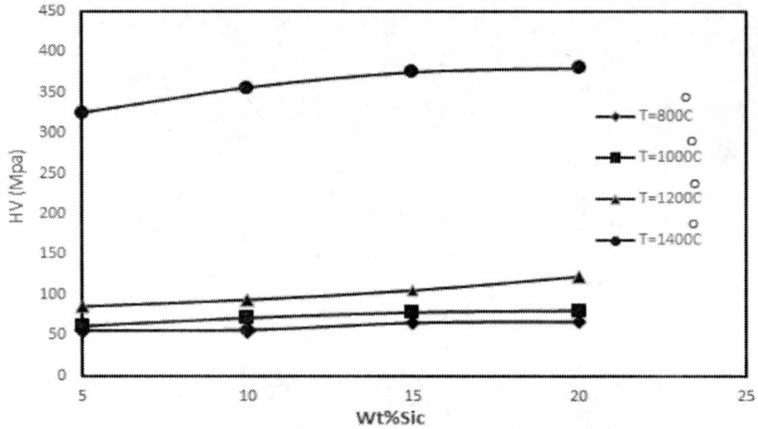

Figure 5: Hardness for the composites ceramics (Y2O3- SiC) ,sintered at different temperatures under static air for 3 hours.

From figure 6 , which represent the fractural strength for all combinations after sintering with respect to the added silicon carbide particles , the increasing in fractural strength started slowly with little increasing at the combination Y_2O_3 95Wt.% -SiC 5Wt.% , and at all the sintering temperatures ,while at the combinations Y_2O_3 90Wt.% , Y_2O_3 85Wt.% and Y_2O_3 80Wt.% the increasing marked as gradually until reach the maximum value of 4.8Mpa at the sintering temperature 1400°C .This behavior is well consistency with the sintered density and hardness results ,which is commonly reflect the effect of the silicon carbide particles dispersion manner within the matrix [16].

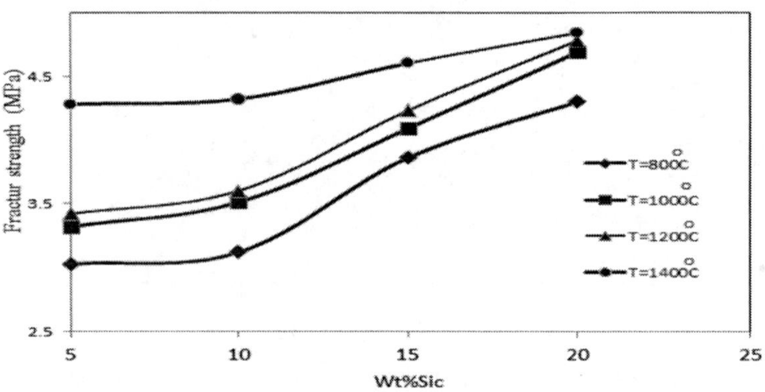

Figure 6: Room temperature flexural strength for the composites ceramics (Y2O3- SiC) ,sintered at different temperatures under static air for 3 hours.

4. Conclusions

The effect of added dispersed silicon carbide particles of (40-60) μm was shown at all the fabricated compacted composites ceramics sintered at different sintering temperatures ranging from(800 ,1000,1200,1400) °C , under static air for 3 hours .The phases presents and the unique one (Y_5Si_3) is indicated too .Clear improvement in sintering density was noticed .More over the damage characterization were achieved by investigated the hardness and fractural strength ,which reflected the improvement for all the fabricated combinations. The best results signed at the combination($Y_2O_3$80-SiC 20) Wt.% ,sintered at 1400 °C ,under static air for 3 hours .

Acknowledgments

Authors wishing to acknowledge assistance from the state commission of research and development chemical research centre ,material lab, ministry of industry , and SABANJA university Istanbul ,Turkey ,for their offering time and instruments facility .

References

[1] G. Magnani, G. Sico, A. Brentari and P. Fabbri 2014 Journal of the European Ceramic Society **34**:4095-409.

[2] R.N.Rao ,S.Das ,P.V.Krishna 2008 Professional Engineering Publishing **222** 1 .

[3] G.Tong ,D.Clark ,C.Bernau ,M.Sanders,O.Rayan 2010 J. Mater .Chem. **20** 6333.

[4] T.Andelman ,S.Gordonov ,G.Busto ,P.V.Moghe, R.E.Riman 2010 Nano scale Res. Lett.**5** 263-273.

[5] C.Lorrette .A .Rean ,L.Briottet 2013 Journal of the European Ceramic Society **32**: 147- 145.

[6] H.Liong ,X.Yao ,J.Zhang ,X.Liu and Z.Hung 2014 Journal of the European Ceramic Society **34**:831-835.

[7] T.T.Issa ,D.K.M Al Nasrawy 2014 Research Journal of Science and IT Management **3** 4 .

[8] Tarik T I , Kawakib J M and Jenan M.H 2015 ,Springer Proceedings in Physics **164** :52-56 DOI : 10 1007/978-3-319-16919-4 -7.

[9] T.T.Issa ,J.M.Haan ,E.T.Abdullah and F.S.Shokr 2016 AIP Conference proceeding **1727** 020012 2016 DOI:10.1063/1.4945967 .

IOP Conf. Series: Materials Science and Engineering **757** (2020) 012064 doi:10.1088/1757-899X/757/1/012064

[10] M. I. Johari, M. Mustapha, O. Mamat, S. Kakooei, T. L. Ginta and M. Z. P Ismadi 2016 ARPN Journal of Engineering and Applied Sciences **11**.

[11] Ulrike Ta¨ffner, Veronika Carle, and Ute Scha¨fer 2004 Metallforschung, ASM Handbook Volume 9 Metallography and Microstructers , Handbook Vol. 9 , ASM International.

[12] Saito, N., Matsuda, S.-I. and Ikegami, T.,. *J. Am. Ceram. Soc.*, 1998, 81, 2023–2028.

[13] Ruchi Nandanwar1, Purnima Singh1 and Fozia Z. Haque 2015 American Chemical Science Journal **5** 1: 1-10.

[14] Wang, X.-H., Chen, P.-L. and Chen, I.-W. J. 2006 Am. Ceram. Soc. **89**:431–437.35.

[15] Huang, Z., Sun, X., Xiu, Z., Chen, Sh. and Tsai, C.-T 2004 Mater. Lett. **58**: 2137–2142.

[16] V. Carle, U. Scha¨fer, U. Ta¨ffner, F. PredelR. Telle, and G. Petzow1991 Pract. Metallogr. **28**: 420–434.

IOP Conf. Series: Materials Science and Engineering **757** (2020) 012065 doi:10.1088/1757-899X/757/1/012065

Structural, Morphological and Optical properties of n-type Porous Silicon-effect of Etching Current Density

H K Abood[1] and F A-H Mutlak[1*]

[1] University of Baghdad, College of Science, Physics Department, Baghdad, Iraq

[*]Corresponding author's e-mail: Falah.mutlak5@gmail.com

Abstract. Porous silicon layers have been prepared from n-type silicon wafers of (111) orientation. XRD, AFM, reflectivity and PL have been used to characterize the structural, morphological, and optical properties of porous silicon. The influence of varying etching current density in the anodizing solution, on structural and optical properties of porous silicon has been investigated. It is observed that pore size increases with etching current density and attain maximum for 56 mA/cm^2 and then decreases. The PL spectrum peak shifts had been slight towards the higher energy side, which supports the quantum confinement effect in porous silicon. The reflectivity shows that the porous silicon surface lower reflectance which is due to the very thin layer of porous silicon and changed refractive index profile at the interface of the bulk silicon and porous silicon material.

Keywords: Porous silicon; Si microstructures; Etching; Solar cell.

1. Introduction

Silicon thin films are used in many devices in various fields of applications such as optics, opto-electronics and microelectronics [1]. A band gap of silicon 1.12 eV leads to efficient detection of visible light and conversion of sun light into electricity. There is a wide range of interest in improving the responsivity of silicon detectors in the visible and the near-infrared (near-IR) regions of the electromagnetic spectrum [2]. In order to increase the efficiency of actual solar cells, many different ways are currently being developed by researchers. Through the focus of work on the nano and micro-structurization of the surface [3]. Nanostructured solar cells offer several advantages for solar cells including; (1) the ability to exceed a single junction solar cell efficiency by implementing new concepts; (2) the ability to overcome practical limitations in existing devices, such as tailoring the material properties of existing materials or using nanostructures to overcome constraints related to lattice matching and; (3) the potential for low cost solar cell structures using self-assembled nanostructures [4]. PS was discovered by Uhlir in 1956 [5], and in 1990, Canham [6] showed that certain PS materials can have large photoluminescence (PL) efficiency at room temperature in the visible region. This result was surprising, since the PL efficiency of bulk silicon is very low, due to its indirect energy band gap and short non-radiative lifetime. The reason for PS giving surprising result was the partial dissolution of silicon, which causes i) the formation of small silicon nanocrystals in the PS material: ii) the reduction of the effective refractive index of PS with respect to bulk silicon, which

increased the light extraction efficiency from PS: and iii) the spatial confinement of the excited carriers in small silicon regions where non-radiative recombination centers are mostly absent. In general, it can be said that the PS is an interconnected network of air holes (pores) in Si. Porous structures are formed with the mechanism proposed by Lehmann and Gosele [7]. There are three different kinds of etching techniques for monocrystalline silicon solar cells: photochemical-etching, electrochemical-etching and photo electro chemical etching [8]. The beneficial effect of the micro texturized Si obtained by femtosecond laser was demonstrated by Kolasinski *et al.* [9, 10] which showed Laser assisted etching in aqueous fluoride solutions allows for selected area formation of nan crystalline porous silicon. Mutlak. [11] Prepared a laser-micro textured Si structure which reduces the reflection of a Si surface. Laser surface texturing method was achieved on Si solar cell by UV femtosecond laser. Drevinskas *et al.* [12] demonstrate that selective KOH etching reveals high contrast nan gratings without affecting the unmodified silica lattice and leads to the control of optical dispersion and enhancement of the retardance exhibited by the nanostructures. Mutlak *et al.* [13] deposited the SnO_2 NPs on chemically etched n-type silicon substrate. The reduction in SnO_2 NPs size accompanied by increasing the surface area to volume ratio and photoreaction of these nanoparticles led to increases of the responsivity, quantum efficiency and detectivity of this UV region detector.
In this paper, we improve the method used to enhance the absorption efficiency of the silicon solar cells by using photo electrochemically anodized, and the formation of nanostructure porous silicon layers is observed and its feature characteristics are studied.

2. Experimental

2. 1 Chemicals

Silicon wafer of n-type (111) with a resistivity of 0.015 Ω.cm and thickness 508 μm was used the following solvents were used as media deionized water, organic ethanol (ScharlauCo., CH_3CH_2OH, 99.9% purity) and HF (48%, SOLVOCHEM, UK).

2. 2 Preparation of PS

The Si wafer cut off into (2×2 cm^2) pieces, rinsed with ethanol to clean the surface from any contamination after that rinsed with 5% Hydrofluoric acid (HF) to remove the native oxide and dried in presence of nitrogen. A thin Aluminum film was deposited on back side of Si wafer to create ohmic contact.
The experimental set – up of laser-assisted etching for produced PS samples is presented in Figure 1. Ecthing cell was made from Teflon with bottom part from stainless steel disk which was placed for contact purposes. Then the silicon wafer was placed and a rubber O-ring was used before placing the upper part which has a center circular of 1cm^2 to allow the electrolyte solution touch the silicon wafer. Two electrodes were used to apply current across the cell, the first one was was the stainless steel disk below silicon wafer and the other was made of gold mesh immersed in the HF solution. The electrical

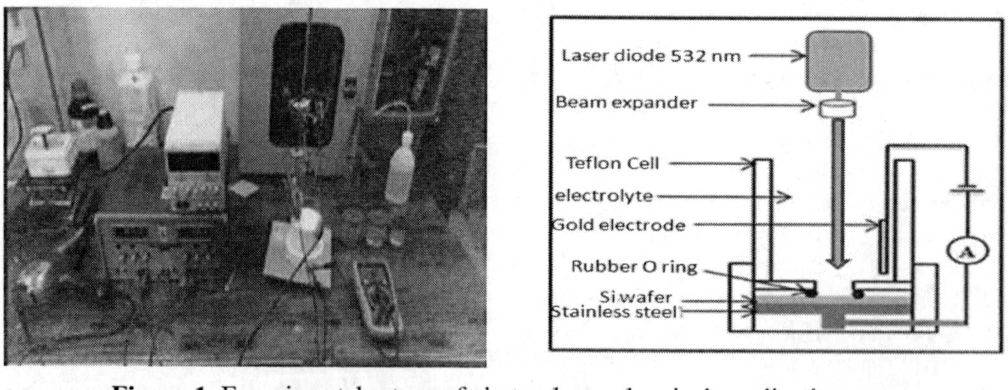

Figure 1. Experimental set-up of photo-electrochemical anodization process.

circuit was completed by QJEDC regulated power supply (QJ3005XII), and and diode laser of wavelength 532 nm and power of 200 mW was used to illuminate the Si surface.
Laser power was obtained after calibration by using (Genetic- EO SOLO2) power meter. The Beam expander was utilized for expand the laser beam to lighting all etching Si area. The PECE process was carried out with different current density (06, 16, 26, 36, 46 and 56 mA/cm^2) and constant HF concentration (15%) for 10 minutes. After etching process, the PS sample was rinsed by pentane, ethanol and dried with stream of air.

2.3 Measurements

The surface morphology and roughness of prepared samples was obtained by atomic force microscopy gave a direct surface image of porous silicon layer. A CSPM-AA3000. AFM instrument was used. The photoluminescence was measured by using a PerkinElmer LS-45 Fluorescence spectrometer. Structural properties of PS have been investigated by X-ray diffraction techniques. A SHIMADZU (XRD-6000, 220/50Hz, JAPAN) of 1.54 A° from Cu-Kα was used. The optical properties of the porous silicon were studied by optical reflectance spectroscopy. The spectra were obtained at room temperature. UV-VIS, Perkin Elmer Lambda 950 spectrophotometer was employed to study the optical Reflectance (R) for PS/Si layer prepared in different conditions for wavelength interval between 400-800 nm.

3. Results and discussion

XRD was used to investigate the size of the crystallites of the PS and to estimate the crystalline degree of etched samples. XRD pattern of bulk silicon become a very sharp peak at 2 = 28.8° showing the single crystalline nature of the wafer. This peak becomes very broad with varying full-width at half maximum (FWHM) for different anodization current densities as shown in Figure 2. Consequently, we can confirm that the PS layer remains crystalline, but it is slightly shifted to a smaller diffraction angle.

The crystallites size obtained for PS samples are shown in Table 1, when estimated by the Scherrer's equation, a significant crystallites size decrease trend can be clearly noted on urrincreasing current density.

Figure 3 shows the 3D AFM image and diameter values distribution of PS samples prepared under different current density of porous silicon in which the irregular and randomly distributed nan crystalline silicon pillars and voids over the entire surface can be seen. However, increasing in current density orders the small pores to exhibit cylindrical shapes giving rise to larger pore diameter.
The morphology characteristics of PS samples are presented in Table 2. The average pore diameter appears in good agreement with classification of mesoporous. At low current density, small pores are formed surrounding by thick columnar structure network of silicon walls. When the current density is increased the pore with large diameter and thin columnar walls is formed.
Figure 4 shows the room temperature PL spectra of various samples. The energy band gap (Eg) values were estimated to be 1.768, 1.771, 1.773, 1.773, 1.776 and 1.779 eV for current densities of 6 mA/cm^2, 16 mA/cm^2, 26 mA/cm^2, 36 mA/cm^2, 46 mA/cm^2 and 56 mA/cm^2, respectively. The observed blue shift in the PL peak with increasing anodization current density was attributed to quantum confinement effect, which led to an increase in the energy band gap.
The FWHM values for different peaks with increasing current density displayed slight narrowing. The decrease in PL intensity with the increase of anodization current density was attributed to the enhanced recombination probability of the carriers that are strongly confined within the nanostructure, where the nature of the band gap is transformed from indirect one to quasi-direct one. The PL peak shift, narrowing of FWHM, and an increase in the peak intensity at 56 mA/cm^2 are consistent with previous findings [14, 15].
It can be seen that the reflectivity of PS is raised with increasing the current density from (6 to 56 mA/cm^2) as shown in Figure 5. Hence, at low current density etching (6 and 16 mA/cm^2), the porosity

of porous silicon layer increased, and refractive index will reduce. Thus the reflectivity of porous silicon is reduced. Reduced reflectivity is also attributed to the light scattering and increased light trapping at wavelength (400-800 nm). The scattering light may be due to the surface roughness at the

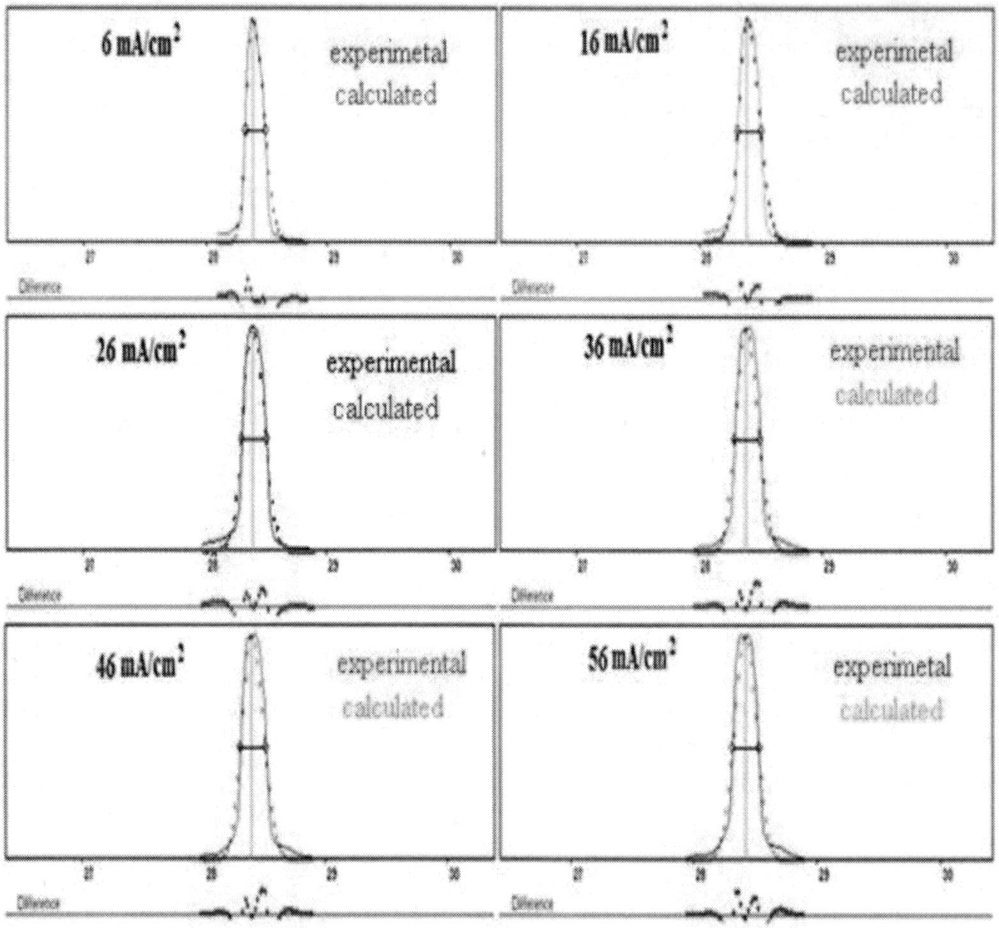

Figure 2. X-ray diffraction patterns for PS layers with different etching current density.

Table 1. Experimental values of X-ray diffraction peaks for PS lyers with different etching current density.

T_S (°C)	x	a(Å)	b (Å)	c (Å)	V $(Å)^3$	c/a (Å)	ρ_m (gm/cm^3)	V2223%	V2212%
	0	5.403	5.477	37.007	1095	6.849	1.428	69.632	29.478
820	0.2	5.406	5.322	36.671	1055	6.783	1.507	41.543	57.567
	0.4	5.403	5.352	36.846	1065	6.819	1.503	68.854	30.256
	0	5.365	5.454	37.212	1088	6.936	1.436	79.554	19.556
830	0.2	5.447	5.361	37.149	1084	6.820	1.452	51.723	47.387
	0.4	5.454	5.465	37.267	1110	6.832	1.456	68.934	30.176

PS-Si interface, whereas roughness is attributed to the etching parameters affect the surface characterization. The high degree of roughness of the PS surface implies the possibility of using the porous layer as an antireflection coating for solar cells because the surface reduces the light reflection. Our results are in consistent with previous studies [16, 17].

Figure 3 Cross-sectional view of AFM images of the PS obtained at different anodization current density of: (A) 06 mA/cm^2, (B) 16 mA/cm^2, (C) 26 mA/cm^2, (D) 36 mA/cm^2, (E) 46 mA/cm^2 and (F) 56 mA/cm^2.

Table 2 The analysis of the PS morphology characteristics prepared at different current densities.

J (mA/cm^2)	Average diameter (nm)	Average roughness (nm)	Root mean square (nm)
06	24.30	14.9	17.1
16	24.79	5.14	5.95
26	34.16	2.71	3.13
36	35.96	1.88	2.17
46	47.86	3.01	3.14
56	55.51	6.35	7.33

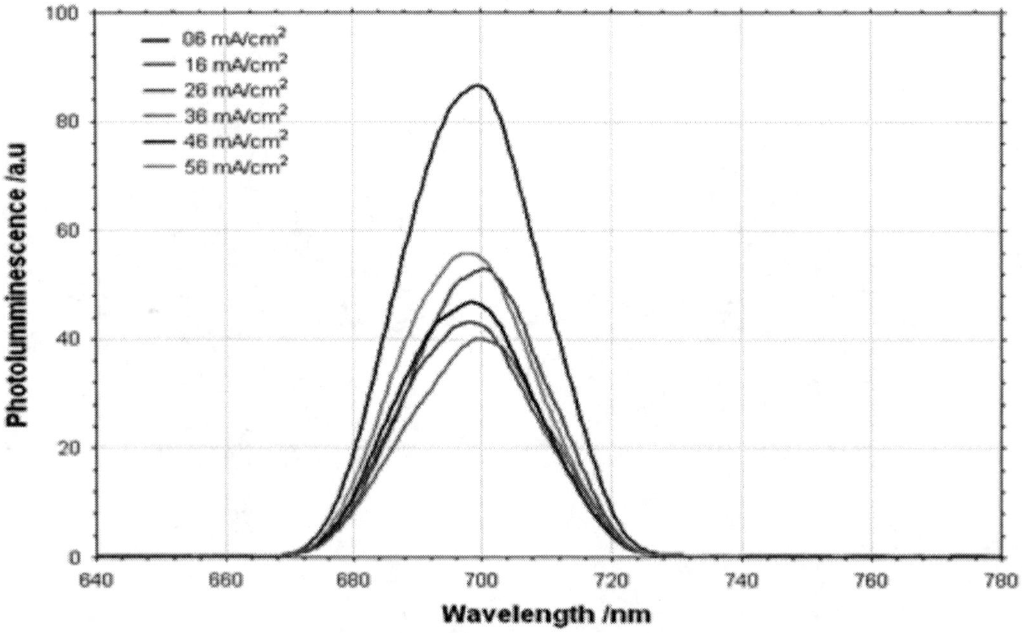

Figure 4 Room temperature PL spectra of PS samples different anodization current density.

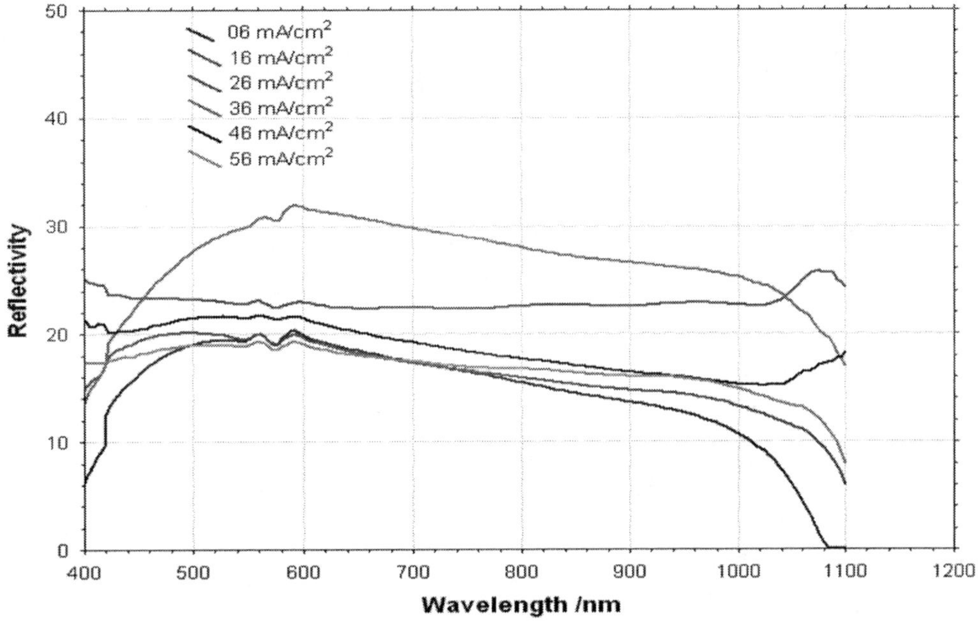

Figure 5 Reflectivity of PS samples with different anodization current density.

4. Conclusions

In this paper, we have prepared micro textured Si structure, which reduces the reflection of a Si surface. Laser-assisted etching method was accomplished on Silicon by laser. The effects of anodization current densities on the morphology and PL spectra of as synthesized PSi were analyzed. It was demonstrated that by varying the anodization current density it is possible to control the porosity and the visible light emitting properties of such PSi. Good quality PSi samples were achieved. PL spectra revealed prominent peaks together with large blue shift, which supported the presence of strong nanoporous structures. Strong surface roughness is established to be responsible for the reduction of the Si crystallite size at higher anodization current density. As a result this approach leads to enhance Si cell efficiency. The current method is much cheaper and simpler than the reactive – ion etching (RIE).

References

[1] Hermann J, Benfarah M, Brunear S, Axente E, Coustillier G, Itina T, Guillemoles J, and Alloncle P 2006 *J. Phys. D: Appl. Phys.* **39** 453.
[2] Maurizio C, Giuseppe C, Mario I, Ivo R and Luigi S 2010 *Sensors* **10** 10571.
[3] Pedraza A, Fowlkes J, Lowndes L 1999 *Appl. Phys. Lett.* **74** 2322.
[4] Honsberg C, Barnett A, Kirkpatrick D, 2006 *4th World Conference on Photovoltaic Energy Conversion*, Hawaii, USA, 7-12 May, Record of the IEEE **2** 2560.
[5] Uhlir A 1956 *Bell Syst. Tech.* J **35** 333.
[6] Canham L 1990 *Appl. Phys. Lett.* **57** 1046.
[7] Bisi O, Stefano O and Pavesi L 2000 *Surface Science Reports* **38** 1.
[8] Wang X, Tebib A, Veschetti Y and Bastide S 2011 *Phys. Status Solidi* A **208** 215.
[9] Kolasinskia K, David M and Mona N 2006 *J. Vac. Sci. Technol.* A **24** 1474.

IOP Conf. Series: Materials Science and Engineering **757** (2020) 012065 doi:10.1088/1757-899X/757/1/012065

[10] Gattass R and Mazur E 2008 *Nat. Photonics* **2** 219.
[11] Mutlak F A-H 2014 *Turk J Phys* **38** 130.
[12] Drevinskas R, Mindaugas G, Beresna M, Bellouard Y and Kazansky P 2015 *Optics Express* **23** 1428.
[13] Mutlak F A-H, Taha A B and Nayef UM 2018 *Silicon* **10** 967.
[14] Behzad K, Yunus W, Talib Z, Zakaria A and Bahrami A 2012 *International Journal of Electrochemical Science* **7** 8266.
[15] Al-Douri Abd H Y, Ahmed N and Hashim M 2013 *International Journal of Electrochemical Science* **8** 11461.
[16] RamizyA, Hassan K, Omar Y Al-Dourib, Mahdi M A 2011 *Applied Surface Science* **257** 6112.
[17] Salman K, Hassan A Z, Omar K 2012 *Int.J.Electrochem.Sci.* **7** 376.

IOP Conf. Series: Materials Science and Engineering **757** (2020) 012066 doi:10.1088/1757-899X/757/1/012066

Effect of sintering temperature on superconducting properties of $Bi_2Pb_2Sr_2Ca_2Cu_{3-x}Co_xO_{10+\delta}$ system

A K Jassim, , M M Abbas[2] and A A Hussien[3]
[1] University of Baghdad, College of Science, Physics Department, Iraq
[2]Al-Hadi University Collage, Baghdad, Iraq
[3] University of Baghdad, College of Science, Physics Department, Iraq
Email:amelalmalki1974@yahoo.com

Abstract. $Bi_2Pb_{0.3}Sr_2Ca_2Cu_{3-x}Co_xO_{10+\delta}$ high temperature superconductors with x=0, 0.2 and 0.4 were prepared by a solid state reaction method. The samples properties have been investigated structurally by X-ray diffraction and morphologically by scanning electron microscopy. Structural analysis showed that two superconducting phases coexist in the samples high temperature Bi-2223 and low-temperature phase 2212 with orthorhombic structure for all samples. Four point probe method was used to study the electrical properties of the samples and all of them showed superconductivity behavior. The transition temperature increases with the increasing Co concentration and sintering temperature. $Bi_2Pb_{0.3}Sr_2Ca_2Cu_{3-x}Co_xO_{10+\delta}$ superconductor sintered at 130 ^{0}C shows highest critical temperature 114 K.

1. Introduction

High temperature superconductors, BiSrCaCuO compounds were discovered in 1988[1]. They have three different phases a Bi-2201 phase, a Bi-2212 phase, and a Bi-2223 phase, according to their compositions; each phase has a different transition temperature to the superconducting state [2,3].The Bi-2223 phase attracted greater interest due to its high critical temperature.

The grain connectivity and the degree of texturing in the Bi-2223 compound depends on various parameters such as starting composition, sintering time and temperature [4].

Azhan *et al.* showed that the heat treatments at 830°C and 860°C; increases the critical temperature of $Bi_{1.6}Pb_{0.4}Sr_2Ca_2Cu_3O_\delta$ system [5].

Muna *et al.* pointed out that increasing of sintering temperature promotes the formation of high- critical temperature (T_C) phase 2223 due to improvement of the nucleation [6].

Ghazala et.al. studied the effect of sintering temperature on BPSCCO Superconductor system prepared by solid state reaction method. The result indicated that there is a decomposition of the high-T_C phase with the increasing of the sintering temperature [7].

On other hand, the superconducting properties can be modified by the addition or substitution of an element with a different ionic radius and bonding characteristics.

In addition the existence of Pb in the started materials promotes the reaction kinetics of the Bi- (2223) phase and enhances the diffusion of the calcium and copper atoms [8].

Jannah *et al.* studied the effect of nano Co_3O_4 addition on the properties of BPCCO system prepared by solid state reaction method. They showed that appropriate amount of Co enhanced the critical temperature of Bi(Pb)SrCaCuO superconductor and the samples have both high-TC phase 2223 and low-T_C phase 2212 [9].

The aim of this work is to prepare and investigate the properties of $Bi_2Pb_{0.3}Sr_2Ca_2Cu_{3-x}Co_xO_{10+\delta}$ superconductors with the different sintering temperature and Co concentration.

2. Experimental

$Bi_2Pb_{0.3}Sr_2Ca_2Cu_{3-x}Co_xO_{10+\delta}$ samples with x=0, 0.2 and 0.4 were prepared by a solid state reaction method. Appropriate weight of high purity powders of materials $Bi_2(CO_3)_3$, Pb_3O_4, $Sr(NO_3)_2$, $CaCo_3$, CuO and CoO were mixed. The mixture was calcined in air at 800°C for 24 h. Then the powders pressed into disk-shaped pellets under pressure 0.5 GPa using hydraulic press type (Specac). The pellets were sintered in air at sintering temperature (Ts) (820, 830, 835 and 840) °C for 140 h.
The electrical resistivity (ρ) of the samples as a function of temperature were obtained by using four point probe DC method at temperature range (77 - 300) K to determine the critical temperature (Tc). The crystal structure of the prepared samples was obtained by using X-ray diffractometer type Philips Source: $Cu_{K\alpha}$ radiation. The lattice parameters were calculated by using a computer program based on Cohen's least square method [3]. The oxygen content (**δ**) in the superconductor samples was determined by using iodometric titration technique
Scanning electron microscopy (SEM) type FEI-SEM model Inspect –S50 were used to study the surface topography of the samples.

3. Results and discussion

XRD patterns of $Bi_2Pb_{0.3}Sr_2Ca_2Cu_{3-x}Co_xO_{10+\delta}$ system for different values of x sintered at 820 and 830°C are shown in Figs.(1-3). The spectra shows an orthorhombic structure of all samples and most of them have two phases Bi-2223 and Bi-2212, in addition these of phases there were some unknown and impurity phase like $Sr_2Ca_2Cu_7O_\delta$ is detected at $2\theta \approx 36.8^\circ$ as reported by Asthana [10] . The appearance of more than two phases could be related to the stacking faults along the c- axis as a result of the displacement of an ion or oxygen defect or to the ordering of cations [11,12].
It could be seen that increase the sintering temperature causes an increase the intensity of the peaks. The most intense peak pattern of samples belongs to the high-2223 phase which also indicates an increase in the volume fraction of this phase [13]. Another feature observed is a present of H(0012) for all samples which is necessary to prove the superconductivity for Bi-compound.
The relative volume fractions of the high Bi-2223 phase (V_{2223}) and low Bi-2212 phase (V_{2212}) were estimated using formulas in reference [14].
The values of the lattice parameters a, b, c, unit cell volumes V, c/a ratio, and density ρ_m of all the samples were calculated and summarized in Table 1. Density of unit cell for all samples was calculated using the following equation in reference [15].

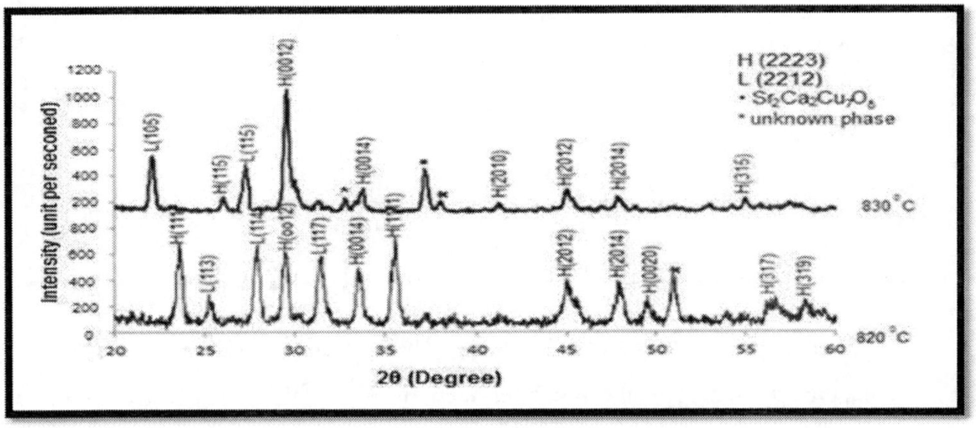

Figure 1. XRD patterns for $Bi_2Pb_{0.3}Sr_2\ Ca_2Cu_3O_{10+\delta}$ sintered at 820 °C and 830 °C for 140h.

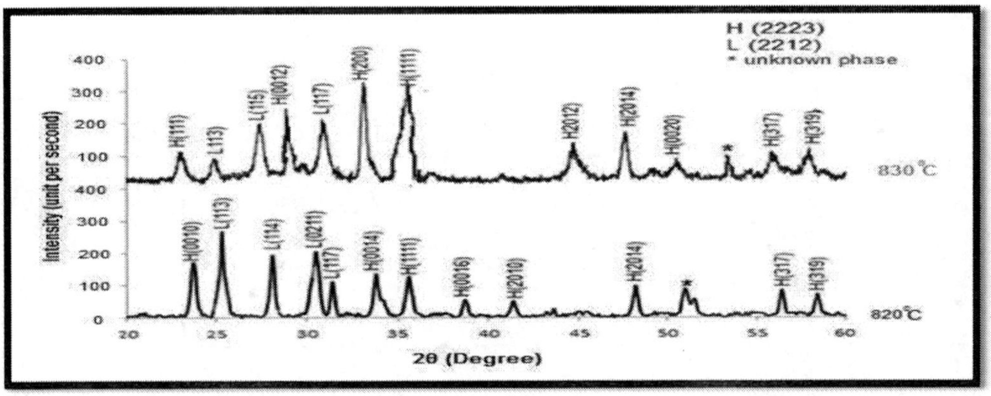

Figure 2. XRD patterns for $Bi_2Pb_{0.3}Sr_2\ Ca_2Cu_{2.8}\ Co_{0.2}\ O_{10+\delta}$ sintered at 820 °C and 830 °C for 140h.

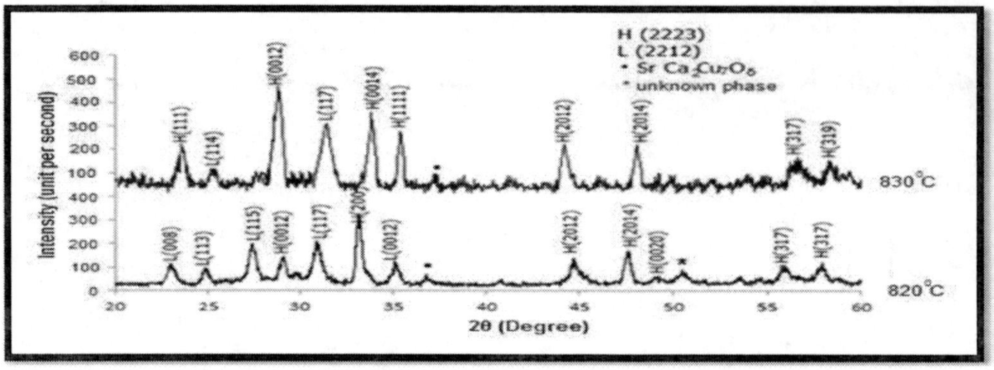

Figure 3. XRD patterns for $Bi_2Pb_{0.3}Sr_2\ Ca_2Cu_{2.8}\ Co_{0.4}\ O_{10+\delta}$ sintered at 820°C and 830°C for 140h.

Table 1. Values of lattice parameters, unit cell volume, c/a ratio, V_{2223} , V_{2212} and density ρ_m for all samples.

Ts (°C)	x	a (Å)	b (Å)	c (Å)	VÅ)^3	c/a	ρ_m(gm /cm^3)	V2223 %	V2212 %
820	0	5.403	5.477	37.007	1095	6.849	1.428	69.632	29.478
	0.2	5.406	5.322	36.671	1055	6.783	1.507	41.543	57.567
	0.4	5.403	5.352	36.846	1065	6.819	1.503	68.854	30.256
830	0	5.365	5.454	37.212	1088	6.936	1.436	79.554	19.556
	0.2	5.447	5.361	37.149	1084	6.820	1.452	51.723	47.387
	0.4	5.454	5.465	37.267	1110	6.832	1.456	68.934	30.176

Measurements of electrical resistivity as a function of temperature showed metallic behavior of the samples as displayed in Figs (4, 5 and 6) respectively. Superconducting transitions to zero resistance had been noticed for the samples sintered at 830 °C. While a complete zero resistance was not observed for the samples sintered at 820 °C, This could be due to the transition within the grains and the presence of Bi-2212 phase [11].

From Table (1) and the see that the transition temperature of the compositions $Bi_2\ Pb_{0.3}Sr_2Ca_2Cu_3O_{10+\delta}$, $Bi_2Pb_{0.3}\ Sr_2\ Ca_2Cu_{2.8}Co_{0.2}O_{10+\delta}$ and $Bi_2Pb_{0.3}Sr_2Ca_2Cu_{2.6}Co_{0.4}O_{10+\delta}$ increases from 103,101 and 108K to 112,113 and 114K respectively as the sintering temperature increasing from 820°C to 830°C.

Enhancement of Tc with increasing Co could mainly due to increasing amounts of Bi-2223 phase as referred in the x-ray analysis, as well as to the increasing the contact areas between the grains which improved the junctions between them during the sintering process with increasing sintering temperature [16 and 17].

Sintering at temperatures 835°C and 840°C produced melted and bended samples; therefore, the sintering temperature is considered critical for the growth of high- T_C phase and the optimum temperature seems to be close to the partial melting point, this agreement with Ghazala et al. [7].

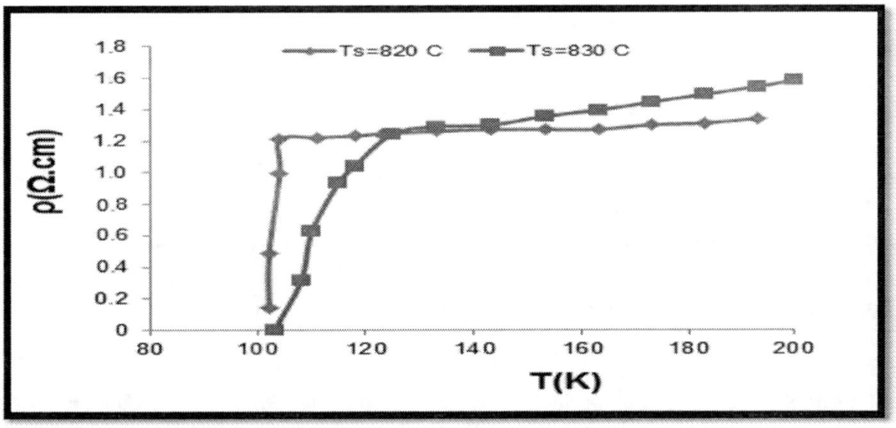

Figure 4. Temperature dependence of resistivity for $Bi_2Pb_{0.3}Sr_2Ca_2Cu_3\,O_{10+\delta}$ sintered at 820°C and 830°C for 140h.

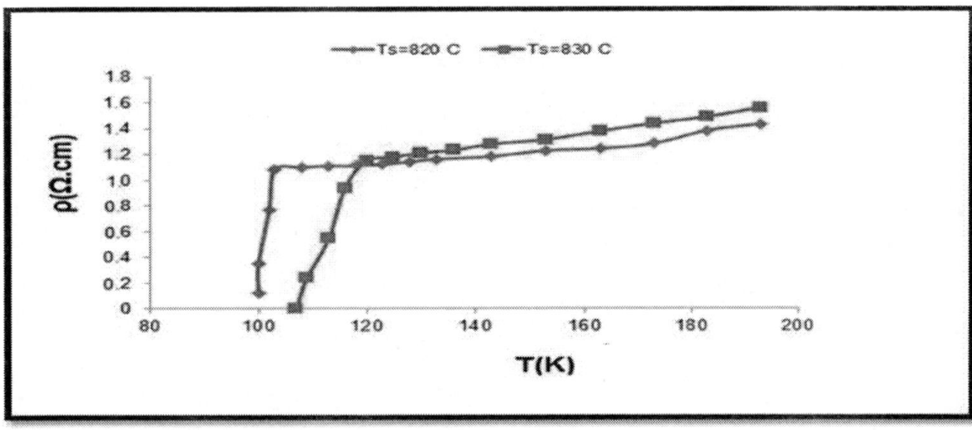

Figure 5. Temperature dependence of resistivity for $Bi_2Pb_{0.3}Sr_2Ca_2Cu_{2.8}Co_{0.2}\,O_{10+\delta}$ sintered at 820°C and 830°C for 140h.

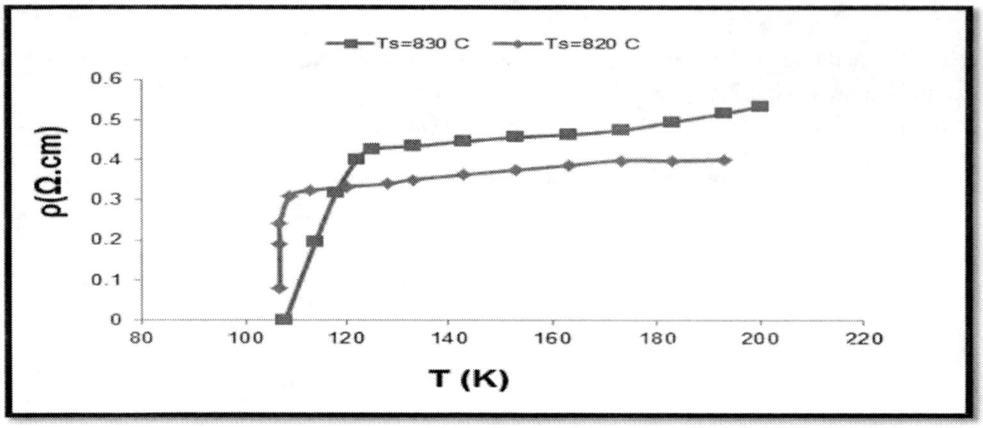

Figure 6. Temperature dependence of resistivity for $Bi_2Pb_{0.3}Sr_2Ca_2Cu_{2.8}Co_{0.4}O_{10+\delta}$ sintered at 820°C and 830°C for 140h.

The oxidation of Bi-2223 superconductor refers to sintering conditions which can occur within the nucleation, growth and reaction process. Bi-2223 compound is very sensitive to oxygen it can be inferred from Table 2 that both the δ and T_C increased, with increasing Co concentration and sintering temperature. The higher T_C value is found for higher value of oxygen content for the samples sintered at 830°C.This could be attributed to the increase of oxygen absorption during crystallization process of the superconductors and this leads to an improvement of the T_C [18].

Table 2. Values of the oxygen content and critical temperature for all samples.

x	T_s (°C)	Oxygen contact(δ)	T_C (K)
0	820	0.0984	103
0.2		0.0965	101
0.4		0.0932	108
0	830	0.2347	112
0.2		0.2434	113
0.4		0.2863	114

Surface topography morphology of the samples with x=0.4 sintered at 820°C and 830°C illustrates in Fig. (7). The images show a composed of different size of stacks irregular plate- like grains.

increasing the sintering temperature leads to form a large grains and enhance the formation of high- T_C phase, each grain grows in random direction through the other gain and creates bigger grains. This coalescence between the grains reveals the growth of the superconducting phase on the account its surrounding phase. However, the enhancement of the samples behavior could be attributed to decreasing the porosity and increasing the contact between the grains during the sintering process time. This confirms our explanation of the higher T_C.

Figure 7. SEM micrographs of fracture surface of samples with x=0.4 sintered at(a)820°C, (b)830°C.

4. Conclusions

The results showed that the properties of $Bi_2Pb_{0.3}Sr_2Ca_2Cu_{3-x}Co_xO_{10+\delta}$ superconductors with x=0, 0.2 and 0.4 depend not only on element composition but it is also much delicate with the sintering process. The sintering temperature is critical for the growth of high- T_C phase and the optimum temperature seems to be close to the partial melting point 830°C. The substitution Co in combination with increase the sintering temperature leads to form large grains and enhance the formation of high- temperature 2223-phase. Highest transit
ion temperature has been determined 114 K for the composition $Bi_2Pb_{0.3}Sr_2Ca_2Cu_{2.6}Co_{0.4}O_{10+\delta}$ accompanied with higher value of oxygen content.

References

[1] Tarascon J ,Page Y, Greene L,.Bagley B, Barboux P, Hwang D, Hull G, Mckinnon W and Giroud M 1988 *Phys.Rev*.B **38** 2504.
[2] Gul I.H., Amin F, Abbasi A.Z., Anis-ur-Rehman M., Maqsood A *2006 Physica C* **449** 139–147.
[3] Truccato M, Cagliero S, Agostino A, Panetta M and Rinaudo G 2006 *Supercond. Sci Technol* **19** 1003–1009.
[4] Muhammad, A.R. and Maqsood, A 2010 *Superconductor.http://www.intechopen.com/ books/superconductor/synthesis- and-thermophysical-characterization-of-bismuth-based-high-tc-superconductors* 249-261

[5] Azhan H, Fariesha F and Khalida S 2011 *J Supercond Nov Magn* **24** 265.
[6] Abbas M ,Oboudi S and, Raoof N 2015 *Materials Sciences and Applications* **6** 310-321.
[7] Hermiz G, Jassim A and AL-Beyaty E 2015 *Almustasirya Education journal* **2** 944.
[8] Abbas M, Abdulridha A,.jassim A 2017 *Digest journal of nanomaterials and biostructures* **12** 1049-1055.
[9] Jannah A , Shukor R, Abdullah H and Agail A 2011 *American Institute of Physics conference series* **1415** 83.
[10] Asthana 1992 "processing–property relationships for high T_c ceramic superconductors in the Bi-Ca-Sr-Cu-O system" *Ph.D. Thesis, University of Illinois, Urbana-Champaign.*
[11] Solunke M B, Sharma P U , Pandya M P, Lakhani V K, Modi K B, Venugopal Reddy P, Shah S S 2005 *Journal of Physics PRAMANA /Indian Academy of Sciences* **65** 481-490
[12] Awad R J. 2012 *J Supercond Nov Magn* **25** 739.
[13] Koyama S, Endo U and Kawai T 1988 *Jap.J.Appl.Phys.* **27** 1861-1863.
[14] Bilgili O., Selamet Y. and Kocabas K. 2008 *J Supercond Nov Magn* **21** 439-449.
[15]. Kresin V Z and.Wolf S A 1990 *Fundamentals of superconductivity* Plenum press, New York.
[16] Heh T S, Chen J R and Tseng T Y 1990 *Jap.J.Appl.Phys* **29** 652-655.
[17] Abbas M, 2015 *International Journal of Current Engineering and Technology* **5** 1908-1910.
[18] Ikeda Y ,Takno , Hiroi Z , Oda K, Kitaguchi H ,Takada J, Miura Y, Takada Y, Yamamoto O and.Mazaki H 1988 *Jap.J.Appl.Phys* **27** 2067-2070.

Cobalt substitution effect on the physical properties of (Bi, Pb)-2223 superconductor

A K Jassim[1,2*], S S Abodi[1] , M M Abbas[2] and A A Hussien[1]

[1] University of Baghdad, College of Science, Physics Department, Baghdad, Iraq

[2]Al-Hadi University Collage, Radiology Techniques department, Baghdad, Iraq

Corresponding author's e-mail: amelalmalki1974@yahoo.com

Abstract. High-temperature superconductors with a nominal composition $Bi_2Pb_{0.3}Sr_2Ca_2Cu_{3-x}Co_xO_{10+\delta}$ for ($0 \leq x \leq 3$) were prepared by solid-state reaction method. Effects of the Co substitution on Cu sites have been investigated to obtain the optimum concentration for the formation and stabilization of the superconducting samples. The results of electrical resistivity measurements of the samples showed that increasing x substituting Co enhance the superconducting properties except for the samples with x= 3 yield semiconductor behavior. The higher critical temperature T_C found at 118 K, which is for the composition $Bi_2Pb_{0.3}Sr_2Ca_2Cu_{2.2}Co_{0.8}O_{10+\delta}$, which has the highest value of excess oxygen content 0.315. The X-ray diffraction analysis for all superconducting samples showed an orthorhombic structure with two phases, high-T_c 2223 phase and low-T_C 2212 phase. X-ray spectroscopy analysis showed that the compositions with x=0.6 and 0.8 samples did not contaminate during the synthesis processes. The scanning electron microscope has been used to identify the morphology of the superconducting phase. The plate-like grains of the high Bi- 2223 phase appeared in most micrographs of the superconducting samples besides changes morphology of samples with increasing dopant concentration.

Keywords: Superconductors; Cobalt; Substitution; Physical properties.

1. Introduction

Superconductor is a material that loses all resistivity due to the flow of electric current when it cooled below a certain temperature, which is called the critical temperature or transition temperature (T_C).

Above this temperature, there is usually little or no indication that the material might be a superconductor [1].BiSrCaCuO (BSCCO) was the first high-temperature superconductor which does not contain any rare earth metals. This family of superconductors can be expressed by the general formula $Bi_2Sr_2Ca_{n-1}Cu_nO_{2n+4+\delta}$, where (n=1, 2, 3), the CuO layers, depend on n values. The Bi-2201, Bi-2212, and Bi-2223 phases have single, double, and triple layers of CuO_2 in the subunit cell, respectively. All three phases have in common both the conduction layer and the charge reservoir layer. The critical temperature of these materials increases with increasing the number of Cu-O planes in the crystalline structure [2, 3]. T_C and lattice parameters of three systems change slightly depending on the cationic substitution, which occurs widely in any of these phases [4]. The T_C between superconductive and nonsuperconductive states strongly depends on the concentration of carriers in the CuO_2 planes, which closely relates to structures in charge reservoirs and the number of CuO_2 planes [5].

$Bi_{1.7}Pb_{0.4}Sr_2Ca_2Cu_{3-x}Ga_xO_\delta$ was prepared with different concentration of Ge by solid-state reaction reaction method [6]. It was shown that all the samples consist of both low-T_C 2212 and high-T_C 2223 phases. The percentage of low-2212 phase was found to increase with the increase Ga-content. Zero resistivity temperature T_C was obtained to be less than 106 K for doped samples. The effects of Ti substitution on the properties of $Bi_{1.7}Pb_{0.3}Sr_2Ca_2Cu_{3-y}Ti_yO_{10+\delta}$ bulk samples were studied for ($0 \leq y \leq 0.4$). [7]. XRD analysis showed an orthorhombic structure with two phases high -2223 phase and Low -2212 phase, in addition, an impure phase was found. The highest critical temperature118 K was found for $Bi_{1.7}Pb_{0.3}Sr_2Ca_2Cu_{2.8}Ti_{0.2}O_{10+\delta}$ compound. This work was performed to improve the high 2223-phase formation and T_C of BiPbSrCaCuO compound by partial substitution of Co in Cu sites. In addition, to clarify the common features of the substitution effect and as there is no report of Co substitution at least to our knowledge.

2. Experimental

The system of $Bi_2Pb_{0.3}Sr_2Ca_2Cu_{3-x}Co_xO_{10+\delta}$ with(x=0, 0.2, 0.4, 0.6, 0.8, 1, 2, 3) was prepared by a solid-state reaction method by mixing oxides, carbonates and nitrates $Bi_2(CO_3)_3$, Pb_3O_4, $Sr(NO_3)_2$, $CaCo_3$, CuO and CoO with molecular weights equivalent to the proportion of high purity powders (99.9%). To get a homogeneous mixture, the powders were mixed using agate mortar, a sufficient quantity of 2-propanol was added to form a paste during the process of grinding for about 1 h. Then the mixture was calcinated in the air at 800°C for 24 h, and after that, the mixture was pressed into pellets (13mm) in diameter and(1.5) mm thick, using hydraulic type SPECAC, under pressure of 0.7 GPa. The pellets were sintered in air at 830°C, for 140 h. The structure of the prepared samples was obtained by using X-ray diffraction (XRD) Philips with CuKα source. The four-point probe method was used to measure the resistivity and to determine the T_C. Scanning electron microscopy (SEM) and energy-dispersive X-ray spectroscopy (EDX) type FEI-SEM model Inspect-S50 were used to study the surface morphology and analysis the composition of the samples.

3. Results and discussion

EDX spectra of the $Bi_2Pb_{0.3}Sr_2Ca_2Cu_{2.4}Co_{0.6}O_{10+\delta}$ and $Bi_2Pb_{0.3}Sr_2Ca_2Cu_{2.2}Co_{0.8}O_{10+\delta}$, superconductors are illustrated in Figures 1 and 2, which show that there is no unwanted element in the sample and implies that the samples are not contaminated during the synthesis process. Moreover, there is no change in the position of Bi, Pb, Sr, Ca, and Cu peaks. In comparison between the two samples, beside of the Co peaks at 0.7 KeV and 6.9 keV for x=0.6 another peak that belongs to Co appeared at 0.9 KeV and 6.6 KeV for x=0.8.

Overlap of some peaks such as Bi and Pb peaks at the same value of energy was observed indicating that Bi ions have partially replaced by Pb ions in the system. Moreover, overlapping is also noticed for Cu and Co. The results confirm that the desired elements in the chemical composition of the synthesized samples are present and well substituted.

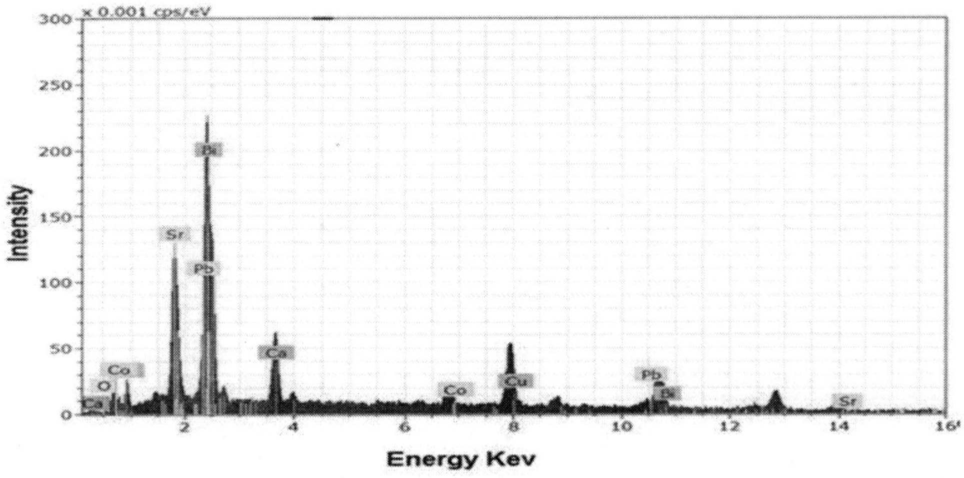

Figure 1. The EDX analysis for $Bi_2Pb_{0.3}Sr_2Ca_2Cu_{2.4}Co_{0.6}O_{10+\delta}$ superconductor

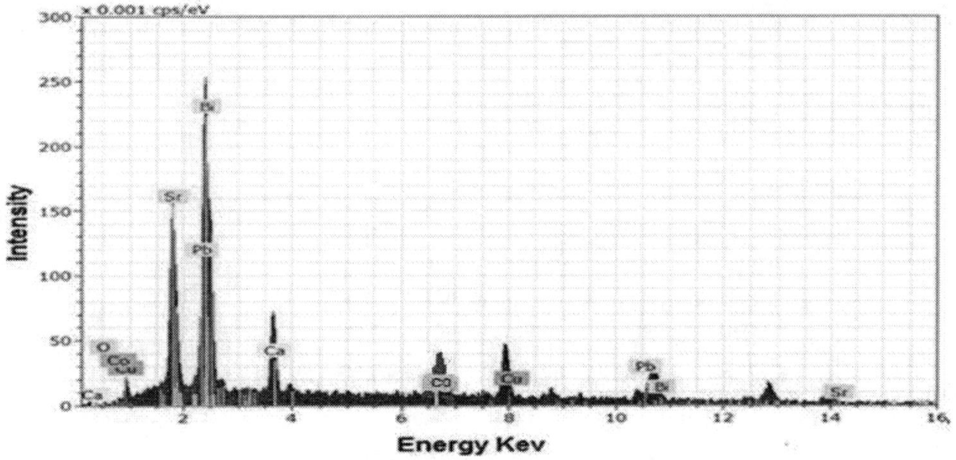

Figure 2. The EDX analysis for $Bi_2Pb_{0.3}Sr_2Ca_2Cu_{2.2}Co_{0.8}O_{10+\delta}$ superconductor

XRD patterns of the nominal composition of $Bi_2Pb_{0.3}Sr_2Ca_2Cu_{3-x}Co_xO_{10+\delta}$ for ($0 \leq x \leq 3$) sintered at 830 °C for 140h are shown in Figure 3. Peaks belong to high and low phases coexistence in the samples; but most of them belong to (Bi, Pb)-2223 phase. Some impurities phases like $Sr_2Ca_2Cu_7O_\delta$ were detected at 2θ around 36° and the other is unknown. Some peaks belong to high T_C-2223 have intensity higher than those of the low-T_C 2212 phase; this case appears more clearly with increasing Co content, and it indicates the increase of the crystalline arrangement degree [8]. Further increase of x= 1-3 decreased the intensity most of the peaks yielded smaller grains (smaller crystallinities) as it will be seen in SEM results.

The relative volume fractions of the high Bi-2223 phase (V_{2223}) and low Bi-2212 phase (V_{2212}) were estimated using formulas in reference [9]. It can be observed in Table 1 and Figure 4 that Co concentration at 0.8 has the highest volume fraction of the Bi-2223 phase, which indicates that Co may act as a phase stabilizer and enhance the nucleation and the formation of high $-T_C$ phase. Further increase of x =1-3 changes the reaction rate and slows down (Bi, Pb)-2223 formation.

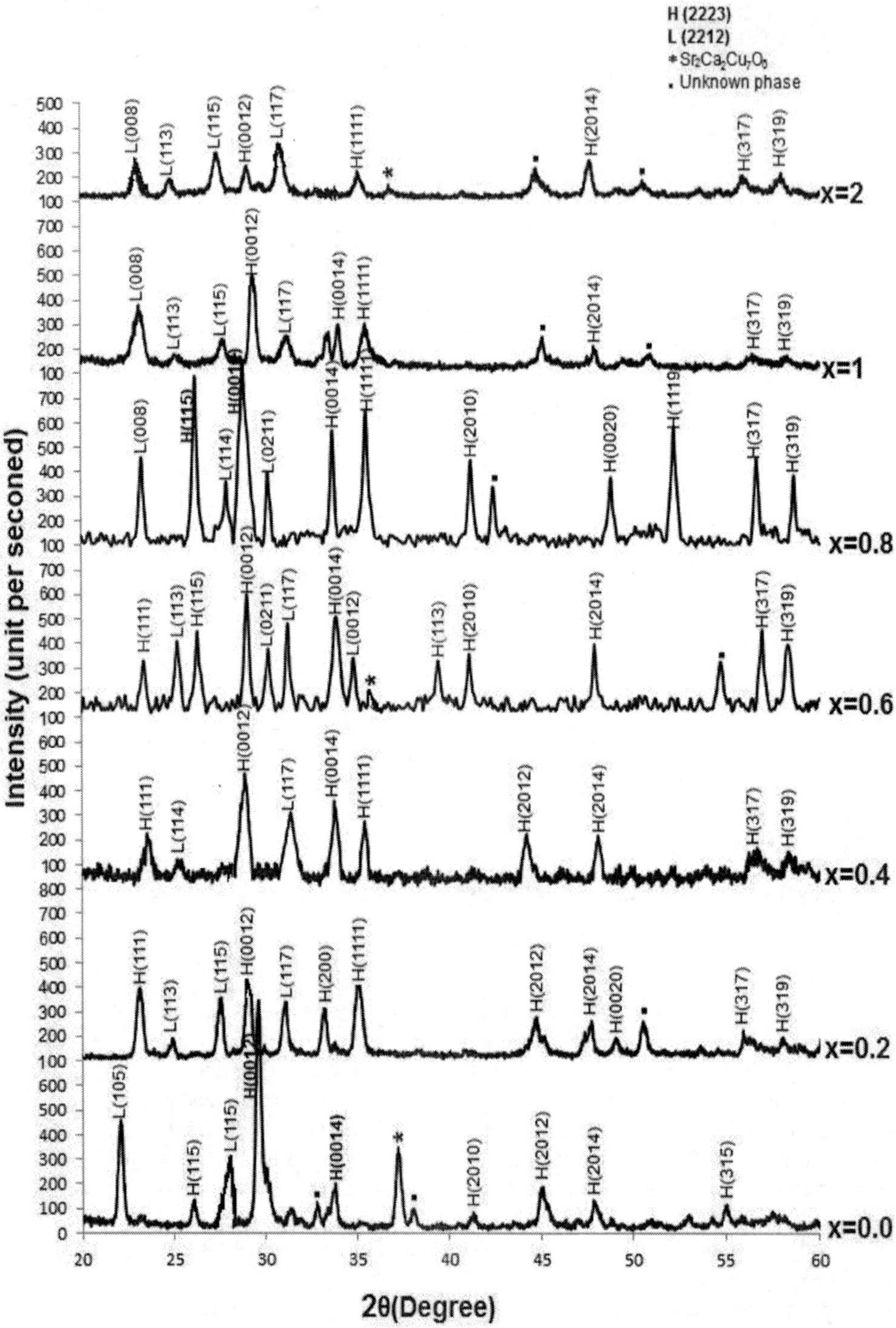

Figure 3. XRD patterns for $Bi_2Pb_{0.3}Sr_2Ca_2Cu_{3-x}Co_xO_{10+\delta}$ samples for x=0, 0.2, 0.4, 0.6, 0.8, 1 and 2 that sintered at 830 °C for 140 h.

The values of the lattice parameters a, b, c, unit cell volumes V, c/a ratio, and density ρ_m of all the samples were calculated and summarized in Table 1. The density of the unit cell for all samples was calculated using the equation in reference [9]. The variation of both ρ_m and c/a ratio as a function of Co content of $Bi_2Pb_{0.3}Sr_2\ Ca_2Cu_{3-x}\ Co_xO_{10+\delta}$ are shown in Figure 5. It can be obtained that there is an increase in both c lattice constant and c/a ratio with the increase of Co content (0<x<1). The reason is due to the differences of the ionic radii for both of Cu^{+3}(0.69Å) and Co^{+2}(0.78Å), which renders c-parameter to be longer or to get deformed. Such results may also be due to the distribution of holes between BiO layers and CuO planes. The doped ions may change the spacing between the CuO layers and thus affect the charge transfer to the CuO layers [10]. On the other side, the gradual decrease of c parameter with increase Co concentration to 1, 2 and 3 decreases this ratio, which may be attributed to the charge ordering phenomenon induced by Co as well as to the change of the oxygen content. Moreover, the interaction between additional bands crosses the Fermi level to grab the holes from CuO band thereafter decrease of the distance between CuO_2 planes that perform the decrease of the c-axis taking into consideration replacement of Bi^{+3} by Pb^{+2} ions this agreement with Muna *et al.* results [11]. The change in a, b, c affects the volume of the unit cell, and so causes a change of density. An increase of the density could be explained by a reduction of the porosity and voids in the sample.

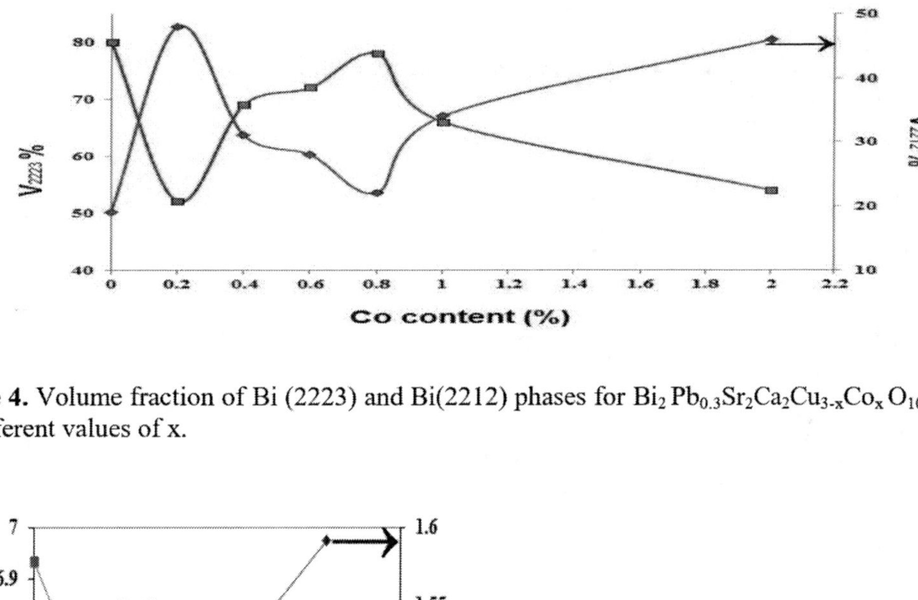

Figure 4. Volume fraction of Bi (2223) and Bi(2212) phases for $Bi_2\ Pb_{0.3}Sr_2Ca_2Cu_{3-x}Co_x\ O_{10+\delta}$ samples for different values of x.

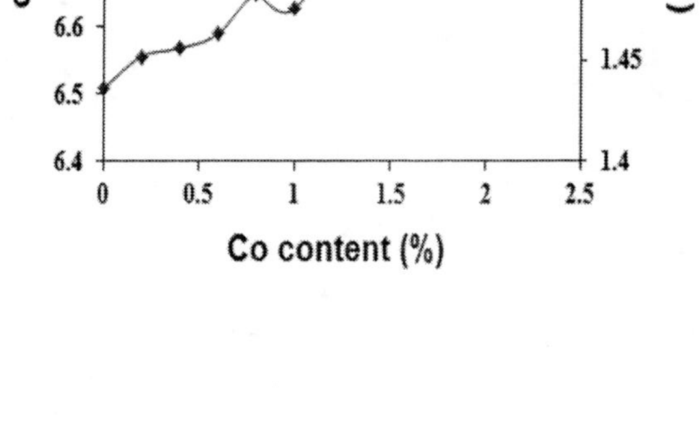

Figure 5. c/a ratio and density as a function of Co content for $Bi_2\ Pb_{0.3}Sr_2Ca_2Cu_{3-x}Co_x\ O_{10+\delta}$ samples.

Table 1. Values of lattice parameters a, b, c, V, c/a ratio, volume fraction, and density for different composition of $Bi_2Pb_{0.3}Sr_2Ca_2Cu_{3-x}Co_xO_{10+\delta}$.

x	a (Å)	b (Å)	c (Å)	c/a	V (Å^3)	ρ_m(gm/cm^3)	V2223%	V2212%
0	5.365	5.454	37.212	6.936	1088	1.436	79.554	19.556
0.2	5.447	5.361	37.149	6.820	1084	1.452	51.723	47.387
0.4	5.454	5.465	37.267	6.832	1110	1.456	68.934	30.176
0.6	5.447	5.456	37.297	6.847	1108	1.463	71.632	27.478
0.8	5.456	5.426	37.354	6.846	1105	1.483	77.762	21.348
1	5.449	5.562	37.112	6.810	1124	1.476	65.578	33.532
2	5.449	5.456	37.086	6.806	1102	1.592	53.712	45.398

Figures 6 and 7 show the resistivity as a function of temperature for $Bi_2Pb_{0.3}Sr_2Ca_2Cu_{3-x}Co_xO_{10+\delta}$ with x ranging from 0 to 3 sintered at 830°C for 140h. All samples are superconductors except the composition with x= 3, which is exhibited like semiconductor resistivity. Samples with x= 1- 2 showed linear behavior in the normal state, which indicates that these samples have optimally doped carriers in the CuO_2 plane, similar results were mentioned by Gul *et al.* [12]. A sharp drop of resistivity was observed for concentration s 0.6,1 and 2 this reveals that the sample consists homogeneity of predominantly Bi-2223 phase [13, 14].

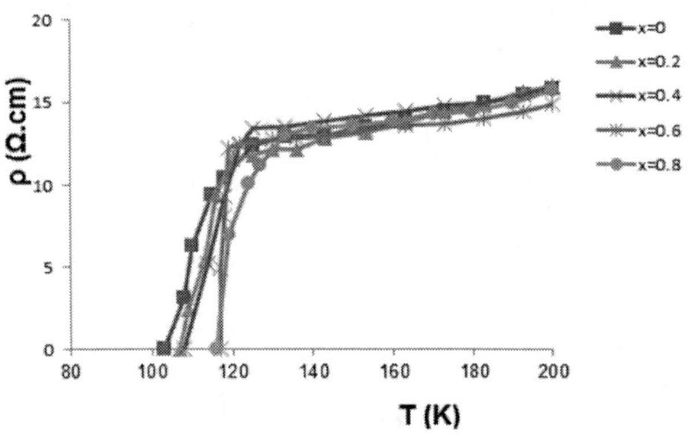

Figure 6. Temperature dependence of resistivity for $Bi_2Pb_{0.3}Sr_2Ca_2Cu_{3-x}Co_xO_{10+\delta}$ with x=0, 0.2, 0.4, 0.6 and 0.8.

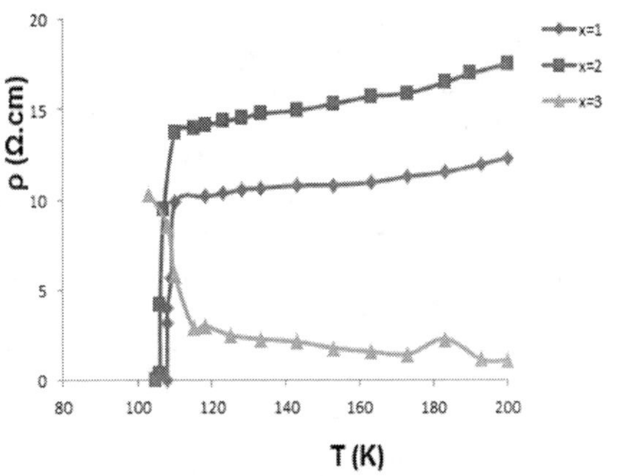

Figure 7. Temperature dependence of resistivity for $Bi_2Pb_{0.3}Sr_2\ Ca_2Cu_{3-x}Co_xO_{10+\delta}$ with x=1, 2 and 3.

It can be noticed from Table 2 that the values of the transition temperature increase from 112K of pure sample to 113,114,117 and 118K depending on the doping concentration 0.2, 0.4, 0.6 and 0.8 respectively. While increasing Co content to x=1 and 2 decreases the transition temperature to 108 and 106K, respectively. Samples with x= 3 yield semiconductor behavior. These values indicate that increasing amounts of Co content causes stabilization and promotes the growth of the Bi-2212 phase at the expense of the Bi-2223 phase [15]. While increasing Co could distort the crystal structure by the extra oxygen intercalated between two adjacent BiO layers which then affected the hole concentration of the copper oxide planes after that decreases T_C this result agrees with that obtained by Man *et al.* [16].

Table 2. Values of T_C and oxygen content for nominal composition $Bi_2\ Pb_{0.3}Sr_2Ca_2Cu_{3-x}Co_xO_{10+\delta}$ with $0 \leq x \leq 3$

x	T_C(K)	δ
0	112	0.2347
0.2	113	0.2434
0.4	114	0.2863
0.6	117	0.2929
0.8	118	0.31519
1	108	0.1592
2	106	0.1011
3	Semi.	-

Upon the results above, it is clear that the relation between T_C and Co concentration is almost bell-like shape, as shown in Figure 8. The superconducting phase may be divided, such as the underdoped, optimally doped, and overdoped regions [17,18]. Most of the samples with the majority of Bi-2223 are in the underdoped state; their superconducting properties tend to optimal values with increasing Co content. Highest T_C was determined at 0.8, which affirms that the sample is in an optimal doping region while a decrease of T_C beyond this concentration appears to be due to the shift of this sample towards the over-doped region. Though increasing Co concentration decreases the T_C, it does not

suppress the formation of (BiPb)-2223 phase, which confirmed by XRD analysis. Similar behavior found by Muna *et al.* [19].
Taple 2 and Figure 8 show the xcess of oxygen content (δ) for the samples $Bi_2Pb_{0.3}Sr_2Ca_2Cu_{3-x}Co_xO_{10+\delta}$ with ($0 \leq x \leq 3$). Both the δ and T_C increased, with increasing x to 0.8 and then reduced for further increasing of substitution, and there is a direct relationship between O content and T_C the higher value is found for a higher value of oxygen content as shown for the sample with x=0.8. This could be attributed to the increase in oxygen absorption during the crystallization process of the superconductors [20]. Thus excess oxygen atoms in the Cu-O_2 layers will create more holes in the perovskite layers, the creation of holes will shorten the Cu-O_2 bond length, and this leads to an improvement of the superconductor properties [21]. Increasing Co beyond 0.8, caused decrease both δ and T_C, this may be accompanied by changes in oxygen content or oxygen order effects which decrease the number of holes in the lattice from the optimum value [22]. Results
suggest that as the hole concentration of the compound increases above a critical value, the superconductor is suppressed Similar behavior which was emphasized by other researchers [19,23]
Furthermore, the addition of Pb to the compounds may relax the modulation by influencing the charge balance, oxygen content, and structural of the relevant layers [24].

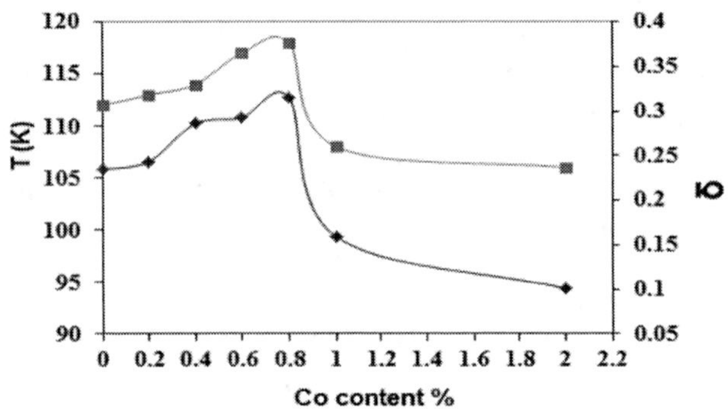

Figure 8. Variation of the T_C and δ as a function of Co content.

Images of scanning electron micrographs of $Bi_2Pb_{0.3}Sr_2Ca_2Cu_{3-x}Co_xO_{10+\delta}$ samples are shown in Figures (9 - 15). All the $Bi_2Pb_{0.3}Sr_2Ca_2Cu_{3-x}Co_xO_{10+\delta}$ samples consisting of packets of plate-like shape and voids expect x=1 and 2. The surface area images of $Bi_2Pb_{0.3}Sr_2Ca_2Cu_3O_{10+\delta}$ show denser microstructure with some porous and irregular sizes of a stack of plate-like (agglomeration of grains) structure which belongs to the high- T_C phase. Samples with x=0.2 show a composed structure of two kinds of particle features namely flaky grains clump of whisker which may be attributed to the cobalt addition (the flaky grains here more obvious) and the different size of stack irregular plate-like grains. This coalescence means the growth of the superconducting phase on account of its surrounding phase. Increasing Co concentration to 0.4, 0.6, and 0.8 cause the growth of the grain in a random direction through the other gain and creates bigger grains. That is in agreement with the results obtained by XRD also decreases the size of voids which enhances the junction between the grains and the formation of high- 2223 phase, Which improves the superconducting properties of the samples and raising the T_C values. On other side, Figures 14 and 15 illustrate how the surface of the samples become smother, and the plate-like structure is destroyed. In other words, it shows very small size grains, indicating the poor formation of the superconducting phase.

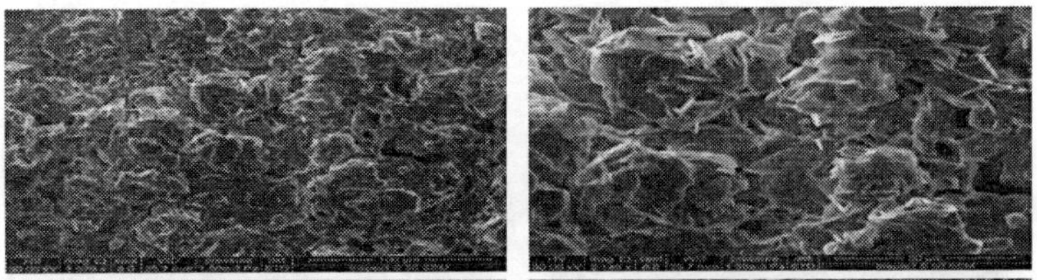

Figure 9. SEM micrographs of the fracture surface of the sample with two magnification and x= 0.1.

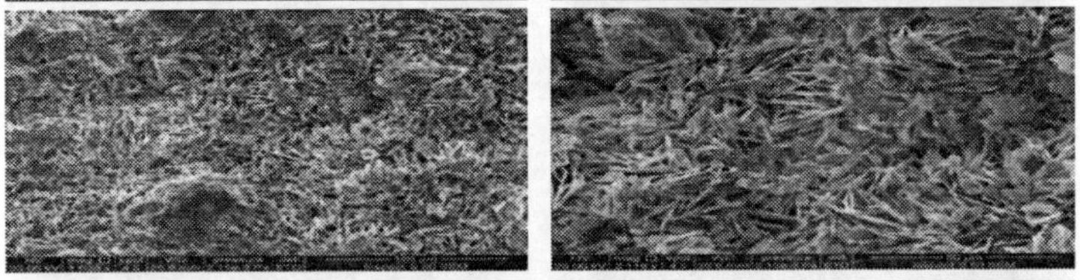

Figure 10. SEM micrographs of the fracture surface of the sample with two magnification and x= 0.2.

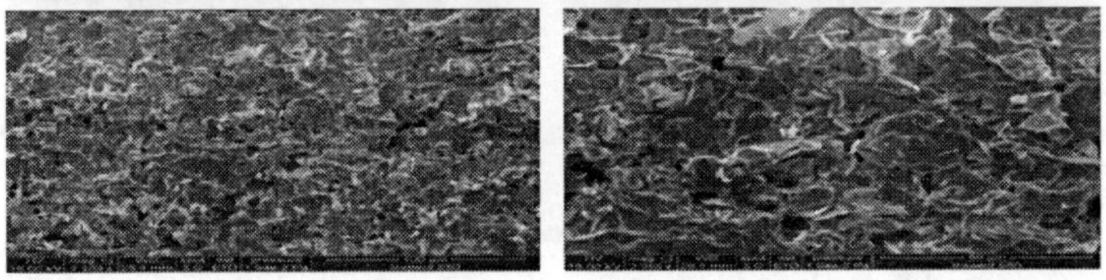

Figure 11. SEM micrographs of the fracture surface of the sample with two magnification and x= 0.4.

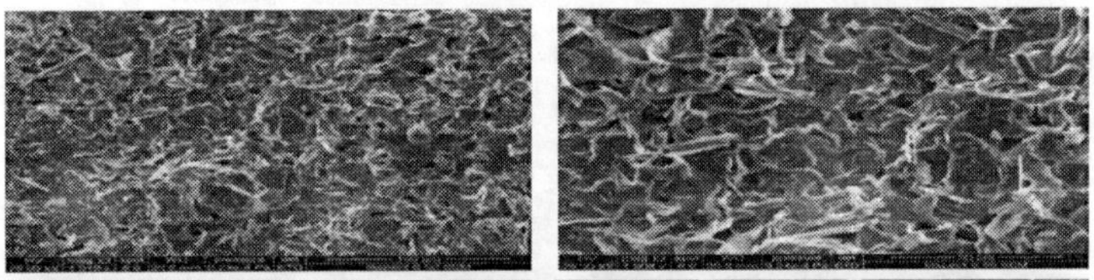

Figure 12. SEM micrographs of the fracture surface of the sample with two magnification and x= 0.6.

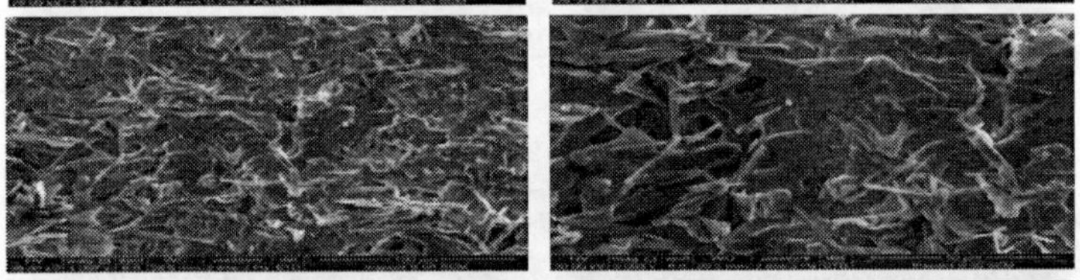

Figure 13. SEM micrographs of the fracture surface of the sample with two magnification and x= 0.8.

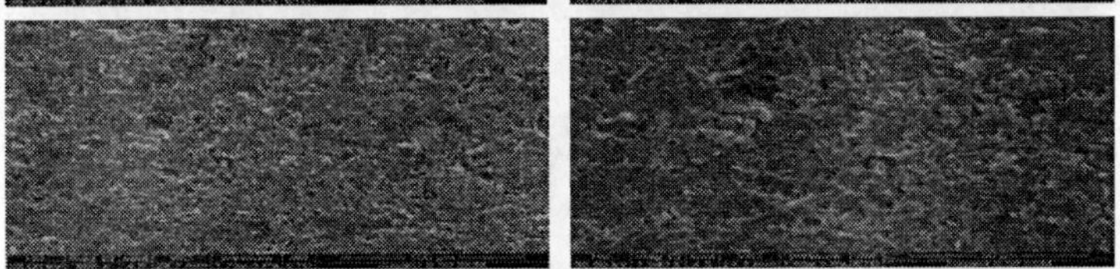

Figure 14. SEM micrographs of the fracture surface of the sample with two magnification and x= 2.

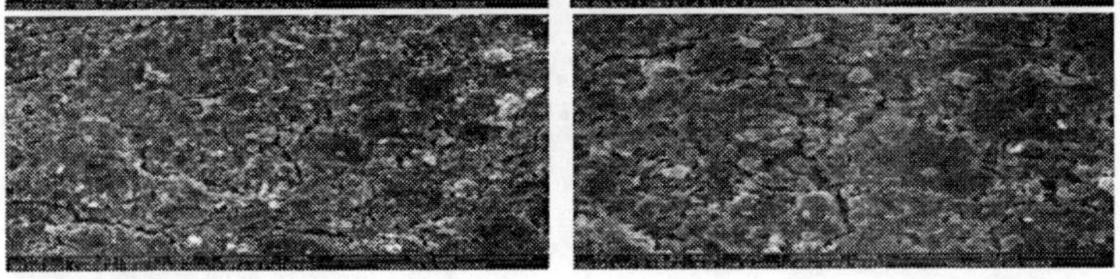

Figure 15. SEM micrographs of the fracture surface of the sample with two magnification and x= 3.

4. Conclusions

The XRD and electrical resistivity results of BPSCCO samples indicate that the variation in lattice parameters and superconductivity behavior are mainly due to the structural distortion and charge ordering phenomenon, which induced by Co substitution in combination with Pb addition, promote the 2223 phase of samples . The maximum T_C= 118 K obtained at the concentration 0.8 in optimal doping regime. The improvement of the superconducting properties of samples is due to the better crystallinity, larger grains, and oxygen incorporation into the sample, which leads to the prominent enhancement of T_C. SEM micrographs show changes morphology of the material with an increasing dopant concentration

References

[1] Fossheim K and Sudbo A 2005 *Superconductivity – Physics and Applications John Willey and Sons Ltd.*
[2] Agarwala P, Srivastava M P, Dheer P N, Pandmanaban P N and Gupat A K 1999 *Physica C* **313** 87
[2] Che G C, Du Y K and Zhao Z X 1994 *Solid-State Commun.* **89** 903.
[4] Hong Z, wang M, Xiong G, and Fan X 1997 *Physica C* **288** 82.
[5] Khan M N and Zakaullah Kh 2006 *J. Rese. Sci.* **17** 59.
[6] Muhammad I and Mahmood R 2012 *World Appl. Sci. J.* **17** 1510.
[7] Hermiz G Y, Jassim A K, Alwan E K 2015 *Asian Academic Rese. J. Multidisciplinary* **2** 281.
[8] Abbas M M, Abbas L K and Salman U 2012 *Energy Procedia* **18** 215.
[9] Bilgili O, Selamet Y and Kocabas K 2008 *J. Super. and Novel Magne.* **21** 439.
[10] AL-Dahash A G 1998 *Effect of Deformation on Electronic Structure and Magnetic Properties of Some High-Temperature Superconductors* Ph.D. Thesis Baghdad University.
[11] Abbas M M, Abdulridha A, Jassim A J, and Hashim F 2018 Citation: *AIP Conference Proceedings* 030009.
[12] Gul I H, Rehman M A, Ali M and Maqsood A 2005 **432** 71.
[13] Iqbal M J, Mehmood R 2009 *J. Alloys and Compound* **477** 386.
[14] Hermiz G Y, Muna M A and Gilioli E 2009, *Attidela Fodazione Giorgio Ronchi Anno Lxiv* **2** 313.
[15] Awad R J 2012 J. *Supercon.and Novel Magnet.* **25** 739.
[16] Man N K, Garg K B, Hien T D, Sinh N H, Phuc N X and enkatesh S V 1998 *J. Magnetism and Magnetic Materials* **527** 177.
[17] Fujii T, Terasaki I, Watanabe T and Matsuda A 2002 *Phys. Rev.* B **66**
[18] Parinov I A 2012 *Microstructure and Properties of High-Temperature Superconductors* Springer-Verlag Berlin.
[19] Abbas M M, Oboudi S F and Raoof N Q 2015 *Materials Scie. and Applica.* **6** 310.
[20] Giri R, Singh H K, Tiwari R S and Srivastava O N 2001 *Bull. Mater. Sci.* **24** 523 .
[21] Hermiz G Y 2001 *Thermal Analysis and phase transformation of HTSC of* $(Bi_{1-x}Pb_x)_2(Sr_{1-y}Ba_y)_2Ca_2Cu_3O10$ *compound* Ph.D. Baghdad University.
[22] Salamati H, Morshedloo T, Kameli P, Baghqi M, Abdolhosseini I, Ahmahvand H and Sohrabi D 2010 *Iranian J. Phy. Rese., Proceedings of the Second National Conference on Advances in Superconductivity,* **10** 2.
[23] Abbas M M, Abbas L K and Bahedh H S 2015 *J. Applied Sci. Rese.* **11** 164.
[24] lkeda Y, Takno M, Hiroi Z, Oda K, Kitaguchi H, Takada J, Miura Y, Takada Y, Yamamoto O, and Mazaki H 1988 *Jap. J .Appl.Phys* **27** 2067.

Calculation Some Atomic Properties For Three and Four Electron Systems by Using Hartree- Fock Method

Wasan A Hussien

Physics department, College of Science, Wasit University.
kutpc19@yahoo.com

Abstract. In this paper many physical properties evaluated like two particles radial density distribution function $D(r1, r2)$, one-particle radial density distribution function$D(r1)$ has been evaluated by using Hartree-Fock type wave function for three and four electron systems (Li , Li^{-1} , Be , Be^{+1} ,B^{+2}) and the inter particle expectation values < *r*> has been evaluated too. Also it was calculation the expectation values for all energies $\langle V_{en}\rangle$, $\langle V_{ee}\rangle$, $\langle V\rangle$, $\langle T\rangle$ and $\langle E_{HF}\rangle$. All the results and the behaviors obtained in this work have been discussed, interpreted and compared with those previously obtained. It was found that the maximum of one-particle density distribution is shifting to the origin by increasing the atomic number (Z), due to the attraction forces between the nucleus and the electron clouds , As atomic number is Z increase, all the total energies are increases because the increasing nuclear charge effect. The results also indicate that attraction energy expectation values $\langle V_{en}\rangle$ are larger than the repulsion energy expectation value $\langle V_{ee}\rangle$.

Keywords: Hartree-Fock, Radial electronic density distribution .

1.Introduction

Any system can be described by a wave function$\Psi(r_1, r_2 \dots r_N)$; it is a mathematical expression to describe the system properties as position, momentum, energy etc. The exact calculation of wave functions for many electron atoms becomes difficult because of the many electron-electron repulsions that should be ignored for simplicity .There are two main kinds of the wave function approximation methods which use in this research, the first one is the uncorrelated wave function approximation Hartree-Fock (HF), and the second one is the correlated wave function approximation Configuration Interaction (CI)[1] Alexei M. Frolov and David M. Wardlaw [2] studied state properties of the four-electron lithium ion Li− in its ground 2S−state and isotope substituted Li^{-6} and Li^{-7} ions in their ground 2^2S−state(s) are determined from the results of accurate. Mohammed A Al-Kaabi [3] Studied configuration interaction method is used to calculate the correlation energy of $(1s)^2$ 2s state for the Beryllium and Boron ions. . The difference between correlated energy and energy (correlation energy) was found to be (-0.01476 a.u) to Beryllium ion,(-0.00787a.u) to Boron ion. The aim of the this work was studied and evaluated some physical properties for three and four electron systems (Li , Li^{-1} , Be , Be^{+1} ,B^{+2}).

2.Theory

Schrödinger's equation cannot be solved exactly for two-electron atoms or ions. So that an approximation method must be used. For atoms, such as Helium and Lithium the variational method is accurate enough, for atoms of higher atomic number, the best approach to find a good wave function lies in first calculation an approximate wave function using the Hartree-Fock procedures which depends on the Central Field approximation. The basic idea of this approximation is that each of the atomic electrons moves in an effective spherically symmetric potential v(r) created by the nucleus and all the other electrons. The problem is the determination of the electron-electron repulsion term $\{\sum_{i<j} \frac{1}{r_{ij}}\}$ [4].

There are two methods to calculate this term; first the simple semi-classical method of Thomas and Fermi, second the more precise Hartree-Fock or self consistent field approach. Hartree wave function for the atoms (ions) is not antisymmetric in the electron coordinates because it uses the simple product wave function .

The generalization of the Hartree method which takes into account this antisymmetry requirement imposed by the Pauli Exclusion Principle was carried out in 1930 by Fock and Slater [4] . This generalization of Hartree theory,is known as the Hartree-Fock method. In the Hartree-Fock approach, it is assumed, in accordance with the independent particle approximation and the Pauli exclusion principle, that the N-electron wave function is a Slater determinant ϕ ,or in other words an antisymmetric product of individual electron spin-orbitals.The optimum Slater determinant is then obtained by using the variational method to determine the best individual electron spin-orbital [5] .

2.1. *Singlet and Triplet States*

A complete description of the state of the atom requires that we take the spin of electrons into account. Each electron has a spin of ½.The spin of an electron may be oriented either parallel or antiparallel to the spin of another electron, giving a total spin of one(S=1) or zero (S=0), spin states with S=0 are called singlets because they can be obtained in only one way. However, when S=1 the resultant spin vector may have three orientations in space, corresponding to m_s=+1,0and -1.Therefore, spin states with S=1 have three spin wave functions and are called triplet .

$$\chi_A=\frac{1}{\sqrt{2}}[\alpha(1)\beta(2)-\alpha(2)\beta(1)] \qquad M_s=0 \qquad ...(1)$$

$$\chi_S=\begin{bmatrix} \alpha(1)\alpha(2) & M_S=+1 \\ \frac{1}{\sqrt{2}}[\alpha(1)\beta(2)+\alpha(2)\beta(1)] & M_S=0 \\ \beta(1)\beta(2) & M_S=-1 \end{bmatrix} \qquad ..(2)$$

where χ_A is an anti symmetric spin wave function, χ_S is a symmetric spin wave function, and $\frac{1}{\sqrt{2}}$ is the normalization factor.

We obtain the total wave function of the atom by combining the orbital wave function given by ψ_S and ψ_A and the spin wave function given by χ_S and χ_A.

ψ_{total} =(Orbital wave function)× (Spin wave function)

According to Pauli Exclusion Principle ψ_{total} must be anti symmetric [6].

We can use any one of the three symmetric two-electron spin functions, so instead of the non degenerate level previously found, we have a triply degenerate level with the three zeroth-order wave functions :

$$\psi_T = \frac{1}{\sqrt{2}}[1s(1)2s(2) - 2s(1)1s(2)]\alpha(1)\alpha(2) \quad \text{...... (3)}$$

$$\psi_T = \frac{1}{\sqrt{2}}[1s(1)2s(2) - 2s(1)1s(2)]\beta(1)\beta(2) \quad \text{..... (4)}$$

$$\psi_T = \frac{1}{\sqrt{2}}[1s(1)2s(2) - 2s(1)1s(2)]\frac{1}{\sqrt{2}}[\alpha(1)\beta(2) + \alpha(2)\beta(1)] \quad \text{....(5)}$$

The requirement of antisymmetry of the overall wave function leads to the zeroth-order wave function.[7]

$$\psi_S = \frac{1}{\sqrt{2}}[1s(1)2s(2) + 2s(1)1s(2)]\frac{1}{\sqrt{2}}[\alpha(1)\beta(2) - \alpha(2)\beta(1)] \quad \text{...(6)}$$

2.2. *Some Atomic Properties*

we demonstrate here some of the atomic properties such as the one-particle radial density distribution function $D(r_1)$, the one particle expectation value $\langle r_1^n \rangle$, Root Mean Square (Standard Deviation) $\Delta r1$ and the energy expectation value $\langle E \rangle$.

2.2.1. *Two Particles Radial Density Distribution Function D(r1, r2) :*

In each individual electronic shell, the two-particle radial density distribution function D(r1 r2) is defined by [8] .

$$D(r_1, r_2) = \iint \Gamma(r_1, r_2) r_1^2 r_2^2 \, d\Omega_1 \, d\Omega_2 \quad \text{.....(7)}$$

where $\Gamma(r_1, r_1)$ tow-particles density $d\Omega_i$ denotes integration over all angular coordinates of the position vector ri .

2.2.2. *One-particle Radial distribution function D(r1) :*

The one particle radial density distribution *D(r₁)* is very important in the study of the electrons in atom, which means the probability of finding electron in each shell and it represents the density distribution of one electron in each shell, and is defined as [9,10] .

$$D(r_1) = \int_0^{\infty} D(r_1, r_2) \, dr_2 \quad \text{... (8)}$$

IOP Conf. Series: Materials Science and Engineering **757** (2020) 012068 doi:10.1088/1757-899X/757/1/012068

2.2.3. *One particle expectation value* $\langle r_1^n \rangle$:

The one particle expectation value $\langle r_1^n \rangle$ is given by using HF approximations [11] :

$$\langle r_1^n \rangle = \int_0^\infty r_1^n D(r_1) dr_1 \quad ....(9)$$

As $n = 0$, the expectation value satisfies the normalization condition

Thus $\langle r_1^0 \rangle = 1$

2.2.4. *Mean Square (Standard Deviation)* $\Delta r1$:

The standard deviation is a useful quantity characterizing certain distribution. It measures the spread of value about the average value which defined as [12]

$$\Delta r_1 = \sqrt{\langle r_1^2 \rangle - \langle r_1 \rangle^2} \quad ....(10)$$

stands for the uncertainty of particle radial position Δr_1 .

2.2.5 ***Energy expectation value*** $< E >$:

The potential energy is simply the sum of the electron-nuclear attraction energy and inter electronic repulsion energy, which are proportional to the expectation values $\langle r_i^{-1} \rangle$ and $\langle r_{ij}^{-1} \rangle$ respectively. Therefore we may write: [13] :

$$\langle V \rangle = -Z \sum_i^N \langle r_i^{-1} \rangle + \sum_{i<j}^N \langle r_{ij}^{-1} \rangle \quad ...(11)$$

The energy expectation value related to the potential energy by [14] :

$$\langle E \rangle = \frac{1}{2} \langle V \rangle$$

Therefore:

$$\langle E \rangle = \frac{1}{2} \left[-Z \sum_i^N \langle r_i^{-1} \rangle + \sum_{i<j}^N \langle r_{ij}^{-1} \rangle \right] \quad ... (12)$$

where Z is the atomic number and energy is in atomic units (a.u.).

3. Result and Discussion.

for the shells (k, L), r1 and r2 are decreased by increased the atomic number Z, as shown the tables (1)and (2) , and the function D (r1, r2) is increased by Z. This is obvious because increasing the atomic number leads to an increase in the effect of the effective charge on the envelopes so approaching the nucleus .

Table (3) show that the results of the one- electron radial density distribution function $D(r_1)$, we can note the maximum values of D(r1) increase as Z increases and decrease their positions towards the nucleus as Z increases according to Coulomb law.

The maximum values for L shell , meane there are two peaks observed, the small peak represented for the probability of finding an electron close to nucleus while the second big peak represented a remote electron from nucleus existing in L shell as shown in fig (1). It was also observed when r1 = 0, then D (r1) =0. This implies the electron is cannot found in the nucleus, and when r1 =∞, then D(r1)=0. That implies the electron within the atom it is not found outside its.Table (5) represent the expectation values of one- particle, the expectation value increments when the atomic number Z increments , when n take negative values -2,-1 and vice-versa when n takes positive values +2, +1. We also note that the values of standard deviation Δr_1 decrease by increasing the atomic number of K,L shell.Note from Table (7) $\langle V_{en} \rangle$ and $\langle V_{ee} \rangle$ increases by increases the atomic number (Z). This behavior can be understood by the fact that each casing shrinks toward the nucleus because of the increased attraction of its nucleus. The distance between the electrons and the nucleus is decreased, and the distance between the two electrons is less, so the potential energy of attraction and contrast increases (the energy between two charges is directly proportional to the amount of the charge and vice versa with the distance between those two charges) .

The attraction energy expectation values $\langle V_{en} \rangle$ are larger than the repulsion energy expectation value $\langle V_{ee} \rangle$, This causes the distance between the electrons and the nucleus to be smaller than the distance between the electrons. This difference increases by increasing the atomic number .

Table 1. Maximum values of the two-particle radial density distribution function $D(r_1, r_2)$ and the location in the Hartree-Fock (HF) wave function for (K,L)Shell, all values are in a.u .

Atom & Ion		Two constants			One constant		
		r_1	r_2	$D(r_1,r_2)$	r_1	r_2	$D(r_1,r_2)$
(K-shell(^{1}s))	Li	0.3764	0.3764	1.884260	0.3938	0.3938	2.097805
	Li^{-1}	0.3768	0.3768	1.627662	0.4187	0.4187	2.095797
	Be	0.2789	0.2789	3.302795	0.2978	0.2978	3.917917
	Be^{+1}	0.2786	0.2786	3.661848	0.2786	0.2786	3.936297
	B^{+2}	0.2172	0.2172	6.028809	0.2208	0.2208	6.379643
(L-shell(^{1}s))	Li^{-1}	0.7958	0.7958	0.014966	0.3187	0.3187	0.002513
		5.4316	0.7958	0.055055	0.3187	2.2288	0.009242
		0.7958	5.4316	0.055055	2.2288	0.3187	0.009242
		5.4316	5.4316	0.202608	2.2288	2.2288	0.033996
	Be	0.3676	0.3676	0.030369	0.2314	0.2314	0.011879
		2.4978	0.3676	0.111669	1.5456	0.2314	0.043767
		0.3676	2.4978	0.111669	0.2314	1.5456	0.043767
		2.4978	2.4978	0.410773	1.5456	1.5456	0.161029

Table 2. Maximum values of the two-particle radial density distribution function $D(r_1, r_2)$ and the location in the Hartree-Fock (HF) wave function for KL-Shell, all values are in a.u .

Atom & Ion	KL(^{3}s)-shell			KL(^{1}s)-shell		
	r_1	r_2	$D(r_1,r_2)$	r_1	r_2	$D(r_1,r_2)$
Li	0.4616	0.4616	0.093813	---------	---------	----------
	3.8608	0.3809	0.189135	3.8608	0.3809	0.189301
	0.3809	3.8608	0.189135	0.3809	3.8608	0.189301
Li^{-1}	0.5326	0.5326	0.105908	---------	---------	----------
	0.3714	5.4248	0.133464	0.3714	5.4248	0.133468
	5.4248	0.3714	0.133464	5.4248	0.3714	0.133468
Be	0.3182	0.3182	0.412309	---------	---------	----------
	0.2768	2.5142	0.396476	0.2768	2.5142	0.398157
	2.5142	0.2768	0.396476	2.5142	0.2768	0.398157
Be^{+1}	0.2967	0.2967	0.249182	---------	---------	----------
	0.2728	2.1538	0.463248	0.2728	2.1538	0.467058
	2.1538	0.2728	0.463248	2.1538	0.2728	0.467058
B^{+2}	0.2178	0.2178	0.455438	---------	---------	----------
	0.2134	1.5228	0.837639	0.2134	1.5228	0.851467
	1.5228	0.2134	0.837639	1.5228	0.2134	0.851467

Table 3. Maximum values of the one-particle radial density distribution function D(r1) and position by using Hartree -Fock wave function for (K,L)Shell, all values are in a.u .

Shell	Atom or ion	Z	r_1	Max.D(r_1)
K(1s)	Li	3	0.3637	1.405855
	Li^{-1}	3	0.3636	1.406211
	Be	4	0.2671	1.945026
	Be^{+1}	4	0.2670	1.945326
	B^{+2}	5	0.2208	2.455357
L(^{1}s)	Li	3	0.2979	0.028822
			3.0998	0.273595
	Li^{-1}	3	0.2989	0.013844
			3.6878	0.169696
	Be	4	0.2151	0.053143
			2.0510	0.395928
	Be^{+1}	4	0.2142	0.075202
			1.8793	0.482236
	B^{+2}	5	0.21854	0.14167
			1.4305	0.69167

IOP Conf. Series: Materials Science and Engineering **757** (2020) 012068 doi:10.1088/1757-899X/757/1/012068

Table 4. values of positions and maximum values of D(r1) by using Hartree -Fock wave function for KL-Shell, all values are in a.u .

Shell	Atom or ion	Z	r_1	Max.D(r_1)
$KL(^1s) \equiv KL(^3s)$	Li	3	0.3615	0.71659
			3.111	0.136866
	Li^{-1}	3	0.3625	0.709700
			3.6867	0.084853
	Be	4	0.2647	0.997413
			2.0446	0.198202
	Be^{+1}	4	0.2637	1.007845
			1.8696	0.241772
	B^{+2}	5	0.22951	1.3312
			1.3934	0.36562

Table 5. The one particle expectation value $\langle r_1^n \rangle$ and standard deviation Hartree -Fock wave function for (K-L) Shell , all values are in a.u .

Shell	Atom or Ion	Results and comparisons	n=-2	n=-1	n=+1	n=+2	Δr_1
$K(^1s)$	Li	Present work	14.8883	2.6850	0.5731	0.4468	0.3440
		Ref.[14]	14.8883	2.6850	0.5731	0.4468	...
	Li^{-1}	Present work	14.8923	2.6855	0.5729	0.4464	0.3437
		Ref.[14]	14.8940	2.685	0.5729	0.4463	0.3436
	Be	Present work	27.7534	3.6819	0.4150	0.2330	0.2464
		Ref.[14]	27.7590	3.6819	0.4149	0.2330	0.2465
	B+2	Present work	44.63224	4.68068	0.32521	0.14267	0.1921
		Ref.[14]	43.5556	4.6667	0.3214	0.1378	0.1856
$L(^1s)$	Li	Present work	0.4354	0.3454	3.8737	17.7384	1.6532
		Ref.[14]	0.4354	0.3454	3.8737	17.7384	...
	Li^{-1}	Present work	0.2255	0.2496	5.7503	43.5989	3.2454
		Ref.[14]	0.1864	0.2420	5.7480	43.5120	3.2357
	Be	Present work	1.0556	0.5225	2.6494	8.4262	1.1861
		Ref.[14]	1.0560	0.5225	2.6498	8.4318	1.1875
	Be^{+1}	Present work	1.4496	0.6074	2.2800	6.0885	0.9434
		Ref.[14]	1.4496	40.6074	2.2801	6.0886	...
	B^{+2}	Present work	2.98160	0.86320	1.6366	3.1303	0.6720
		Ref.[14]	2.98160	0.86320	1.6366	3.1303	....

Table 6. The one particle expectation value $\langle r_1^n \rangle$ and standard deviation, in Hartree -Fock wave function for (KL) Shell , all values are in a.u .

Shell	Atom or Ion	Results and comparisons	n=-2	n=-1	n=+1	n=+2	Δr_1
	Li	Present work	7.6619	1.5152	2.2234	9.0926	2.0369
		Ref.[14]	7.6619	1.5152	2.2231	9.0874	...
	Li^{-1}	Present work	7.5589	1.4675	3.1616	22.0226	3.4679
		Ref.[14]	7.559	1.4670	3.1600	21.978	3.4626
KL(^{1}s) ≡ KL(^{3}s)	Be	Present work	14.4045	2.1022	1.5322	4.3296	1.4078
		Ref.[14]	14.4080	2.1022	1.5324	4.3324	1.4086
	Be^{+1}	Present work	14.6065	2.1449	1.3475	3.1608	1.1598
		Ref.[14]	...	2.1449	1.3483	3.1670	...
	B^{+2}	Present work	23.28609	2.7719	1.02435	1.80924	0.8211
		Ref.[14]	22.29265	2.67860	1.04596	1.85058	0.86980

Table 7. The attraction energy expectation values $\langle V_{en} \rangle$ and The repulsion energy expectation values$\langle V_{ee} \rangle$ in the HF wave functions and the total energy , all values are in a.u.

Atom or Ion	$\langle V \rangle$			$-\langle E \rangle = -1/2\langle V \rangle$		E_{corr}
	$-\langle V_{en} \rangle$	$\langle V_{ee} \rangle$	$-\langle V \rangle$	Present work	Ref.[15]	
Li	17.14637	2.28092	14.86545	7.43273	7.43272	0.04534
Li^{-1}	17.61005	2.75354	14.856506	7.42826	7.42800	0.06735
Be	33.63518	4.48915	29.146034	14.5730	14.5730	0.0944
Be^{+1}	31.88935	3.33456	28.55479	14.2774	14.2774	0.0474
B^{+2}	51.01327	4.433070	46.5802	23.2901	23.4787	-----

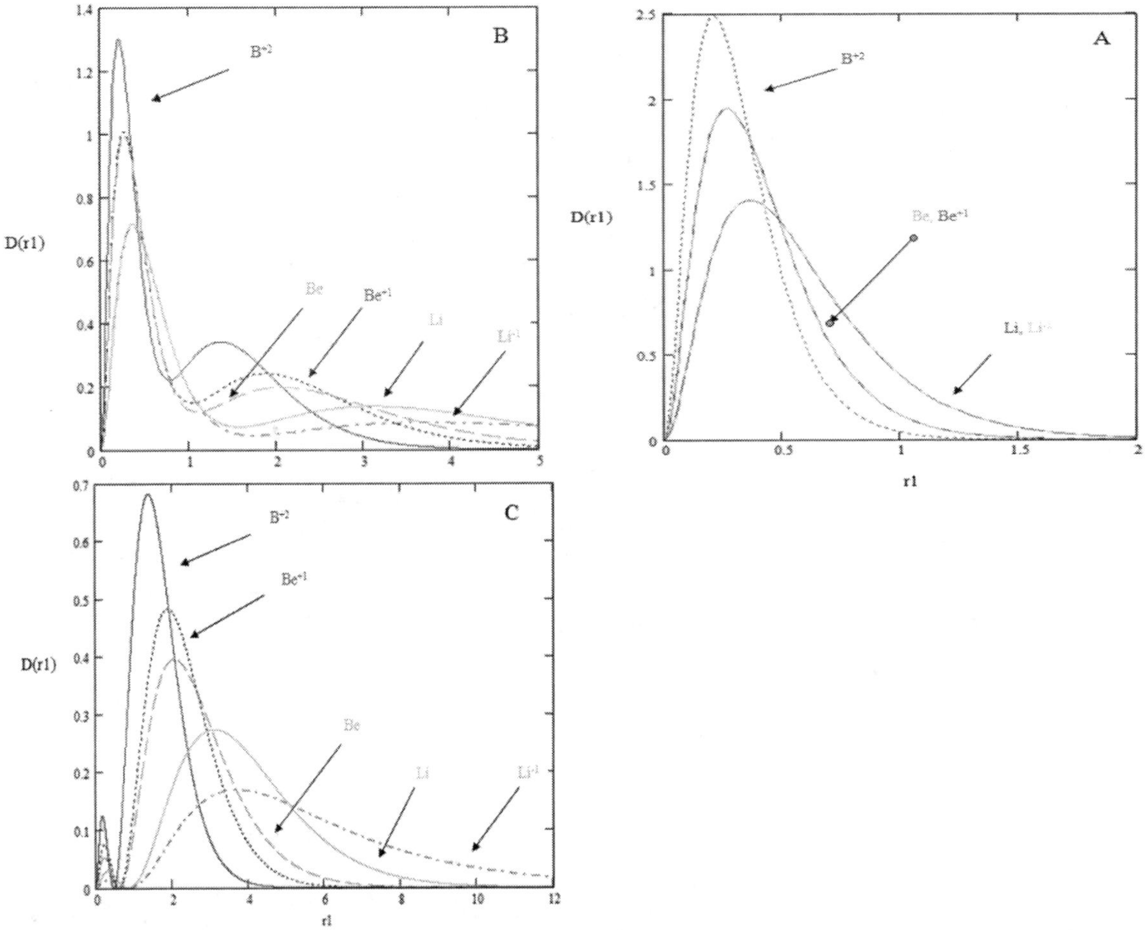

Figure 1. The relation between one-electron radial density distribution function with the position for (Li , Li^{-1} , Be, Be^{+1},B) : (A) for K shell , (B) for KL shell , (C) for L shell .

4. Conclusion

1. As Z increases not only $D(r_1, r_1)$ becomes more diffuse but it also decreases in magnitude these results are due to the Coulomb attraction force between nucleus and electrons which will be greater when Z goes from 2 to 4 or when Z goes from 3 to 5.
2. The maximum of one-particle density distribution is shifting to the origin by increasing the atomic number (Z), due to the attraction forces between the nucleus and the electron clouds.
3. The maximum probability of the function $D(r_1)$ is directly proportional to the atomic number (Z).
4. For each Z, the one –particle expectation values for HF wave functions are increasing when n goes from 1 to 2 and decreasing when n goes from -2 to 1.This means that the expectation values weight different regions of space.
5. As atomic number is Z increase, all the total energies are increases because the increasing nuclear charge effect.
6. The attraction energy expectation values $\langle V_{en} \rangle$ are larger than the repulsion energy expectation value $\langle V_{ee} \rangle$.

IOP Conf. Series: Materials Science and Engineering **757** (2020) 012068 doi:10.1088/1757-899X/757/1/012068

5. Reference

[1] Bransden B H and Joachain C J, 2003 *Physics of Atoms and Molecules,* ,(England: second edition).
[2] Alexei M F and David M W 2015 *physics . atom-ph* ,**v4** pp 1411.1138 .
[3] Mohammed A A, 2009 *Journal of Kerbala University*, **Vol. 7** No.1 Scientific.
[4] Donald A and John D, 1998 A Molecular Approach, *Physical Chemistry.*
[5] Bransden B H and Charles J J, 2003 *Physics of Atoms and Molecules*, (England: second edition).
[6] Marcelo A and Edward J F, 1968 *Fundamental University Physics* **vol.III**, Hardcover.
[7] Snow R L and Bills J L, 1974, *J Chem.Educ.***vol.51**, No.9, p 585.
[8] Roothaan C C, Lester M S and Wiess A W, 1960, *Rev.Modern Phys.*, **vol. 32**, p 186.
[9] Banyard K E and Baker C C, 1969 *J.Chem.Phys.,* **vol.51**,N.6.
[10] King F W and Dressel P R, 1989 *J.Chem. Phys.,* **vol.90**, No.1, p 6449.
[11] Russell J B, 1975 *Can. J. Phys.,* **vol. 53**, p. 592.
[12] Banyard K E 1971,*Publicado en Anales de Fisica*, **vol.67**, p 409.
[13] Levine I N, 1976 *Quantum Chemistry*, (New York: second edition).
[14] Bunge C F, Barrientos J A and Bunge A V, 1993 *Atomic Data and Nuclear Data Tables*, **Vol. 53**, Issue 1, p. 113-162.
[15] Hartree W, Hartree D R and Manning M F, 1941 *Phys.Rev.* **vol. 59**, p. 299.

Effect of Cold Atmospheric Pressure Plasma on Fertility Hormones for Female Rats.

N Yasoob A [1] , K J Khaleel[2] and H H Murbat [1]
[1]University of Baghdad, College of Science for Women, Department of Physics, Baghdad, Iraq.
[2] University of Al-Mustansiriyah, Iraqi Center for Cancer and Medical Genetic Research, Baghdad, Iraq.
Corresponding author email: nooryasoob@gmail.com

Abstract.

The direct application of cold atmospheric plasma (CAP) is the main scope of plasma medicine in or on the organism for curative purposes. Cold plasma is both effective in disrupting a wide range of microorganisms including multiple drug resistant ones (MDRs) and to stimulate proliferation of mammalian cells. It has obtained by Floating Electrode Dielectric Barrier Discharge (FE-DBD) system . The present study aimed to show the effected of cold plasma on the fertility hormones LH, Prolactin, Estrogen, and Testosterone hormones for healthy adult female rats (Albino) / bulb c). There are divided into many groups according to time exposure of plasma (15, 30, 60, and 90 second) and a reference group (without plasma exposure), with different time intervals after exposure (1, 2, 3, 7, 14, 28 days). The concentration of hormones was measured using Enzyme Linked Immunosorbent Assay (ELISA). The results showed that, the CAP effected slightly on the Luteinizing Hormone (LH) and Estradiol Hormone (E2) the concentration levels slightly increased and stay near to the reference value for all times. Prolactin hormone (PRL) was effected significantly from the first day after exposure to CAP for all times, it was increased significantly ($P < 0.0001$) for short time of exposure (15, 30 sec) than control group value, and it was decreased significantly ($P < 0.0001$) than control value for time of exposure (60 , 90 sec). The result of testosterone hormone is decreased significantly from the first day after exposure to CAP for all time exposure (15, 30, 60, and 90 sec).

Keywords: FE-DBD, CAP, fertility hormones, LH, Prolactin, Estrogen, and Testosterone.

1.Introduction

Plasma is referred to a partially or wholly ionized gas [1] . Plasma is a quasi neutral gas contains a mixture of reactive components including charged species (positive and negative ions, electrons , neutrals, visible, ultraviolet (UV), infrared (IR) radiation, and other electromagnetic fields, reactive oxygen species (ROS) and reactive nitrogen species (RNS) (2).

It is classified into two main groups according to its temperature: high temperature (fusion plasma) and low temperature plasma (gas discharge), low temperature plasma is subdivided into thermal and non-thermal plasma (3). Thermal plasma can be used in many different areas such as preserving the environmental from pollution, to process industrial waste (plasma gasification) (4), reducing environmental pollution by destroying toxic substances (5), chemical vapour deposition(6), and it was using in surgery for coagulation and tissue cutting, on the other hand, low temperature plasmas(< 40°C) is already prefer in plasma medicine to avoid damage the living tissues (2).

temperature of cold plasma devices normally operate below 40°C, $T_e >> T_i \sim T_g \sim 300$ K (3). This property can be used in cases that are sensitive to high temperatures such as living tissue and some biological applications, it is effective and safety for medical application have individual requirements, whose inhibits tissues to heating in some applications. Such as manageable, creditable, and operation at low temperature and open atmospheres (2) , painless *in vivo* applications and without damaging the surrounding tissue (7) .

In recent years cold plasma become interest in biological and medical applications, there are a broadband of implementations such as, sterilization surface and living tissues and inactivation of bacteria (8-10), the other characteristic of plasma that used frequently in medicine is blood coagulation (11, 12), wound healing (13), proliferation of adipose tissue derived stem cells (14), cancer treatment(15, 16), clinical application has started in the field of wound healing and treatment of infective skin diseases (2) and aesthetic medicine. Otherwise, cold plasma used in farming called plasma farming such as used to development chicken embryonic (17, 18), variety of advantages in the promotion of plant growth (19, 20), and using to water purification (21, 22).

A number of sources of non thermal plasma with low temperature close to room temperature (27~40 °C) that enhanced to be safe in biological implementation including , corona discharge (23-25), glow discharge (26, 27) , and dielectric barrier discharge DBD (28-30), that employed for serviceable biological and medical fields, all these atmospheric pressure plasmas are small scale and filamentary and are generated inside small discharge gaps (p*d-scaling of breakdown voltage by Paschen curve) (3).

Floating electrode dielectric barrier discharge (FE-DBD) is a type of DBD, which is operate under atmospheric pressure. DBD consists of two electrodes, the plasma is ignited in the gap between an isolated high voltage electrode and the target, when one of electrode is floated, it is called FE-DBD, the second electrode is the desired target (such as living tissues). The biological mechanisms of direct plasma interaction with living tissue caused by the change acquires in the liquid environment of cells by plasma, reactive oxygen and nitrogen species (ROS, RNS) generated in, or channeled to liquid by cold plasma considered a main and special character in biological plasma activity (2).

When cells or living tissue surfaces are exposed to cold DBD plasma a different biological reactive species are generated (ROS, RNS), DBD plasma begins selectively and amplifies ROS signals to promote cell proliferation and propagation. However, intracellular ROS formation after excessive plasma treatment of DBD results in a deficiency of cells. ROS balance plays an important role in physiological processes of sperm, while high concentrations of ROS lead to sperm morphological diseases (19).

Zhank J.J. et al. (2017) were investigated in reproductive capacity for chicken and development of embryonic in chicken eggs, there results are shown when exposure for 1 min of plasma at voltages of 11.7 kV to 27.6 kV enhanced chicken embryonic development, but when exposure to more duration for 4 min of plasma embryonic death (17). The aim of this study to investigate the effective of cold plasma on reproductive hormones.

2.Material and method

One hundred twenty five (125) healthy adult female Albino rats were selected for the study, the aged of them (2-3) months, and average weight 225±1.417g they were breeding in the animal house of Iraqi Center for Cancer and Medical Genetic Research. Animals are housed in a (special cages) with well ventilated, room temperature between 22-25 °C with light-controlled (12 h light, 12 h dark) room. They were provided with standard rat chow pellets. It is divided into multi groups, the first group is considered as reference group (control), and the remnant are divided according to time exposure to CAP (15, 30, 60, and 90 second) each group consist of 30 rats. Then, each group subdivided to six groups (n=5) according to interval after exposure to CAP the effect of CAP on the fertility hormones is measured after (1, 2, 3, 7, 14, 28) days. The blood was drawn from the heart of each rat by cardiac puncture, then it is putting in gel tube at room temperature, and centrifuged at a speed of 2500 rpm for 10 minutes.

The collected serum is stored at (- 20 °C) for measurement fertility hormones by (ELISA) technique.

Animals are directly exposed to cold plasma (CAP) on the skin from the dorsal side after hair removal for an area of approximately 2 cm^2, it is exposing for different time exposure to CAP and blood was drawn after time periods (1, 2, 3, 7, 14, 28) days.

Floating electrode dielectric barrier discharge (FE-DBD) system that is made and set up by the authors for producing cold plasma for biomedical applications.

The concentration of LH, Prolactin, Estrogen, and Testosterone hormones are measured by Enzyme Linked Immunosorbent Assay (ELISA) technique micro plate reader (Awareness USA) according to instruction of (Rapid Labs Limited, UK, monobind ink. USA, bioactive diagnostic GmbH.).

For statistical analysis using one way (ANOVA) analysis of variance, to verify the least significant difference between groups, and t-test analysis of variance to verify significant difference between controls with each group separately, using Prism version 7.04 for Windows, GraphPad Software, La Jolla California USA.

3.Result and discussion

In this study the effects of CAP on the fertility hormones for female rats are different from each other, the result of the CAP effect on the Luteinizing Hormone (LH) and Estradiol Hormone (E_2) indicates these hormones are slightly effected by cold plasma for all time intervals (15, 30, 60, and 90 sec).

When using long time of plasma exposure 3 min, all hormones are inhibited and arrived near to zero.

For LH hormone, the results show significant increased ($P<0.05$) for time exposure (30, 60, and 90 sec), but for smaller than 30 sec the result shown that non significant increased as shown in tables (1) and figures (1) . At the same time as the result of E_2 hormone showed that, significant increased for time of exposure (30, 60 sec) and non significant increased for time (15, 90 sec), as shown in tables (2) and figures (2), this result agree with J.J. Zhang et al. (31) that improved had no significant changes in estradiol level for female chicken.

Table 1. Effect of CAP on the LH (mIU/ml) concentration for female rats

Exposure time	1 day	2 day	3 day	7 day	14 day	28 day
	Mean ±SD	*Mean ±SD*	*Mean ±SD*	*Mean ±SD*	*Mean ±SD*	*Mean ±SD*
control	4.488±0.016	4.488±0.016	4.488±0.016	4.488±0.016	4.488±0.016	4.488±0.016
15 s	4.566±0.100	4.584±0.081	4.596±0.089	4.561±0.066	4.611±0.051	4.577±0.069
30 s	4.534±0.036	4.531±0.039	4.673±0.007	4.661±0.014	4.667±0.014	4.657±0.020
60 s	4.622±0.042	4.709±0.014	4.615±0.029	4.677±0.036	4.674±0.056	4.620±0.128
90 s	4.666±0.048	4.646±0.023	4.634±0.012	4.638±0.007	4.626±0.024	4.669±0.024

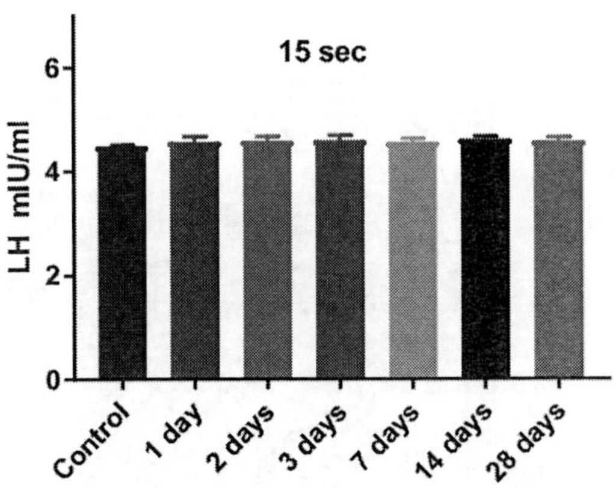

(a)

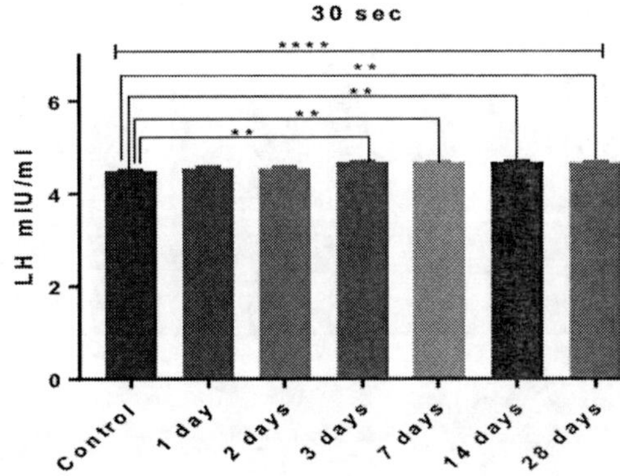

(b)

IOP Conf. Series: Materials Science and Engineering **757** (2020) 012069 doi:10.1088/1757-899X/757/1/012069

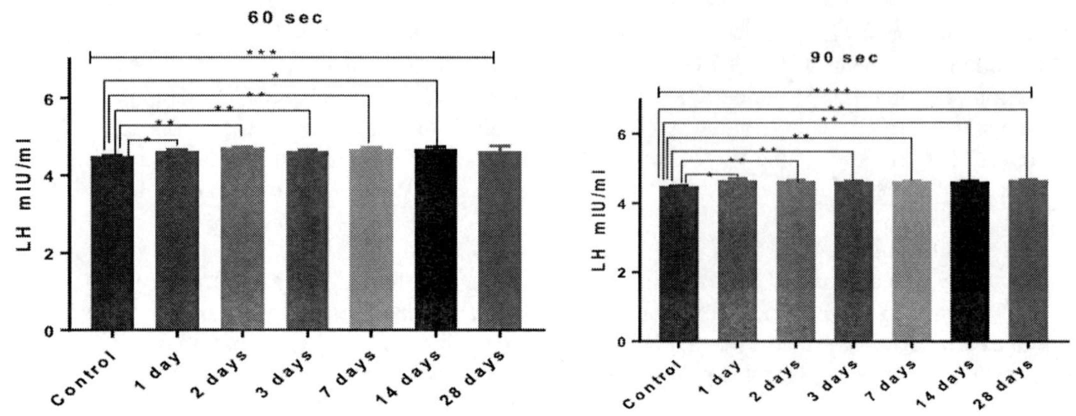

(c) (d)

Figure 1. Effect of CAP on the LH concentration for female rats with time after exposure (1, 2, 3, 7, 14, and 28) days for different time exposure to CAP (a) 15 sec. (b) 30 sec. (c) 60 sec. (d) 90 sec. Data are expressed as means ±SD,*p<0.05, **p<0 .01, ***p<0 .001, **** p<0.0001 (*) compared with control group.

Table 2. Effect of CAP on the E2 (pg/ml) concentration for female rats

Exposure time	**1 day**	**2 day**	**3 day**	**7 day**	**14 day**	**28 day**
	Mean ±SD	*Mean ±SD*	*Mean ±SD*	*Mean ±SD*	*Mean ±SD*	*Mean ±SD*
control	56.901±0.326	56.901±0.326	56.901±0.326	56.901±0.326	56.901±0.326	56.901±0.326
15 s	56.902±0.462	56.965±0.649	56.958±0.522	56.960±0.523	57.489±0.173	56.707±0.502
30 s	57.692±0.085	57.826±0.219	57.658±0.177	57.670±0.100	57.631±0.127	57.810±0.122
60 s	57.901±0.360	57.469±0.191	57.346±0.155	57.784±0.039	57.424±0.108	57.441±0.232
90 s	56.710±0.386	57.197±0.266	57.041±0.542	56.588±0.323	56.700±0.331	56.763±0.313

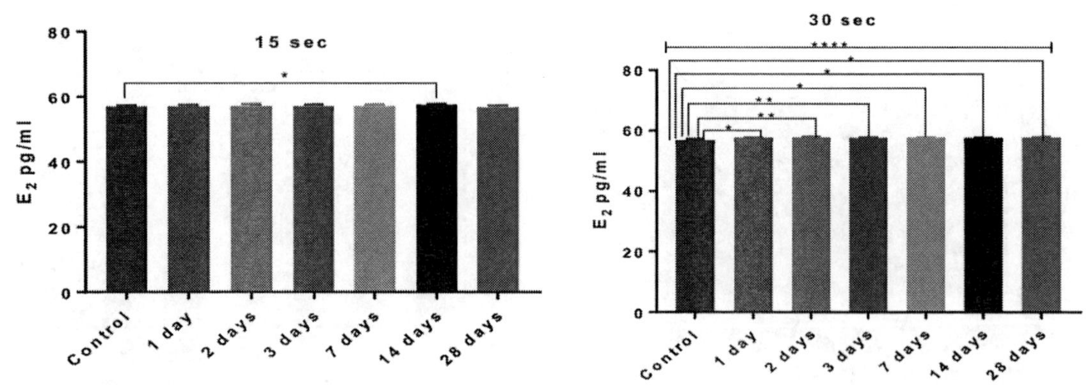

(a) (b)

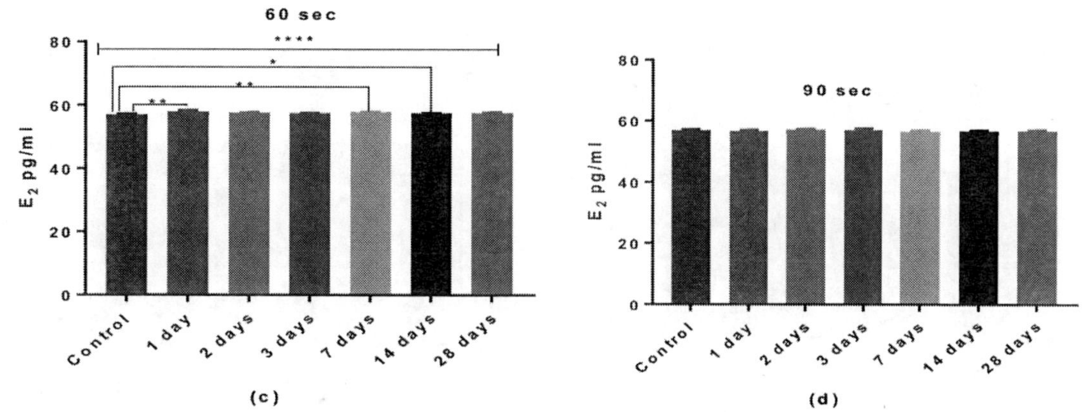

Figure 2. Effect of CAP on the Estradiol concentration hormone for female rats with time after exposure (1, 2, 3, 7, 14, and 28) days for different time exposure to CAP (a) 15 sec. (b) 30 sec. (c) 60 sec. (d) 90 sec. Data are expressed as means ±SD,*p<0.05, **p<0 .01, ***p<0 .001, **** p<0.0001 (*) compared with control group.

Prolactin hormone (PRL) are significant effected from the first day after exposure to CAP for all exposure times (15, 30, 60, and 90 sec) as shown in table (3) , but for short time of exposure (15, 30 sec) the concentration of PRL is increased significantly ($P < 0.0001$) than control value, and decreased significantly ($P < 0.0001$) than control value for time exposure (60 , 90 sec) as shown if fig (3)

Table 3. Effect of CAP on the PRL (ng/ml) concentration for female rats

Exposure time	1 day	2 day	3 day	7 day	14 day	28 day
	Mean ±SD	Mean ±SD	Mean ±SD	Mean ±SD	Mean ±SD	Mean ±SD
control	2.218±0.058	2.218±0.058	2.218±0.058	2.218±0.058	2.218±0.058	2.218±0.058
15 s	2.671±0.106	2.632±0.070	2.613±0.074	2.817±0.323	2.627±0.040	2.549±0.018
30 s	2.672±0.174	2.658±0.060	2.645±0.071	2.602±0.016	2.789±0.089	2.530±0.037
60 s	1.003±0.303	1.851±0.020	1.238±0.139	0.844±0.304	0.987±0.205	1.486±0.301
90 s	0.829±0.199	1.203±0.370	1.259±0.513	0.437±0.140	0.444±0.066	0.599±0.060

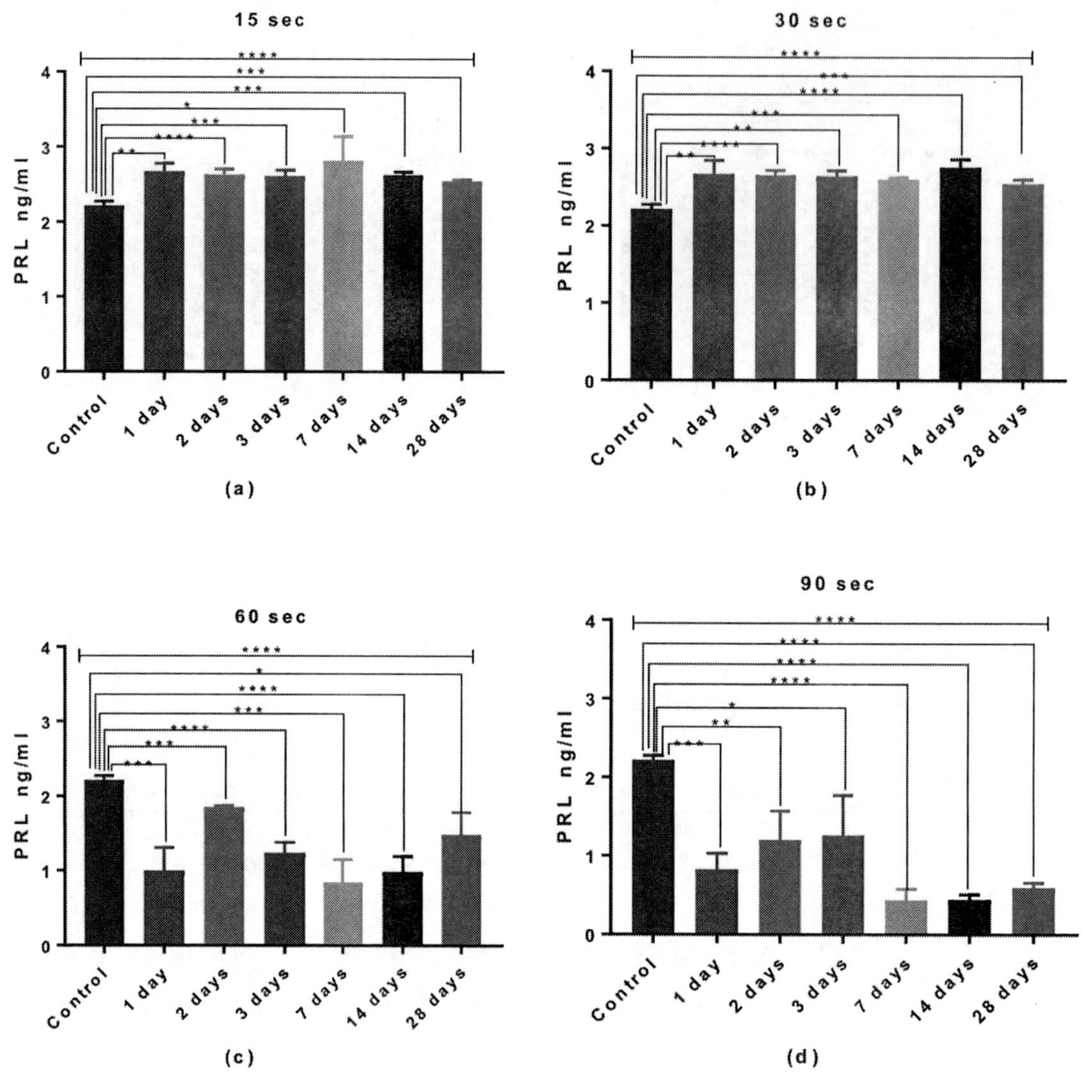

Figure 3. Effect of CAP on the Prolactin concentration hormone for female rats with time after exposure (1, 2, 3, 7, 14, and 28) days for different time exposure to CAP (a) 15 sec. (b) 30 sec. (c) 60 sec. (d) 90 sec. Data are expressed as means ±SD,*p<0.05, **p<0 .01, ***p<0 .001, **** p<0.0001 (*) compared with control group.

The result of testosterone hormone is shown in table (4), it is decreased significantly from the first day after exposure to CAP for all exposure times (15, 30, 60, and 90 sec), it's value still less than control even for long periods after exposure (28 days) as shown in fig (4).

These results be through with J.J. Zhang et al. (18) that improved high time of exposure 4 min caused embryonic death because high concentrations of ROS that lead DNA damage to sperm, Weak chromium integrity, low mobility and fertilization capacity, but when exposed to 1 min soybean sprout growth and chicken development.

Table 4. Effect of CAP on the Testosterone (ng/ml) concentration for female rats

Exposure time	1 day	2 day	3 day	7 day	14 day	28 day
	Mean ±SD	Mean ±SD	Mean ±SD	Mean ±SD	Mean ±SD	Mean ±SD
control	0.594±0.175	0.594±0.175	0.594±0.175	0.594±0.175	0.594±0.175	0.594±0.175
15 s	0.275±0.018	0.307±0.022	0.307±0.013	0.290±0.004	0.311±0.003	0.288±0.027
30 s	0.280±0.041	0.387±0.138	0.259±0.026	0.266±0.034	0.379±0.018	0.305±0.014
60 s	0.398±0.064	0.218±0.002	0.316±0.011	0.324±0.071	0.321±0.018	0.328±0.041
90 s	0.309±0.015	0.311±0.007	0.322±0.056	0.311±0.009	0.295±0.024	0.321±0.067

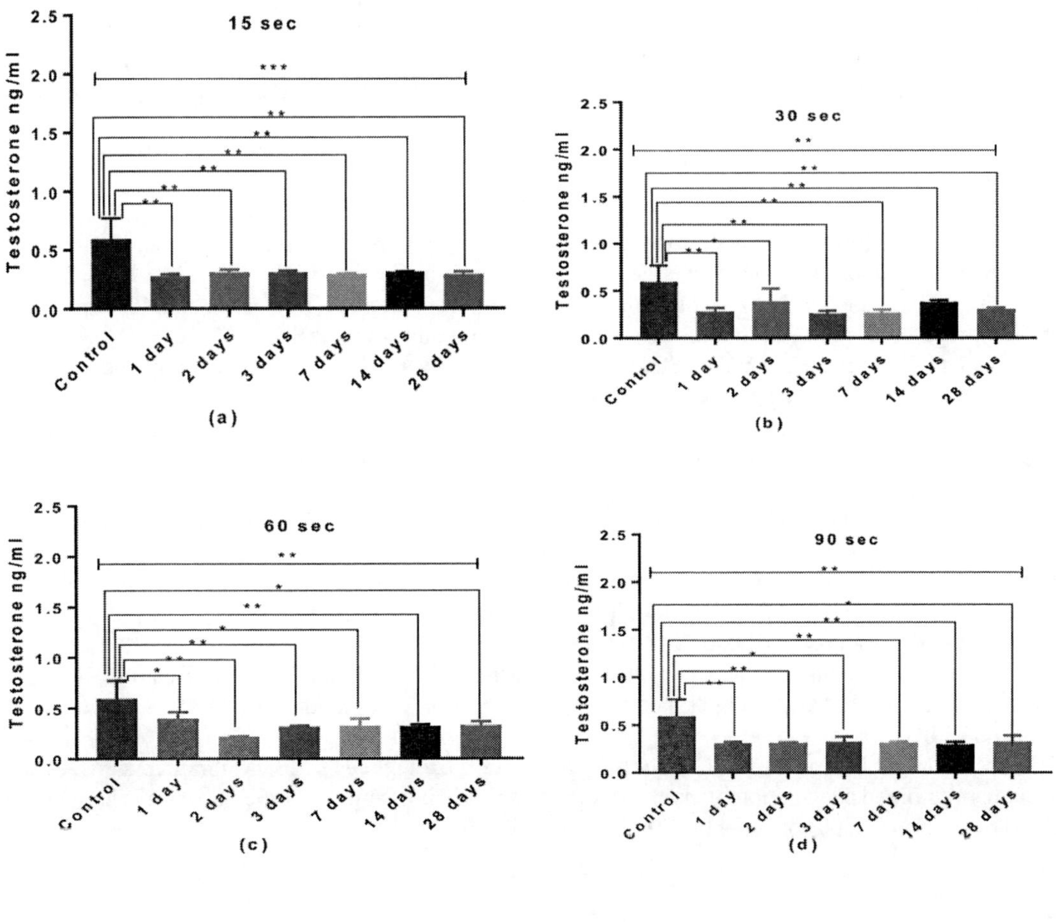

Figure 4. Effect of CAP on the Testosterone concentration for female rats with time after exposure (1, 2, 3, 7, 14, and 28) days for different time exposure to CAP (a) 15 sec. (b) 30 sec. (c) 60 sec. (d) 90 sec. Data are expressed as means ±SD,*p<0.05, **p<0 .01, ***p<0 .001, **** p<0.0001 (*) compared with control group.

4. Conclusion

In the present study non thermal plasma was obtained from FE-DBD, and applied directly on the live organism (female rats) for different time of exposure (15, 30, 60, and 90 sec). The effect was shown after (1, 2, 3, 7, 14, 28 days). The CAP have maximum effected on the testosterone hormone, it is decreased significantly for all time of exposure (15, 30, 60, and 90 sec). Prolactin hormone is effected by CAP too, it is increased for time of exposure (15, 30 sec), and decreased significantly for long time of exposure (60, 90 sec). On the other hand, the effected of CAP on the concentration of LH and E_2 hormones are slightly for all time of exposure (15, 30, 60, and 90 sec).
We suggested to use cold atmospheric plasma for exposure time (60 and 90 sec) for hyperprolactinemia with non pregnant females, and (15, 30, 60, and 90 sec) for decreasing testosterone hormone in females that suffering increased in testosterone hormone and showing whether it could be inhibited by plasma.

5.References

1. Chen FF. Introduction to plasma physics and controlled fusion: Springer; 1984.
2. Weltmann K-D, Metelmann H-R, von Woedtke T. Low temperature plasma applications in medicine. Europhysics News. 2016;47(5-6):39-42.
3. Nehra V, Kumar A, Dwivedi H. Atmospheric non-thermal plasma sources. International Journal of Engineering. 2008;2(1):53-68.
4. Anyaegbunam Felix N. Plasma Gasification for waste management and sustainable renewable clean energy generation. Proceedings of Nigerian Academy of Science. 2013;6(1).
5. Aziz KHH, Miessner H, Mahyar A, Mueller S, Kalass D, Moeller D, et al. Removal of dichloroacetic acid from aqueous solution using non-thermal plasma generated by dielectric barrier discharge and nano-pulse corona discharge. Separation and Purification Technology. 2019;216:51-7.
6. Yasoob N, j. N, H. N, H. R. study the morphological and optical properties of CdO doped ZnO thin film prepared by (PLD) technique. journal of college of education. 2018;1:147-56.
7. Isbary G, Shimizu T, Li Y-F, Stolz W, Thomas HM, Morfill GE, et al. Cold atmospheric plasma devices for medical issues. Expert review of medical devices. 2013;10(3):367-77.
8. Vaze ND, Gallagher MJ, Park S, Fridman G, Vasilets VN, Gutsol AF, et al. Inactivation of bacteria in flight by direct exposure to nonthermal plasma. IEEE Transactions on Plasma Science. 2010;38(11):3234-40.
9. Att RR, Al Haideri H, Murbat HH. Influence of Cold Atmospheric Plasma on Acinetobacter baumannii. Baghdad Science Journal. 2019;16(1 61-151:(ملحق.
10. Fridman G, Peddinghaus M, Ayan H, Fridman A, Balasubramanian M, Gutsol A, et al. Blood Coagulation and Living Tissue Sterilization by Floating-Electrode Dielectric Barrier Discharge in Air. Plasma Chem Plasma Process. 2006;26:425–42.
11. Ayan H, Fridman G, Staack D, Gutsol AF, Vasilets VN, Fridman AA, et al. Heating effect of dielectric barrier discharges for direct medical treatment. IEEE Transactions on Plasma Science. 2009;37(1):113-20.
12. Kalghatgi SU, Fridman G, Cooper M, Nagaraj G, Peddinghaus M, Balasubramanian M, et al. Mechanism of blood coagulation by nonthermal atmospheric pressure dielectric barrier discharge plasma. IEEE Transactions on plasma science. 2007;35(5):1559-66.

13. Naderi N, Zaefizadeh M. Expression of growth factors in re-epithelialization of diabetic foot ulcers after treatment with non-thermal plasma radiation. Biomed Res India. 2017;28:3402-7.
14. Park J, Lee H, Lee HJ, Kim GC, Kim S-S, Han S, et al. Non-thermal atmospheric pressure plasma is an excellent tool to activate proliferation in various mesoderm-derived human adult stem cells. Free Radical Biology and Medicine. 2019;134:374-84.
15. Pereira S, Pinto E, Ribeiro P, Sério S. Study of a Cold Atmospheric Pressure Plasma jet device for indirect treatment of Squamous Cell Carcinoma. Clinical Plasma Medicine. 2019;13:9-14.
16. Balzer J, Demir E, Kogelheide F, Fuchs PC, Stapelmann K, Opländer C. Cold atmospheric plasma (CAP) differently affects migration and differentiation of keratinocytes via hydrogen peroxide and nitric oxide-related products. Clinical Plasma Medicine. 2019;13:1-8.
17. Zhang JJ, Jo JO, Do Luong Huynh MG, Kim N, Lee SB, Lee HK, et al. Lethality of inappropriate plasma exposure on chicken embryonic development. Oncotarget. 2017;8(49):85642.
18. Zhang J, Kwon T, Kim S, Jeong D. Plasma Farming: Non-Thermal Dielectric Barrier Discharge Plasma Technology for Improving the Growth of Soybean Sprouts and Chickens. Plasma. 2018;1(2):285-96.
19. Zhang JJ, Jo JO, Mongre RK, Ghosh M, Singh AK, Lee SB, et al. Growth-inducing effects of argon plasma on soybean sprouts via the regulation of demethylation levels of energy metabolism-related genes. Scientific reports. 2017;7:41917.
20. Volin JC, Denes FS, Young RA, Park SMT. Modification of Seed Germination Performance through Cold

Plasma Chemistry Technology. CROP SCIENCE. 2000;40:1706–18.
21. Mahyar A, Miessner H, Mueller S, Aziz KHH, Kalass D, Moeller D, et al. Development and Application of Different Non-thermal Plasma Reactors for the Removal of Perfluorosurfactants in Water: A Comparative Study. Plasma Chemistry and Plasma Processing. 2019:1-14.
22. Iervolino G, Vaiano V, Palma V. Enhanced removal of water pollutants by dielectric barrier discharge non-thermal plasma reactor. Separation and Purification Technology. 2019.
23. Chang J-S, Lawless PA, Yamamoto T. Corona Discharge Processes

IEEE TRANSACTIONS ON PLASMA SCIENCE. 1991;19(6).
24. Arola K, Kallioinen M, Reinikainen S-P, Hatakka H, Mänttäri M. Advanced treatment of membrane concentrate with pulsed corona discharge. Separation and Purification Technology. 2018;198:121-7.
25. Ayrton ST, Jones R, Douce DS, Morris MR, Cooks RG. Uncatalyzed, regioselective oxidation of saturated hydrocarbons in an ambient corona discharge. Angewandte Chemie International Edition. 2018;57(3):769-73.
26. Romani VP, Olsen B, Collares MP, Oliveira JRM, Prentice-Hernández C, Martins VG. Improvement of fish protein films properties for food packaging through glow discharge plasma application. Food hydrocolloids. 2019;87:970-6.
27. Pardeshi H, Patil P, Bange J. Development of atmospheric pressure glow discharge plasma assisted CVD system for the deposition of SiOx coatings. Journal of Instrumentation. 2019;14(06):P06026.
28. Kostov KG, Honda RY, Alves LMS, Kayama ME. Characteristics of Dielectric Barrier Discharge Reactor for Material Treatment

Brazilian Journal of Physics. 2009;39(2).

29. Rad ZS, Davani FA. Non-thermal atmospheric pressure dielectric barrier discharge plasma source construction and investigation on the effect of grid on wound healing application. Clinical Plasma Medicine. 2016;4(2):56-64.
30. Tran ND, Harada N, editors. Developing of Aging Wine Model by Dielectric Barrier Discharge (DBD) at Atmospheric Pressure. 2018 4th International Conference on Green Technology and Sustainable Development (GTSD); 2018: IEEE.
31. Zhang JJ, Chandimali N, Kang TY, Kim N, Mok YS, Kwon T, et al. Growth and male reproduction improvement of non-thermal dielectric barrier discharge plasma treatment on chickens. Journal of Physics D: Applied Physics. 2018;51(20):205201.

IOP Conf. Series: Materials Science and Engineering **757** (2020) 012070 doi:10.1088/1757-899X/757/1/012070

Anomalies in the magnetic properties of some terbium-yttrium iron garnets at low temperatures

M Lahoubi[1] **and A Boutaba**[1]

[1]Department of Physics, Laboratory L.P.S., Faculty of Sciences, Badji-Mokhtar Annaba University, Annaba 23000, Algeria

Email: mlahoubi@hotmail.fr & mlahoubi@gmail.com

Abstract. This work is aimed to study the influence of the replacement of the magnetic cations Tb^{3+} by diamagnetic cations Y^{3+} on the 'low temperature point', (T_B) predicted by Belov for ferrimagnets with and without magnetic compensation point, (T_{comp}). Isothermal magnetizations, $M_T(H)$ are carried out on single crystals of two mixed terbium-yttrium ferrite garnets $Tb_xY_{3-x}Fe_5O_{12}$ (x = 1.98 and 0.37). The measurements are performed in the 1.5-300 temperature range and under DC magnetic field, (H) up to 80 kOe applied along the <111> and <100> directions. Using the approximation to the second order with respect to H for the expansion of $M_T(H)$, the specific features are obtained on the basis of the temperature dependences of the spontaneous, (M_S) and saturation, (M_{Sat}) magnetizations, the paraprocess magnetic susceptibility, (χ_p) and the parameter, (|b|) of the bH^2 term. For x = 1.98, M_S with H // <111> (the easy axis (EA) of magnetization) and M_{Sat} with H // <100> (the hard axis of magnetization) exhibit Néel's curves (N-type) with T_{comp} = 137 K and the ferromagnetic-type low-temperature paraprocess. In the area of the T_B^{TbIG} point (58 K) found before for x = 3, anomalies occur in the temperature dependences of χ_p and the derivatives, (dM_S/dT) and (dM_{Sat}/dT). Whereas for x = 0.37 with H // <100> which is the EA at 1.5 K up to 40 K certainly, the $M_S(T)$ dependence with a Néel's P-type curve without a T_{comp} point is revealed. As for x = 1.98, there appear likewise in the region of the T_B^{TbIG} point, anomalies in the temperature variation of the corresponding physical quantities but with the manifestation of the antiferromagnetic paraprocess phenomenon. The results obtained for x = 0.37 in the case of the free sphere show that |b| undergoes changes at three spin reorientation temperatures, T_{SR1} = 40 K, T_{SR2} = 140 K, and T_{SR3} = 232 K. We found that as T increases, the total spontaneous phase transition occurs in a wide T-range from the <100> axis via stable and intermediate low-symmetry angular phases of <uuw>-type towards the high symmetry phase of the <111> axis.

1. Introduction

Rare-earth iron garnets (REIGs) are substances with the general formula $RE_3Fe_5O_{12}$ where RE^{3+} is a rare-earth ion or the yttrium (cubic structure with space group $Ia\bar{3}d$ - (O_h^{10}) N° 230). They constitute so far an important class of magnetic compounds whose fundamental properties are based on three-sublattice ferrimagnets. There is at first a strong negative exchange interaction between the Fe^{3+} ions located on the tetrahedral (d) and octahedral [a] sites. Consequently, the magnetic moments of these two sublattices are oriented antiparallel to each other, and in many cases the [a]-(d) sublattices can be treated as a single resultant iron YIG subsystem for which the temperature dependence of the spontaneous magnetization, ($M_S^{YIG}(T)$) curve exhibits a Weiss-type behavior. In the end of the 50s, the magnetic properties of these materials were studied by Pauthenet [1]. This author has found that the magnetic order in this subsystem gives rise to a Néel temperature (T_N^{YIG} = 560 K [1]). The third sublattice is formed by the RE^{3+} ions which are located on the dodecahedral {c} sites. The magnetic order in this sublattice is produced by the relatively much weaker negative exchange interaction between the {c} and (d) sublattices than the [a]-(d) interaction responsible of T_N^{REIG} (nearly identical for all REIGs) which provide undoubtedly the best illustration of ferrimagnetism. Then, the resultant spontaneous magnetization, ($M_S^{REIG}(T)$) may vanish at some magnetic compensation temperature, (T_{comp}) lower than T_N, in the heavy REIGs only, such GdIG, TbIG and DyIG for example for which T_{comp} is equal respectively to 290, 246 and 220 K [1]. In the early 60s, Belov proposed a theory based in the molecular-field approximation [2] to explain that the appearance of this particular T_{comp} point is due to the 'weakly ordered' RE^{3+} sublattice (later named 'weak sublattice'), i.e., subject to a weak exchange field H_{exc}^{cd} compared to the effective exchange field H_{exc}^{ad} induced by the YIG subsystem (mainly by the (d) sublattice). Starting from the lowest temperatures, the spontaneous magnetization of the {c} sublattice, ($M_S^{RE}(T)$) exhibits an exponential decay with temperature. This behavior differs sharply from that of the YIG subsystem which was later named 'strong sublattice'. In addition, based on the anomalous temperature dependences of $M_S^{REIG}(T)$, Belov and Nikitin [3] have predicted in their seminal 1965 paper with GdIG as a prototype, that in ferrimagnets namely ones with a T_{comp} point with N-type curve (in Néel's notation) must give rise to a 'low-temperature point', (T_B) and 'low-temperature paraprocess' phenomenon. These predictions were proposed subsequently for a second group of ferrimagnets with M-, and P-type curves, namely ones without a T_{comp} point [2]. The transition at T_B is, in a sense, 'induced' by the exchange field H_{exc}^{cd} of the neighbouring 'strong sublattice'. The existence of this T_B point has been discovered in a number of REIGs with heavy RE^{3+} ions, those possessing a T_{comp} point such as GdIG where Gd^{3+} is in $^8S_{7/2}$ ground state [3] or the group of REIGs with TbIG and DyIG as an example, for which the magnetic moments of their RE^{3+} ions have nonzero orbital angular momentum [4-7]. Based on the magnetization of polycrystalline samples, Pauthenet [1] revealed that the magnetic moments at 0 K of these RE^{3+} ions of this group are much less than the moments of the free ions. Instead of looking for explanation of these differences in the representation of a magnetic model at low temperatures which differs from the ferrimagnetic model, e.g., a Yafet-Kittel model, the results were interpreted using Néel' suggestion that "there is a quenching of the moment of the RE^{3+} ions due to the influence of a strong asymmetric crystalline field (CF)". It is worth keeping in mind that some other concomitant effects are well-known to appear at low temperatures where significant deviations from the collinear ferrimagnetic ordering have been interpreted as evidence of a high magnetic anisotropy due to a strong competition between CF and {c}-(d) exchange interactions. As a result, a more complex magnetic structure of the RE^{3+} moments, such as the so-called "double umbrella", appears at low temperature and was shown by neutron diffraction studies on polycrystalline TbIG and DyIG samples [8-11]. However, the lack of data on T_B for most RE-substituted-YIG compounds, $RE_xY_{3-x}IG$ is presumably due, in our knowledge, to the fact that no relevant measurements are available. This is the case for the mixed $Tb_xY_{3-x}IG$ system which has not received considerable attention. Since the first reported paper which was based on the magnetostriction properties of polycrystalline samples with x varying from 3 to 0 [12], high magnetic field measurements have been performed on single crystals [8, 13, 14] for the determination of the magnetic anisotropy and the change of the easy direction of magnetization from <111> to <100> when x decreases from 3 to 0.37. Then, subsequent papers have been devoted only to the magnetic and magneto-optical properties of single crystals with a low concentration ($x < 0.65$) at low temperatures and under high magnetic fields to study characteristic features such the jumps in the magnetization and Faraday rotation due to the Jahn-Teller effect which induces an instability of the magnetic structure and the crossing of the levels of Tb^{3+} ions [15-17 and refs. Therein].

2. Experimental

Single crystals of $Tb_xY_{3-x}IG$ have been grown from a $PbO/PbF_2/B_2O_3$ solvent using the standard spontaneous nucleation technique (details concerning the preparation method are already described elsewhere [18]) and three compositions have been synthesized: x = 1.98, 1, and 0.37. They were checked carefully by the specific magnetization method in the low magnetic field range and from the spontaneous magnetization $M_S(T)$ dependences, it appears that there is a T_{comp} point except for x = 0.37. The reduction of $T_{comp}(x)$ with x is found quasi linear, and its value which is constantly lower than $T_N \sim 560$ K (independently of x) vanishes at the critical content $x_{cr} = 0.725 \pm 0.02$ with $M_S = 0$ and $T_{comp} = 0$. The data reported in this paper concern only compounds with the two following concentrations: x = 1.98 and 0.37. The specimens are in the form of a sphere (diameter of about 3.1 mm) and their respective weights are equal to 0.0413 and 0.08143 g. The samples were encapsulated in special sample holders for the isothermal magnetization measurements, ($M_T(H)$) as a function of the DC external magnetic field, (H). The internal field, (H_i) can be easily obtained by correcting the field H of the amount of the demagnetizing field, (H_D) calculated with the demagnetization factor of a sphere. The collected $M_T(H)$ curves are overall reported in Bohr magneton, (μ_B) per two ($Tb_xY_{3-x}YIG$) formula units (2F.u. = one mole). Using the extraction technique of a magnetometer, the $M_T(H)$ curves were obtained with H up to 80 kOe from 1.5 K to 300 K. The field H was applied parallel to the principal crystallographic directions successively, <111> and <100>. These orientations were checked with an X-ray Laüe technique with an accuracy being about one degree and the total misalignment with H in all the experiments is estimated to be less than two degrees in angle. Specific measurements were also performed in the case where the sample is allowed to rotate freely on itself.

3. Results and discussion

The present work gives the principal features. From the temperature dependences of some magnetic characteristics, low-T anomalies occur in the region of the T_B^{TbIG} point estimated at 58 K for the pure TbIG [2, 4, 5]. The ferrites with x = 1.98 and 0.37 are stated to belong to the respective classes of 'strongly anisotropic ferrimagnet' and 'weakly anisotropic ferrimagnet' [19]. The results found for each sample are analyzed separately and if a comparison is required, they will be presented together.

3.1. Low temperature point and paraprocess phenomenon in $Tb_{1.98}Y_{1.02}IG$

In Fig. 1, are reported the $M_T(H)$ curves measured with H applied along <111> axis which was considered as the easy axis of magnetization [13]. While no measureable hysteresis can be observed, they show a non-linear field dependence up to 80 kOe with a significant negative curvature above and below $T_B^{TbIG} = 58$ K, with the exception of values of temperatures in the neighbourhood of $T_{comp} = 137$ K. It should be remembered that in the vicinity of T_{comp}, field induced phase transitions (FIPTs) between collinear and canted magnetic phases should occur when H falls a lower H limit [19]. At 137 K, $M_T(H)$ does not exhibits any measurable anomaly and is almost linearly field dependent and behaves as the straight line passing through the origin. That means that Tb and Fe sublattices are no more collinear but canted, as it was observed close $T_{comp} = 243.5$ K in the recent study of TbIG [20].

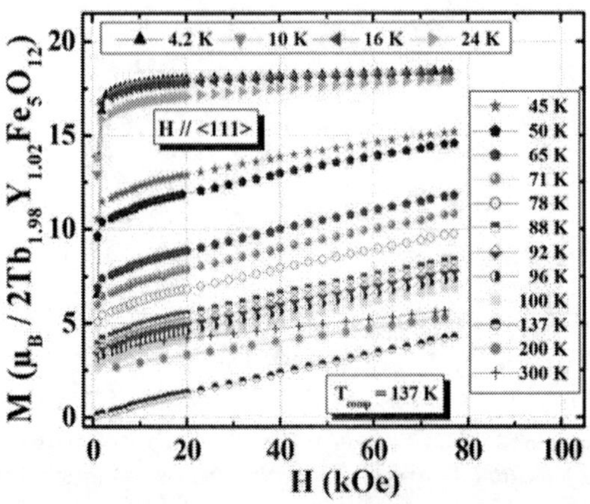

Figure 1. $Tb_{1.98}Y_{1.02}IG$. $M_T(H)$ measured in the 4.2-300 K temperature range with H // <111>.

The variation of $M_T(H)$ with H can be represented in a good approximation by the following quadratic expression [5]: $M_T(H) = M_S(T) + \chi_P(T)H + b(T)H^2$, where $\chi_P(T) = \partial M / \partial H\,(T)$ and b(T) are identified respectively with the paraprocess susceptibility and the parameter which should be able to take into account the curvature of the curves. The two magnetic phases (collinear-or coaxial and canted) may appear each one separately or together. The $M_T(H)$ curves present an initial curved portion of approach to technical saturation which can extend up to 15 kOe below 100 K. Based on the least-squares refinement technique, M_S is defined by extrapolation to H = 0 from the H > 15 kOe part. For H // <111>, M_S is well matched at 4.2 K (17.77 μ_B/mol) with previous value (17.90 μ_B/mol) [13].

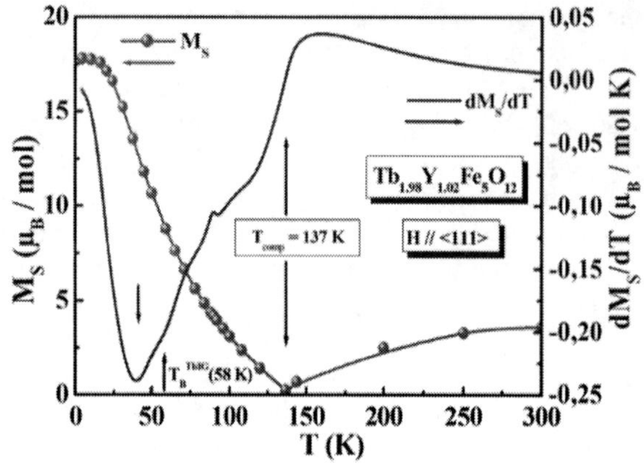

Figure 2. $Tb_{1.98}Y_{1.02}IG$. Temperature dependences of M_S and its derivative dM_S/dT with H // <111>.

The temperature dependences of M_S and its derivative, dM_S/dT are displayed in Fig. 2. As T increases, the $M_S(T)$ curve decreases from 4.2 K and a N-type dependence arises with a T_{comp} point at 137 K.

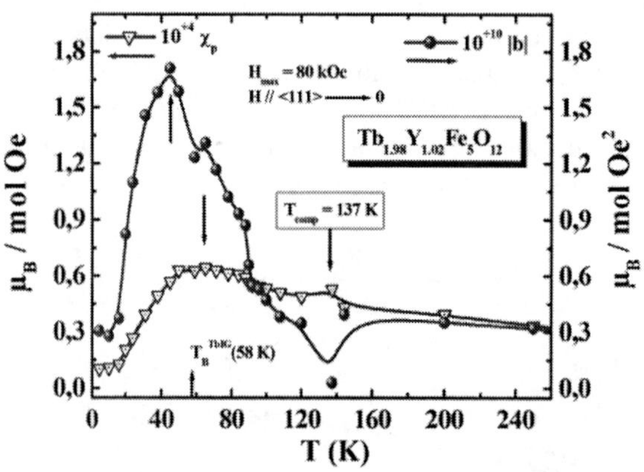

Figure 3. $Tb_{1.98}Y_{1.02}IG$. $\chi_P(T)$ and $|b|(T)$ curves with H // <111>.

Knowing that $M_S(T)$ changes sign on passing through T_{comp}, the dM_S/dT variation shows a discontinuous change because of the effect of the paraprocess. Below T_{comp}, the curve decreases abruptly with decreasing T and reaches a pronounced minimum at ~42 K i.e., in the low-part region of T_B^{TbIG}. Then, the curve increases with decreasing T up 4.2 K. Shown in Fig. 3, are the temperature dependences of both $10^{+4}\chi_P(T)$ and $10^{+10}|b|(T)$ curves. As is evident, and starting below room temperature, they increase when T decreases showing discontinuities near T_{comp} due to FIPTs [19], in agreement with that reported previously in high DC magnetic fields [13]. Below T_{comp}, each curve reaches a maximum respectively, at 65 and 45 K, then, with further decrease of T, diminishes but does not attains zero at T = 4.2 K ($10^{+4}\chi_P(T)$ = 0.1041 μ_B/mol Oe and $10^{+10}|b|(T)$ = 0.3040 μ_B/mol Oe^2).

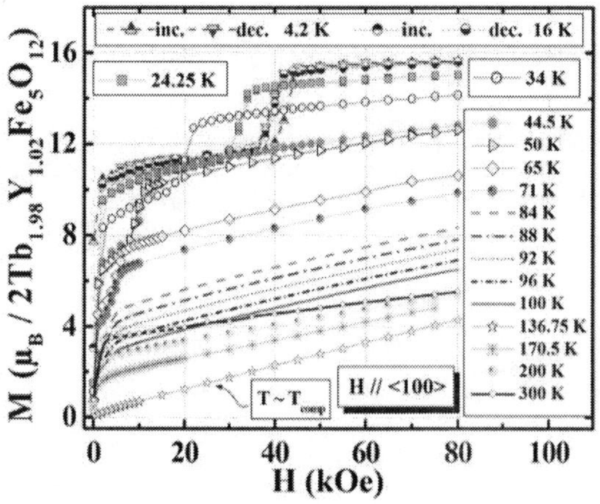

Figure 4. $Tb_{1.98}Y_{1.02}IG$. $M_T(H)$ measured in the 4.2-300 K temperature range with H // <100>.

The $M_T(H)$ curves displayed in Fig. 4 are obtained when H is applied along the <100> axis which was found as the hard axis of magnetization in the whole temperature range [13, 18].

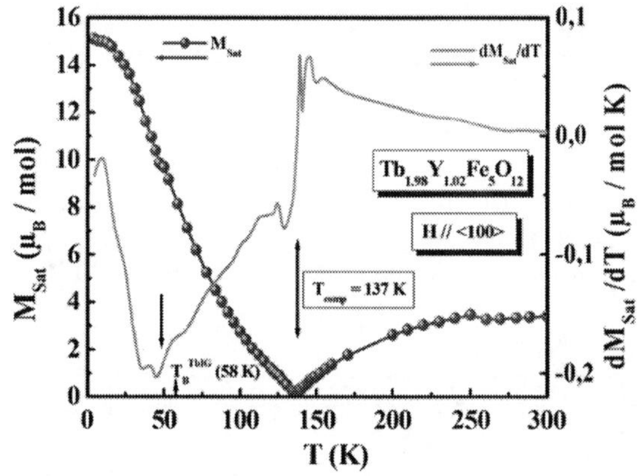

Figure 5. $Tb_{1.98}Y_{1.02}IG$. Temperature dependences of M_{Sat} and its derivative dM_{Sat}/dT with H // <100>.

One FIPT possessing a spin-reorientation phase transition (SRPT) nature from <111> to <100> occurs with a magnetization transition, at a transition field, (H_{C1}) in the 4.2-100 K range. The values of H_{C1} are well determined by the anomalies (magnetization jump (ΔM_1) in this case) on differentiated $M_T(H)$ curves with respect to H with ΔM_1 equal to 3.6 μ_B at 4.2 K. Then, the transition line, $H_{C1}(T)$ is separated in two regions by a critical point, (R_1) at $T_1 = 101 \pm 2$ K: the FIPT has a first order nature for $T < T_1$ and should be of second order above T_1 ($\Delta M_1 = 0$) with H_{C1} being defined up to 170 K. Within the relative experimental errors, the observed hysteresis of about 4 kOe at 4.2 K with $H_{C1} = 40 \pm 2$ kOe disappears completely above 16 K. Figs. 5, and 6 present respectively, the temperature dependences of magnetization saturation, (M_{Sat}) and its derivative, (dM_{Sat}/dT) as well as both curves $10^{+4}\chi_p$ and $10^{+10}|b|$ determined at each T for H > H_{C1}. It is seen that M_{Sat} shows a similar N-type dependence and dM_{Sat}/dT (Fig. 5) displays a minimum at 44.5 K, i.e. few Kelvin below T_B^{TbIG}, whereas it is seen from a comparative analysis of the $10^{+4}\chi_p$ and $10^{+10}|b|$ data (Fig. 6) that a maximum occurs at 78 and 71 K, respectively, i.e., values higher than T_B^{TbIG}. The 71 K-maximum of the low-T anomaly is due to the fact that in this direction of the hard axis of magnetization <100>, our high field extrapolation method up to H = 0 depends highly from the anisotropy of magnetization, (M_S<111> – M_{Sat}<100>)(T) which presents an inflection point at 45 K (curve not shown). The second 78 K-maximum, can be explained by the following arguments. At T << T_{comp}, where the magnetization of the Tb sublattice, (M_{Tb}) is oriented along H, an intensive ferromagnet-type paraprocess occurs, which decreases the magnetic component of the entropy. The magnetic order increases upon application of H since the thermal motion only partially destroys magnetic ordering in the 'weak' sublattice at the T_B point. As the spontaneous phase transition at the point T_B appears as an incomplete transition by comparison to the transition at T_N, at T > T_B, the high magnetic field H continues to act on the M_{Tb} sublattice and comes into play an effect in the shift of the low-T anomaly at higher values of T.

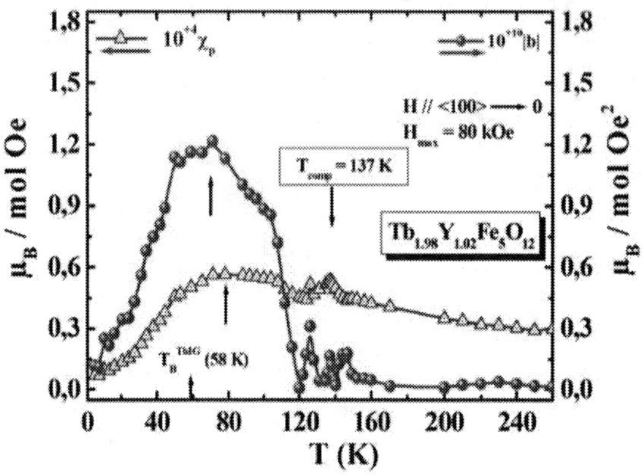

Figure 6. $Tb_{1.98}Y_{1.02}IG$. χ_p(T) and |b|(T) curves with H // <100>.

3.2. Low temperature point, paraprocess phenomenon and spin reorientation in $Tb_{0.37}Y_{2.63}IG$

Figure 7 illustrates the M_T(H) curves measured with H applied up to 60 kOe along the <100> direction which was found as the easy axis of magnetization in the 1.5-40 K range certainly and the results presented here are in a natural continuation of those previously published [8, 21].

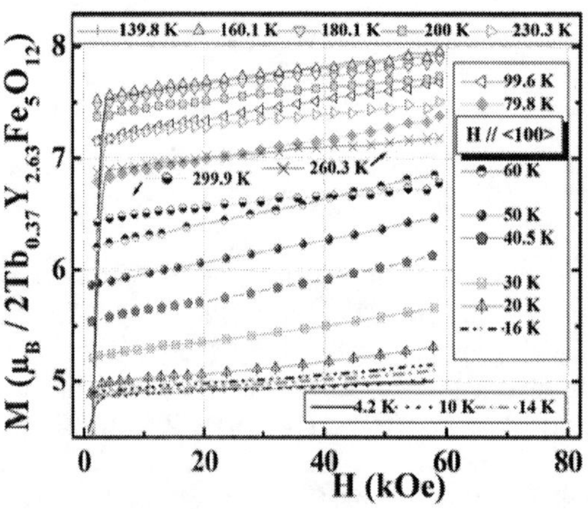

Figure 7. $Tb_{0.37}Y_{2.63}IG$. M_T(H) measured at various values of T from 4.2 K to 300 K with H // <100>.

They are free of hysteresis, without any anomaly or FIPT and show a slight positive curvature which increases near the T_B^{TbIG} point. The temperature dependences of M_S and dM_S/dT are plotted in Fig. 8.

IOP Conf. Series: Materials Science and Engineering **757** (2020) 012070 doi:10.1088/1757-899X/757/1/012070

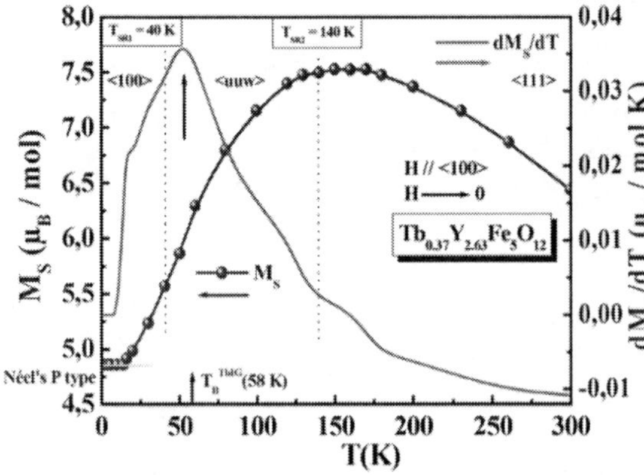

Figure 8. $Tb_{0.37}Y_{2.63}IG$. Temperature dependences of M_S and its derivative dM_S/dT with H // <100>.

As T increases, the $M_S(T)$ curve grows from 1.5 K up to the 150-170 K range and shows a P-type dependence. At the contrary, Belov and co-workers [2] have reported that $M_S(T)$ curve has the M-type from the interpolation between 4.2 and 60 K of the data obtained on $Tb_xY_{3-x}IG$ powder samples with x = 0.1 and 0.3. Such dependence of our $M_S(T)$ curve arises when close to 0 K, the 'weak sublattice' of the Tb^{3+} moments is 'well' ordered. The $M_S(T)$ curve has a nearly 'horizontal' portion between 1.5 and 10 K with M_S = 4.88 μ_B/mol, and the increase of M_S starts at 14 K with M_S = 4.90 μ_B/mol and reaches a slightly curved flat maximum (~ 7.52 μ_B/mol) in the 130-180 K range. Then, the dM_S/dT versus T shows an abrupt maximum at ~ 53 K due to the effect of the low-T paraprocess and a small anomaly in the region of the flat maximum. As in the case of $Tb_{1.98}Y_{1.02}IG$ with T_{comp} = 137 K, the Tb^{3+} moments which are considered as the 'weak sublattice' must give rise to a Néel's P-type curve without a T_{comp} point. They also account for a low-T_B point and the paraprocess phenomenon. The later must be of an antiferromagnet-type paraprocess because the field vector **H** acts to orient the $\mathbf{M_{Fe}}$ vector of the YIG sublattice and hence that of the 'weak sublattice' will be oriented opposite to **H**.

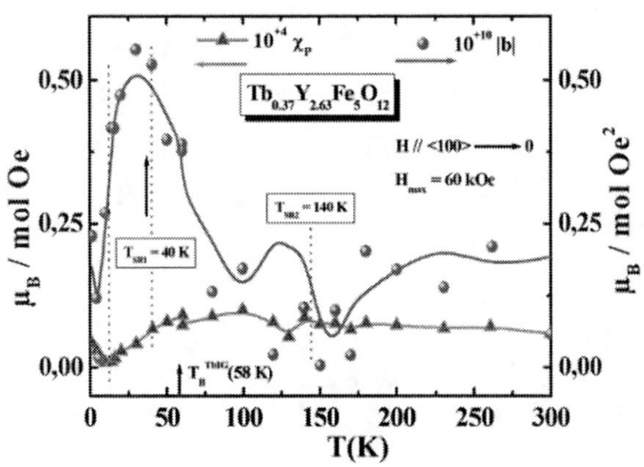

Figure 9. $Tb_{0.37}Y_{2.63}IG$. $\chi_P(T)$ and |b|(T) curves with H // <100>.

Shown in Fig. 9, are the $10^{+4}\chi_P(T)$ and $10^{+10}|b|(T)$ dependences for x = 0.37. It is clearly seen that a minimum appears at 14 K in the $\chi_P(T)$ curve which in fact is dictated by the nature itself of the P-type $M_S(T)$ dependence. There is a no well-defined paraprocess susceptibility maximum which is obscured by other concomitant effects over a wide temperature interval beyond and below T_B^{TbIG}. The |b|(T) curve displays a great maximum in the low-T part region of the T_B^{TbIG} point and close T_{SR1} = 40 K, the temperature of the spontaneous SRPT1 which was revealed in earlier works when this sample was involved [8, 21]. The SRPT1 is related with the lower continuous change from the easy axis of magnetization <100> to the low-symmetry angular phases of <uuw>-type. In the 130-180 K temperature range, the small minimum and the inflection point near T_{SR2} = 140 K observed in the $\chi_p(T)$ and |b|(T) curves respectively, are due to the effect of the SRPT2 [8, 21]. This upper continuous change from <uuw> to <111> is found in agreement with the similar one observed at 135 and 160 K respectively for $Tb_xY_{3-x}IG$ samples with x = 0.26 and 0.40 [22].

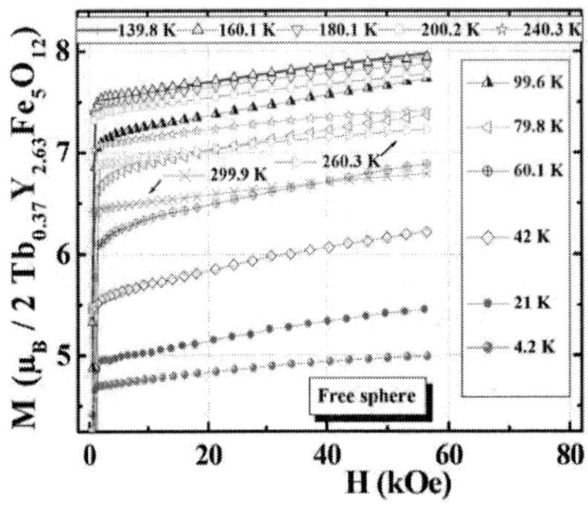

Figure 10. $Tb_{0.37}Y_{2.63}IG$. $M_T(H)$ measured at various values of T from 4.2 K to 300 K in the case of the free rotating sample.

The $M_T(H)$ curves in the case of the free rotating sample are displayed in Fig. 10. They are expected to exhibit similar behavior than that observed for <100> in Fig. 7. As T increases, the M_S curve plotted in Fig. 11 grows up to the same flat maximum in the 120-180 K range whereas the P-type dependence is not observed again due to the limited data from 4.2 K up to 21 K. The dM_S/dT curve displays the same 53 K-maximum shows in Fig. 8 whereas the small anomaly observed in the 120-180 K range is absent.

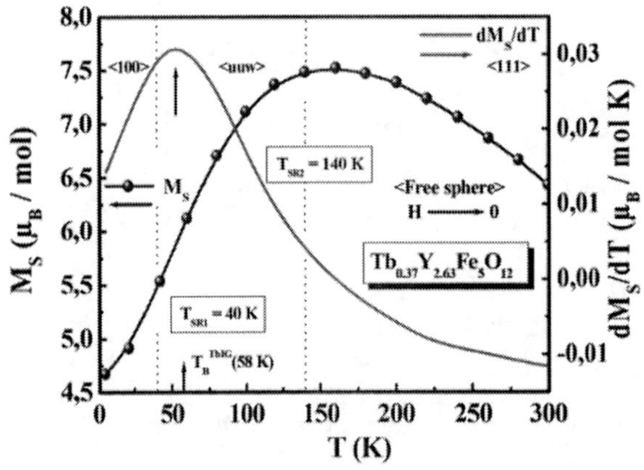

Figure 11. $Tb_{0.37}Y_{2.63}IG$. Temperature dependences of M_S and its derivative dM_S/dT in the case of the free sphere.

In Fig. 12, are reported the temperature dependences of both $10^{+4}\chi_p(T)$ and $10^{+10}|b|(T)$ curves in the case of the spherical crystal of $Tb_{0.37}Y_{2.63}IG$ is allowed to rotate freely on itself when H is applied.

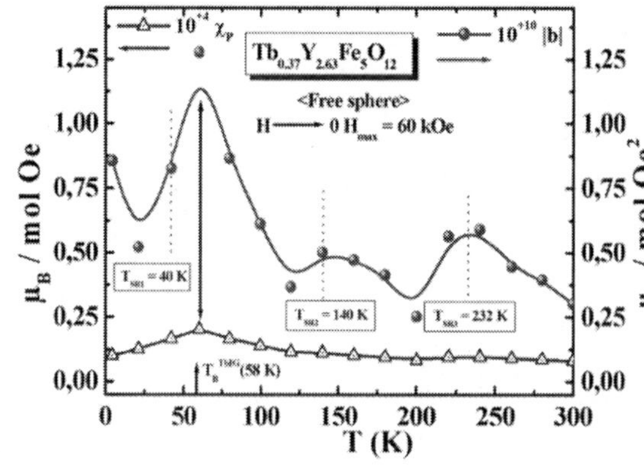

Figure 12. $Tb_{0.37}Y_{2.63}IG$. $\chi_p(T)$ and |b|(T) curves in the case of the free sphere.

Clearly, the situation is notably different with that observed for H parallel to <100>. Maxima are found in the vicinity of the T_B^{TbIG} point for both $\chi_p(T)$ and |b|(T) curves. The |b|(T) data give another experimental confirmation for the SRPT1 at T_{SR1} and SRPT2 at T_{SR2}. Furthermore, there is evidence that another upper change occurs at the 232 K-maximum. We still assume that T_{SR1} is defined as the transition temperature when the spontaneous magnetization M_S<111> (deduced for H < H_C applied along the hard axis of magnetization <111>) separates from the value of the projection of M_S<100> on the <111> direction, (M_S<100>/$\sqrt{3}$) [21]. In that case, T_{SR2} is defined again as the transition temperature when the difference M_S<100> – M_S<111> vanishes [8]. However, it had been reported earlier [22 and refs. Therein] that in $Tb_xY_{3-x}IG$ sample with a nearly similar composition (x = 0.40), a third SRPT3 was observed with another upper change at T_{SR3} = 240 K, in addition with the second SRPT2 at T_{SR2} = 160 K. The same phenomenon was also revealed for samples with x = 0.2, and 0.26 with T_{SR2} = 135 K and T_{SR3} = 227 K. Consequently, mention should be made for this experimental fact in support of the occurrence of an SRPT3 at T_{SR3} = 232 K. It was assumed at first that the $\mathbf{M_{Fe}}$ vector may be reoriented to the intermediate difficult axis <110> at T_{SR3} [23] whereas it was argued in [22] that the SRPT3 occurs at T_{SR3} from another <uuw> phase to an <111> collinear one. We agree with this second interpretation which was found valid on the basis of the magnetic phase diagram (MPD) of cubic materials allowing the first three magnetocrystalline anisotropy constants, (K_1, K_2, K_3). From the MPD plotted as function of the reduced anisotropy constants, ($K_1' = K_1/K_3$) and ($K_2' = K_2/K_3$) with $K_3 > 0$ [24], the estimated required conditions ($-0.17 \leq K_1' < 0$ and $K_2' = 0$) have allowed the existence of the SRPT3 [22]. Concerning the stability of the intermediate <uuw> phases which take place between T_{SR1} and T_{SR2}, we use the predictions of the model based on the analytical expressions of K_1, K_2, K_3 as function of the non-zero diagonal tensor components $G_{\alpha = x,y,z}$ representing the Fe-Tb anisotropic interactions in a high temperature approximation [21]. They provided a good explanation on the basis of the estimated values of the ratios (G_x/G_z) and (G_y/G_z) which were found equivalent with the specific required conditions ($-0.75 \leq K_1' \leq 0$ and $0 < K_2' \approx 0$) [21]. The $\mathbf{M_{Fe}}$ vector, being in an arbitrary direction in one of the cubic $\{(0\bar{1}0)\}$-type planes, is reoriented continuously in the 40-140 K range. So, the first intermediate <uuw> phase takes place up T_{SR2} = 140 K on the basis of a PO-trajectory situated inside the upper CFH area of the MPD near the CF-line [24]. Whereas in the 140-232 K range, another PO-trajectory close the line described by $K_2' = 0$ is selected inside the lower CFH area for the second intermediate <uuw> phase. We note that the lack of data on T_{SR1} for samples with x = 0.2, 0.26, and 0.40 is presumably due to the fact that no relevant measurements are available [22, 23]. This could be due to the reduced investigated 80-300 K range which was used before in their experiments. We must always remember some several phenomena which occur in the wide T-region between T_{SR1} and T_{SR3}: fast variation of the Tb moment because of the concomitant effects due to the estimated exchange splitting (40 cm^{-1}) and the T_B point; the opening of the 'square pyramid' umbrella of the Tb moments predicted at 4.2 K is suggested to start below T_a = 80 K, the temperature of the appearance of the anisotropy of magnetization which reaches a maximum value of one μ_B/mol at 4.2 K [8]; the higher levels states of the Tb^{3+} ions are populated beyond 100 K which change the nature of the anisotropy; during the increase of T, a small but significant variation of $M_{Fe}(T)$ needs to be considered. All these considerations imply that the trajectories become much more complicated than is predicted in the MPD. So, with $K_3 > 0$ varying with T, a large variety of spontaneous SRPTs (first and second order) can occur in the mixed $Tb_xY_{3-x}IG$ system with $x < x_{cr}$.

4. Conclusion

Experimental data based on magnetization $M_T(H)$ measurements carried out on two $Tb_xY_{3-x}IG$ single crystals (x = 1.98 and 0.37) are reported. The spontaneous M_S and saturation M_{Sat} magnetizations, the paraprocess susceptibility χ_p and the parameters, dM_S/dT, dM_{Sat}/dT and |b| are obtained using the approximation to the second order with respect to H for the expansion of $M_T(H)$. From the temperature dependences of these physical quantities, it was established that anomalies occur for both Tb content in the same region of the low-temperature point T_B^{TbIG} estimated at 58 K for TbIG. The existence of the ferromagnetic- and antiferromagnetic-type low-temperature paraprocess as well as some other concomitant effects is confirmed. For x = 0.37 and H // <100>, a P-type curve $M_S(T)$ dependence showing a nearly 'horizontal' portion in the 1.5-14 K range is observed. As T increase in the 40-232 K range, three SRPTs are involved in the total spontaneous spin reorientation <100>→<uuw>→<111>.

5. References

[1] Pauthenet R 1958 Les propriétés magnétiques des ferrites d'yttrium et de terres rares de formule $5Fe_2O_3.3M_2O_3$ *Thesis University of Grenoble France* Order N°. **81** pp.1–38
[2] Belov K P 1996 *Phys. Usp.* **39** 623
[3] Belov K P and Nikitin S A 1965 *Phys. Status Solidi* **12** 453
[4] Lahoubi M Wang W Varazashvili V 2017 *J. Phys.: Conf. Ser.* **827** 012014
[5] Lahoubi M 2018 *Physica* B **536** 96
[6] Boutaba A Lahoubi M Varazashvili V Pu S 2019 *J. Magn. Magn. Mater.* **476** 551
[7] Boutaba A Lahoubi M Varazashvili V Pu S 2019 *J. Supercond. Novel Magn.* **32** 3087.
[8] Lahoubi M 2012 *J. Physics: Conf. Ser.* **340** 012068
[9] Lahoubi M 2012 Temperature evolution of the double umbrella magnetic structure in terbium iron garnet *Neutron Diffraction* (Ed. Khidirov I) IntechOpen Chapter 10 pp. 203−230
[10] Lahoubi M 2015 *J. Phys.: Conf. Ser.* **574** 012099
[11] Lahoubi M Ouladdiaf B 2015 *J. Magn. Magn. Mater.* **373** 108
[12] Kiryukhin V P Sokolov V I 1967 *Sov. Phys.* JETP **23** (2) 287
[13] Guillot M Marchand A Desvignes J M Le Gall H 1989 *Z. Phys. Condens. Matter* **74** 335
[14] Lahoubi M Wang W Pu S 2015 *Physica* B **476** 129
[15] Lagutin A S Druzhinina R F 1992 *Sov. Phys.* JETP **75** (6) 1001
[16] Valiev U V 1994 *Phys. Solid State* **36** (2) 299
[17] Lahoubi M G Fillion 1997 *J. Phys. IV France* **7** C1-287
[18] Lahoubi M 1986 Structures magnétiques non colinéaires et transitions de phases dans les grenats ferrimagnétiques de terbium et d'yttrium *Thesis Doctorat of State in Physics, University of Grenoble France* pp.1–145
[19] Zvezdin A K 1995 Field Induced Phase Transitions in Ferrimagnets *Handbook of Magnetic Materials* Buschow K H J (Ed.), Elsevier Science, Amsterdam Chapter 4 pp. 405–543
[20] Lahoubi M Wang W 2015 *J. Magn. Magn. Mater.* **393** 437
[21] Lahoubi M Kihal A Fillion G 2000 *Physica* B **284-288** 1503
[22] Tsitskishvili K F *et al.* 1982 *Sov. Phys. Solid State* **24** (11) 1964
[23] Baazov N G *et al.* 1980 *Sov. Phys. Solid State* **22** (1) 162
[24] Borodin V A Doroshev V D Tarasenko T N 1983 *Fiz. Metal. Metalloved.* **56** (2) 220

The effect of the spinning speed variation on the perovskite solar cell efficiency

Mohammed S. Mehde[1], Abdulnasser M. Al-Gebori[2], Ali K. Hantoosh[2*]

[1]Al-karkh University of Science, Baghdad, Iraq

[2]Department of Applied Science, University of Technology, Baghdad, Iraq

[*]Corresponding author email: alialmousoi2006@yahoo.com

Abstract

In this work, compact titanitum dioxide (c-TiO_2), amorphous titanitum dioxide (mp-TiO_2) and methylammonium lead iodide ($CH_3NH_3PbI_3$) films were prepared by spin coating method under different different rotational speed of the $CH_3NH_3PbI_3$ films. Many characterization measurements were achieved on the deposited $CH_3NH_3PbI_3$ films in order to study the change that resulted from different parameters. The X-ray diffraction (XRD) measurements revealed that the lattice constants of the fabricated $CH_3NH_3PbI_3$ films were close to the slandered values. The atomic force microscopy (AFM) measurements were conducted in order to study the morphology of the fabricated $CH_3NH_3PbI_3$ films at different scale, these results displayed that the grain size of the $CH_3NH_3PbI_3$ films in the range between about 160nm to 210 nm. The field emiision scanning electron microscopy (FE-SEM) results of the fabricated $CH_3NH_3PbI_3$ films were well agreed to the AFM result especially in the grains shape and distribution. Transmittance measurements have been done to determine the variation of the energy gap and absorption coefficient the $CH_3NH_3PbI_3$ films. The structure form of the perovskite solar cells were FTO/c-TiO_2/mp-TiO_2/$CH_3NH_3PbI_3$/Au, the C-V measurements were achieved to determine the built-in potential, which found to have the highest value when using 4000 rpm during the synthesis of the perovskite layer. In addition, the I-V measurements in dark conditions have been conducted to compute the ideality factor and the saturation current, which are considered important parameters that influences the characteristics of the solar cells. The best solar cell efficiency was obtained at spinning speed of 4000 rpm.

1. Introduction

The perovskite solar cells (PSCs) that involved with organic-inorganic halide materials, have gained a great deal of attention of solar cells research society this is because an incredible devices efficiency improvement from 3.8% to 22.1% since 2009 until the writing of this thesis [1,2]. The perovskite already obtained much interest as a potential replacement of the silicon solar cells, which yet to take the main predominant position in the present-day PV markets, with recorded power conversion efficiency around 26% [3]. This kind of efficiency of solar cell that have small gap, attracted the current attention particularly from the researchers who work in the dye-sensitized solar cells or solar cells based on organic materials, because some materials can be utilized in both perovskite solar cell and organic photovoltaic. The materials of the perovskite have been observed with hugely variable band

IOP Conf. Series: Materials Science and Engineering **757** (2020) 012071 doi:10.1088/1757-899X/757/1/012071

gap (for instance, $CH_3NH_3PbX_3$ possess band gaps of values from 1.5 eV to 2.3 eV) [4], and high value of light absorption coefficient (more than 10^4 cm^{-1}) [5], which is similar to other thin film solar cell materials such as CdTe [6] and copper zinc tin sulfide (CZTS) [7]. Its low-cost and easy fabrication techniques also serve as the possible advantages compared to silicon-based devices that require complicated and costly high-vacuum deposition methods. Reports of successful cell fabrication on flexible substrates even indicated a greater possibility to the large-scale roll-to-roll manufacturing of PSCs that can be used in the industries [8]. The perovskite materials have a common formulation of ABX_3, where A is an organic cation (for instance methyl-ammonium $CH_3NH_3^+$), B referred to a metallic cation (e.g., Pb^{2+}) and X attributed to the halide anion (e.g., I^-). Miyasaka et al. [1] only less than one decade ago reported an efficiency of 3.8% based on a DSSC structure. Due to the application of liquid electrolyte in the hole-transporting material (HTM), the stability of solar cell was very weak and did not attract much attention. Similar trial was done by Park et al. [9] with the increased efficiency of 6.5% but stability was still the main problem because of the instability of HTM layer due to the liquid medium. The quick enhancement of the efficiency of PSCs make perovskite being predictable to be similar with the stable effectiveness of c-Si solar cells while all other types of non-silicon solar cells experienced great obstacles in more developments. Rendering of the theoretic calculations built on the famous Shockley-Queisser limit, the perovskite solar cells, which have the formula of ($CH_3NH_3PbI_{3-x}Cl_x$), might attain efficiency about 25–27% [10]. In the present study, the thin film of c-TiO_2, mp-TiO_2, and $CH_3NH_3PbI_3$ were prepared by spin coating method and investigation of the characteristics of the fabricated solar cell.

2. Experimental

2.1 FTO glass preparation

The substrates used in this work were FTO glass in size of 1.5cm × 1cm with surface resistivity of 14Ω sq^{-1}, these FTO glass substrates were etched by a distance of 3mm along the substrate length. The etching process was done by using zinc powder and HCl (2M). The zinc powder was placed on the edge of the substrate while the rest of the substrate was covered by insulator tape then HCl was poured on it and waiting for 1 min. then the resulting paste was removed and the FTO glass soaked in deionized water and removing any remaining paste. The processes of substrates cleaning are extremely important since even small quantity of dust or any other material present on the FTO glass will affect the quality of the fabricated film. The cleaning was done by these steps: 1- the substrate were put in mixed of solution of soap and water in 100 watt ultrasonic device for 15min. 2- Immersing the substrate in deionized water and run the ultrasonic for 15min. 3- The substrates were washed in acetone ultrasonically for a period of 15min. 4- The substrates were putting in ethanol and ultrasonically were washed for 15 min. 5-Repeating the point number 2 for the same time. 6-Ultrasonically washing the substrates by immersing it in isopropanol for 15 min. 7-Drying the substrate by using compressed air. 8- Putting the substrate in oven at 100 °C for 15min. 9-Applying UV light to the substrate for 30 min.

2.2Materials preparation

Firstly we prepared a 2M of HCl from 12M HCl by using the relation M1V1=M2V2. Secondly absolute ethanol was put in two glass container each one have 5ml, then 0.635 ml of Titanium isopropoxide (TIP) was added to one container and 0.07 ml of HCl (2M) was added to the other container after that these two container were placed in an ice bath magnetic stirrer for 30 min. and then these two solution were mixed and run the ice bath magnetic stirrer for another 30min then the final solution was placed in the refrigerator. The final solution of compact TiO_2 has a colorless appearance. To prepare the meso-porous TiO_2, 5ml of absolute ethanol was used to solve 1g of TiO_2 paste then the solution was placed on the magnetic stirrer at room temperature for 12 hours. The final solution was seemed to have a milky like appearance. The $CH_3NH_3PbI_3$ solution was synthesized by dissolving PbI_2 of mass 461 mg in 0.635 ml of DMF and 0.071ml of DMSO solution and placed all these materials on magnetic stirrer at 70°C for 30 min., after leaving this solution to cool to the ambient temperature, 159 mg of MAI was added to the solution and stirred at room temperature for 30 min. lastly we obtained a very clear yellow solution.

2.3 Device fabrication and characterization

The compact TiO_2 was deposited by using spin coater when the rotational speed is (2000, 3000, 4000 rpm) and the time was sit to be 60 s. After the spin coating completed, the samples were put in the oven, at 100 °C for 30 min and then annealed at 500 °C for 1 hour and leaved to cool to the ambient temperature for about 4 hours. The deposition of the meso-porous TiO_2 was done in the same condition of the compact TiO_2.

After the completion of the processes, the samples were left to cool to the ambient temperature gradually. The process of deposition of $CH_3NH_3PbI_3$ film was achieved by spin coating with pouring of anti-solvent (diethyl ether) in order to crystalize the perovskite material. After finishing this process the films were annealed in order to improve the crystallization. The deposition of Au was achieved by sputtering. The morphology of the films was examined using MIRA3 TESCAN scanning electron microscope (SEM). ARA AFM was used to study the topography of the films. The x-ray diffraction (XRD) patterns of substrate and perovskite films were obtained using Philips PW 1840 X – ray diffractometer of wavelength λ = 1.5406°A from Cu - Kα. A Cecile CE 7200 Spectrophotometer supplied by Aquarius company was used to record the optical transmission for the films (375-900 nm). Dark current-voltage measurements carried out by applying voltage supplied to the sample from a stabilized D.C. fine power supply, type L 30-2 Farnell of range (0.1-5)V. The current passing through the device was measured using a Keithley (2400) sourcemeter. The I-V measurements of the solar cells were done under 100mW/cm^{-2} light illumination power densities supplied by a Halogen lamp which is connected to a variable AC power supply and calibrated by Keithley 2400.

3. Results and Discussion

3.1 Structural Measurements

Fig. 1 represents three X-ray patterns of $CH_3NH_3PbI_3$ that fabricated in spin coating method with different rotation speed one can notice that all peaks have nearly the same diffraction planes. The standard lattice parameters (a and c) of the $CH_3NH_3PbI_3$ in tetragonal phase are very close to the sample prepared in rotation speed of 4000 rpm. Table. 1 shows the lattice constants and the grain size and Fig. 2 shows the variation of the lattice parameters and grain size with the rotation speed. The duration of spin coating process will affect the properties of the fabricated films due to the removal more solution from substrate that result in decreasing the film thickness, surface morphology and crystallization degree.

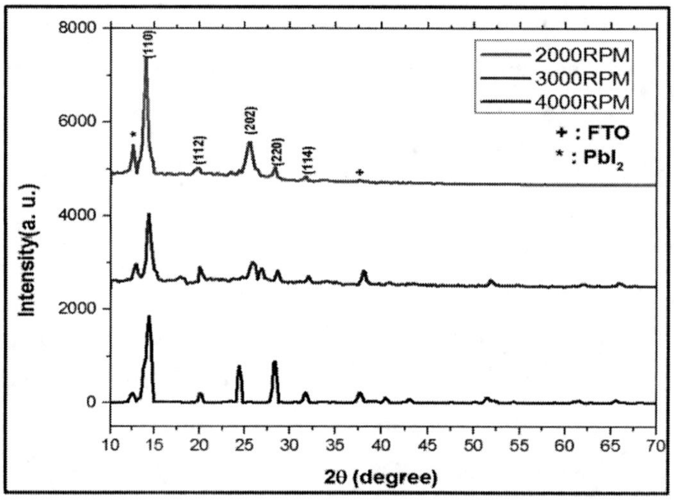

Figure1. XRD patterns of the $CH_3NH_3PbI_3$ thin films deposited on FTO glass substrate at different speed rotation: 2000 rpm, 3000 rpm and 4000 rpm.

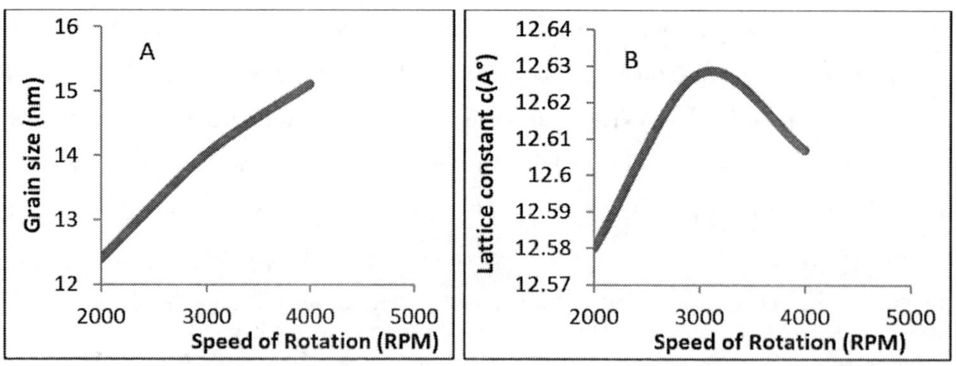

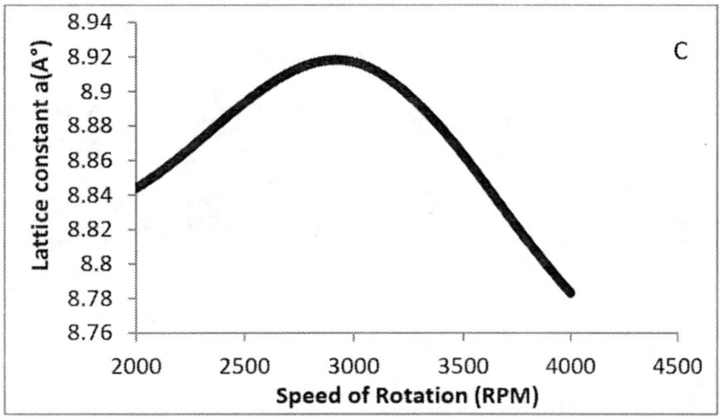

Figure. 2 the effect of the rotation speed of the deposited $CH_3NH_3PbI_3$ films on A: grain size, B: lattice constant c and C:lattice constant a.

Table 1. the result of XRD data of the $CH_3NH_3PbI_3$ thin films deposited on FTO glass substrate at different speed rotation.

Rotation speed(RPM)	FWHM(^{0}) at (110)	Grain size(nm)	a(A°)	c(A°)
2000	0.678	11.8018	8.852	12.577
3000	0.658	12.1675	8.917	12.563
4000	0.695	11.5095	8.783	12.607

3.2 Morphology analysis

The surface morphology of films has been investigated using AFM images, which produces images of surfaces at very high magnification. Fig. 3 displays the AFM images of samples fabricated in different spinning speed, the average diameter, of sample prepared at spinning speed of 2000 rpm, is about 210nm, with increasing the speed of spinning the average diameter decreases and reach the values of 195nm and 170 nm for samples fabricated at spinning speed of 3000 rpm and 4000 rpm respectively.

The surface morphology of samples fabricated with different spinning speed is shown in Fig. 4, the FE-SEM images of samples are found to have almost the same morphology which may mean the spinning speed does not affect morphology of the fabricated films, the average grain size of samples prepared with spinning speed 2000rpm, 3000rpm and 4000 rpm are 176 nm, 184 nm and 187 nm respectively.

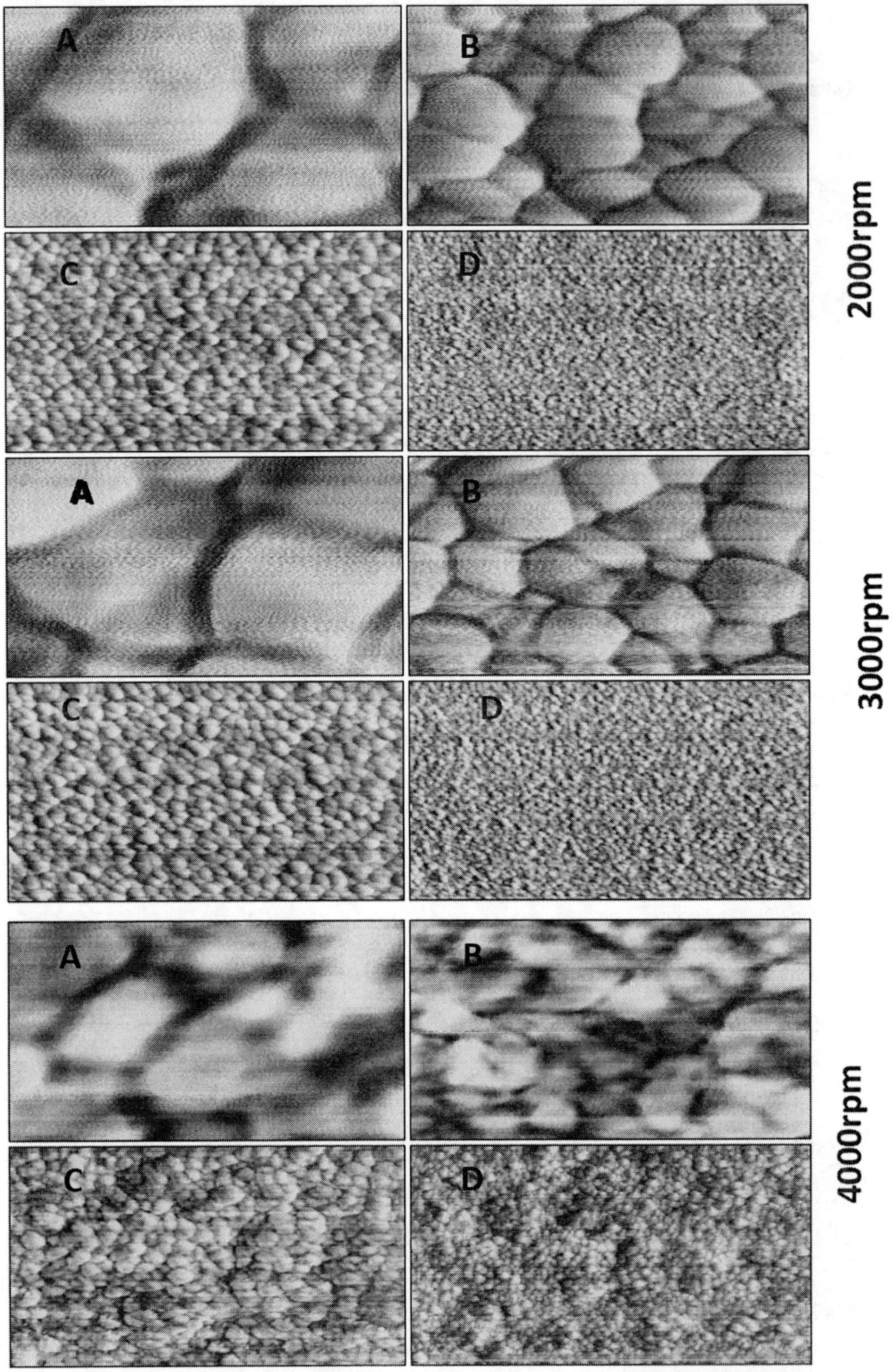

Figure. 3. AFM images of $CH_3NH_3PbI_3$ film prepared with different spinning speed (20000, 3000, 4000), the letters here refer to the image scale A:0.5 μm , B:1μm ,C:5 μm and D:10 μm

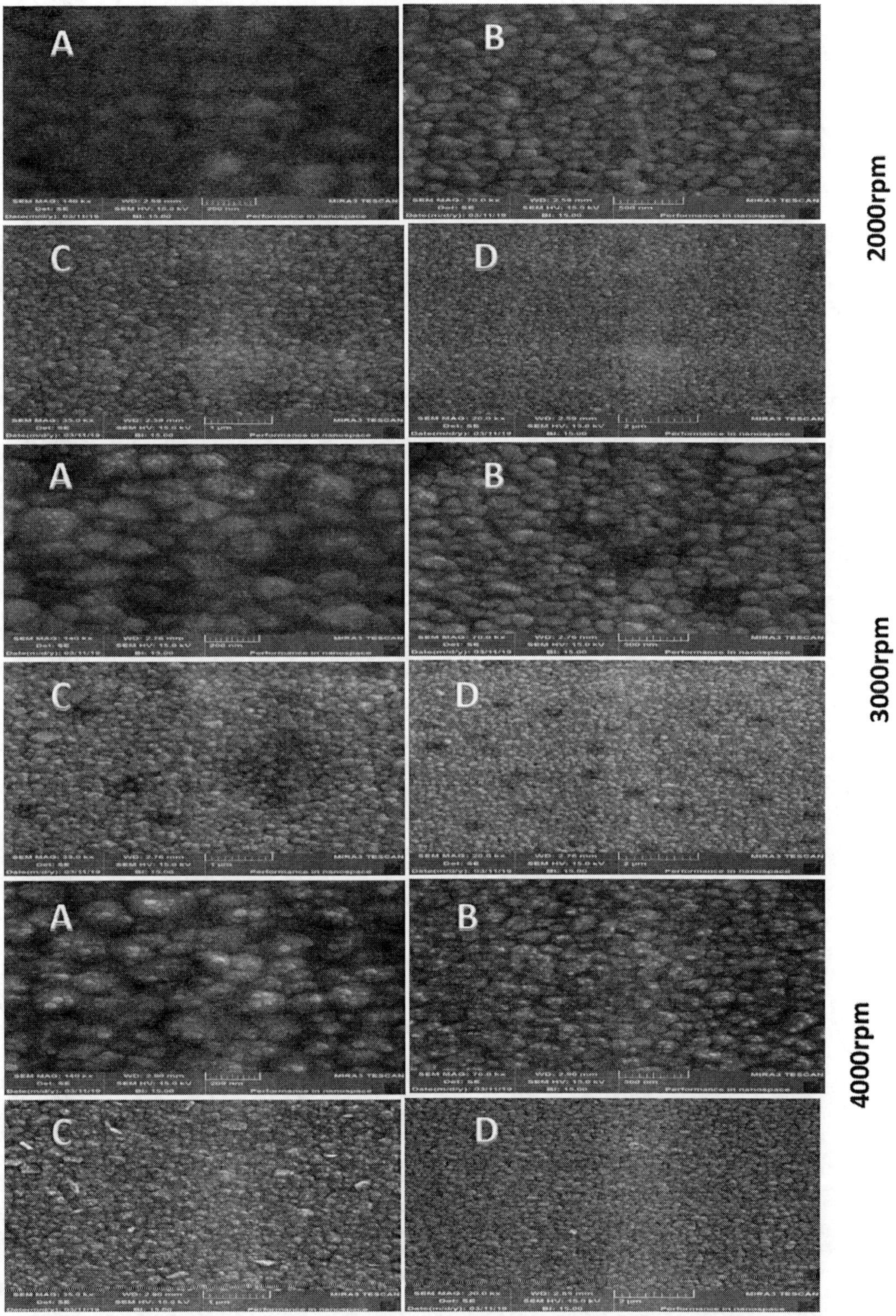

Figure 4. FE-SEM images of $CH_3NH_3PbI_3$ film prepared with different spinning speed (2000,3000,4000), the letters here refer to the image scale A:0.2 μm , B:0.5μm ,C:1 μm and D:2 μm

IOP Conf. Series: Materials Science and Engineering **757** (2020) 012071 doi:10.1088/1757-899X/757/1/012071

3.3 Optical Measurements

To evaluate the optical properties of the perovskite films prepared by spin coating method at different conditions, we measured the UV-visible to calculate the transmittance, absorption coefficient and the energy gap. The speed of rotation in the spin coating process has influence in the resultant film morphology, this is due to the effect of diffusion of solution through the substrate that vary depending on the rotation speed, which would change the centrifugal force that responsible on the spread of solution. It can be clearly observed from Fig .5 that transmittance is higher in speed rotation of 4000 rpm, and has maximum transmittance (about 65%) at long wavelength and then the transmittance begin to decrease at about 750nm wavelength to 35% then remain constant in short wavelengths. This result of lower transmittance at high rotational speed (4000 rpm) explains that the thickness and the morphology of thin films prepared by spin coating method are changed significantly due to small solution amount remained during the process.

The absorption coefficient versus wavelength, of samples fabricated at different rotational speed, is shown in Fig. 6. At the short range wavelength the absorption coefficient, for sample prepared at 4000 rpm spinning, appears to have lower value compared to the other samples, in the other hand, as the wavelength increased, the 4000 rpm sample has higher absorption coefficient (in the range between 600 nm and 1100 nm) relative to samples prepared with lower speed. The energy gap of $CH_3NH_3PbI_3$ thin films deposited at different spinning speed are almost equal value of about 1.53 eV, this result may be explained as the spinning speed has no big influence on the optical energy gap of the fabricated samples as shown in Fig. 7.

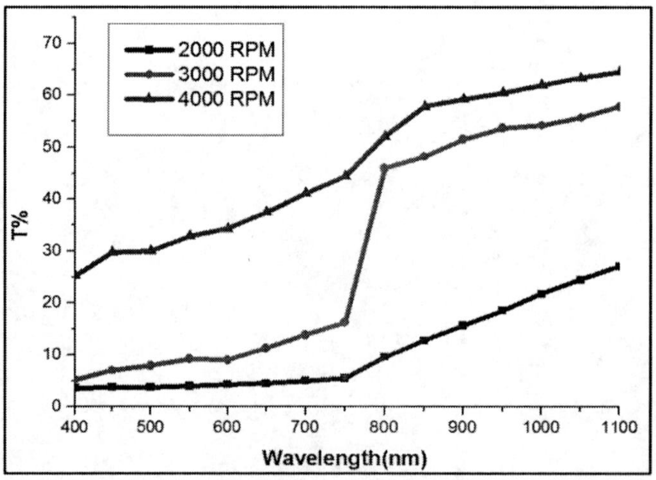

Figure 5. Transmittance as a function of wavelength for $CH_3NH_3PbI_3$ thin films deposited at different rotational speed.

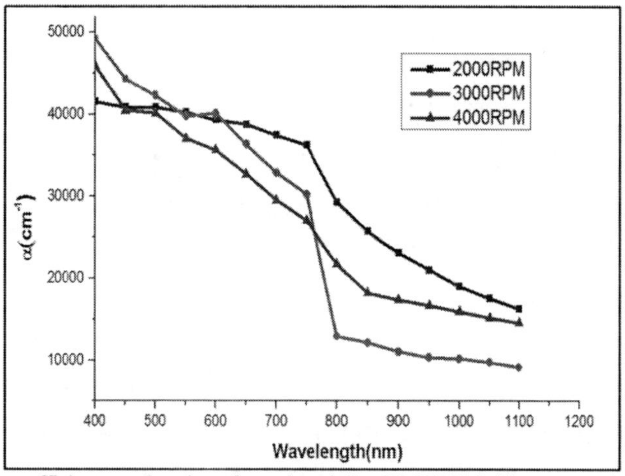

Figure 6. Absorption coefficient as a function of wavelength for $CH_3NH_3PbI_3$ films prepared at different rotational speed.

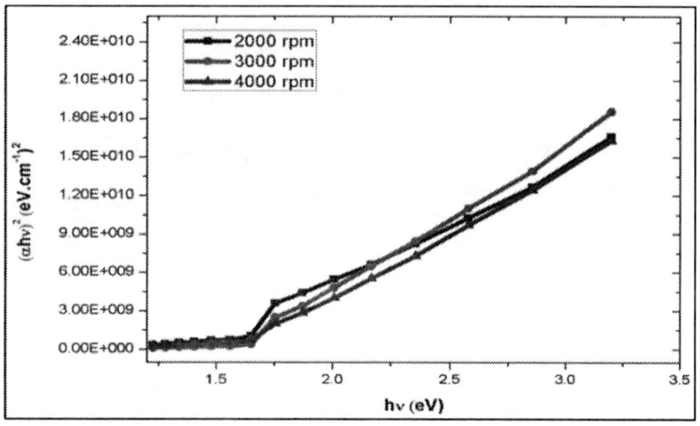

Figure 7. Shows the estimated energy gap of $CH_3NH_3PbI_3$ thin films deposited at different rotational speed.

3.4 Electrical Measurements

3.4.1 Capacitance –Voltage (C-V) Measurement

In order to obtain efficient energy conversion in solar cells with low mobility LHM, a high build in potential (V_{bi}) is necessary to prevent significant losses due to carrier recombination processes competing with charge extraction processes [1]. The capacitance equation is given by

$$\frac{1}{C^2} = \frac{2}{qA^2\varepsilon_0}\left(\frac{1}{N\varepsilon_r}\right)\left(V_{bi} - V_{bias} + \frac{k_BT}{q}\right) \quad (1)$$

Where C is the junction capacitance, A is the junction area, ε is the vacuum permittivity, and N is the activated dopant density in semiconductor and ε_r is the relative permittivity. Fig. 8 shows the change in junction capacitance with applied voltage for FTO/compactTiO_2 /mp TiO_2/$CH_3NH_3PbI_3$/Au solar cells prepared at different parameters of the absorber layer ($CH_3NH_3PbI_3$). It's clear from this figure that there is an increase in capacitance with increasing voltage bias and this variation of capacitance, as the bias change, is caused by changes of the width of the depletion region, As the bias reversely increases, the extension of the depletion region starts at the $CH_3NH_3PbI_3$/Au interface, then goes through the $CH_3NH_3PbI_3$ and finally into the TiO_2 because of the different N and εr values in both $CH_3NH_3PbI_3$ and TiO_2, C-V curves under reverse bias are bent into two stages. Similar phenomena were observed for III-V semiconductors [11]. The roughly constant capacitance at stage II is due to the heavy doping level in TiO_2, in comparison with the smaller slope related to the small doping level in $CH_3NH_3PbI_3$. The sharp edge of the curves in Fig. 9 corresponds to the transition point of the depletion region from stage I to stage II. According to equation (1), the Vbi of a finite $CH_3NH_3PbI_3$ layer is extracted from the intersection. The fluctuations in some curves found in Fig. 9 of stage II can be explained by non-uninform doping in the TiO_2 layer. Because of the huge effective surface area of nano-porous TiO_2 and the rough surface of $CH_3NH_3PbI_3$ layer. The diffusion potential or built in potential (V_{bi}) can be estimated by plotting between V and $1/C_2$ (Mott-Shottky plot) and intercept the linear portion of the line with x-axis. Table 2 shows the summary results obtained from C-V measurements basically included the built in potential values.

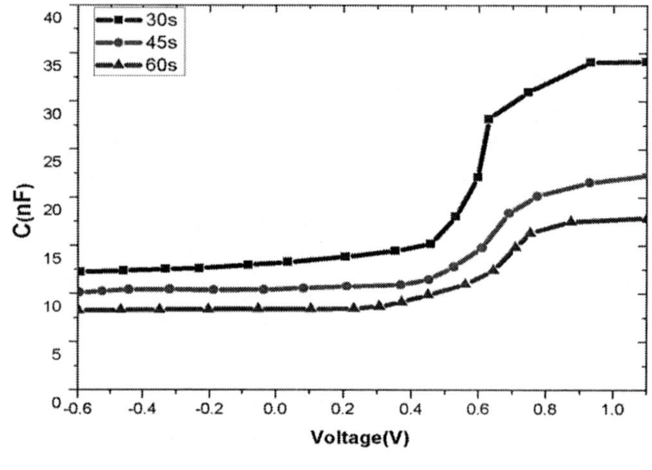

Figure 8. Capacitance as a function of applied voltage for FTO/c TiO_2 /mp TiO_2/$CH_3NH_3PbI_3$/Au solar cell at different synthesis Speed of spinning.

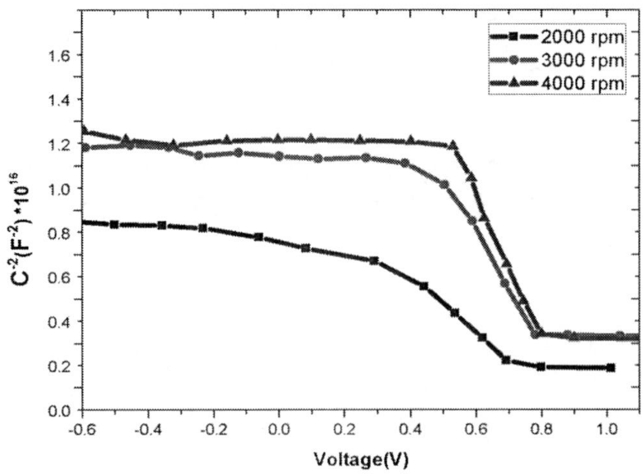

Figure 9. Estimation of built in potential from the relationship of voltage and $1/C^2$ for FTO/c TiO2 /mp $TiO_2/CH_3NH_3PbI_3$/Au solar cell at different Speed of spinning of $CH_3NH_3PbI_3$ film.

Table 2. Values of built in potential of FTO/c TiO_2 /mp TiO_2 / $CH_3NH_3PbI_3$ / Au solar cells prepared at spinning speed of the absorber layer ($CH_3NH_3PbI_3$).

Parameters	Values	Built –in voltage (diffusion potential) (V)
Spinning Speed(rpm)	2000	0.85
	3000	0.9
	4000	0.91

3.4.2 Dark Current-Voltage Measurements (I-V)

It can be noticed from the Fig. 10 that the junction exhibits rectifying behavior, this rectifying behavior is attributed to heterojunction potential barrier at the $TiO_2/CH_3NH_3PbI_3$ interface. The formation of the heterojunction structure is referred to the difference in energy gap between the TiO_2 and $CH_3NH_3PbI_3$. Some of electrical characteristics of $TiO_2/CH_3NH_3PbI_3$ sandwich structure can be calculated from dark I-V measurements, such as ideality factor and saturation current that were calculated according to the equations in previous work [10].Both Is and n provide important information on the predominant current transport mechanism in a device. There are a number of ways to obtaining these parameters, although the most consistent method involves their determination using the dark J-V curves. The slope of the J-V curve gives the n, whereas the y-intercept gives the J_o. They are usually considered together as there is great deal of correlation between these two parameters. It is desirable that Js be as low as possible, so that higher voltages are required for the dark current to equal the light generated current. This increases the Voc of the devices. A lower value of n will lead to reduction in Voc value but that will help to increase the fill factors of devices. Table 3 describes the ideality factor and the saturation current of $TiO_2/CH_3NH_3PbI_3$/Au solar cells prepared at different rotation speed of the absorber layer ($CH_3NH_3PbI_3$).

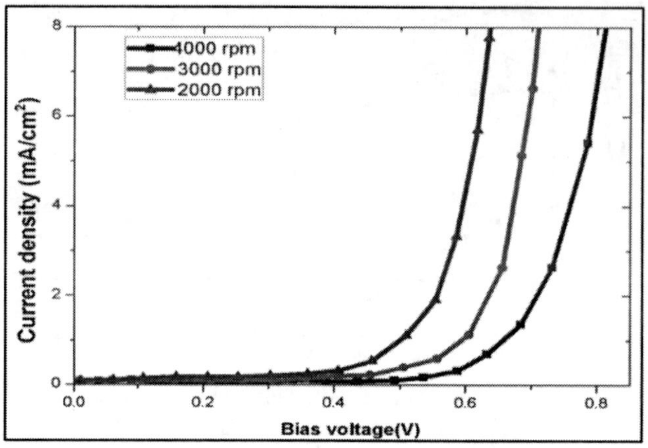

Figure 10 . Dark I-V characteristics of FTO/c TiO2 /mp $TiO_2/CH_3NH_3PbI_3$/Au solar cell at different Speed of spinning of $CH_3NH_3PbI_3$ film.

Table 3. shows the values of parameters of the solar cell extracted from the dark (J-V).

Parameters	Values	Ideality Factor	Saturation current (μA)
Spinning Speed(rpm)	2000	2.86	1.7
	3000	2.58	1.26
	4000	2.153	0.69

4. Solar Cell Measurements

Fig. 11 presents the (J-V) characteristics of FTO/c TiO_2 /mp $TiO_2/CH_3NH_3PbI_3$/Au structure solar cells with different spinning speed of the prepared $CH_3NH_3PbI_3$ film. The detailed photovoltaic parameters such as open-circuit voltage (Voc), FF, short-circuit current density (Jsc), PCE, Rs and Rsh are given in Table 4. It is obvious from the table that the solar cell, which prepared with spinning speed of 2000 rpm, had higher Rs and lower Rsh, this result will lead to lower the photocurrent which lowers the efficiency of the solar cell. By increasing the spinning speed to 3000 rpm and 4000 rpm the value of Jsc is drastically increasing to higher values as well as the other parameters will improve, this may be due to the good interaction between the spinning speed and the anti-solvent diffusion process that occurs during the formation of the perovskite layer. On the other hand, the FF of the device prepared at 4000 rpm has the higher value, this may be attributed to good quality formation of perovskite at this spinning speed as presented in Table 4.

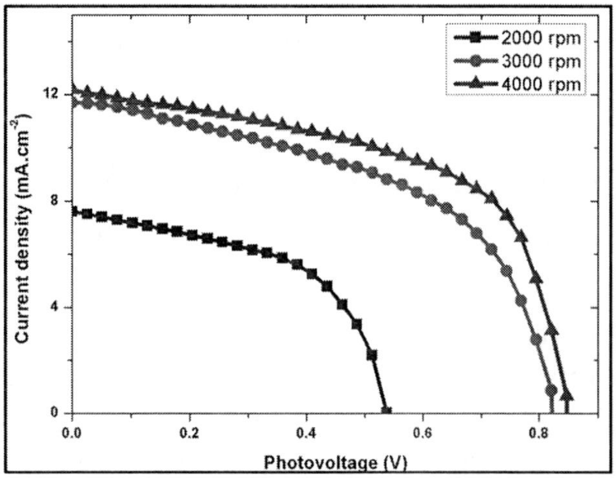

Figure 11. J-V characteristics of the solar cells under illumination of 100 mW/cm^{-2} with different spinning speed.

Table 4. Illustrates Parameters of the fabricated solar cells that prepared with different spinning speed.

Spinning Speed (rpm)	V_{oc}(V)	J_{sc}(mA/cm^2)	FF	PCE (%)	R_s (Ω.cm^2)	R_{sh} (Ω.cm^2)
2000	0.551	7.612	0.514	2.156	18.53	233
3000	0.834	11.708	0.508	4.96	12.2	347
4000	0.859	12.188	0.577	5.857	10.04	380.17

5. Conclusions

The different rotation speeds play an important role on the quality of the perovskite films and lead to different influence of the fabricated solar cells. The best efficiency of the solar cells found at speed of 4000 rpm. AFM and SEM analysis showed that there is big effect on the structure and morphology of the perovskite film that prepared with different speeds, it is found that the grain size becomes smaller when the rotation speed increases. From the dark I-V measurements, the ideality factor and the saturation current will decrease with increase the rotation speed.

References

[1] Kojima, A.; Teshima, K.; Shirai, Y.; Miyasaka, T. . *J. Am. Chem. Soc.* **2009,** *131*, 6050-6051.

[2] Saliba, M.; Matsui, T.; Seo, J.Y.; Domanski, K.; Correa-Baena, J.P.; Nazeeruddin, M.K.; Zakeeruddin, S.M.; Tress, W.; Abate, A.; Hagfeldt A.. *Energy Environ. Sci.* **2016,** *9,* 1989-1997.

[3] Yoshikawa, K.; Kawasaki. H.; Yoshida, W.; Irie, T.; Konishi, K.; Nakano, K.; Uto, T.; Adachi, D.; Kanematsu, M.; Uzu, H.. *Nat. Energy* **2017,** *2,* 17032.

[4] Jeon, N.J.; Noh, J.H.; Yang, W.S.; Kim, Y.C.; Ryu, S.; Seo, J.; Seok, S.I.. *Nature* **2015,** *517,* 476-480.

[5] Park, N.G. PSCs: An emerging photovoltaic technology. *Mater. Today* **2015,** *18,* 65-72.

[6] Green, M.A.; Baillie, A.H.; Snaith, H.J.. *Nat. Photonics* **2014**, *8,* 506-514.

[7] Rangel-Cárdenas, J.; Sobral, H. *Materials* **2017,** *10* ,607.

[8] Shi, Z.; Jayatissa, A.H. *Prog. Nat. Sci. Mater. Int.* **2017**, *27,* 550-555.

[9] Leo, K. Perovskite photovoltaics: signs of stability. *Nat. Nanotechnol.* **2015** *10* 574-575.

[10] Baikie, T.; Fang, Y.; Kadro, J.M.; Schreyer, M.; Wei, F.; Mhaisalkar, S.G. *J. Mater. Chem. A.* **2013,** *1*, 5628.

[11] Thjeel, H.A.; Suhail, A.M.; Naji, A.N.; Al-Zaidi, Q.G.; Muhammed, G.S.; Naum, F.A. Adv. Mater. Phy. Chem. **2012**, *l01*, 70-77.

Removing atmospheric effects for Multi spectral images (OLI 8) using ATCOR model

Asmaa Maher [1], Nawal K. Ghazal [2]
[1]Department of Medical Physics, College of Applied Science, University of Fallujah, Iraq.
[2] Department of Remote Sensing & GIS, Science College, University of Baghdad .

Abstract. The atmospheric correction of satellite images is an important first step for different remote sensing applications such as estimation of vegetation indices. In atmospheric corrections, most uncertainties arise from temporal and spatial variations in aerosol types and quantities. Thus, considered validation estimate Aerosol is an essential step in the validation atmospheric correction algorithms. In the current study, two models of atmospheric correction algorithms ATCOR (ATCOR 3 and ATCOR 2) were applied to remove atmospheric effects of test sites in Middle part of Iraq. Statistical results of ATCOR 2 was shown to be successful in the urban parts to remove effective type of aerosol which could be chosen to process satellite images in areas under investigation.

Keywords: ATCOR 2 , ATCOR3, Atmospheric correction , OLI.

1. Introduction:

Before reaching to the earth's surface, radiation used for remote sensing should travel through some distance of the earth's atmosphere. Atmospheric gases and particles can affect the incoming lights and radiations. Mechanisms of absorption and scattering can cause these effects [1,2]. Theoretically, depending upon the particle size which lead to scattering and the wavelength of scattered radiation, scattering can be classified into three categories [3]: There is a predominance of Rayleigh scattering, as there is interaction between electromagnetic radiation and particles which are smaller than the incoming light's wavelength, where there is an inverse relationship between the wavelength's forth power and the incoming light's wavelength. There is more scattering of the shorter wavelengths than the longer wavelengths. In the aspect of remote sensing, the most important kind of scattering is Rayleigh scattering. It leads to the distortion of the spectral characteristics of reflected lights in comparison with the measurements performed on the ground [4,5].

Atmospheric particles having the same wavelength size show the dispersion of mi. The emergence of multiple scattering occurs when the size is similar to the wavelength of the radiation contained with a particle size in the atmosphere. This type results from the scattering of aerosols, a mixture of dust, gases and water vapour. It is usually limited to lower atmosphere, where there is huge amounts of large particles that dominate at overcast cloudy conditions. It affects the total spectral part which begins from ultra violet to almost near the infrared region [3,5]. Aerosols exhibit non-selective scatterings, and these aerosols are water droplets with particle size much greater than the wavelength energy (visible and almost mid infrared). Scatterings performed by atmospheric particles are dominant mechanisms which result in radiometric distortions of image data, as well as the probable sensor malfunction impacts [5].

When there is an interaction between electromagnetic radiation and the atmosphere, absorption is the other major mechanism at work. This phenomenon causes the atmospheric molecules to absorb energy at different wavelengths. The three major atmospheric components which absorb radiation are ozone, water and carbon dioxide [1,2].In several canopies, most non-photosynthetic vegetation (NPVs) are obscured under a tightly-closed leaf canopy; the shortwave red wavelengths used for measuring (NPVs) usually can not penetrate the upper canopy for interaction with these (NPVs). Thus, only exposed (NPVs) have important influence on the spectral reflectance of the vegetated ecosystem, and during their exposure, (NPVs) scatter photons so effectively to infrared range shortwaves, in contrast with the green vegetation which are highly absorbed in that range[6].

The potentially ideal source to detect changes and analyze trends is provided by time series of remote-sensed image that is acquired at high temporal and spatial resolutions [7]. Methods of correcting sensor sensitivity changes, and topographic with viewing angle impacts involved in the category of radiometric correction which is well- established. Removal of atmospheric effect is the most complicated step in the radiometric correction processes [8]. The intensity of electromagnetic radiation intensity is recorded by digital sensors of each spot and shown on the surface of earth as digital number (DN) to all spectral bands. The exact DN range that is utilized by the sensor relies on its radiometric resolution. The true units are recorded as W m-2 ster-1 mm-1, though DN values recorded by the sensors are proportional to radiances. Most of image processing depend upon raw values of DN where the true spectral radiance is not a concern [9,10].

Modified darkest pixel atmospheric correction is considered as a darkest pixel DP. The dark target of the surface radiance is proposed to possess approximately a zero surface reflectance or radiance [11,12]. If the images stem from the same sensor, then we can use a simple sun elevation correction. The trick is to normalize the images as if they are taken with the sun at the zenith [2]. Haze is composed of atmospheric aerosols and molecules which scatters and absorbs solar radiations, thus rendering the downward and upward solar radiance to be recorded with remote-sensing sensor. The aim of Haze correction is to remove these effects from raw data, which can be beneficial to several space borne RS applications. Therefore, for each band of a RS image, reducing haze should be done independently [2].

1.1 Study Area

Baghdad province is the first land area, which is the capital and the main educational, cultural and economic city in Iraq. Geographically, it is situated in the right center of Iraq as shown in figure (1). Baghdad province is located between a number of Iraqi provinces (latitudes 33.312805° to the north and longitude 44.361488° to the east in the plain area).

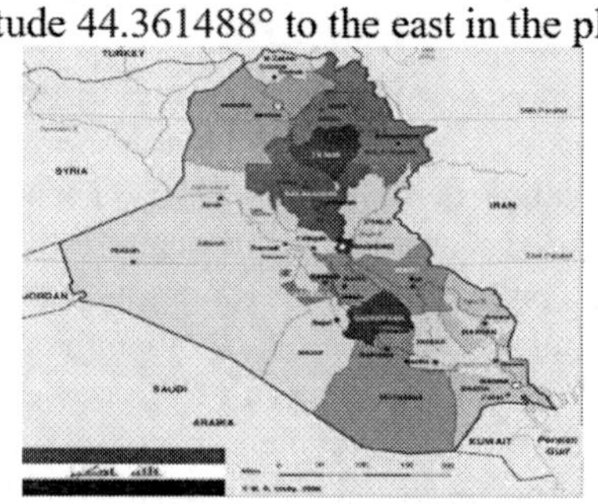

a

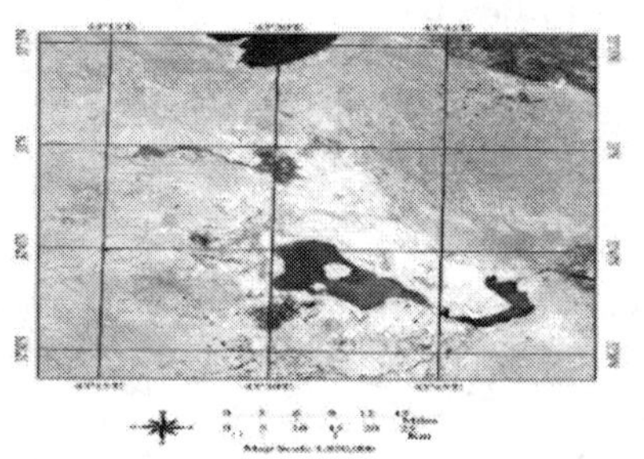

b

Figure (1): Administrative map of Iraq and the photomap of study area.(a) Administrative map of Iraq showing the international and local borders between the provinces. Iraq is situated between the geographic coordinates latitude (37.38°→28.5°N), and longitude. (38.70°→48.75°E).and (b) The photomap of study area of Baghdad province in Iraq, Lat. 33.312805° N, and Long. 44.361488° E

2. Methodology

2.1 ATCOR atmospheric correction model

The atmospheric and topographic correction ATCOR is a broadly method used to correct the atmospheric satellite data [13]. ATCOR is applied to compute the reflectances image ground of the reflective spectral band, and the thermal band emissivity image. For nearly flat terrain or horizontal surface, ATCOR (2) is considered as the spatially-adaptive fast atmospheric corrections algorithms, while ATCOR (3) is used for rugged topographic surfaces, therefore, the digital elevation model (DEM) is applied in ATCOR (3) algorithms for atmospheric corrections purpose [14,15]. ATCOR algorithm, also depends upon (MODTRAN-4 RTC), which is a surface reflectance, without considering the adjacency impact, and it is derived by the following equation [16]:

$$\rho = \frac{1}{a_1}\left(\frac{d^2 \ \pi L^*}{E_{TOA}\cos\theta_i} - a_0\right) \ \ldots\ldots\ldots\ldots(1)$$

Where: E_{TOA} is a solar spectral radiance on the surface that is perpendicular to the sun rays outside the atmosphere. d^2: is a direct distance to sun. a_0 and a_1 coefficients derived from the main atmospheric parameter estimation: the column of water vapor, type of aerosol and the optical thicknesses.

In brief, ATCOR is beneficial for processing solar region bands between 400 –2500 nanometers. ATCOR model was split into separated codes optimized for airborn and satellite sensors. ATCOR (ATCOR-2/3) satellite reinforces the small to moderate FOV sensors only, and it is composed of separate codes to rugged and flat terrain [17]. On applying ATCOR 3, the next set of empirical functions G Eqs. 2&3 serve in decreasing higher values of reflectance in extreme geometry areas to obtain reflectance values that are close to those of nearest areas with a moderate incident angle.

$$G = cos(i)/\cos(t) \quad ....(2)$$

$$G = \sqrt{\cos(i)/\cos(t)} \quad ....(3)$$

Where G is the empirical function, t is the threshold angle ,i is the incident angle and g lower boundary. G function is ranging between specified lower boundary (g) and (1), which means that the specified lower boundary is g < G < 1. Only the area of extreme incidence and / or existence angle is included, initiating with the threshold angle (t). Values of G more than (1) are set to the figure (1), while values lower than g boundary are set to (g), indicating that the process functions in a geometric order from (t) to (90°). The majority of satellite sensor with high spatial resolution possess viewing angle closer to nadir, but the sun is often at the farthest distance from nadir. Thereby, the incident angles mostly penetrate into the 60–90° critical regions. Equations 4 and 5 are also based on the existence angle (e), and such options are involved for tilting sensor. When the scene is recorded along with greater off-nadir view angles (i.e. 20–30°), where (e) could be nearer to solar zenith. In this condition, angle (i) and angle (e) may get access into critical region (60–90°) together.

$$\mathrm{G} = \cos(\mathrm{i}) * \cos(\mathrm{e})/\cos(\mathrm{t}) \quad ....(4)$$

$$\mathrm{G} = \sqrt{\cos(\mathrm{i}) * \cos(\mathrm{e})/\cos(\mathrm{t})} \quad ....(5)$$

3. Results and discussion

To be sure result of atmospheric correction in this project are accurate, the researchers have made multiple visits to the region of study area (Baghdad) .During these trips , the researcher carried with them a GPS device (GPS 72) show in figure (2-a). The researchers conducted a field visit and samples were taken from the area of the study area (cover, "water, vegetation and soil"), as shown in Figure (2-b) to measure the spectral reflection of each sample using ASD.

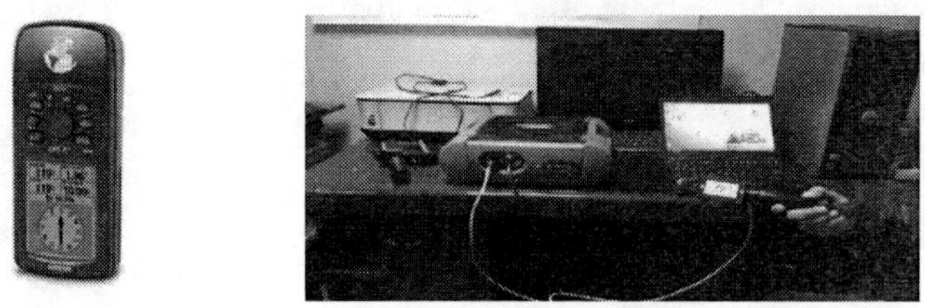

a b

Figure (2): (a) GPS – 72 devices, (b) Analytical Spectral Devices (ASD)

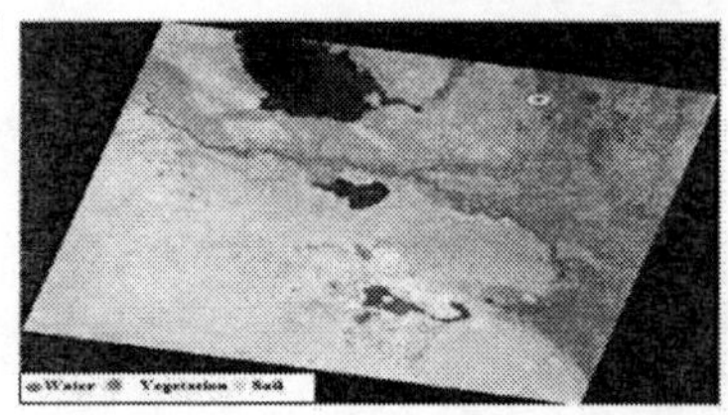

Figure (3): The location of samples (water, vegetation and soil)

Three different areas were identified and processed in software ENVI by region of interest (ROI) and applied FLAASH and QUAC atmospheric correction methods as shown in (Figure 4):

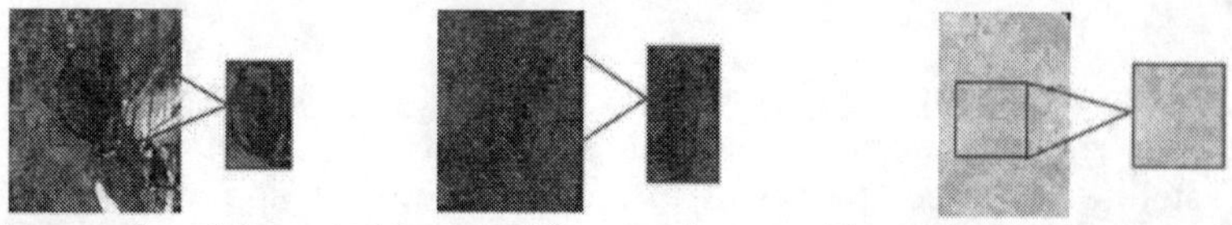

Figure(4): The image of multi-spectral image of first steady area . The region inside the red square has\been processed with atmospheric correction models.(a) first region selected has processed Includes water ,(b) second region selected has processed Includes vegetation and (c) third region selected has processed Includes soil.

Figure(5)represents the spectral fingerprint measured by an ASD device, and this fingerprint is free of atmospheric effects.

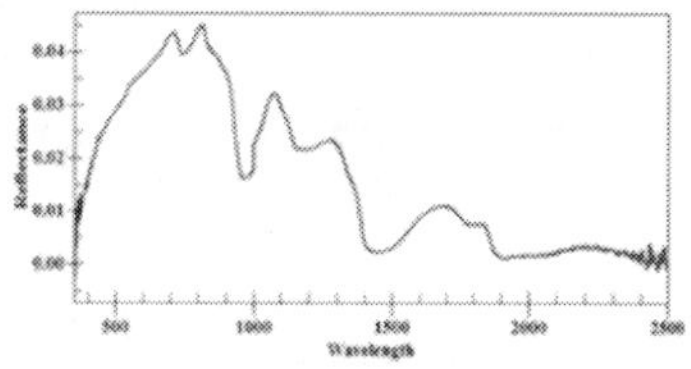

a

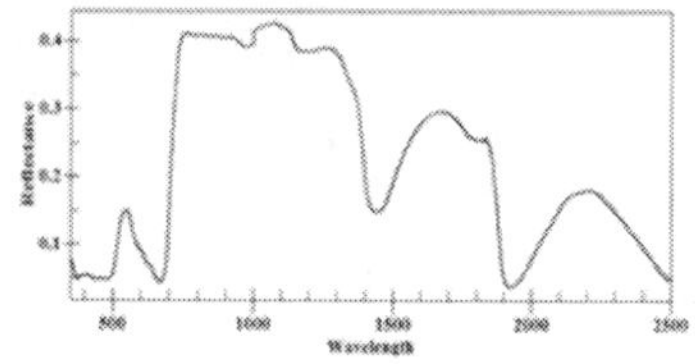

b

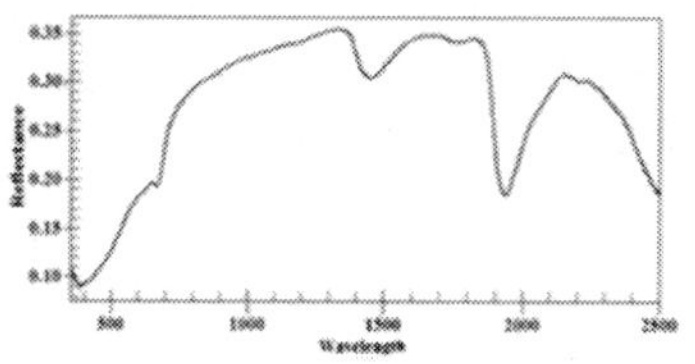

c

Figure(5): Spectral profile of ROI land cover ASD spectral without distortion,(a) water,(b) vegetation, and (c) soil.

ATCOR processing is implemented in OLI 8 products. These top-of-atmosphere (TOA) data are radiometrically and geometrically corrected. The Major ATCOR output is atmospherically corrected Bottom-of-atmosphere (BOA) reflectance for all sensor spectral bands. Figure (6) and (7) show the result of ATCOR 3. First of all, there was a visual comparison between applying various radiometric corrections algorithm and the landsats image No. (8) of the first studying area. The same type of "urban " aerosol was used to parametrize all atmospheric correction algorithms, in a visibility greater than 15 km. Compared to the genuine DN image, brightnesses of other images vary because of the parameterization that is utilized in every algorithm for the correction of the atmosphere and radiometric impacts. Remote sensing in 2015, Nov., Raw 37 and path169.

$$G = cos(i)/\cos(t)$$

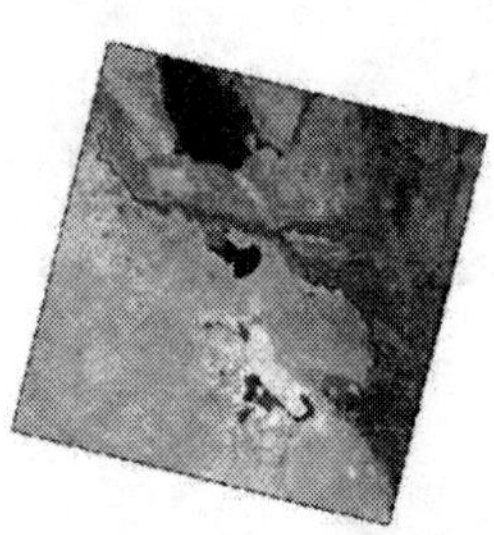

Before ATCOR3

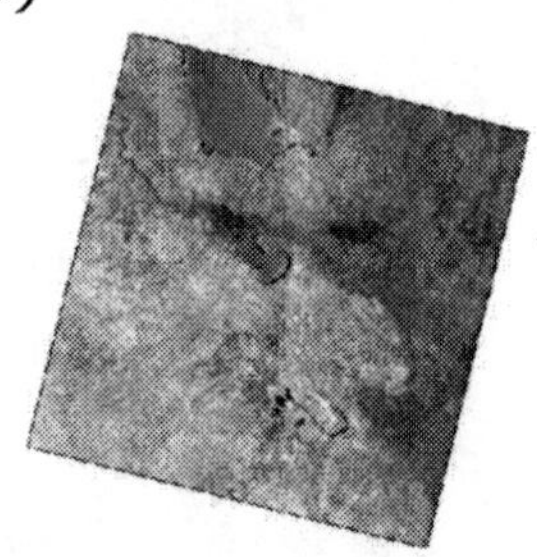

After Atcor3

$$G = \sqrt{\cos(i)/\cos(t)}$$

Before ATCOR3 **After Atcor3**

Figure (6): Empirical functions applied Eqs. 2and3 for multispectral image

$$G = \cos(i) * \cos(e)/\cos(t)$$

Before ATCOR3 **After Atcor3**

$$G = \sqrt{\cos(i) * \cos(e)/\cos(t)}$$

Before ATCOR3 **After Atcor3**

Figure (7): Empirical functions applied Eqs. 4and5for multispectral image

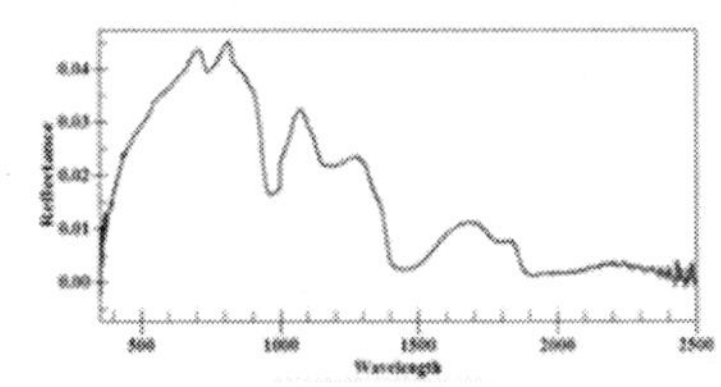

a

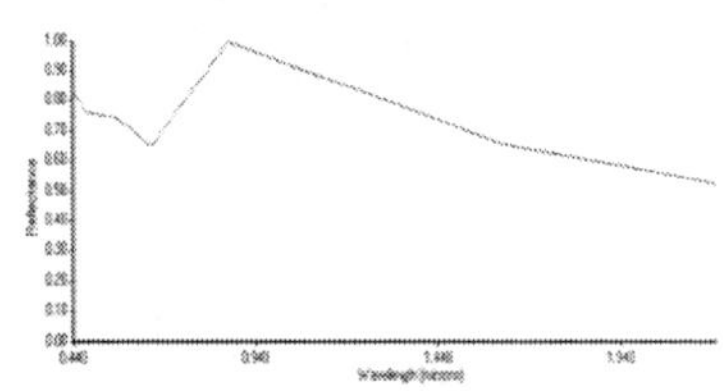

b

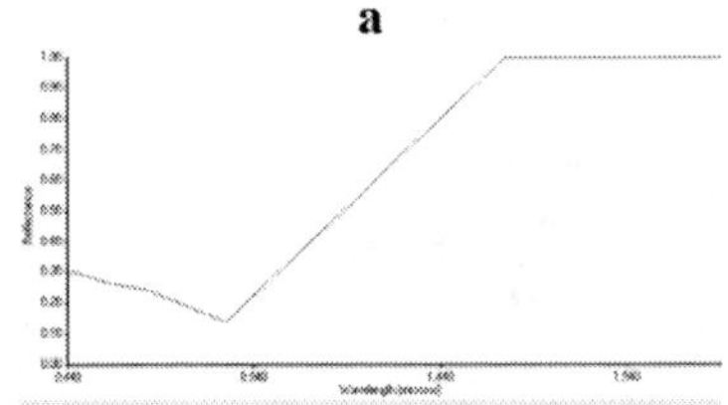

c

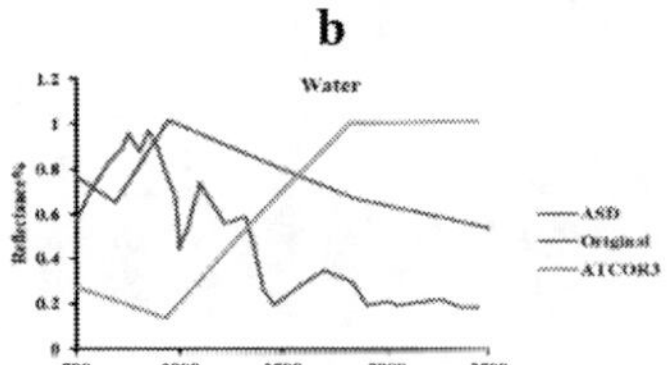

d

Figure(8): The spectral profile for applied Equ.(2) for study area for land cover of water ,(a) ASD spectral without distortion, (b): Before ATCOR3, (c): After ATCOR3, (d) comparison between a, b and c.

IOP Conf. Series: Materials Science and Engineering **757** (2020) 012072 doi:10.1088/1757-899X/757/1/012072

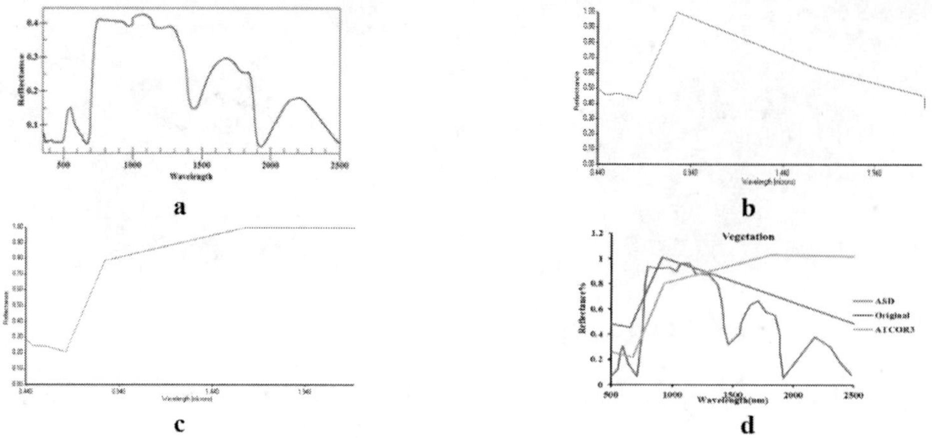

Figure(9): The spectral profile for applied Equ.(2) for study area for land cover of Vegetation ,(a) ASD spectral without distortion, (b): Before ATCOR3, (c): After ATCOR3, (d) comparison between a, b and c.

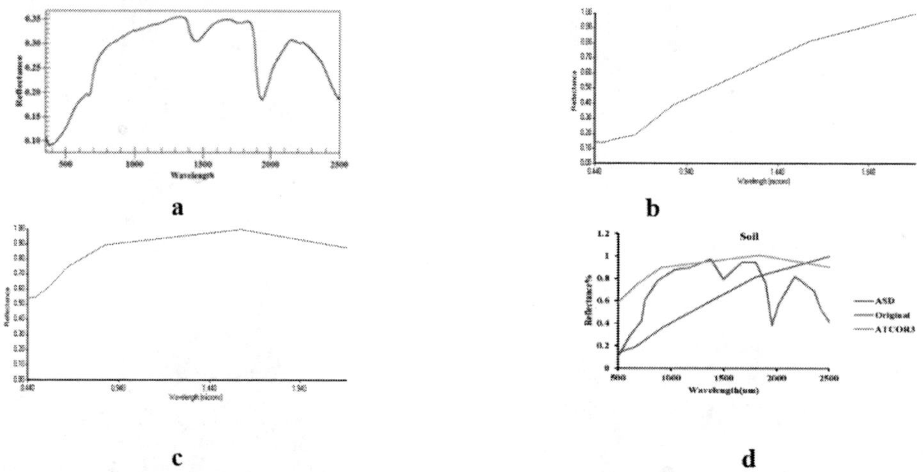

Figure(10): The spectral profile for applied Equ.(2) for study area for land cover of Soil ,(a) ASD spectral without distortion, (b): Before ATCOR3, (c): After ATCOR3, (d) comparison between a, b and c .

Figures (8,9 and 10,-a) representing the spectral fingerprint with distortion. In the range of the wavelength (0.473–0.658), micrometer ATCOR(3) to water land cover indication rating of water vapor absorption. For spectral vegetation, diagnostic dips at (0.866) micrometer due to high absorption because of water vapor, which matches the presence of chlorophyll-b in the healthful leaves as shown in figure (9):

c d

Figure(11): The spectral profile for applied Equ.(4) for the study area for land cover of Water ,(a) ASD spectral without distortion, (b): Before ATCOR3, (c): After ATCOR3, (d) comparison between a, b and c .

a b

c d

Figure(12): The spectral profile for applied Equ.(4) for the study area for land cover of Vegetation ,(a) ASD spectral without distortion, (b): Before ATCOR3, (c): After ATCOR3, (d) comparison between a, b and c .

a b

c d

Figure(13): The spectral profile for applied Equ.(4) for the study area for land cover of Soil ,(a) ASD spectral without distortion, (b): Before ATCOR3, (c): After ATCOR3, (d) comparison between a, b and c .

In the range of the wavelength (0.89-1.54) micrometer, there is a very strong correction, because weak atmospheric attenuation land area of study, while the range (0.46–0.77) micrometer is highly strong atmospherically attenuation, because of water vapor, as shown in figures (11,12 and 13).

Table 1. Delta standard deviation (SD) of the image bands before and after ATCOR3

No. of bands	(B_1)	(B_2)	(B_3)	(B_4	(B_5)	(B_6)	(B_7)
ΔSD(Eqe.1)	3034.46	-1681.34	-993.27	370.88	1017.71	2201.40	1130.16
ΔSD(Eqe.2)	2910.47	-1791.97	-1079.21	292.05	960.74	2201.40	1130.16
ΔSD(Eqe.3)	3827.42	-908.72	-427.21	778.67	1211.79	2201.40	1130.16
ΔSD(Eqe.4)	3548.08	-1238.40	-727.35	630.58	1152.15	2201.40	1130.16

Figure (14)show the result of ATCOR 2. Atmospheric conditions occurring during each image acquisition date are re-evaluated, and each thermal region model is run for all bands of Landsat image to determine the sensitivity of ATCOR2's different atmospheres.

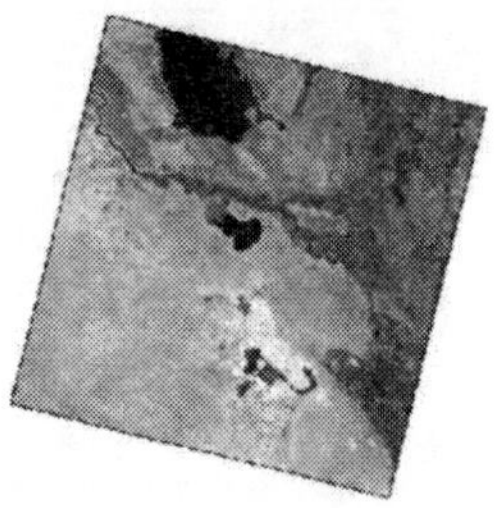

Before ATCOR2

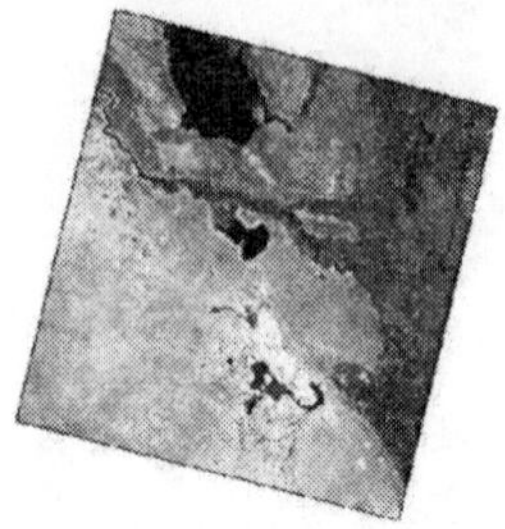

After ATCOR2

Figure (14): Applied ATCOR 2 algorithm for study area

Table 2. Delta standard deviation (SD) of the image bands before and after ATCOR2 for study area

No. of bands	(B_1)	(B_2)	(B_3)	(B_4	(B_5)	(B_6)	(B_7)
ΔSD	2550.54	-2205.95	-2304.07	-2358.71	-6195.32	-22531.52	-24491.19

Figure (15) illustrate convergence spectra of the water after correction when compared with ASD, we note that the reflective is reduced in the visible range (425 – 700)nm. We observe the highest wavelength reversal value in the 750 nm. There is a similarity between ASD and the resulting spectral profile ATCOR2 at wavelengths(426 – 1500) nm .

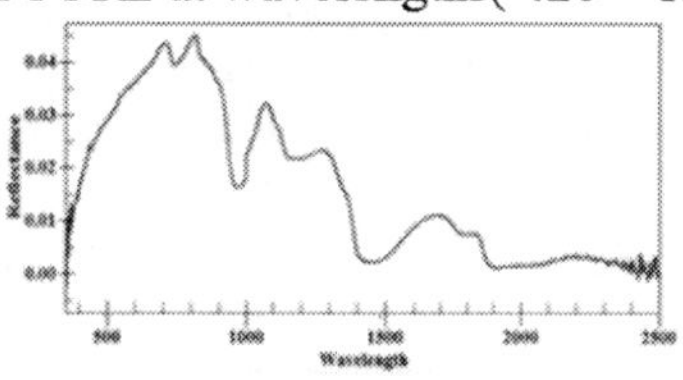

a

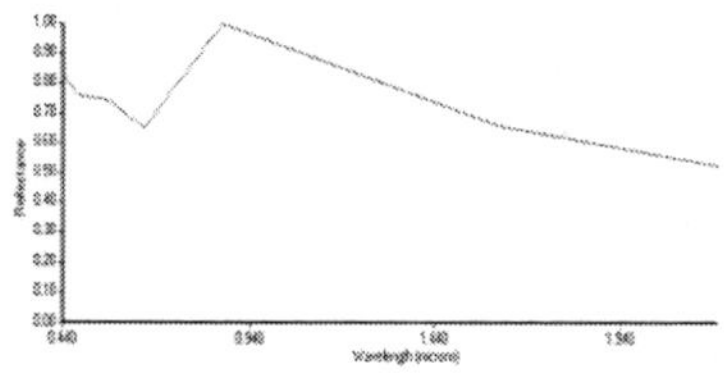

b

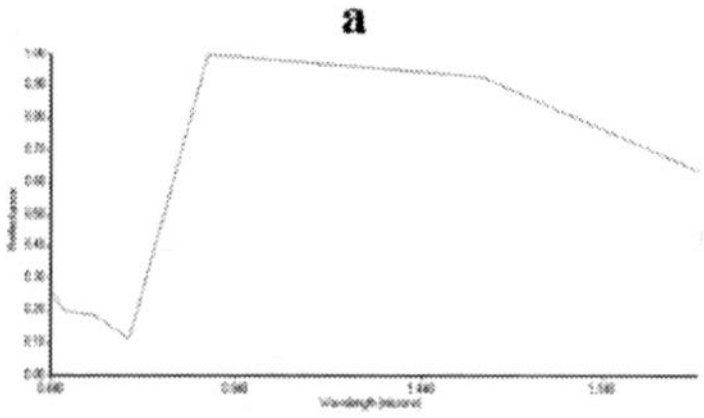

c

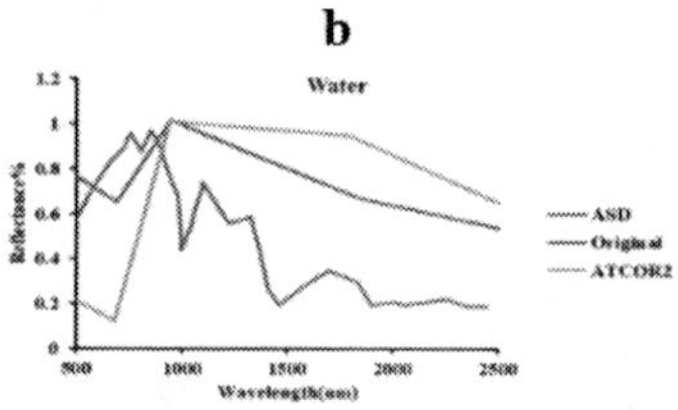

d

Figure (15): Spectral profile of the study area of Water ,(a) ASD spectral without distortion, (b): Before ATCOR2, (c): After ATCOR2,(d) comparison between a, b and c.

Figure (16) illustrate at the vegetation deduction area we notice a similarity between the ASD spectra with ATCOR2 technique at the range(1430 – 1450)nm indicating the success of the technique in removing the effects of atmospheric gases . Absorption in the visible area is an indication of the absorption of oxygen gases and water, as well as increased absorption at the wavelength 1470 indicates the presence of CO_2 and H_2O.

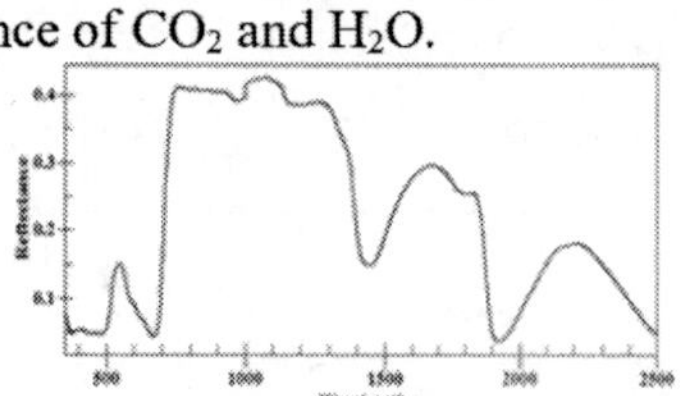

a

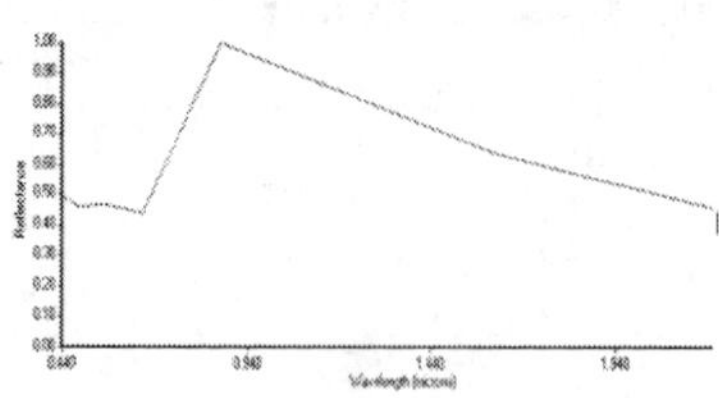

b

c d

Figure (16): Spectral profile of the study area of Vegetation ,(a) ASD spectral without distortion, (b): Before ATCOR2, (c): After ATCOR2, (d) comparison between a, b and c

Figure (16) illustrate in the Soil spectra we notice a similarity between the ASD spectra with ATCOR2 technique at the range(1430 – 1480)nm indicating the success of the technique in removing the effects of atmospheric gases.

a b

c d

Figure (17): Spectral profile of the study area of Soil ,(a) ASD spectral without distortion, (b): Before ATCOR2, (c): After ATCOR2, (d) comparison between a, b and c

4. Conclusions:

This work included 2 ATCOR modules, which are (ATCOR(2) and ATCOR(3) on OLI 8 images with their accuracy were evaluated means of ground truth measurement. In deep blue channel, ATCOR's empirical approach for reflectance difference STD computing before and after ATCOR was negative. Water spectra were compared to ATCOR2 illustrate convergence spectra of the water after correction when compared , we noted that the reflective is reduced in the visible range (425 – 700)nm. Spectra vegetation of ATCOR2 and ASD A were comparable in both results indicating that the spectra of both images showed similarity in the behaviour we notice a similarity between the ASD spectra with ATCOR2 technique at the range(1430 – 1450)nm indicating the success of the technique in removing the effects of atmospheric gases . Absorption in the visible area is an indication of the absorption of oxygen gases and water, as well as increased absorption at the wavelength 1470 indicates the presence of Carbon dioxide gas and water vapour. The soil spectra of ATCOR2 and ASD were compared with each other and the results indicated that the spectra of both images showed a similarity in the technique at the range(1430 – 1480)nm indicating the success of the technique in removing the effects of atmospheric gases. The value of difference STD for the bands (b1, b4, b5, b6 and b7) before and after ATCOR3 (Flat region) were positive. water land cover after ATCOR 3 indication rating of water vapour absorption in the range of the wavelength (0.473–0.658), micrometer. For spectral vegetation, diagnostic dips at (0.866) micrometer due to high absorption because of water vapour, which matches the presence of chlorophyll-b in the healthful leaves .The algorithm ATCOR 2 yielded results closer to the ASD in the sedimentary plain region.

IOP Conf. Series: Materials Science and Engineering **757** (2020) 012072 doi:10.1088/1757-899X/757/1/012072

References

[1] Canada C. for Remote sensing 2007, Taylor and Francis 8-14
[2]Tempfli K.,Norman K., Gerrit C. H. and Lucas L. F. J. 2009,The International Institute for Geo-Information Science and Earth Observation (ITC)158-188.
[3]Shefali A., Kishtawal C.M.,and Minakshi K.r. 2005 ,world meteorological organisation 23-38.
[4]Thomas L., Ralph W. K., and Jonathan C. 2015, Wiley and Sons 7th Edition.
[5]Prashant K. 2007, Thesis of Master of Science in Geo-information Science and Earth Observation, the International Institute for Geo-information Science and Earth.
[6]Tappan G. 1980,KARS, Univ. of Kansas, Lawrence.
[7]Mahmoud E. H. , Agnès B. , Bruno L. , Olivier H., Gérard D. and Matthieu R. 2008, Sensors journal8 2774-2791.
[8]Hadjimitsis D.G. and Themistocleous K. 2008, Digital Heritage – Proceedings of the 14th International Conference on Virtual Systems and Multimedia.
[9]A. Athos, G. Diofantos, D. Dimitrios and P. George. 2011The Case Studies of Palaepaphos and Nea Paphos Sites in Cyprus Remote Sensn 32605-2629.
[10]Chrysoulakis N., Abrams M., Feidas H., Arai K. , International Journal of Remote Sensing, 31 6347-6385.
[11]Hadjimitsis D.G., and Clayton C.R.I. 2004, International Journal of Remote Sensing 25 3651-3674.
[12]Hadjimitsis D.G., Clayton C.R.I., Retalis A. 2004, A revised procedure for environmental applications of satellite remotely sensed imagery 5239, 464-471.
[13]Bringfried P., Magdalena M. 2014, Proc. of SPIE 9242 92420.
[14]San a B. T., and Suzen M. L. 2010, International Archives of the Photogrammetry, Remote Sensing and Spatial Information Science, XXXVIII, Part 8, .
[15]themistocleous K. 2012;phd. Thesis, Cyprus University of Technology.
[16]Francisco E., Javier M., Javier M. and Dionisio R2017, Sensors journal 17 2639.
[17]Richter, R. 1996, International Journal of Remote Sensing, 17 1201–1214.

STUDY THE EFFECT OF TIO_2 NANOPARTICLES ON PHYSICAL PROPERTIES OF BIOPOLYMER BLEND

Mohammed K. Jawad, Noor K. Abid

Department of Physics, College of Science, University of Baghdad, Baghdad, Iraq

E-mail: mohamedkadhom@yahoo.com

Abstract. Chitosan (CH) / Poly (1-vinylpyrrolidone-co-vinyl acetate) (PVP-co-VAc) (7:3) blend [CH/(PVP-co-VAc)] reinforced with two types of nanoparticle of titanium oxide TiO_2 (type I & II) were prepared by solution casting method. The nanocomposites prepared with different concentrations (1, 2, 3.5, 5, 7.5, and 10) wt%. The tests carried out were AFM, FTIR, Tensile strength, Thermal conductivity measurement, and Weight gain. The results of the AFM test showed that the roughness of the surface decreased with the addition of a small amount of nanoparticles (1 wt.%) to the CH/(PVP-co-VAc) blend. FTIR results for blend film indicates a significant degree of composition between (PVP-co-VAc) and CH molecules. Nanocomposite films indicate that the composition among polymer blends and Nano are associated with changes in intensity, shape, and position of modes. The mechanical properties of chitosan were enhanced by the addition of TiO_2 nanoparticle with significant increases in tensile strength (from 47.87 MPa to 64.95 ~ 70 MPa). Because interfacial bonding and homogenous distribution between the nanoparticles and the blend were supportive of markedly improved mechanical strength .The results of thermal conductivity and water absorption showed that the highest values were achieved in the polymer blend (CH/ [PVP-co-VAc]), whereas its decrease with the increase of the nanoparticle incorporation.

Keywords: Natural polymer, nanocomposites, Physical properties.

1. Introduction

Chitosan (CH) is a natural carbohydrate biopolymer derived by deacetylation (DA) of chitin, a major component of the shells of a crustacean such as a crab, shrimp, and crawfish. After cellulose, chitin is the second most abundant natural biopolymer found in nature like cellulose; chitosan is a fiber. However, unlike plant fiber, chitosan possesses unique properties, including the ability to form films, optical structural characteristics and much more [1,2]. Chitosan shows first pledge for employ as a scaffold in tissue engineering, wound dressing applications, the antimicrobial therapy of textiles as well as water disinfection and microbial control [3-5]. Enforcement like food packaging and wound dressings frequently demand processing of chitosan into films. That is not straightforward as chitosan is insoluble in most common solvents (including water) but can be beat by dissolving chitosan in dilute aqueous acidic solutions [6]. It has been confirmed that aqueous acetic acid is one of the most suitable solvents in terms of the resulting film. The aim of this work is to report the effect of TiO_2 nanoparticle on the AFM, FTIR, tensile properties of polymer blend [CH/(PVP-co-VAc)] in addition to the thermal conductivity and weight gain. Maxing of two or more polymers is an important technique to improve the cost-effectiveness of commercial products. Chitosan films are brittle and not suitable for use in its dry state. Poly (1-vinylpyrrolidone-co-vinyl acetate) was select to blend with chitosan to enhance the mechanical properties with degradability of chitosan films [7].

IOP Conf. Series: Materials Science and Engineering **757** (2020) 012073 doi:10.1088/1757-899X/757/1/012073

2. Experimental Part

Chitosan (CH) with a degree of deacetylation of 85% and average molecular weight of 345,500 g·mol^{-1}, Titanium dioxide TiO_2 type (I) was about 65 nm, type (II) 42 nm anatase nanopowder, 99.7% trace metal basis, and Poly (1-vinylpyrrolidone-co-vinyl acetate) [PVP-co-VAc] with a molecular weight of ~5.0 × 10^4 g. mol^{-1} was obtained from Sigma-Aldrich.

2.1 Film Preparation

The blend of (CH) and (PVP-co-VAc) with the fixed ratio (7:3) have been prepared by dissolving 0.4 g of CH in 30 mL acetic acid (1% v/v), and 0.4 g of (PVP-co-VAc) in 10 mL acetic acid (1% v/v). The two solutions were put separately under continuous stirring for 2 h at room temperature, then mixed for about 1 hour. Finally, cast into a petri dish with 8 cm in diameter at room temperature for 24 h in a desiccator to ensure the removal of bubbles, after that the samples dried at 50 °C in an electric oven at for 6 h, then the films were peeled off. The nano-composites preparation was the same as the blend but adding the TiO_2 nanoparticles slowly to the (PVP-co-VAc) solution with continuous stirring for 30 min (by weight relative to CH) are 1, 2, 3.5, 5, 7.5 and 10 wt. % before mixing it with the CH solution to have proper distribution as tabulated in Table (1).

Table 1. The samples assingning and composition of blend{CP}named [CH+(PVP-co-VAc)] and the (TiO_2 [I], TiO_2 [II]) nanoparticles.

Sample type	**CP**	T1	T2	T3	T4	T5	T6	P1	P2	P3	P4	P5	P6
Nano TiO_2 wt.%	0	1	2	3.5	5	7.5	10	1	2	3.5	5	7.5	10
Type of NanTiO_2	-	(I)	(I)	(I)	(I)	(I)	(I)	(II)	(II)	(II)	(II)	(II)	(II)

2.2 Characterizations of films

1. Atomic Force Microscopy

Surface morphology and roughness of Titanium dioxide TiO_2 with two nanoparticle size (65& 42 nm), blend and nanocomposite films were studied by Atomic Force Microscope (AFM) analysis type (Angstrom AA3000, USA) for the measurement the roughness of the surface.

2. FTIR spectroscopy

FTIR characterization was done on the polymer electrolytes using the Shimadzu FT-IR 8000 series spectrum over the range of 400-4000 cm^{-1} with resolution 4 of the spectra obtained at room temperature and recorded in the transmittance mode.

3. Tensile strength

The tensile strength of [CH/(PVP-co-VAc)] blend and nanocomposite specimens were characterized for their properties in the longitudinal direction by ASTM D882-02. The films were cut into slices with a dimension of 10 mm x 80 mm and with a thickness of 0.17 -0.24 mm. These specimens were all tested at a gauge length of 50 mm; tensile tests were performed at a crosshead speed of 5 $mm.min^{-1}$.

4. Thermal conductivity

The thermal conductivity of the blend [CH/(PVP-co-VAc)] film and TiO_2 nanocomposite films was studied by using Lee's disc method, homemade by the worker in material Laboratory Department of Physics. It's designed to calculate the thermal conductivity of primarily thermal insulators such as. A thin, large surface area sample of 3 cm in diameter is used to determination of thermal conductivity to reach steady-state quickly.

5. Water absorption

Swelling behavior of the blend and $CaCo_3$ nanocomposite films was studied on rectangular specimens with (10×10 mm). The dry weight (W_1) of each sample was measured, and then the sample was immersed in distilled water of pH (6.5) at the time interval of 1, 2, and 3 h. The measurement was taken every 10 min, then, the sample was taken out, and the excess water on its surface was wiped with a filter paper. The swollen films eventually weighed. Weight gain of the samples was registered by periodic removal of the specimens from immersion and weighted. The weight gain % calculated, according to Coarse Aggregate AS per IS 2386 part 3, in Eq. 1 for blend and nanocomposite films.

$$\text{Weight gain\%} = [(W_2 - W_1) / W_1] \times 100\% \quad ..... (1)$$

where W_2 is the weight of the wet sample at time (t) after swelling, and W_1 is the weight of the dry sample before expanding.

3. Results and Discussion

3.1 Atomic Force Microscopy

The surface topology of the synthesize of Titanium dioxide TiO_2 with two nanoparticle size (65& 42 nm), blend and nanocomposite films have characterized by AFM. The result showed that the roughness of the surface decreased with the addition of the nanoparticles to the CH/(PVP-co-VAc) blend as shown in the Table (2) and Figure (1).

Table 2. The Roughness average of AFM test result for the blend and nanocomposite (T1, T6, P1, P6, C1, and C6) films.

Sample	Nano type	Ave. Grain size (nm)	Nano wt. %	Ave. Roughness (nm)
CP	-	-	0 %	12.6
T1	TiO_2 (I)	65	1	0.87
T6	TiO_2 (I)	65	10	18
P1	TiO_2(II)	42	1	1.63
P6	TiO_2(II)	42	10	9.2

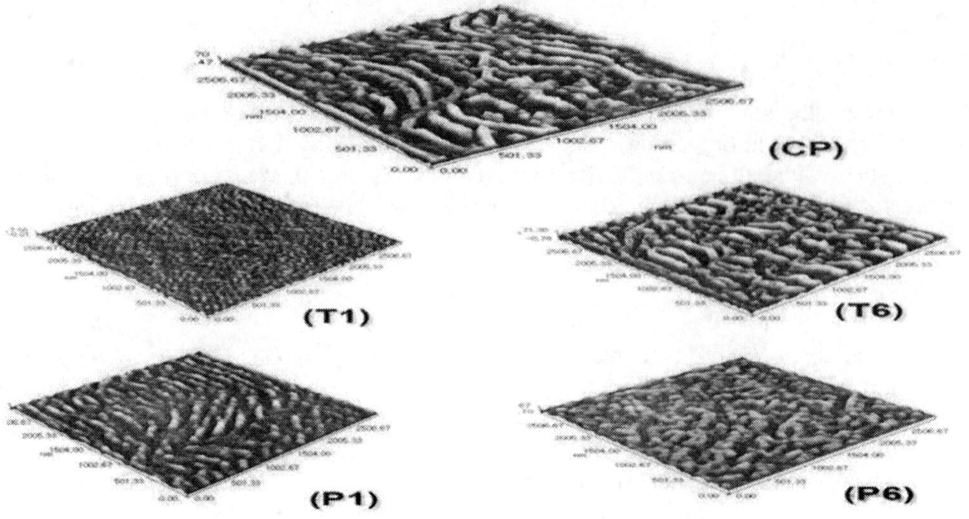

Figure 1. 3D view of the AFM image of the blend (CP) and nanocomposites (T1, T6, P1, and P6) films.

3.2 FTIR Spectroscopy

Figure (2) shows the FTIR spectra of raw materials, as for CH and (PVP-co-VAc) the prominent peak observed at 3431.84 cm^{-1} corresponds to the intermolecular hydrogen bonding. A peak obtained at 2925.53cm^{-1}, 2855.66cm^{-1} indicates the asymmetric and symmetric C-H stretching in CH2 respectively. A peak observed at 1718.20cm^{-1}, 1623.22 cm^{-1} indicates the presence of C=O is stretching in secondary amide and N-H bending, respectively. These peaks confirm the blending of (PVP-co-VAc) and chitosan [8]. From Figure (2) the FTIR spectrum of TiO_2 nanoparticles it can noticed clearly that the bands for Ti-O and Ti-O-Ti are present in the 800 - 400 cm^{-1} region, the former observed in a higher wavenumber than the latter [9,10]. The peak about (800) cm^{-1} is allocated to a (Ti-O) vibration, where the oxygen atom is in the non-binding condition. The bands at 760 cm^{-1}, 680 cm^{-1}, 560 cm^{-1}, 500 cm^{-1} and 468 cm^{-1} are associated to (Ti-O-Ti) vibrations [11,12]. These assignments of vibrational modes of the blend and nanocomposites bands were tabulated in Table (3).

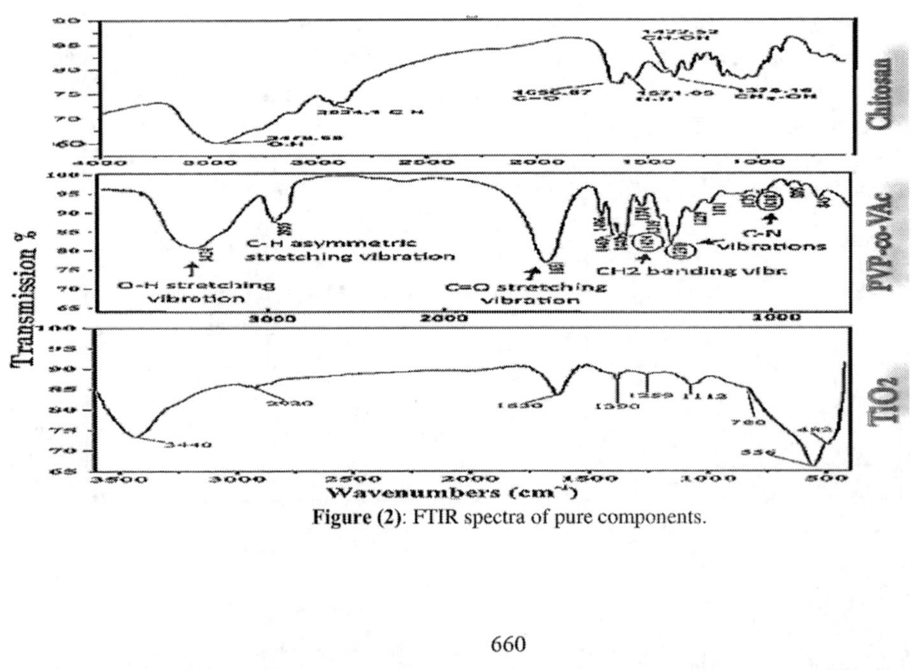

Figure (2): FTIR spectra of pure components.

Table 3. Assignments of vibrational modes and list of the observed peak for blend and nanocomposites.

Assignments	Wavenumbers (cm^{-1})				
	CP (0%)	T4 (5%)	T6 (10%)	P4 (5%)	P6 (10%)
O-H & N-H	3298 3369 3436 3583	3242 3355 3450 3506	3234 3330 3444 3743	3413 3423 3747	3232 3301 3365 3456
C=O	1670 1737	1645 1660 1733	1647 1670 1741	1660 1733	1676 1735
CH3 & C-H	1080 1157 1245 1373	1072 1157 1251 1379	1081 1157 1249	1029 1159 1247 1375	1074 1155 1244 1373
Ti-O &Ti-O-Ti	435 486 649 734	460 540 653 775	443 532 617 665	422 576 649 796	459 559 651 732
CH2 & C–H	842 898 946	873	838 929	840 898	842 896 946

Figures (3) shows the FTIR spectra for polymer blend and nanocomposites. The peaks (3436 ~ 3460) for (T4, T6, P4 & P6) became narrower and had high intensity.

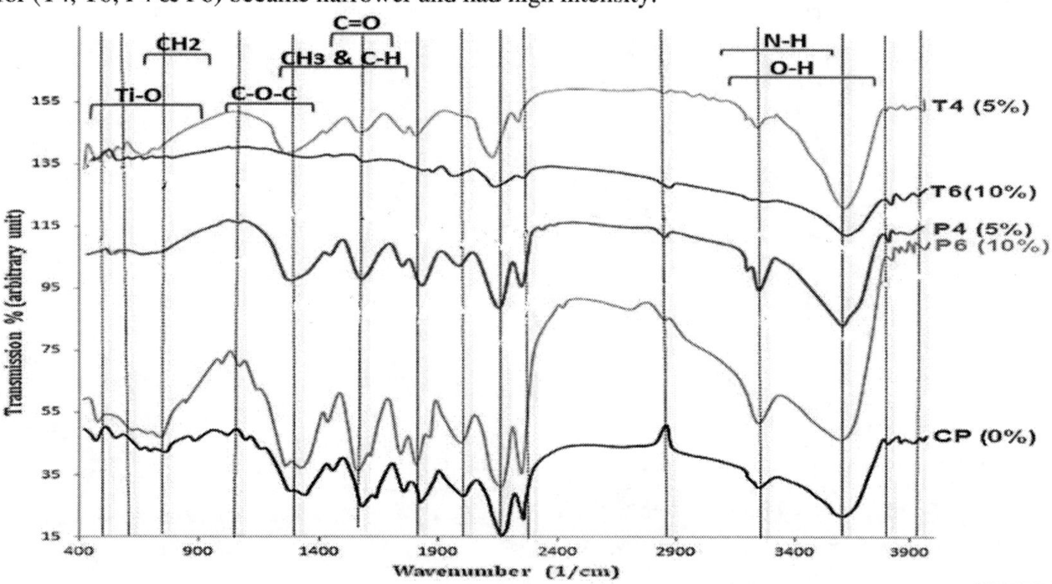

Figure (3). FT-IR spectra for the polymer blend (CP) and nanocomposite (T4, T6, P4, P6) films.

The low intensity of blend peaks (1245, 1670 & 3436) as shown in the table (3.4) may be due to that the incorporated actively interact with (PVP-co-VAc) molecules, and then disturb the interactions among (PVP-co-VAc) molecules. The introduction of nanoparticle makes the intensity of the band dramatically enhanced, which may be related to an overlapping effect, from the introduction of a large number of Nano bonds [13].

3.3 Tensile strength

From Figure (4-a) and Table (4), the result shows that the nanocomposites with type (I) of TiO_2 have the higher tensile strength value (64 MPa) for T2 and the higher elongation value (3.8) was for T5. While the nanocomposites reinforced with TiO_2 (II) shows that the tensile strength increase with increasing the weight ratio of TiO_2 except for sample P5 with (7.5 wt.%) TiO_2 as seen in Figure (4-b) and Table (4). The decrease in tensile properties can be related to the poor distribution or agglomeration of nanoparticles.

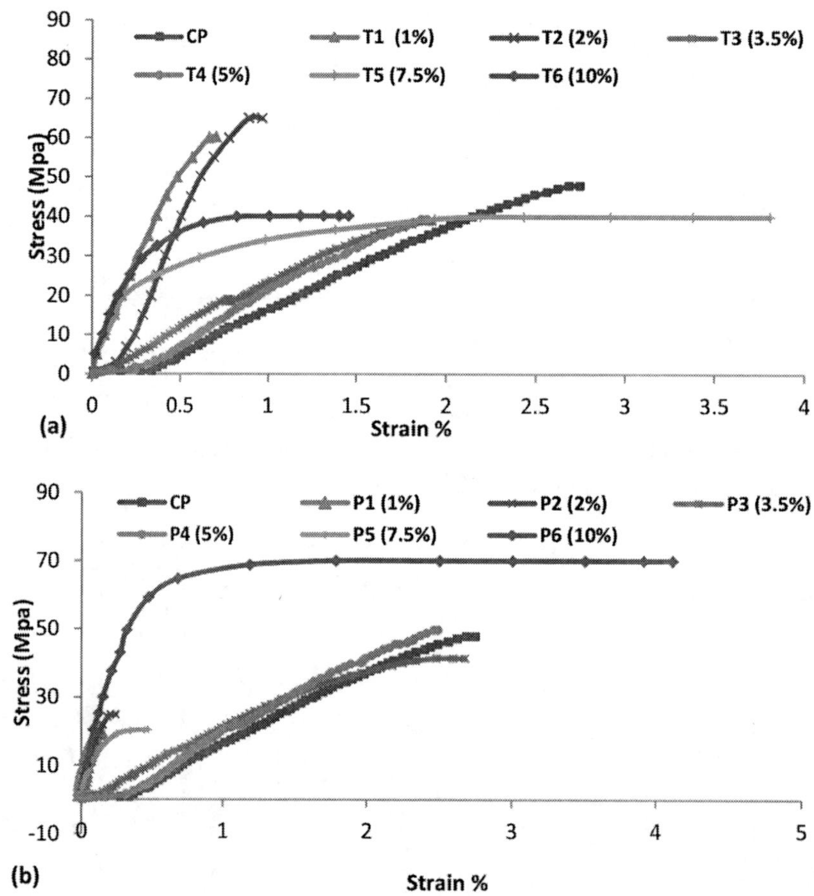

Figure 4. Stress-strain curves for blend and nanocomposite films. (a) (CP) and (T1~T6) films, (b) (CP) and (P1~P6) films.

The various amount of TiO_2 (1, 2, 3.5, 5, 7.5 and 10 wt. %) was conventionally added to [CH/(PVP-co-VAc)] blend. From Table (4), it was found that the addition nanocomposite could promote the physical properties and mechanical properties of the films because nanoparticles have significant surface and size effects. The higher nanoparticle content maybe lead to poor allocation of the nanoparticle across the matrix which caused the stress in the continuous phase to increase and promote poor tensile strength. The imperfect distribution of the nanoparticle can explain the reduction in tensile strength through the matrix [14,15].

Table 4. Tensile properties of the blend and the nanocomposite with TiO_2 (type I & II).

Sample	Nano-type & percent wt.%		Tensile strength (MPa)	Elongation at Max. %
CP	0		47.87	2.689
T1	TiO_2 (I)	1	60.22	0.7
T2	TiO_2 (I)	2	64.95	0.96
T3	TiO_2 (I)	3.5	38.67	1.920
T4	TiO_2 (I)	5	39.33	1.891
T5	TiO_2 (I)	7.5	40	3.8
T6	TiO_2 (I)	10	40.1	1.45
P1	TiO_2(II)	1	19.9	0.12
P2	TiO_2(II)	2	24.87	0.23
P3	TiO_2(II)	3.5	41.39	2.680
P4	TiO_2(II)	5	49.65	2.489
P5	TiO_2(II)	7.5	20.4	0.46
P6	TiO_2(II)	10	70	4.1

The mechanical properties of the blend and nanocomposites are a result of the polymer conformation and the attraction energies between the [(PVP-co-VAc)/CH] chains [16,17].

3.4 Thermal conductivity

The thermal conductivity of blend (CP) and TiO_2, nanocomposite films result showed that the highest values of thermal conductivity were accomplished for the polymer blend (CP), (CH/ [PVP-co-VAc]), while it decrees with increasing nanoparticle incorporation, as shown in Figure (5) & Table (5). Small particle size has a large surface area that means more considerable thermal interface resistance.

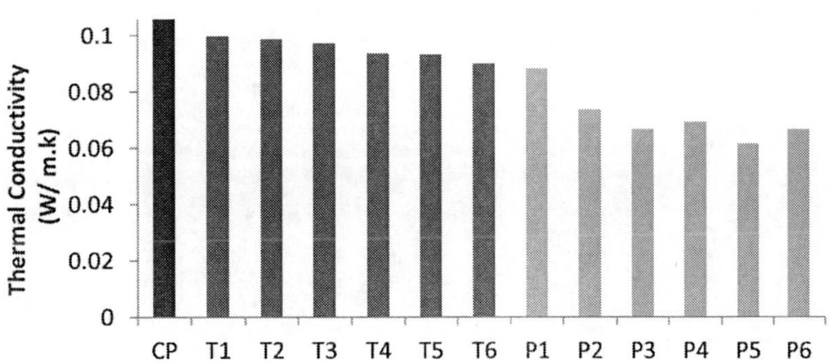

Table 5. The thermal conductivity of the blend and nanocomposites films.

Sample type	Nano wt.%	Type of nano TiO_2	Thermal conductivity (W/m.k)
CP	-	-	0.106761
T1	1	TiO_2 (I)	0.099552
T2	2	TiO_2 (I)	0.098553
T3	3.5	TiO_2 (I)	0.097307
T4	5	TiO_2 (I)	0.093597
T5	7.5	TiO_2 (I)	0.093316
T6	10	TiO_2 (I)	0.090003
P1	1	TiO_2 (II)	0.088163
P2	2	TiO_2 (II)	0.073552
P3	3.5	TiO_2 (II)	0.066779
P4	5	TiO_2 (II)	0.069047
P5	7.5	TiO_2 (II)	0.061471
P6	10	TiO_2 (II)	0.066767

Figure 5. Thermal conductivity of (CP) blend and nanocomposite films.

3.5 Water absorption

Swelling behavior of blend [CH/(PVP-co-VAc)] film and [CH/(PVP-co-VAc)/TiO_2], [CH/(PVP-co-VAc)/$CaCo_3$] nanocomposite films were done by using rectangular samples with (10×10 mm). The result shows that the water uptake for the blend is higher than the other nanocomposite films, and the weight gain decrease with the increase of the nanoparticle as shown in Figure (6). The water uptake occurred due to the formation of micro-cracks and voids in the surface of the samples during the longer degradation stages. That may be expounded as the voids and flaws existing on the surface and inside the samples have been attacked and filled with the solution. This process occurred first, leading to swelling, the following step separates polymers chains apart, and the swollen polymer generates sufficient pressure to destroy the intermolecular chains, leading to polymer degradation. It is worth mentioning that the interfaces are prone to chemical attack, and this can be true for the blends, and after attacking the interfaces, liquids will diffuse into the material [18,19].

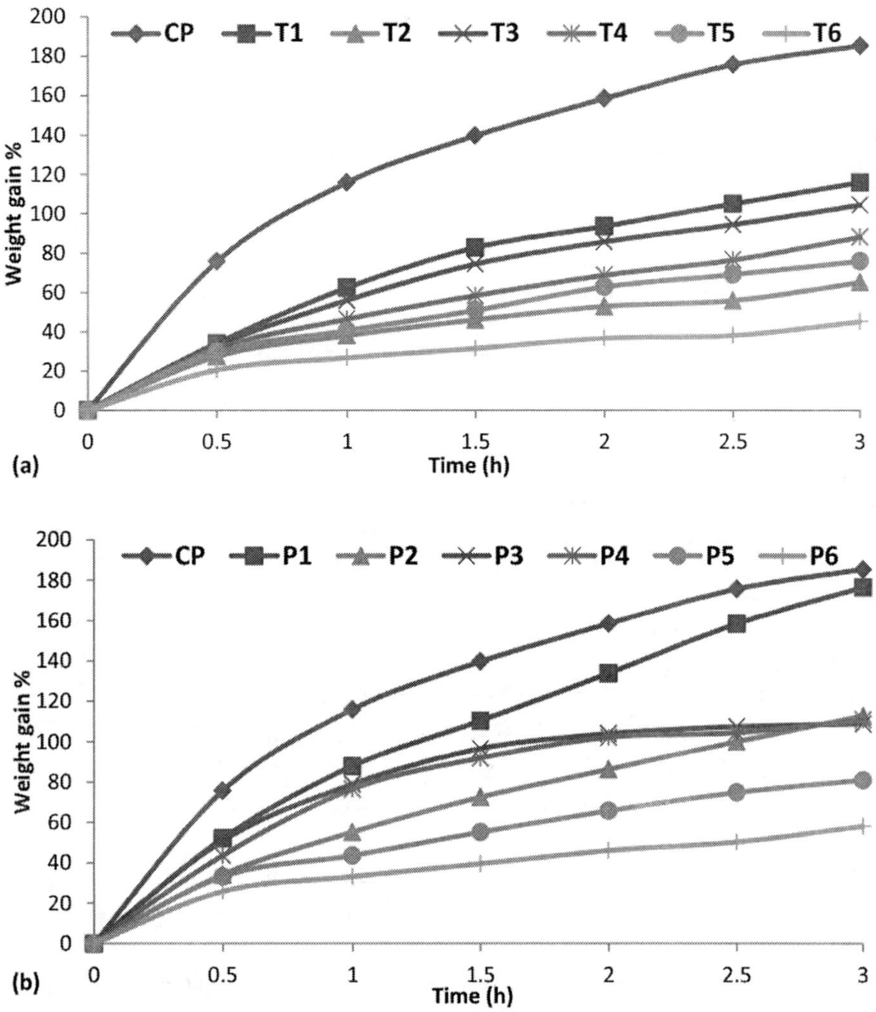

Figure 6. Effect of TiO_2 content on swelling behavior of blend (CP) at various interval time; (0.5,1, 1.5, 2,2.5, & 3 h), at RT for (a) (CP) and (T1-T6) and (b) (CP) and (P1-P6).

The water absorption of the blend film was the highest among the other films because chitosan has increased the hydrophilicity due to its hydroxyl groups (OH) and amino groups (NH_2) while that of (PVP-co-VAc) depends on upon only hydroxyl groups (OH). Therefore, hydroxyl group can form stronger hydrogen bonds with water (H_2O) than an amino group [20], while the water gain of the nanocomposites is lower than the (CH/ PVP-co-VAc) blend because it contains TiO_2 the nanoparticles decreases and spread throughout the plastic and be able to block oxygen, carbon dioxide, and moisture [21].

4. Conclusions

Incorporation of the TiO_2 nanoparticle to the blend [CH/[PVP-co-VAc)] via solution casting method produced the TiO_2 nanocomposite films. The immobilized TiO_2 exhibited an excellent distribution in chitosan film when adding a small amount of TiO_2. The results show that the tensile strength of these nanocomposites was enhanced compared to the blend [CH/(PVP-co-

VAc)] film. The addition of nanoparticle to the blend reduced its absorption ability to the water as well as its thermal conductivity. The TiO_2 nanocomposites will be promising sources for photocatalyst materials for various organic compounds in the environment.

References

[1] Meyers, S. P., Chen, H. M., No, H. K. and Lee, K. S, 1990 International By-Products Conference. Anchorage, Alaska, 161-171.
[2] Sun-Ok Fernandez-Kim, Ph.D., 2004 Seoul National University, Department of Food Science.
[3] Chung, Y.-C.; Su, Y.-P.; Chen, C.-C.; Jia, G.; Wang, H.-L.; Wu, J.C.W.and Lin, J.-G., 2004 Acta Pharm, **25**, 932–936.
[4] Li, Q.; Mahendra, S.; Lyon, D.Y.; Brunet, L.; Liga, M.V.; Li, D. and Alvarez, P.J.J. 2008 Water Res, **42**, 4591–4602.
[5] Khairul Anuar Mat Amin and Marc Soft Materials Group, 2o12 School of Chemistry, University of Wollongong, **4**, 590-599.
[6] Rhim, J.W.; Weller, C.L.and Ham, K.S., 1998, Food Sci. Biotechnol, **7**, 263–268.
[7] POKHREL S., ADHKAR R., AND YADAV P. N., 2017, ASIAN JOURNAL OF CHEMISTRY, VOL.**29**(7), 1602-1606.
[8] Nasreen K., Vijayalakshmi K., Gomathi T., Anbalagan S. and Sudha P.N., 2013 Der Pharmacia Lettre, **5**, 145-150.
[9] Burgos M. and Langlet M., 1999 Thin Solid Films, **349**(1-2), 19-23.
[10] Castellano F.N., Stipkala J.M., Friedman L.A. and Meyer G.L.,1994 Chemistry of Materials, **6**(11), 2123-2129.
[11] Djaoued Y., Badilescu S., Ashrit P.V. and Robichaud J., 2001 The Internet Journal of Vibrational Spectroscopy, **5**(4).
[12] Phillippi C. M. and Lyon S.R., 1971 Physical Review B, **3**(6), 2086-2087.
[13] Huawen H., John H. X., Hong H., Allan C. and Liang H., 2013 Carbohydrate Polymers, **91**, 305– 313.
[14] Torres F.G., Arroyo O.H. and Gómez C., 2007 Journal of Thermoplastic Composite Materials, **20**, 207-223.
[15] Asmaa S. K., Mohammed K. J., Abdulkareem M.A. Alsammarraie and Harith I. J.,2017 Asian J. Chem.,**29**(3) 675-678.
[16] Herald T.J., Obuz E., Twombly W.W., and Rausch K.D., 2002 Cereal Chem. **79**(2), 261-264.
[17] Khairul A. M. A. and Marc, 2012 Soft Materials Group, School of Chemistry, University of Wollongong, **4**, 590-599.
[18] Rola A-K A., 2001 M.Sc. thesis, Department of Applied Science, University of Technology.
[19] Bilqis M. Z., 1996 PhD thesis, Department of Applied Science, University of Technology.
[20] Tang C., Chen N., Zhang Q., Wang K., Fu Q. and Zhang X., 2009 Polymer. Degrade. Stab, **94**, 124–131.
[21] Lu S., Chenzhong M., Yifeng H., Shangang D., Xiao L., Yongping B. and Yudong H.,2011 Iranian Polymer Journal, **20**(12) 955-967.

The effect of polyaniline addition on ZnO for the fabrication of high performance photodetection

Dalal K Thbayh[1*], Rawnaq A Talib[1], Kareema M Ziadan[2]

[1]Polymer Research Centre, University of Basra, Iraq

[2]University of Basra, College of Science, Department of Physics, Basra, Iraq

[*]Corresponding authors: E-mail address: dalalkarad2014@gmail.com

E-mail address: Rounq4@gmail.com
E-mail address: kmziadan16@gmail.com

Abstract The effect of PAni loading (5 and 15%) on the properties (PAni-ZnO) composite was investigated. X-ray diffraction analysis confirms the hexagonal wurtzite-type ZnO crystal structure, while ZnO peaks' intensity decreases with increasing PAni content. Field emission scanning microscopy observations indicate, in the absence of PAni, ZnO crystallizes as twinned hexagonal rod nanostructure; the size of crystals of ZnO-15%PAni is shorter along ZnO-5% PAni while the width of the crystals is larger with increasing concentration of PAni. The composites deposited onto glass, PET, and Si substrates revealed interesting characteristics for the fabricated photodetector devices. The photodetector based on ZnO-15%PAni / Si substrate demonstrates the best performance with a high gain of 43 and a sensitivity of 4962.

Keywords: ZnO; Polyaniline; Composite; Photodetector; Twin hexagonal.

1. Introduction

Composites of polymer and inorganic material with specific properties are gaining more attention for various applications. There are three main preparations procedures which can be employed to obtain polymer-inorganic composites, namely direct mixing of two or more components in solvent, in situ polymerization of monomer units in the presence of inorganic particles and dissolve mixing of polymer with inorganic particles [1]. The incorporation of inorganic component such as ZnO into polymer matrices could result in the modification of optical, electrical and mechanical properties of the respective components [2]. Polyaniline (PAni) is one of the most studied organic based conducting polymer, with ideal properties for various applications because of its low cost, simple preparation procedure, good environmental stability, and can be easily doped. It is a p-type conducting polymer which can be produced as powder[3], thin film[4], or fiber[5].

Zinc oxide (ZnO) as n-type semiconductor is one of the versatile inorganic compound that can be used in wide range of applications because of its low cost and can be readily produced in various forms such as nanowires [6], nanoflowers [7] nanosheets [8], and nanorods [9].

Ultraviolet (UV) photodetectors based on semiconductor have found applications in water sterilization, human traffic safety, flame sensing, and pollution monitoring systems [10]. Composites and p-n Junction from conducting polymer and metal oxide have been produced for numerous applications such as gas sensors [11], humidity sensors [12], biosensors [13], and photo sensors [14]. However, conducting polymer/ metal oxide composites-based sensors are relatively new and more studies are needed to evaluate the effects of preparation conditions on the evolution of the physical properties towards the development of high-performance photodetectors. Several researchers have reported on the preparation of ZnO-PAni composite thin films [15-17] and the preparation of twin-structure ZnO with PVP, PEO and PVA [18-20]. More investigation should be conducted on the preparation of twin-structure ZnO with PAni to elucidate its characteristics for sensor applications.

The aim of this research work consists on low temperature growth of twin ZnO - PAni composites. A low cost and simple polymer-ZnO based photodetectoris fabricated. The use of glass and silicon as substrates will further reduce the cost of the device. In addition, the use of polyethylene terephthalate (PET) as substrate could provide on advantage due to its flexibility, light weight, transparency, and high resistivity.

2.Experimental and methods

1.0 g zinc acetate hexahydrate ($Zn(CH_3COO)_2.6H_2O$) powder was dissolved in 50 ml of distilled-water then stirred for 10 min at 30°C. After that, 5 ml of (1M) ammonium hydroxide was added to the solution under vigorous stirring at 70°C for 2 h. Subsequently, a certain amount of PAni (M_w=15,000 from Sigma-Aldrich) was added to the mixture; 5 and 15 wt.%, then the mixture was kept under stirring for 3h at 50°C. The as-obtained precipitate was washed several times with distilled water to remove any residual reactants and finally dried in oven at 80°C for 1h.The glass substrates, PET, and Si wafer were cut into small square shapes, then cleaned in a beaker containing isopropyl alcohol solution and ethanol alcohol solution in an ultrasonic bath for 15 min. Si substrate was cleaned by Corporation of America (RCA) method[21].The composites films were deposited onto cleaned substrates by casting technique in formic acid to obtain homogeneous solution.

Morphological observations were performed by field emission scanning electron microscopy (FESEM) using FEI Nova Nano SEM 450. EDX. Structural analysis was checked by high-resolution X-ray diffraction using PANalyticalX'Pert Pro MRD system equipped with Cu-Kα radiation source (λ=0.1541 nm). The optical measurements were studied at the room temperature by Raman spectroscopy using JobinYvon HR 800 UV spectrometer. Photoconductivity (UV-sensing) measurements were performed using a current source 2400 source Meter Keithely .

3.Results and discussions

3.1 Morphological observations

Figure 1 Illustrates FESEM images of the fabricated powder ZnO / (5,15%) PAni composites. It can be seen obviously the twining hexagonal rod-type ZnO-PAni composite nanostructures. The crystals in ZnO-15% PAni are shorter than ZnO-5% PAni, while the width is greater at a higher PAni concentration. It is important to highlight that PAni is preferentially adsorbed onto ZnO (0001) face, which will retard the crystal growth perpendicular to this face. Hence, it is suggested that twin-like ZnO has been capped with polymer chain. As it is known that the hexagonal ZnO has the $^{+}$c [0001] direction by zinc-terminated and the $^{-}$c [000$1^{-}$] also direction by oxygen- terminated, so that the positive ion $Zn(NH_3)_4^{+2}$ adsorption happen towards $^{-}$c-axis and the negative ion $Zn(OH)_4^{-2}$ will be adsorbed toward $^{+}$c-axis. After then, the two rods are formed. The as-obtained results are consistent with some previous research works using other kinds of the polymers, such as PVP, PEO-b-PMMA and polyethylenimine (PEI) [18, 19, 22].

The EDX spectrum (Figure1) reveals the existence of O and Zn elements, in addition to C peaks arising from PAni. It can be seen that the presence of C in ZnO-PAni composite results from an interaction of PAni with ZnO particles. moreover besides, one can see that the observed percentage of C increases with increasing the weight ratio PAni/ZnO. Hence, it is evident that the addition of PAni has a direct influence on the increase of the diameter of crystal rods in the composites.

IOP Conf. Series: Materials Science and Engineering **757** (2020) 012074 doi:10.1088/1757-899X/757/1/012074

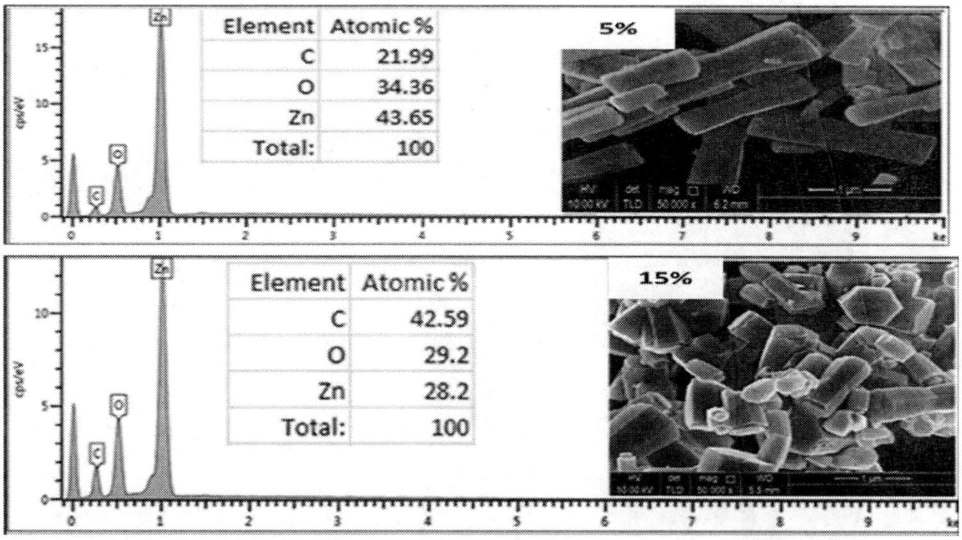

Figure 1. FESEM images and EDX data of (ZnO-5%) PAni, and (ZnO-15%) PAni

Figure 2 shows the surface morphology of the films after dissolving the powder composites in formic acid. It can be noticed that the shape changes from semi-hexagonal large grain distribution (ZnO-5%PAni) to a layered-shaped nanostructure (ZnO-15%PAni). This result indicates that increasing the concentration of PAni in the composite improves its solubility in formic acid, given that the formic acid is a good solvent of polyaniline. Thus, increasing the ratio of PAni to 15 wt% subsequently increases the number of PAni chains between the spaces of ZnO crystalline structure [23], resulting in the formation of a thick composite layer meanwhile semi-hexagonal particles of ZnO-PAni gradually dissipates.

IOP Conf. Series: Materials Science and Engineering **757** (2020) 012074 doi:10.1088/1757-899X/757/1/012074

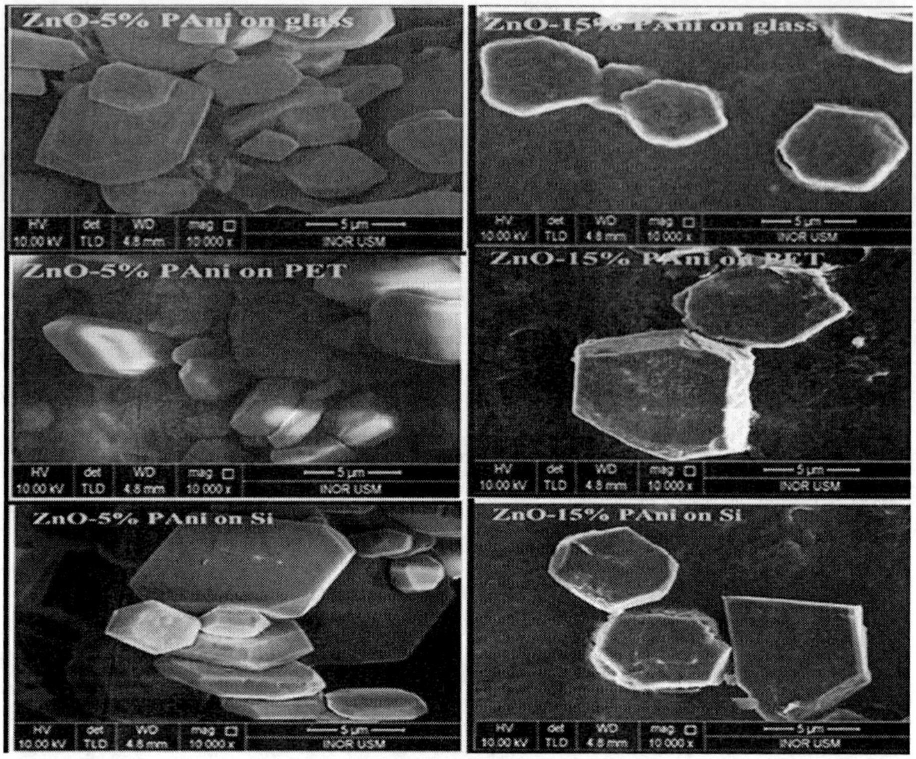

Figure 2: FESEM images of microparticle ZnO-PAni composites deposited on different substrates: glass, PET, and Si.

XRD diffraction patterns of the fabricated ZnO/PAni composites were shown in Figure3. All that was observed diffraction peaks can be identified within the hexagonal wurtzite-type structure of ZnO phase, in the agreement with JCPDS card No. 80-0075. The minor remaining peak located at 2θ = 24.42° can be assigned to the reflection of PAni. The intense and sharp diffraction peaks indicate the good crystallinity of the as-prepared ZnO-PAni composites. This confirms that the presence of PAni did not modify the original hexagonal structure of ZnO phase without deterring its crystalline quality.

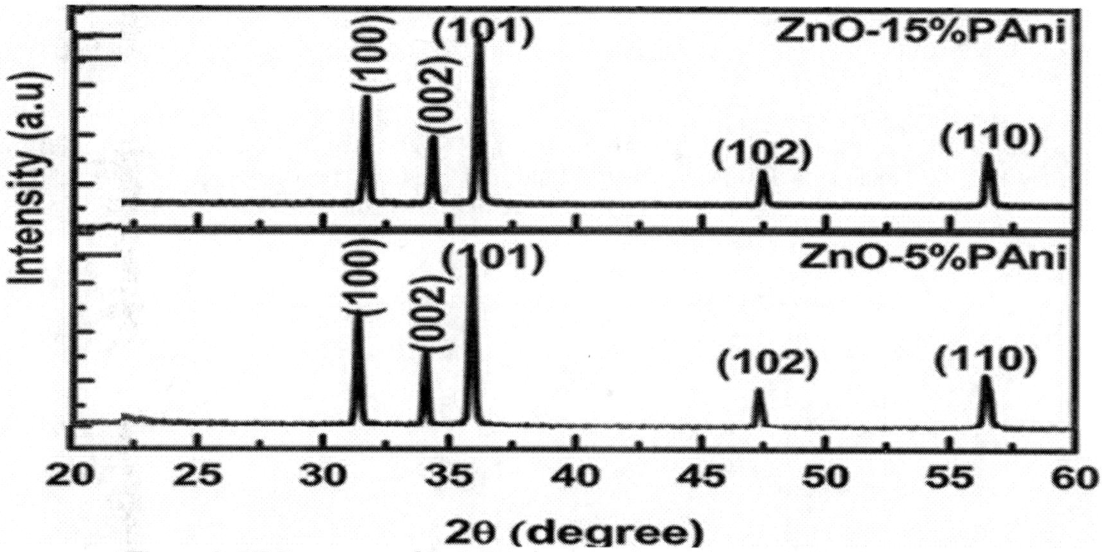

Figure 3: XRD patterns of ZnO-PAni composites at (5-15wt %) ratio of PAni.

The average size of the crystals (G) of ZnO-PAni composites structure has been estimated by measure the full width at half maxima (β) of (101) reflection (Table 1) and using the Debye Scherrer's formula [24, 25]:

$$G = \frac{0.9\,\lambda}{B\cos\theta} \qquad \text{..........................} (1)$$

Where the wavelength(λ) of the X-ray source is equal = 0.15405(nm) , and θ is the angle of the diffraction peak. The crystallite size of ZnO phase increases significantly from 34 nm of (ZnO-5%) PAni to 52 nm of (ZnO-15%) PAni. This might be attributed to a higher surface along the prism plan when compared to the basal plane in the hexagonal close packed crystal structure; thereby more of PAni chain can be expected to become attached to the prism surface of the unit cell rather than that of the basal surface. When the compressive strain is in one direction, there will be a tensile strain on the opposite direction. Thus, the lattice constant is increased along a-axis and caused the unit cell to expand [26-28].

Table 1: Structural properties of ZnO-PAni composites.

Sample	2θ (degree) (101)	2θ (degree) (100)	2θ (degree) (002)	FWHM (degree) (101)	D (nm)	ε_z(%)	ε_a(%)
ZnO-5% PAni	36.216	31.733	34.333	0.2458	34.00	0.0885	-0.0034
ZnO-15% PAni	36.205	31.730	34.375	0.1604	52.11	-0.0298	0.0057

3.2 Ultraviolet (UV) photodetectors

Figure 4 (a, b and c) displays the repeatability performance of ZnO-PAni composites conducted by applyed a varying voltage (1, 3, and 5 V) by using thick film onto glass, PET, and Si substrates. The device has potential to repeat the same behavior although it has been exposed to the light and dark process over and over. Hence, it could be concluded from that, all the fabricated sensors possess the stability was good during the measurement for 20 sec under 400 nm with an intensity of 0.958 mW/ cm^2, but the sensor based on 15% ZnO-PAni composite onto Si substrate exhibits the best performance. The mechanism of light sensing of ZnO-PAni composites are discussed as follows. Both ZnO and PAni can be excited by light to produce photo-generated carriers and excited electrons, meanwhile the conducting PAni absorbs the light and produces (e^-) that transfers to the CB of ZnO[29]. Simultaneously, the photo-generated holes in VB of ZnO are able to move freely to the composites surface through HOMO PAni.

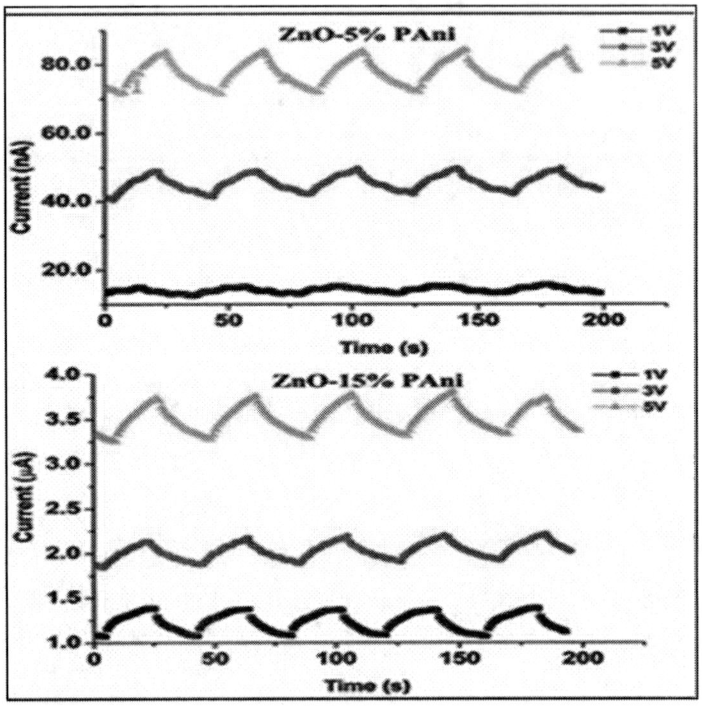

Figure 4: The repeatability characteristic (ON/OFF) of Pt-(ZnO-PAni) composites-Pt on (a) glass substrate under pulsed UV- light (400 nm) at various bias voltages.

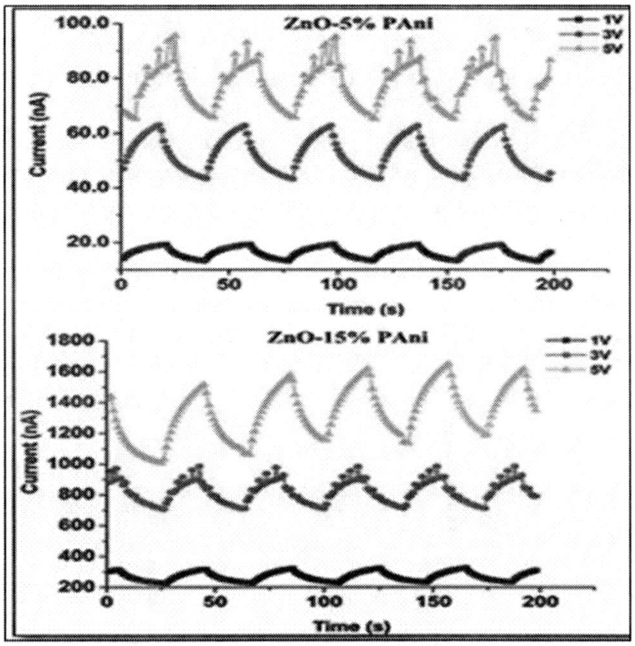

Figure 4: The repeatability characteristic (ON/OFF) of Pt-(ZnO-PAni) composites-Pt on (b) PET substrate under pulsed UV- light (400 nm) at various bias voltages.

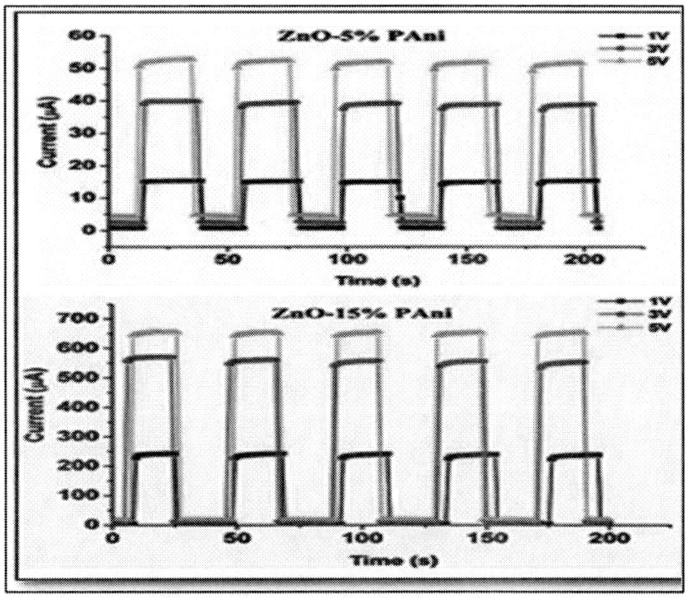

Figure 4: The repeatability characteristic (ON/OFF) of Pt-(ZnO-PAni) composites-Pt on (C) Si substrate under pulsed UV- light (400 nm) at various bias voltages.

So, the photo-generated electrons and holes will move in opposite directions, reducing the recombination probability and making the charge separation more efficient leading to a higher photocurrent [30, 31] as shown in Figure 5. A huge photoconduction sensitivity of the composite deposited onto Si substrate could be due to the higher photon absorption in ZnO-PAni composite layer, whereby some of photo-generated electrons may diffuse to ZnO-PAni/Si junction and subsequently swept into Si side. In general, both layers may contribute to the photocurrent [32], which led to higher sensitivity of ZnO-PAni/Si photodetector as compared to ZnO-PAni composite on insulating substrates (glass and PET). Table 2 summarizes the response time, fall time, sensitivity, and gain measured at various bias voltages for ZnO-PAni composites deposited onto different substrates.

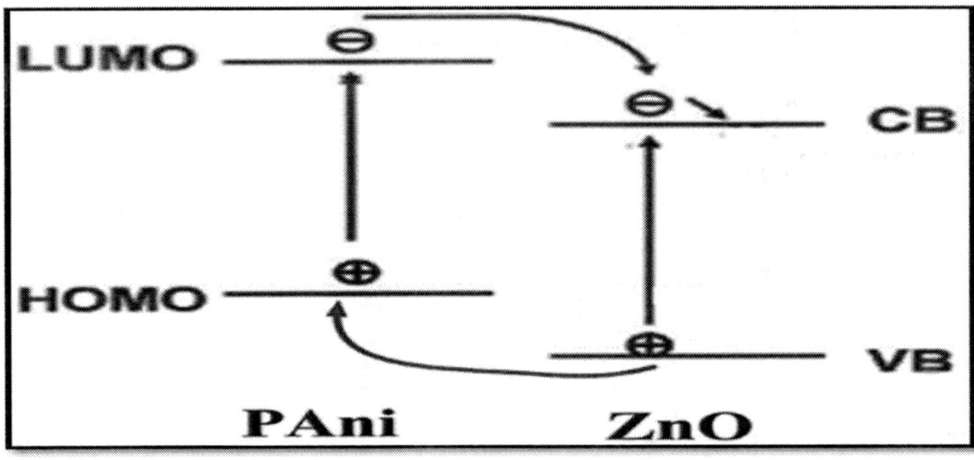

Figure 5: Schematic representation of energy levels and processes involved under UV illumination.

Table 2: The response time, fall time, sensitivity, and gain of the Pt –(ZnO-15%PAni) composites Pt at various bias voltages onto different substrates.

Sample	Substrate	Voltage (V)	Response time (s)	Fall time (s)	Sensitivity %	Gain
ZnO-5%PAni	glass	1	10.93	17.75	15.30	1.18
		3	11.75	15.29	17.39	1.27
		5	13.52	15.98	17.22	1.17
	PET	1	14.64	11.25	43.70	1.45
		3	12.48	14.40	44.30	1.46
		5	8.92	10.82	30.90	1.30
	Si	1	1.052	0.959	2157	25.3
		3	1.067	0.967	1561	16.9
		5	1.072	0.967	991.1	12.3

Sample	Substrate	Voltage (V)	Response time (s)	Fall time (s)	Sensitivity %	Gain
ZnO-15%PAni	glass	1	13.77	11.90	35.30	1.30
		3	15.42	17.40	17.60	1.15
		5	12.68	16.55	16.90	1.13
	PET	1	13.08	13.89	40	1.40
		3	15.69	13.51	28.6	1.31
		5	13.82	14.76	47.1	1.45
	Si	1	0.936	0.852	5925	79.7
		3	1.050	0.956	5450	71.8
		5	0.802	0.885	4962	43

4.Conclusions

ZnO-PAni nanocomposites with enhanced sensing characteristics have been fabricated through cost-effective chemical method starting from the mixture of Zn $(CH_3COO)_2.6H_2O$ and PAni. XRD analysis revealed sharp and defined peaks, implying the high crystallinity of the as-prepared ZnO-PAni composites. FESEM observations revealed twinned-hexagonal rod nanostructure; the crystals of ZnO-15%PAni possess smaller length and larger width compared to ZnO-5% PAni, suggesting that ZnO was capped with polymer. The surface morphology of the as-fabricated composites thick film changed from a nearly hexagonal large grain distribution to a layer shape. The photoconductive properties of the UV thick film deposited onto glass, PET, and Si slides were investigated. The photodetector fabricated from 15% PAni composite onto with Si substrate showed a high gain of 43 and high sensitivity 4962 at voltage of 1 V.

Acknowledgments

The authors would like to thanks Nano Optoelectronics Research and Technology (NOR) of the School of Physics, University Sains Malaysia. Also, they have research support from University of Basrah according to its policy on objectivity in research.

References

[1] C. S. Wu, Y. J. Huang, T. H. Hsieh, P. T. Huang, B. Z. Hsieh, Y. K. Han, *et al.*, "Studies on the conducting nanocomposite prepared by in situ polymerization of aniline monomers in a neat (aqueous) synthetic mica clay," *Journal of Polymer Science Part A: Polymer Chemistry,* vol. 46, pp. 1800-1809, 2008.

[2] E. Tang, G. Cheng, X. Pang, X. Ma, and F. Xing, "Synthesis of nano-ZnO/poly (methyl methacrylate) composite microsphere through emulsion polymerization and its UV-shielding property," *Colloid and Polymer Science,* vol. 284, pp. 422-428, 2006.

[3] L. Ding, X. Wang, and R. Gregory, "Thermal properties of chemically synthesized polyaniline (EB) powder," *Synthetic Metals,* vol. 104, pp. 73-78, 1999.

[4] N. Agbor, M. Petty, and A. Monkman, "Polyaniline thin films for gas sensing," *Sensors and Actuators B: Chemical,* vol. 28, pp. 173-179, 1995.

[5] H. L. Wang, R. J. Romero, B. R. Mattes, Y. Zhu, and M. J. Winokur, "Effect of processing conditions on the properties of high molecular weight conductive polyaniline fiber," *Journal of Polymer Science Part B: Polymer Physics,* vol. 38, pp. 194-204, 2000.

[6] Y. Zhang, M. K. Ram, E. K. Stefanakos, and D. Y. Goswami, "Synthesis, characterization, and applications of ZnO nanowires," *Journal of Nanomaterials,* vol. 2012, p. 20, 2012.

[7] Z. Wang, X.-f. Qian, J. Yin, and Z.-k. Zhu, "Large-scale fabrication of tower-like, flower-like, and tube-like ZnO arrays by a simple chemical solution route," *Langmuir,* vol. 20, pp. 3441-3448, 2004.

[8] H. Kou, X. Zhang, Y. Du, W. Ye, S. Lin, and C. Wang, "Electrochemical synthesis of ZnO nanoflowers and nanosheets on porous Si as photoelectric materials," *Applied Surface Science,* vol. 257, pp. 4643-4649, 2011.

[9] W. I. Park, D. Kim, S.-W. Jung, and G.-C. Yi, "Metalorganic vapor-phase epitaxial growth of vertically well-aligned ZnO nanorods," *Applied Physics Letters,* vol. 80, pp. 4232-4234, 2002.

[10] J. Sun, F.-J. Liu, H.-Q. Huang, J.-W. Zhao, Z.-F. Hu, X.-Q. Zhang, *et al.*, "Fast response ultraviolet photoconductive detectors based on Ga-doped ZnO films grown by radio-frequency magnetron sputtering," *Applied Surface Science,* vol. 257, pp. 921-924, 2010.

[11] H. Tai, Y. Jiang, G. Xie, and J. Yu, "Preparation, characterization and comparative NH 3-sensing characteristic studies of PANI/inorganic oxides nanocomposite thin films," *Journal of Materials Science & Technology,* vol. 26, pp. 605-613, 2010.

[12] P.-G. Su and L.-N. Huang, "Humidity sensors based on TiO2 nanoparticles/polypyrrole composite thin films," *Sensors and Actuators B: Chemical,* vol. 123, pp. 501-507, 2007.

[13] A. A. Ansari, G. Sumana, R. Khan, and B. Malhotra, "Polyaniline-cerium oxide nanocomposite for hydrogen peroxide sensor," *Journal of Nanoscience and Nanotechnology,* vol. 9, pp. 4679-4685, 2009.

[14] G. Paul, A. Bhaumik, A. Patra, and S. Bera, "Enhanced photo-electric response of ZnO/polyaniline layer-by-layer self-assembled films," *Materials Chemistry and Physics,* vol. 106, pp. 360-363, 2007.
[15] S. P. Ansari and F. Mohammad, "Studies on Nanocomposites of Polyaniline and Zinc Oxide Nanoparticles with Supporting Matrix of Polycarbonate," *ISRN Materials Science,* 2012.
[16] P. SL, C. MA, and S. Shashwati, "Effect of camphor sulfonic acid doping on structural, morphological, optical and electrical transport properties on polyaniline-ZnO nanocomposites," *Soft Nanoscience Letters,* 2012.
[17] T. Pandiyarajan, R. Mangalaraja, and B. Karthikeyan, "Enhanced ultraviolet fluorescence in surface modified ZnO nanostructures: Effect of PANI," *Spectrochimica Acta Part A: Molecular and Biomolecular Spectroscopy,* vol. 147, pp. 280-285, 2015.
[18] M. Öner, J. Norwig, W. H. Meyer, and G. Wegner, "Control of ZnO crystallization by a PEO-b-PMAA diblock copolymer," *Chemistry of materials,* vol. 10, pp. 460-463, 1998.
[19] T. Thirugnanam, "Effect of polymers (PEG and PVP) on sol-gel synthesis of microsized zinc oxide," *Journal of Nanomaterials,* vol. 2013, p. 43, 2013.
[20] M. Distaso, M. Mačković, E. Spiecker, and W. Peukert, "Early Stages of Oriented Attachment: Formation of Twin ZnO Nanorods under Microwave Irradiation," *Chemistry-A European Journal,* vol. 18, pp. 13265-13268, 2012.
[21] A. Uzum, T. Ashikaga, T. Noguchi, H. Kanda, T. Matsuoka, T. Nakanishi, *et al.*, "Water Soluble Aluminum Paste Using Polyvinyl Alcohol for Silicon Solar Cells," *International Journal of Photoenergy,* vol. 2015, 2015.
[22] X. Hu, Y. Masuda, T. Ohji, and K. Kato, "Polyethylenimine-guided self-twin zinc oxide nanoarray assemblies," *Crystal Growth and Design,* vol. 9, pp. 3598-3602, 2009.
[23] R. Nosrati, A. Olad, and R. Maramifar, "Degradation of ampicillin antibiotic in aqueous solution by ZnO/polyaniline nanocomposite as photocatalyst under sunlight irradiation," *Environmental Science and Pollution Research,* vol. 19, pp. 2291-2299, 2012.
[24] P. Prepelita, R. Medianu, B. Sbarcea, F. Garoi, and M. Filipescu, "The influence of using different substrates on the structural and optical characteristics of ZnO thin films," *Applied surface science,* vol. 256, pp. 1807-1811, 2010.
[25] H. S. Al-Salman and M. Abdullah, "Structural, optical, and electrical properties of Schottky diodes based on undoped and cobalt-doped ZnO nanorods prepared by RF-magnetron sputtering," *Materials Science and Engineering: B,* vol. 178, pp. 1048-1056, 2013.
[26] S. B. Kondawar, P. T. Patil, and S. P. Agrawal, "Chemical vapour sensing properties of electrospun nanofibers of polyaniline/ZnO nanocomposites," *Advanced Material Letters,* vol. 5, pp. 389-395, 2014.
[27] D. Sharma, B. Kaith, and J. Rajput, "Single Step In Situ Synthesis and Optical Properties of Polyaniline/ZnO Nanocomposites," *The Scientific World Journal,* vol. 2014, 2014.
[28] L. Geng, Y. Zhao, X. Huang, S. Wang, S. Zhang, and S. Wu, "Characterization and gas sensitivity study of polyaniline/SnO2 hybrid material prepared by hydrothermal route," *Sensors and Actuators B: Chemical,* vol. 120, pp. 568-572, 2007.
[29] S. Ameen, M. S. Akhtar, Y. S. Kim, O.-B. Yang, and H.-S. Shin, "An effective nanocomposite of polyaniline and ZnO: preparation, characterizations, and its photocatalytic activity," *Colloid and Polymer Science,* vol. 289, pp. 415-421, 2011.
[30] H.-G. Li, G. Wu, M.-M. Shi, L.-G. Yang, H.-Z. Chen, and M. Wang, "ZnO/poly (9, 9-dihexylfluorene) based inorganic/organic hybrid ultraviolet photodetector," *Applied Physics Letters,* vol. 93, p. 153309, 2008.
[31] L. Wang, D. Zhao, Z. Su, F. Fang, B. Li, Z. Zhang, *et al.*, "High spectrum selectivity organic/inorganic hybrid visible-blind ultraviolet photodetector based on ZnO nanorods," *Organic Electronics,* vol. 11, pp. 1318-1322, 2010.
[32] M. R. Esmaeili-Rad and S. Salahuddin, "High performance molybdenum disulfide amorphous silicon heterojunction photodetector," *Scientific Reports,* vol. 3, 2013.

IOP Conf. Series: Materials Science and Engineering **757** (2020) 012075 doi:10.1088/1757-899X/757/1/012075

Comparison Study of Some Mechanical Properties of Epoxy Composites usinguncarbonized Orange peel and Carbonized Orange peel

H K Hameed*, W H. Jassim
Department of Physics, College of Education for Pure Science IbnAlHaitham, University of Baghdad, Baghdad, Iraq

*Corresponding author email:**roaahashim@yahoo.com**

Abstract.Inthis investigation the epoxy was reinforced by orange peel and carbonized orange peel particles with percentages (5%, 10%, 15% and 20% by weight).Mechanical tests like:Tensile, flexural,Hardness, impact and compression were carried out on these natural epoxy composites. The results showed the tensile strength have a higher value by adding (15% by weight)of orange peel and carbonized orange peel particles to epoxy,while the value (10% by weight) of addition is suitable to get improvement in the other mechanical properties as flexural strength, Hardness, impact and compressive strength .The epoxy / carbonized orange peel powder have the best valuesin all mechanical properties than that of epoxy /orange peel powder.

Keywords: Natural fiber, Orange peel, carbonized orange peel, Carbonized, Mechanical properties.

1.Introduction

Natural fiber-reinforced compound composite encompasses a niceinterest among material scientistsassociated engineers in recent years because of the necessity for developing an environmentally friendly material**[1]**When compound materials strengthened with artificial fibers like glass and carbon give benefits of high stiffness and strength to weight magnitude relation as compared tostandardconstruction materials[2-4] natural filler materials appropriate alternatives to artificial materials for reinforcing polymers like epoxy because oftheir environmental friend lines and cost- effectiveness. Many analysis efforts are place to checkthe effectiveness of natural fiber based materials on the mechanical behavior of epoxy composites, focusing primarily on fibers and their weight percent'inside the composites.

The researchers Senthil J and Madan Raj. P in (2015) were used the eggshell powder with (15 and 20 wt %)to resin system like epoxies and polyesters, they found that addition of eggshell powder to the polymer leads to increase in hardness and flexural strength [5].In the same year the researcherWidadHamdi J. used both white and Broun chicken eggshell with (1,2,3,4 and 5%) volume fraction , she found that the value 3% of addition is the best to improvement in the hardness[6].

Orange is one among the foremostcommon fruits within the world. The waste generated from the fruits has to be place into helpfuluse. Some of the workers like Raghad[7] studied the mechanical properties of polyester/orange peel composite. The weight fraction of orange peel (2, 4, 6, 8 and10%)was added to polyester using hand –lay technique, she found that the utmost hardness, tensile and flexural isobtainingfor the composite with the 10% reinforced orange peel.

The present investigation has been used the orange peel and carbonized orange peel with different weight fraction of loading to epoxy matrix to improvement some mechanical properties like: flexural strength, Hardness, impact and compressive strength.

2. Experimental work

2.1Preparation of composites

The orange peels were collected regionally and were preserved for five days to get rid of the wetness from the peels. Theun carbonized orange peel and carbonizedorange peel was grinded into fine powder with (150μm) as shown in figure1,.A glass mold of 100x10x4mm as shown infigure2,was used for casting the composite sheet. For fast and simpleremoval of the specimen a Teflon sheet was connected to the inner and outer surfaces of the mold. The weight percent of orange peel powder for orange peel and carbonized (5,10, 15and 20wt %), were mixed with the matrix material consisting of epoxy glue andwithin the magnitude relation of2:1. Then the resin was mixed uniformly with hardener, the mixture was poured fastidiously into the molds and left in the mold for 72 hours, care was taken to avoid formation of air bubbles throughout running.

A hardness test (Shore D) the specimens cut according to (ASTM D-2242). Impact tester wont to confirmthe energy needed to interrupters rupture samples by victimization testing methodology for Charpy with sample dimensions (100 x 10 x 4 mm) according to international system (ISO-179).Compression tester carried out according to the test specification of ASTM (D695-12) specimens with a dimension of (4 x 4 x 8) mm was used in this work.

The tensile testing machine model H50KT factory-made by (Tinius Olsen/UK). The samples dimensions (100 x 10 x 4 mm) in accordance with ASTM-D638.Flexural tester using instrument model H50KT factory-madeby (Tinius Olsen/UK). The samples dimensions (100 x 10 x 4 mm) in accordancewith ASTM-D790 were used for the check,as shown in figure (2-b,c).

Figure1. Orange peel samples (a) orange peel (b) orange peel powder
(c) Carbonized (d)Carbonizedorange peel powder

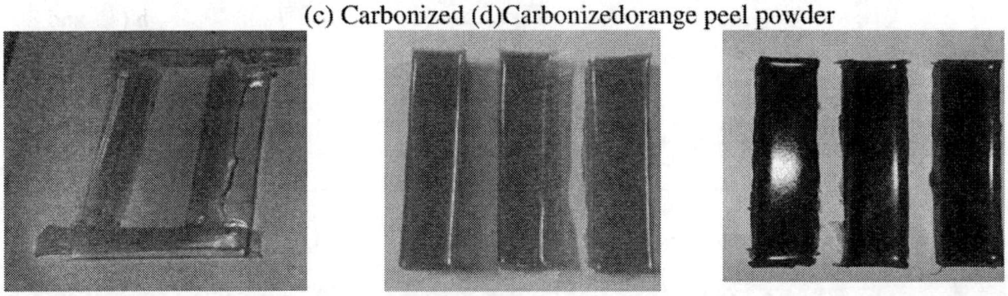

a b c

Figure2.a. Glass mold b. epoxy/orange peel specimens ,c. epoxy/ Carbonized orange peelspecimens

3.Result and Discussion

Figure 3,shows that the (15wt %) of orange peel and carbonized orange peel powder is the best ratio of addition to epoxy to improve the tensile strength of epoxy composite.The epoxy/ carbonized orange peel powder shows the a higher values of tensile strength by comparison with epoxy/ orange peel powder composites using the ratio of loading (10,15 and 20wt%), this may be due to that the carbonized orange peel will increasing the mechanical tensile properties ,which is agood agreement with [8].

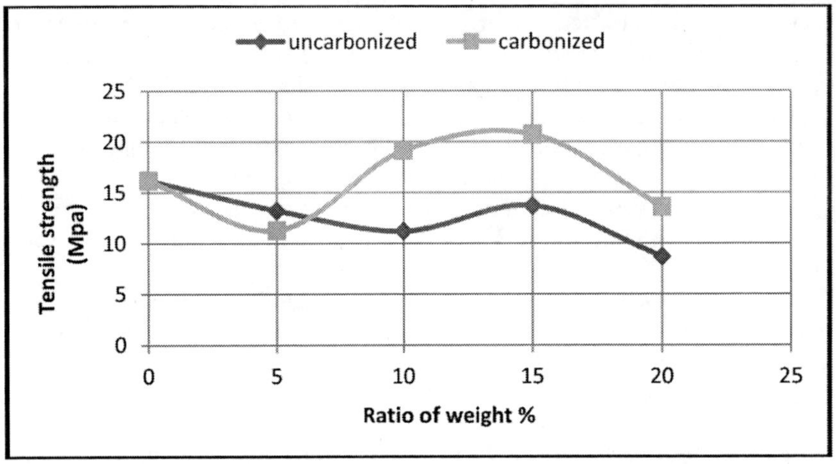

Figure3. Variation of tensile strength of epoxy reinforced with Wt.% orange peel particles addition

Figure4,shows that epoxy/ carbonized orange peel powder composite have a higher value of flexural strength at (10wt%) of loading but the composite epoxy/ orange peel powder have the lower value ,this because the carbonized particles have a good dispersion in epoxy matrix and also the good bonding between the powder particles and epoxy chains [5,9].

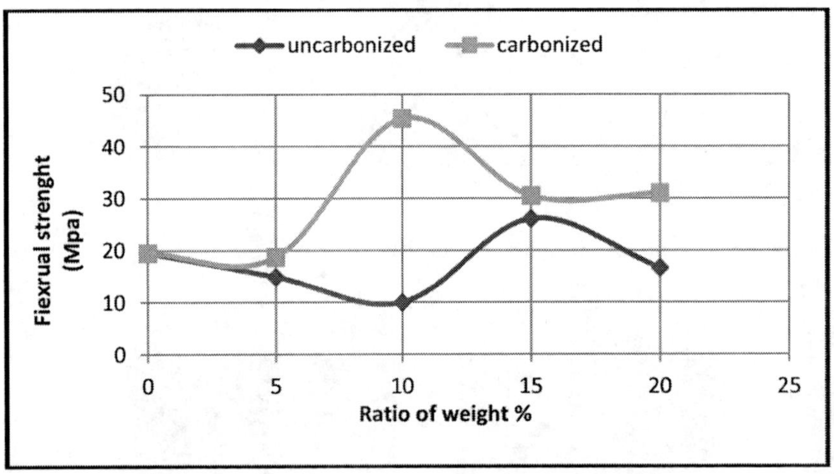

Figure 4.Variation of flexural strength of epoxy reinforced with Wt.% orange peel particles addition

Figure5 and figure 6 shows that the hardness and impact values respectively ,which are increase with increase the loading with orange peel and carbonized orange peel powder reaching to (10wt%) of addition , after this value the hardness and impact will decrease for both kind of composites . The epoxy / carbon orange peel powder have a higher values of hardness and impact than that of epoxy / orange peel powder, this because the carbonized particles play a good role in strengthening the epoxy matrix [5].

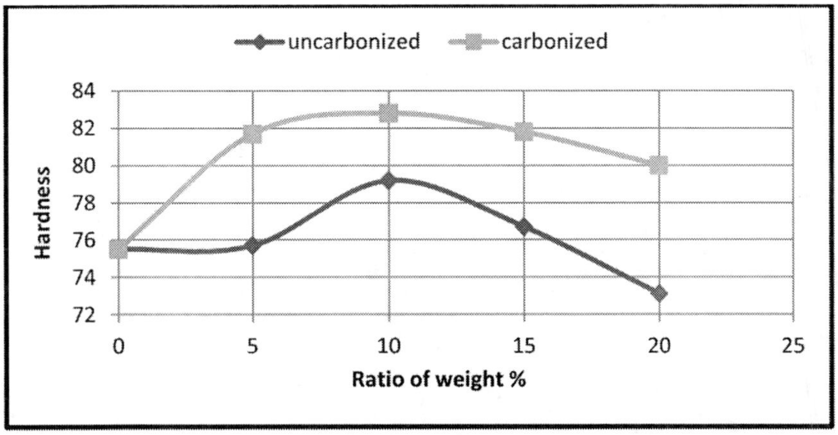

Figure5.Variance the hardness values of epoxy composites with ratio Wt.%

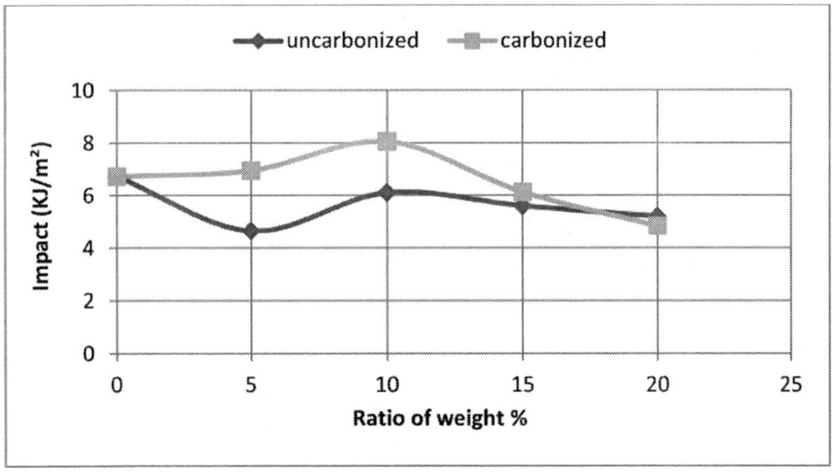

Figure 6.Variance the impact values of epoxy composites with ratio Wt.%.

Compressive testcould be a basic style of check accustomedcharacterize materials. Static compression tests apply a compressive loadtill failure [10, 11].Figures (7-10) shows that epoxy/ carbonized orange peel powder shows higher bearing to load than that of neat epoxy and of the epoxy/ orange peel powder this may be due to the uniform dispersion of micro-sized fillercarbonized particles in the composites. These micro-sized particles have greater interaction with matrix, which directly increased the load transferring behavior of the epoxy resin; this result is similar to report by [12].

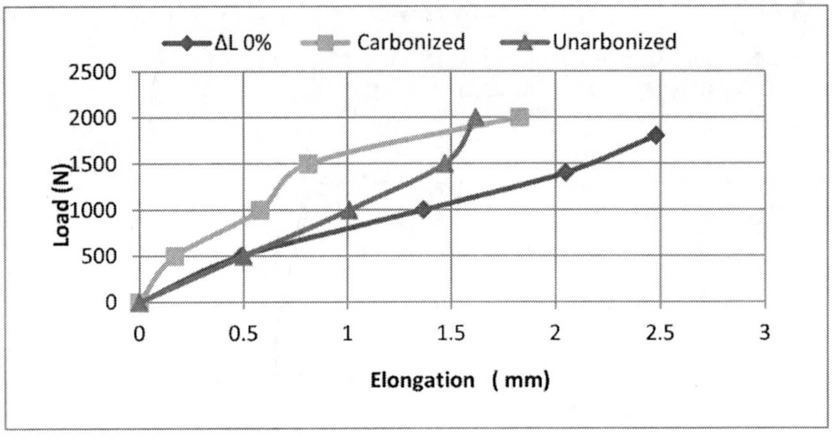

Figure 7. Variation of load of neat epoxy and epoxy/ 5% Wt. orange peel addition Composites with elongation

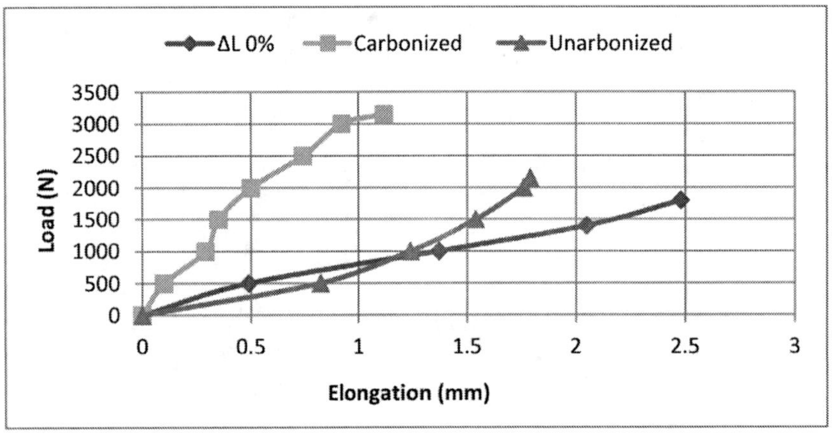

Figure 8.Variation of load of neat epoxy and epoxy/ 10% Wt. orange peel addition Composites with elongation

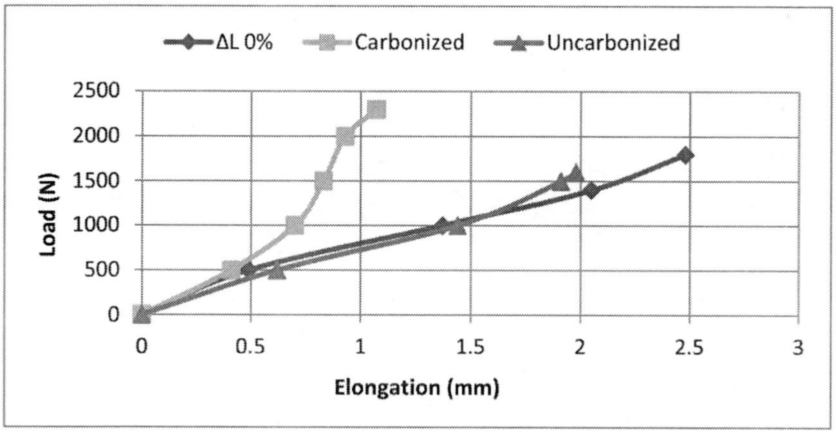

Figure 9. Variation of load of neat epoxy and epoxy/ 15% Wt. orange peel addition Composites with elongation

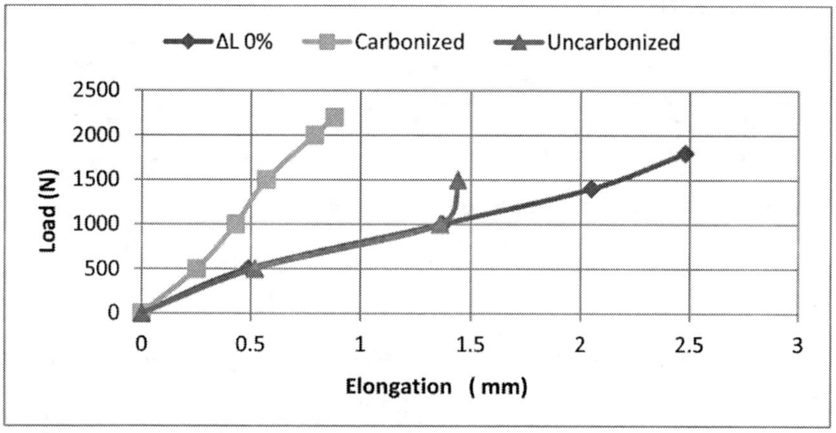

Figure 10.Variation of load of neat epoxy and epoxy/ 20% Wt. orange peel addition Composites with elongation .

Figure11, shows that the epoxy with (10Wt.%)of carbonized orange peel powder have the higher values of compressive strength, this may be due to a good bonding between the particles powder and epoxy matrix at this value of loading, where the orange peel powder act as a bridge between the epoxy chains, this result is a good agreement with [5].

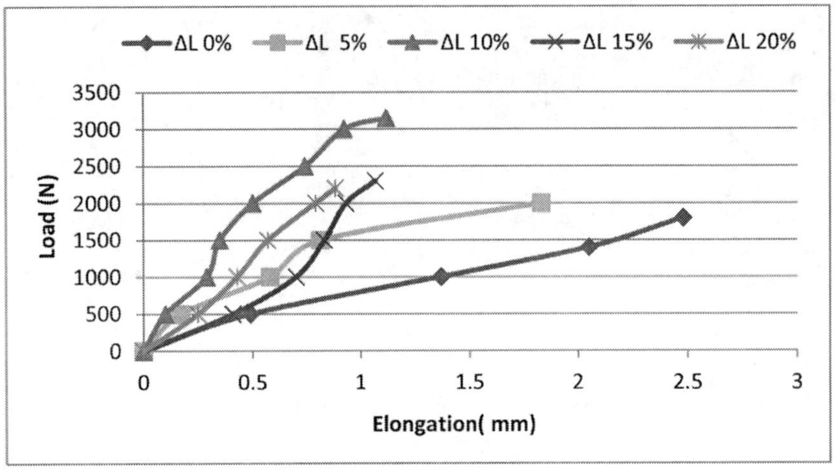

Figure11. Variation in load with elongation for epoxy /carbonized orange peel powder Composites .

4.Conclusions

The addition of orange peel powder to epoxy will improve the mechanical properties, the best value of addition is at (15Wt. %)for tensile test, but it is found (10Wt. %)for the other mechanical tests like flexural, Hardness, impact and compression strength. The carbonized orange peel powder act as reinforcing centers to improve the mechanical properties of neat epoxy.

5. Reference

[1] Hann H. S, 2014, Ibn Al-HaithamJ. for Pure and Appl. Sci., Vol. 27,N. 1.
[2] Katla..P .K ,Chetty.N., 2016, IOSR Journal of Mechanical and Civil Engineering.
[3] Ahmed .M .A., Sameer.H.N . and Mohammed .H.M, 2016, Int. J. Chem. V.14, N.4.
[4] Shashir S.K, Apsarraj A.D, Shashidhar A.L, Basavaroodh A.B, 2017, Journal of Emerging Technologies and Innovative Research (JETIR),V.4, N.07.
[5] Senthil J. ,MadanRaj.P, 2015, International Journal on Mechnical Engineering and Robtics (IJMER)V.3.

[6] Widad H. J, 2015,Advances in physics Theories and applications , Vol. 44.

[7] PRAVEEN. K, 2015, International Journal of Engineering Sciences & Research Technology, V. 4, N.1.

[8]Sukhdeep .S, Dharmpal. D, Lakshya .A,V.K.Gupta, 2014,Slsevier, V.6 .

[9] Nikhil, B.Singh, L. Aggarwal, 2017, International Journal of Engineering Research & Technology (IJERT), V.6, N.11.
[10] Jauhar. F, 2016, info teknik, V.17, N.1.

[11] F.A.Hummdai , Engineering and Technology Journal, V.30 N.12,(2012).

[12] B.D. Jeevan, K. V. Arun, S. J. Prashanth, 2016, International Journal of Engineering Science and Computing,V. 6, N. 6.

The Role of New Technologies for Envelops of Buildings In Improving The Internal Environment Of University Buildings

Luma Mohammed yehia , Maysaa Moffeq Alobaidi , Ebtesam Sameer

Department of Architecture Engineering, College of Engineering, University of Mosul, Mosul/Iraq

Luma.m.yahya @uomosul.edu.iq, maysaa.moffeq @uomosul.edu.iq

Abstract. The impact of the progress in the field of technology on the development of engineering sciences, including architecture and provided the modern technologies of the architect designed enormous possibilities in the formation and design of buildings and thus improve the environmental performance of those buildings, providing them with comfort for their occupants, and recent studies have proved the importance of the quality of the internal environment on the scientific performance in the educational buildings from here The importance of research in the role and importance of modern techniques in improving the internal environment of educational buildings, including university, where the research assumes the existence of a range of techniques employed by the designer in the design of building covers to improve internal environmental performance and The research problem was based on a discussion of the factors influencing the quality of the internal environment and a series of studies that dealt with the design of the modern buildings and their environmental impact, in order to derive the main variables and apply them in the study The process, which analyzed a range of contemporary university buildings, led to conclusions and recommendation .

Keywords: Technologies, Envelops, Environment, University Buildings.

1.Introduction.

The educational buildings of the English language colleges in the world are designed to provide educational services in their ages, because they contain a large number of occupants and for long periods that require the greatest comfort for their occupants. And environmental comfort in buildings for the first degree in heat and audio and visual comfort [1]. Therefore, the good design of the building cortex improves students' performance and increases the level of thermal, audio and visual comfort [2], The climatic environment is the most influential force on the internal environment of the building, especially when the environmental control devices are limited, and the techniques are weak so must adapt to them, and the human body to expend energy to adapt to environmental conditions and when the exchange of energy is a little environment is within the limits of rest [3]. Thermal comfort means that the occupants of the building do not feel the need to adapt space during their stay. Visual comfort means the quality of light that makes visual visual tasks (reading, writing, etc.) easier. Audio comfort enables the occupants of the building (teachers and students) to hear each other easily and reduce the noise that affects the internal environment [4].

2. Previous literature.

1.1. Study Kamuna and Raouf("Integration Of The Work Of The Building As An Employee System Of Advanced Technology In The Face Of Climatic Conditions ForeignAffairs"):

The study examined the development of societies and their adaptability to the necessary and continuous changes to develop Its response to the surrounding climatic environment, and the use of modern technology in the construction of buildings characterized by enveloping High dynamic and responsive to the surrounding climatic conditions and adapt the design (form and function) to work As a whole integrated to provide comfort to the occupant and the style preserves the versatility and

IOP Conf. Series: Materials Science and Engineering **757** (2020) 012076 doi:10.1088/1757-899X/757/1/012076

provides. The study reviewed Arab experiences In this field for the purpose of drawing ideas and employing them in buildings in line with the climate of Iraq [5].

1.2. Study Al – Essawi ("The effect of the exterior design of the building on the thermal acquisition and thermal comfort of the users"):

The study examined the concept of the effect of the outer shell of the building on thermal acquisition as a means of understanding the conditions And the effect on the thermal behavior of the building, finding the best solutions for the outer shell Which directly helps to achieve the required thermal comfort for building users has been studied evolution Architecture through the ages has the most important influences on buildings, and the concept of climate design has been confirmed by research on The need to adopt the principle of climate design for the outer shell of the building through climate analysis, study Climatic aspects of the facades according to solar radiation Thermal loads and the exchange process by conductivity [6].

1.3. Study Halawa and Others("A review on energy conscious designs of building façades in hot and humid climates: Lessons for (and from) Kuala Lumpur and Darwin")):

This study argues that there is a lack of a systematic and comprehensive analysis of the available literature regarding the energy and thermal performance of building façades based on the various possible design and technical configurations, especially in hot and humid climates. Important decisions should be made by architects and engineers during the early design stages of buildings with viewpoints to the ultimate impacts of building physics on the overall energy performance and indoor comfort conditions of buildings [7].

1.4. Study Barnaś ("Double-Skin Façades – The Shaping Of Modern Elevations – Technology And Materials"):

The study showed the importance and role of design in balancing the challenges of maintaining the correct internal atmosphere of the building and providing the highest possible energy efficiency. Considering that the facade of the building is an integral part of it, and that the use of modern technologies in the design of the facades [8]. All studies confirmed the importance of employing technology in architecture to reach the best performance of buildings,see Figure 1:

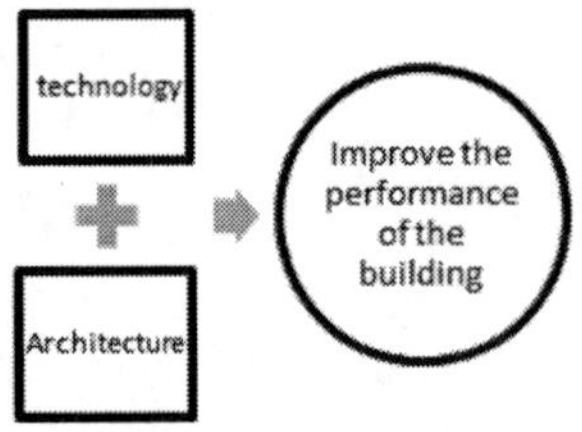

Figure 1. Employing technology in architecture to reach the best performance of buildings(researchers)

2. Building envelope.

It is a building material that surrounds the internal spaces of the building with certain building techniques and is the boundary between the environment The interior of the building and the external environment, it is also the most important element to get a suitable environment for the convenience of users The buildings and the protection of the building.

3.1. Roof

The most exposed to climate variables in the building's envelope , for example, exposed to the sun Throughout the day, unlike facades that are exposed to hours only during the day, and there are techniques used Treatment of ceilings:

1.Thermal Insulation Roofs: There are many insulating materials that can be used for insulation Roofs Low density materials are used to achieve the best thermal insulation.
2.The use of a reflective material for the sun: a coating of ceilings with light colors, reflective materials are considered The heat is effective in limited areas.
3.Insulation by providing air space (secondary ceilings): is to leave an air vacuum between the structural roof of the building And the secondary roof to reduce heat transfer inside the buildings as a good thermal insulation.
4.The use of the roof consists of two separate layers: leading to the movement of air between them where it works The upper layer of the umbrella leads to lower air temperature below and thus the heat is not permeable to The inner layer of the roof.
5.Shading the roof by different means, such as the surface gardens: the soil is a heat insulating material in addition to that Plants shade the ceiling and cool the air that touches it and the ceiling is built in a special way so that the weight bearing And prevents leakage of water [9].
6.The use of tilted or curved ceilings, such as vaulted or cylindrical ceilings, is part of the ceiling Often shaded .

3.2. Facades

The outer face of the building or the group of waves that we see according to the optical axis of the eye The Observer [10]. High technical interfaces:

1. Kinetic Facades These interfaces are flexible and able to change shape, configuration and routing to respond automatically For surrounding environmental factors and their changes this includes temperature, humidity, wind ... etc, (one building Ocean in Korea) so that the solar blades made of fibrous plates are used to be Strong and flexible allow light to pass through[11]

2. Using a smart and moving shading system (eg Abu Dhabi Towers Building), providing heat reduction Acquired by up to 50% which helps to reduce energy consumption up to 40% [11]

3. Solar Facades (green pix project in China) The ability to generate electricity provides up to 40% energy, thus reducing thermal bridges and providing comfort Interior [12].

4. Use of assembled facades: Trombe Well is one of the most common walls Simply consists of a single or double glass layer, an air gap and a thick wall with a single layer Dark absorbs the radiation and converts it into long wavelengths of radiation and stores them in the wall and moves Heat in the space that exists into the interior space [13].

5. The use of double skin facades: a system consisting of two shells Glass is placed in a way that allows the movement of air in the middle vacuum to be natural or mechanical With the help of fans (eg the swissre tower project). The width of the middle corridor ranges from 20 cm to 2 m. The double facade can secure the hardness of the building Reduce heat loss in winter. It may reduce energy consumption to 50% [14].

6. The use of corridor corridors (eg the Stadthur building in Düsseldorf) This interface is airflow on each floor separately where the vertical spacing is placed between the spaces to reduce From the

IOP Conf. Series: Materials Science and Engineering **757** (2020) 012076 doi:10.1088/1757-899X/757/1/012076

spread of fire and sounds that can spread through the vacuum between the inner and outer layer, The corridor is so wide that it can be used as a service platform.

7. The use of adaptive interfaces: It uses two types of static systems and mobile interfaces Mechanically responsive to surrounding environmental conditions to provide a comfortable internal environment for building users.

4. Construction Material.

The angle of rotation of the horizontal axis (330 °) of solar panel, it is moving from the extreme right to the far left to trace the motion of the sun from east to west. The angle of the vertical axis is (from 0 to 90 degrees) with any horizon moving from a level parallel to the horizon to a vertical level with the horizon to track the movement of the sun from north to south .

1-Using a Smart Generator: Materials that have the ability to receive the information caused by the catalysts External and internal environment and take appropriate response .These materials are either [15](Variable properties: substances that respond by changing one of their properties (chemical, thermal , Mechanical and magnetic) and are either variable color or streamlined) [16].It contains energy-transforming materials: substances that convert energy from one form to another directly and indirectly [16],and be either (converted to light or thermal or photoelectric or light or Electric or magnetic) Types of smart materials ;(Concrete,Carbon fiber concrete ,Light concrete, Transparent concrete, Aerial concrete ,porous concrete. , Bricks Intelligent, Cement reduces pollution)[16][18].

2-Glass: Glass develops to become an advanced construction material and plays an important role in building ,like(Color-born glass, Self-cleaning glass)[19][20]

3- Aerogel material: A highly transparent gel material, the majority of which is air-like and is glass-like Of heat insulation materials are characterized by their light weight It is not possible Flammable and used in windows and transparent walls[17].

4- Photonics: Materials that change their color properties by exposure to light and are used to paint facades. In order to interact with the movement of the sun and light dominated [7].

5- Aluminum: Used since long periods and used as a cortex (Lawrence London Center) or cladding of facades Theology in Cambridge) .

6- Plastic: It is an organic mixture with the ability to form and be on the following types: (Poly vinyl chloride .,PVC - U (Basquepool Stadium London 2012).,Poly Tetra Fluoro Ethylene (roof of Jeddah airport for pilgrims).

7- Ceramic Material The materials improve the internal environment by:(Control the inside of the mini.,Control of natural lighting entering the building.,Refreshing thermal comfort by,reducing heat loss and thermal gain. ,Humidity control. ,Control the natural ventilation of the building to provide good air for the interior spaces.

5. Local materials. .

Traditional architecture is characterized by the use of materials available in the surrounding environment such as bricks and vaulting Wood, marble and tiles. These materials (bricks and stone) are of high thermal capacity The ability to store heat for long hours during the day and then return it to

IOP Conf. Series: Materials Science and Engineering **757** (2020) 012076 doi:10.1088/1757-899X/757/1/012076

the outer spaces during Night (after the absence of energy source) and these materials are characterized by thickness and thickness of the wall in the architecture The traditional between (0 , 36 - 0 , 75) according to its location and has the advantage of having a large thermal delay time ranging from (12 - 15) hours. These materials are environmentally friendly because they do not emit any toxic gases or effects Negative for human health such as dyes and chemical decomposition of certain substances, as it is recyclable materials And use again.

6. The use of traditional techniques and elements.

Traditional architecture was to meet human needs and adapt them to the environment Surrounding it is an environmentally efficient architecture so many modern studies have emerged demanding a return to use Elements used by traditional architecture in adapting to the climate environment These elements are:

1.Using the inner courtyard: a medium or lateral space open to the sky helps to provide lighting Natural ventilation and the use of renewable energy .

2. The picker or pedicure: A channel that rises above the building and contains at its summit an opening corresponding to the direction of the wind to pull the air passing over the building (which is cooler than the internal air) and a push into the building.

3.The Badkir is an open air channel from the four sides and inside it, two vertical poles along the channel to pull the air in any direction from it.Technology can be used in the design of the building's envelope .

This is evidenced by the existence of a set of technical, formal and technological techniques that have been used in the design of the building's envelope to improve the internal environmental performance of the building as shown in Table1.

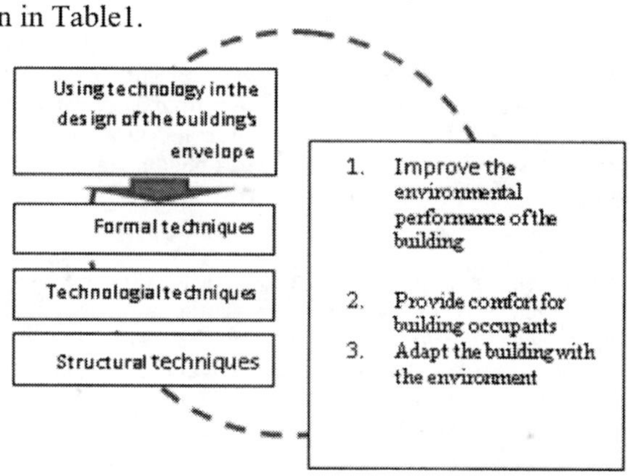

Figure 2. Impact Of Using Technology In The Design Of The Building's Envelope(researchers)

Table 1. techniques using in design of Envelope (researchers)

Technics used	**Envelope of building**			
	Elevation		roof	
Formal technics	Type of Shapes	Constance flexibility traditional	Type of Shapes	flate slop curve
	Opens	Variety Unity size		

IOP Conf. Series: Materials Science and Engineering **757** (2020) 012076 doi:10.1088/1757-899X/757/1/012076

		shape		
Technological Technics		smart sustainability Traditional		smart sustainability Traditional
Construction Technics	Structure	Double skin Trombe wall	Structure	
	Materials	Corridor Change characters Adaptive Transformer of energy Traditional	materials	

7.Practical study.

The paper in his practical study select a group of different university buildings As shown in Table 2, and the research examined the vocabulary presented by the theoretical framework, which was ascertained in Tables 3,4

Table 2. groups of university projects

projects	Subhead
A  Tsinghua University (2006)	The building is designed as a "demonstration" of the possibility of reducing carbon dioxide emissions in China. The design incorporates negative and active strategies to control the external environment in order to improve internal environmental conditions, and the terraced gardens are designed from shaded panels that produce energy for the building.
B  American University in Cairo	Environmental treatments suitable for desert areas, such as directing the openings towards the northern wind and the use of internal pavements, gardens and water that contribute to air cooling, use a vegetative barrier to protect from dusty wind, use sandstone to construct 80% of external walls, Heat, energy-efficient materials, and all American university buildings generally have all the features of a smart building (automation, responsiveness, compatibility with the environment and sustainability).
C  King Abdullah University of Science and Technology (2009)	Project designers have inspired Bedouin tents to create a huge roof system that runs across the building blocks on the campus to prevent the sun from building facades and pedestrians. This helps to facilitate natural ventilation. And the design of wind towers that encourage air flow in the pedestrian paths, the wind towers used are solar wind tower also known as solar chimnies, the design of shading devices on the campus using Mashrabia, and the University employed many clean energy technologies, and the use of natural daylight with the presence of sensors Automatic light in addition to solar thermal arrays to heat local water. And the use of building materials that reduce any environmental effects
D  RMIT Design Hub University (2012)	The front of the building is envelope ed with disks that can rotate toward the sun. The building also has a number of ESD features and includes strategies for water, waste and recycling management. The exterior of the building includes automatic solar shading which reduces energy consumption and lighting is controlled by a BMS device to reduce the need for artificial lighting

E	 NationalTaiwan University (2013)	The rectangular building has large openings with two or three floors. These openings reduce the size of the building block and connect it to nature. The facades of the building have balconies and concrete concrete that provide shadows for the building. The roof units took free formations resembling lotus blossoms touching at the edges. Natural, the large glass facades of the library were surrounded by water bodies that soften the heat radiation of the earth around it.
F	 SDU Campus Kolding University (2014)	The building envelope is equipped with a dynamic solar shade, which adapts to specific climatic conditions and user patterns and provides optimal daylight with sensors that continuously measure light and heat levels and are mechanically controlled
G	 Emerson college (2014)	The exterior of the building has a traditional appearance and is equipped with screens, media links, audio, integrating lighting with the structure, making the upper deck of the building a flexible engine for outdoor displays, and transforming the wavy screens into a dynamic visual background. In response to local weather conditions, the automatic parachute system opens and closes horizontal panels of outdoor curtain walls installed on glass to reduce heat absorption while allowing increased daylight and visibility. And the use of emerging renewable energy systems
H	 University of Southern – Denmark (SDU) (2015)	The building was designed like a glass house envelope ed with a glass-paneled shade. The outer layer of the building is composed of two layers. The first is made of transparent glass panels, envelope ed by a second layer consisting of high-performance special white concrete slabs. These concrete panels consist of circular openings of seven different sizes, protect the building from direct radiation and allow natural and air lighting . And the use of low impact materials on the environment.
I	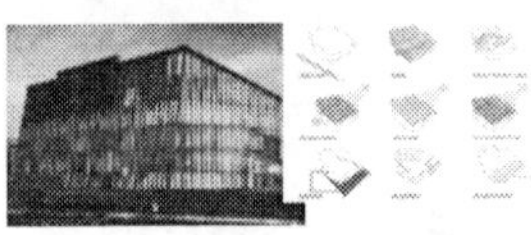 Energy Academy Europe University (2016)	Sustainability has been the rule of design from the very first moment, a low-tech approach has been chosen to take advantage of the four natural elements to their fullest potential. The land is used for cooling or pre-cooling fresh air. The solar chimney at the top of the surface leads to normal airflow through the structure,. The innovative 3-D roof also allows for maximum daylight and energy harvesting
J	 Brick Passive Designed University(2017)	Use the porous brick screens in front of the glass façade to allow for an appropriate amount of natural light, and put in the front of the building details of the local environment, air outlet is like a chimney to allow the movement of air. The adoption of the designer on the quality of materials and the location of screens and openings and size was intended to filter and distribute light and air to Appropriate level.
K	 Silpakorn Universit (2017)	The building envelope is designed to allow clean air in built space and allows natural light to enter the interior spaces, thus reducing the energy consumption and filtering of the lighting entering the building and thus getting better performance of the lighting. In the process of construction, it was built on prefabricated parts and construction techniques to speed up the construction process
L	 Life Sciences Building for the University of Washington (2018)	The building's cover is composed of vertical glass solar fins from the outside and is expected to generate enough electricity to light more than 12,400 square feet of offices throughout the year. Other sustainable features include operable windows to cool natural ventilation, cooled beams and waves, a water reclamation system to irrigate the greenhouse, radioactive floors and solar panels on the roof [31].

Table 3. techniques using in design of Elavation of building(researchers)

Technics	Elevation			Projects											
				A	B	C	D	E	F	G	H	I	J	K	L
Formal	Types of shapes	Constance		1				1		1				1	
		Flexibility					1		1	1	1	1			1
		traditional			1	1				1			1		
	opens	Varity		1					1						
		Unity	Size				1	1				1			1
			shape				1	1	1	1	1	1	1	1	
Techno logy	Smart			1		1	1		1	1	1				
	Sustainablity			1	1	1				1		1	1		1
	Traditional			1	1	1			1				1		
constructional	Structure	Double skin					1		1	1	1		1		1
		Thromb wall													
		Corrider													
	Materials	Change charechter							1	1					
		Adaptive					1		1	1	1	1		1	
		Transe formation		1		1	1								1
		Traditional			1								1		

Table 4. techniques using in design of roof of building(researcher)

Technics	Roofs			Projects											
				A	B	C	D	E	F	G	H	I	J	K	L
Formal	Types of shapes	flat		1	1	1	1	1	1	1	1				1
		slop										1			
		curve			1										
	opens	Varity													
		Unity	Size												
			shape												
Technol ogy	Smart					1						1			1
	Sustainablity			1				1		1		1	1		1
	Traditional				1				1		1				
constructional	Structure	One layer slab		1	1	1	1	1			1		1		1
		two layer slab							1			1		1	
	Materials	insulation					1	1		1	1	1	1	1	1
		Translation													
		photovolic				1						1			1

8. Discussion of results.

The results of applying the effect of the use of modern technologies in the envelops of university buildings and for the projects elected in the practical study, which is shown in Table 5.

Table 5. the effect of the use of modern technologies in the envelops of university(researcher)

Effect of using technices on envelops	Projects											
	A	**B**	**C**	**D**	**E**	**F**	**G**	**H**	**I**	**J**	**K**	**L**
Reduce convection	1	1	1	1	1	1		1		1	1	1
Provide natural lighting	1	1	1	1	1	1	1	1	1	1	1	1
Control of ventilation	1	1	1			1	1	1	1		1	1
Noise control							1					
Reduce environmental pollution	1			1				1				1
Energy saving	1	1	1	1	1	1	1	1	1	1	1	1

The proportion of flexible forms used by elected projects 54.5% the proportion of traditional and fixed forms by 36% At the level of openings, the percentage of the unit in the form was 72% and the percentage of the unit in size was 36%. The percentage of diversity in the use of openings was 18% of the total projects selected,The use of smart technologies was 54.5%, while the use of sustainability techniques was 63.6%, while the traditional was 45.4% of the total projects,At the level of structural structure, the ratio of double skin use was 54.5%, Projects used variable building materials by 18.1%, while adaptive materials were 54.5% and energy converters 36.3%, while traditional materials were 18.1%. and Results on the level of roof: flat shape were used by 81.1%, while sloping and curved ceilings were 18.1%,The use of smart technologies was 27.3% or sustainable 63.6% and Photovoltaic cells used 27.3%.

9. Conclusions:

The new techniques used in the university buildings to improve the internal environment of these buildings through a set of mechanisms used by the designers in the building's envelope (the façade,roofs) discussed by the paper . The results of the application showed the concentration of designers to employ formal techniques through the use of flexible forms and the adoption of the unit in the formation Openings. Technological techniques included the use of sustainable mechanisms in the design of facades and ceilings, such as the use of solar cells, moving vents and green elements in the roof of buildings, The technology influenced the internal environment of the university buildings through its effect on the heat transfer through the building envelope and by 91.6% while it provided natural lighting for the buildings where 100% of the total buildings were used and thus reduced the amount of energy and its availability. This confirms the trend of designers to achieve environmental sustainability. For their university buildings as an important and necessary direction in the design of university buildings through the techniques of the design of the facades and ceilings of these buildings. The envelpe is the direct part that plays the main role in balancing and providing comfort in the internal environment of occupants of university buildings while designers adopted the use of construction techniques by separating the structural structure of The building side to find a corridor to move the air and the use of smart building materials and adapted to obtain the best environmental performance and improve the internal environment.

IOP Conf. Series: Materials Science and Engineering **757** (2020) 012076 doi:10.1088/1757-899X/757/1/012076

10. References

[1] E Douglas 2010 *Green Schools As High Performance Learning Facilities National Clearinghouse For Educational Facilities*, September

[2] S Olson & C Jenny C 2006 *Sustainable K-12 Schools*, Leonardo Academy Inc ,January 31.

[3] S Alrawi 1998 *Climate Design For Building Sites: Applications For Model Designs Of Schools*, University Of Technology, Faculty Of Engineering, Master Thesis.

[4] C Eley 2001 *High Performance Schools Best Practices Manual*",Chps.

[5] G Kamuna and L Raouf 2011 *Integrating the Work of the Building as an Employee System for Advanced Technology in the Face of External Climate Conditions*, published in Baghdad University.

[6] M El-Essawy 2003 *Effect of the exterior design of the building on the thermal acquisition and thermal comfort of users,* Unpublished Master Thesis, Faculty of Engineering, Cairo University.

[7] E Halawa & others 2017 *A review on energy conscious designs of building façades in hot and humid climates: Lessons for (and from) Kuala Lumpur and Darwin,* Renewable and Sustainable Energy Reviews , http://dx.doi.org/10.1016/j.rser.2017.08.061

[8] J Barnaś *2014 ,Double-Skin Façades – The Shaping Of Modern Elevations – Technology And Materials Technical Transactions*, Architecture,7-A.

[9] The world of electricity and energy *,Design of thermal buildings to rationalize the introduction of electricity*, 2013 (0nline).

[10] A Ahmed , 2012 *Simulation of the technical developments of architectural facades*, Damascus University Journal of Engineering Sciences

[11] F Russell 2014 *Kinetic Architecture: Designs For Active Envelopes* , Images , UK

[12] P Munari and G,Roecher 2012 *Solor Energy Systems In Architecture* ,International Energy Agency, Sweden

[13] P Torcellini,2004 *Trombe Walls In Low-Energy Buildings :* Practical Experiences Nrel , USA.

[14] P Harris 2004 *Duble Skin Facades For Office Building* Land University ,Sweden.

[15] S bohidr And Others 2015 *Smart Materials For Future* , International Journal Of Advance Research In Scipence And Engineering ,IJARSE, Special Issue(02),February.

[16] M Addington.,& D Schodek. 2005 *Smart Materials And New Technologies For The Architecture And Design Professions.*Architectural Press,An Imprint Of Eleservier, Oxford, UK:Linacre House,Jordan Hill.

[17] F Aladdin and others 2015 *Smart and Nanomaterial Building Materials Introduction to Increasing the Efficiency and Integration of Smart Buildings*, Scientific Publishing House and Printing Press at Jazan University, Jazan University Journal, Saudi Arabia.

IOP Conf. Series: Materials Science and Engineering **757** (2020) 012077 doi:10.1088/1757-899X/757/1/012077

Effect Annealing Temperature on the Physical Properties of Cadmium Sulphide thin Films Deposited by using Thermal Evaporation Technique

Jumaa Mohammed[1], Jasim M. Abbas[2], Falah I. Mustafa[3*]
[1,2]University of Anbar, College of Education for Pure Sciences,
[3]Solar Energy Research Center, Renewable Energy Directorate,
[1,2,3] Higher Education and Scientific Research Ministry, Baghdad, Iraq

falah_im@yahoo.com

Abstract. This study measured some of the structural, electrical and optical properties of cadmium sulfide (CdS) thin films with thickness of 150 nm deposited on glass substrate by using thermal evaporation system under vacuum pressure $1x10^{-5}$mbar. X-ray diffraction technology was used to analyses the structural properties of the CdS films. Results showed that the CdS films type cube polycrystalline structure and the preferred growth path was [111].In evaluating DC conductivity, we observed two mechanism of activation energy. In the measuring Hall effect shows that the electrical conductivity of the CdS films is negative carriers (n-type). The result of the (AFM) measurements showed that the grain size and average roughness increased with annealing temperature increased. In optical properties it were found that transmittance 80% at RT that decreases with increasing annealing temperature. It also found that absorption coefficient, extinction coefficient and refraction index increased with increasing annealing temperature. Optical energy gap were found decreased with increasing annealing temperature as (RT, 150 and 300 C°) the results (2.4, 2.38 and 2.32 eV) respectively. The results electrical properties showed that the CdS thin films had two activation energy and Hall effect were found that the mobility increase with increasing annealing temperature.

1. Introduction

Cadmium sulfide (CdS) is one of the most suitable II – VI semiconductor materials for applications in solar cells technology. It is a direct band gap semiconductor having energy of 2.42 eV for single crystalline material at room temperature [1, 2]. It is n-type in electrical conduction and is widely used in the fabrication of electronic devices due to its unique optoelectronic features. Some of the applicable areas where CdS semiconductors have found optimum use fullness are: gas sensors [3–5], thin film field effect transistors [6], photo resistors [7], photo sensors [8–10].The deposition methods and characteristics of cadmium sulfide semiconducting thin films has been received as considerable amount of interest due to their potential application in the area of electronic and opto-electronic devices fabrication such as solar cells [11, 12] thin film FET transistors, Light emitting diodes and photonic devices, CdS is employed to make nanocrystals, optical filters, and all optical switches [13, 14]. CdS can exist in two case of crystalline medications: the hexagonal (wurtzite) phase and the cubic (zincblende) phase [15]. Heat treatment is often used to tune the structure and properties of thin films such as CdS. It is because crystalline quality of the film plays an important role in the utilization of the CdS films for solar cell applications and thermal annealing leads to improve the crystalline quality of the films by the removal of strains, which can lead to the phase transition and thereby changing the band gap [16]. Also, such an annealing leads to phase transition from the metastable cubic phase to a stable hexagonal phase of CdS [17].

2. Experimental procedure

A Thermal evaporation technique was used to prepare CdS samples with high purity(99.999%)on the glass substrates, before deposition, the samples were cleaned with alcohol and distilled water, dried with filtration paper and subjected to dry air current from a blower. Before deposition of the thin film, deposition electrodes from the aluminum material and using appropriate masks on the substrate using (Edward 306, coating unit) device under pressure4*10^{-5} Torr, The distance between the samples and the boat was 20 cm. The thickness required (150 nm) for the prepared CdS films was calculated using the approximate weight method in accordance with the following

$m = 2\pi r^2 \rho t$........... (1)

Where

That m is the mass of the material, r^2 is the distance between the boat(source) and the Sample, ρ is the density of the material and t is the thickness of the film

Then, the samples put under oven with annealing temperature at 150C^0 and 300C^{0}for 2 h.

3. Results and discussion

The results include the x-ray diffraction (XRD) of the CdS thin films, the findings of the electrical measurements (continuous conductivity and the Hall Effect) and the optical properties.

Structural properties

XRD

Studying the X-ray diffraction of the prepared CdS films with a thickness of 150 nm showed that the CdS films exhibit a cube-type polycrystalline structure. Moreover, grain size increases with rising annealing temperature because of the reduction in grain boundaries [18]. . Figures (1) show the X-ray spectrum of CdS thin films.

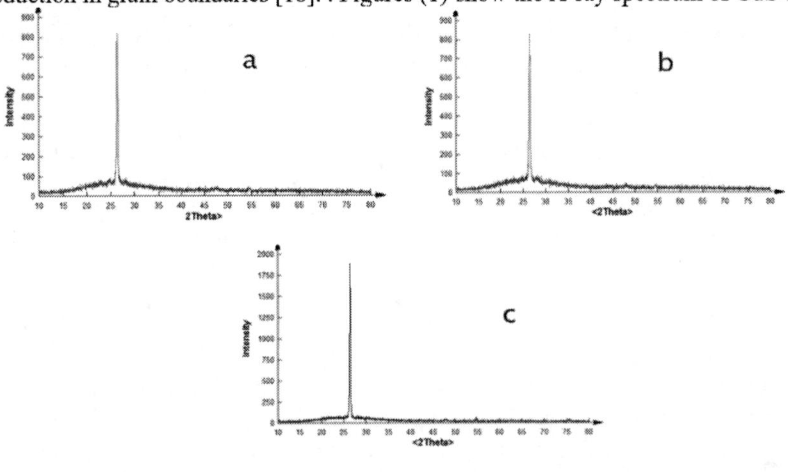

Fig (1): X-Ray spectrum of CdS thin film (a) at RT, (b) annealing at 150C^0 and (c) annealing at 300C^0

Atomic force microscopy (AFM)

The morphology of the surface of thin film must be investigated because of its influence on the efficiency of manufactured solar cells. This test determines the rates of granular size, surface roughness and square root roughness at different annealing temperature, as shown in Table (1) .The images in two and three dimensions are shown in Figures (2). The AFM analysis indicates that grain size, root mean square (RMS), and average roughness increase with annealing temperature. This result is consistent with the study [19].

IOP Conf. Series: Materials Science and Engineering **757** (2020) 012077 doi:10.1088/1757-899X/757/1/012077

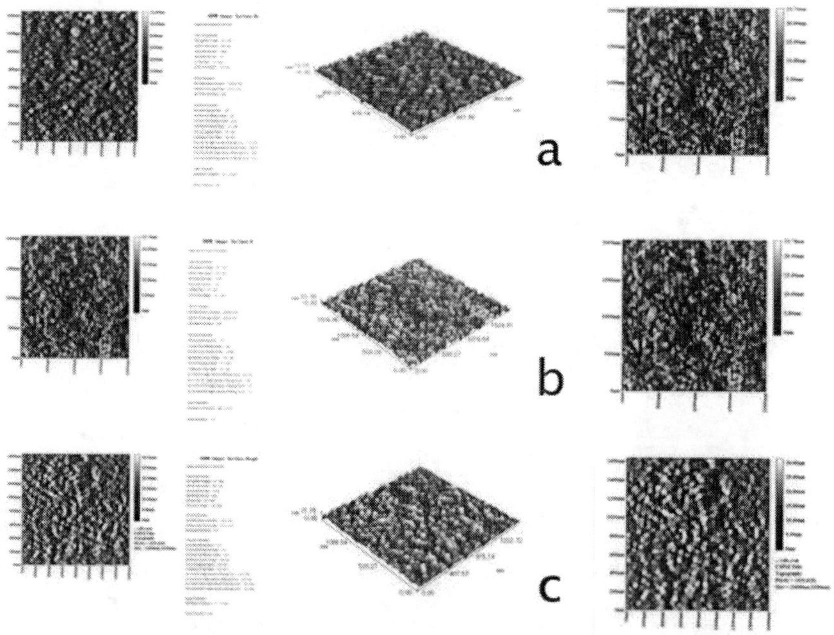

Fig (2): AFM photos CdS thin film (a) at RT, (b) annealing at 150C^0 and (c) annealing 300C^{0}with two dimensions and three dimensions

The table (1) shows the measurements of the atomic force microscope

sample	Annealing Temperature	Average Diameter nm	RMS Roughness nm	Ave Roughness nm
CdS	RT	40.61	2.68	2.22
	150 C	50.83	6.19	5.27
	300 C	60.03	6.95	5.7

Electric properties D.C electric conductivity

The continuous conductivity of CdS films was measured by changing resistance with temperature to obtain information regarding the nature of the conductivity mechanism of the CdS films within the thermal range of (293k-433k). Then, the activation energy was calculated. Figures (3) show the relationship between the $ln\sigma_{d.c}$ of reverse temperature (1000/T) and the increase in continuous electrical conductivity with increasing temperature. This trend is one of the most important characteristics of semiconductors [20]. Evidently, the two types of activation energy corresponds to the results of the X-ray tests, that is, the films exhibit a polycrystalline structure and the activation energy can be calculated using the following equation[21]: The conductivity measurement indicates that continuous conductivity increases with annealing temperature as shown in Table (2)

$\sigma_{d.c}=\sigma_0 \exp(-Ea/kBT)$............................ (2)

IOP Conf. Series: Materials Science and Engineering **757** (2020) 012077 doi:10.1088/1757-899X/757/1/012077

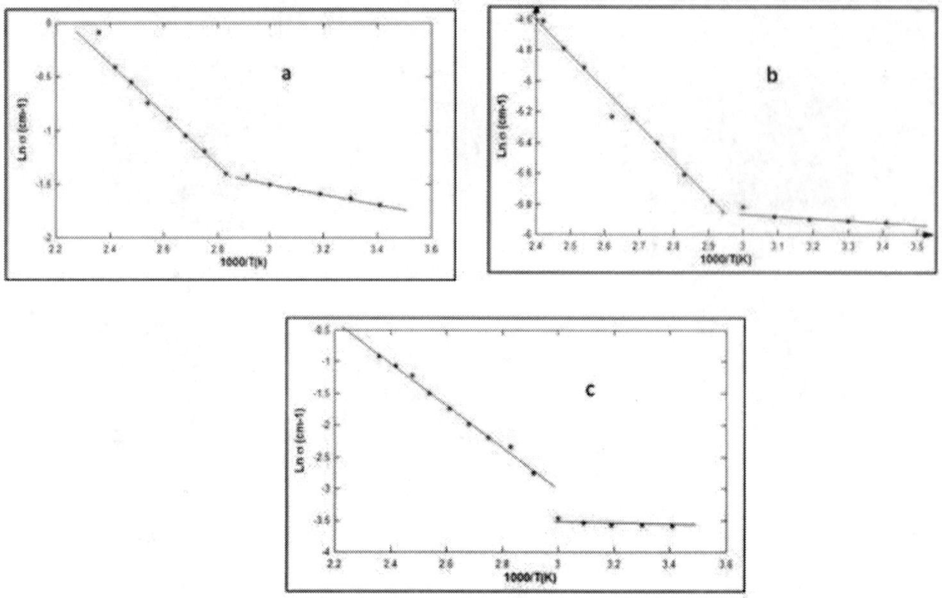

Fig (3): Change (Lnϭ) with (1000/T) of CdS (a) at RT, (b) annealing at 150C⁰ and (c) annealing at 300C⁰

Table (2) activation energy of CdS thin film for various annealing temperatures

sample	Annealing Temperature	E_{a1} eV	Range K	E_{a2} eV	Range K	σ_{RT} $\Omega^{-1}.cm^{-1}$
CdS	R.T	0.039	273-323	0.288	333-433	4.34 E-3
	°150 C	0.013	273-323	0.208	333-433	0.138
	°300 C	0.0107	273-323	0.188	333-433	0.31

Hall Effect

Hall Effect measurements were performed at room temperature for pure CdS thin films to determine the type, concentration, and mobility of the majority of the charge carriers of these films. The pure CdS films are n-type, there by indicating that the majority of the charge carriers are the electrons, whereas the minority of the charge carriers are holes. This result may be attributed to the density of the Cd atoms, which act as a donors with a higher density than the S voids. From the measurement of the hall Effect, we observe that Mobility increases with annealing temperature as shown in Table (3) .This trend is ascribed to an increase in grain size, which decreases grain boundaries.

Table (3): measurement of the Hall Effect of the CdS thin film for various annealing temperatures

sample	Annealing Temperature	R_H cm2/c	N_H $(cm)^{-3}$	$\sigma_{(\Omega.cm)^{-1}}$	Type	μ_H $cm^2/v.sec$
CdS	RT	-6.636 E+1	9.406 E+16	9.854 E-1	n	6.539 E+1
	150C⁰	-3.295E+4	1.895E+14	2.235E-3	n	7.363E+1
	300C⁰	-1.352E+7	4.616E+11	3.000E-5	n	4.056E+2

Optical properties

The transmittance and absorbance of CdS films were recorded within the wavelength of (300-1100nm). Figure (10) shows transmittance as a function of the wavelength, it increases by increasing the wavelength. Pure CdS films exhibit good transparency in the visible and near-ultraviolet region where transmittance is approximately 80%. However, transmittance decreases with increasing annealing temperature as shown in figure (4).This result is consistent with researcher's result (Ragavendar et al) [22]

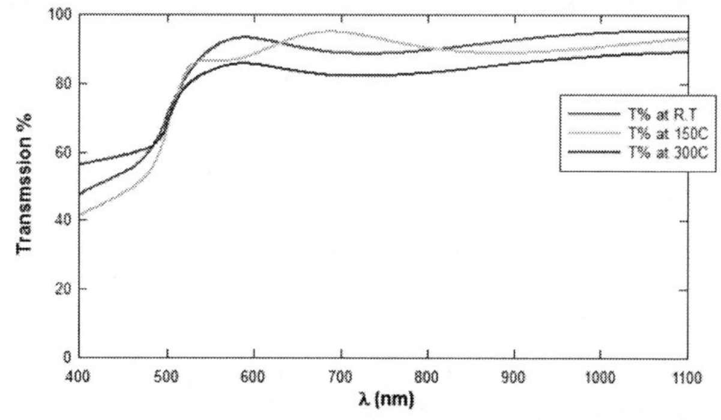

Fig (4): Transmission of CdS thin film

Energy gap is one of the most important optical constants that depends on the physical properties of semiconductors to manufacture electronic devices, such as solar cells, optical detectors and optical diodes .The energy gap of the pure CdS films was Calculated by plotting the relationship between$(\alpha h\upsilon)^2$and $(h\upsilon)$as shown in figure (13), and from the straight line extension of the curve and its intersection with the x-axis. The energy gap was found to be (2.4) ev at room temperature. When the annealing temperature increases to 150 C^0 and 300 C^0, the energy gap of the CdS thin film decreased to 2.38 ev and 2.32 ev, respectively. These values are close to those in a previous study [23].

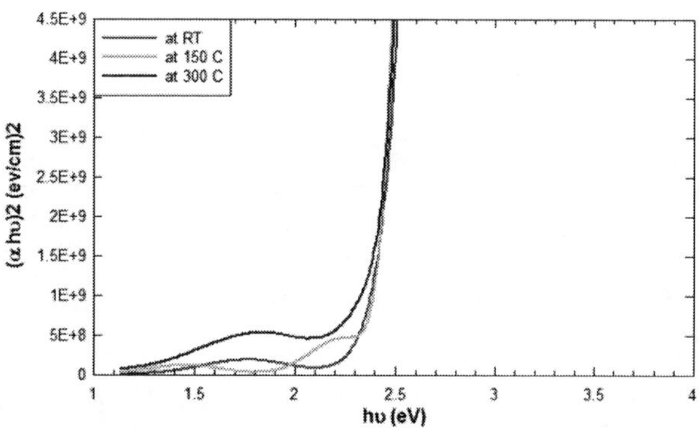

Fig (5): optical energy gap of CdS thin film with relationship between$(\alpha h\upsilon)^2$and $(h\upsilon)$

Equation (3) shows that the coefficient of extinction is related to the absorption coefficient. Through this equation, the coefficient of extinction of the CdS films can be calculated as follows.

$$k = \frac{\alpha\lambda}{4\pi} \quad \text{.................(3)}$$

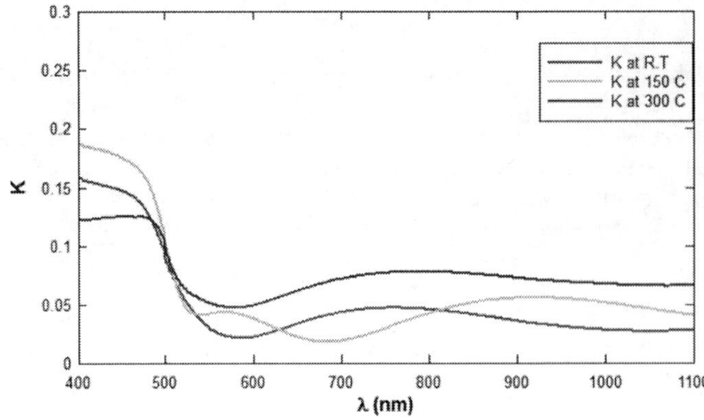

Figure (6) shows the change in the coefficient of extinction as a function of wavelength

The refractive index was calculated using equation (4). The refractive index is associated with the coefficient of extinction and reflectivity, there, the refractive index increases with annealing temperature as shown in figure (7).

.... (4)$n = [(\frac{4R}{(R-1)^2})^{1/2} - k^2 - \frac{(R+1)}{(R-1)}]$

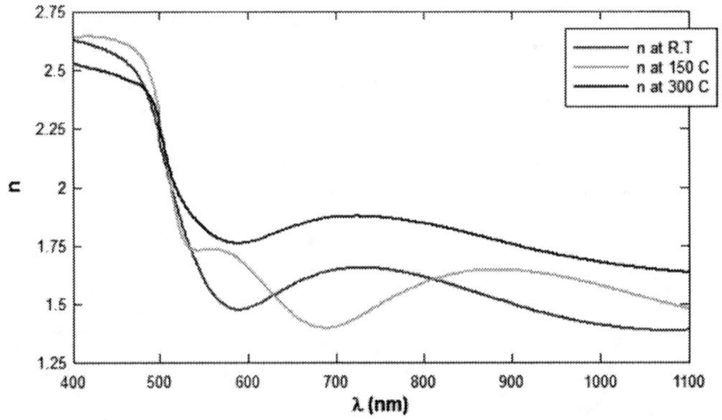

Fig (7): The refractive index as a function of the wavelength

The interaction between light and medium charges is due to the absorption of energy in the material and the subsequent polarization of medium charges. This polarization is typically described as the complex dielectric constant of the medium [24] the real and imaginary dielectric constant of the CdS films were calculated using equations (5) and (6)

$\varepsilon_r = n^2 - k^2$..........(5)
$\varepsilon_i = 2nk$............(6)
Figures (8) indication that the dielectric constant in the real and imaginary parts generally decreases with increasing wavelength

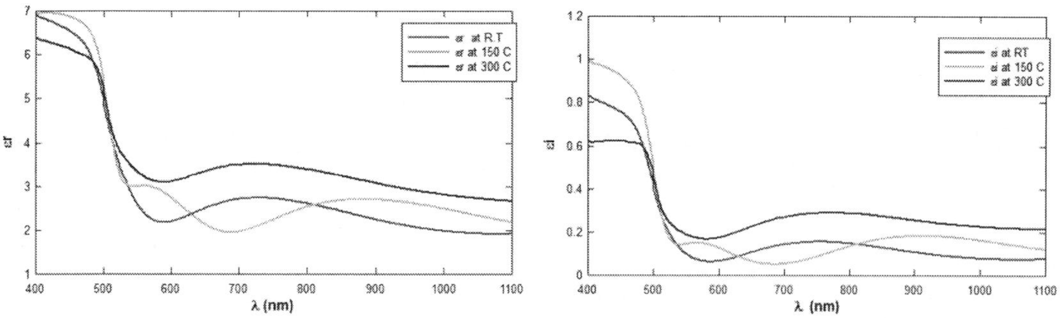

(a) (b)

Fig. 8(a): The real dielectric constant as a function of the wavelength,(b): The imaginary dielectric constant as a function of the wavelength

4. Conclusions

The effect of annealing on the structure .optical and electrical properties of CdS thin films deposited by thermal evaporation technique were studied. . The polycrystalline structure is a cubic type, and the dominant orientation of growth is [111] The AFM images of CdS thin films that fabricated at different substrate temperatures shows that the smoothness and uniform thickness. The average grain sizes were CdS films were increased with increasing of substrate temperatures. The films show a direct optical energy gap (Eg opt), and it decreases with the increase of annealing temperature. . The optical transmittance varied with the annealing temperature extinction coefficient, refractive index and absorption coefficient show an increase when the annealing temperature increases. CdS films exhibit high continuous electrical conductivity that increases with annealing temperature The Continuous conductivity of CdS film is negative type

References

[1] O. toma, R. Pascu, M. Dinescu, C. Besleaga, T.L. Mitran, N. Scarisoreanu, S. Antohe, 2011, *Chalcogenide Letters*, Vol. 8, No. 9, p. 541 – 548 .

[2] C. Wu, J. Jie, L. Wang, Y. Yu, Q. Peng, X. Zhang, J. Cai, H. Guo, D. Wu, Y. Jiang, 2010, *Nanotechnology*, Vol. 21, 505203.

[3] A. Giberti, D. Casotti, G. Cruciani, B. Fabbri, A. Gaiardo, V. Guidi, C. Malagu, G. Zonta, S. Gherardi,2015, *Sensors Actuators, B Chem.*, Vol. 207, 504–510.

[4] X. Jiao, L. Zhang, Y. Lv, Y. Su,2013, *Sensors Actuators*, B Chem., Vol. 186, 750–754.

[5] W. Kim, M. Baek, K. Yong, 2016, *Sensors Actuators, B Chem.*, Vol. 223, 599–605.

[6] B. Mereu, G. Sarau, E. Pentia, V. Draghici, M. Lisca, T. Botila, L. Pintilie, 2004, *Mater. Sci. Eng. B*, Vol.109, 260–263.

[7] H. Sezen, A. A. Rockett, S. Suzer, 2012, *Anal. Chem.* 84, 2990–2994.

[8] B. G. An, H. R. Kim, M. J. Kang, J. G. Park, Y. W. Chang, J. C. Pyun, 2016, *Anal. Chim. Acta*, Vol. 927, 99–106.

[9] X. Li, C. Hu, Z. Zhao, K. Zhang, H. Liu,2013, Sensors Actuators, B Chem., Vol.182, 461–466.

[10] X. Li, Q. Yang, H. Hua, L. Chen, X. He, C. Hu, Y. Xi, J.2015, *Alloys Compd.*, Vol.630, 94–99.

[11] Sung-Gi Hur and Eui-T K, Ji-Hong L, Geun-Hong K. 2008, *Characterization of photoconductive CdS Thin films prepared on glass substrates for photoconductive-sensor applications. J. Vac. Sci. Technol. B*, Vol. 26, 4: 1334-1337.

[12] Yasmeen Z D, Salah Q H, Saba J H, Najiba A H. 2015, *Effect of Substrate Temperature on the Morphological and Optical Properties of Nanocrystalline ZnO Films Formed By DC Magnetron Sputtering. Journal of Applied Physics,* Vol. 7, 1: 59-63.

[13] Demir R, Gode F. 2015, *Structural, optical and electrical properties of nanocrystalline CdS thin films Grown by chemical bath deposition method. Chalcogenide Letters*; Vol. 12, 2: 4350.

[14] Ashok Ch, Venkateswara K R, Shilpa Ch Ch, Rajendar V Lakshmi R N.2014, *Fabrication and Characterization of CdS Thin Films for the Solar Cell Applications. International Journal of Chem.Tech. Research*; Vol. 6, 6:3367-3370.

[15] Oladiran A A, Oluwaseun A, Kolawole S Y.2012, *Study of optical and crystallographic porperties of CBD grown CdS thin films. IJRRAS*; Vol. 12, 3

[16] Han, Junfeng, et al. 2011*"Optimized chemical bath deposited CdS layers for the improvement of CdTe solar cells." Solar Energy Materials and Solar Cells,* Vol. 95, 3: 816-820.

[17] Mishra S, Ingale A, Roy UN, Gupta A. *2007, Study of annealing-induced changes in CdS thin films Using X-ray diffraction and Ramanspectroscopy. Thin Solid Films*, Vol. 516:91-98.

[18] Metin, Hulya, Mehmet Ari, Selma Erat, Semra Durmuş, Mehmet Bozoklu, and Artur Braun. 2010, *"The Effect of annealing temperature on the structural, optical, and electrical properties of CdS films." Journal of Materials Research*, Vol.25, no. 1 : 189-196.

[19] Lee S, Lee ES, Kim TY, Cho JS, Eo YJ, Yun JH, Cho A. 2015, *Effect of annealing treatment on CdS/CIGS thin film solar cells depending on different CdS deposition temperatures. Solar Energy Materials and Solar Cells.* Vol. 1; 141:299-308.

[20] Choubey, Ravi Kant, Dipti Desai, S. N. Kale, and Sunil Kumar. 2016, *"Effect of annealing treatment and deposition temperature on CdS thin films for CIGS solar cells applications." Journal of Materials Science: Materials in Electronics*, Vol. 27, no. 8: 7890-7898.

[21] Mao, Ling-Feng, H. Ning, Changjun Hu, Zhaolin Lu, and Gaofeng Wang. 2016,*"Physical modeling of activation energy in organic semiconductor devices based on energy and momentum conservations." Scientific reports*, Vol. 6 : 24777.

[22] Mariappan, R., Ponnuswamy, V., Ragavendar, M., Krishnamoorthi, D., & Sankar, C. 2012, *The effect of annealing temperature on structural and optical properties of undoped and Cu doped CdS thin films. Optik-International Journal for Light and Electron Optics,* 123(12) ,1098-1102.

[23] Metin, H., and R. Esen. 2003, *"Annealing studies on CBD grown CdS thin films." Journal of Crystal Growth,* Vol. 258.1-2 : 141-148.

[24] NAFTALY, M., et al. 2016, *Dielectric constants of bulk ferroelectric PZT measured by terahertz time-domain Spectroscopy .Advances in Applied Ceramics*, Vol. 115. 5: 260-263

AUTHOR INDEX